MEDICAL INFRARED IMAGING
PRINCIPLES AND PRACTICES

MEDICAL INFRARED IMAGING

PRINCIPLES AND PRACTICES

Edited by

Mary Diakides
Joseph D. Bronzino
Donald R. Peterson

CRC Press
Taylor & Francis Group
Boca Raton London New York

CRC Press is an imprint of the
Taylor & Francis Group, an **informa** business

CRC Press
Taylor & Francis Group
6000 Broken Sound Parkway NW, Suite 300
Boca Raton, FL 33487-2742

First issued in paperback 2017

© 2013 by Taylor & Francis Group, LLC
CRC Press is an imprint of Taylor & Francis Group, an Informa business

No claim to original U.S. Government works

Version Date: 20121023

ISBN 13: 978-1-138-07229-9 (pbk)
ISBN 13: 978-1-4398-7249-9 (hbk)

Library of Congress Cataloging-in-Publication Data

Medical infrared imaging : principles and practices / edited by Mary Diakides, Joseph D. Bronzino, and Donald R. Peterson.
    p. ; cm.
    Includes bibliographical references and index.
    Summary: "This book explores new ideas, concepts, and technologies such as dynamic thermal imaging, thermal texture mapping, and thermal multispectral imaging commonly used in the clinical environment. The coverage ranges from historical background, concepts, clinical applications, standards, and infrared technology. Of interest to the medical and biomedical engineering communities, the book can provide many opportunities for developing and conducting multidisciplinary research in many areas of medical infrared imaging"--Provided by publisher.
    ISBN 978-1-4398-7249-9 (hardback : alk. paper)
    I. Diakides, Mary. II. Bronzino, Joseph D., 1937- III. Peterson, Donald R. (Donald Russell), 1970-
    [DNLM: 1. Infrared Rays--diagnostic use. 2. Thermography--methods. WN 205]

616.07'54--dc23   2012024644

Visit the Taylor & Francis Web site at
http://www.taylorandfrancis.com

and the CRC Press Web site at
http://www.crcpress.com

Contents

Preface

The evolution of technological advances in infrared sensor technology, image processing, "smart" algorithms, knowledge-based databases, and their overall system integration has resulted in new methods of research and use in medical infrared imaging. The development of infrared cameras with focal plane arrays no longer requiring cooling added a new dimension to this modality. New detector materials with improved thermal sensitivity are now available, and production of high-density focal plane arrays (640×480) has been achieved. Advance read-out circuitry using on-chip signal processing is now commonly used. These breakthroughs permit low-cost and easy-to-use camera systems with thermal sensitivity less than 50 mK, as well as spatial resolution of 25–50 μm, given the appropriate optics. Another important factor is the emerging interest in the development of smart image processing algorithms to enhance the interpretation of thermal signatures. In the clinical area, new research addresses the key issues of diagnostic sensitivity and specificity of infrared imaging. Increased efforts are underway to achieve quantitative clinical data interpretation in standardized diagnostic procedures. For this purpose, clinical protocols and appropriate training are emphasized.

New concepts such as dynamic thermal imaging and thermal texture mapping (thermal tomography) and thermal multispectral imaging are being commonly used in clinical environments. Other areas such as three-dimensional infrared are being investigated.

These new ideas, concepts, and technologies are covered in this book. We have assembled a set of chapters that range in content from historical background, concepts, clinical applications, standards, and infrared technology.

Chapter 1 deals with worldwide advances in and a guide to thermal imaging systems for medical applications. Chapter 2 presents a historical perspective and the evolution of thermal imaging. Chapters 3 through 5 are comprehensive chapters on technology and hardware including detectors, detector materials, un-cooled focal plane arrays, high-performance systems, camera characterization, electronics for on-chip image processing, optics, and cost-reduction designs.

Chapter 6 deals with the physiological basis of the thermal signature and its interpretation in a medical setting. It discusses the physics of thermal radiation theory and the physiology as related to infrared imaging. Chapters 7 and 8 cover innovative concepts such as dynamic thermal imaging and thermal tomography that enhance the clinical utility leading to improved diagnostic capability. Chapter 9 presents thermal texture mapping as used in a clinical environment. Chapters 10 and 11 expose the fundamentals of infrared breast imaging, equipment considerations, early detection, and the use of infrared imaging in a multimodality setting. Chapters 12 through 16 are on innovative image processing techniques for the early detection of breast cancer. Chapter 17 discusses the practical value of inspection and measurements of thermal imaging during surgical interventions. Chapter 18 presents biometrics, a novel method for facial recognition. Today, this technology is of utmost importance in the area of homeland security and other applications. Chapter 19 deals with infrared monitoring of therapies using multispectral optical imaging in Kaposi's sarcoma investigations at the National Institutes of Health (NIH). Chapters 20 through 30 deal with the use of infrared in various clinical applications: fever, surgery, dental, skeletal and neuromuscular diseases, as well as the quantification of the TAU image technique in the relevance and stage of a disease. Chapter 31 is on infrared imaging in veterinary medicine. Chapter 32 discusses the complexities and importance of standardization, calibration, and protocols for effective and reproducible results. Chapter 33 deals with databases and primarily with the storage and retrieval of thermal images. Chapter 34 addresses the ethical obligations in infrared research and clinical practice.

This book will be of interest to both the medical and biomedical engineering communities. It could provide many opportunities for developing and conducting multidisciplinary research in many areas of medical infrared imaging. These range from clinical quantification to intelligent image processing for enhancement of the interpretation of images and for further development of user-friendly high-resolution thermal cameras. These would enable the wide use of infrared imaging as a viable, noninvasive, low-cost, first-line detection modality.

This book benefitted from the input of an editorial advisory board consisting of

Professor Hairong Qi, University of Tennessee, Knoxville, Tennessee
Dr. Moinuddin Hassan, U.S. Food and Drug Administration (FDA), Silver Spring, Maryland
Dr. Pradeep Buddharaju, University of Houston, Houston, Texas
Professor Ioannis Pavlidis, University of Houston, Houston, Texas
Professor Francis J. Ring, University of Glamorgan, Pontypridd, Wales
Professor Boguslaw Wiecek, Technical University of Lodz, Lodz, Poland

Hairong Qi deserves accolades for the tremendous effort she expended, as does Moinuddin Hassan, for the organization of papers, distribution to reviewers, follow-up, and so on. All the reviewers are acknowledged and accorded our thanks for their timely and excellent reviews.

Select color figures are available at the Download section of this book's web page on the CRC Press website (http://crcpress.com/product/isbn/9781439872499).

MATLAB® is a registered trademark of The MathWorks, Inc. For product information, please contact:

The MathWorks, Inc.
3 Apple Hill Drive
Natick, MA 01760-2098 USA
Tel: 508-647-7000
Fax: 508-647-7001
E-mail: info@mathworks.com
Web: www.mathworks.com

Editors

Nicholas A. Diakides received his DSc in electrical engineering (biomedical engineering, telecommunications, and computer science) from George Washington University in 1979. He devoted the majority of his professional career to the development and assessment of sensor systems, biomedical technology, medical infrared (IR) imaging, and bioinformatics, engaging this work in a variety of governmental positions under the Office of the Secretary of Defense (OSD-S&T, DARPA, ARO, and ONR). As the director of the survivability enhancement division, U.S. Army Laboratory Command (1984–1989) and as program manager for various areas of IR technology and electro-optics at the Army Night Vision and Electro-Optics Laboratory (1962–1983), he studied and developed several novel applications of IR imaging, including "smart" image processing, computer-aided detection, knowledge-based databases, IR-linked information-technology, and telemedical systems. For this work, Dr. Diakides received the Department of the Army Research and Development Achievement Award in 1973, and he was presented the Commander's Award for Civilian Service by the Department of the Army for his significant contribution in military programs ranging from the creation of vision aids for night fighting to the improvement of antitank munitions' effectiveness in 1989. Moving from the government to the private sector, he established and served as president of Advanced Concepts Analysis, Inc. (1989–2009), a corporation dealing with advanced biomedical technology and innovative defense research on sensors.

Dr. Diakides was an active member of the Institute of Electrical and Electronics Engineers—Engineering in Medicine and Biology Society (IEEE-EMBS). He served as the publicity chair and as a member of the conference and technical program committees at the 16th annual IEEE-EMBS International Conference, Baltimore, Maryland, in 1994. He organized IR imaging-related workshops and symposia for IEEE-EMBS international conferences from 1994 to 2006, and he served on a number of IEEE-USA committees, including the R&D Policy Committee (1994–2009), Healthcare Engineering Policy (1989–1994), and EMBS Technical Committee (on imaging and imaging processing, 2005–2009).

Dr. Diakides authored book chapters on "Phosphorous Screens" in *Electronics Engineers Handbook* (2nd Edition, McGraw-Hill Book Company, 1982) and "Advances in Medical Infrared Imaging" in *Medical Infrared Imaging* (CRC Press, 2007). He served as a guest editor for *IEEE EMDB Magazine* special issues on medical infrared imaging (July/August 1999, May/June 2000, November/December 2002) and was the section editor of "Infrared Imaging" in the *Biomedical Engineering Handbook* (3rd Edition, CRC Press, 2006). He was the co-editor of *Medical Infrared Imaging* (CRC Press, 2007) and published more than 50 papers in peer-reviewed scientific literature.

Dr. Diakides was the inventor of the MedATR concept that led to the first IR-CAD for the early detection of breast abnormalities and other applications. A pioneer in the development of knowledge-based databases with standardized IR signatures validated by pathology, he led by 1994 and for the remainder of his exceptional career an international effort to establish the use of advanced digital infrared imaging in medicine, and he championed this effort as a member of the Executive Committee of the American Academy of Thermology (1998–2009). In recognition of his lifelong achievement and commitment to the field, he was named a Fellow of the American Institute of Medical and Biological Engineering.

Dr. Nicholas A. Diakides unexpectedly passed away on August 9, 2009, at his home in Falls Church, Virginia, after recovering from a severe infection. He is survived by his soul mate and wife of 36 years, Mary, brother Michael Diakides, and daughter Anastasia (Tasha) Diakides.

Joseph D. Bronzino earned a BSEE from Worcester Polytechnic Institute, Worcester, Massachusetts, in 1959, an MSEE from the Naval Postgraduate School, Monterey, California, in 1961, and a PhD in electrical engineering from Worcester Polytechnic Institute in 1968. He is presently the Vernon Roosa Professor of Applied Science, an endowed chair at Trinity College, Hartford, Connecticut, and president of the Biomedical Engineering Alliance and Consortium (BEACON), which is a nonprofit organization consisting of academic and medical institutions as well as corporations dedicated to the development of new medical technology. To accomplish this goal, BEACON facilitates collaborative research, industrial partnering, and the development of emerging companies.

Dr. Bronzino is the author of over 200 journal articles and 15 books, including *Technology for Patient Care* (C.V. Mosby, 1977), *Computer Applications for Patient Care* (Addison-Wesley, 1982), *Biomedical Engineering: Basic Concepts and Instrumentation* (PWS Publishing Co., 1986), *Expert Systems: Basic Concepts* (Research Foundation of State University of New York, 1989), *Medical Technology and Society: An Interdisciplinary Perspective* (MIT Press and McGraw-Hill, 1990), *Management of Medical Technology* (Butterworth/Heinemann, 1992), *The Biomedical Engineering Handbook* (CRC Press, 1st Edition, 1995; 2nd Edition, 2000; 3rd Edition, 2006), *Introduction to Biomedical Engineering* (Academic Press, 1st Edition, 1999; 2nd Edition, 2006), *Biomechanics: Principles and Applications* (CRC Press, 2002), *Biomaterials: Principles and Applications* (CRC Press, 2002), *Tissue Engineering* (CRC Press, 2002), and *Biomedical Imaging* (CRC Press, 2002).

Dr. Bronzino is a fellow of IEEE and the American Institute of Medical and Biological Engineering (AIMBE), an honorary member of the Italian Society of Experimental Biology, past chairman of the Biomedical Engineering Division of the American Society for Engineering Education (ASEE), a charter member of the Connecticut Academy of Science and Engineering (CASE), a charter member of the American College of Clinical Engineering (ACCE), a member of the Association for the Advancement of Medical Instrumentation (AAMI), past president of the IEEE-Engineering in Medicine and Biology Society (EMBS), past chairman of the IEEE Health Care Engineering Policy Committee (HCEPC), and past chairman of the IEEE Technical Policy Council in Washington, DC. He is a member of Eta Kappa Nu, Sigma Xi, and Tau Beta Pi. He is also a recipient of the IEEE Millennium Medal for "his contributions to biomedical engineering research and education" and the Goddard Award from WPI for Outstanding Professional Achievement in 2005. He is presently editor-in-chief of the Academic Press/Elsevier BME Book Series.

Donald R. Peterson is an associate professor of medicine and the director of the biodynamics laboratory in the School of Medicine at the University of Connecticut (UConn). He serves jointly as the director of the biomedical engineering undergraduate program in the School of Engineering and recently served as the director of the graduate program and as the BME Program chair. He earned a PhD in biomedical engineering and an MS in mechanical engineering at UConn and a BS in aerospace engineering and a BS in biomechanical engineering from Worcester Polytechnic Institute. Dr. Peterson has 16 years of experience in biomedical engineering education and offers graduate-level and undergraduate-level courses in BME in the areas of biomechanics, biodynamics, biofluid mechanics, and ergonomics, and he teaches in medicine in the subjects of gross anatomy, occupational biomechanics, and occupational exposure and response. Dr. Peterson's scholarly activities include over 50 published journal articles, 3 textbook chapters, and 12 textbooks, including his new appointment as Co-Editor-in-Chief for *The Biomedical Engineering Handbook* by CRC Press.

Dr. Peterson has over 21 years of experience in biomedical engineering research and has been recently focused on measuring and modeling human, organ, and/or cell performance, including exposures to various physical stimuli and the subsequent biological responses. This work also involves the investigation of human–device interaction and has led to applications on the design and development of tools and various medical devices. Dr. Peterson is faculty within the occupational and environmental medicine group at the UConn Health Center, where his work has been directed toward the objective analysis of the anatomic and physiological processes involved in the onset of musculoskeletal and neuromuscular diseases, including strategies of disease mitigation. Recent applications of his research include human interactions with existing and developmental devices such as powered and non-powered tools, spacesuits and space tools for NASA, surgical and dental instruments, musical instruments, sports equipment, and computer-input devices. Other overlapping research initiatives focus on cell mechanics and cellular responses to fluid shear stress, the acoustics of hearing protection and communication, human exposure and response to vibration, and the development of computational models of biomechanical performance.

Dr. Peterson is also the co-executive director of the Biomedical Engineering Alliance and Consortium (BEACON; www.beaconalliance.org), which is a nonprofit entity dedicated to the promotion of collaborative research, translation, and partnership among academic, medical, and industry people in the field of biomedical engineering to develop new medical technologies and devices.

Tributes to
Dr. Nick Diakides

It is not possible in a few sentences to pay full tribute to my friend Nick Diakides. I have known Nick as a friend and professional associate for 40 years, and its difficult to put that in the past tense. Nick and I first met in 1971 as employees of the Night Vision Laboratory at Fort Belvoir, Virginia. Nick was already a pioneer in the biomedical engineering field with his doctoral thesis on interpretation of human brain waves with the goal of automating man–machine interfaces for the military. This is currently a hot research topic around the world. In the 1980s Nick was working on optical countermeasures and provided me with a powerful study for my smart sensors work at the Defense Advanced Research Projects Agency (DARPA). In the 1990s Dr. Diakides focused on biomedical endeavors with a study on the application of military technology to early detection of breast cancer. The study was done for me in my capacity as Director for Research (and later Sensors) in the Office of the Director, Defense Research and Engineering. The Army Research Office and DARPA were co-sponsors. Nick was again a pioneer, hoping to provide women around the world with a painless, noninvasive approach to early screening. He inspired many of his sponsors with this idea and the IEEE published Tanks to Tumors. He continued this important work until he died at the young age of 79. He and his wife became internationally known for this and began working with the IEEE to publish dedicated handbooks in the rapidly expanding field of thermal imaging for biomedical applications. I was working with him until 2009 to help fulfill this vision. He can be thankful that his wife, Mary Diakides, and many others continue to be active in promoting the use of thermal imaging for detection of tumors. It is an honor to have known and worked with Dr. Nick Diakides over the past decades. He has made a great contribution to applications of thermal imaging in biomedicine.

Dr. Jasper C. Lupo
Physicist

Nick Diakides was the soul of the infrared imaging community for a quarter century. He has been the mentor of an entire generation of infrared imaging researchers. A great scientist and heart. We are missing him dearly.

Ioannis Pavlidis
University of Houston

The Egyptians realized the importance of temperature measurements in understanding human physiology. Today, accurate measurement of skin surface temperature has evolved thanks to the development of sophisticated electronic detectors, computers, and software. Initially, only the military had access to the best technology. Nick was one of the key persons who recognized the importance of bringing together the highly developed equipment available to the military in order to improve research and medical practices associated with thermography. This concept started with the first report on the clinical use of IR imaging as a diagnostic aid when, in 1956, Lawson discovered that the skin temperature over a cancer in a breast was higher than that of normal tissue. This was in accordance to the recently developed program called "Tanks to Tumors." Those of us who have had the pleasure of knowing Nick will particularly remember him for his endless enthusiasm in promoting the medical use of thermography. His contributions to this area of science, his hard work, perseverance, and dreams live on in the chapters of this book, including ours, which we dedicate to his memory.

David D. Pascoe
Auburn University

James B. Mercer
University of Tromsø, Norway

Louis de Weerd
University Hospital of North Norway

My first contact with Nick was over the research we were doing on infrared breast imaging. I was taken back by his kindness and the level of knowledge he had in this area. Over the years Nick and I had many discussions regarding the direction and advancements in this technology. He always encouraged us to keep up with our research and to join him in presenting our work through the IEEE-EMBS and textbooks.

We have all lost a great man, for Nick gave so much to the advancement of medical infrared imaging. I will miss our discussions and his kind, gentlemanly manner.

Farewell friend,

William C. Amalu, DC, DABCT, FIACT
President, International Academy of Clinical Thermology
Medical Director, International Association of Certified Thermographers

Dr. Diakides made a great contribution to medical thermography. He knew and made contact with so many people in different parts of the world. We remember his visit to Poland, where he showed great interest in the way thermography was developing in Europe. This book is one indication of his many international contacts.

Professor Anna Jung
Warsaw
Vice President, European Association of Thermology

This book and the now third edition is a tribute to Dr. Nicholas Diakides, who has done so much for infrared imaging and for its application in medicine. His visionary outlook, which often gave him a practical view "over the wall" when others sometimes failed to see a realistic view of this technology was always valuable. He brought many people around the world into contact, and this was a personal catalyst in research and development of infrared imaging. Through the years this has certainly added to the growth of applications for thermal imaging in medical and biological fields. His contribution to science especially through his inspiration remains and hopefully will continue through the readers of this latest volume.

Professor Francis J. Ring MSc, DSc
Fellow of the Institute of Physics and Engineering in Medicine, Fellow of the Royal Society of Medicine
Former President of the European Association of Thermology
Faculty of Advanced Technology, University of Glamorgan

Dr. Nicholas Diakides was instrumental in shepherding my work in the utilization of functional IR imaging in evaluating patients with presumptive CRPS. I will forever be grateful to Nick for the mentorship he so kindly and without reservation provided over many years.

Timothy D. Conwell, DC
Medical Director, Colorado Infrared Imaging Center

Good-bye, Nick! I will remember our long telephone discussions, USA–Poland, usually at night. You always have been full of energy, planning sessions at IEEE-EMBS conferences, workshops, visits, and other activities. I was impressed by your bright idea of the importance of IR-thermal imaging in medical diagnostics. I remember also your telling me how important it was to perform objective, high-quality research, real scientific work, to be able to publish the results in well-recognized, impact journals. I always appreciated your efforts to help us in finding money for participation in the events you organized and the challenging atmosphere of "your" workshops. Meetings with you and Mary in different places around the world, such as in Zakopane—the trip to Dolina Koscieliska, on a shaking horse wagon, and many others. Of most importance was your positive attitude and ability to bring together people working in medical thermal IR-imaging from all parts of the world in order to get the best synergetic results from such meetings. Without your energy, it would not be possible to promote so effectively and to revitalize IR-thermal imaging in medical diagnostics. Now, we witness important progress in this field. This would not be possible without your personal impact.

Nick, thank you for all you have done for us!

Antoni Nowakowski
Department of Biomedical Engineering, Gdansk University of Technology,
Gdansk, Poland

We are thankful to have this opportunity to write a small tribute to Nick, our dear friend and colleague for the last decade. Words are insufficient to capture our appreciation and respect for his always-present encouragement and nonyielding dedication to our work (thermal texture mapping or TTM) and to the field of infrared imaging. No one could have any doubt about his contribution and single-minded devotedness to this field. It is his pursuit of his beliefs and dreams about infrared imaging that allowed him to arrive at the present evolution in the field.

Why do we only pay tribute and say kind words to people who have passed away? Why wouldn't we do so to our colleagues now who are still striving to make this field alive and thrive in medicine. We know this truth deep in our heart, just like Nick—that this is a tool that will fundamentally change medicine and how it will be practiced. Our action will be the real tribute to Nick. Let us work together and make it better for men and women who deserve to have better health and greater knowledge of the secret of health. Overall, this is what infrared imaging is about, and this is what Nick and Mary are working on, their dream and our dream together.

Our love to Nick—may he be in peace in his continuing journey. Whenever we think of him, he is with us, ever present in our heart.

Professor Zhong Qi Liu and Dr. H. Helen Liu
Bioyear Medical Systems, Beijing, China
TTM Wellness Center, Houston, Texas

I have no words to thank Nick for his energetic, prompt, and strong support to the research in infrared imaging. His advice was always effective and wise. But what I remember most is the friend with whom to share ideas and opinions. Thanks, Nick.

Arcangelo Merla
Department of Neurosciences and Imaging, G. d'Annunzio University, Chieti-Pescara, Italy. ITAB—
Institute for Advanced Biomedical Technology, Foundation G.d'Annunzio, Chieti, Italy

I have known Nick since I was a PhD student at North Carolina State University. He has since become a dear friend and mentor in both career and life. It is always delightful to talk to him at conferences and workshops. His enthusiasm, his vision, and his solid belief toward IR imaging in breast cancer study have revitalized the field. We miss you so much, Nick!

Hairong Qi
University of Tennessee

Friend's wedding

Air Force service

Nick's brothers and sisters

Father–daughter tender moments

New Year's party

Infrared image station—TTM

Contributors

Kim Abramson
Institute for BioTechnology Futures
New York, New York

P.D. Ahlgren
Ville Marie Medical and Women's Health Center
Montreal, Quebec, Canada

William C. Amalu
Pacific Chiropractic and Research Center
Redwood City, California

Kurt Ammer
Ludwig Boltzmann Research Institute for
    Physical Diagnostics
Vienna, Austria

and

University of Glamorgan
Pontypridd, Wales, United Kingdom

Michael Anbar
University at Buffalo, State University of
    New York
Buffalo, New York

Raymond Balcerak
(deceased)
RSB Consulting LLC

Normand Belliveau
Ville Marie Medical and Women's Health Center
Montreal, Quebec, Canada

Reinhold Berz
German Society of Thermography and
    Regulation Medicine (DGTR)
Waldbronn, Germany

Pradeep Buddharaju
Department of Computer Science
University of Houston
Houston, Texas

Paul Campbell
Ninewells Hospital
Dundee, Scotland, United Kingdom

Victor Chernomordik
National Institutes of Health
Bethesda, Maryland

Timothy D. Conwell
Colorado Infrared Imaging Center
Denver, Colorado

Mary Diakides
Advanced Concepts Analysis, Inc.
Falls Church, Virginia

Nicholas A. Diakides
(deceased)
Advanced Concepts Analysis, Inc.
Falls Church, Virginia

C. Drews-Peszynski
Laser Diagnostics and Therapy Center
Technical University of Lodz
Lodz, Poland

Ronald G. Driggers
U.S. Army CERDEC Night Vision
    and Electronic Sensors Directorate
Fort Belvoir, Virginia

Robert L. Elliot
Elliot-Elliot-Head Breast Cancer Research and
    Treatment Center
Baton Rouge, Louisiana

M. Etehadtavakol
Medical Image and Signal Processing Research
    Center
Isfahan University of Medical Science
Isfahan, Iran

Amir H. Gandjbakhche
National Institutes of Health
Bethesda, Maryland

Israel Gannot
Department of Biomedical Engineering
Tel Aviv University
Tel Aviv, Israel

N. Gheissari
Electrical and Computer Engineering Department
Isfahan University of Technology
and
Medical Image and Signal Processing Research
    Center
Isfahan University of Medical Science
Isfahan, Iran

James Giordano
Center for Neurotechnology Studies
Potomac Institute for Policy Studies
Arlington, Virginia

and

Krasnow Institute for Advanced Studies
George Mason University
Fairfax, Virginia

and

Wellcome Centre for Neuroethics
University of Oxford
Oxford, United Kingdom

Barton M. Gratt
School of Dentistry
University of Washington
Seattle, Washington

Michael W. Grenn
U.S. Army CERDEC Night Vision
    and Electronic Sensors Directorate
Fort Belvoir, Virginia

Moinuddin Hassan
Center for Devices and Radiological Health
U.S. Food and Drug Administration (FDA)
Silver Spring, Maryland

Jonathan F. Head
Elliot-Elliot-Head Breast Cancer Research and
    Treatment Center
Baton Rouge, Louisiana

William B. Hobbins
Women's Breast Health Center
Madison, Wisconsin

Stuart B. Horn
U.S. Army CERDEC Night Vision
    and Electronic Sensors Directorate
Fort Belvoir, Virginia

T. Jakubowska
Laser Diagnostics and Therapy Center
Technical University of Lodz
Lodz, Poland

Bryan F. Jones
Medical Imaging Research Group
School of Computing
University of Glamorgan
Pontypridd, Wales, United Kingdom

A. Jung
Department of Pediatrics
Military Institute of Medicine
Szaserow, Warsaw, Poland

Mariusz Kaczmarek
Department of Biomedical Engineering
Gdansk University of Technology
Gdansk, Poland

Jana Kainerstorfer
National Institutes of Health
Bethesda, Maryland

B. Kalicki
Department of Pediatrics
Military Institute of Medicine
Szaserow, Warsaw, Poland

John R. Keyserlingk
Ville Marie Medical and Women's Health Center
Montreal, Quebec, Canada

Andrey Kondyurin
Institute for Laser and Information Technologies
    of Russian Academy of Sciences
Troitsk, Moscow Region

Phani Teja Kuruganti
Oak Ridge National Laboratory
Oak Ridge, Tennessee

Joshua E. Lane
Departments of Surgery and Internal Medicine
Mercer University School of Medicine
Macon, Georgia

and

Department of Dermatology
Emory University School of Medicine
Atlanta, Georgia

Richard F. Little
National Institutes of Health
Bethesda, Maryland

H. Helen Liu
Flow-Of-Light Natural Health
Institute of Holistic Health and Science
Houston, Texas

Zhong Qi Liu
Academy of TTM Technologies and Bioyear
  Medical Instrument
Beijing, China

Caro Lucas
Electrical and Computer Engineering
  Department
University of Tehran
Tehran, Iran

Jasper Lupo
Applied Research Associates, Inc.
Falls Church, Virginia

Alessandro Mariotti
Department of Neurosciences and Imaging
G. d'Annunzio University
Chieti-Pescara, Italy

James B. Mercer
University of Tromsø
Tromsø, Norway

Arcangelo Merla
University of Chieti-Pescara
Pescara, Italy

E.Y.K. Ng
School of Mechanical and Aerospace Engineering
and
College of Engineering
Nanyang Technological University
Nanyang, Singapore

Paul R. Norton
U.S. Army CERDEC Night Vision
  and Electronic Sensors Directorate
Fort Belvoir, Virginia

Antoni Nowakowski
Department of Biomedical Engineering
Gdansk University of Technology
Gdansk, Poland

David D. Pascoe
Auburn University
Auburn, Alabama

Jeffrey Paul
Applied Research Associates, Inc.
Alexandria, Virginia

Ioannis Pavlidis
Department of Computer Science
University of Houston
Houston, Texas

Joseph G. Pellegrino
U.S. Army CERDEC Night Vision
  and Electronic Sensors Directorate
Fort Belvoir, Virginia

Philip Perconti
U.S. Army CERDEC Night Vision
  and Electronic Sensors Directorate
Fort Belvoir, Virginia

Ram C. Purohit
Auburn University
Auburn, Alabama

Hairong Qi
Department of Electrical Engineering and
  Computer Science
University of Tennessee
Knoxville, Tennessee

Francis J. Ring
Medical Imaging Research Unit
University of Glamorgan
Pontypridd, Wales, United Kingdom

Jan Rogowski
Department of Cardiosurgery
Gdansk University of Medicine
Gdansk, Poland

Gian Luca Romani
University of Chieti-Pescara
Pescara, Italy

A. Rustecka
Department of Pediatrics
Military Institute of Medicine
Szaserow, Warsaw, Poland

S. Sadri
Electrical and Computer Engineering
    Department
Isfahan University of Technology
and
Medical Image and Signal Processing Research
    Center
Isfahan University of Medical Science
Isfahan, Iran

Gerald Schaefer
Department of Computer Science
Loughborough University
Loughborough, United Kingdom

Claus Schulte-Uebbing
German Society of Thermography and
    Regulation Medicine (DGTR)
Munich, Germany

Wesley E. Snyder
Department of Electrical and Computer
    Engineering
North Carolina State University
Raleigh, North Carolina

Robert Strakowski
Institute of Electronics
Technical University of Lodz
Lodz, Poland

M. Strzelecki
Institute of Electronics
Technical University of Lodz
Lodz, Poland

Alexander Sviridov
Institute for Laser and Information Technologies
    of Russian Academy of Sciences
Troitsk, Moscow Region

Roderick Thomas
Swansea Institute of Technology
Swansea, Wales, United Kingdom

Tracy A. Turner
Private Practice
Minneapolis, Minnesota

R. Vardasca
Computer Science and Communications
    Research Centre
School of Technology and Management
Polytechnic Institute of Leiria
Leiria, Portugal

Jay Vizgaitis
U.S. Army CERDEC Night Vision
    and Electronic Sensors Directorate
Fort Belvoir, Virginia

Abby Vogel
Georgia Institute of Technology
Atlanta, Georgia

Louis de Weerd
University Hospital of North Norway
Tromsø, Norway

Sven Weum
Auburn University
Auburn, Alabama

Boguslaw Wiecek
Institute of Electronics
Technical University of Lodz
Lodz, Poland

Maria Wiecek
Institute of Electronics
Technical University of Lodz
Lodz, Poland

M. Wysocki
Laser Diagnostics and Therapy Center
Technical University of Lodz
Lodz, Poland

Robert Yarchoan
National Institutes of Health
Bethesda, Maryland

Mariam Yassa
Ville Marie Medical and Women's Health Center
Montreal, Quebec, Canada

E. Yu
Ville Marie Medical and Women's Health Center
Montreal, Quebec, Canada

Jason Zeibel
U.S. Army CERDEC Night Vision
    and Electronic Sensors Directorate
Fort Belvoir, Virginia

J. Zuber
Department of Pediatrics
Military Institute of Medicine
Szaserow, Warsaw, Poland

1

Advances in Medical Infrared Imaging: An Update

Nicholas A. Diakides
Advanced Concepts Analysis, Inc.

Mary Diakides
Advanced Concepts Analysis, Inc.

Jasper Lupo
Applied Research Associates, Inc.

Jeffrey Paul
Applied Research Associates, Inc.

Raymond Balcerak
RSB Consulting LLC

1.1 Introduction

Since the last publication of the *Medical Infrared Imaging* handbook in 2008, our dear friend and medical infrared (IR) pioneer, Dr. Nick Diakides passed away. His wife, Mary Diakides, Advanced Concepts Analysis, continues the quest to bring thermal imaging into the forefront for medical imaging, especially the early detection of breast cancer. It is an honor to have been Nick's friends and to continue efforts to realize his goals. If we succeed, someday women and their doctors will have a new screening option to complement existing tools such as mammography.

During the intervening few years, there have been steady improvements in low cost, high-resolution thermal cameras that operate at room temperature. With the continued investment of military R&D, it is now possible to buy uncooled cameras with nearly 1.0 megapixel arrays at prices below $15,000 in a competitive industrial base. These cameras can resolve temperature differences as low as 0.035°K, thus providing exquisite detail for medical diagnostics.

IR imaging in medicine has been used in the past but without the advantage of twenty-first century technology. In 1994, under the Department of Defense (DOD) grants jointly funded by the Office of the Secretary of Defense (S&T), the Defense Advanced Research Projects Agency (DARPA) and the Army Research Office (ARO), a concerted effort was initiated to re-visit this subject. Specifically, it was to explore the potential of integrating advanced IR technology with "smart" image processing for use in medicine. The major challenges for acceptance of this modality by the medical community were investigated. It was found that the following issues were of prime importance: (1) standardization and quantification of clinical data, (2) better understanding of the pathophysiological nature of thermal signatures, (3) wider publication and exposure of medical IR imaging in conferences and leading journals, (4) characterization of thermal signatures through an interactive web-based database, and (5) training in both image acquisition and interpretation.

Over the last 10 years, significant progress has been made internationally by advancing a thrust for new initiatives worldwide for clinical quantification, international collaboration, and providing a forum for coordination, discussion, and publication through the following activities: (1) medical IR imaging symposia, workshops, and tracks at IEEE/Engineering in Medicine and Biology Society (EMBS) conferences from 1994 to 2004, (2) three EMBS, Special Issues dedicated to this topic [1–3], (3) the DOD "From Tanks to Tumors" Workshop [4]. The products of these efforts are documented in final government technical reports [5–8] and IEEE/EMBS Conference Proceedings (1994–2004).

Likewise, great technological progress has been made in camera development over the last 15 years. Early IR cameras used a small, linear array of 1×180 detectors, which used a mechanically scanned mirror to form an image, and required cryogenic cooling in order to produce a clear image. Electrical contact was made to each individual detector by means of a manually connected wire—a very laborious and expensive process. The images formed by these cameras were limited in spatial and thermal resolution. The 1990s saw the advent of 2-D thermal imaging arrays and the emergence of un-cooled thermal imaging technology. Developers focused on producing larger and larger arrays, capitalizing in large measure on the same technology that drives the commercial digital camera boom. Thermal cameras with 2-D arrays produce images without mechanical scanning in much the same way as a modern digital camera. Electrical scanning and lithographic techniques provide an elegant and simplified image readout that produces standard analog and digital video with very little external processing. New cameras mentioned above can feed imagery directly into a laptop computer or a video monitor.

Presently, IR imaging is used in many different medical applications. The most prominent of these are oncology (breast, skin, etc.), vascular disorders (diabetes DVT, etc.), pain, surgery, tissue viability, monitoring the efficacy of drugs and therapies; respiratory (recently introduced for testing of SARS).

There are various methods used to acquire IR images: Static, Dynamic, Passive and Active—Dynamic Area Telethermometry (DAT), subtraction, and so on [9], Thermal Texture Mapping (TTM), Multispectral/Hyperspectral, Multimodality, and Sensor Fusion. A list of current applications and IR imaging methods are listed in Table 1.1. Figures 1.1 and 1.2 illustrate thermal signatures of breast screening and Kaposi sarcoma.

1.2 Worldwide Use of IR Imaging in Medicine

1.2.1 United States of America and Canada

IR imaging is beginning to be reconsidered in the United States, largely due to new IR technology, advanced image processing, powerful, high-speed computers, and the promotion of research. This is evidenced by the increased number of publications available in open literature and national databases such as *Index Medicus* and *Medline* (National Library of Medicine) on this modality. Currently, there are several academic institutions with research initiatives in IR imaging. Some of the most prominent are the following: NIH, John's Hopkins University, University of Houston, University of Texas. NIH has several ongoing programs: vascular disorders (diabetes, deep-venous thrombosis), monitoring angiogenesis activity—Kaposi sarcoma, pain-reflex sympathetic dystrophy, monitoring the efficacy of radiation therapy, organ transplant—perfusion, multispectral imaging.

Johns Hopkins University carries out research on microcirculation, monitoring angiogenic activity in Kaposi sarcoma and breast screening, laparoscopic IR images—renal disease.

TABLE 1.1 Medical Applications and Methods

Applications	IR Imaging Methods
Oncology (breast, skin, etc.)	Static (classical)
Pain (management/control)	Dynamic (DAT, subtraction, etc.)
Vascular disorders (diabetes, DVT)	Dynamic (active)
Arthritis/rheumatism	TTM
Neurology	Multispectral/hyperspectral
Surgery (open heart, transplant, etc.)	Multimodality
Ophthalmic (cataract removal)	Sensor fusion
Tissue viability (burns, etc.)	
Dermatological disorders	
Monitoring efficacy of drugs and therapies	
Thyroid	
Dentistry	
Respiratory (allergies, SARS)	
Sports and rehabilitation medicine	

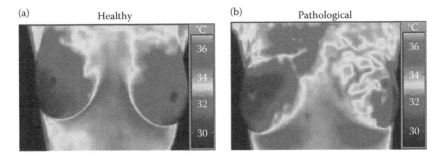

FIGURE 1.1 An application of infrared technique for breast screening: (a) healthy; (b) pathological breast. (Courtesy of Prof. Reinhold Berz, MD. Informatics, Germany.)

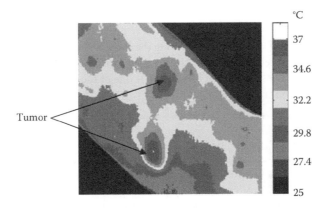

FIGURE 1.2 An application of infrared technique for cancer research. (Adapted from Hassan, M. et al., *Technology in Cancer Research and Treatment*, 3(5), 451–457, 2004.)

University of Houston just created an IR imaging laboratory to investigate using IR the facial thermal characteristics for such applications as lie detection and other behavioral issues (fatigue, anxiety, fear, etc.).

There are two medical centers specializing in breast cancer research and treatment which use IR routinely as part of their first-line detection system, which also includes mammography and clinical exam. These are: EHH Breast Cancer and Treatment Center, Baton Rouge, LA and Ville Marie Oncology Research Center, Montreal, Canada. Their centers are fully equipped with all state-of-the-art imaging equipment. These centers are members of the coalition team for the development of a "knowledge-based" database of thermal signatures of the breast with "ground-truth" validation.

1.2.2 China

China has a long-standing interest in IR imaging. More recently, the novel method Thermal Texture Mapping (TTM) has added increased specificity to static imaging. It is known that this method is widely used in this country, but unfortunately there is no formal literature about this important work. This is urgently needed in order for TTM to be exposed and accepted as a viable, effective method by the international community. The clinical results obtained through this method should be published in open literature of medical journals and international conference proceedings. Despite the lack of the availability of this documentation, introduction of TTM has been made to The National Institutes of Health (NIH). They are now using this method and its camera successfully in detection and treatment in Kaposi sarcoma (associated with AIDS patients). China has also shown interest in breast cancer detection and associated angiogenesis. There are further possibilities for high-level research for this method in the United States and abroad.

1.2.3 Japan

IR imaging is widely accepted in Japan by the government and the medical community. More than 1500 hospitals and clinics use IR imaging routinely. The government sets the standards and reimburses clinical tests. Their focus is in the following areas: blood perfusion, breast cancer, dermatology, pain, neurology, surgery (open-heart, orthopedic, dental, cosmetic), sports medicine, oriental medicine. The main research is performed at the following universities: University of Tokyo—organ transplant; Tokyo Medical and Dental University (skin temperature characterization and thermal properties); Toho University—neurological operation); Cancer Institute hospital (breast cancer). In addition, around 40 other medical institutions are using IR for breast cancer screening.

1.2.4 Korea

The involvement in IR imaging in Korea began during the early 1990s. More than 450 systems are being used in hospitals and medical centers. Primary clinical applications are neurology, back pain/treatment, surgery, oriental medicine. Yonsei College of Medicine is one of the leading institutions in medical IR imaging research along with three others.

1.2.5 United Kingdom

The University of Glamorgan is the center of IR imaging; the School of Computing has a thermal physiology laboratory which focuses in the following areas: medical IR research, standardization, training (degree offered), "SPORTI" Project funded by the European Union Organization. The objective of this effort is to develop a reference database of normal, thermal signatures from healthy subjects.

The Royal National Hospital of Rheumatic Diseases specializes in rheumatic disorders, occupational health (Raynaud's Disease, Carpal Tunnel Syndrome, and Sports Medicine).

The Royal Free University College Medical School Hospital specializes in vascular disorders (diabetes, DVT, etc.), optimization of IR imaging techniques and Raynaud's Phenomenon.

1.2.6 Germany

University of Leipzig uses IR for open-heart surgery, perfusion and micro-circulation. There are several private clinics and other hospitals that use IR imaging in various applications.

EvoBus-Daimler Chrysler uses IR imaging for screening all their employees for wellness/health assessment (occupational health).

InfraMedic, AG, conducts breast cancer screening of women from 20 to 85 years old for the government under a two-year grant. IR is the sole modality used. Their screening method is called Infrared Regulation Imaging (IRI).

1.2.7 Austria

Ludwig Bolzman Research Institute for Physical Diagnostics has carried out research in IR for many years and it publishes the *Thermology International* (a quarterly journal of IR clinical research and instrumentation). This journal contains papers from many Thermology Societies. A recent issue contains the results of a survey of 2003 international papers dedicated to Thermology [10].

The General Hospital, University of Vienna, does research mainly in angiology (study of blood and lymph vessels) diabetic foot (pedobarography).

1.2.8 Poland

There has been a more recent rapid increase in the use of IR imaging for medicine in Poland since the Polish market for IR cameras was opened up. There are more than 50 cameras being used in the following medical centers: Warsaw University, Technical University of Gdansk, Poznan University, Lodz University, Katowice University, and the Military Clinical Hospital. The research activities are focused on the following areas: Active IR imaging, open-heart surgery, quantitative assessment of skin burns, ophthalmology, dentistry, allergic diseases, neurological disorders, plastic surgery, thermal image database for healthy and pathological cases and multispectral imaging (IR, visual, x-ray, ultrasound).

In 1986 the Eurotherm Committee was created by members of the European Community to promote cooperation in the thermal sciences by gathering scientist and engineers working in the area of Thermology. This organization focuses on quantitative IR thermography and periodically holds conferences and seminars in this field [11].

1.2.9 Italy

Much of the clinical use of IR imaging is done under the public health system, besides private clinics. The ongoing clinical work is in the following areas: dermatology (melanoma), neurology, rheumatology, anesthesiology, reproductive medicine, sports medicine. The University of G. d'Annunzio, Chieti, has an imaging laboratory purely for research on IR applications. It collaborates on these projects with other universities throughout Italy.

There are other places, such as Australia, Norway, South America, Russia, and so on that have ongoing research as well.

1.3 IR Imaging in Breast Cancer

In the United States, breast cancer is a national concern. There are 192,000 cases a year; it is estimated that there are 1 million women with undetected breast cancer; presently, the figure of women affected is

1.8 million; 45,000 women die per year. The cost burden of the U.S. healthcare is estimated at $18 billion per year. The cost for early-stage detection is $12,000 per patient and for late detection it is $345,000 per patient. Hence, early detection would potentially save $12B annually—as well as many lives. As a result, the U.S. Congress created "The Congressionally Directed Medical Research Program for Breast Cancer." Clinical IR has not as yet been supported through this funding. Effort is being directed toward including IR. Since 1982 FDA has approved IR imaging (Thermography) as an adjunct modality to mammography for breast cancer as shown in Table 1.2.

Ideal characteristics for an early breast cancer detection method as defined by the Congressionally Directed Medical Research Programs on Breast Cancer are listed in Table 1.3. IR imaging meets these requirements with the exception of the detection of early lesions at 1000–10,000 cells which has not yet been fully determined.

A program is underway in the United States to develop a prototype web-based database with a collection of approximately 2000 patient thermal signatures to be categorized into three categories: normal, equivocal, suspicious for developing algorithms for screening and early detection of tumor development.

The origin of this program can be traced back to a 1994 multiagency DoD grant sponsored by the Director for Research in the Office of the Director, Defense Research and Engineering, the Defense Advanced Research Projects Agency, and the Army Research Office. This grant funded a study to determine the applicability of advanced military technology to the detection of breast cancer—particularly thermal imaging and automatic target recognition. The study produced two reports, one in 1995 and another in 1998; these studies identified technology, concepts, and ongoing activity that would have direct relevance to a rigorous application of IR. Rigor was the essential ingredient to further progress. The US effort had been dormant since the 1970s because of the limitations imposed by poor sensors, simplistic imaging processing, lack of automatic target recognition, and inadequate computing power. This was complicated by the fact that the virtues of IR had been overstated by a few developers.

TABLE 1.2 Imaging Modalities for Breast Cancer Detection Approved by FDA

Film-screen mammography
Full-field digital mammography
Computer-aided detection
Ultrasound
Magnetic resonance imaging (MRI)
Positron emission tomography (PET)
Thermography
Electrical impedance imaging

Source: Mammography and Beyond, Institute of Medicine, National Academy Press, 2001.

TABLE 1.3 Ideal Characteristics for an Early Breast Cancer Detection Method

Detects early lesion
Available to population (48 million U.S. women age 40–70 y)
High sensitivity/high specificity (in all age groups)
Inexpensive
Noninvasive
Easily trainable and with high quality assurance
Decreases mortality
Infrared imaging meets all the above requirements

In 1999, the director for research in the Office of the Director, Defense Research and Engineering and the deputy assistant secretary of the Army for Installations and Environment; Environmental Safety and Occupational Health formulated a technology transfer program that would facilitate the use of advanced military technology and processes to breast cancer screening. Funds were provided by the Army and the project was funded through the Office of Naval Research (ONR).

A major milestone in the US program was the Tanks to Tumors workshop held in Arlington, VA, Dec. 4–5, 2001. The workshop was co-sponsored by Office of the Director, Defense Research and Engineering, Space and Sensor Technology Directorate; the deputy assistant secretary of the Army for Environment, Safety and Occupational Health; the Defense Advanced Research Projects Agency; and the Army Research Office. The purpose was to explore means for exploiting the technological opportunities in the integration of image processing, web-based database management and development, and IR sensor technology for the early detection of breast cancer. A second objective was to provide guidance to a program. The government speakers noted that significant military advances in thermal imaging, and automatic target recognition coupled with medical understanding of abnormal vascularity (angiogenesis) offer the prospect of automated detection from one to two years earlier than other, more costly and invasive screening methods.

There were compelling reasons for both military and civilian researchers to attend: (1) recognition of breast cancer as a major occupational health issue by key personnel such as Raymond Fatz, deputy assistant secretary of the Army for Installations and Environment; Environmental Safety and Occupational Health; (2) growing use of thermal imaging in military and civilian medicine (especially abroad); (3) maturation of military technology in automatic target recognition (ATR), ATR evaluation, and low-cost thermal imaging; (4) emerging transfer opportunities to and from the military. In particular, ATR assessment technology has developed image data management, dissemination, collaboration, and assessment tools for use by government and industrial developers of ATR software used to find military targets in thermal imagery. Such tools seem naturally suited for adaptation to the creation and use of a national database for IR breast cancer imagery and the evaluation of screening algorithms that would assist physicians in detecting the disease early. Finally, recent IR theories developed by civilian physicians indicate that the abnormal vascularity (angiogenesis) associated with the formation of breast tumors may be detected easily by IR cameras from one to five years before any other technique. Early detection has been shown to be the key to high survival probability.

The workshop involved specialists and leaders from the military R&D, academic, and medical communities. Together they covered a multidisciplinary range of topics: military IR sensor technology, automatic target recognition (ATR), smart image processing, database management, interactive web-based data management, IR imaging for screening of breast cancer, and related medical topics. Three panels of experts considered: (1) Image Processing and Medical Applications; (2) Website and Database; (3) Sensor Technology for Medical Applications. A subject area expert led each. The deliberations of each group were presented in a briefing to the plenary session of the final day. Their outputs were quite general; they still apply to the current program and are discussed below for the benefit of all future US efforts.

1.3.1 Image Processing and Medical Applications

This group focused on the algorithms (ATR approaches) and how to evaluate and use them. It advised that the clinical methods of collection must be able to support the most common ATR approaches, for example, single frame, change detection, multilook, and anomaly detection. They also provided detailed draft guidelines for controlled problem sets for ATR evaluation. Although they thought a multisensor approach would pay dividends, they stressed the need to quantify algorithm performance in a methodical way, starting with approaches that work with single IR images.

1.3.2 Website and Database

This panel concerned itself with the collection and management of an IR image database for breast cancer. It looked particularly at issues of data standards and security. It concluded that the OSD supported Virtual Distributed Laboratory (VDL), created within the OSD ATR Evaluation Program, is a very good model for the medical data repository to include collaborative software, image management software, evaluation concepts, data standards, security, bandwidth, and storage capacity. It also advised that camera calibration concepts and phantom targets be provided to help baseline performance and eliminate unknowns. It noted that privacy regulations would have to be dealt with in order to post the human data but suggested that this would complicate but not impede the formation of the database.

1.3.3 Sensor Technology for Medical Applications

The sensor panel started by pointing out that, if angiogenesis is a reliable early indicator of risk, then thermal imaging is ideally suited to detection at that stage. Current sensor performance is fully adequate. The group discussed calibration issues associated with hardware design and concluded that internal reference is desirable to insure that temperature differences are being measured accurately. However, they questioned the need for absolute temperature measurement; the plenary group offered no counter to this. This group also looked at the economics of thermal imaging, and concluded that recent military developments in uncooled thermal imaging systems at DARPA and the Army Night Vision and Electronic Sensing Division would allow the proliferation of IR cameras costing at most a few thousand dollars each. They cited China's installation of over 60 such cameras. The panel challenged ATR and algorithm developers to look at software methods to help simplify the sensor hardware, for example, frame-to-frame change detection to replace mechanical stabilization.

IR imaging for medical uses is a multidisciplinary technology and must include experts from very different fields if its full potential is to be realized. The Tanks to Tumors workshop is a model for future US efforts. It succeeded in bringing several different communities together—medical, military, academic, industrial, and engineering. These experts worked together to determine how the United States might adopt thermal imaging diagnostic technology in an orderly and demonstrable way for the early detection of breast cancer and other conditions. The panel recommendations will serve to guide the transition of military technology developments in ATR, the VDL, and IR sensors, to the civilian medical community. The result will be a new tool in the war against breast cancer—a major benefit to the military and civilian population. Detailed proceedings of this workshop are available from ACA, Falls Church, VA.

1.4 Guide to Thermal Imaging Systems for Medical Applications

1.4.1 Introduction

The purpose of this section is to provide the physician with an overview of the key features of thermal imaging systems and a brief discussion of the marketplace. It assumes that the reader is somewhat familiar with thermal imaging theory and terminology as well as the fundamentals of digital imaging. It contains a brief, modestly technical guide to buying sensor hardware, and a short list of active websites that can introduce the buyer to the current marketplace. It is intended primarily to aid the newcomer; however, advanced workers may also find some of these websites useful in seeking custom or cutting-edge capabilities in their quest to better understand the thermal phenomenology of breast cancer.

1.4.2 Background

As discussed elsewhere, the last decade has seen a resurgence of interest in thermal imaging for the early detection of breast cancer and other medical applications, both civilian and military. There was a brief period in the 1970s when thermal imaging became the subject of medical interest. That interest waned due to the combination of high prices and modest-to-marginal performance. Dramatic progress has been made in the intervening years; prices have dropped thanks to burgeoning military, domestic, and industrial use; performance has improved significantly; and new technology has emerged from Defense investments. Imaging electronics, digitization, image manipulation software, and automatic detection algorithms have emerged. Cameras can be had for prices that range from about $10,000 on up. These cameras can provide a significant capability for screening and data collection. The camera field is highly competitive; it is possible to rent, lease, or buy cameras from numerous vendors and manufacturers.

1.4.3 Applications and Image Formats

Currently, thermal imaging is being used for research and development into the phenomenology of breast cancer detection, and for screening and cuing in the multimodal diagnosis and tracking of breast cancer in patients. The least stressful and most affordable is the latter. Here, two types of formats can be of general utility: uncalibrated still pictures and simple uncalibrated video. Such formats can be stored and archived for future viewing. Use of such imagery for rigorous research and development is not recommended. Furthermore, there may be legal issues associated with the recording and collection of such imagery unless it is applied merely as a screening aid to the doctor rather than as a primary diagnostic tool. In other words, such imagery would provide the doctor with anecdotal support in future review of a patient's record. In this mode, the thermal imagery has the same diagnostic relevance as a stethoscope or endoscope, neither of which is routinely recorded in the doctor's office. Imagery so obtained would not carry the same diagnostic weight as a mammogram. Still cameras and video imagers of this kind are quite affordable and compact. They can be kept in a drawer or cabinet and be used for thermal viewing of many types of conditions including tumors, fractures, skin anomalies, circulation, and drug affects, to name a few. The marketplace is saturated with imagers under $10,000 that can provide adequate resolution and sensitivity for informal "eyeballing" the thermal features of interest. Virtually any image or video format is adequate for this kind of use.

For medical R&D, in which still imagery is to be archived, shared, and used for the testing of software and medical theories, or to explore phenomenology, it is important to collect calibrated still imagery in lossless archival formats (e.g., the so-called "raw" format that many digital cameras offer). It is thus desirable to purchase or rent a radiometric still camera with uncompressed standard formats or "raw" output that preserves the thermal calibration. This kind of imagery allows the medical center to put its collected thermal imagery into a standard format for distribution to the Virtual Distributed Laboratory and other interested medical centers. There are image manipulation software packages that can transform the imagery if need be. On the other hand, the data can be transmitted in any number of uncompressed formats and transformed by the data collection center. The use of standard formats is critical if medical research centers are to share common databases. There is no obvious need yet for video in R&D for breast cancer, although thermal video is being studied for many medical applications where dynamic phenomena are of interest.

1.4.4 Dynamic Range

The ability of a camera to preserve fine temperature detail in the presence of large scene temperature range is determined by its dynamic range. Dynamic range is determined by the camera's image digitization and formation electronics. Take care to use a camera that allocates an adequate number of bits to the digitization of the images. Most commercially available cameras use 12 bits or more per pixel. This is quite adequate to preserve fine detail in images of the human body. However, when collecting images, make sure there is nothing in the field of view of the camera that is dramatically cooler or hotter than the subject; that is, avoid scene temperature differences of more than roughly 30°C (e.g., lamps, refrigerators, or radiators in the background could cause trouble). This is analogous to trying to use a visible digital camera to capture a picture of a person standing next to headlights—electronic circuits may bloom or sacrifice detail of the scene near the bright lights. Although 12-bit digitization should preserve fine temperature differences at a 30°C delta, large temperature differences generally stress the image formation circuitry, and undesired artifacts may appear. Nevertheless, it is relatively easy to design a collection environment with a modest temperature range. A simple way to do this is to simply fill the camera field of view with the human subject. Experiment with the imaging arrangement before collecting a large body of imagery for archiving.

1.4.5 Resolution and Sensitivity

The two most important parameters for a thermal sensor are its sensitivity and resolution. The sensitivity is measured in degrees Celsius. Modest sensitivity is on the order of a tenth of a degree Celsius. Good sensitivity sensors can detect temperature differences up to 4 times lower or 0.025 degrees. This sensitivity is deemed valuable for medical diagnosis, since local temperature variations caused by tumors and angiogenesis are usually higher than this. The temperature resolution is analogous to the number of colors in a computer display or color photograph. The better the resolution, the smoother the temperature transitions will be. If the subject has sudden temperature gradients, those will be attributable to the subject and not the camera.

The spatial resolution of the sensor is determined primarily by the size of the imaging chip or pixel count. This parameter is exactly analogous to the world of proliferating digital photography. Just as a 4 megapixel digital camera can make sharper photos than a 2 megapixel camera, pixel count is a key element in the design of a medical camera. There are quite economical thermal cameras on the market with 320 × 240 pixels, and the images from such cameras can be quite adequate for informal screening; imagery may appear to be grainy if magnified unless the viewing area or field of view is reduced. By way of example, if the image is of the full chest area, about 18 inches, then a 320 pixel camera will provide the ability to resolve spatial features of about a 16th of an inch. If only the left breast is imaged, spatial features as low as 1/32 inch can be resolved. On the other hand, a 640 × 480 camera can cut these feature sizes in half. Good sensitivity and pixel count ensures that the medical images will contain useful thermal and spatial detail. In summary, although 320 × 240 imagery is quite adequate, larger pixel counts can provide more freedom for casual use, and are essential for R&D in medical centers. Although the military is developing megapixel arrays, they are not commercially available. Larger pixel counts have advantages for consumer digital photography and military applications, but there is no identified, clear need at this time for megapixel arrays in breast cancer detection. Avoid the quest for larger pixel counts unless there is a clear need. Temperature resolution should be a tenth of a degree or better.

1.4.6 Calibration

Another key feature is temperature calibration. Many thermal imaging systems are designed to detect temperature differences, not to map calibrated temperature. A camera that maps the actual surface temperature is a radiographic sensor. A reasonably good job of screening for tumors can be accomplished by only mapping local temperature differences. This application would amount to a third eye for the physician, aiding him in finding asymmetries and temperature anomalies—hot or cold spots. For example, checking circulation with thermal imaging amounts to looking for cold spots relative to the normally warm torso. However, if the physician intends to share his imagery with other doctors, or use the imagery for research, it is advisable to use a calibrated camera so that the meaning of the thermal differences can be quantified and separated from display settings and digital compression artifacts. For example, viewing the same image on two different computer displays may result in different assessments. But, if the imagery is calibrated so that each color or brightness is associated with a specific temperature, then doctors can be sure that they are viewing relevant imagery and accurate temperatures, not image artifacts.

It is critical that the calibration be stable and accurate enough to match the temperature sensitivity of the camera. Here caution is advised. Many radiometric cameras on the market are designed for industrial applications where large temperature differences are expected and the temperature of the object is well over 100°C; for example, the temperature difference may be 5°C at 600°C. In breast cancer, the temperature differences of interest are about a tenth of a degree at about 37°C. Therefore, the calibration method must be relevant for those parameters. Since the dynamic range of the breast cancer application is very small, the calibration method is simplified. More important are the temporal stability, temperature resolution, and accuracy of the calibration. Useful calibration parameters are: of 0.1°C resolution at 37°C, stability of 0.1°C per hour (drift), and accuracy of ±0.3°C. This means that the camera can measure a temperature difference of 0.1°C with an accuracy of ±0.3°C at body temperature. For example, suppose the breast is at 36.5°C; the camera might read 36.7°C.

Two methods of calibration are available—internal and external. External calibration devices are available from numerous sources. They are traceable to NIST and meet the above requirements. Prices are under $3000 for compact, portable devices. The drawback with external calibration is that it involves a second piece of equipment and more complex procedure for use. The thermal camera must be calibrated just prior to use and calibration imagery recorded, or the calibration source must be placed in the image while data are collected. The latter method is more reliable but it complicates the collection geometry.

Internal calibration is preferable because it simplifies the entire data collection process. However, radiometric still cameras with the above specifications are more expensive than uncalibrated cameras by $3000–$5000.

1.4.7 Single Band Imagers

Today there are thermal imaging sensors with suitable performance parameters. There are two distinct spectral bands that provide adequate thermal sensitivity for medical use: the medium wave IR band (MWIR) covers the electromagnetic spectrum from 3 to 5 μm in wavelength, approximately; the long-wave infrared band (LWIR) covers the wavelength spectrum from about 8 to 12 μm. There are advocates for both bands, and neither band offers a clear advantage over the other for medical applications, although the LWIR is rapidly becoming the most economical sensor technology. Some experimenters believe that there is merit to using both bands.

MWIR cameras are widely available and generally have more pixels, hence higher resolution for the same price. Phenomenology in this band has been quite effective in detecting small tumors and temperature asymmetries. MWIR sensors must be cooled to cryogenic temperatures as low as 77 K. Thermoelectric coolers are used for some MWIR sensors; they operate at 175–220 K depending on the design of the imaging chip. MWIR sensors not only respond to emitted radiation from thermal sources but they also sense radiation from broadband visible sources such as the sun. Images in this band can contain structure caused by reflected light rather than emitted radiation. Some care must be taken to minimize reflected light from broadband sources including incandescent light bulbs and sunlight. Unwanted light can cause shadows, reflections, and bright spots in the imagery. Care should be taken to avoid direct illumination of the subject by wideband artificial sources and sunlight. It is advisable to experiment with lighting geometries and sources before collecting data for the record. Moisturizing creams, sweat, and other skin surface coatings should also be avoided.

The cost of LWIR cameras has dropped dramatically since the advent of uncooled thermal imaging arrays. This is a dramatic difference between the current state of the art and what was available in the 1970s. Now, LWIR cameras are being proliferated and can be competitive in price and performance to the thermoelectrically cooled MWIR. Uncooled thermal cameras are compact and have good resolution and sensitivity. Cameras with 320 × 240 pixels can be purchased for well under $10,000. The trend in uncooled IR cameras is toward larger format arrays with smaller pixel size. Sensors with 640 × 480 pixels, individual pixel size of 25 μm and sensitivity of less than 50 mk, are on the market. These larger format arrays have thermal sensitivity equal to or greater than the previous generation of smaller format arrays, indicative of advances in the pixel design and manufacturing technology. Sensors in this band are far less likely to be affected by shadows, lighting, and reflections. Nevertheless, it is advisable to experiment with viewing geometry, ambient lighting, and skin condition before collecting data for the record and for dissemination.

Although large format arrays are becoming available, there is a cost advantage in smaller format arrays, either 320 × 240 or 160 × 120. The smaller format arrays integrated with signal processing algorithms can produce excellent images for short-term applications encountered in medical imaging. Also, in medical applications, where the targets are static, and data can be collected over several frames, frame integration and image enhancement software can provide excellent imaging, with moderate investment in camera equipment.

1.4.8 Emerging and Future Camera Technology

There are emerging developments that may soon provide for a richer set of observable phenomena in the thermography for breast cancer. Some researchers are already simultaneously collecting imagery in both the MWIR and LWIR bands. This is normally accomplished using two cameras at the same time. Developers of automatic screening algorithms are exploring schemes that compare the images in the two bands and emphasize the common elements of both to get greater confidence in detecting tumors. More sophisticated software (based on neural networks) learns what is important in both bands. Uncooled detector arrays have been demonstrated that operate in both bands simultaneously. It is likely that larger or well-endowed medical centers can order custom imagers with this capability this year.

Spectroscopic (hyperspectral) imaging in the thermal bands is also an important research topic. Investigators are looking for phenomenology that manifests itself in fine spectral detail. Since flesh is a thermally absorptive and scattering medium, it may be possible to detect unique signatures that help detect tumors. Interested parties should ask vendors if such cameras are available for lease or purchase.

In addition, research is underway to integrate tunable spectral filters with the imaging sensor. Since the filters are integrated with the sensor, the overall camera size and weight will be similar to conventional cameras. The spectral tuning range can be flexible, and depending upon filter design, tuning for either the 3–5 or 8–11 μm bands is feasible. A narrow spectral bandwidth, approximately 0.1 μm, will provide a comprehensive set of spectral data for in-depth analysis. Dual band imagery and multispectral data arguably provide physicians with a richer set of observables.

1.4.9 Summary Specifications

Table 1.4 summarizes the key parameters and their nominal values to use in shopping for a camera.

1.4.9.1 How to Begin

Those who are new to thermal phenomenology should carefully study the material in this handbook. Medical centers, researchers, and physicians seeking to purchase cameras and enter the field may wish to contact the authors or leading investigators mentioned in this handbook for advice before looking for sensor hardware. The participants in the Tanks to Tumors workshop and the MedATR program may already have the answers. If possible, compare advice from two or more of these experts before moving on; the experts do not agree on everything. They are currently using sensors and software suitable for building the VDL database. They may also be aware of public domain image screening

TABLE 1.4 Summary of Key Camera Parameters

	Application: Recording	Application: Informal
Format	Digital stills	Video or stills
Compression	None	As provided by mfr
Digitization (dynamic range)	12 bits or more	12 bits nominal
Pixels (array size)	320×240 up to 640×480	320×240
Sensitivity	0.04°C, 0.1°C max	0.1°C
Calibration accuracy	±0.3°C	Not required
Calibration range	Room and body temperature	Not required
Calibration resolution	0.1°C	Not required
Spectral band	MWIR or LWIR	MWIR or LWIR

software. Once advice has been collected, the potential buyer should begin shopping at the one or more of the websites listed in this section. Do not rely on the website alone. Most vendors provide contact information so that the purchaser may discuss imaging needs with a consultant. Take advantage of these advisory services to shop around and survey the field. Researchers may also wish to contact government and university experts before deciding on a camera. Finally, many of the vendors below offer custom sensor design services. Some vendors may be willing to lease or loan equipment for evaluation. High-end, leading-edge researchers may need to contact component developers at companies such as Raytheon, DRS, BAE, or SOFRADIR to see if the state of the art supports their specific needs.

1.4.9.2 Supplier Websites

The reader is advised that all references to brand names or specific manufacturers do not imply an endorsement of the vendor, producer, or its products. Likewise, the list is not a complete survey; we apologize for any omissions. Since the last printing, FLIR Systems Inc. acquired Indigo Systems and L-3 acquired the Raytheon thermal imaging business.

http://www.thermal-eye.com/
http://medicalir.com/
http://www.infrared-camera-rentals.com/
http://www.electrophysics.com/Browse/Brw_AllProductLineCategory.asp
http://www.cantronic.com/ir860.html
http://www.nationalinfrared.com/Medical_Imaging.php
http://www.flirthermography.com/cameras/all_cameras.asp
http://www.mikroninst.com/
http://www.baesystems.com/ProductsServices/bae_prod_s2_mim500.html
http://www.flir.com/US/
http://x26.com/articles.html
http://www.infraredsolutions.com/
http://www.isgfire.com/
http://www.infrared.com/
http://www.sofradir.com/
http://www.drs.com/Products/RSTA/MX2A.aspx

1.5 Summary, Conclusions, and Recommendations

Today, medical IR is being backed by more clinical research worldwide where state-of-the-art equipment is being used. Focus must be placed on the quantification of clinical data, standardization, effective training with high-quality assurance, collaborations, and more publications in leading peer-reviewed medical journals.

For an effective integration of twenty-first century technologies for IR imaging we need to focus on the following areas:

- IR camera systems designed for medical diagnostics
- Advanced image processing
- Image analysis techniques
- High-speed computers
- Computer-aided detection (CAD)
- Knowledge-based databases
- Telemedicine

Other areas of importance are

- Effective clinical use
- Protocol-based image acquisition
- Image interpretation
- System operation and calibration
- Training
- Continued research in the pathophysiological nature of thermal signatures
- Quantification of clinical data

In conclusion, this noninvasive, nonionizing imaging modality can provide added value to the present multi-imaging clinical setting. A thermal image measures metabolic activity in the tissue and thus can noninvasively detect abnormalities very early. It is well known that early detection leads to enhanced survivability and great reduction in health care costs. With these becoming exorbitant, this would be of great value. Besides its usefulness at this stage, a second critical benefit is that it has the capability to noninvasively monitor the efficacy of therapies [12].

References

1. Diakides, N.A. (Guest Editor): Special issue on medical infrared imaging, *IEEE/Engineering in Medicine and Biology*, 17(4), Jul/Aug 1998.
2. Diakides, N.A. (Guest Editor): Special issue on medical infrared imaging, *IEEE/Engineering in Medicine and Biology*, 19(3), May/Jun 2000.
3. Diakides, N.A. (Guest Editor): Special issue on medical infrared imaging, *IEEE/Engineering in Medicine and Biology*, 21(6), Nov/Dec 2002.
4. Paul, J.L., Lupo, J.C., From tanks to tumors: Applications of infrared imaging and automatic target recognition image processing for early detection of breast cancer, *Special Issue on Medical Infrared Imaging, IEEE/Engineering in Medicine and Biology*, 21(6), 34–35, Nov/Dec 2002.
5. Diakides, N.A., Medical applications of IR focal plane arrays, Final Progress Report, U.S. Army Research Office, Contract DAAH04–94-C-0020, Mar 1998.
6. Diakides, N.A., Application of army IR technology to medical IR imaging, Technical Report, U.S. Army Research Office Contract DAAH04-96-C-0086 (TCN 97–143), Aug 1999.
7. Diakides, N.A., Exploitation of infrared imaging for medicine, Final Progress Report, U.S. Army Research Office, Contract DAAG55-98-0035, Jan 2001.
8. Diakides, N.A., Medical IR imaging and image processing, Final Report U.S. Army Research Office, Contract DAAH04-96-C-0086 (TNC 01041), Oct 2003.
9. Anbar, M., *Quantitative Dynamic Telethermometry in Medical Diagnosis and Management*, CRC Press, Boca Raton, FL, 1994.
10. Ammer, K. (Ed. In Chief), *Journal of Thermology, Intl.*, 14(1), Jan 2004.

11. Balageas, D., Busse, G., Carlomagno, C., Wiecek, B. (Eds.), *Proceedings of Quantitative Infrared Thermography 4*, Technical University of Lodz, Poland, 1998.
12. Hassan, M. et al., Quantitative assessment of tumor vasculature and response to therapy in Kaposi's sarcoma using functional noninvasive imaging, *Technology in Cancer Research and Treatment*, 3(5), 451–457, Oct 2004.

Historical Development of Thermometry and Thermal Imaging in Medicine

Francis J. Ring
University of Glamorgan

Bryan F. Jones
University of Glamorgan

Fever was the most frequently occurring condition in early medical observation. From the early days of Hippocrates, when it is said that wet mud was used on the skin to observe fast drying over a tumorous swelling, physicians have recognized the importance of a raised temperature. For centuries, this remained a subjective skill, and the concept of measuring temperature was not developed until the sixteenth century. Galileo made his famous thermoscope from a glass tube, which functioned as an unsealed thermometer. It was affected by atmospheric pressure as a result.

In modern terms we now describe heat transfer by three main modes. The first is conduction, requiring contact between the object and the sensor to enable the flow of thermal energy. The second mode of heat transfer is convection where the flow of a hot mass transfers thermal energy. The third is radiation. The latter two led to remote detection methods.

Thermometry developed slowly from Galileo's experiments. There were Florentine and Venetian glassblowers in Italy who made sealed glass containers of various shapes, which were tied onto the body surface. The temperature of an object was assessed by the rising or falling of small beads or seeds within the fluid inside the container. Huygens, Roemer, and Fahrenheit all proposed the need for a calibrated scale in the late seventeenth and early eighteenth century. Celsius did propose a centigrade scale based on ice and boiling water. He strangely suggested that boiling water should be zero, and melting ice 100 on his scale. It was the Danish biologist Linnaeus in 1750 who proposed the reversal of this scale, as it is known today. Although International Standards have given the term Celsius to the 0 to 100 scale today, strictly speaking it would be historically accurate to refer to degrees Linnaeus or centigrade [1].

The clinical thermometer, which has been universally used in medicine for over 130 years, was developed by Dr. Carl Wunderlich in 1868. This is essentially a maximum thermometer with a limited scale around the normal internal body temperature of 37°C or 98.4°F. Wunderlich's treatise on body temperature in health and disease is a masterpiece of painstaking work over many years. He charted the progress of all his patients daily, and sometimes two or three times during the day. His thesis was written in German for Leipzig University and was also translated into English in the late nineteenth century [2]. The significance of body temperature lies in the fact that humans are homeotherms who are capable of maintaining a constant temperature that is different from that of the surroundings. This is essential to the preservation of a relatively constant environment within the body known as homeostasis. Changes in temperature of more than a few degrees either way is a clear indicator of a bodily dysfunction; temperature variations outside this range may disrupt the essential chemical processes in the body.

Today, there has been a move away from glass thermometers in many countries, giving rise to more disposable thermocouple systems for routine clinical use.

Liquid crystal sensors for temperature became available in usable form in the 1960s. Originally the crystalline substances were painted on the skin that had previously been coated with black paint. Three of four colors became visible if the paint was at the critical temperature range for the subject. Micro-encapsulation of these substances, that are primarily cholesteric esters, resulted in plastic sheet detectors. Later these sheets were mounted on a soft latex base to mold to the skin under air pressure using a cushion with a rigid clear window. Polaroid photography was then used to record the color pattern while the sensor remained in contact. The system was reusable and inexpensive. However, sensitivity declined over 1–2 years from the date of manufacture, and many different pictures were required to obtain a subjective pattern of skin temperature [3].

Convection currents of heat emitted by the human body have been imaged by a technique called Schlieren photography. The change in refractive index with density in the warm air around the body is made visible by special illumination. This method has been used to monitor heat loss in experimental subjects, especially in the design of protective clothing for people working in extreme physical environments.

Heat transfer by radiation is of great value in medicine. The human body surface requires variable degrees of heat exchange with the environment as part of the normal thermo-regulatory process. Most of this heat transfer occurs in the infrared, which can be imaged by electronic thermal imaging [4]. Infrared radiation was discovered in 1800 when Sir William Herschel performed his famous experiment to measure heat beyond the visible spectrum (see Figure 2.1). Nearly 200 years before, Italian observers

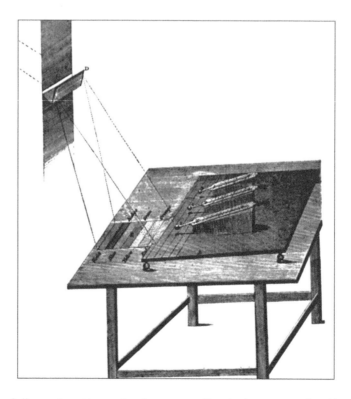

FIGURE 2.1   Herschel's experiment to examine the presence of heat in the spectrum, found beyond the visible red.

had noted the presence of reflected heat. John Della Porta in 1698 observed that when a candle was lit and placed before a large silver bowl in church, that he could sense the heat on his face. When he altered the positions of the candle, bowl, and his face, the sensation of heat was lost.

William Herschel, in a series of careful experiments, showed that not only was there a "dark heat" present, but that heat itself behaved like light, it could be reflected and refracted under the right conditions. William's only son, John Herschel, repeated some experiments after his father's death, and successfully made an image using solar radiation. This he called a "thermogram," a term still in use today to describe an image made by thermal radiation. John Herschel's thermogram was made by focusing solar radiation with a lens onto to a suspension of carbon particles in alcohol. This process is known as evaporography [5].

A major development came in the early 1940s with the first electronic sensor for infrared radiation (see Figure 2.2). Rudimentary night vision systems were produced toward the end of World War II for use by snipers. The electrons from near-infrared cathodes were directed onto visible phosphors which converted the infrared radiation into visible light. Sniperscope devices, based on this principle, were provided for soldiers in the Pacific in 1945, but found little use.

At about the same time, another device was made from indium antimonide; this was mounted at the base of a small Dewar vessel to allow cooling with liquid nitrogen. A cumbersome device such as this, which required a constant supply of liquid nitrogen, was clearly impractical for battlefield use but could be used with only minor inconvenience in a hospital. The first medical images taken with a British prototype system, the "Pyroscan" (see Figure 2.3) were made at The Middlesex Hospital in London and The Royal National Hospital for Rheumatic Diseases in Bath between 1959 and 1961. By modern standards, these thermograms were very crude.

In the meantime, the cascade image tube, that had been pioneered during World War II in Germany, had been developed by RCA into a multialkali photocathode tube whose performance exceeded expectations. These strides in technology were motivated by military needs in Vietnam; they were classified and, therefore, unavailable to clinicians. However, a mark 2 Pyroscan was made for medical use in 1962,

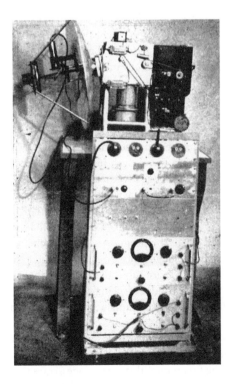

FIGURE 2.2 First English IR prototype camera.

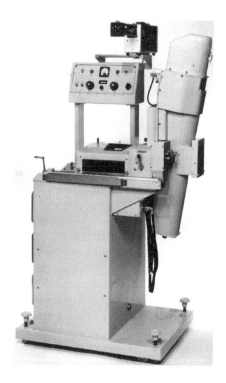

FIGURE 2.3 1960s development from the prototype in 1942. Pyroscan made for medical trials.

with improved images. The mechanical scanning was slow and each image needed from 2 to 5 min to record. The final picture was written line by line on electro-sensitive paper. In the 1970s, the U.S. Military sponsored the development of a multielement detector array that was to form the basis of a real-time framing imager. This led to the targeting and navigation system known as Forward Looking InfraRed (FLIR) systems which had the added advantage of being able to detect warm objects through smoke and fog.

During this time the potential for thermal imaging in medicine was being explored in an increasing number of centers. Earlier work by the American physiologist J. Hardy had shown that the human skin, regardless of color, is a highly efficient radiator with an emissivity of 0.98 which is close to that of a perfect black body. Even so, the normal temperature of skin in the region of 20–30°C generated low intensities of infrared radiation at about 10 μm wavelength [6]. The detection of such low intensities at these wavelengths presented a considerable challenge to the technology of the day. Cancer detection was a high priority subject and hopes that this new technique would be a tool for screening breast cancer provided the motivation to develop detectors. Many centers across Europe, the United States, and Japan became involved. In the United Kingdom, a British surgeon, K. Lloyd Williams showed that many tumors are hot and the hotter the tumor, the worse the prognosis. By this time, the images were displayed on a cathode-ray screen in black and white. Image processing by computer had not arrived, so much discussion was given to schemes to score the images subjectively, and to look for hot spots and asymmetry of temperature in the breast. This was confounded by changes in the breast through the menstrual cycle in younger women. The use of false color thermograms was only possible by photography at this time. A series of bright isotherms were manually ranged across the temperature span of the image, each being exposed through a different color filter, and superimposed on a single frame of film.

Improvements in infrared technology were forging ahead at the behest of the U.S. Military during the 1970s. At Fort Belvoir, some of the first monolithic laser diode arrays were designed and produced with a capability of generating 500 W pulses at 15 kHz at room temperature. These lasers were able to image objects at distances of 3 km. Attention then turned to solid state, gas, and tunable lasers which were used in a wide range of applications.

By the mid-1970s, computer technology made a widespread impact with the introduction of smaller mini and microcomputers at affordable prices. The first "personal" computer systems had arrived. In Bath, a special system for nuclear medicine made in Sweden was adapted for thermal imaging. A color screen was provided to display the digitized image. The processor was a PDP8, and the program was loaded every day from paper-tape. With computerization many problems began to be resolved. The images were archived in digital form, standard regions of interest could be selected, and temperature measurements obtained from the images. Manufacturers of thermal imaging equipment slowly adapted to the call for quantification and some sold thermal radiation calibration sources to their customers to aid the standardization of technique. Workshops that had started in the late 1960s became a regular feature, and the European Thermographic Association was formed with a major conference in Amsterdam in 1974. Apart from a range of physiological and medical applications groups were formed to formulate guidelines for good practice. This included the requirements for patient preparation, conditions for thermal imaging and criteria for the use of thermal imaging in medicine and pharmacology [7,8]. At the IEEE EMBS conference in Amsterdam some 20 years later in 1996, Dr. N. Diakides facilitated the production of a CD-ROM of the early, seminal papers on infrared imaging in medicine that had been published in *ACTA Thermographica* and the *Journal of Thermology*. This CD was sponsored by the U.S. Office of Technology Applications, Ballistic Missile Defence Organisation and the U.S. National Technology Transfer Center Washington Operations and is available from the authors at the Medical Imaging Research Group at the University of Glamorgan, UK [9]. The archive of papers may also be searched online at the Medical Imaging Group's website [9].

A thermal index was devised in Bath to provide clinicians with a simplified measure of inflammation [10]. A normal range of values was established for ankles, elbows, hands, and knees, with raised values obtained in osteoarthritic joints and higher values still in rheumatoid arthritis. A series of clinical trials with nonsteroid, anti-inflammatory, oral drugs, and steroid analogues for joint injection was published using the index to document the course of treatment [11].

Improvements in thermal imaging cameras have had a major impact, both on image quality and speed of image capture. Early single-element detectors were dependent on optical mechanical scanning. Both spatial and thermal image resolutions were inversely dependent on scanning speed. The Bofors and some American imagers scanned at 1–4 frames/s. AGA cameras were faster at 16 frames/s, and used interlacing to smooth the image. Multielement arrays were developed in the United Kingdom and were employed in cameras made by EMI and Rank. Alignment of the elements was critical, and a poorly aligned array produced characteristic banding in the image. Professor Tom Elliott FRS solved this problem when he designed and produced the first significant detector for faster high-resolution images that subsequently became known as the Signal Processing In The Element (Sprite) detector. Rank Taylor Hobson used the Sprite in the high-resolution system called Talytherm. This camera also had a high specification Infrared zoom lens, with a macro attachment. Superb images of sweat pore function, eyes with contact lenses, and skin pathology were recorded with this system.

With the end of the cold war, the greatly improved military technology was declassified and its use for medical applications was encouraged. As a result, the first focal plane array detectors came from the multielement arrays, with increasing numbers of pixel/elements, yielding high resolution at video frame rates. Uncooled bolometer arrays have also been shown to be adequate for many medical applications. Without the need for electronic cooling systems these cameras are almost maintenance free. Good software with enhancement and analysis is now expected in thermal imaging. Many commercial systems use general imaging software, which is primarily designed for industrial users of the technique. A few dedicated medical software packages have been produced, which can even enhance the images from the older cameras. CTHERM is one such package that is a robust and almost universally usable program for medical thermography [9]. As standardization of image capture and analysis becomes more widely accepted, the ability to manage the images and, if necessary, to transmit them over an intranet or Internet for communication becomes paramount. Future developments will enable the operator of thermal imaging to use reference images and reference data as a diagnostic aid. This, however, depends on the level of standardization that can be provided by the manufacturers, and by the operators themselves in the performance of their technique [12].

Modern thermal imaging is already digital and quantifiable, and ready for the integration into anticipated hospital and clinical computer networks.

References

1. Ring, E.F.J., *The History of Thermal Imaging in the Thermal Image in Medicine and Biology*, eds. Ammer, K. and Ring, E.F.J., pp. 13–20. Uhlen Verlag, Vienna, 1995.
2. Wunderlich, C.A., *On the Temperature in Diseases, a Manual of Medical Thermometry*. Translated from the second German edition by Bathurst Woodman, W., The New Sydenham Society, London, 1871.
3. Flesch, U., Thermographic techniques with liquid crystals in medicine. In *Recent Advances in Medical Thermology*, eds. Ring, E.F.J. and Phillips, B., pp. 283–299. Plenum Press, New York, 1984.
4. Houdas, Y. and Ring, E.F.J., *Human Body Temperature, Its Measurement and Regulation.* Plenum Press, New York, 1982.
5. Ring, E.F.J., The discovery of infrared radiation in 1800. *Imaging Science Journal*, 48, 1–8, 2000.
6. Jones, B.F., A reappraisal of infrared thermal image analysis in medicine. *IEEE Transactions on Medical Imaging*, 17, 1019–1027, 1998.
7. Engel, J.M., Cosh, J.A., Ring, E.F.J. et al., Thermography in locomotor diseases: Recommended procedure. *European Journal of Rheumatology and Inflammation*, 2, 299–306, 1979.
8. Ring, E.F.J., Engel, J.M., and Page-Thomas, D.P., Thermological methods in clinical pharmacology. *International Journal of Clinical Pharmacology*, 22, 20–24, 1984.
9. CTHERM website www.medimaging.org.
10. Collins, A.J., Ring, E.J.F., Cash, J.A., and Brown, P.A., Quantification of thermography in arthritis using multi-isothermal analysis: I. The thermographic index. *Annals of the Rheumatic Diseases*, 33, 113–115, 1974.
11. Bacon, P.A., Ring, E.F.J., and Collins, A.J., Thermography in the assessment of antirheumatic agents. In *Rheumatoid Arthritis*, eds. J.L. Gordon and B.L. Hazleman, pp. 105–110. Elsevier/North-Holland Biochemical Press, Amsterdam, 1977.
12. Ring, E.F.J. and Ammer, K., The technique of infrared imaging in medicine. *Thermology International*, 10, 7–14, 2000.

Infrared Detectors and Detector Arrays

Paul R. Norton
U.S. Army CERDEC Night Vision and Electronic Sensors Directorate

Stuart B. Horn
U.S. Army CERDEC Night Vision and Electronic Sensors Directorate

Joseph G. Pellegrino
U.S. Army CERDEC Night Vision and Electronic Sensors Directorate

Philip Perconti
U.S. Army CERDEC Night Vision and Electronic Sensors Directorate

There are two general classes of detectors: *photon* (or quantum) and *thermal* detectors [1,2]. Photon detectors convert absorbed photon energy into released electrons (from their bound states to conduction states). The material bandgap describes the energy necessary to transition a charge carrier from the valence band to the conduction band. The change in charge carrier state changes the electrical properties of the material. These electrical property variations are measured to determine the amount of incident optical power. Thermal detectors absorb energy over a broad band of wavelengths. The energy absorbed by a detector causes the temperature of the material to increase. Thermal detectors have at least one inherent electrical property that changes with temperature. This temperature-related property is measured electrically to determine the power on the detector. Commercial infrared imaging systems suitable for medical applications use both types of detectors. We begin by describing the physical mechanism employed by these two detector types.

3.1 Photon Detectors

Infrared radiation consists of a flux of photons, the quantum-mechanical elements of all electromagnetic radiation. The energy of the photon is given by

$$E_{\text{ph}} = h\nu = hc/\lambda = 1.986 \times 10^{-19}/\lambda \text{ J}/\mu\text{m} \tag{3.1}$$

where h is the Planck's constant, c is the speed of light, and λ is the wavelength of the infrared photon in micrometers (μm).

Photon detectors respond by elevating a bound electron in a material to a free or conductive state. Two types of photon detectors are produced for the commercial market:

- Photoconductive
- Photovoltaic

3.1.1 Photoconductive Detectors

The mechanism of photoconductive detectors is based upon the excitation of bound electrons to a mobile state where they can move freely through the material. The increase in the number of conductive electrons, n, created by the photon flux, Φ_0, allows more current to flow when the detective element is used in a bias circuit having an electric field E. The photoconductive detector element having dimensions of length L, width W, and thickness t is represented in Figure 3.1.

Figure 3.2 illustrates how the current–voltage characteristics of a photoconductor change with incident photon flux (Chapter 4).

The response of a photoconductive detector can be written as

$$R = \frac{\eta q R E \tau (\mu_n + \mu_p)}{E_{ph} L} (V/W) \tag{3.2}$$

where R is the response in volts per watt, η is the quantum efficiency in electrons per photon, q is the charge of an electron, R is the resistance of the detector element, τ is the lifetime of a photoexcited electron, and μ_n and μ_p are the mobilities of the electrons and holes in the material in volts per square centimeter per second.

Noise in photoconductors is the square root averaged sum of terms from three sources:

- Johnson noise
- Thermal generation–recombination
- Photon generation–recombination

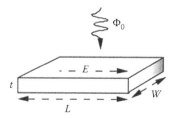

FIGURE 3.1 Photoconductive detector geometry.

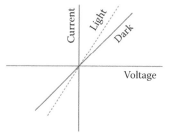

FIGURE 3.2 Current–voltage characteristics of a photoconductive detector.

Expressions for the total noise and each of the noise terms are given in Equations 3.3 through 3.6:

$$V_{noise} = \sqrt{V_{Johnson}^2 + V_{ph\,g-r}^2 + V_{th\,g-r}^2} \tag{3.3}$$

$$V_{Johnson} = \sqrt{4kTR} \tag{3.4}$$

$$V_{ph\,g-r} = \frac{\sqrt{\eta\phi(WL)2qRE\tau(\mu_n + \mu_p)}}{L} \tag{3.5}$$

$$V_{th\,g-r} = \sqrt{\frac{np}{n+p}\tau\left(\frac{Wt}{L}\right)2qRE(\mu_n + \mu_p)} \tag{3.6}$$

The figure of merit for infrared detectors is called D^*. The units of D^* are cm $(Hz)^{1/2}/W$, but are most commonly referred to as Jones. D^* is the detector's signal-to-noise ratio (SNR), normalized to an area of 1 cm^2, to a noise bandwidth of 1 Hz, and to a signal level of 1 W at the peak of the detectors response. The equation for D^* is

$$D_{peak}^* = \frac{R}{V_{noise}}\sqrt{WL} \text{ (Jones)} \tag{3.7}$$

where W and L are defined in Figure 3.1.

A special condition of D^* for a photoconductor is noted when the noise is dominated by the photon noise term. This is a condition in which the D^* is maximum.

$$D_{blip}^* = \frac{\lambda}{2hc}\sqrt{\frac{\eta}{E_{ph}}} \tag{3.8}$$

where "blip" denotes background-limited photodetector.

3.1.2 Photovoltaic Detectors

The mechanism of photovoltaic detectors is based on the collection of photoexcited carriers by a diode junction. Photovoltaic detectors are the most commonly used photon detectors for imaging arrays in current production. An example of the structure of detectors in such an array is illustrated in Figure 3.3

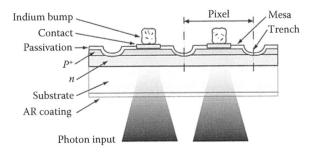

FIGURE 3.3 Photovoltaic detector structure example for mesa diodes.

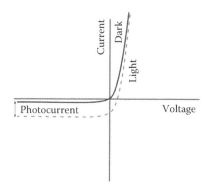

FIGURE 3.4   Current–voltage characteristics of a photovoltaic detector.

for a mesa photodiode. Photons are incident from the optically transparent detector substrate side and are absorbed in the *n*-type material layer. Absorbed photons create a pair of carriers, an electron and a hole. The hole diffuses to the *p*-type side of the junction creating a photocurrent. A contact on the *p*-type side of the junction is connected to an indium bump that mates to an amplifier in a readout circuit where the signal is stored and conveyed to a display during each display frame. A common contact is made to the *n*-type layer at the edge of the detector array. Adjacent diodes are isolated electrically from each other by a mesa etch cutting the *p*-type layer into islands.

Figure 3.4 illustrates how the current–voltage characteristics of a photodiode change with incident photon flux (Chapter 4).

The current of the photodiode can be expressed as

$$I = I_0(e^{qV/kT} - 1) - I_{photo} \tag{3.9}$$

where I_0 is reverse-bias leakage current and I_{photo} is the photoinduced current. The photocurrent is given by

$$I = I_0(e^{qV/kT} - 1) - I_{photo} \tag{3.10}$$

where Φ_0 is the photon flux in photons/cm²/s and A is the detector area.

Detector noise in a photodiode includes three terms: Johnson noise, thermal diffusion generation and recombination noise, and photon generation and recombination. The Johnson noise term is written in terms of the detector resistance $dI/dV = R_0$ at zero bias as

$$i_{Johnson} = \sqrt{4kT/R_0} \tag{3.11}$$

where k is the Boltzmann's constant and T is the detector temperature. The thermal diffusion current is given by

$$i_{diffusion\ noise} = q\sqrt{2I_s\left[\exp\left(\frac{eV}{kT}\right) - 1\right]} \tag{3.12}$$

where the saturation current, I_s, is given by

$$I_s = q n_i^2 \left[\frac{1}{N_a} \sqrt{\frac{D_n}{\tau_{n_0}}} + \frac{1}{N_d} \sqrt{\frac{D_p}{\tau_{p_0}}} \right] \tag{3.13}$$

where N_a and N_d are the concentration of p- and n-type dopants on either side of the diode junction, τ_{n0} and τ_{p0} are the carrier lifetimes, and D_n and D_p are the diffusion constants on either side of the junction, respectively.

The photon generation–recombination current noise is given by

$$i_{\text{photon noise}} = q \sqrt{2 \eta \Phi_0} \tag{3.14}$$

When the junction is at zero bias, the photodiode D^* is given by

$$D_\lambda^* = \frac{\lambda}{hc} \eta e \frac{1}{\left[(4kT/R_0 A) + 2e^2 \eta \right]} \tag{3.15}$$

In the special case of a photodiode that is operated without sufficient cooling, the maximum D^* may be limited by the dark current or leakage current of the junction. The expression for D^* in this case, written in terms of the junction-resistance area product, $R_0 A$, is given by

$$D_\lambda^* = \frac{\lambda}{hc} \eta e \sqrt{\frac{R_0 A}{4kT}} \tag{3.16}$$

Figure 3.5 illustrates how D^* is limited by the $R_0 A$ product for the case of dark-current-limited detector conditions.

For the ideal case where the noise is dominated by the photon flux in the background scene, the peak D^* is given by

$$D_\lambda^* = \frac{\lambda}{hc} \sqrt{\frac{\eta}{2 E_{\text{ph}}}} \tag{3.17}$$

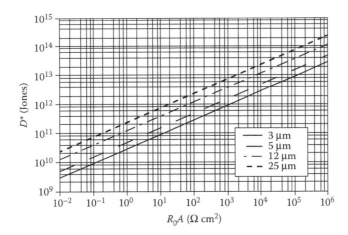

FIGURE 3.5 D^* as a function of the detector resistance–area product, $R_0 A$. This condition applies when detector performance is limited by dark current.

Comparing this limit with that for a photoconductive detector in Equation 3.8, we see that the background-limited D^* for a photodiode is higher by a factor of square root of 2 ($\sqrt{2}$).

3.2 Thermal Detectors

Thermal detectors operate by converting the incoming photon flux to heat [3]. The heat input causes the thermal detector's temperature to rise and this change in temperature is sensed by a bolometer. A bolometer element operates by changing its resistance as its temperature is changed. A bias circuit across the bolometer can be used to convert the changing current to a signal output.

The coefficient α is used to compare the sensitivity of different bolometer materials and is given by

$$\alpha = \frac{1}{R_d}\frac{dR}{dT} \tag{3.18}$$

where R_d is the resistance of the bolometer element, and dR/dT is the change in resistance per unit change in temperature. Typical values of α are 2–3%.

Theoretically, the bolometer structure can be represented as illustrated in Figure 3.6. The rise in temperature due to a heat flux ϕ_e is given by

$$\Delta T = \frac{\eta P_0}{G(1 + \omega^2\tau^2)^{1/2}} \tag{3.19}$$

where P_0 is the radiant power of the signal in watts, G is the thermal conductance (K/W), h is the percentage of flux absorbed, and ω is the angular frequency of the signal. The bolometer time constant, τ, is determined by

$$\tau = \frac{C}{G} \tag{3.20}$$

where C is the heat capacity of the detector element.

The sensitivity or D^* of a thermal detector is limited by variations in the detector temperature caused by fluctuations in the absorption and radiation of heat between the detector element and the background. Sensitive thermal detectors must minimize competing mechanisms for heat loss by the element, namely, convection and conduction.

Convection by air is eliminated by isolating the detector in a vacuum. If the conductive heat losses were less than those due to radiation, then the limiting D^* would be given by

$$D^*(T,f) = 2.8 \times 10^{16}\sqrt{\frac{\varepsilon}{T_2^5 + T_1^5}} \text{ Jones} \tag{3.21}$$

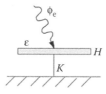

FIGURE 3.6 Abstract bolometer detector structure, where C is the thermal capacitance, G is the thermal conductance, and o is the emissivity of the surface. ϕ_e represents the energy flux in W/cm^2.

where T_1 is the detector temperature, T_2 the background temperature, and o the value of the detector's emissivity and equally its absorption. For the usual case of both the detector and background temperature at normal ambient, 300 K, the limiting D^* is 1.8×10^{10} Jones.

Bolometer operation is constrained by the requirement that the response time of the detector be compatible with the frame rate of the imaging system. Most bolometer cameras operate at a 30 Hz frame rate—33 ms frame. Response times of the bolometer are usually designed to be on the order of 10 ms. This gives the element a fast-enough response to follow scenes with rapidly varying temperatures without objectionable image smearing.

3.3 Detector Materials

The most popular commercial cameras for thermal imaging today use the following detector materials [4]:

- InSb for 5 μm medium-wavelength infrared (MWIR) imaging
- $Hg_{1-x}Cd_xTe$ alloys for 5 and 10 μm long-wavelength infrared (LWIR) imaging
- Quantum well detectors for 5 and 10 μm imaging
- Uncooled bolometers for 10 μm imaging

We will now review a few of the basic properties of these detector types.

Photovoltaic InSb remains a popular detector for the MWIR spectral band operating at a temperature of 80 K [5,6]. The detector's spectral response at 80 K is shown in Figure 3.7. The spectral response cutoff is about 5.5 μm at 80 K, a good match to the MWIR spectral transmission of the atmosphere. As the operating temperature of InSb is raised, the spectral response extends to longer wavelengths and the dark current increases accordingly. It is thus not normally used above about 100 K. At 80 K, the R_0A product of InSb detectors is typically in the range of 10^5–10^6 Ω cm²—see Equation 3.16 and Figure 3.5 for reference.

Crystals of InSb are grown in bulk boules up to 3 in. in diameter. InSb materials is highly uniform and combined with a planar-implanted process in which the device geometry is precisely controlled, the resulting detector array responsivity is good to excellent. Devices are usually made with a *p/n* diode polarity using diffusion or ion implantation. Staring arrays of backside illuminated, direct hybrid InSb detectors in 256 # 256, 240 # 320, 480 # 640, 512 # 640, and 1024 # 1024 formats are available from a number of vendors.

HgCdTe detectors are commercially available to cover the spectral range from 1 to 12 μm [7–13]. Figure 3.8 illustrates representative spectral response from photovoltaic devices, the most commonly used type. Crystals of HgCdTe today are mostly grown in thin epitaxial layers on infrared-transparent CdZnTe crystals. Short-wavelength infrared (SWIR) and MWIR material can also be grown on Si substrates with CdZnTe buffer layers. Growth of the epitaxial layers is by liquid-phase melts, molecular

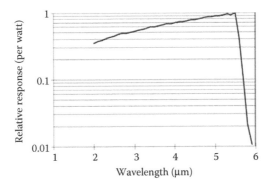

FIGURE 3.7 Spectral response per watt of an InSb detector at 80 K.

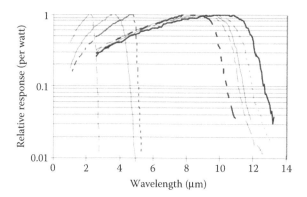

FIGURE 3.8 Representative spectral response curves for a variety of HgCdTe alloy detectors. Spectral cutoff can be varied over the SWIR, MWIR, and LWIR regions.

beams, or by chemical vapor deposition. Substrate dimensions of CdZnTe crystals are in the 25–50 cm^2 range and Si wafers up to 5–6 in. (12.5–15 cm) in diameter have been used for this purpose. The device structure for a typical HgCdTe photodiode is shown in Figure 3.3.

At 80 K, the leakage current of HgCdTe is small enough to provide both MWIR and LWIR detectors that can be photon-noise dominated. Figure 3.9 shows the R_0A product of representative diodes for wavelengths ranging from 4 to 12 μm.

The versatility of HgCdTe detector material is directly related to being able to grow a broad range of alloy compositions in order to optimize the response at a particular wavelength. Alloys are usually adjusted to provide response in the 1–3 μm SWIR, 3–5 μm MWIR, or the 8–12 μm LWIR spectral regions. Short-wavelength detectors can operate uncooled, or with thermoelectric coolers that have no moving parts. Medium- and long-wavelength detectors are generally operated at 80 K using a cryogenic cooler engine. HgCdTe detectors in 256 # 256, 240 # 320, 480 # 640, and 512 # 640 formats are available from a number of vendors.

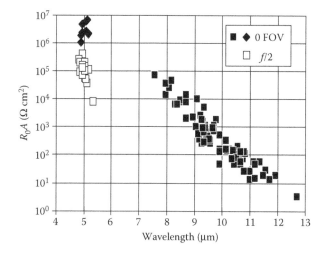

FIGURE 3.9 Values of R_0A product as a function of wavelength for HgCdTe photodiodes. Note that the R_0A product varies slightly with illumination—0° field-of-view compared with $f/2$—especially for shorter-wavelength devices.

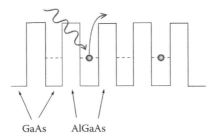

GaAs   AlGaAs

FIGURE 3.10 Quantum wells generate bound states for electrons in the conduction band. The conduction bands for a QWIP structure are shown consisting of $Al_xGa_{1-x}As$ barriers and GaAs wells. For a given pair of materials having a fixed conduction band offset, the binding energy of an electron in the well can be adjusted by varying the width of the well. With an applied bias, photoexcited electrons from the GaAs wells are transported and detected as photocurrent.

Quantum well infrared photodetectors (QWIPs) consist of alternating layers of semiconductor material with larger and narrower bandgaps [14–20]. This series of alternating semiconductor layers is deposited one layer upon another using an ultrahigh vacuum technique such as molecular beam epitaxy (MBE). Alternating large- and narrow-bandgap materials give rise to quantum wells that provide bound and quasi-bound states for electrons or holes [1–5].

Many simple QWIP structures have used GaAs as the narrow-bandgap quantum well material and $Al_xGa_{1-x}As$ as the wide bandgap barrier layers as shown in Figure 3.10. The properties of the QWIP are related to the structural design and can be specified by the well width, barrier height, and doping density. In turn, these parameters can be tuned by controlling the cell temperatures of the gallium, aluminum, and arsenic cells as well as the doping cell temperature. The quantum well width (thickness) is governed by the time interval for which the Ga and As cell shutters are left opened. The barrier height is regulated by the composition of the $Al_xGa_{1-x}As$ layers, which are determined by the relative temperature of the Al and Ga cells. QWIP detectors rely on the absorption of incident radiation within the quantum well and typically the well material is doped n-type at an approximate level of 5×10^{17}.

The QWIP detectors require that an electric field component of the incident radiation be perpendicular to the layer planes of the device. Imaging arrays use diffraction gratings as shown in Figure 3.11. In particular, the latter approach is of practical importance in order to realize two-dimensional detector arrays. The QWIP focal plane array is a reticulated structure formed by conventional photolithographic techniques. Part of the processing involves placing a two-dimensional metallic grating over the focal plane pixels. The grating metal is typically angled at 45° patterns to reflect incident light obliquely so as to couple the perpendicular component of the electric field into the quantum wells thus producing the photoexcitation. The substrate material (GaAs) is backside thinned and a chemical/mechanical polish is used to produce a mirrorlike finish on the backside. The front side of the pixels with indium bumps are

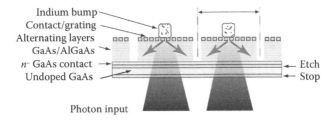

Indium bump
Contact/grating
Alternating layers
GaAs/AlGaAs
n^- GaAs contact — ← Etch
Undoped GaAs — ← Stop

Photon input

FIGURE 3.11 Backside illuminated QWIP structure with a top side diffraction grating/contact metal. Normally incident light is coupled horizontally into the quantum wells by scattering off a diffraction grating located at the top of the focal plane array.

flip-chip bonded to a readout IC. Light travels through the back side and is unabsorbed during its first pass through the epilayers; upon scattering with a horizontal propagation component from the grating, some of it is then absorbed by the quantum wells, photoexciting carriers. An electric field is produced perpendicular to the layers by applying a bias voltage at doped contact layers. The structure then behaves as a photoconductor.

The QWIP detectors require cooling to about 60 K for LWIR operation in order to adequately reduce the dark current. They also have comparatively low quantum efficiency, generally less than 10%. They thus require longer signal integration times than InSb or HgCdTe devices. However, the abundance of radiation in the LWIR band in particular allows QWIP detectors to still achieve excellent performance in infrared cameras.

The maturity of the GaAs-technology makes QWIPs particularly suited for large commercial focal plane arrays with high spatial resolution. Excellent lateral homogeneity is achieved, thus giving rise to a small fixed-pattern noise. QWIPs have an extremely small $1/f$ noise compared to interband detectors (like HgCdTe or InSb), which is particularly useful if long integration times or image accumulation are required. For these reasons, QWIP is the detector technology of choice for many applications where somewhat smaller quantum efficiencies and lower operation temperatures, compared to interband devices, are tolerable. QWIPs are finding useful applications in surveillance, night vision, quality control, inspection, environmental sciences, and medicine.

Quantum well infrared detectors are available in the 5 and 10 μm spectral region. The spectral response of QWIP detectors can be tuned to a wide range of values by adjusting the width and depth of quantum wells formed in alternating layers of GaAs and GaAlAs. An example of the spectral response from a variety of such structures is shown in Figure 3.12. QWIP spectral response is generally limited to fairly narrow spectral bandwidth—approximately 10–20% of the peak response wavelength. QWIP detectors have higher dark currents than InSb or HgCdTe devices and generally must be cooled to about 60 K for LWIR operation.

The quantum efficiencies of InSb, HgCdTe, and QWIP photon detectors are compared in Figure 3.13. With antireflection coating, InSb and HgCdTe are able to convert about 90% of the incoming photon flux to electrons. The QWIP quantum efficiencies are significantly lower, but work at improving them continues to occupy the attention of research teams.

We conclude this section with a description of Type-II superlattice detectors [21–26]. Although Type-II superlattice detectors are not yet used in arrays for in commercial camera system, the technology is briefly reviewed here because of its potential future importance. This material system mimics an intrinsic detector material such as HgCdTe, but is "bandgap engineered." Type-II superlattice structures

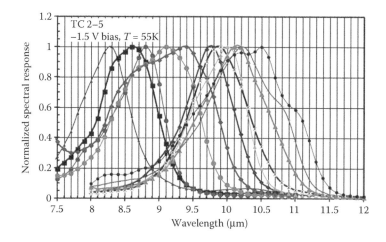

FIGURE 3.12 Representative spectral response of QWIP detectors.

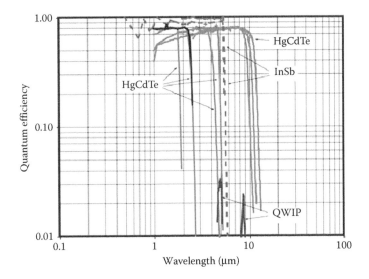

FIGURE 3.13 Comparison of the quantum efficiencies of commercial infrared photon detectors. This figure represents devices that have been antireflection coated.

are fabricated from multilayer stacks of alternating layers of two different semiconductor materials. Figure 3.14 illustrates the structure. The conduction band minimum is in one layer and the valence band minimum is in the adjacent layer (as opposed to both minima being in the same layer as in a Type-I superlattice).

The idea of using Type-II superlattices for LWIR detectors was originally proposed in 1977. Recent work on the MBE growth of Type-II systems by [7] has led to the exploitation of these materials for IR detectors. Short-period superlattices of, for example, strain-balanced InAs/(Ga,In)Sb lead to the

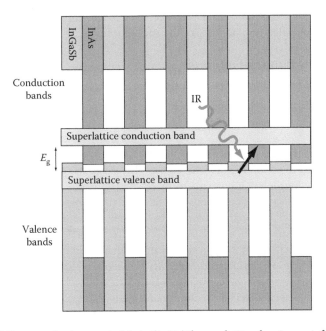

FIGURE 3.14 Band diagram of a short-period InAs/(In,Ga)Sb superlattice showing an infrared transition from the heavy hole (hh) miniband to the electron (e) miniband.

formation of conduction and valence minibands. In these band states heavy holes are largely confined to the (Ga,In)Sb layers and electrons are primarily confined to the InAs layers. However, because of the relatively low electron mass in InAs, the electron wave functions extend considerably beyond the interfaces and have significant overlap with heavy-hole wave functions. Hence, significant absorption is possible at the minigap energy (which is tunable by changing layer thickness and barrier height).

Cutoff wavelengths from 3 to 20 μm and beyond are potentially possible with this system. Unlike QWIP detectors, the absorption of normally incident flux is permitted by selection rules, obviating the need for grating structures or corrugations that are needed with QWIPs. Finally, Auger transition rates, which place intrinsic limits on the performance of these detectors and severely impact the lifetimes found in bulk, narrow-gap detectors, can be minimized by judicious choices of the structure's geometry and strain profile.

In the future, further advantages may be achievable by using the InAs/Ga(As,Sb) material system where both the InAs and Ga(As,Sb) layers may be lattice matched to InAs substrates. The intrinsic quality obtainable in these structures can be in principle superior to that obtained in InAs/(Ga,In)Sb structures. Since dislocations may be reduced to a minimum in the InAs/Ga(As,Sb) material system, it may be the most suitable Type-II material for making large arrays of photovoltaic detectors.

Development efforts for Type-II superlattice detectors are primarily focused on improving material quality and identifying sources of unwanted leakage currents. The most challenging problem currently is to passivate the exposed sidewalls of the superlattice layers where the pixels are etched in fabrication. Advances in these areas should result in a new class of IR detectors with the potential for high performance at high operating temperatures.

3.4 Detector Readouts

Detectors themselves are isolated arrays of photodiodes, photoconductors, or bolometers. Detectors need a readout to integrate or sample their output and convey the signal in an orderly sequence to a signal processor and display [27].

Almost all readouts are integrated circuits (ICs) made from silicon. They are commonly referred to as readout integrated circuits, or ROICs. Here, we briefly describe the functions and features of these readouts, first for photon detectors and then for thermal detectors.

3.4.1 Readouts for Photon Detectors

Photon detectors are typically assembled as a hybrid structure, as illustrated in Figure 3.15. Each pixel of the detector array is connected to the unit cell of the readout through an indium bump. Indium bumps allow for a soft, low-temperature metal connection to convey the signal from the detector to the readout's input circuit.

Commercial thermal imagers that operate in the MWIR and LWIR spectral regions generally employ a direct injection circuit to collect the detector signal. This is because this circuit is simple and works well with the relatively high photon currents in these spectral bands. The direct injection transistor feeds the signal onto an integrating capacitor where it stored for a time called the integration time. The integration time is typically around 200 μs for the LWIR spectral band and 2 ms for the MWIR band, corresponding to the comparative difference in the photon flux available. The integration time is limited by the size of the integration capacitor. Typical capacitors can hold on the order of 3×10^7 electrons.

For cameras operating in the SWIR band, the lower flux levels typically require a more complicated input amplifier. The most common choice employs a capacitive feedback circuit, providing the ability to have significant gain at the pixel level before storage on an integrating capacitor.

Two readout modes are employed, depending upon the readout design:

- Snapshot
- Rolling frame

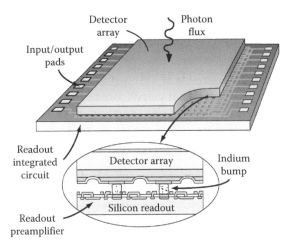

FIGURE 3.15 Hybrid detector array structure consists of a detector array connected to a readout array with indium metal bumps. Detector elements are usually photodiodes or photoconductors, although photocapacitors are sometimes used. Each pixel in the readout contains at least one addressable switch, and more often a preampflifier or buffer together with a charge storage capacitor for integrating the photosignal.

In the snapshot mode, all pixels integrate simultaneously, are stored, and then read out in sequence, followed by resetting the integration capacitors. In the rolling frame mode, the capacitors of each row are reset after each pixel in that row is read. In this case, each pixel integrates in different parts of the image frame. A variant of the rolling frame is an interlaced output. In this case, the even rows are read out in the first frame and the odd rows in the next. This corresponds to how standard U.S. television displays function.

It is common for each column in the readout to have an amplifier to provide some gain to the signal coming from each row as it is read. The column amplifier outputs are then fed to the output amplifiers. Commercial readouts typically have one, two, or four outputs, depending upon the array size and frame rate. Most commercial cameras operate at 30 or 60 Hz.

Another common feature found on some readouts is the ability to operate at higher frame rates on a subset of the full array. This ability is called windowing. It allows data to be collected more quickly on a limited portion of the image.

3.4.2 Thermal Detector Readouts

Bolometer detectors have comparatively lower resistance than photon detectors and relatively slow inherent response times. This condition allows readouts that do not have to integrate the charge during the frame, but only need to sample it for a brief time. This mode is frequently referred to as pulse-biased.

The unit cell of the bolometer contains only a switch that is pulsed on once per frame to allow current to flow from each row in turn to the column amplifiers. Bias is supplied by the row multiplexer. Sample times for each detector are typically on the order of the frame time divided by the number of rows. Many designs employ differential input column amplifiers that are simultaneously fed an input from a dummy or blind bolometer element in order to subtract a large fraction of the current that flows when the element is biased.

The nature of bolometer operation means that the readout mode is rolling frame. Some designs also provide interlaced outputs for input to TV-like displays.

3.4.3 Readout Evolution

Early readouts required multiple bias supply inputs and multiple clock signals for operation. Today, only two clocks and two bias supplies are typically required. The master clock sets the frame rate. The integration clock sets the time that the readout signal is integrated, or that the readout bias pulse is applied. On-chip clock and bias circuits generate the additional clocks and biases required to run the readout. Separate grounds for the analog and digital chip circuitry are usually employed to minimize noise.

Current development efforts are beginning to add on-chip analog-to-digital (A/D) converters to the readout. This feature provides a direct digital output, avoiding significant difficulties in controlling extraneous noise when the sensor is integrated with an imaging or camera system.

3.5 Technical Challenges for Infrared Detectors

Twenty-five years ago, infrared imagining was revolutionized by the introduction of the Probeye Infrared camera. At a modest 8 pounds, Probeye enabled handheld operation, a feature previously unheard of at that time when very large, very expensive IR imaging systems were the rule. Infrared components and technologies have advanced considerably since then. With the introduction of the Indigo Systems Omega camera, one can now acquire a complete infrared camera weighing less than 100 g and occupying 3.5 in.[3]

Many forces are at play enabling this dramatic reduction in camera size. Virtually all of these can be traced to improvements in the silicon IC processing industry. Largely enabled by advancements in photolithography, but additionally aided by improvements in vacuum deposition equipment, device feature sizes have been steadily reduced. It was not too long ago that the minimum device feature size was just pushing to break the 1-μm barrier. Today, foundries are focused on production implementation of 65–90 nm feature sizes.

The motivation behind such significant improvements has been the high-dollar/high-volume commercial electronics business. Silicon foundries have expended billions of dollars in capitalization and R&D aimed at increasing the density and speed of the transistors per unit chip area. Cellular telephones, personal data assistants (PDAs), and laptop computers are all applications demanding smaller size, lower power, and more features—performance—from electronic components. Infrared detector arrays and cameras have taken direct advantage of these advancements.

3.5.1 Uncooled Infrared Detector Challenges

The major challenge for all infrared markets is to reduce the pixel size while increasing the sensitivity. Reduction from a 50-μm pixel to a 25-μm pixel, while maintaining or even reducing noise equivalent temperature difference (NETD), is a major goal that is now being widely demonstrated (see Figure 3.16).

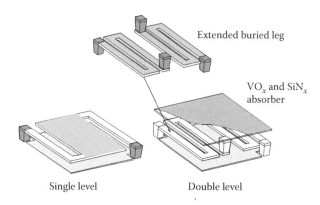

FIGURE 3.16   Uncooled microbolometer pixel structures having noise-equivalent temperature difference (NEΔT) values <50 mK: single level for 2 mil (50 μm) pixels in a 240 × 320 format and double level for 1 mil (25 μm) pixels in a 480 × 640 format. (Courtesy of Raytheon Vision Systems.)

The trends are illustrated by a simple examination of a highly idealized bolometer: the DC response of a detector in which we neglect all noise terms except temperature fluctuation noise, and the thermal conductance value is not detector area dependent (i.e., we are not at or near the radiation conductance limit). Using these assumptions, reducing the pixel area by a factor of four will reduce the SNR by a factor of eight as shown below:

$$\Delta T_{\text{signal|DCresponse}} = \frac{P_{\text{signalDC}}}{G_{\text{th}}} = \frac{\gamma A_{\text{D}} I_{\text{light}}}{G_{\text{th}}} \tag{3.22}$$

where P_{signalDC} is the DC signal from IR radiation (absorbed power) [W], A_{D} is the detector area [m^2], I_{light} is the light intensity [W/m^2], G_{th} is the thermal conductance [W/K], and γ is a constant that accounts for reflectivity and other factors not relevant to this analysis.

For a detector in the thermal fluctuation limit, the root mean square temperature fluctuation noise is a function of the incident radiation and the thermal conductance of the bolometer bridge.

$$\Delta T_{\text{noise}} \sqrt{\langle \Delta T^2 \rangle} = \sqrt{\frac{kT^2}{C_{\text{th}}}} \tag{3.23}$$

where T is the operating temperature in Kelvin, k is the Boltzmann's constant, and C_{th} is the total heat capacity of the detector in joules per Kelvin [J/K].

The total heat capacity can be written as $C_{\text{th}} = c_{\text{p}} A_{\text{d}} Z_{\text{bridge}}$, where Z_{bridge} is the bolometer bridge thickness in meters and c_{p} is the specific heat of the detector in J/K-m^3.

The SNR is then

$$\frac{\Delta T_{\text{signal}}}{\Delta T_{\text{noise}}} = \frac{\gamma A_{\text{D}} I_{\text{light}}}{G_{\text{th}}} \sqrt{\frac{c_{\text{p}} A_{\text{D}} Z_{\text{bridge}}}{kT^2}} = \frac{\gamma A_{\text{D}} I_{\text{light}}}{G_{\text{th}}} A_{\text{D}}^{3/2} \sqrt{\frac{c_{\text{p}} Z_{\text{bridge}}}{kT^2}} \tag{3.24}$$

It can be seen that the SNR goes as the area to the three halves. Therefore, a 4× reduction in detector area reduces the SNR by a factor of eight for this ideal bolometer case. Thermal conductance is assumed constant, that is, the ratio of leg length to thickness remains constant as the detector area is reduced. In practical constructions, reducing the pixel linear dimensions by 2× also reduces the leg length by 2×, thus the thermal conductance increases and aggravates the problem. In order to improve the SNR caused by the 4× loss in area, one may be tempted to reduce the thermal conductance G_{th} by 8×. To accomplish this, the length of the legs must be increased and their thickness reduced. By folding the legs under the detector, as seen in Figure 3.10, one can achieve this result. However, an 8× reduction in thermal conductance would result in a detrimental increase in the thermal time constant.

The thermal time constant is given by $\tau_{\text{thermal}} = C_{\text{th}}/G_{\text{th}}$. The heat capacity is reduced by 4× because of the area loss. If G_{th} is reduced by a factor of 8×, then $\tau_{\text{thermal}} = 2C_{\text{th}}/G_{\text{th}}$ is increased by a factor of two. This image smear associated with this increased time constant would prove problematic for practical military applications.

In order to maintain the same time constant, the total heat capacity must be reduced accordingly. Making the detector thinner may achieve this result except that it also increases the temperature fluctuation noise. From this simple example, one can readily see the inherent relationship between SNR and the thermal time constant.

We would like to maintain both an equivalent SNR and thermal time constant as the detector cell size is decreased. This can be achieved by maintaining the relationships between the thermal conductance, detector area, and bridge thickness as shown in the following.

The thermal time constant is given by

$$\tau_{\text{thermal}} = \frac{C_{\text{th}}}{G_{\text{th}}} = \frac{c_{\text{p}} A_{\text{D}} Z_{\text{bridge}}}{G_{\text{th}}} \tag{3.25}$$

Equating the thermal time constant of the large and small pixels and doing the same with the SNR leads to the following relationships, where the primed variables are the parameters required for the new detector cell:

$$\tau_{\text{thermal}} = \frac{c_{\text{p}} A_{\text{D}} Z_{\text{bridge}}}{G_{\text{th}}} = \frac{c_{\text{p}} A_{\text{D}}' Z_{\text{bridge}}'}{G_{\text{th}}'} \tag{3.26}$$

$$\frac{\Delta T_{\text{signal}}}{\Delta T_{\text{noise}}} = \frac{\gamma A_{\text{D}} I_{\text{light}}}{G_{\text{th}}} \sqrt{\frac{c_{\text{p}} A_{\text{D}} Z_{\text{bridge}}}{kT^2}} = \frac{\gamma A_{\text{D}}' I_{\text{light}}}{G_{\text{th}}'} \sqrt{\frac{c_{\text{p}} A_{\text{D}}' Z_{\text{bridge}}'}{kT^2}} \tag{3.27}$$

Rearranging τ_{thermal} to find the ratio $G_{\text{th}}/G_{\text{th}}'$ and substituting into the SNR, we obtain

$$\frac{Z_{\text{bridge}}'}{Z_{\text{bridge}}} = \frac{A_{\text{D}}'}{A_{\text{D}}}, \quad \text{and it follows that} \quad \frac{G_{\text{th}}'}{G_{\text{th}}} = \left(\frac{Z_{\text{bridge}}'}{Z_{\text{bridge}}}\right)^2 \tag{3.28}$$

So, it becomes evident that a 4× reduction in pixel cell area requires a 16× , and not an 8× , reduction in thermal conductance to maintain equivalent SNR and thermal time constant. This gives some insight into the problems of designing small pixel bolometers for high sensitivity. It should be noted that in current implementations, the state-of-the-art sensitivity is about 10× from the thermal limits.

3.5.2 Electronics Challenges

Specific technology improvements spawned by the commercial electronics business that have enabled size reductions in IR camera signal processing electronics include

- Faster digital signal processors (DSPs) with internal memory 1 MB
- Higher-density field-programmable gate arrays (FPGAs) (>200 K gates and with an embedded processor core)
- Higher-density static (synchronous?) random access memory >4 MB
- Low-power, 14-bit differential A/D converters

Another enabler, also attributable to the silicon industry, is reduction in the required core voltage of these devices (see Figure 3.17). Five years ago, the input voltage for virtually all-electronic components was 5 V. Today, one can buy a DSP with a core voltage as low as 1.2 V. Power consumption of the device is proportional to the square of the voltage. So a reduction from 5- to 1.2-V core represents more than an order of magnitude power reduction.

The input voltage ranges for most components (e.g., FPGAs and memories) are following the same trends. These reductions are not only a boon for reduced power consumption, but also these lower power devices typically come in much smaller footprints. IC packaging advancements have kept up with the higher-density, lower-power devices. One can now obtain a device with almost twice the number of I/Os in 25% of the required area (see Figure 3.18).

All of these lower-power, smaller-footprint components exist by virtue of the significant demand created by the commercial electronics industry. These trends will continue. Moore's law (logic density

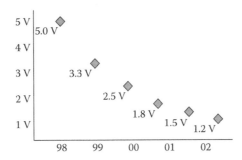

FIGURE 3.17　IC device core voltage versus time.

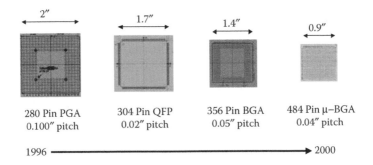

FIGURE 3.18　Advancements in component packaging miniaturization together with increasing pin count that enables reduced camera volume.

in bits/in.[2] will double every 18 months) nicely describes the degree by which we can expect further advancements.

3.5.3 Detector Readout Challenges

The realization of tighter design rules positively affects reduction in camera size in yet another way. Multiplexers, or ROICs, directly benefit from the increased density. Now, without enlarging the size of the ROIC die, more functions can be contained in the device. On-ROIC A/D conversion eliminates the need for a dedicated, discrete A/D converter. On-ROIC clock and bias generation reduces the number of vacuum dewar feedthroughs to yield a smaller package as well as reducing the complexity and size of the camera power supply. Putting the nonuniformity correction circuitry on the ROIC reduces the magnitude of the detector output signal swing and minimizes the required input dynamic range of the A/D converter. All of these increases in ROIC functionality come with the increased density of the silicon fabrication process.

3.5.4 Optics Challenges

Another continuing advancement that has helped reduced the size of IR cameras is the progress made at increasing the performance of the uncooled detectors themselves. The gains made at increasing the sensitivity of the detectors has directly translated to reduction in the size of the optics. With a sensitivity goal of 100 mK, an *F*/1 optic has traditionally been required to collect enough energy. Given the recent sensitivity improvements in detectors, achievement of 100 mK can be attained with an *F*/2 optic. This reduction in required aperture size greatly reduces the camera size and weight. These improvements

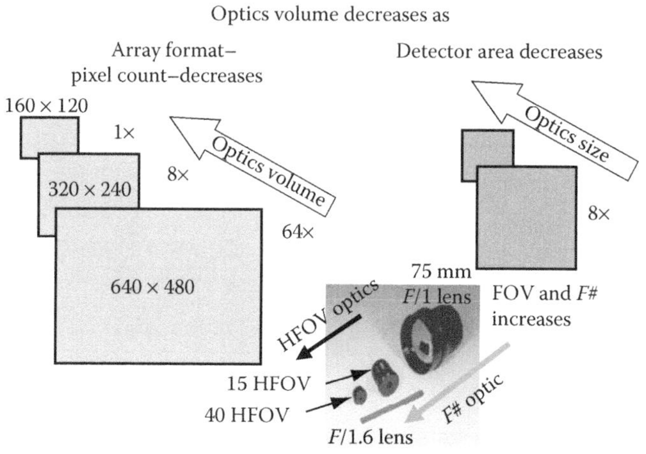

FIGURE 3.19 Trade-off between optics size and volume and *f/#*, array format, and pixel size.

in detector sensitivity can also be directly traceable to improvements in the silicon industry. The same photolithography and vacuum deposition equipments used to fabricate commercial ICs are used to make bolometers. The finer geometry line widths translate directly to increased thermal isolation and increased fill factor, both of which are factors in increased responsivity.

Reduction in optics' size was based on a sequence of NEDT performance improvements in uncooled VO_x microbolometer detectors so that faster optics F/1.4 to F/2 could be utilized in the camera and still maintain a moderate performance level. As indicated by Equations 3.29 through 3.33, the size of the optics is based on the required field-of-view (FOV), number of detectors (format of the detector array), area of the detector, and F# of the optics (see Figure 3.19). The volume of the optics is considered to be approximately a cylinder with a volume of $\pi r^2 L$. In Equations 3.29 through 3.33, FL is the optics focal length equivalent to L, D_o is the optics diameter and $D_o/2$ is equivalent to r, A_{det} is the area of the detector, F# is the *f*-number of the optics, and HFOV is the horizontal FOV.

$$\text{FL} = \frac{\#\text{ horizontal detectors}}{\tan(\text{HFOV}/2)} = \frac{\sqrt{A_{det}}}{2} \tag{3.29}$$

$$D_o = \frac{\#\text{ horizontal detectors}}{\tan(\text{HFOV}/2)} = \frac{\sqrt{A_{det}}}{2F\#} \tag{3.30}$$

$$F\# = \frac{\text{FL}}{D_o} \tag{3.31}$$

$$\text{Volume}_{\text{optics}} = \pi \left[\frac{D_o}{2}\right]^2 = \text{FL}$$

$$= \pi \left[\frac{(\#\text{ horizontal detectors}/(\tan(\text{HFOV}/2))) = (\sqrt{A_{det}}/2F\#)}{2}\right]^2 = \text{FL} \tag{3.32}$$

$$\text{Volume}_{\text{optics}} = \pi \left[\frac{(\#\text{ horizontal detectors}/(\tan(\text{HFOV}/2))) = \sqrt{A_{det}}}{32F\#^2}\right]^3 \tag{3.33}$$

Uncooled cameras have utilized the above enhancements and are now only a few ounces in weight and require only about 1 W of input power.

3.5.5 Challenges for Third-Generation Cooled Imagers

Third-generation cooled imagers are being developed to greatly extend the range at which targets can be detected and identified [28–30]. The U.S. Army rules of engagement now require identification prior to attack. Since the deployment of first- and second-generation sensors, there has been a gradual proliferation of thermal imaging technology worldwide. Third-generation sensors are intended to ensure that the U.S. Army forces maintain a technological advantage in night operations over any opposing force.

Thermal imaging equipment is used to first detect an object, and then to identify it. In the detection mode, the optical system provides a wide FOV (WFOV—*f*/2.5) to maintain robust situational awareness [31]. For detection, LWIR provides superior range under most Army fighting conditions. MWIR offers higher spatial resolution sensing, and a significant advantage for long-range identification when used with telephoto optics (NFOV—*f*/6).

3.5.5.1 Cost Challenges: Chip Size

Cost is a direct function of the chip size since the number of detector and readout die per wafer is inversely proportional to the chip area. Chip size in turn is set by the array format and pixel size. Third-generation imager formats are anticipated to be in a high-definition 16×9 layout, compatible with future display standards, and reflecting the soldier's preference for a wide FOV. An example of such a format is 1280×720 pixels. For a 30-μm pixel, this format yields a die size greater than 1.5×0.85 in. (22×38 mm). This will yield only a few die per wafer, and will also require the development of a new generation of dewar-cooler assemblies to accommodate these large dimensions. A pixel size of 20 μm results in a cost saving of more than 2×, and allows the use of existing dewar designs.

3.5.5.1.1 Two-Color Pixel Designs

Pixel size is the most important factor for achieving affordable third-generation systems. Two types of two-color pixels have been demonstrated. Simultaneous two-color pixels have two indium–bump connections per pixel to allow readout of both color bands at the same time. Figure 3.20 shows an example of a simultaneous two-color pixel structure. The sequential two-color approach requires only one indium bump per pixel, but requires the readout circuit to alternate bias polarities multiple times during each frame. An example of this structure is illustrated in Figure 3.21. Both approaches leave very little area available for the indium bump(s) as the pixel size is made smaller. Advanced etching technology is being developed in order to meet the challenge of shrinking the pixel size to 20 μm.

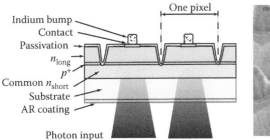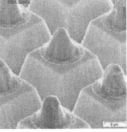

FIGURE 3.20 Illustration of a simultaneous two-color pixel structure—cross section and SEM. Simultaneous two-color FPAs have two indium bumps per pixel. A 50-μm simultaneous two-color pixel is shown.

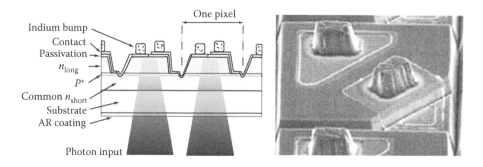

FIGURE 3.21   Illustration of a sequential two-color pixel structure—cross section and SEM. Sequential two-color FPAs have only one indium bump per pixel, helping to reduce pixel size. A 20-μm sequential two-color pixel is shown.

3.5.5.2 Sensor Format and Packaging Issues

The sensor format was selected to provide a wide FOV and high spatial resolution. Target detection in many Army battlefield situations is most favorable in LWIR. Searching for targets is more efficient in a wider FOV, in this case, $f/2.5$. Target identification relies on having 12 or more pixels across the target to adequately distinguish its shape and features. Higher magnification, $f/6$ optics combined with MWIR optical resolution enhances this task.

Consideration was also given to compatibility with future standards for display formats. Army soldiers are generally more concerned with the width of the display than the height, so the emerging 16:9 width to height format that is planned for high-definition TV was chosen.

A major consideration in selecting a format was the packaging requirements. Infrared sensors must be packaged in a vacuum enclosure and mated with a mechanical cooler for operation. Overall array size was therefore limited to approximately 1 in. so that it would fit in an existing standard advanced dewar assembly (SADA) dewar design. Figure 3.22 illustrates the pixel size/format/FOV trade within the design size constraints of the SADA dewar.

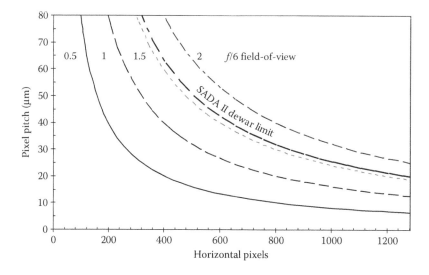

FIGURE 3.22   Maximum array horizontal format is determined by the pixel size and the chip size limit that will fit in an existing SADA dewar design for production commonality. For a 20-μm pixel and a 1.6° FOV, the horizontal pixel count limit is 1280. A costly development program would be necessary to develop a new, larger dewar.

3.5.5.3 Temperature Cycling Fatigue

Modern cooled infrared focal plane arrays are hybrid structures comprising a detector array mated to a silicon readout array with indium bumps (see Figure 3.15).

Very large focal plane arrays may exceed the limits of hybrid reliability engineered into these structures. The problem stems from the differential rates of expansion between HgCdTe and Si, which results in large stress as a device is cooled from 300 K ambient to an operating temperature in the range of 77–200 K. Hybrids currently use mechanical constraints to force the contraction of the two components to closely match each other. This approach may have limits—when the stress reaches a point where the chip fractures.

Two new approaches exist that can extend the maximum array size considerably. One is the use of silicon as the substrate for growing the HgCdTe detector layer using MBE. This approach has shown excellent results for MWIR detectors, but not yet for LWIR devices. Further improvement in this approach would be needed to use it for third-generation MWIR/LWIR two-color arrays.

A second approach that has proven successful for InSb hybrids is thinning the detector structure. HgCdTe hybrids currently retain their thick, 500 µm, CdZnTe epitaxial substrate in the hybridized structure. InSb hybrids must remove the substrate because it is not transparent, leaving only a 10-µm-thick detector layer. The thinness of this layer allows it to readily expand and contract with the readout. InSb hybrids with detector arrays over 2 in. (5 cm) on a side have been successfully demonstrated to be reliable.

Hybrid reliability issues will be monitored as a third-generation sensor manufacturing technology and is developed to determine whether new approaches are needed.

In addition to cost issues, significant performance issues must also be addressed for third-generation imagers. These are now discussed in the following section.

3.5.5.4 Performance Challenges

3.5.5.4.1 *Dynamic Range and Sensitivity Constraints*

A goal of third-generation imagers is to achieve a significant improvement in detection and ID range over second-generation systems. Range improvement comes from higher pixel count, and to a lesser extent from improved sensitivity. Figure 3.23 shows relative ID and detection range versus pixel size in the MWIR and LWIR, respectively. Sensitivity (D^* and integration time) have been held constant, and the format was varied to keep the FOV constant.

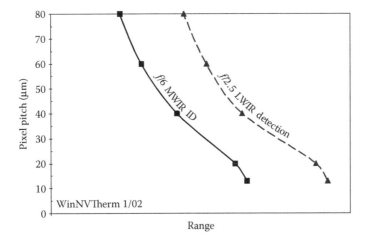

FIGURE 3.23 Range improves as the pixel size is reduced until a limit in optical blur is reached. In the examples above, the blur circle for the MWIR and LWIR cases are comparable since the *f*/number has been adjusted accordingly. D^* and integration time have been held constant in this example.

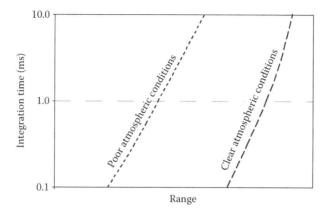

FIGURE 3.24 Range in a clear atmosphere improves only modestly with increased sensitivity. The case modeled here has a 20-μm pixel, a fixed D', and variable integration time. The 100× range of integration time corresponds to a 10× range in SNR. Improvement is more dramatic in the case of lower-atmospheric transmission that results in a reduced target signal.

Sensitivity has less effect than pixel size for clear atmospheric conditions, as illustrated by the clear atmosphere curve in Figure 3.24. Note that here the sensitivity is varied by an order of magnitude, corresponding to two orders of magnitude increase in integration time. Only a modest increase in range is seen for this dramatic change in SNR. In degraded atmospheric conditions, however, improved sensitivity plays a larger role because the signal is weaker. This is illustrated in Figure 3.24 by the curve showing range under conditions of reduced atmospheric transmission.

Dynamic range of the imager output must be considered from the perspective of the quantum efficiency and the effective charge storage capacity in the pixel unit cell of the readout. Quantum efficiency and charge storage capacity determine the integration time for a particular flux rate. As increasing number of quanta are averaged, the SNR improves as the square root of the count. Higher-accuracy A/D converters are therefore required to cope with the increased dynamic range between the noise and signal levels. Figure 3.25 illustrates the interaction of these specifications.

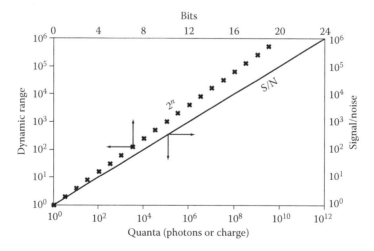

FIGURE 3.25 Dynamic range (2^n) corresponding to the number of digital bits (n) is plotted as a discrete point corresponding to each bit and referenced to the left and top scales. SNR, corresponding to the number of quanta collected (either photons or charge) is illustrated by the solid line in reference to the bottom- and right-hand scales.

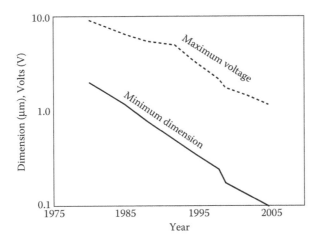

FIGURE 3.26 Trends for design rule minimum dimensions and maximum bias voltage of silicon foundry requirements.

System interface considerations lead to some interesting challenges and dilemmas. Imaging systems typically specify a noise floor from the readout on the order of 300 μV. This is because system users do not want to encounter sensor signal levels below the system noise level. With readouts built at commercial silicon foundries now having submicrometer design rules, the maximum bias voltage applied to the readout is limited to a few volts—this trend has been downward from 5 V in the past decade as design rules have shrunk, as illustrated in Figure 3.26. Output swing voltages can only be a fraction of the maximum applied voltage, on the order of 3 V or less.

This means that the dynamic range limit of a readout is about 10,000—80 db in power—or less. Present readouts almost approach this constraining factor with 70–75 db achieved in good designs. In order to significantly improve sensitivity, the noise floor will have to be reduced.

If sufficiently low readout noise could be achieved, and the readout could digitize on chip to a level of 15–16 bits, the data could come off digitally and the system noise floor would not be an issue. Such developments may allow incremental improvement in third-generation imagers in the future. Figure 3.27 illustrates an example of an on-chip A/D converter that has demonstrated 12 bits on chip.

FIGURE 3.27 Focal planes with on-chip A/D converters have been demonstrated. This example shows a 900 × 120 TDI scanning format array. (Photo supplied by Lester Kozlowski of Rockwell Scientific, Camarillo, CA.)

A final issue here concerns the ability to provide high charge storage density within the small pixel dimensions envisioned for third-generation imagers. This may be difficult with standard CMOS capacitors. Reduced oxide thickness of submicrometer design rules does give larger capacitance per unit area, but the reduced bias voltage largely cancels any improvement in charge storage density. Promising technology in the form of ferroelectric capacitors may provide much greater charge storage densities than the oxide-on-silicon capacitors now used. Such technology is not yet incorporated into standard CMOS foundries. Stacked hybrid structures[*] [32] may be needed as at least an interim solution to incorporate the desired charge storage density in detector–readout–capacitor structures.

3.5.5.4.2 High Frame Rate Operation

Frame rates of 30–60 fps are adequate for visual display. In third-generation systems, we plan to deploy high frame rate capabilities to provide more data throughput for advanced signal processing functions such as automatic target recognition (ATR), and missile and projectile tracking. An additional benefit is the opportunity to collect a higher percentage of available signal. Higher frame rates pose two significant issues. First, output drive power is proportional to the frame rate and at rates of 480 Hz or higher; this could be the most significant source of power dissipation on the readout. Increased power consumption on chip will also require more power consumption by the cryogenic cooler. These considerations lead us to conclude that high frame rate capabilities need to be limited to a small but arbitrarily positioned window of 64 × 64 pixels, for which a high frame rate of 480 Hz can be supported. This allows for ATR functions to be exercised on possible target locations within the full FOV.

3.5.5.4.3 Higher Operating Temperature

Current tactical infrared imagers operate at 77 K with few exceptions—notably MWIR HgCdTe, which can use solid-state thermoelectric (TE) cooling. Power can be saved, and cooler efficiency and cooler lifetime improved if focal planes operate at temperatures above 77 K.

Increasing the operating temperature results in a major reduction of input cryogenic cooler power. As can be seen from Figure 3.28, the coefficient of performance (COP) increases by a factor of 2.4 from 2.5% to 6% as the operating temperature is raised from 80 to 120 K with a 320 K heat sink. If the operating temperature can be increased to 150 K, the COP increases fourfold. This can have a major impact on input power, weight, and size.

Research is underway on an artificial narrow-bandgap intrinsic-like material—strained-layer superlattices of InGaAsSb—which have the potential to increase operating temperatures to even higher

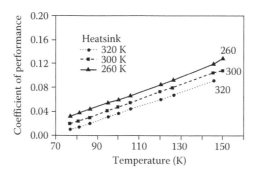

FIGURE 3.28 Javelin cooler coefficient of performance versus temperature.

---

[*] It should be noted that the third-generation imager will operate as an on-the-move wide-area step-scanner with automated ATR versus second-generation systems that rely on manual target searching. This allows the overall narrower field of view for the third-generation imager.

levels [33]. Results from this research may be more than a decade away, but the potential benefits are significant in terms of reduced cooler operating power and maintenance.

The above discussion illustrates some of the challenges facing the development of third-generation cooled imagers. In addition to these are the required advances in signal processing and display technologies to translate the focal plane enhancements into outputs for the user. These advances can be anticipated to not only help to increase the range at which targets can be identified but also to increase the rate of detection and identification through the use of two-color cues. Image fusion of the two colors in some cases is anticipated to help find camouflaged targets in clutter. Improved sensitivity and two-color response is further anticipated to minimize the loss of target contrast now encountered because of diurnal crossover. Future two-color imagers together with novel signal processing methods may further enhance the ability to detect land mines and find obscured targets.

3.6 Summary

Infrared sensors have made major performance strides in the last few years, especially in the uncooled sensors area. Cost, weight, and size of the uncooled have dramatically been reduced, allowing a greater proliferation into the commercial market. Uncooled sensors will find greater use in the medical community as a result. High-performance cooled sensors have also been dramatically improved, including the development of multicolor arrays. The high-performance sensors will find new medical applications because of the color discrimination and sensitivity attributes now available.

References

1. D.G. Crowe, P.R. Norton, T. Limperis, and J. Mudar, Detectors, in *Electro-Optical Components*, W.D. Rogatto, Ed., Vol. 3, ERIM, Ann Arbor, MI; J.S. Accetta and D.L. Schumaker, Executive Eds., *Infrared and Electro-Optical Systems Handbook*, SPIE, Bellingham, WA, 1993, revised 1996, Chapter 4, pp. 175–283.
2. P.R. Norton, Detector focal plane array technology, in *Encyclopedia of Optical Engineering*, R.G. Driggers, Ed., Vol. 1, Marcel Dekker, New York, 2003, pp. 320–348.
3. P.W. Kruse, and D.D. Skatrud, Uncooled infrared imaging arrays and systems, in *Semiconductors and Semimetals*, R.K. Willardson and E.R. Weber, Eds., Academic Press, New York, 1997.
4. P. Norton, Infrared image sensors, *Opt. Eng.*, 30, 1649–1663, 1991.
5. T. Ashley, I.M. Baker, T.M. Burke, D.T. Dutton, J.A. Haigh, L.G. Hipwood, R. Jefferies, A.D. Johnson, P. Knowles, and J.C. Little, *Proc. SPIE*, 4028, 398–403, 2000.
6. P.J. Love, K.J. Ando, R.E. Bornfreund, E. Corrales, R.E. Mills, J.R. Cripe, N.A. Lum, J.P. Rosbeck, and M.S. Smith, Large-format infrared arrays for future space and ground-based astronomy applications, *Proceedings of SPIE*; *Infrared Spaceborne Remote Sensing IX*, Vol. 4486–38; pp. 373–384, 29 July–3 August, 2001; San Diego, USA.
7. The photoconductive and photovoltaic detector technology of HgCdTe is summarized in the following references: D. Long and J.L. Schmidt, Mercury-cadmium telluride and closely related alloys, in *Semiconductors and Semimetals* 5, R.K. Willardson and A.C. Beer, Eds., Academic Press, New York, pp. 175–255, 1970; R.A. Reynolds, C.G. Roberts, R.A. Chapman, and H.B. Bebb, Photoconductivity processes in 0.09 eV bandgap HgCdTe, in *Proceedings of the 3rd International Conference on Photoconductivity*, E.M. Pell, Ed., Pergamon Press, New York, pp. 217, 1971; P.W. Kruse, D. Long, and O.N. Tufte, Photoeffects and material parameters in HgCdTe alloys, in *Proceedings of the 3rd International Conference on Photoconductivity*, E.M. Pell, Ed., Pergamon Press, New York, pp. 233, 1971; R.M. Broudy and V.J. Mazurczyk (HgCd) Te photoconductive detectors, in *Semiconductors and Semimetals*, 18, R.K. Willardson and A.C. Beer, Eds., Chapter 5, Academic Press, New York, pp. 157–199, 1981; M.B. Reine, A.K. Sood, and T.J. Tredwell, Photovoltaic infrared detectors, in *Semiconductors and Semimetals*, 18, R.K. Willardson and A.C. Beer, Eds., Chapter 6, pp. 201–311;

D. Long, Photovoltaic and photoconductive infrared detectors, in *Topics in Applied Physics* 19, *Optical and Infrared Detectors*, R.J. Keyes, Ed., Springer-Verlag, Heidelberg, pp. 101–147, 1970; C.T. Elliot, infrared detectors, in *Handbook on Semiconductors* 4, C. Hilsum, Ed., Chapter 6B, North Holland, New York, pp. 727–798, 1981.

8. P. Norton, Status of infrared detectors, *Proc. SPIE*, 2274, 82–92, 1994.
9. I.M., Baker, Photovoltaic IR detectors, in *Narrow-Gap II–VI Compounds for Optoelectronic and Electromagnetic Applications*, P. Capper, Ed., Chapman and Hall, London, pp. 450–473, 1997.
10. P. Norton, Status of infrared detectors, *Proc. SPIE*, 3379, 102–114, 1998.
11. M. Kinch, HDVIP® FPA technology at DRS, *Proc. SPIE*, 4369, 566–578, 1999.
12. M.B. Reine., Semiconductor fundamentals—Materials: Fundamental properties of mercury cadmium telluride, in *Encyclopedia of Modern Optics*, Academic Press, London, 2004.
13. A. Rogalski., HgCdTe infrared detector material: History, status and outlook, *Rep. Prog. Phys.*, 68, 2267–2336, 2005.
14. S.D. Guanapala, B.F. Levine, and N. Chand, *J. Appl. Phys.*, 70, 305, 1991.
15. B.F. Levine, *J. Appl. Phys.*, 47, R1–R81, 1993.
16. K.K. Choi., *The Physics of Quantum Well Infrared Photodetectors*, World Scientific, River Edge, New Jersey, 1997.
17. S.D. Gunapala, J.K. Liu, J.S. Park, M. Sundaram, C.A. Shott, T. Hoelter, T.-L. Lin, S.T. Massie, P.D. Maker, R.E. Muller, and G. Sarusi, 9 μm Cutoff 256×256 GaAs/AlGaAs quantum well infrared photodetector hand-held camera, *IEEE Trans. Elect. Dev.*, 45, 1890, 1998.
18. S.D. Gunapala, S.V. Bandara, J.K. Liu, W. Hong, M. Sundaram, P.D. Maker, R.E. Muller, C.A. Shott, and R. Carralejo, Long-wavelength 640×480 GaAs/AlGaAs quantum well infrared photodetector snap-shot camera, *IEEE Trans. Elect. Dev.*, 44, 51–57, 1997.
19. M.Z. Tidrow et al., Device physics and focal plane applications of QWIP and MCT, *Opto-Elect. Rev.*, 7, 283–296, 1999.
20. S.D. Gunapala and S.V. Bandara, Quantum well infrared photodetector (QWIP) focal plane arrays, in *Semiconductors and Semimetals*, R.K. Willardson and E.R. Weber, Eds., 62, Academic Press, New York, 1999.
21. G.A. Sai-Halasz, R. Tsu, and L. Esaki, *Appl. Phys. Lett.*, 30, 651, 1977.
22. D.L. Smith and C. Mailhiot, Proposal for strained type II superlattice infrared detectors, *J. Appl. Phys.*, 62, 2545–2548, 1987.
23. S.R. Kurtz, L.R. Dawson, T.E. Zipperian, and S.R. Lee, Demonstration of an InAsSb strained-layer superlattice photodiode, *Appl. Phys. Lett.*, 52, 1581–1583, 1988.
24. R.H. Miles, D.H. Chow, J.N. Schulman, and T.C. McGill, Infrared optical characterization of InAs/GaInSb superlattices, *Appl. Phys. Lett.*, 57, 801–803, 1990.
25. F. Fuchs, U. Weimar, W. Pletschen, J. Schmitz, E. Ahlswede, M. Walther, J. Wagner, and P. Koidl, *J. Appl. Phys. Lett.*, 71, 3251, 1997.
26. Gail J. Brown, Type-II InAs/GaInSb superlattices for infrared detection: An overview, *Proc. SPIE*, 5783, 65–77, 2005.
27. J.L. Vampola, Readout electronics for infrared sensors, in *Electro-Optical Components*, Chapter 5, Vol. 3, W.D. Rogatto, Ed., *Infrared and Electro-Optical Systems Handbook*, J.S. Accetta and D.L. Schumaker, Executive Eds., ERIM, Ann Arbor, MI and SPIE, Bellingham, WA, pp. 285–342, 1993, revised 1996.
28. D. Reago, and S. Horn, J. Campbell, and R. Vollmerhausen, Third generation imaging sensor system concepts, *SPIE*, 3701, 108–117, 1999.
29. P. Norton, J. Campbell III, S. Horn, and D. Reago, Third-generation infrared imagers, *Proc. SPIE*, 4130, 226–236, 2000.
30. S. Horn, P. Norton, T. Cincotta, A. Stoltz, D. Benson, P. Perconti, and J. Campbell, Challenges for third-generation cooled imagers, *Proc. SPIE*, 5074, 44–51, 2003.

31. S. Horn, D. Lohrman, P. Norton, K. McCormack, and A. Hutchinson, Reaching for the sensitivity limits of uncooled and minimally cooled thermal and photon infrared detectors, *Proc. SPIE*, 5783, 401–411, 2005.
32. W. Cabanskia, K. Eberhardta, W. Rodea, J. Wendlera, J. Zieglera, J. Fleißnerb, F. Fuchsb, R. Rehmb, J. Schmitzb, H. Schneiderb, and M. Walther, 3rd gen focal plane array IR detection modules and applications, *Proc. SPIE*, 5406, 184–192, 2004.
33. S. Horn, P. Norton, K. Carson, R. Eden, and R. Clement, Vertically-integrated sensor arrays—VISA, *Proc. SPIE*, 5406, 332–340, 2004.
34. R. Balcerak and S. Horn, Progress in the development of vertically-integrated sensor arrays, *Proc. SPIE*, 5783, 384–391, 2005.

4

Infrared Camera Characterization

Joseph G. Pellegrino
U.S. Army CERDEC Night Vision and Electronic Sensors Directorate

Jason Zeibel
U.S. Army CERDEC Night Vision and Electronic Sensors Directorate

Ronald G. Driggers
U.S. Army CERDEC Night Vision and Electronic Sensors Directorate

Philip Perconti
U.S. Army CERDEC Night Vision and Electronic Sensors Directorate

Many different types of infrared (IR) detector technology are now commercially available and the physics of their operation has been described in an earlier chapter. IR imagers are classified by different characteristics such as scan type, detector material, cooling requirements, and detector physics. Prior to the 1990s, thermal imaging cameras typically contained a relatively small number of IR photosensitive detectors. These imagers were known as *cooled scanning systems* because they required cooling to cryogenic temperatures and a mechanical scan mirror to construct a two-dimensional (2D) image of the scene. Large 2D arrays of IR detectors, or staring arrays, have enabled the development of *cooled staring systems* that maintain the sensitivity over a wide range of scene–flux conditions, spectral bandwidths, and frame rates. Staring arrays consisting of small bolometric detector elements, or microbolometers, have enabled the development of *uncooled staring systems* that are compact, lightweight, and of low power (see Figure 4.1).

The sensitivity, or thermal resolution, of uncooled microbolometer focal plane arrays (FPA) has improved drastically over the past decade, resulting in IR video cameras that can resolve temperature differences under nominal imaging conditions as small as 20 millidegrees Kelvin using $f/1.0$ optics. Advancements in the manufacturing processes used by the commercial silicon industry have been instrumental in this progress. Uncooled microbolometer structures are typically fabricated on top of the silicon-integrated circuitry (IC) designed to readout the changes in resistance for each pixel in the array. The silicon-based IC serves as an electrical and mechanical interface for the IR microbolometer.

The primary measures of IR sensor performance are sensitivity and resolution. When measurements of the end-to-end or human-in-the-loop (HITL) performance are required, the visual acuity of an observer through a sensor is included. The sensitivity and resolution are both related to the hardware and software that comprises the system, while the HITL includes both the sensor and the observer. Sensitivity is determined through radiometric analysis of the scene environment and the quantum electronic properties of the detectors. Resolution is determined by an analysis of the physical and optical properties, the detector array geometry, and other degrading components of the system in the same manner as complex electronic circuit/signal analysis. The sensitivity of cooled and uncooled staring

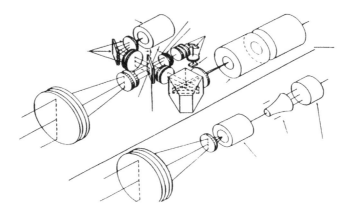

FIGURE 4.1   Scanning and staring system designs.

IR video cameras has improved by more than a factor of 10 compared to scanning systems that were commercially available in the 1980s and the early 1990s [1,2].

Sensitivity describes how the sensor performs with respect to the input signal level. It relates noise characteristics, responsivity of the detector, light gathering of the optics, and the dynamic range of the sensor. Radiometry describes the amount of light leaving the object and background and the amount which is collected by the detector. Optical design and detector characteristics are of considerable importance in sensor-sensitivity analysis. In IR systems, noise-equivalent temperature difference (NETD) is often a first-order description of the system sensitivity. The three-dimensional (3D) noise model [1] describes more detailed representations of the sensitivity parameters. The sensitivity of the scanned long-wave infrared (LWIR) cameras operating at video frame rates is typically limited by very short detector integration times on the order of tens or hundreds of microseconds. The sensitivity of staring IR systems with high quantum efficiency detectors is often limited by the charge integration capacity, or well capacity, of the readout integrated circuit (ROIC). The detector integration time of staring IR cameras can be tailored to optimize the sensitivity for a given application and may range from microseconds to tens of milliseconds.

The second type of measure is resolution. Resolution is the ability of the sensor to image small targets and to resolve fine detail in large targets. Modulation transfer function (MTF) is the most widely used resolution descriptor in IR systems. Alternatively, it may be specified by a number of descriptive metrics such as the optical Rayleigh criterion or the instantaneous field-of-view of the detector. These metrics are component-level descriptions and the system MTF is an all-encompassing function that describes the system resolution. Sensitivity and resolution can be competing system characteristics and they are the most important issues in initial studies for a design. For example, given a fixed sensor aperture diameter, an increase in focal length can provide an increase in resolution, but may decrease sensitivity [2]. A more detailed consideration of the optical design parameters is included in the next chapter.

Quite often metrics such as NETD and MTF, are considered as separable. However, in an actual sensor, sensitivity and resolution performances are interrelated. As a result, the minimum resolvable temperature difference (MRT or MRTD) has become a primary performance metric for IR systems.

This chapter addresses the parameters that characterize a camera's performance. A website advertising IR camera would in general contain a specification sheet that contains some variation of the terms that follow. A goal of this section is to give the reader working knowledge of of these terms so that it will enable them better to obtain the correct camera for their application:

- 3D noise
- NETD
- Dynamic range

- MTF
- MRT and Minimum detectable temperature (MDT)
- Spatial resolution
- Pixel size

4.1 Dimensional Noise

The 3D noise model is essential for describing the sensitivity of an optical sensor system. Modern imaging sensors incorporate complex focal plane architectures and sophisticated postdetector processing and electronics. These advanced technical characteristics create the potential for the generation of complex noise patterns in the output imagery of the system. These noise patterns are deleterious and therefore need to be analyzed to better understand their effects upon performance. Unlike classical systems where "well-behaved" detector noise predominates, current sensor systems have the ability to generate a wide variety of noise types, each with distinctive characteristics temporally, as well as along the vertical and horizontal image directions. Earlier methods for noise measurements at the detector preamplifier port that ignored other system noise sources are no longer satisfactory. System components following the stage that include processing may generate additional noise and even dominate total system noise.

Efforts by the Night Vision and Electronic Sensor Directorate to measure the second-generation IR sensors uncovered the need for a more comprehensive method to characterize noise parameters. It was observed that the noise patterns produced by these systems exhibited a high degree of directionality. The data set is 3D with the temporal dimension representing the frame sequencing and the two spatial dimensions representing the vertical and horizontal directions within the image (see Figure 4.2).

To acquire this data cube, the field of view of a camera to be measured is flooded with a uniform temperature reference. A set number n (typically around 100) of successive frames of video data are then collected. Each frame of the data consists of the measured response (in Volts) to the uniform temperature source from each individual detector in the 2D FPA. When many successive frames of data are "stacked" together, a uniform source-data-cube is constructed. The measured response may be either analog (RS-170) or digital (RS-422, Camera Link, Hot Link, etc.) in nature depending on the camera interface being studied.

To recover the overall temporal noise, first the temporal noise is calculated for each detector in the array. A standard deviation of the n measured voltage responses for each detector is calculated. For an h by v array, there are separate hv values where each value is the standard deviation of n-voltage measurements. The median temporal noise among these hv values is stated as the overall temporal noise in Volts.

Following the calculation of temporal noise, the uniform source-data-cube is reduced along each axis according to the 3D noise procedure. There are seven noise terms as part of the 3D noise definition. Three components measure the average noise present along each axis (horizontal, vertical, and

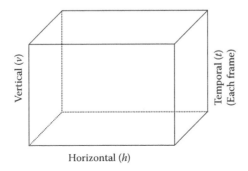

FIGURE 4.2 An example of a uniform source-data-cube for 3D noise measurements. The first step in the calculation of 3D noise parameters is the acquisition of a uniform source-data-cube.

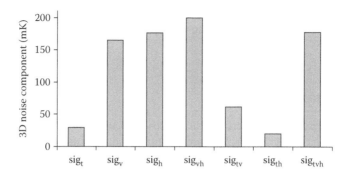

FIGURE 4.3 The 3D noise values for a typical data cube. The spatial nonuniformity can be seen in the elevated values of the spatial 3D noise components σ_h, σ_v, and σ_{vh}. The white noise present in the system (σ_{tvh}) is roughly the same magnitude as the spatial 3D noise components.

temporal) of the data cube (σ_h, σ_v, and σ_t). Three terms measure the noise common to any given pair of the axes in the data cube (σ_{tv}, σ_{th}, and σ_{vh}). The final term measures the uncorrelated random noise (σ_{tvh}). To calculate the spatial noise for the camera, each of the 3D noise components that are independent of time (σ_v, σ_h, and σ_{vh}) are added in quadrature. The result is quoted as the spatial noise of the camera in Volts.

To represent a data cube in a 2D format, the cube is averaged along one of the axes. For example, if a data cube is averaged along the temporal axis, then a time-averaged array is created. This format is useful for purely visualizing spatial noise effects as three of the components are calculated after temporal averaging (σ_h, σ_v, and σ_{vh}). These are the time-independent components of 3D noise. The data cube can also be averaged along both the spatial dimensions. The full 3D noise calculation for a typical data cube is shown in Figure 4.3.

Figure 4.4 shows an example of a data cube that has been temporally averaged. In this case, many spatial noise features are present. Column noise is clearly visible in Figure 4.4; however, the dominant spatial noise component appears to be the "salt and pepper" fixed pattern noise. The seven 3D noise components are shown in Figure 4.5. σ_{vh} is clearly the dominant noise term, as it is expected due to the high fixed pattern noise. The column noise σ_h and the row noise σ_v are also dominant. In this example,

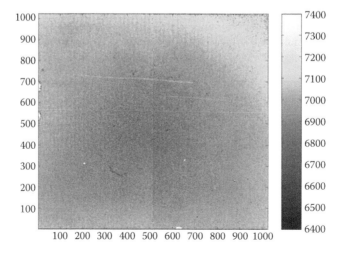

FIGURE 4.4 An example of a camera system with high spatial noise components and very low temporal noise components.

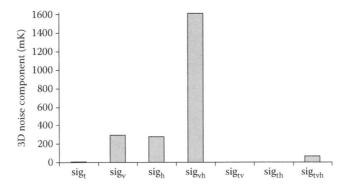

FIGURE 4.5 The 3D noise components for the data cube used to generate the dominant noise as it is expected due to the high fixed pattern noise. Note that the amount of row and column noise is significantly smaller than the fixed pattern noise. All the 3D noise values with a temporal component are significantly smaller than the purely spatial values.

the overall bulls-eye variation in the average frame dominates the σ_v and σ_h terms. Vertical stripes present in the figure add to σ_v, but this effect is negligible in comparison, leading to similar values for σ_v and σ_h. In this example, the temporal components of the 3D noise are two orders of magnitude lower than σ_{vh}. If this data cube were to be plotted as individual frames, we would see that the successive frames would hardly change and the dominant spatial noise would be present (and constant) in each frame.

4.2 Noise-Equivalent Temperature Difference

In general, imager sensitivity is a measure of the smallest signal that is detectable by a sensor. For IR imaging systems, NETD is a measure of sensitivity. Sensitivity is determined using the principles of radiometry and the characteristics of the detector. The system intensity transfer function (SITF) can be used to estimate the NETD. NEDT is the system noise rms voltage over the noise differential output. It is the smallest measurable signal produced by a large target (extended source), in other words, the minimum measurable signal.

The equation below describes NETD as a function of noise voltage and the SITF. The measured NETD values are determined from a line of video-stripped image of a test target, as depicted in Figure 4.10. A square test target is placed before a blackbody source. The delta *T* is the difference between the blackbody temperature and the mask. This target is then placed at the focal point of an off-axis parabolic mirror. The mirror serves the purpose of a long optical path length to the target and yet relieves the tester from the concerns over atmospheric losses to the difference in temperature The image of the target is shown in Figure 4.6. The SITF slope for the scan line in Figure 4.6 is the $\Delta\Sigma/\Delta T$, where $\Delta\Sigma$ is the signal measured for a given ΔT. The N_{rms} is the background signal on the same line.

$$\text{NETD} = \frac{N_{rms}[\text{volts}]}{\text{SITF\_Slope}[\text{volts} / \text{K}]}$$

After calculating both the temporal and spatial noise, a signal transfer function (SiTF) is measured. The field of view of the camera is again flooded with a uniform temperature source. The temperature of the source is varied over the dynamic range of the camera's output while the response of the mean array voltage is recorded. The slope of the resulting curve yields the SiTF responsivity in volts per degree Kelvin change in the scene temperature. Once both the SiTF curve and the temporal and spatial noise in volts are known, the NETD can be calculated. This is accomplished by dividing the temporal and spatial

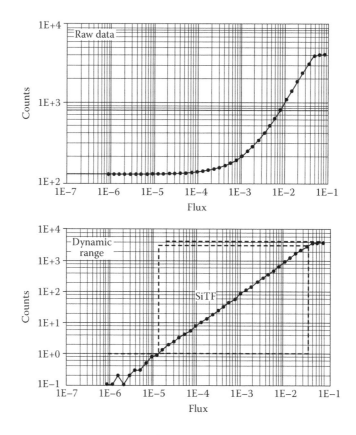

FIGURE 4.6 Dynamic range and system transfer function.

noise in volts by the responsivity in volts per degree Kelvin. The resulting NETD values represent the minimum discernable change in scene temperature for both spatial and temporal observation.

The SiTF of an electro-optical (EO) or IR system is determined by the signal response once the dark offset signal has been subtracted. After subtracting the offset signal due to nonflux effects, the SiTF plots the counts output relative to the input photon flux. The SiTF is typically represented in response units of voltage, signal electrons, digital counts, and so on versus units of the source: blackbody temperature, flux, photons, and so on. If the system behaves linearly within the dynamic range then the slope of the SiTF is constant. The dynamic range of the system, which may be defined by various criteria, is determined by the minimum (i.e., signal-to-noise ratio = 1) and maximum levels of operation.

4.3 Dynamic Range

The responsivity function also provides information on the dynamic range and linearity. The camera dynamic range is the maximum measurable input signal divided by the minimum measurable signal. The NEDT is assumed to be the minimum measurable signal. For AC systems, the maximum output depends on the target size and therefore the target size must be specified if the dynamic range is a specification. Depending upon the application, the maximum input value may be defined by one of the several methods. One method for specifying the dynamic range of a system involves having the ΔV_{sys} signal reach some specified level, say 90% of the saturation level as shown in Figure 4.7. Another method to assess the maximum input value is based on the signal's deviation from linearity. The range of data points that fall within a specified band is designated as the dynamic range. A third approach involves specifying the minimum SiTF of the system.

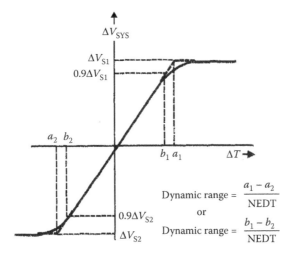

FIGURE 4.7 Dynamic range defined by linearity.

For most systems, the detector output signal is adjusted both in gain and offset system so that the dynamic range of the A/D converter is maximized. Figure 4.8 shows a generic detector system that contains an 8-bit A/D converter. The converter can handle an input signal between 0 and 1 V and an output between 0 and 255 counts. By selecting the gain and offset system, any detector voltage range can be mapped into the digital output. Figure 4.9 shows three different system gains and offsets. When the

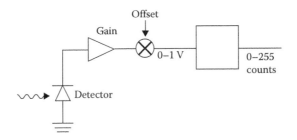

FIGURE 4.8 A system with an 8-bit A/D converter.

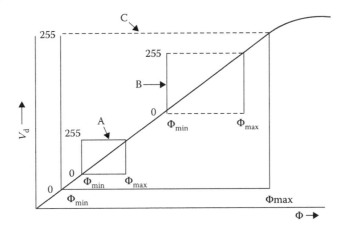

FIGURE 4.9 Different gains and voltage offsets that affect the input-to-output transition.

source flux level is less than Φ_{min}, the source will not be seen (i.e., it will appear as 0 counts). When the flux level is greater than Φ_{max}, the source will appear as 255 counts, and the system is said to be saturated. The gain parameters, Φ_{min} and Φ_{max} are redefined for each gain and offset level setting.

Output A below occurs with the maximum gain. Point B occurs with moderate gain and C with minimum gain. For the various gains, the detector output gets mapped into the full dynamic range of the A/D converter.

4.4 Modulation Transfer Function

The MTF of an optical system measures a system's ability to faithfully image a given object. Consider for example the bar pattern shown in Figure 4.10, with the cross section of each bar being a sine wave. Since the image of a sine wave light distribution is always a sine wave, although the aberrations may be bad, the image is always a sine wave. Therefore, the image will have a sine wave distribution with its intensity shown in Figure 4.10.

When the bars are coarsely spaced, the optical system has no difficulty in faithfully reproducing them. However, when the bars are more tightly spaced, the contrast,

$$\text{Contrast} = \frac{\text{bright} - \text{dark}}{\text{bright} + \text{dark}}$$

begins to fall off as shown in panel c. If the dark lines have an intensity = 0, the contrast = 1, and if the bright and dark lines are equally intense, contrast = 0. The contrast is equal to the MTF at a specified spatial frequency. Furthermore, it is evident that the MTF is a function of spatial frequency and position within the field.

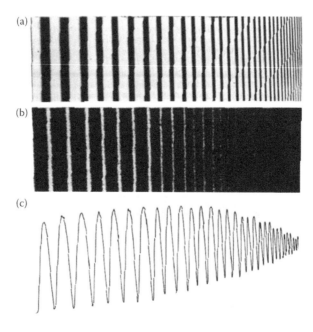

FIGURE 4.10 This figure shows the falloff in MTF as the spatial frequency increases. Panel (a) is a sinusoidal test pattern, panel (b) is the optical system's (negative) response, and panel (c) shows the contrast as a function of spatial frequency.

4.5 Minimum Resolvable Temperature

Whenever a camera is turned on, the observer subconsciously makes a judgment about the image quality. The IR community uses the MRT and the MDT as standard measures of image quality. The MRT and MDT depend upon the IR imaging system's resolution and sensitivity. MRT is a measure of the ability to resolve detail and it is inversely related to the MTF, whereas the MDT is a measure to detect something. The MRT and MDT deal with an observer's ability to perceive low-contrast targets which are embedded in noise.

MRT and MDT are not absolute values rather they are temperature differentials relative to a given background. They are sometimes referred to as the MRTD and the MDTD.

The theoretical MRT is

$$\text{MRT}(f_x) = \frac{k \cdot (\text{NEDT})}{\text{MTF}_{\text{perceived}}(f_x)} \cdot \sqrt{\{\beta_1 + \cdots + \beta_n\}}$$

where $\text{MTF}_{\text{perceived}} = \text{MTF}_{\text{SYS}}\text{MTF}_{\text{MONITOR}}\,\text{MTF}_{\text{EYE}}$. The $\text{MTF}_{\text{system}}$ is defined by the product $\text{MTF}_{\text{sensor}}$ $\text{MTF}_{\text{optics}}\,\text{MTF}_{\text{electronics}}$. Each β_i in the equation is an eye filter that is used to interpret the various components of noise. As certain noise sources increase, the MRT also increases. MRT has the same ambient temperature dependence as the NEDT; as the ambient temperature increases, MRT decreases. Because the MTF decreases as the spatial frequency increases, the MRT increases with the increasing spatial frequency. The overall system response depends on both sensitivity and resolution. The MRT parameter is bounded by sensitivity and resolution. Figure 4.10 shows that different systems may have different MRTs. System A has a better sensitivity because it has a lower MRT at low spatial frequencies. At mid-range spatial frequencies, the systems are approximately equivalent and it can be said that they provide equivalent performance. At higher frequencies, System B has better resolution and can display finer detail than system A. In general, neither sensitivity, resolution, nor any other single parameter can be used to compare systems and many quantities must be specified for a complete system-to-system comparison.

4.6 Spatial Resolution

The term resolution applies to two different concepts with regard to the vision systems. Spatial resolution refers to the image size in pixels—for a given scene, more pixels means higher resolution. The spatial resolution is a fixed characteristic of the camera and cannot be increased by the frame grabber of postprocessing techniques. For example, the zooming techniques, merely interpolate between pixels to expand an image without adding any new information to what the camera provided. However, it is easy to decrease the resolution, by simply ignoring part of the data. The National Instruments frame grabbers provide a "scaling" feature that instructs the frame grabber to sample the image to return a 1/2, 1/4, 1/8, and so on, scaled image. This is convenient when system bandwidth is limited and you do not require any precision measurements of the image.

The other use of the term "resolution" is commonly found in data acquisition applications and refers to the number of quantization levels used in A/D conversions. In this sense, higher resolution means that you would have improved the capability of analyzing low-contrast images. This resolution is specified by the A/D converter; the frame grabber determines the resolution for analog signals whereas the camera determines digital signals (the frame grabber must have the capability of supporting whatever resolution the camera provides).

4.6.1 Pixel Size

Camera pixel size consists of tiny dots that make up a digital image. So let us say that a camera is capable of taking images at 640×480 pixels. A little math shows us that such an image would contain

3,07,200 pixels or 0.3 megapixels. Now let us say the camera takes 1024×768 images. This gives us 0.8 megapixels. Larger the number of megapixels, the more the image detail . Each pixel can be one of 16.7 million colors.

The detector pixel size refers to the size of the individual sensor elements that make up the detector part of the camera. If we had two charge-coupled device (CCDs) detectors with equal quantum efficiency (QEs), one has 9 μm pixels and the other has 18 μm pixels (i.e., the pixels on CCD#2 are twice the linear size of those on CCD #1) and we put both of these CCDs into cameras that operate identically, then the image taken with CCD#1 will require 4× exposure of the image taken with CCD#2. This seeming discrepancy is due to its entirety to the area of the pixels in the two CCDs and could be compared to the effectiveness of the rain-gathering gauges with different rain collection areas: A rain gauge with a 2-in. diameter throat will collect 4× as much rain water as a rain gauge with a 1-in. diameter throat.

References

1. J. D'Agostino and C. Webb, 3-D analysis framework and measurement methodology for imaging system noise. *Proc. SPIE*, 1488, 110–121, 1991.
2. R.G. Driggers, P. Cox, and T. Edwards, *Introduction to Infrared and Electro-Optical Systems*, Artech House, Boston, MA, 1998, p. 8.

5

Infrared Camera and Optics for Medical Applications

Michael W. Grenn
U.S. Army CERDEC Night Vision and Electronic Sensors Directorate

Jay Vizgaitis
U.S. Army CERDEC Night Vision and Electronic Sensors Directorate

Joseph G. Pellegrino
U.S. Army CERDEC Night Vision and Electronic Sensors Directorate

Philip Perconti
U.S. Army CERDEC Night Vision and Electronic Sensors Directorate

The infrared radiation emitted by an object above 0 K is passively detected by infrared imaging cameras without any contact with the object and is nonionizing. The characteristics of the infrared radiation emitted by an object are described by Planck's blackbody law in terms of spectral radiant emittance.

$$M_\lambda = \varepsilon(\lambda) \frac{c_1}{\lambda^5 (e^{c_2/\lambda T} - 1)} \, \text{W/cm}^2 \, \mu\text{m}$$

where c_1 and c_2 are constants of 3.7418×10^4 W $\mu\text{m}^4/\text{cm}^2$ and 1.4388×10^4 W μm K. The wavelength, λ, is provided in micrometers and $o(\lambda)$ is the emissivity of the surface. A blackbody source is defined as an object with an emissivity of 1.0, so that it is a perfect emitter. Source emissions of blackbodies at nominal terrestrial temperatures are shown in Figure 5.1. The radiant exitance of a blackbody at a 310 K, corresponding to a nominal core body temperature of 98.6°F, peaks at approximately 9.5 µm in the long-wave infrared (LWIR). The selection of an infrared camera for a specific application requires consideration of many factors including sensitivity, resolution, uniformity, stability, calibratability, user controllability, reliability, object of interest phenomenology, video interface, packaging, and power consumption.

Planck's equation describes the spectral shape of the source as a function of wavelength. It is readily apparent that the peak shifts to shorter wavelengths as the temperature of the object of interest increases. If the temperature of a blackbody approaches that of the sun, or 5900 K, the peak of the

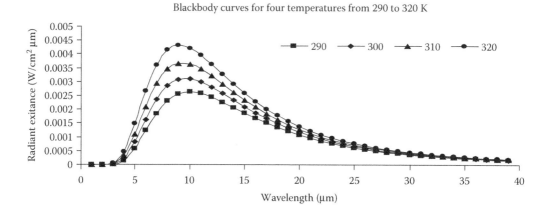

FIGURE 5.1   Planck's blackbody radiation curves.

spectral shape would shift to 0.55 μm or green light. This peak wavelength is described by Wien's displacement law

$$\lambda_{\text{max}} = 2898/T \ \mu m$$

Figure 5.2 shows the radiant energy peak as a function of temperature in the LWIR. It is important to note that the difference between the blackbody curves is the "signal" in the infrared bands. For an infrared sensor, if the background temperature is 300 K and the object of interest temperature is 302 K, the signal is the 2 K difference in flux between these curves. Signals in the infrared ride on very large amounts of background flux. This is not the case in the visible. For example, consider the case of a white object on a black background. The black background is generating no signal, while the white object is generating a maximum signal assuming the sensor gain is properly adjusted. The dynamic range may be fully utilized in a visible sensor. For the case of an IR sensor, a portion of the dynamic range is used by the large background flux radiated by everything in the scene. This flux is never a small value; hence sensitivity and dynamic range requirements are much more difficult to satisfy in IR sensors than in visible sensors.

A typical infrared imaging scenario consists of two major components, the object of interest and the background. In an IR scene, the majority of the energy is emitted from the constituents of the scene. This emitted energy is transmitted through the atmosphere to the sensor. As it propagates through the atmosphere it is degraded by absorption and scattering. Obscuration by intervening objects and additional energy emitted by the path also affect the target energy. This effect may be very small in short-range imaging applications under controlled conditions. All these contributors, which are not the object of interest, essentially reduce one's ability to discriminate the object. The signal is further degraded by the optics of the sensor. The energy is then sampled by the detector array and converted to electrical signals. Various electronics amplify and condition this signal before it is presented to either a display for human interpretation or an algorithm like an automatic target recognizer for machine interpretation. A linear systems approach to modeling allows the components' transfer functions to be treated separately as contributors to the overall system performance. This approach allows for straightforward modifications to a performance model for changes in the sensor or environment when performing tradeoff analyses.

The photon flux levels (photons per square centimeter per second) on Earth is 1.5×10^7 in the daytime and around 1×10^{10} at night in the visible. In the MWIR, the daytime and nighttime flux levels are 4×10^{15} and 2×10^{15}, respectively, where the flux is a combination of emitted and solar-reflected flux. In the LWIR, the flux is primarily emitted where both day and night yield a 2×10^{17} level. At first look, it

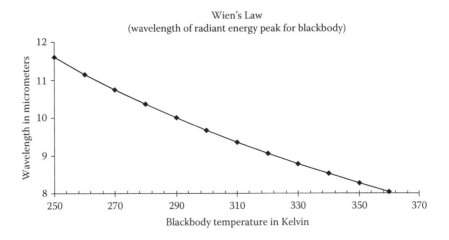

FIGURE 5.2   Location of peak of blackbody radiation, Wien's law.

appears that the LWIR flux characteristics are as good as a daytime visible system, however, there are two other factors limiting performance. First, the energy bandgaps of infrared sensitive devices are much smaller than in the visible, resulting in significantly higher detector dark current. The detectors are typically cooled to reduce this effect. Second, the reflected light in the visible is modulated with target and background reflectivities that typically range from 7% to 20%.

In the infrared, where all terrestrial objects emit, a two-degree equivalent blackbody difference in photon flux between object and background is considered high contrast. The flux difference between two blackbodies of 302 K compared to 300 K can be calculated in a manner similar to that shown in Figure 5.1. The flux difference is the signal that provides an image, hence the difference in signal compared to the ambient background flux should be noted. In the LWIR, the signal is 3% of the mean flux and in the MWIR it is 6% of the mean flux. This means that there is a large flux pedestal associated with imaging in the infrared.

There are two major challenges accompanying the large background pedestal in the infrared. First, the performance of a typical infrared detector is limited by the background photon noise and this noise term is determined by the mean of the pedestal. This value may be relatively large compared to the small signal differences. Second, the charge storage capacity of the silicon input circuit mated to each infrared detector in a staring array limits the amount of integrated charge per frame, typically around 10^7 charge carriers. An LWIR system in a hot desert background would generate 10^{10} charge carriers in a 33 ms integration time. The optical *f*-number, spectral bandwidth, and integration time of the detector are typically tailored to reach half well for a given imaging scenario for dynamic range purposes. This well capacity-limited condition results in a sensitivity, or noise equivalent temperature difference (NETD) of 10–30 times below the photon-limited condition. Figure 5.3 shows calculations of NETD as a function of background temperature for MWIR- and LWIR-staring detectors dominated by the photon noise of the incident IR radiation. At 310 K, the NETD of high-quantum efficiency MWIR and LWIR focal plane arrays (FPAs) is nearly the same, or about 3 millidegrees K, when the detectors are permitted to integrate charge up to the frame time, or in this case about 33 ms. The calculations show the sensitivity limits from the background photon shot noise only and do not include the contribution of detector and system temporal and spatial noise terms. The effects of residual spatial noise on NETD are described later in the chapter. The well capacity assumed here is 10^9 charge carriers to demonstrate sensitivity that could be achieved under large well conditions. The MWIR device is photon limited over the temperature range and begins to reach the well capacity limit near 340 K. The 24 μm pitch 9.5 μm cutoff LWIR device is well capacity limited over the entire temperature range. The 18 μm pitch 9.5 μm cutoff LWIR device becomes photon limited around 250 K. Various on-chip signal processing techniques, such as

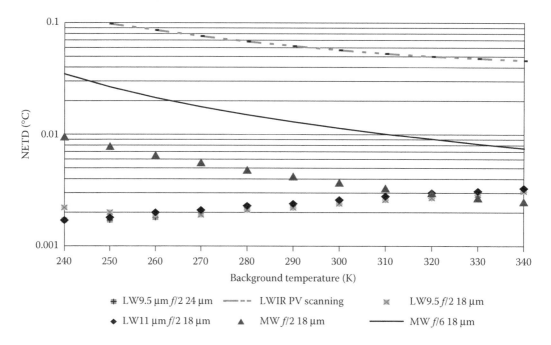

FIGURE 5.3 Background-limited NETD for high quantum efficiency MWIR and LWIR detectors.

charge skimming and charge partitioning, have been investigated to increase the charge capacity of these devices. In addition, as the minimum feature sizes of the input circuitry decreases, more real estate in the unit cell can be allocated to charge storage.

Another major difference between infrared and visible systems is the size of the detector and diffraction blur. Typical sizes for MWIR and LWIR detectors, or pixels, range from 20 to 50 μm. Visible detectors less than 6 μm are commercially available today. The diffraction blur for the LWIR is more than ten times larger than the visible blur and MWIR blur is eight times larger than visible blur. Therefore, the image blur due to diffraction and detector size is much larger in an infrared system than a visible system. It is very common for infrared staring arrays to be sampling limited where the sample spacing is larger than the diffraction blur and the detector size. Dither and microscanning are frequently used to enhance performance. A more detailed discussion of the optical considerations of infrared sensors is provided later in the chapter.

Finally, infrared staring arrays consisting of cooled photon detectors or uncooled thermal detectors may have responsivities that vary dramatically from pixel to pixel. It is common practice to correct for the resulting nonuniformity using a combination of factory preset tables and user inputs. The nonuniformity can cause fixed pattern noise in the image that can limit the performance of the system even more than temporal noise and these effects are demonstrated in the next section.

5.1 Infrared Sensor Calibration

Significant advancement in the manufacturing of high-quality FPAs operating in the short wave infrared (SWIR), MWIR, and LWIR has enabled industry to offer a wide range of affordable camera products to the consumer. Commercial applications of infrared camera technology are often driven by the value of the information it provides and price points set by the marketplace. The emergence of uncooled microbolometer FPA cameras with sensitivity less than 0.030°C at standard video rates has opened many new applications of the technology. In addition to the dramatic improvement in sensitivity over the past several years, uncooled microbolometer FPA cameras are characteristically compact, lightweight, and low power. Uncooled cameras are commercially available from a variety of domestic and foreign vendors including Agema, BAE Systems, CANTRONIC Systems, Inc., DRS and DRS Nytech, FLIR Systems, Inc., Indigo Systems, Inc., Electrophysics Corp., Inc., Infrared Components Corp., IR Solutions, Inc., and Raytheon, Thermoteknix Systems Ltd., ompact, low power. The linearity, stability, and repeatability of the system intensity transfer function (SiTF) may be measured to determine the suitability of an infrared camera for accurate determination of the apparent temperature of an object of interest. LWIR cameras are typically preferred for imaging applications that require absolute or relative measurements of object irradiance or radiance because emitted energy dominates the total signal in the LWIR. In the MWIR, extreme care is required to ensure the radiometric accuracy of data. Thermal references may be used in the scene to provide a known temperature reference point or points to compensate for detector-to-detector variations in response and improve measurement accuracy. Thermal references may take many forms and often include temperature-controlled extended area sources or uniformly coated metal plates with contact temperature sensors. Depending on the stability of the sensor, reference frames may be required in intervals from minutes to hours depending on the environmental conditions and the factory presets. Many sensors require an initial turn-on period to stabilize before accurate radiometric data can be collected. An example of a windows-based graphical user interface (GUI) developed at NVESD for an uncooled imaging system for medical studies is shown in Figure 5.4. The system allows the user to operate in a calibrated mode and display apparent temperature in regions of interest or at any specified pixel location including the pixel defined by the cursor. Single frames and multiple frames at specified time intervals may be selected for storage. Stability of commercially available uncooled cameras is provided earlier.

The LTC 500 thermal imager had been selected as a sensor to be used in a medical imaging application. Our primary goal was to obtain from the imagery calibrated temperature values within an accuracy of approximately a tenth of a degree Celsius. The main impediments to this goal consisted of several sources of spatial nonuniformity in the imagery produced by this sensor, primarily the spatial variation of radiance across the detector FPA due to self heating and radiation of the internal camera components, and to a lesser extent the variation of detector characteristics within the FPA. Fortunately, the sensor provides a calibration capability to mitigate the effects of the spatial nonuniformities.

We modeled the sensor FPA as a 2D array of detectors, each having a gain G and offset K, both of which are assumed to vary from detector to detector. In addition we assumed an internally generated radiance Y for each detector due to the self-heating of the internal sensor components (also varying from

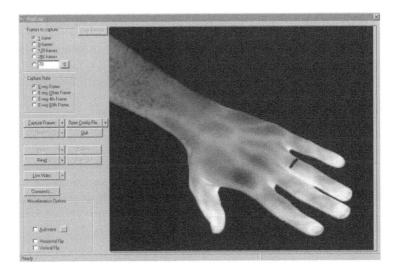

FIGURE 5.4 Windows-based GUI developed at NVESD for an uncooled medical imaging system.

detector to detector, as well as slowly with time). Lastly, there is an internally programmable offset C for each detector which the sensor controls as part of its calibration function. Therefore, given a radiance X incident on some detector of the FPA from the external scene, the output Z for that detector is given by

$$Z = GX + GY + K + C$$

5.2 Gain Detector

Individual detector gains were calculated by making two measurements. First, the sensor was allowed to run for several hours in order for the internal temperatures to stabilize. A uniform blackbody source at temperature T_1 (20°C) was used to fill the field of view (FOV) of the sensor and an output image Z_1 was collected. Next the blackbody temperature was set to T_2 (40°C) and a second output image Z_2 was collected. Since the measurement interval was small (<1–2 min) we assume the Y values remain constant, we have (for each detector):

$$Z_1 = GX_1 + GY + K + C$$
$$Z_2 = GX_2 + GY + K + C$$

where X_1 and X_2 refer to the external scene radiance corresponding to temperatures T_1 and T_2 incident on the detector and were calculated by integrating Planck's blackbody function over the 8–12 μm spectral band of the sensor. Taking the difference, we have (for each detector):

$$G = \frac{Z_2 - Z_1}{X_2 - X_1}$$

where the numbers here refer to measurements at different temperatures and, again, the value G (as well as Z and X) are assumed to vary from detector to detector (detector subscripts were omitted for clarity).

5.2.1 Nonuniformity Calibration

The LTC 500 provides the capability for nonuniformity calibration that allows the user to remove nonuniformities across the FPA assuming they do not change too rapidly with time. The procedure involves placing a uniform blackbody source across the FOV of the sensor and pressing the calibrate button. At this point, the sensor internally adjusts the value of a programmable offset for each detector so that the output Z of each detector is equal to a constant that we will denote Z_{CAL} (which the sensor sets to the midpoint of the digital pixel range, i.e., 16384).

Let D_1 and D_2 be two detectors selected from the 2D FPA, the outputs Z_1, Z_2 are then given by

$$Z_1 = G_1 X_1 + G_1 Y_1 + K_1 + C_1$$
$$Z_2 = G_2 X_2 + G_2 Y_2 + K_2 + C_2$$

where now the numbers refer to different detectors, and as before G is gain, X is the incident radiance from the external scene, Y is the internal self-heating radiance on the detectors, K is a possible offset variation from detector to detector, and C represents the programmable calibration offset for each detector. If we fill the FOV with a uniform blackbody at some temperature (T_{CAL}) producing a uniform radiance X_{CAL} on the FPA and activate the calibration function, we have

$$Z_1 = Z_{CAL} = G_1 X_{CAL} + G_1 Y_1 + K_1 + C_1$$
$$Z_2 = Z_{CAL} = G_2 X_{CAL} + G_2 Y_2 + K_2 + C_2$$

so

$$C_1 = Z_{CAL} - G_1 X_{CAL} - G_1 Y_1 - K_1$$
$$C_2 = Z_{CAL} - G_2 X_{CAL} - G_2 Y_2 - K_2$$

where X_{CAL} is calculated by spectrally integrating Planck's function from 8 to 12 μm at $T = T_{CAL}$. C_1 and C_2 will now retain these values until the sensor is either recalibrated or powered down. Now, for some arbitrary externally supplied radiance X_1, X_2 on the FPA we have

$$Z_1 = G_1 X_1 + G_1 Y_1 + K_1 + C_1$$
$$Z_1 = G_1 X_1 + G_1 Y_1 + K_1 + Z_{CAL} - G_1 X_{CAL} - G_1 Y_1 - K_1$$
$$Z_1 = G_1 X_1 + Z_{CAL} - G_1 X_{CAL}$$
$$Z_1 = G_1 (X_1 - X_{CAL}) + Z_{CAL}$$

and, similarly

$$Z_2 = G_2 (X_2 - X_{CAL}) + Z_{CAL}$$

therefore, the output of each detector depends only on the individual detector gain (which we know) and the external radiance incident on the detector. The spatially varying components (Y and K) have been removed.

Rearranging to solve for radiance input X as a function of the output intensity Z we have

$$X = \frac{(Z - Z_{CAL})}{G} + X_{CAL}$$

Given a precomputed lookup table RAD2TEMP of T, X pairs we can take the radiance value X and look up the corresponding temperature T for any pixel in the image. Hence we have computed temperature T_C as a function of radiance X on any detector:

$$T_C = \text{RAD2TEMP}[X]$$

5.3 Operational Considerations

Upon testing the system in a scenario that more accurately reflected the operational usage anticipated (i.e., with up to 10 ft between the sensor and the measured object), we encountered an unexpected discrepancy between the computed temperature values and the actual values as reported by the blackbody temperature display. We decided to assume that actual temperature values would vary linearly with the values computed by the above method. Therefore, we added a second step to the calibration procedure that requires the user to collect an image at a higher temperature than the calibration temperature T_{CAL}. Also, this second measurement would be made at a sensor to blackbody distance of approximately 10 ft. So now, we have two computed temperatures and two actual corresponding temperatures. Then we compute a slope and y-intercept describing the (assumed) linear relationship between the computed and actual temperatures.

$$T_A = MT_C + B$$

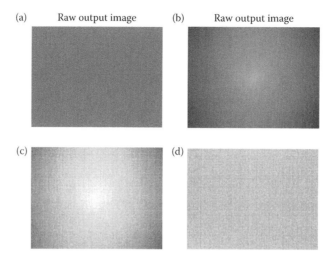

FIGURE 5.5 Gain calculation. (a) *Low temperature*: uniform 30°C black body source, calibrated at 30°, $\mu = 16381$, $\sigma = 1.9$ (raw counts). (b) *High temperature*: uniform 40°C black body source, calibrated at 30°, $\mu = 16842$, $\sigma = 11.3$ (raw counts). (c) *Processed using uniform gain*: uniform 35°C black body source, calibrated at 30°, $\mu = 34.74$, $\sigma = 0.12$ (°C). (d) *Processed using computed gain*: uniform 35°C black body source, calibrated at 30°, $\mu = 34.74$, $\sigma = 0.04$ (°C).

where

$$M = \frac{T_{A_2} - T_{A_1}}{T_{C_2} - T_{C_1}}$$

and

$$B = T_{A_1} - MT_{C_1}$$

where T_A is the adjusted temperature, T_C is the computed temperature from the previously described methodology. M and B are recomputed during each nonuniformity calibration.

From a camera perspective, the SiTF in digital counts for a $DC = A + BeT^4$.

$$T = \left(\frac{DN - A}{B\varepsilon} \right)^{1/4}$$

By adding this second step to the calibration process, we were able to improve the accuracy of the computed temperature to within a tenth of a degree for the test data set (Figure 5.5).

5.4 Infrared Optical Considerations

This section focuses mainly on the MWIR and LWIR since optics in the near infrared (NIR) and SWIR and very similar to that of the visible. This area assumes a basic knowledge of optics, and applies that knowledge to the application of infrared systems.

5.4.1 Resolution

Designing an IR optical system is first initiated by developing a set of requirements that are needed. These requirements will be used to determine the focal plane parameters and desired spectral band.

These parameters in turn drive the first-order design, evolving into the focal length, entrance pupil diameter, FOV, and *f*/number.

If we start with the user inputs of target distance, size, cycle criteria (or pixel criteria), and spectral band, we can begin designing our sensor. First, we calculate the minimum resolution angle (α, in radians). This parameter is also known as the instantaneous field of view (IFOV), and can be calculated by

$$\alpha = \frac{\text{Size}_{\text{tar}}}{(\text{Range})(2 \times \text{Cycles})}$$

or

$$\alpha = \frac{\text{Size}_{\text{tar}}}{(\text{Range})(\text{Pixels})}$$

On the basis of the wavelength, we can determine the minimum entrance pupil diameter that is necessary to distinguish between the two blur spots.

$$\text{EPD} = \frac{1.22\lambda}{\alpha}$$

Knowing the detector size and pixel pitch we can then determine the minimum focal length based on our IFOV that is required to meet our resolution requirements. Longer focal lengths will provide better spatial resolution.

$$\text{EFL} = \frac{\text{Pitch}}{\alpha}$$

Once we know our focal length, we can determine our FOV based on the height of the detector and the focal length. The vertical and horizontal fields of view are calculated separately based on their respective dimensions.

$$\theta = 2\tan\left[\frac{0.5h}{\text{EFL}}\right]$$

where *h* is the full detector height (or width). Depending on the system requirements, the size of the FPA may want to be scaled to match the desired FOV. Arrays with more pixels provide for greater resolution for a given FOV. However, smaller arrays cost less. Scaling an optical system to match the FOV for a different array format results in the scaling of the focal length, and thus the resolution.

The *f*/number is then calculated as the ratio of the focal length to the entrance pupil diameter.

$$f/\text{number} = \frac{\text{EFL}}{\text{EPD}}$$

The *f*/number of the system can be further optimized based on two parameters: the sensitivity and the blur circle. The minimum *f*/number is already set based on the calculated minimum entrance pupil diameter and focal length. The *f*/number can be adjusted to improve sensitivity by trying to optimize the blur circle to match the diagonal dimension of the detector pixel. This method provides a way to maximize the amount of energy on the pixel while minimizing aliasing.

$$f/\text{number} = \frac{\text{Pixel}_{\text{diagonal}}}{2.44\lambda}$$

A faster f/number is good in many ways as it can improve the resolution by reducing the blur spot and increasing the optics cutoff frequency. It also allows more signals to the detector and gives a boost in the signal to noise ratio. A fast f/number is absolutely necessary for uncooled systems because they have to overcome the noise introduced from operation at warmer temperatures. Faster f/numbers also help in environments with poor thermal contrast. However, a faster f/number also means that the optics will be larger, and the optical designs will be more difficult. Faster f/numbers introduce more aberrations into each lens making all aberrations more difficult to correct, and a diffraction limited system harder to achieve. A cost increase may also occur due to larger optics and tighter tolerances. A tradeoff has to occur to find the optimal f/number for the system. The table below shows the tradeoffs between optics diameter, focal length, resolution, and FOV.

5.5 Spectral Requirement

The spatial resolution is heavily dependent on the wavelength of light and the f/number. Diffraction limits the minimum blur size based on these two parameters.

$$d_{\text{spot}} = 2.44\lambda(f/\text{number})$$

The table given below compares the blur sizes for various wavelengths and f/numbers:

	Spot Size (μm)			
f/Number	$\lambda = 0.6$	$\lambda = 2$	$\lambda = 4$	$\lambda = 10$
1	1.5	4.9	9.8	24.4
2.5	3.7	12.2	24.4	61.0
4	5.9	19.5	39.0	97.6
5.5	8.1	26.8	53.7	134.2
7	10.2	34.2	68.3	170.8

First-Order Parameters Resolution 7.510

	f/Number	Focal Length	Field of View	Entrance Pupil Diameter
Impact resolution	A faster f/number results in a smaller optics blur due to diffraction. The result is an improved diffraction limit, and thus better spatial resolution. However, two things can adversely impact this improvement. Aberrations increase with faster f/number, potentially moving a system out of being diffraction limited, in which case the faster f/number can potentially hurt you. Also, a fixed	Increasing the focal length will increase spatial resolution. However, the amount of improvement may be limited by the size of the allowed aperture, as having to go to a slower f/number can reduce some of the gains. Longer focal lengths also result in narrower FOVs, which can	Narrower FOVs are the direct result of longer focal lengths. Longer focal lengths result in improved spatial resolution. The FOV can also vary by changing the size of the FPA. If all other parameters are maintained, and the FPA size is increased merely through the addition of more pixels, then the FOV increases without impacting resolution. If the number of pixels stay the same, but the pixel size is increased, then	The entrance pupil diameter (EPD) can impact the resolution in three ways. Increasing the EPD while maintaining focal length improves resolution by utilizing a faster f/number. Increasing the EPD while maintaining a constant f/number results in a longer focal length, and thus increase spatial resolution. Maintaining a

	f/Number	Focal Length	Field of View	Entrance Pupil Diameter
	front aperture system will have to reduce its focal length to accommodate the faster f/number, thus reducing spatial resolution through a change in focal length	be limited by stabilization issues	resolution is decreased. If pixel size is constant, number of pixels is increased, and focal length is scaled to maintain a constant FOV, then the resolution scales with the focal length	constant EPD while increasing focal length results in an improved spatial resolution due to the focal length, but a reduced resolution due to diffraction. The point where there is no longer significant improvement is dependent on the pixel size

5.5.1 Depth of Field

It is often desired to image targets that are located at different distances in object space. Two targets that are separated by a distance will both appear to be equally in focus if they fall within the depth of field (DOF) of the optics. A target that is closer to the optics than the DOF will appear defocused. In order to bring the out-of-focus target back in focus, it is required to refocus the optics so that the image plane shifts back to the location of the detector. Far targets focus shorter than near targets. If it is assumed that the optics are focused for an infinite target distance, the near DOF, known as the hyperfocal distance (HFD), can be-1pt found by

$$\text{HFD} = \frac{D^2}{2\lambda}$$

where D is the entrance pupil diameter, and λ is the wavelength in the same units as the diameter. This approximation can be made in the infrared because we can assume that we are utilizing a diffraction-limited system.

This formula is dependent only on the aperture diameter and wavelength. It is easily seen that shorter wavelengths and larger apertures have larger HFDs. This relationship is based on the Rayleigh limit which states that as long as the wavefront error is within a 1/4 wavelength of a true spherical surface, it is essentially diffraction limited. The depth of field can then be improved by focusing the optics to be optimally focused at the HFD. The optics are then in focus from that point to infinity based on the 1/4 wave criteria, but also for a 1/4 wave near that target distance. The full DOF of the system then becomes half the HFD to infinity. This is-1pt approximated by

$$\text{DOF} = \frac{D^2}{4\lambda}$$

If the region of interest is not within these bounds, it is possible to approximate the near and far focus points for a given object-1pt distance with

$$Z_{\text{near}} = \frac{\text{HFD} \times Z_\text{o}}{\text{HFD} + (Z_\text{o} - f)}$$

$$Z_{\text{far}} = \frac{\text{HFD} \times Z_\text{o}}{\text{HFD} - (Z_\text{o} - f)}$$

where X is the object distance that the objects are focused for and f is the focal length of the optics.

Work has been done with digital image-processing techniques to improve the DOF by applying a method known as Wavefront Coding, developed by Cathey and Dowksi at the University of Colorado.

This effectiveness of this technique has been well documented and demonstrated for the visible spectral band. Efforts are underway to demonstrate the effectiveness with LWIR uncooled cameras.

5.6 Selecting Optical Materials

The list of optical materials that transmit in the MWIR and LWIR spectrum is very short compared to that found in the visible spectrum. There are 21 crystalline materials and a handful chalcogenide glasses that transmit radiation at these wavelengths. Of the 21 crystalline materials, only 6 are practical to use in the LWIR, and 9 are usable in the MWIR. The remaining possess poor characteristics such as being hygroscopic, toxic to the touch, and so on, making them impractical for use in a real system. The list grows shorter for multispectral applications as only four transmit in the visible, MWIR, and LWIR. The chalcogenide glasses are an amorphous conglomerate of two or three infrared transmitting materials. Table 5.1 lists practical infrared materials along with a chart of their spectral bands. Transmission losses due to absorption can be calculated from the absorption coefficient of the material at the specified wavelength.

$$T_{abs} = e^{-at}$$

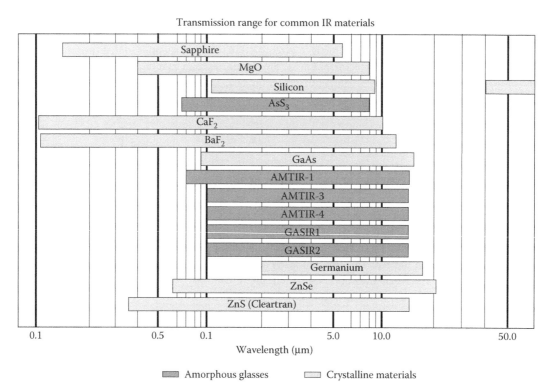

5.6.1 Special Considerations

In the LWIR, germanium is by far the best material to use for color correction and simplicity of design due to its high index of refraction and low dispersion. It is possible to design entire systems with germanium, but there are some caveats to this choice. Temperature plays havoc on germanium in two ways. It has a very high dn/dT (0.000396 K^{-1}), defocusing a lens that changes temperature of only a few degrees. It also has an absorption property in the LWIR for temperatures greater than 57°C. As the optic temperature rises above this point, the absorption coefficient increases, reducing the transmission. The high cost of germanium can also be a factor. It is not always good choice for a low-cost sensor, as

TABLE 5.1 Infrared Optical Materials

| Material | Refractive Index | | dµ/dT (K⁻¹) | Spectral Range (µm) |
	4 µm	= 10 µ		
Germanium	4.0243	4.0032	0.000396	2.0–17.0
Gallium arsenide	3.3069	3.2778	0.000148	0.9–16.0
ZnSe	2.4331	2.4065	0.000060	0.55–20.0
ZnS (cleartran)	2.2523	2.2008	0.000054	0.37–14.0
AMTIR-1	2.2514	2.4976	0.000072	0.7–14.0
AMTIR-3	2.6200	2.6002	0.000091	1.0–14.0
AMTIR-4	2.6487	2.6353	−0.000030	1.0–14.0
GASIR 1	2.5100	2.4944	0.000055	1.0–14.0
GASIR 2	2.6038	2.5841	0.000058	1.0–14.0
Silicon	3.4255	N/A	0.000160	1.2–9.0
Sapphire	1.6753	N/A	0.000013	0.17–5.5
BaF_2	1.4580	1.4014	−0.000015	0.15–12.5
CaF_2	1.4097	1.3002	−0.000011	0.13–10.0
As_2S_3	2.4112	2.3816	−0.0000086	0.65–8.0
MgO	1.6679	N/A	0.000011	0.4–8.0

the lens may end up costing more than the FPA. A good thermal match to compensate for the dn/dT of germanium is AMTIR-4. Its negative dn/dT provides for an excellent compensator for the germanium. It is possible to design a two lens germanium/AMTIR-4 lens system that does not require refocusing for over a 60°C temperature range.

The low-cost optics are silicon, ZnS, and the chalcogenide glasses. Silicon is only usable in the MWIR, and although the material is very inexpensive, its hardness can make it very difficult to diamond turn, and thus expensive. Although it does not diamond turn well, it does grind and polish easily, providing a very inexpensive solution when complex surfaces such as aspheres and diffractives are not used. ZnS is relatively inexpensive to germanium and ZnSe, but is relatively expensive to silicon and the chalcogenide glasses. The chalcogenide glasses are by far the least expensive to make and manufacture making them an excellent solution for low-cost system design. There are three types of the chalcogenide glasses that are moldable. AMTIR-4, a product of Amorphous Materials, Inc., has a lower melting point that the other chalcogenides making it the easiest to mold. Another material GASIR1 and GASIR2, products of Umicore, have also been demonstrated as being moldable. They have similar optical properties to that of AMTIR-1 and AMTIR-3.

5.6.2 Coatings

The high indices of refraction of most infrared materials lead to large fresnel losses, and thus require AR coatings. The transmission for a plane uncoated surface is shown below. In air, $n_1 = 1$.

$$T = 1 - \left(\frac{n_1 - n_2}{n_1 + n_2} \right)^2$$

The total reflectance off both sides of an uncoated plate is the multiplication of the two surfaces. This is in turn multiplied by the transmission of the material due to absorption. An example is given below:

Example

Uncoated zinc selenide flat, $n = 2.4$ in air, $t = 1.5$ cm thickness, absorption = 0.0005.
Total transmission through both sides = (0.83)(0.999)(0.83) = 0.688.

Standard AR coatings are readily available for all of the materials previously listed. Generally, better than 99% can be expected for an AR-coated lens for either the MWIR or LWIR. If dual band operation is required, expect this performance to drop to 96%, and for the price to go up. The multispectral coatings are more difficult to design, and result in having many more layers. Infrared beamsplitter coatings can be difficult and expensive to manufacture. Care should be taken in specifying both the transmission and reflection properties of the beamsplitter. Specifications that are too stringent often lead to huge costs and schedule delays. Also, it is very important to note that the choice of which wavelength passes through and which wavelength is reflected can make a significant impact on the performance of a beamsplitter. In general, transmitting the longer wavelength and reflecting the shorter wavelength will boost performance and reduce cost.

5.6.3 Reflective Optics

Reflective optics can be a very useful and effective design too in the infrared. Reflective optics have no chromatic aberrations, and allow for diffraction-limited solutions for very wide spectral bands. However, the use of reflective optics is somewhat limited to narrow FOVs and have difficulty with fast *f*/numbers. In addition, the type of reflective system can impact the performance fairly significantly for longer wavelengths. The most common type of design is the Cassegrain, which has two mirrors that are aligned on the same optical axis. The secondary mirror acts as an obscuration to the primary, which results in a degraded MTF due to diffraction around the obscuration. This effect is not apparent in most wavebands because the MTF loss occurs after the Nyquist frequency of the detector. However, this is not the case for the LWIR where the MTF drop occurs before Nyquist. To overcome this effect, most reflective systems used in the LWIR are off-axis reflective optics. These optics will provide diffraction-limited MTF as long as the *f*/numbers do not get too fast, and the FOVs do not get too large. The off-axis nature makes these reflective systems hard to align, and expensive to manufacture.

Acknowledgments

The authors thank John O'Neill, Jason Zeibel, Tim Mikulski, and Kent McCormack for the data; EO-IR Measurements, Inc. for developing the graphical user interface for the medical infrared imaging camera; and Leonard Bonnell, Vipera Systems, Inc., for the development of the infrared endoscope.

References

1. J. D'Agostino and C. Webb, 3-D analysis framework and measurement methodology for imaging system noise. *Proc. SPIE*, 1488, 110–121 (1991).
2. R.G. Driggers, P. Cox, and T. Edwards, *Introduction to Infrared and Electro-Optical Systems*. Artech House, Boston, MA, p. 8 (1998).
3. G.C. Holst, *Electro-Optical Imaging System Performance*. JCD Publishing, Winter Park, FL, p. 347 (1995).
4. M.W. Grenn, Recent advances in portable infrared imaging cameras. *Proc. IEEE–EMBS*, Amsterdam (1996).
5. M.W. Grenn, Performance of portable staring infrared cameras. *Proc. IEEE/EMBS* Oct. 30–Nov. 2, Chicago, IL, USA (1997).

Further Information

Holst, G.C. *Electro-Optical Imaging System Performance*. JCD Publishing, Winter Park, FL, p. 432 (1995).
Johnson, J. Analysis of image forming systems. *Proc. IIS*, 249–273 (1958).

Ratches, J.A. *NVL Static Performance Model for Thermal Viewing Systems*, USA Electronics Command Report ECOM 7043, AD-A011212 (1973).

Schade, O.H. Electro-optical characteristics of television systems. *RCA Review*, IX(1–4) (1948).

Sendall, R. and Lloyd, J.M. Improved specifications for infrared imaging systems. *Proc. IRIS*, 14, 109–129 (1970).

Vollmerhausen, R.H. and Driggers, R.G. NVTHERM: Next generation night vision thermal model. *Proc. IRIS Passive Sensors*, 1 (1999).

6

Physiology of Thermal Signals

David D. Pascoe
Auburn University

James B. Mercer
University of Tromsø

Louis de Weerd
*University Hospital of
North Norway*

6.1 Overview

William Herschel first recognized heat emitted in the infrared (IR) wave spectrum in the 1800s. Medical IR, popularly known as IR-thermography has utilized this heat signature since the 1960s to measure and map skin temperatures. Our understanding of the regulation of skin blood flow, heat transfers through the tissue layers, and skin temperatures has radically changed during these past 40 years, allowing us to better interpret and evaluate these thermographic measurements. During this same period of time, improved camera sensitivity coupled with advances in focal plan array technology and new developments in computerized systems with assisted image analysis have improved the quality of the noncontact, noninvasive thermal map or thermogram [1–3]. In a recent electronic literature search in Medline using the keywords "thermography" and "thermogram" more than 5000 hits were found [4]. In 2003 alone, there were 494 medical references, 188 basic science, 148 applied science (14 combined with Laser Doppler and 28 combined with ultrasound research), and 47 in biology including veterinary medicine [5]. Further databases and references for medical thermography since 1987 are available [6].

This review will highlight some of the literature and applications of thermography in medical and physiological settings. More specifically, IR thermography and the structure and functions of skin thermal microcirculation can provide a better understanding of (1) thermoregulation and skin thermal patterns (e.g., comfort zone, influences of heat, cold, and exercise stressors, etc.), (2) assess skin blood perfusion (e.g., skin grafts), (3) observe and diagnose vascular pathologies that manifest thermal disturbances in the cutaneous circulation, (4) evaluate thermal therapies, and (5) monitor patient/subject/athlete's recovery as evidenced by the resumption of normal thermal patterns during the rehabilitation process for some musculoskeletal injuries.

At the outset, it needs to be stressed that an IR image is a visual map of the skin surface temperature that can provide accurate thermal measurement but cannot quantify measurements of blood flow to the skin tissue. It is also important to stress that recorded skin temperatures may represent heat transferred from within the core through various tissue layers to the skin that may be the result of conductive or radiant heat provided from an external thermal stressor. To interpret thermographic images and thermal measurement, a basic understanding of physiological mechanisms of skin blood flow and factors that influence heat transfers to the skin must be considered to evaluate this dynamic process. With this understanding, objective data from IR-thermography can add valuable information and complement other methodologies in the scientific inquiry and medical practices.

6.2 Skin Thermal Properties in Response to Stress

The thermal properties of the skin surface can change dramatically to maintain our core temperature within a narrowly defined survival range (cardiac arrest at 25°C to cell denaturation at 45°C) [7]. This remarkable task is accomplished despite a large variability in temperatures both from the hostile environment and internal production of heat from metabolism. Further perturbations to core and skin temperature may result from thermal stressors associated with injury, fever, hormonal milieu, and disease. The skin responds to these thermal challenges by regulating skin perfusion.

During heat exposure or intense exercise, skin blood flow can be increased to provide greater heat dissipating capacity. The thermal properties of the skin combined with increased cutaneous circulation operate as a very efficient radiator of heat (emitted radiant heat of 0.98 compared to a blackbody source of 1.0) [8]. Evaporative cooling of sweat on the skin surface further enhances this heat dissipating process. Under hyperthermic conditions, the skin masterfully combines anatomical structure and physiological function to protect and defend the organism from potentially lethal thermal stressors by regulating heat transfers between core, skin, and environment. When exposed to a cold environment, the skin surface nearly eliminates blood flow and becomes an excellent insulator. Under these hypothermic conditions, our skin functions to conserve our body's core temperature. It accomplishes this by reducing convective heat transfers, minimizing heat losses from the core and lessening the possibility of excessive cooling from the environment.

The ability of the skin to substantially increase blood flow, far in excess of the tissue's metabolic needs, alludes to the tissue's role and potential in heat transfer mechanisms. The nutritive needs of skin tissue has been estimated at 0.2 mL/min per cubic centimeter of skin [9], which is considerably lower than the maximal rate of 24 mL/min per cubic centimeter of skin (estimated from total forearm circulatory measurement during heat stress) [10]. If one were to approximate skin tissue as 8% of the forearm, then skin blood would equate to 250–300 mL/100 mL of skin per minute [11]. Applying this flow rate to an estimated skin surface of 1.8 m² (average individual), suggests that approximately 8 L of blood flow could be diverted to the skin to dissipate heat at rate of 1750 W to the environment [12,13]. This increased blood flow required for heat transfers from active muscle tissue and skin blood flow for thermoregulation is made available through the redistribution of blood flow (splanchnic circulatory beds, renal tissues) and increases in cardiac output [14]. The increased cardiac output has been suggested to account for two-thirds of the increased blood flow needs, while redistribution provides the remaining one-third [15]. Several good reviews are available regarding cutaneous blood flow, cardiovascular function, and thermal stress [14–19].

6.3 Regulation of Skin Blood Flow for Heat Transfers

In the 1870s, Goltz, Ostromov, and others injected atropine and pilocaprine into the skin to help elucidate the sympathetic neural innervation of the skin tissue for temperature, pressure, pain, and the activation of the sweat glands [20]. The reflex neural regulation of skin blood flow relies on both sympathetic vasoconstrictor and vasodilator controls to modulate internal heat temperature transfers to the skin. The vasoconstrictor system is responsible for eliminating nearly all of the blood flow to the skin during exposure to cold. When exposed to thermal neutral conditions, the vasoconstrictor system maintains the vasomotor tone of the skin vasculature from which small changes in skin blood flow can elicit large changes in heat dissipation. Under these conditions, the core temperature is maintained; mean skin temperature is stable, but dependent upon the extraneous influences of radiant heat, humidity, forced convective airflow, and clothing. A naked individual in a closed room with ambient air temperature between 27°C and 28°C can retain thermal equilibrium without shivering or sweating [21]. Slightly dressed individuals are comfortable in a neutral environment with temperatures between 20°C and 25°C. This thermoneutral zone provides the basis for the clinical testing standards for room temperatures being set at 18–25°C during IR thermographic studies. Controlling room test conditions is important when measuring skin temperature responses as changes in ambient temperatures can alter the fraction of flow shared between the musculature and skin [19]. Under the influence of whole-body hyperthermic conditions, removal of the vasomotor vasoconstrictor tone can account for 10–20% of cutaneous vasodilation, while the vasodilator system provides the remaining skin blood flow regulation [10]. Alterations in the threshold (onset of vascular response and sweating) and sensitivity (level of response) in vasodilation blood flow control can be related to an individual's level of heat acclimation [22], exercise training [22], circadian rhythm [23,24], and women's reproductive hormonal status [10]. Recent literature suggests that some observed shifts in the reflex control are the result of female reproductive hormones. Both estrogen and progesterone have been linked to menopausal hot flashes [10].

Skin blood flow research has identified differences in reflex sympathetic nerve activation for various skin surface regions. In 1956, Gaskell demonstrated that sympathetic neural activation in the acral regions (digits, lips, ears, cheeks, and palmer surfaces of hands and feet) is controlled by adrenergic vasoconstrictor nerve activity [20]. In contrast, I.C. Roddie in 1957 demonstrated that in nonacral regions, the adrenergic vasoconstrictor activity accounts for less than 25% of the control mechanism [25]. In the nonacral region, the sympathetic nervous system has both adrenergic (vasoconstriction) and nonadrenergic (vasodilator) components. While the vasodilator activity in the nonacral region is well accepted, the vasoconstrictor regulation is not fully understood and awaits the identification of neural cotransmitters that mediate the reductions in blood flow. For a more in-depth discussion of regional sympathetic reflex regulation, see Charkoudian [10] and Johnson and Proppe [13].

A further distinction in blood flow regulation can be found in the existence of arteriovenous anastamoses (AVA) that are principally found in acral tissues but not commonly found in nonacral tissues of the legs, arms, and chest regions [10,26,27]. The AVAs are thick-walled, low-resistance vessels that provide a direct blood flow route from arterioles to the venules. The arterioles and AVA, under sympathetic adrenergic vasoconstrictor control, modulate and substantially control flow rates to the skin vascular plexuses in these areas. When constricted, blood flow into subcutaneous venous plexus is reduced to a very low level (minimal heat loss); while, when dilated, extremely rapid flow of warm blood into the venous plexus is allowed (maximal heat loss). The skin sites where these vessels are found are among those where skin blood flow changes are discernible to the IR-thermographer. While the AVA are most active during heat stress, their thick walls and high-velocity flow rates do not support their significant role in heat transfers to adjoining skin tissue [12].

Localized cooling of the skin surface can cause skin blood flow vasoconstriction induced by the stimulation of the nonadrenergic sympathetic nerves. This localized cooling or challenge can be used as a diagnostic tool by IR thermographers to identify clinically significant alterations in skin blood flow response. Using a cold challenge test, Francis Ring developed a vasoplasticity test for Raynaud's syndrome based on the temperature gradient in the hand following a cold water immersion (20°C for 60 s) [28]. Challenge testing and IR imaging has also been used to evaluate blood flow thermal patterns in patients with carpal tunnel syndrome pre- and post-surgery [29]. Skin blood flow vasodilation in response to local heating is stimulated by the release of sensory nerve neuropeptides or the nonneural stimulation of the cutaneous arteriole by nitric oxide. Thermographic imagers have exploited this localized warming response to provide a skin blood flow challenge [10].

As stated earlier, conditions of thermal stress from heat exposure and increased metabolic heat from exercise necessitate increases in skin blood flow to transfer the heat to the environment. This could have serious blood pressure consequences if it were not for baroreflex modulation of skin blood flow in regulating both sympathetic vasoconstriction and vasodilation [30,31]. With mild heat stress for 1 h (38°C and 46°C, 42% relative humidity), cardiac output was not significantly impacted [32–34]. When exposing the individual to longer duration bouts and higher temperatures, significant changes in cardiac output have been observed. During these hyperthermic bouts, skin blood flow will withdraw vasodilation in nonacral regions in response to situations that displace blood volumes to the legs (lower body negative pressure or upright tilting) [35]. In contrast, under normothermic conditions, skin blood flow will demonstrate a sympathetic vasoconstriction to these same blood volume situations. Thus, it appears that the baroreceptor response can activate either sympathetic pathway. Withdrawing vasodilation under normothermic conditions was not an option in this inactive system [36].

In summary, skin blood flow during whole-body thermal stress is regulated by neural reflex control via sympathetic vasoconstriction and vasodilation. There are structural (AVA) and neural mechanisms (vasoconstriction vs. vasodilation) differences between the acral and nonacral regions. During local thermal stress, stimulation of the sensory afferent nerve, nitric oxide stimulation of cutaneous arteriole, and inhibition of sympathetic vasoconstrictor system regulate changes in local blood flow. During thermal stress and increased heat from exercise metabolism, the skin blood flow can be dramatically increased. This increase in skin blood flow is matched to increased cardiac output and peripheral resistance to maintain blood pressure.

6.4 Heat Transfer Modeling Equations for Microcirculation

The capacity and ability of blood to transfer heat through various tissue layers to the skin can be predicted from models. These models are based on calculations from tissue conductivity, tissue density, tissue specific heat, local tissue temperature, metabolic or externally derived heat sources, and blood velocity. Many current models were derived from the 1948 Pennes model of blood perfusion, often referred to as the "bioheat equation" [9]. The bioheat equation calculates volumetric heat that is equated to the proportional volumetric rate of blood perfusion. The Pennes model assumed that thermal equilibrium occurs in the capillaries and venous temperatures were equal to local tissue temperatures. Both of these assumptions have been challenged in more recent modeling research. The assumption of thermal equilibrium within capillaries was challenged by the work of Chato [37] and Chen and Holmes in 1980 [38]. Based on this vascular modeling for heat transfer, thermal equilibrium occurred in "thermally significant blood vessels" that are approximately 50–75 μm in diameter and located prior to the capillary plexus. These thermally significant blood vessels derive from a tree-like structure of branching vessels that are closely spaced in countercurrent pairs. For a historical perspective of heat transfer bioengineering, see Chato in Reference 37 and for a review of heat transfers and microvascular modeling, see Baish in [39].

This modeling literature provides a conceptual understanding of the thermal response of skin when altered by disease, injury, or external thermal stressors to skin temperatures (environment or application of cold or hot thermal sources, convective airflow, or exercise). From tissue modeling, we know that tissue is only slightly influenced by the deep tissue blood supply but is strongly influenced by the cutaneous circulation. With the use of IR-thermography, the skin temperatures can be accurately quantified and the thermal pattern mapped. However, these temperatures cannot be assumed to represent thermal conditions at the source of the thermal stress. Furthermore, the thermal pattern only provides a visual map of the skin surface in which heat is dissipated throughout the skin's multiple microvascular plexuses.

6.5 Cutaneous Circulation Measurement Techniques

A brief review of some of the techniques and procedures that have been employed to reveal skin tissue structure, rates and variability of perfusion, and factors that provide regulatory control of blood flow are provided. This serves to inform the reader as to how these measurement techniques have molded our understanding of skin structure and function. It is also important to recognize the advantages and disadvantages each technique brings to our experimental investigations. It is the opinion of the authors that IR-thermography can provide complimentary data to information obtained from these other methodologies.

6.5.1 Procedures and Techniques

Visual and microscopic views of skin have provided scientists with a structural layout of skin layers and blood flow. Despite our understanding of the structural organization, we still struggle to understand the regulation and functioning of the skin as influenced by the multitude of external and internal stressors. Since ancient times, documents have recorded visible observations made regarding changes to skin color and temperature that underscore some of the skin's functions. These observations include increased skin color and temperature when someone is hyperthermic, skin flushing when embarrassed, and the appearance of skin reddening when the skin is scratched. In contrast, decreases of skin color and temperature are observed during hypothermia, when blood flow is occluded, or during times of circulatory shock. In addition to observations, testing procedures and techniques have been employed in search of an understanding of skin function. Skin blood flow has provided one of the greatest challenges and has been notoriously difficult to record quantitatively in terms of mL/min per gram of skin tissue.

6.5.2 Dyes and Stains

Stains and dyes have been a useful tool to investigate the structure of various histological tissue preparations. In 1889 Overton used Florescin, a yellow dye that glows in visible light, to visualize the lipid bilayer membrane structure [40]. Florescin is still used today to illuminate the blood vessel in various tissues. In the early 1900s, August Krogh was able to identify perfused capillaries by staining them with writer's ink before the tissue was excised and observed under the microscope. Using a different approach to observe the structure of skin blood flow, Michael Salmon in the 1930s developed a radio-opaque preservative mixture that was injected into fresh cadavers and produced detailed pictures of the arterial blood flow to skin tissue regions which were mapped for the entire skin surface [41]. Scientists have also used Evans Blue Dye to investigate changes in blood volume by calculating the dilution factor of pre- and post-samples. Evans Blue and Pontamine Sky Blue dyes have also been used to investigate skin microvascular permeability and leakage as induced by various stimuli [42]. A more recent staining technique involves the use of an IR absorbing dye, indocyanine green (ICG). The dye ICG fluoresces with invisible IR light when captured by special cameras sensitive to these light wavelengths. Recent publications have suggested that ICG video angiographies provide qualitative and quantitative data from which clinicians may assess tissue blood flow in burn wounds [43,44].

6.5.3 Plethysmography

The rationale for venous occlusion plethysmography (VOP) is that a partially inflated cuff around a limb exceeds the pressure of venous blood flow but does not interfere with arterial blood flow to the limb. This can be effectively accomplished when the cuff is inflated to a pressure just below diastolic arterial pressure. Consequently, portions of the limb distal to the cuff will swell as blood accumulates. The original VOP technology relied on measuring the rate of swelling as indicated by the displacement volume of the water-filled chamber around the limb. Later, gauges were used to record changes in limb circumference as the limb expanded. Either way, the geometric forces allows one to express the changes in terms of a starting volume, thus scales are labeled as mL/min per 100 mL in the illustrations of VOP data. Unfortunately, the 100 mL reference quantity refers to the whole limb (not the quantity of the skin) and the VOP technique provides discontinuous measurement, usually four measurements per minute. Currently, VOP represents the most reliable quantitative measure of skin blood flow. More recently, a modified plethysmography technique has been developed that equates changes in blood flow to the changes in electrical impedance, when a mild current is introduced into the blood flow and recorded by serially placed electrodes along the limb.

6.5.4 Doppler Measurement

The Doppler effect describes the shift in wavelength frequency or pitch that result when sound or light from any portion of the electromagnetic spectrum are influenced by the distance and directional movement of the object. Both ultrasound and laser Doppler techniques rely on this physical principle to make the skin blood flow measurement. A continuous-wave Doppler ultrasound emits a high-frequency sound wave from which reflected or echo sounds can be used to calculate the direction of flow and velocity of moving objects (circulating blood cells). When ultrasound equipment is pulsated, the depth of the circulatory flow vessel can be identified. The blood flow is assessed in Doppler units or volts and can be continuously monitored over a small surface area. Similar measurements can be made with the laser Doppler technique, but neither technique is able to yield quantitative blood flow values except through reference data obtained by VOP.

6.5.5 Clearance Measurement

In the 1930s, scientists relied on the thermal conductance method for determining blood flow in tissues. This methodology was based on measurement of heat dissipation in blood flow downstream from a known heat source (thermocouple). The accuracy of the methodology was assumed through strong correlations made with direct drop counting from isolated sheep spleens. However, these blood flow measurements had problems related to trauma associated with the obstruction and insertion of the thermocouple. During the 1950s and 1960s, injection of small amounts of radioactive isotopes was used in an attempt to better quantify skin blood flow [45]. In order to accurately measure skin blood flow, the isotopes had to be freely diffusible and able to cross the capillary endothelium to enter the blood stream. The blood flow measurement (mL blood flow per 100 g of tissue) obtained through this methodology was not reproducible and subjects were exposed to injected radioactive isotope substances.

6.5.6 Thermal Skin Measurement

Physicians and parents have often relied upon touching a child's forehead to identify a fever or elevated temperature. While this is a crude measure of skin temperature, it has demonstrated practical importance. The development of the thermistor has provided quantifiable measurements of the skin surface. This has been extensively used in research related to thermoregulation and in research in which the skin temperature for a particular physiological perturbation must be quantified. Mean skin temperature formulas have been developed in which various regional temperatures (3–15 sites) are combined or a weighted mean based on the DuBois formula for surface area is calculated from the regions measured. When measuring skin temperatures under the influence of colder environments, skin temperature distribution is heterogeneous, especially in the acral regions [46]. Under these conditions, the various placements of the thermistor will provide different temperature measurement that may not be indicative of the mean skin temperature for that region. In contrast, the skin surface temperature is more homogeneous under warm environmental conditions. To provide an accurate measurement, physiologists investigating mean skin temperatures have modified the number of thermistors required during testing as dictated by environmental temperatures [46]. Thermistor attachment to the skin surface is problematic as it creates a microenvironment between the probe, skin, and adhering tape and exerts a pressure on the site. Furthermore, one site placement of the thermistor cannot represent variable responses under conditions and within regions that demonstrate heterogeneous temperature distributions.

IR thermography provides a thermal map of the skin surface area by measuring the radiant heat that is emitted. Current IR-thermography machines provide accurate skin surface temperatures (<0.05°C) that are noninvasive, noncontact measurements through the use of stable detectors. These systems produce high-speed–high-resolution images from which thermal data can be pictorially and quantitatively stored and analyzed. While IR-radiation begins at wavelengths of 0.7 μm, current IR-thermographic imagers are operating at either the mid-range (3–5 μm) or long range (8–14 μm) wavelengths [1–3]. At these wavelengths the skin's ability to emit the radiant heat is 0.98 on a scale of 1.0 for a perfect radiator, blackbody surface [8]. IR-thermographers sometimes rely on "challenge tests" in which the skin thermal response is evaluated after cold or hot water immersion, convective airflow, or an exercise bout that alters skin blood flow. Under these thermally challenging conditions, the clinical importance of abnormalities of skin blood flow may be more apparent. However, one must also recognize that as the heat is being transferred through the various layers of skin, some of the heat is dissipated into the adjoining tissues. The heat decay as the blood traverses the layers of tissue and its dispersion pattern within the circulatory plexus of the skin may disguise the origin of the tissue producing the abnormal thermal response.

6.5.7 Measurement Techniques and Procedures Summary

Since ancient times, significant progress has been made in the quest to understand the anatomy and physiology of skin blood flow and the thermal heat transfers that are dissipated to the skin surface. Current measurement technologies are unable to quantify skin blood flow per skin tissue area (mL/min per 100 mL skin). At this time, our best estimations of skin blood flow come from VOP. VOP provides discontinuous blood flow measurements based on tissue perfusion of a whole limb extremity. In recent years, laser Doppler blood flow measures have been popular when investigating more localized tissue areas, but this technique must be calibrated through the data obtained by plethysmography. In the quest to understand skin surface heat transfers, noncontact IR-thermography provides researchers with accurate measures of skin temperature for specific locations and thermal maps of regions of interest. Thermographic images provide spatial and temporal changes in skin temperature that are representative of spatial and temporal changes in perfusion. However, the scientist and clinician should be aware of blood–skin heat transfer properties prior to interpreting and evaluating the thermal responses of skin perfusion or investigating pathologies that are thermally transferred to the skin.

6.6 Objective Thermography

The main focus of this section is to present some specific examples to show that IR thermal imaging is a useful, objective tool for medical and physiological-based investigations. Although IR thermal imaging cannot determine quantitative measurements of blood flow, this imaging can quantify skin thermal measurement that can be correlated to qualitative evaluations of skin blood flow. As with any technique, it is imperative for the IR-thermographer to understand the anatomical and dynamic physiological features (e.g., vasomotor activity) of the blood vessels involved in skin circulation in order to interpret skin temperature responses. This is particularly important in situations where skin blood flow and temperature are responding to some external or internal thermal stress (e.g., clinician's cold challenge testing, exercise, environment, etc.) or nonthermal stimuli (e.g., disease, injury, medications, etc.). When properly assessed, skin temperature can provide researchers, scientists, and physicians with valuable information about the blood flow and thermal regulation of organs and tissues (see Examples 6.6.1 through 6.6.7).

6.6.1 Efficiency of Heat Transport in the Skin of the Hand (Example 6.1)

The AVA are specialized vascular structures within acral regions. At normal body temperature, sympathetic vasoconstrictor nerves keep these anastomoses closed. However, when the body becomes overheated, sympathetic discharge is greatly reduced so that the anastomoses dilate allowing large quantities of warm blood to flow into the subcutaneous venous plexus, thereby promoting heat loss from the body (Figure 6.1a). In thermal physiology, hands and feet are well recognized as effective thermal windows. By controlling the amount of blood perfusing through these extremities, the temperature of the skin surface can change over a wide range of temperatures. It is at such peripheral sites that the effect of changes in blood flow, and therefore skin temperature are most clearly discernable.

The following example demonstrates how efficient skin blood flow in the hand is dissipating a large incoming radiant heat load. The radiant heat load was provided by a Hydrosun® type 501 water-filtered infrared-A (wIRA) irradiator. The Hydrosun allows a local regional heating of skin tissue with a higher penetration depth than that of conventional IR therapy. The unique principle of operation involves the use of a hermetically sealed water filter in the radiation path to absorb those IR wavelengths emitted by conventional IR lamps which would otherwise harm the skin (OH-group at 0.94, 1.18, 1.38, and 1.87 μm). The total effective radiation from the Hydrosun lamp was 400 mW/cm², which is about three times the intensity of the IR-A radiation from the sun. Throughout the time course of the experiment (Figure 6.1a), sequential digital IR-thermal images were taken every 2 s. With the image analysis software, five points within the radiation field of each image were selected for temperature measurement and used to construct temperature curves shown in Figure 6.1b.

In the experiment (Figure 6.1b) a rubber mat with an emissivity close to 1.0 was irradiated with the Hydrosun wIRA lamp for a period of 25 min at a standard distance of 25 cm. At this distance the circular irradiation field had a diameter of about 16 cm. Since the rubber material has a high emissivity it rapidly absorbs heat from the lamp. As can be seen from the time course of the five temperature curves in Figure 6.2 (five fixed measuring points within the radiation field), the center of the irradiated rubber mat rapidly reached temperatures over 90°C during the first 10 min of irradiation. After the first 10 min of irradiation, the hand of a healthy 54-year-old male subject was abruptly placed on the irradiated rubber mat and kept there for a further 10 min, before being removed. When the hand was placed onto the irradiation field, three of the five fixed temperature-measuring sites now measured skin temperature on the dorsal skin surface. Based on the temperature curves in Figure 6.1b, skin temperature on the surface of the hand directed toward the irradiator never increased higher than 40°C, while the two remaining temperature-measuring points on the rubber mat remained at their original high values. The patient suffered no thermal discomfort from placing his hand on the "hot" rubber mat, even though the temperature of the center of the rubber mat was more than 90°C before the hand was placed on it. This was presumably due to a combination of the low heat capacity of the rubber material combined with a high rate of skin blood flow that is capable of rapidly dissipating heat. The visual data created from the IR-thermal images coupled with the temperature curves which were calculated from sequential IR-thermal images clearly demonstrates the large heat transporting capacity of the skin.

6.6.2 Effect of a Reduced Blood Flow (Example 6.2)

During exposure to cold, heat loss from the extremities is minimized by reducing blood flow (vasoconstriction). In cool environments, this vasomotor tone results in a heterogeneous distribution of skin surface temperatures; while in a warm environment the skin surface becomes more homogeneous (see

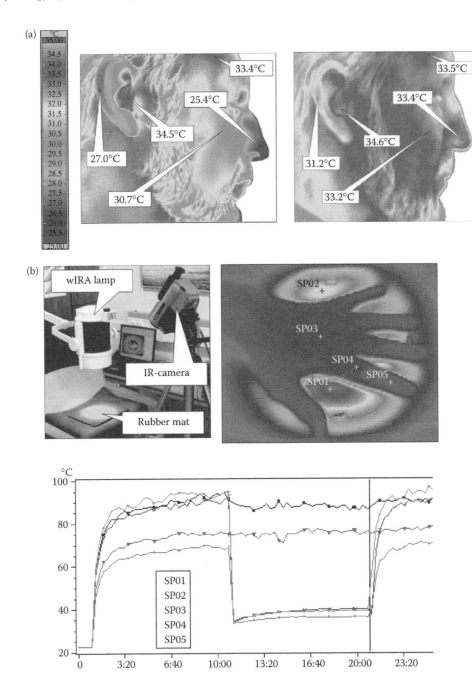

FIGURE 6.1 (a) A thermogram of a healthy 52-year-old male subject in a cold environment (ca. 15°C; left panel) and in a warm environment (ca. 25°C; right panel). In the cold environment arteriovenous anastomoses in the nose and the auricle region of the ears are closed, resulting in low skin temperatures at these sites. Also note reduced blood flow in the cheeks but not on the forehead, where the blood vessels in the skin are unable to vasoconstrict. (b) Efficiency of skin blood flow as a heat transporter. The photograph in the upper left panel shows a rubber mat being heated with a water-filtered IR-A irradiator at high intensity (400 mW/cm²) for a period of 20 min. In the lower panel the time course of surface temperature of the mat at five selected spots as measured by an IR-camera is given. During the last 10 min of the 20-min heating period, the left hand of a healthy 54-year-old male subject was placed on the mat. Note that skin surface temperature of the hand remains below 39°C.

Figure 6.2a). The importance of the integrity of blood flow in a limb as a whole can also be demonstrated at room temperature by totally cutting off the blood supply to the limb with the aid of a pressure cuff. This is demonstrated in the experiment shown in Figure 6.2b. In this experiment, skin surface on the back of the left hand of a 60-year-old healthy male subject was irradiated for a 10-min period with a wIRA lamp (see description in Section 6.6.1). The fingers were not heated. The heating was repeated twice. In the upper panel (Figure 6.2b) intact blood supply is demonstrated. In the lower panel (Figure 6.2b), thermal response to heating of the limb while blood flow was totally restricted using a blood

(a)   30 min equilibration in cool environment (20°C, 30% rh)

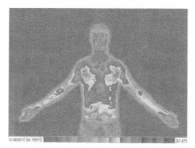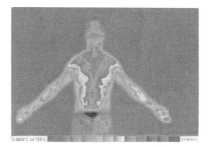

30 min equilibration in warm environment (41°C, 30% rh)

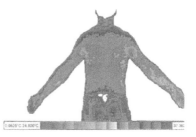

Range of temperature (°C) within regions at two climatic conditions.

	Anterior torso	Posterior torso	Anterior arms	Posterior arms	Palmer hands	Dorsal hands
Cool 20°C	5.0	5.1	6.3	4.5	5.2	4.0
Warm 41°C	3.0	3.2	3.2	3.6	4.0	3.5

FIGURE 6.2 (a) A thermoregulatory response of the torso when exposed to a cool or warm environment. Note the heterogeneous temperature distribution in the cool environment as opposed to the more homogeneous temperature distribution in the warm environment. Different shades of grey represent different temperatures. One might recognize the difficulty related to choosing one point (thermocouple data) as the reference value in a region. (b) Skin heating with and without intact circulation in a 65-year-old healthy male subject. The two panels on the right show the time course of six selected measuring sites (four spot measurements and two area measurements) of skin temperatures as determined by IR-thermography before, during, and after a period in which the skin surface on the back of the left hand was heated with a water-filtered IR-A irradiator at high intensity (400 mW/cm²) in two separate experiments. The results shown in the upper right panel were performed under a situation with a normal intact blood circulation, while the results shown in the lower left panel were made during total occlusion of circulation in the right arm by means of an inflated pressure cuff placed around the upper arm. The time of occlusion is indicated. The IR-thermogram on the left was taken just prior to the end of the heating period in the experiment described in the upper panel. The location of the temperature measurement sites on the hand (four spot measurements [Spot 1 to Spot 4] and the average temperature within a circle [Areas 5 and 6]) are indicated on the thermogram.

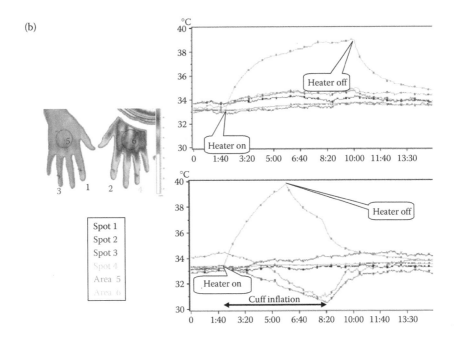

FIGURE 6.2 (Continued).

pressure cuff is presented. The time course of the temperature curves shown in the upper panel of Figure 6.2b demonstrates a gradual increase in skin temperature, eventually stabilizing just below 39°C. The finger skin temperature on the heated hand only showed a minor increase during the heating period. When the lamp was switched off the accumulated heat was rapidly dissipated and skin temperature of the heated area returned to the preheating level. In the lower panel the experiment was repeated, but as the heater was turned on, a pressure cuff placed around the upper arm was inflated to above systolic pressure to totally cut off blood supply to the arm. As can be seen, the rate of rise of skin temperature over the heated skin area was quite rapid, soon reaching a level deemed to be uncomfortably hot, after which the heater was turned off (at this stage the pressure cuff was still inflated). During the heating period finger skin temperature on the same hand steadily decreased (i.e., the fingers were passively cooling due to lack of circulation). After the heater was turned off the temperature of the back of the left hand also began to decrease indicating a passive cooling. When the pressure cuff was reopened and blood flow to the arm reestablished, the rate of cooling on the back of the hand increased (active cooling) and the skin temperature of the cooled fingers also rapidly returned to normal.

While the same result could have been gained by using other methods to measure skin temperature, such as thermocouples, the visual effect gained by using IR-thermography provides much more information than a single-point measurement. Various points or areas of interest can be investigated during and after the experimental procedures. Modern digital image processing software provides endless possibilities, especially with systems allowing the recording of sequential IR-images. Today modern firewire technology permits one to record IR-images at very high frequencies (100 Hz and greater).

6.6.3 Median Nerve Block in the Hand (Example 6.3)

The nervous supply to the hand is via three main nerves: radial, ulnar, and median nerve. The approximate distribution of the sensory innervation by these nerves is shown in Figure 6.3a. With a sudden removal (or disruption) of the sympathetic discharge to one of the main nerves, a maximal vasodilatation of blood vessels in the area supplied by the nerve is observed. Such a vasodilatation will result in a significant increase in skin blood flow to that area and therefore in a rise in skin temperature. The IR-thermogram in Figure 6.3b shows such a response. The subject is a 40-year-old female suffering from excessive sweating in the palms of her hands, a condition known as hyperhidrosis palmaris. Excessive sweating causes cooling of the skin surface. The blue color in Figure 6.3b represents areas of excessive sweating as well as areas of lowest skin temperature. Hyperhidrosis palmaris is treated by subcutaneous injections of botulinum toxin. The skin in the palm of the hand is very sensitive and anesthesia is therefore required. A very effective way to provide adequate anesthesia is the median nerve block. At the level of the wrist, a local anesthetic is injected around the median nerve and blocks the nervous activity in the median nerve. The resulting increase in skin temperature on the palmar side of the left hand after such an injection is clearly seen in Figure 6.3b. One sees that part of the thumb and half of the fourth finger show no increase in temperature, closely matching the predicted area of distribution for this nerve in Figure 6.3a. Although a median nerve block was applied to the right hand, neither vasodilatation nor a change in skin temperature was seen. The botulinum toxin injections on this side were painful. The placement of the anesthetic was therefore incorrect, and as a result, normal nervous activity in the right hand was observed. In Figure 6.3c,d, the thermograms show the hands of a 36-year-old female patient. Her left wrist (middle of the red circle) was punctured with a sharp object, resulting in partial nerve damage. The strong vasodilatory response, due to the nerve damage, results in an increased skin temperature as shown in Figure 6.3d.

The diagnosis of acute nerve damage can be a challenge for a surgeon. In the acute situation, pain makes a proper physical examination often impossible. To evaluate the extent of the nerve injury, cooperation of the patient is necessary and the use of local anesthesia can mask the extent of the injury. IR-thermography proves to be a helpful, noninvasive diagnostic tool by visualizing changes in skin temperature and, indirectly blood flow to the skin, due to the nerve injury.

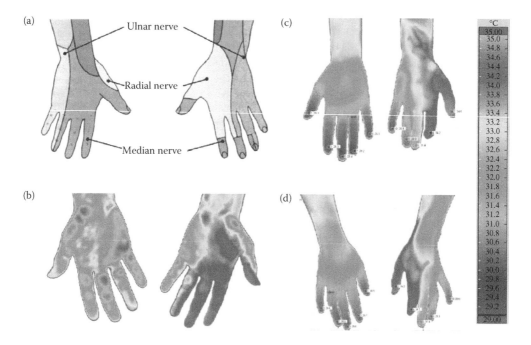

FIGURE 6.3 (a) The distribution of cutaneous nerves to the hand. (b) IR-thermogram of a 40-year-old female patient following a successful nerve block of the left median nerve. (c) and (d) IR-thermograms of the hands of a 36-year-old female patient whose left wrist (middle of the red circle) was punctured with a sharp object resulting in partial nerve damage (motor and sensory loss). The strong vasodilatory response resulting from partially severed nerves can be easily seen.

6.6.4 IR-Thermography and Laser Doppler Mapping of Skin Blood Flow (Example 6.4)

As mentioned in the introduction above, one of the drawbacks of using IR-thermography to indirectly indicate changes in skin blood flow is being able to decide, for example, if an increase in skin temperature is due to a parallel increase in skin blood flow. One way to verify whether observed skin temperature changes are related to changes in skin blood flow is to combine IR-thermography with a more direct measurement technique of skin blood flow. Laser Doppler can ascertain the direction of skin blood flow but these measurements are restricted to small surfaces and blood flow is not quantified but reported as changes in Doppler units or volts [45]. With IR-thermography rapid changes in skin temperature over a large area can be easily measured. Laser-Doppler mapping involves a scanning technique using a lower power laser beam in a raster pattern and requires time to complete a scan (up to minutes depending on the size of the scanned area). During the time it takes to complete a scan it is possible that the skin blood flow has changed in the skin area examined at the start compared to the skin area being examined at the end of each scan. This is most likely to happen in dynamic situations where skin blood flow is rapidly changing. Despite this draw back, the use of IR-thermography and laser Doppler mapping can provide a complimentary investigative view of this dynamic process.

In Figure 6.4, both IR-thermography and scanning laser Doppler mapping have been simultaneously used to examine a healthy 44-year-old female subject. In this experiment the subject was submitted to a 20 min heating of the right side of the abdomen using a water-filtered IR-A irradiation lamp (see Section 6.6.1). During the period of heating, the left side of the abdomen was covered with a drape to prevent this side of the body from being heated. At predetermined intervals prior to, during, and after the heating period both IR-thermal images and laser Doppler mapping images were recorded, the latter with a MoorLDI™ laser Doppler imager (LDI-1), Moor Instruments, England. As can be seen in Figure 6.5, the wIRA heating lamp causes a large increase in skin temperature on the irradiated side of the abdomen. After the end of the heating period, the temperature of the heated area gradually decreases. These changes nicely correspond with the laser Doppler mapping images. A semi quantitative value of blood flow (perfusion units) for each LDI image was also calculated (right panels in Figure 6.5). Each blood flow profile corresponds well with their respective temperature profiles.

6.6.5 Use of Dynamic IR-Thermography to Highlight Perforating Vessels (Example 6.5)

The procedure of dynamic thermography involves promoting changes in skin blood flow by local heating or cooling. In the example shown in Figure 6.5, a mild cooling (2 min period of fan cooling) of the skin overlying the abdominal area of a 36-year-old female patient prior to undergoing breast reconstruction surgery was performed in order to highlight perforating blood vessels [41]. These blood vessels originate in deeper lying tissue and course their way toward the skin surface, although they may not necessarily reach the skin surface. By invoking a local cooling, a temporary vasoconstrictor response is initiated in skin blood vessels. Blood flow in the perforating blood vessels is little affected and following the end of the cooling period they rapidly contribute to the skin rewarming. During the rewarming process the localization of the "hot-spots" caused by these perforating vessels becomes more diffuse as heat from them spreads into neighboring tissue. This technique allows one to more easily identify so-called perforator vessels of the medial and lateral branches of the deep inferior epigastric artery, one or more of which will be selected for reconnection to the internal mammary artery (the usual recipient vessel) during reconstruction of a new breast from this abdominal tissue.

6.6.6 Reperfusion of Transplanted Tissue (Example 6.6)

With the development of microvascular surgery, it has become possible to connect (anastomose) blood vessels with a diameter as small as 0.5 mm to each other. Transplantation of tissue from one place

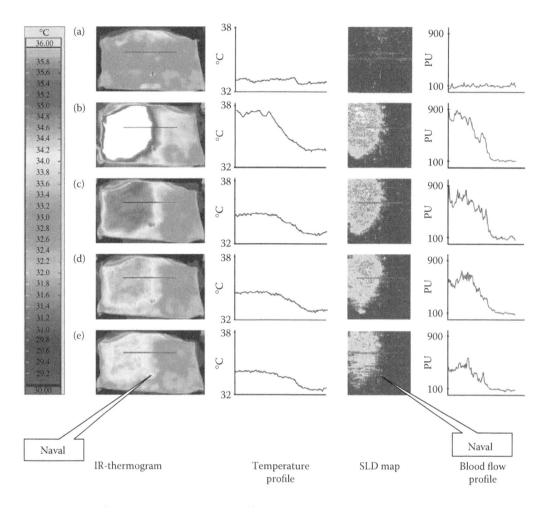

FIGURE 6.4 IR-thermograms, temperature profiles, scanning laser doppler (SLD) scans, and blood flow profiles (perfusion units PU) of the abdominal area of a 44-year-old healthy female subject before (a), immediately after (b), and 5 min (c), 10 min (d), and 20 min (e) after a 20 min heating of the right side of the abdomen with a water-filtered IR-A irradiation lamp. In IR-thermogram (b) the white color indicates skin temperatures greater than 36°C.

to another place in the same patient (autologous transplantation) or from one individual to another (allologous transplantation) has become daily practice in medicine. The technique requires high surgical skills. During the operation, blood supply to the transplant is completely severed and the blood vessels to the transplant are later connected to recipient vessels. These recipient blood vessels will supply blood to the transplant at the new site. Kidney transplantation is a well-known example of allologous transplantation. Here, a patient with a nonfunctional kidney receives a donor kidney to provide normal renal function. Nowadays, autologous tissue transplantation is an integrated part in trauma surgery and cancer surgery. The successful reperfusion of the transplant is obviously essential for its survival. A nonsuccessful operation causes psychological stress for the patient and is indeed an inefficient use of resources stress. IR-thermography provides an ideal noninvasive and rapid method for monitoring the reperfusion status of transplanted tissue.

Autologous breast reconstruction is a critical part of the overall care plan for patients faced with a diagnosis of breast cancer and a plan that includes removal of the breast. The deep inferior epigastric perforator (DIEP) flap is the state of the art in autologous breast reconstruction today. The technique

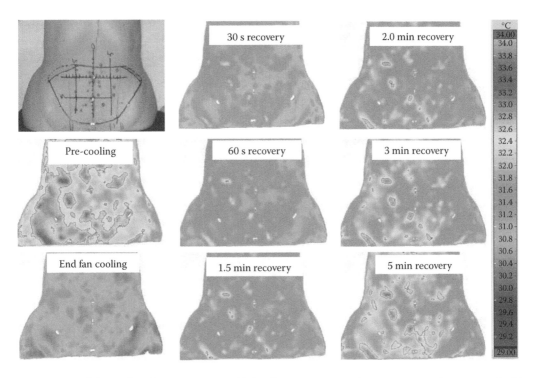

FIGURE 6.5 Abdominal cooling in a 32-year-old female patient prior to breast reconstruction surgery. In this procedure skin and fat tissue from the abdominal area was used to reconstruct a new breast. The localization of suitable perforating vessels for reconnection to the mammary artery is an important part of the surgical procedure. In this sequence of thermograms the patient was first subjected to a mild cooling of the abdominal skin (2 min fan cooling). IR-thermograms of the skin over the entire abdominal area were taken prior to, immediately after, and at various time intervals during the recovery period following the cooling period. To help highlight the perforating vessels, an outline function has been employed in which all skin areas having a temperature greater than 32.5°C are enclosed in solid lines.

uses skin and fatty tissue from the lower abdomen for reconstruction of a new breast (Figure 6.6). The fatty tissue and skin on the lower abdomen are supplied by many blood vessels. By using this tissue as a DIEP flap in autologous breast reconstruction, the blood supply to this tissue is reduced to one single artery and veins, each with a diameter of 1.0–1.5 mm. A critical period during the operative procedure is the connection (anastomosing) of the flap to the recipient vessels. The blood vessels of the DIEP flap are anastomosed to the internal mammary vessels to provide blood supply to the flap. Blood circulation to the newly reconstructed breast is dependent on the viability of the microvascular anastomosis. The series of IR-thermograms shown in Figure 6.6 were taken at various time intervals after reestablishing the blood supply to the flap in a patient undergoing autologous breast reconstruction with a DIEP flap.

During the operation, there is a period (ca. 50 min) with no blood supply to the dissected flap, and as a result, it cools down. The rate and pattern of rewarming of the flap after anastomoses to the recipient vessels provides the surgeon with important information on the blood circulation in the transplanted flap. With a successful and adequate outcome of the anastomosis, the rewarming response was found to be rapid and well distributed (Figure 6.6). A poor rewarming response often made an extra venous anastomosis necessary to improve venous drainage (the most common problem). In such cases, IR-thermography provides an excellent method to quickly verify improvement in the blood flow status in the flap. In addition, in the days following surgery, examination of the skin temperatures using IR-thermography of the newly reconstructed breast was found to be a quick and easy way to monitor its

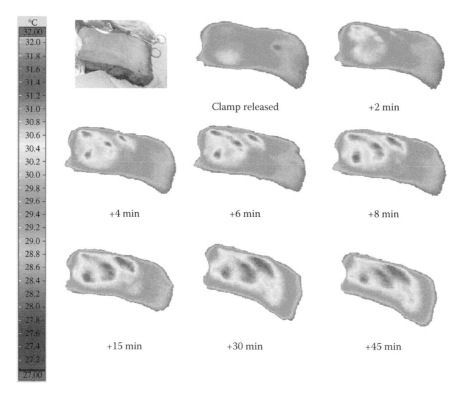

FIGURE 6.6 IR thermal images of an abdominal skin flap during breast reconstruction surgery. The sequence of IR-thermal images demonstrates the return of heat to the excised skin flap following reestablishing its blood supply (anastomizing of a mammary artery and vein to a single branch of the deep inferior epigastric artery and a concomitant vein). Prior to this procedure the excised skin flap had been without a blood supply for about 50 min and consequently cooled down. The photograph in the upper left panel shows the skin flap in position on the chest wall prior to being shaped into a new breast.

blood flow status. The peripheral areas of the new breast can suffer from diminished blood circulation. Improvement in blood circulation could be seen during post surgery.

6.6.7 Sports Medicine Injury (Example 6.7)

Thermal images of an injury sustained by an 18-year-old American football player after a helmet impacted the superior portion of his right shoulder. X-ray and MRI immediately post game showed no structural damage. IR images were taken 12 h post-injury in a controlled environment (22°C, 30% rh, after 20 min equilibration). Upon examination, the athlete was unable to lift his right arm above shoulder level. Note the disruption in the thermal pattern in the right side torso view (T4 spinal region on left side of image) and the very cold temperatures and pattern extending down the left arm. After 3 weeks of rehabilitation, the weakness experienced in the shoulder and arm was resolved. IR imaging confirmed a return to normal symmetrical thermal pattern in the back torso and warmer arm temperatures.

6.6.8 Conclusions

As pointed out in the introduction to this section the main objective of this chapter is to try and persuade those who are not familiar with IR-thermography that this technology can provide objective data, which makes clinical sense. The reader has to be aware that the examples presented above only represent a tiny fraction of the possibilities that this technology provides. It is important to realize that an IR-thermal image of a skin area under examination will more often not provide the examiner with a satisfactory result. It is important to keep in mind that blood flow is a very dynamic event and in many situations useful clinical information can only be obtained by manipulating skin blood flow, for example by local cooling. Thus, the need to have a basic understanding of the physiological mechanisms behind the control of skin blood flow cannot be overemphasized. One also has to realize that this technology should be thought of as an aid to making a clinical diagnosis and in many cases should, if possible, be combined with other complementary techniques. For example, IR-thermography does not have the ability to pinpoint the location of a tumor in the breast. Consequently, IR-thermography's role is in addition to mammography, ultrasound and physical examination, not in lieu of. IR-thermography does not replace mammography and mammography does not replace IR-thermography, the tests complement each other.

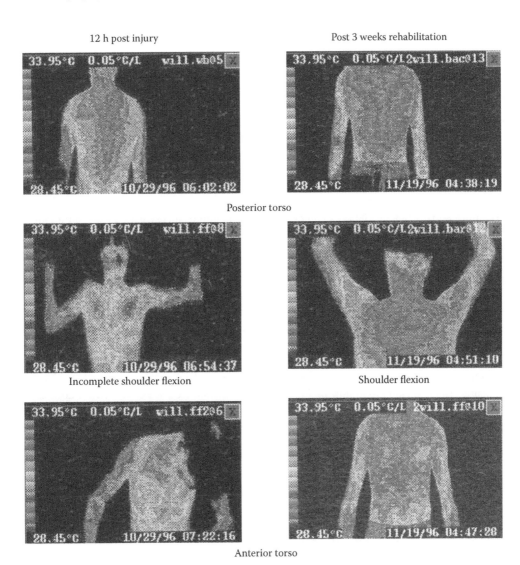

FIGURE 6.7 An American football player who suffered a helmet impact to the right shoulder (left side of image). The athlete was unable to flex his arm above shoulder level. Note the asymmetrical torso pattern in the posterior torso.

Acknowledgments

Examples of IR images from patients and volunteers (Figures 6.1, 6.2b, 6.3 through 6.7) from the Department of Medical Physiology, Faculty of Medicine, University of Tromsø, Norway and the Department of Clinical Physiology, Hillerød Hospital, Hillerød, Denmark. The IR images in these examples were taken using two different IR-cameras: A Nikon Laird S270 cooled camera and a FLIR ThermaCAM® PM695 uncooled bolometer. The respective image analysis software used was ThermaCAM researcher ver. 2.1 and PicWin-IRIS (EBS Systemtechnik GmbH, Munich, Germany). The studies described in Examples 6.1 and 6.2 were carried out at the Department of Clinical Physiology, Hillerød Hospital, Hillerød, Denmark in collaboration with consultant Dr. Stig Pors Nielsen. The assistance of Master student Lise Bøe Setså in the studies described in Examples 6.5 and 6.6 is also acknowledged. Examples of IR images (Figures 6.2a and 6.7) come from the Thermal Lab, Auburn University, Alabama, United States, utilizing a Computerized Thermal Imager 2000. We would like to acknowledge the assistance of John Eric Smith and Estevam Strecker in these projects.

References

1. Jones B.F. and Plassmann P. Digital infrared thermal imaging of human skin. *IEEE Eng. Med. Biol.* 21, 41–48, 2002.
2. Otsuka K., Okada S., Hassan M., and Togawa T. Imaging of skin thermal properties with estimation of ambient radiation temperature. *IEEE Eng. Med. Biol.* 21, 65–71, 2002.
3. Head J.F. and Elliott R.L. Infrared imaging: Making progress in fulfilling its medical promise. *IEEE Eng. Med. Biol.* 21, 80–85, 2002.
4. Park J.Y., Kim S.D., Kim S.H., Lim D.J., and Cho T.H. The role of thermography in clinical practice: Review of the literature. *Thermol. Int.* 13, 77–78, 2003.
5. Ammer K. Thermology 2003—A computer-assisted literature survey with a focus on nonmedical applications of thermal imaging. *Thermol. Int.* 14, 5–36, 2004.
6. Abernathy M. and Abernathy T.B. *International Bibliography of Medical Thermology.* Vol. 2, Washington, DC, American College of Thermology, Georgetown University Medical Center, 1987.
7. Pascoe D.D., Bellinger T.A., and McCluskey B.S. Clothing and exercise II: Influence of clothing during exercise/work in environmental extremes. *Sports Med.* 18, 94–108, 1994.
8. Flesch, U. Physics of skin-surface temperature. In *Thermology Methods.* Engel J.M., Flesch U., and Stüttgen G. (eds), translated by Biederman–Thorson M.A. Federal Republic of Germany Weinheim, pp. 21–37, 1985.
9. Pennes H.H. Analysis of tissue and arterial blood temperatures in resting human forearm. *J. Appl. Physiol.* 1, 93–102, 1948.
10. Charkoudian N. Skin blood flow in adult thermoregulation: How it works, when it does not, and why. *Mayo Clin. Proc.* 78, 603–612, 2003.
11. Greenfield A.D.M. The circulation through the skin. In *Handbook of Physiology—Circulation.* Hamilton W.P. (eds), Washington DC, American Physiological Society, Sec. 3, Vol. 2 (Chapter 39), pp. 1325–1351, 1963.
12. Johnson J.M., Brenglemann G.L., Hales J.R.S., Vanhoutte M., and Wenger C.B. Regulation of the cutaneous circulation. *Fed. Proc.* 45, 2841–2850, 1986.
13. Johnson J.M. and Proppe D.W. Cardiovascular adjustments to heat stress. In *Handbook of Physiology—Environmental Physiology.* Fregly M.J. and Blatteis C.M. (eds), Oxford, Oxford University Press/American Physiological Society, pp. 215–243, 1996.
14. Rowell L.B. Human cardiovascular adjustments to exercise and thermal stress. *Physiol. Rev.* 54, 75–159, 1974.
15. Rowell L.B. Cardiovascular adjustments in thermal stress. In *Handbook of Physiology. Section 2 Cardiovascular System, Vol. 3 Peripheral Circulation and Organ Flow*, J.T. Shepard and F.M. Abboud (eds), Bethesda, MD: American Physiological Society, pp. 967–1024, 1983.

16. Rowell L.B. *Human Circulation: Regulation during Physiological Stress*. New York: Oxford University Press, 1986.

17. Sawka M.N. and Wenger C.B. Physiological responses to acute exercise–heat stress. In *Human Performance Physiology and Environmental Medicine at Terrestrial Extremes*. Pandolf K.B., Sawka M.N., and Gonzales R.R. (eds), Indianapolis, IN: Benchmark Press, pp. 97–151, 1988.

18. Johnson J.M. Circulation to the skin. In *Textbook of Physiology*. Patton H.D., Fuchs A.F., Hille B., Scher A.M., and Steiner R. (eds), Philadelphia: W.B. Saunders Co., Vol. 2 (Chapter 45), 1989.

19. Johnson J.M. Exercise and the cutaneous circulation. *Exercise Sports Sci. Rev.* 20, 59–97, 1992.

20. Garrison F.H. *Contributions to the History of Medicine*. New York: Hafner Publishing Co., Inc., p. 311, 1966.

21. Kirsh K.A. Physiology of skin-surface temperature. In *Thermology Methods*. Engel J.M., Flesch U., and Stüttgen G. (eds), translated by Biederman–Thorson M.A., Federal Republic of Germany/Weinheim, pp. 1–9, 1985.

22. Roberts M.P., Wenger C.B., Stölwik J.A.J., and Nadel E.R. Skin blood flow and sweating changes following exercise training and heat acclimation. *J. Appl. Physiol.* 43, 133–137, 1977.

23. Stephenson L.A. and Kolka M.A. Menstrual cycle phase and time of day alter reference signal controlling arm blood flow and sweating. *Am. J. Physiol.* 249, R186–R192, 1985.

24. Aoki K., Stephens D.P., and Johnson J.M. Diurnal variations in cutaneous vasodilator and vasoconstrictor systems during heat stress. *Am. J. Physiol. Regul. Integr. Comp. Physiol.* 281, R591–R595, 2001.

25. Rodie I.C., Shepard J.T., and Whelan R.F. Evidence from venous oxygen saturation that the increase in arm blood flow during body heating is confined to the skin *J. Physiol.* 134, 444–450, 1956.

26. Gaskell P. Are there sympathetic vasodilator nerves in the vessels of the hands? *J. Physiol.* 131, 647–656, 1956.

27. Fox R.H. and Edholm O.G. Nervous control of the cutaneous circulation. *J. Appl. Physiol.* 57, 1688–1695, 1984.

28. Ring E.E.J. Cold stress testing in the hand. In *The Thermal Image in Medicine and Biology*. Ammer K. and Ring E.E.J. (eds), Wein: Uhlen-Verlag, pp. 237–240, 1995.

29. Pascoe D., Purohit R., Shanley L.A., and Herrick R.T. Pre and post operative thermographic evaluations of CTS. In *The Thermal Image in Medicine and Biology*. Ammer K. and Ring E.E.J. (eds), Wein: Uhlen-Verlag, pp. 188–190, 1995.

30. Faithfull N.S., Reinhold P.R., van den Berg A.P., van Roon G.C., Van der Zee J., and Wike-Hooley J.L. Cardiovascular challenges during whole body hyperthermia treatment of advanced malignancy. *Eur. J. Appl. Physiol.* 53, 274–281, 1984.

31. Finberg J.P.M., Katz M., Gazit H., and Berlyne G.M. Plasma rennin activity after acute heat exposure in non-acclimatized and naturally acclimatized man. *J. Appl. Physiol.* 36, 519–523, 1974.

32. Carlsen A., Gustafson A., and Werko L. Hemodynamic influence of warm and dry environment in man with and without rheumatic heart disease. *Acta Med. Scand.* 169, 411–417, 1961.

33. Damato A.N., Lau S.H., Stein E., Haft J.I., Kosowsky B., and Cohen S.J. Cardiovascular response to acute thermal stress (hot dry environment) in unacclimatized normal subjects. *Am. Heart J.* 76, 769–774, 1968.

34. Sancetta S.M., Kramer J., and Husni E. The effects of "dry" heat on the circulation of man. I. General hemodynamics. *Am. Heart J.* 56, 212–221, 1958.

35. Crandall C.G., Johnson J.M., Kosiba W.A., and Kellogg D.L. Jr. Baroreceptor control of the cutaneous active vasodilator system. *J. Appl. Physiol.* 81, 2192–2198, 1996.

36. Kellogg D.L. Jr, Johnson J.M., and Kosiba, W.A. Baroreflex control of the cutaneous active vasodilator system in humans. *Cir. Res.* 66, 1420–1426, 1990.

37. Chato J.C. A view of the history of heat transfer in bioengineering. In *Advances in Heat Transfer*. Cho Y.J. (eds), Boston: Academic Press, Inc./Harcourt Brace Jovanovich Publishers, Vol. 22, pp. 1–19, May 1981.

38. Chen M.M. and Holmes K.R. Microvascular contributions in tissue heat transfer. In *Thermal Characteristics of Tumors: Applications in Detection and Treatment*. Jain R.K. and Guillino P.M. (eds), *Ann. N.Y. Acad. Sci.* 335, 137, 1980.

39. Baish J.W. Microvascular heat transfer. In *The Biomedical Engineering Handbook*. Bronzino J.D. (eds), Boca Raton, FL: CRC Press/IEEE Press, Vol. 2, 98, pp. 1–14, 2000.

40. Hille B. Membranes and ions: Introduction to physiology of excitable cells. In *Textbook of Physiology*. Patton H.D., Fuchs A.F., Hille B., Scher A.M., and Steiner R. (eds), Philadelphia: W.B. Saunders Co., pp. 2–4, 1989.

41. Taylor G.I. and Tempest M.N. (eds), *Michael Salmon: Arteries of the Skin*. London: Churchill Livingstone, 1988.

42. He S. and Walls A.F. Human mast cell trypase: A stimulus of microvascular leakage and mast cell activation. *Eur. J. Pharmacol.* 328, 89–97, 1997.

43. Kalmolz L.P., Haslik A.H., Donner A., Winter W., Meissl G., and Frey M. Indocyanine green video angiographics help identify burns requiring operating. *Burns* 29, 785–791, 2003.

44. Flock S.T. and Jacques S.L. Thermal damage of blood vessels in a rat skin-flap window chamber using indocyanine green and pulsated alexandrite laser: A feasibility study. *Lasers Med. Sci.* 8, 185–196, 1993.

45. Ryan T.J., Jolles B., and Holti G. *Methods in Microcirculation Studies*. London: HK Lewis and Co., Ltd, 1972.

46. Olsen B.W. How many sites are necessary to estimate a mean skin temperature? In *Thermal Physiology*. Hales J.R.S. (eds), New York, Raven Press, pp. 33–38, 1984.

7

Quantitative Active Dynamic Thermal IR-Imaging and Thermal Tomography in Medical Diagnostics

Antoni
Nowakowski
*Gdansk University of
Technology*

7.1 Introduction

Static infrared (IR) thermal imaging has a number of attractive properties for its practical applications in industry and medicine. The technique is noninvasive and harmless as there is no direct contact of the diagnostic tool to an object under test, the data acquisition is simple and the equipment is transportable and well adapted to mass screening applications. There are also some fundamental limitations concerning this technique. Only processes characterized by changes in temperature distribution on external surfaces directly accessible by an IR-camera can be observed. The absolute value of temperature measurement is usually not very accurate due to generally limited knowledge of the emission coefficient. Many harmful processes are not inducting any changes in surface temperature, for example, in industrial applications material corrosion or cracks are usually not visible in IR thermographs, in medicine in mammography inspection a cyst may mask cancer, and so on. For such cases active dynamic thermal imaging methods with sources of external excitations are helpful. Therefore, the nondestructive evaluation (NDE) of materials using active dynamic thermal IR-imaging is extensively studied; see, for example, proceedings of QIRT [1]. The method is already well developed in some of industrial applications but in medicine this technique is almost unknown.

Active dynamic thermal (ADT) IR-imaging, known in industry either as *infrared-nondestructive testing* (IR-NDT), or as thermographic nondestructive testing (TNDT or just NDT), is under intensive development during at least the past 25 years [1,2]. The concept of material testing by active thermography is based on delivery of external energy to a specimen and observation of the thermal response. In ADT imaging only thermal transients are studied. Such approach allows visualization of material subsurface abnormalities or failures and has already gained high recognition in technical applications [1,2]. In medicine it was applied probably for the first time also around 25 years ago [3,4]. Microwave excitation was applied to evidence breast cancer. Unfortunately, after early experiments the research was suspended, probably due to poor control of microwave energy dissipation. Again some proposals to use ADT in medicine have been published at the end of the last years of the twentieth century [5–7].

The visualization of affected regions needs some extra efforts therefore the role and practical importance of the use of synthetic pictures in ADT for medical applications should be underlined [8–10]. In this case equivalent thermal model parameters are defined and calculated for objective quantitative data visualization and evaluation.

Potential role of medical applications of ADT is clearly visible from experiments on phantoms, and *in vivo* on animals as well as in clinical applications [11,12].

7.1.1 Thermal Tomography

Another concept based on ADT is *thermal tomography* (TT). Vavilow et al. [13–15] dealing with IR-NDTs, proposed this term almost 25 years ago. Even today, the concept is not new this modality may be regarded as being still at the early development stage in technical applications, especially taking into account the limited number of practical applications published up to now. The first proposals to apply the concept of thermal tomography in medicine are just under intensive development in the Department of Biomedical Engineering TUG [16–18]. The main differences comparing TT to simple ADT arise due to necessity of advanced reconstruction procedures applied for determination of internal structure of tested objects.

Tomography is known in medical diagnostics as the most advanced technology for visualization of tested object internal structures. X-rays—CT (computed tomography), NMR—MRI (nuclear magnetic resonance imaging), US—ultrasound imaging, SPECT (single-photon emission computed tomography), PET (positron-emission tomography) are the modalities of already established position in medical diagnostics. There are three additional tomography modalities, still in the phase of intensive research and development—optical tomography (OT), electroimpedance tomography (EIT), and TT. Here we concentrate our notice on ADT and TT. The aim of this chapter is discussion of potential applications of both modalities in medical diagnostics and analysis of existing limitations.

General concept of tomography requires collection of data received in the so-called projections—measurements performed for a specific direction of applied excitation; data from all projections form a scan. Having a model of a tested object the inverse problem is solved showing internal structure of the object. In the oldest tomographic modality—CT—the measurement procedure requires irradiation of a tested object from different directions (projections), all possible projections giving one scan. Then, using a realistic model of a tested object, a reconstruction procedure based on solving the inverse problem allows visualization of external structure of the object. In early systems, a 2D picture of internal organs was shown; nowadays more complex acquisition systems allow visualization of 3D cases. More or less similar procedures are applied in all other tomography modalities, including TT, OT, and EIT. The main advantages of TT, OT, and EIT technologies are: fully safe interaction with tested objects (organisms) and relatively low cost of instrumentation.

The concept and problems of validity in medical diagnostics of ADT imaging as well as of TT are here discussed. In both cases practically the same instrumentation is applied and only the data treatment and object reconstruction differs. Also, the sources of errors influencing quality of measurements and limiting accuracy of reconstruction data are the same. Main limitations are due to the necessity of using proper thermal models of living tissues, which are influenced by physiological processes from one side and are not very accurate due to hardly controlled experiment conditions from the other side.

The main element allowing quantitative evaluation of tested objects is the use of realistic thermal equivalent models. The measurement procedure requires use of heat sources, which should be applied to the object under test (OUT). Thermal response at the surface of OUT to external excitation is recorded using IR-camera. Usually the natural recovery to initial conditions gives reliable data for further analysis. The dynamic response recorded as a series of IR pictures allows reconstruction of properties of the OUT equivalent thermal model. Either thermal properties of the model elements (for defined structure of OUT) or the internal structure (for known material thermal properties) may be determined and recognized. The main problem is correlation of thermal and physiological properties of living tissues, which may strongly differ for *ex vivo* and *in vivo* data. Practical measurement results from phantoms and *in vivo* animal experiments as well as from clinical applications of ADT performed in the Department of Biomedical Engineering Gdansk University of Technology (TUG), Poland and in co-operating clinics of the Medical University of Gdansk are taken into account for illustration of this chapter. The studies in this field are concentrated on applications of burns, skin transplants, cancer visualization, and open-heart surgery, evaluation and diagnostics [19–26].

In the following subchapters all elements of the ADT and TT procedures are described. First, thermal tissue properties are defined. Then basic elements of thermal model construction are described. Following is description of the experiment of active dynamic thermography based on OUT excitation (pulse) and IR recording. Finally the procedures of model identification are described based on phantom and *in vivo* experiments. Some clinical applications illustrate practical value of the described modalities.

7.2 Thermal Properties of Biological Tissues

What is the basic difference between IR-thermography (IRT) and ADT and TT?

In IRT, main information is the absolute value of temperature, T, and distribution of thermal fields, $T(x,y)$. Therefore, regions of high temperature (hot spots) or low temperature (cold spots) are determined giving usually data of important diagnostic value. Unfortunately the accuracy of temperature measurements is limited. It is mainly due to limited accuracy of IR-cameras; due to limited knowledge of the emission coefficient, as individual features of living tissues may be strongly diversified; finally due to hardly controlled conditions of the environment, what may strongly influence temperature distribution at the surface and its absolute values. Additionally surface temperature does not always properly reflect complicated processes, which may exist underneath or in the tissue bulk or which may be masked by fat or other biological structures.

In ADT and TT basic thermal properties of tissues are quantitatively determined; the absolute value of temperature practically is not interesting. Only thermal flows are important. We ask the question—how fast are thermal processes? In ADT, specific equivalent parameters are defined and visualized. In TT directly either spatial distribution of thermal conductivity is of major importance or for known tissue properties the geometry of the structure is determined.

Basic thermal properties (Figures of Merit) of materials and tissues important in ADT and in TT are following:

- k—Thermal conductivity—(W m^{-1} K^{-1}), it describes ability of a material (tissue) to conduct heat in the steady-state conditions

- c_p—Specific heat—(J kg^{-1} K^{-1}), describes ability of a material to store the heat energy. It is defined by the amount of heat energy necessary to raise the temperature of a unit mass by 1 K
- ρ—Density of material—(kg m^{-3})
- ρc_p—Volumetric specific heat—(J m^{-3} K^{-1})
- α—Thermal diffusivity—(m^2 s^{-1}); is defined as

$$\alpha = \frac{k}{\rho \cdot c_p} \tag{7.1}$$

Volumetric specific heat and thermal conductivity are responsible for thermal transients described by the equation

$$\frac{\partial T}{\partial t} = \alpha \nabla^2 T \tag{7.2}$$

For one-directional heat flow (what describes the case of infinite plate structures composed of uniform layers and uniformly excited at the surface) this may be rewritten in the form:

$$\frac{\partial T}{\partial t} = \frac{k}{\rho \cdot c_p} \cdot \frac{\partial^2 T}{\partial x^2} \tag{7.3}$$

Thermal diffusivity describes heat flow and is equivalent to the reciprocal of the time constant τ describing the electrical *RC* circuit:

$$\alpha = \frac{k}{\rho \cdot c_p} \leftrightarrow \frac{1}{\tau} = \frac{1}{RC} \tag{7.4}$$

Based on this analogy, materials are frequently described by equivalent thermal model composed of thermal resistivity R_{th} and thermal capacity C_{th}, which are responsible for the value of the thermal time constant, τ_{th},

$$\tau_{th} = R_{th} C_{th} \tag{7.5}$$

Those are the most frequently used equivalent parameters. Determination of such parameters is easy based on measurements of thermal transients and using fitting procedures for determination of the applied model parameters.

Additionally one may define thermal inertia as $k \rho c_p$.

Useful may be also introducing the definition of thermal effusivity β, defined as the root-square of the thermal inertia—(J m^{-2} K^{-1} s$^{-1/2}$) or (W s$^{1/2}$ m^{-2} K^{-1}).

$$\beta^2 = k \cdot \rho \cdot c_p \tag{7.6}$$

Importance of IRT and ADT/TT in medical diagnostics is complementary. Regions of abnormal vascularization are detected in IRT as *hot spots* in the cases of intensive metabolic processes or as *cold spots* for regions of affected vascularization or necrosis. The same places may represent higher or lower thermal conductivity, but this is not a rule. Thermal properties of tissues may differ significantly depending on the vascularization, physical structure, water content, and so on. Knowledge of thermal properties and geometry of tested objects may be used in medical diagnostics if correlation of thermal properties and specific physiological features are known. The character of the data allowing objective quantitative description of tissues or organs is the most important.

The main advantage of thermal tissue parameter characterization comparing to absolute temperature measurements is relatively low dependence on external conditions, for example, ambient temperature, what usually is of great importance in IRT. The main disadvantage is still limited knowledge of thermal tissue properties and very complicated structure of biological organs. Although the first results of thermal tissue properties have been published at the end of the nineteenth century, the main data, still broadly cited, were collected around 30 years ago [27–30]. Unfortunately, literature data of thermal tissue properties should be treated as not very reliable, because measurement conditions are not always known [31], *in vivo* and *ex vivo* data are mixed, and so on. Table 7.1 illustrates this problem. Detailed description of measurement methods of biological tissue thermal parameters is given in References 18, 31, and 32.

In Table 7.2 basic thermal tissue properties—mean values of data given by different studies and for different tissues taken *in vitro* at 37°C—are collected [18].

It has to be underlined that thermal tissue properties are temperature dependent (typically around 0.2–1%/°C). In most of ADT and TT measurements this effect may be neglected as being not significant for data interpretation, as possible temperature differences are usually limited. The differences of temperature in *in vivo* experiments may be usually neglected as not important at all. Also there is a strong influence of water content and blood perfusion, see, for example, results of Valvano et al. (1985) in Reference [33]. Usually effective thermal conductivity and diffusivity are applied to overcome this problem in modeling of thermal processes. Additionally, in subtle analysis one should remember that some tissues might be anisotropic [18,31].

For mixture of N substances, each characterized by thermal conductivity $k_1,...,k_N$, the effective thermal conductivity is the mean value of the volumetric content of components:

$$k_{tk} = \frac{\sum_{i=1}^{N} k_i \cdot V_i}{\sum_{i=1}^{N} V_i} = \rho_{tk} \sum_{i=1}^{N} k_i \frac{m_i}{m_{tk}} \cdot \frac{1}{\rho_i} \tag{7.7}$$

TABLE 7.1 Thermal Properties of Muscle Tissue

Muscle	Conditions	Thermal Conductivity (W m⁻¹ K⁻¹)	Diffusivity (m² s⁻¹)
Heart	37°C	0.492–0.562	No data
Heart	37°C	0.537	1.47×10^{-7}
Heart	5–20°C	No data	$1.47–1.57 \times 10^{-7}$
Heart (dog)	*In vivo*	0.49	No data
Heart (dog)	37°C	0.536	1.51×10^{-7}
Heart (dog)	21°C	No data	$1.47–1.55 \times 10^{-7}$
Heart (swine)	38°C	0.533	No data
Skeletal	37°C	0.449–0.546	No data
Skeletal (dog)	No data	No data	1.83×10^{-7}
Skeletal (cow)	30°C	No data	1.25×10^{-7}
Skeletal (cow)	24–38°C	0.528	No data
Skeletal (swine)	30°C	No data	1.25×10^{-7}
Skeletal (ship)	21°C	No data	$1.51–1.67 \times 10^{-7}$

Source: From Hryciuk M., PhD dissertation—Investigation of layered biological object structure using thermal excitation (in Polish), *Politechnika Gdanska*, 2003. Shitzer A. and Eberhart R.C., *Heat Transfer in Medicine and Biology*, Plenum Press, New York, London, 1985.

TABLE 7.2 Mean Values of Thermal Tissue Properties Taken *In Vitro*

Parameter	Density	Thermal Conductivity	Specific Heat	Volumetric Specific Heat	Diffusivity	Effusivity
Unit of Measure	(kg m^{-3})	[W/(m K)]	[J/(kg K)]	[J/(m^3 K)]	(m^2 s^{-1})	[J/(m^2 K s$^{1/2}$)]
Symbol	ρ	K	c	ρc ($\times 10^6$)	α ($\times 10^{-7}$)	β
Soft Tissues						
Heart muscle	1060	0.49–0.56	3720	3.94	1.24–1.42	1390–1490
Skeletal muscle	1045	0.45–0.55	3750	3.92	1.15–1.4	1330–1470
Brain	1035	0.50–0.58	3650	3.78	1.32–1.54	1375–1480
Kidney	1050	0.51	3700	3.89	1.31	1410
Liver	1060	0.53	3500	3.71	1.27–1.43	1320–1400
Lung	1050	0.30–0.55	3100	3.26	0.92–1.69	990–1340
Eye	1020	0.59	4200	4.28	1.38	1590
Skin superficial layer	1150	0.27	3600	4.14	0.62	1060
Fat under skin	920	0.22	2600	2.39	0.92	725
Hard Tissues						
Tooth-enamel	3000	0.9	720	2.16	4.17	1400
Tooth-dentine	2200	0.45	1300	2.86	1.57	1130
Cancellous bone	1990	0.4	1330	2.65	1.4–1.89	990–1150
Trabecular bone	1920	0.3	2100	4.03	0.92–1.26	1220–1430
Marrow	1000	0.22	2700	2.70	0.82	770
Fluids						
Blood; 44% HCT	1060	0.49	3600	3.82	1.28	1370
Plasma	1027	0.58	3900	4.01	1.45	1520

Source: From Hryciuk M., PhD dissertation—Investigation of layered biological object structure using thermal excitation (in Polish), *Politechnika Gdanska*, 2003. Shitzer A. and Eberhart R.C., *Heat Transfer in Medicine and Biology*, Plenum Press, New York, London, 1985.

where the index *tk* concerns all tissue, and the index *i* – *i*th component. Analogically, one may calculate effective specific heat either as the mean value of mass components

$$c_{wtk} = \frac{\sum_{i=1}^{N} c_{wi} \cdot m_i}{\sum_{i=1}^{N} m_i} \tag{7.8}$$

or as the mean value of volumetric components

$$\rho_{tk} c_{wtk} = \frac{\sum_{i=1}^{N} \rho_i c_{wi} \cdot V_i}{\sum_{i=1}^{N} V_i} \tag{7.9}$$

Marks and Burton [34] propose for the skin a thermal model, composed of three components—water, fat, and proteins—using the values collected in Table 7.3. Though, it was suggested that more realistic values of thermal conductivity are given by Bowman [18,28].

TABLE 7.3 Thermal Properties for the Equivalent Three-Component Model

Quantity	Density	Thermal Conductivity [W/(mK)]		Specific Heat	Specific Heat (volumetric)
Unit	(kg/m³)	K		[J/(kg K)]	[J/(m³K)]
Symbol	ρ	see Reference 28		c_w	$\rho \cdot c_w \, (\times 10^6)$
Water	1000	0.628	0.6	4200	4.2
Fat	815	0.231	0.22	2300	1.87
Proteins	1540	0.117	0.195	1090	1.68

Source: From Marks R.M. and Bartan S.P., *The Physical Nature of the Skin*, MTP Press, Boston, 1998.

7.3 Thermal Models and Equivalent/Synthetic Parameters

Basic analysis of existing thermal processes in correlation with physiological processes is necessary to understand significance of thermal flows in medical diagnostics. Thermal models are very useful to study such problems. It should be underlined that ADT and TT are based on analysis of thermal models of tested objects. Solution of the so-called direct problem while external excitation (determination of temperature distribution in time and space for assumed boundary conditions and known model parameters) involves simulation of temperature distribution in defined object under test. It allows study on thermal tomography concept, simulation of heat exchange and flows using optimal excitation methods, analysis of theoretical, and practical limitations of proposed methods, and so on.

Direct problems may be properly solved only for realistic thermal models. Having such a model, one can compare the results of experiments and can try to solve the reverse problem. This responds to the question—what is the distribution of thermal properties in the internal structure of a human body? Such knowledge may be directly used in clinical diagnostics if the correlation of thermal and physiological properties is high.

In biologic applications solution of the direct problem is basic to see relationship between excitation and temperature distribution at the surface of a tested object, what is a measurable quantity. This requires solution of the heat flow in 3D space. Equation 7.10, representing the general parabolic heat flow [35] describes this problem mathematically:

$$\mathrm{div}\left(k \cdot \mathrm{grad}\, T\right) - c_p \rho \frac{\partial T}{\partial t} = -q\,(P,t) \tag{7.10}$$

where T is the temperature in K; k the thermal conductivity in W m^{-1} K^{-1}, c_p the specific heat in J g^{-1} K^{-1}; ρ the material (tissue) density in g m^{-3}, t the time in seconds, and $q(P,t)$ is the volumetric density of generated or dissipated power in W m^{-3}. For biologic tissues, Pennes [36] defined "the biologic heat flow equation":

$$c_p \rho \frac{\partial T(x,y,z,t)}{\partial t} = k\nabla^2 T(x,y,z,t) + q_b + q_m + q_{ex} \tag{7.11}$$

where $T(x,y,z,t)$ is the temperature distribution at the moment t, q_b (W/m³) the heat power density delivered or dissipated by blood, q_m the heat power density delivered by metabolism, and q_{ex} is the heat power density delivered by external sources. Solving of this equation, including all processes influencing tissue temperature, is very complicated and analytically even impossible. Generally there are three approaches in analysis and solving of heat transfers and distribution of temperature.

The first one is analytic, usually very simplified description of an object, typically using the Fourier series method. Analytical solutions of heat flows are known only for very simple structures of well-defined shapes, what usually is not the case acceptable for analysis of biological objects. The simplest solution assumes one-directional flow of energy, as it is for multilayer, infinite structures, thermally uniformly excited. Distribution of temperature, if several sources exist, may be solved assuming the superposition method.

The second option is the use of numerical methods, usually based on the finite element method (FEM); to model more complicated structures. There are several commercial software packages broadly known and used for solving problems of heat flows, usually combining a mechanical part, including generator of a model mesh as well as modules solving specific thermal problems, see, for example, References [37] and [38]. Also general mathematical programs, as MATLAB® [39] or Mathematica [40], contain proper tools allowing thermal analysis. FEM methods allow solution of 3D heat flow problems in time. Functional variability of thermal properties and nonlinear parametric description of tested objects may be easily taken into consideration. Again, there are basic limitations concerning the dimension of a problem to be solved (complexity and number of model elements), which should be taken into account from the point of view of the computational costs.

The solution of Equations 7.10 and 7.11 may be given in the explicit form:

$$T_n^{i+1} = \frac{\Delta t}{\rho c_p V_n} \left[\sum_m k_{nm} T_m^i + \left(\frac{\rho c_w V_n}{\Delta t} - \sum_m k_{nm} T_m^i \right) + q_n V_n \right] \tag{7.12}$$

or in the implicit form:

$$T_n^{i+1} - T_n^i = \frac{\Delta t}{\rho c_p V_n} \left[\sum_m k_{nm} \left(T_m^{i+1} - T_n^i \right) + q_n V_n \right] \tag{7.13}$$

where i indicates moment of time, Δt is the time step, n the position in space (node number), V_n the volume of the n node, m indicates neighboring nodes, k_{nm} is thermal conductivity between the nodes n and m, T_n^i the temperature of a node at the beginning of a time step, T_n^{i+1} the temperature of a node at the end of the time step, and q_n is the power density representing generation of heat per unit volume at the nth node. Each node represents a specific part of the modeled structure defined by mass, dimensions, and physical properties. The node is assumed to be thermally uniform and isotropic. Boundary conditions may be assumed discretional, depending on real conditions of experiments to be performed (under analysis). Usually adiabatic or isothermic conditions and a value of excitation power are assumed. The boundary conditions, the value of excitation power as well as individual properties of any node may be modified with time!

In the explicit method the temperature of each node is determined at the end of the time step, taking into account the heat balance based on temperature, heat generation, and thermal properties at the beginning of the time step. Assuming choice of a not proper time step or model geometry may result in lack of stable solution or in poor accuracy of analysis.

In the implicit method, the increase of temperature in each node is calculated from the heat balance resulting from temperatures, heat generation, and thermal properties during the time step. The solution is stable. The duration of a time step may be modified in the adaptation procedure assuming several conditions, for example, maximum rise of temperature in any of the nodes. Increase of iteration steps followed by the need of many possible modifications of parameters during iterations may lead to unacceptable rise in computation time. Still, the mesh geometry may influence accuracy of analysis.

The numerical modeling is preferable in cases where solution of the direct problem may be sufficient, for example, in analysis of methods of investigation, because it allows easy and accurate modifications of important factors influencing measurements. The forward problem may be solved for any configuration of mesh and tissue as well as excitation parameters. Unfortunately, limited knowledge of living tissue properties is reducing possibility of using simulation methods for reliable *in vivo* case analysis.

The third approach is to build simple equivalent models of tested tissues or organs based on such synthetic parameters as thermal resistivity (conductivity) and thermal capacity. This solution, as the simplest one, seems to be the most useful practically. A medical doctor may accept relatively simple, still reliable description, based on synthetic parameters easily determined for such simple models. As an example, thermal structure of the skin may be represented by three equivalent layers described by the $R_{th}(1-3)C_{th}(1-3)$ model. Values of such simple model parameters may be relatively easily determined from experimental data using fitting procedures and well correlated to physical phenomena, for example, depth of burns. This is also probably the easiest method to be applied for solving the inverse problem, and therefore, the best offer for modern computer technology-based diagnostic tools. Also, correlation of model parameters coefficients is not as complicated as in the case of the FEM approach.

7.4 Measurement Procedures

Study of heat transfer enables quantification of thermo-physical material properties such as thermal diffusivity or conductivity and finally detection of subsurface defects. The main drawback of thermal measurement methods is limitation in number of contact sensors distributed within a tested object or application of IR-methods of temperature measurements, allowing observation of the surface temperature distribution only, what in both cases results in limited accuracy of analysis.

The ADT and TT are based on IR technology. The general concept of measurements performed in such applications is shown in Figure 7.1. External thermal excitation source (heating or cooling) is applied to a tested object (TO). First, steady-state temperature distribution on TO surface is registered using the IR-camera. The next step requires thermal excitation application, and registration of temperature transients on the TO surface. The control unit synchronizes the processes of excitation and registration. All data are stored in the data acquisition system (DAS) for further computer analysis.

To allow quantitative evaluation of TO properties, the procedures of both, ADT and TT, are based on determination of thermal models of TO. Typical procedure of model parameter estimation is shown in Figure 7.2. For the measurement structure shown in Figure 7.1 an equivalent model of TO must be assumed (developed and applied in the procedure). In this case usually a simple model, for example, the three layers $R_{th}(1-3)C_{th}(1-3)$, is applied. Thermal excitation (from a physical source as well as simulated for the assumed model), applied simultaneously to TO and to its model, results in thermal transients, registered at the surface of TO by the IR-camera (applied and simulated for the model). For each pixel, the registered temperature course is identified (typically the least-squares approximation—LSA and exponential model for thermal transients are applied). The result (measured thermal course) is compared with the simulated transient and if necessary the thermal model is modified to fit to experimental results.

The registration process is illustrated in Figure 7.3. An example for a pulse excitation and registration of temperature only during the recovery phase (in this case—cooling, after heating excitation) is shown. This is a typical case of practical measurements using optical excitation. Each pixel showing the temperature distribution at the TO surface is represented by an equivalent model, therefore single excitation

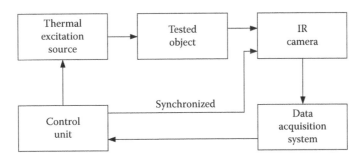

FIGURE 7.1 Schematic diagram of ADT and TT instrumentation.

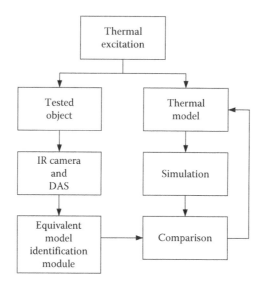

FIGURE 7.2   Model-based identification of a tested structure (object).

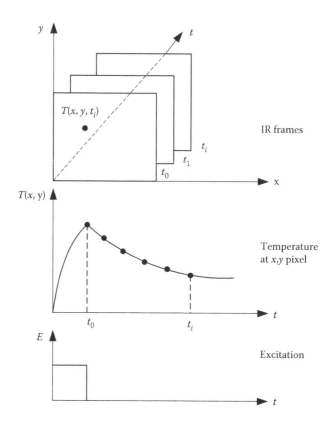

FIGURE 7.3   Registration procedure—after thermal excitation (here heating), a series of IR frames are recorded in controlled moments, starting just at the moment of switching the excitation off; temperature of each pixel is recorded to calculate time constants specific in each pixel.

and registration of transient temperatures at the position *x,y* allows identification of the structure *in depth*, resulting in 3D picture of TO.

There are several procedures of NDE of defects by observation of temperature changes of an inspected sample surface using IR-imaging [2]:

1. Continuous heating of a tested object and observation of surface temperature changes during thermal stimulation (step heating, long pulse); the main drawback of the method is limited possibility of quantitative measurements. The information may be similar to classical static thermography with enhancement of specific defects.
2. Sinusoidal heating and synchronized observation of temperature distribution during stimulation (lock-in technique); this is the most accurate NDT–ADT method but not practical in medical applications as the experiment is rather difficult and time consuming. Additionally data analysis is not simple as thermal biofeedback may be influencing values of tissue thermal properties.
3. Pulse excitation (e.g., heating using optical excitation, air fan for cooling, etc.) and observation during the heating and/or the cooling phases. Several procedures are possible, as pulsed ADT, multi-frequency binary sequences (MFBS) technique, pulse phase thermography (PPT), and other. The description in the following text is concentrating on a single pulse excitation, because this experiment is relatively simple, fast and of accuracy acceptable in medical applications. The time of excitation may be set short enough to eliminate biofeedback interactions, which are otherwise difficult in interpretation.

7.5 Measurement Equipment

Typical measurement set is shown in Figure 7.4. The main elements of the set are shown: an IR-camera; a fast data acquisition system; a set of excitation sources with a synchronized driving unit.

The better are the camera properties the higher accuracy measurements are possible, what is a condition for proper reconstruction of thermal properties of TO. Minimal technical requirements for an IR-camera are: repetition rate 30 frames/s; MRTD better than 0.1°C; LW—long wavelengths range preferable; FOV dependent on application—typically in clinical conditions measurements taken from around 1 m distance to a patient are advisable. Still application of a cooled FPA camera with higher

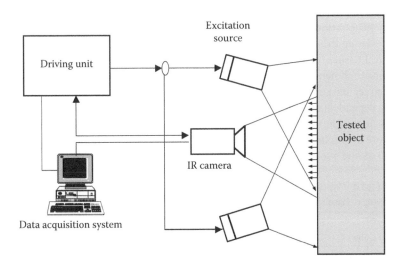

FIGURE 7.4   Measurement set—tested object is exposed to external thermal excitation; the IR-camera, synchronically with excitation records, surface temperature distribution in time.

registration speed and better MRTD would be advisable. SW cameras may be used for other than optical excitations or for analysis of the recovery phase only (time when the excitation is switched off).

Very important is the use of proper excitation sources. Application of optical halogen lamps, laser sources, microwave applicators, even ultrasound generators for heating and air fans or cold sprays for cooling, and some other has been noted [41–44]. As a heating source, an optical or IR lamp, a laser beam, a microwave generator, or an ultrasonic generator can be used. Electromagnetic or mechanical irradiation generated by a source (usually distant) is illuminating a tested object and generating a heating wave proportional to the irradiation energy and to the absorption rate, specific for a tested object material and varying with wavelength of excitation. Microwave irradiation, ultrasonic excitation or electric current flow might generate heat inside a specimen proportional to its dielectric or mechanical properties. For the microwave excitation the main role in the heating process plays the electrical conductivity of a tested object. For the ultrasonic excitation the thermo-elastic effect and mechanical hysteresis are responsible for temperature changes proportional to the applied stress tensor. For the optical irradiation the absorption coefficient is describing the rate of energy converted into heat.

In ADT, the basic information is connected with time-dependent reaction of tested structures therefore it is very important to know the switching properties of the heating or cooling sources, especially the raising and the falling time of generated pulses. Dynamic properties of heating or cooling sources must be taken into account during modeling the different shapes of excitation. Some extra measures should be practically adopted, for example, to avoid the interaction between a lamp, which is self-heated during experiments, and a tested object; sometimes a special shutter is needed to assure proper shape of a heating signal. For microwave excitation the applicator must be directly connected to a tested object therefore a time for mechanical removing of an applicator from the heated object to allow thermographic observation is additionally needed what practically eliminates application of such sources in medical applications.

There are several conclusions regarding application of different sources of thermal excitation:

1. We are dealing with biological objects; therefore, heating should be limited to a safe level, not exceeding 42°C; while cooling temperature of a living tissue should not be lower than 4°C.
2. Thermal excitation should be as uniform as possible; nonuniform excitation results in decrease of accuracy and misinterpretation of measurement data. It is relatively easy to fulfil this condition for optical excitation but very difficult for ultrasonic or microwave methods, even using specially developed applicators.
3. The other important factor is the depth of heat penetration. For different sources of excitation the energy penetration can vary from superficial only absorption to millimeters for some laser beams or even up to a few centimeters for microwave excitation. Ultrasounds are not applicable for biological tissues but of rigid structures only. As an example the data concerning microwave penetration are shown in Table 7.4. High heat penetration may be very interesting for discrimination of deep regions; unfortunately control of microwave energy absorption practically is impossible being dependent on tissue electrical properties. This feature makes use of microwave sources in medical applications very problematic.
4. Temperature interactions should be limited to periods not affected by biological feedback; additionally to assure proper data treatment special care should be devoted to fast switching the

TABLE 7.4 Wavelengths and Penetration Depths of the Microwave Source 2.45 GHz for Typical Tissue with Low and High Water Content

Water Content	λ—In Air (cm)	λ—In Tissue (cm)	σ—Tissue Conductivity (S/m)	Penetration Depth (cm)
High (muscle, skin)	12.2	1.76	2.21	1.70
Low (fat, bones)	12.2	5.21	0.96–2.13	11.2

Source: From Duek A.F., *Physical Properties of Tissue*, Academic Press, London, 1990.

excitation power on and off. For optical sources mechanical shutters are advisable, as electrical switching is effective in the visible range but also other elements as, for example, housing may be heated to a level influencing sensitive IR cameras.

5. Noncontact methods, as optical, seem to be especially appreciated, as aseptic conditions are easy to be secured.

6. Concluding—optical excitation for heating and air fan or cryo-therapy CO_2 devices for cooling seem to be the best options for daily practice in medical diagnostics.

Even ADT is not so sensitive to external conditions as the classical IRT still there are several conditions to be assured. Generally, patients should be prepared for experiments and all "golden standards" valid for classical thermography here should also be applied [46].

7.6 Procedures of Data and Image Processing

Depending on the method of excitation several procedures may be adopted to process the measurement data and then to extract diagnostic information. Here we concentrate only on pulse excitation as it is shown in Figure 7.3. Measurements of transients during the heating and the cooling phases allow further use of procedures developed for pulsed thermography (PT), ADT IR-imaging, pulse phase thermography (PPT), and TT. In all four cases, excitation and registration of thermal transients is performed in the same way using IR-camera. Visualization data differ as it is expressed respectively by thermal contrast images in PT; thermal time constants in ADT and phase shift images in PPT, finally thermal conductivity distribution in TT.

7.6.1 Pulsed Thermography

Pulsed thermography applies calculation of *the maximal thermal contrast index*—$C(t)$ [2].

$$C(t) = \frac{T_\mathrm{d}(t) - T_\mathrm{d}(0)}{T_\mathrm{s}(t) - T_\mathrm{s}(0)} \tag{7.14}$$

where T is the temperature, t the time; for $t = 0$ the sample temperature is maximal, d,s are the subscripts indicating the defect and the sample. With given number of temperature images acquired during PT, the thermal contrast image may be expressed also as

$$C(x,y,t) = \frac{T(x,y,t) - T(x_0,y_0,t)}{T(x_0,y_0,t)} \tag{7.15}$$

where x,y is the pixel coordinates and x_0, y_0 the pixel coordinates of the reference point in the chosen defect area.

To calculate contrast images, a defect template or a reference point should be defined. In medical applications, it is almost impossible to indicate such templates therefore to overcome this limitation estimation of the behavior of living tissues in PT conditions is possible. As an example, heat transfer caused by blood flow is different in normal and affected (e.g., cancerous) tissues. During the heating process the affected tissues usually are heated more than the normal tissues, what is caused by reduced thermoregulation. Further image processing (segmentation) allows discrimination of the so-called *hot spots* and *cold spots* images. Another possibility is to define *the normalized differential PT index* (NDPTI) [47]:

$$\mathrm{NDPTI}(x,y,t) = \frac{T(x,y,0) - T(x,y,t)}{T(x,y,0) + T(x,y,t)} \tag{7.16}$$

(a) 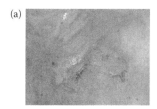 (b) 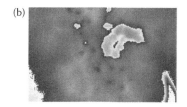 (c)

FIGURE 7.5 Patient with burn wound, first day after the accident—photograph of the burn (a), NDPTI image of the same field (b), and the NDPTI image at the second day (c).

Automatic processing requires definition of thresholds, which can extract only "hot spots" or "cold spots" in the image (i.e., pixels with a value lower than the threshold will compose an image background, while pixels with a value equal or greater than the threshold will compose—"hot spots"). Because popular indexes are constructed to indicate higher probability of detected elements by a higher index value (low values—"cold spots"; high values—"hot spots") it is possible to calculate a negative NDPTI image, too.

$$\mathrm{NPTI}(x,y,t) = 1 - \mathrm{NDPTI}(x,y,t) = \frac{2 * T(x,y,t)}{T(x,y,0) + T(x,y,t)} \qquad (7.17)$$

The signal-to-noise ratio (SNR) of this technique is rather small. It can be improved by averaging images, repeating the pulse excitation with time interval long enough for proper cooling.

As an example, Figure 7.5a is a photograph of the burns caused by fire. Figure 7.5b,c shows fields of IIa/b degree burn evidenced at the first and the second day after the accident. The effect of treatment is also very well visible comparing the results of measurements done day by day. In the related example the burn area was small enough to avoid grafting.

7.6.2 Active Dynamic Thermal IR-Imaging

The ADT is based on comparison of measurements with a simple multilayer thermal model described by Equations 7.4 and 7.5 giving a simplified description of parametric images of thermal time constants, whose values are correlated with internal structure of tested objects [11,48]. Recorded in time thermal transients at a given pixel are described by exponential models. Here, the two exponential models are describing the process of natural cooling after switching the heating pulse off.

$$T(t) = T_{\min} + \Delta T_1 e^{-t/\tau_{1c}} + \Delta T_2 e^{-t/\tau_{2c}} \qquad (7.18)$$

Model-based identification of a tested structure (object) is performed as it is shown in Figure 7.2. The recorded thermal response is fitted to a model. In many cases the two exponential description is fully sufficient for practical applications. As an example, typical transient in time for a single pixel (e.g., from Figure 7.5) is shown in Figure 7.6, with fitting procedure applied to one and two exponential models.

The two exponential models, as Equation 7.18, are fully sufficient for the data of accuracy available in the performed experiment. Some more pictures and examples of the ADT procedure are shown in the following paragraphs, especially in skin burn evaluation [49,50]. It should be underlined that time constants for the heating phase τ_h and for the cooling phase τ_c usually are different due to different heat flow paths.

7.6.3 Pulse Phase Thermography

The procedure of PPT [2,51] is based on Fourier transform. The sequence of IR images is obtained as in conventional PT experiments, Figure 7.7. Next, for each pixel (x,y) in every of N images the

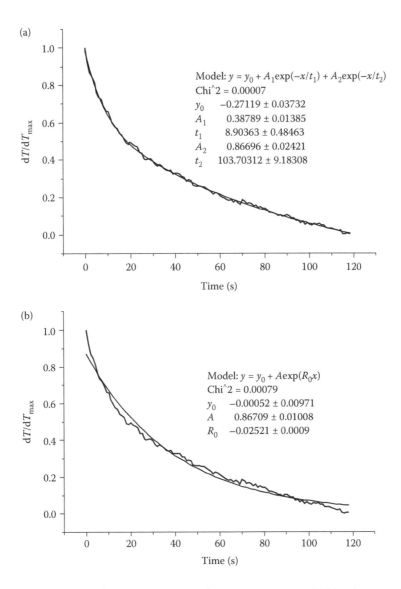

FIGURE 7.6 Fitting of the models to the measurement data—5-parameter model (a) and 3-parameter model (b).

temporal evolution $g_i(t)$ is extracted and then the discrete Fourier transform is performed using the formula:

$$F_i(f) = \frac{1}{N} \sum_{n=0}^{N-1} g_i(t) \exp\left(\frac{-j2\pi f t}{N}\right) = \mathrm{Re}_i(f) + j\mathrm{Im}_i(f)$$

$$\varphi_i(f) = \mathrm{atan}\left(\frac{\mathrm{Im}_i(f)}{\mathrm{Re}_i(f)}\right); \qquad A_i(f) = \sqrt{\mathrm{Re}_i(f)^2 + \mathrm{Im}_i(f)^2}$$

(7.19)

More valuable diagnostic information is given in the phase shift; therefore, this parameter is of major interest.

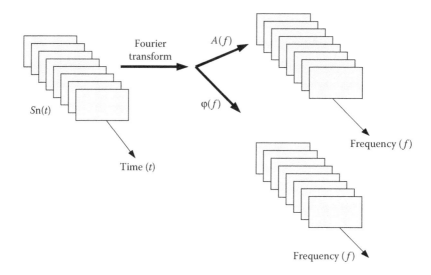

FIGURE 7.7   Procedure of calculation of synthetic pictures in PPT.

Interesting characteristics of PPT are evident since it combines the speed of PT with the features of lock-in approach (the phase and the magnitude image). The condition to get reliable results is application of fast and very high-quality IR-cameras.

As this method was not applied in medicine, yet, we do not show any examples, but extensive discussion of the validity of the procedure may be found in Reference [51].

7.6.4 Thermal Tomography

Thermal tomography should allow tomographic visualization of 3D structures, solving the real inverse problem—find distribution of thermal conductivity inside a tested object. This attempt is fully successful in the case of skin burn evaluation [16–18], giving a powerful tool in hands of medical doctors allowing reliable quantitative evaluation of the burn depth [52–54]. Other applications are under intensive research [55–57].

In TT the first step is determination of thermal properties of a tested object, necessary for development of its thermal equivalent model, to be applied for further data treatment. A typical procedure allowing the model parameters determination is shown in Figure 7.8. The forward problem is solved

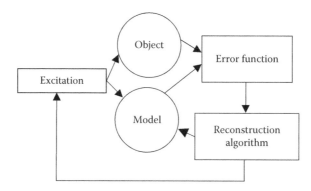

FIGURE 7.8   Determination of model parameters, based on comparison of measurement and simulation data, valid for all kinds of tomography.

and assumed model parameters are modified to satisfy a chosen criterion for model and measurement data comparison. The second step is organization of an active dynamic thermography experiment and registration of thermal transients forced by external excitation.

The next step requires advanced calculations to solve the inverse problem. The information of the correlation of the calculated figure of merit with a specific diagnostic feature is necessary to take a proper diagnostic decision based on the performed measurements. The decision function is directed to determine the internal thermal structure of the tested object based on thermal transients at its surface. In this case, a real inverse problem has to be solved. We perform this procedure using MATLAB scripts [17,18]. As a result the multilayer structure of different thermal properties may be reconstructed. To solve this problem additional assumptions, as number of layers, values of thermal properties or dimensions, are necessary because in general the problem is mathematically ill posed and nonlinear.

The method allows, under some conditions, not only to calculate the thermal effusivity, but also to reconstruct thermal diffusivity distribution in a layered structure. It is done using the nonlinear least-squares optimization algorithm in inversion of the direct problem:

$$
\left.\begin{array}{c}
T_0(x) \\
T_{\text{exc}} \\
\Phi(0,t) \\
k(0)
\end{array}\right\} \Rightarrow \alpha(x) \tag{7.20}
$$

where $T_0(x)$ is the initial temperature distribution in the object, T_{exc} the temperature of the heater, $\Phi(0,t)$ the heat flux at the surface, and $k(0)$ is the thermal conductivity of the outermost layer. In medical diagnostics of burn depth determination the problem is defined in a discrete form allowing determination of the parameter D, describing the depth of specific thermal properties, for example, depth of the burned tissue (the unknown parameter is local thickness of the tested object $d = D\,\Delta x$):

$$
\left.\begin{array}{c}
T_m^1, \quad m = 1,\ldots,M \\
\Phi_{\text{exc}} \\
T_1^p, \quad p = 1,\ldots,P \\
[k_m,\rho c_m]\big|_{m\le D} = [k_{\text{rub}},\rho c_{\text{rub}}] \\
[k_m,\rho c_m]\big|_{D<m<M} = [k_{\text{air}},\rho c_{\text{air}}]
\end{array}\right\} \Rightarrow D \tag{7.21}
$$

where the upper indexes describe the time domain and lower—the space domain. The inverse problem allows solution for which the direct simulation of temperature distribution is closest to the experiment results. This is by solving the problem, which is set by the goal function $F(D)$:

$$
F(D) = \sum_{p=1}^{P} (T_{\text{dir}}^p(D) - T_{\text{meas}}^p)^2 \xrightarrow{\quad D \quad} \min \tag{7.22}
$$

where T_{dir}^p is the surface temperature for simulation performed for assumed value of the parameter D and T_{meas}^p the values measured. In the following examples, the MATLAB *lsqnonlin* function was adopted to solve this problem based on modified interior-reflective Newton method [18]. Description of multilayer structures is also possible and may be defined in the form illustrated in Figure 7.9 [18,54]. One-directional flow of heat energy is assumed.

What should be underlined is that the TT procedure requires advanced calculations and very reliable data, both from measurements as well as for modeling. Comparing with ADT TT is a much more

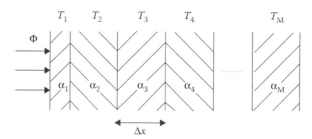

FIGURE 7.9 Spatial discretization of the properties of a multilayer object under test.

demanding method; therefore, it seems that typical for ADT description using thermal time constants at present stage of development seems to be easily accepted by medical staff.

7.7 Experiments and Applications

There are three groups of study performed:

1. On phantoms, for evaluation of technical aspects of instrumentation to be applied for *in vivo* experiments
2. On animals, *in vivo* experiments, for reference data
3. Clinical, for evaluation of the practical meaning of applied procedures

As the ADT as well as TT methods are new in medical diagnostics here we show some results of animal *in vivo* experiments and a few clinical cases. Described applications concern burns diagnostics and evaluation of cardiosurgery interventions. Results of phantom experiments and some other clinical applications were described in other cited publications, for example, References [22,26,43].

7.7.1 Organization of Experiments

The clinical trials are done in *the Department of Plastic Surgery and Burns*, and in *the Department of Cardiac Surgery and Cardiology* of the Medical University of Gdansk. The leading medical personnel had all necessary legal rights for carrying the experiments, approved by the local ethical commission. All clinical experiments were done during normal diagnostic or treatment procedures, as the applied PT, ADT, and TT are noninvasive, aseptic, and safe. The excess of evoked surface temperature rise is around 2–4°C and the observation by the IR-camera is taken from a distance of 1 m. For skin temperature measurements illuminated area usually is kept wet using a thin layer of ointment to satisfy condition of constant evaporation from a tested surface and a constant value of emissivity coefficient. In cardiac surgery experiments with open chest, the surface of the heart is observed where the conditions of emissivity are regarded as constant. The IR-camera is located in a way not disturbing any activity of a surgeon.

The *in vivo* experiments on domestic pigs were performed in *the Department of Animal Physiology, Gdansk University*. The research program received a positive opinion of the institutional board of medical ethics. Housing and care of animals were in accordance with the national regulations concerning experiments on living animals. The choice of animals was intentional. No other animals have skin, blood, or heart properties so close to the human organs than pigs do [45]. Therefore, the experimental data carrying important medical information are of direct relationship to human organ properties.

Our burn experiments were based on the research of Singer et al. [58] and his methodology on standardized burn model. Following his experiment more than 120 paired sets of burns were inflicted on the back skin of 14 young domestic anesthetized pigs using aluminum and copper bars preheated in water

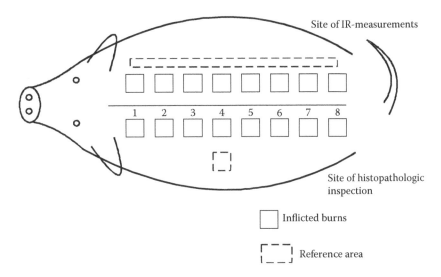

FIGURE 7.10 Location of experimental burns.

in the range from 60°C to 100°C. Typical distribution of burns is shown in Figure 7.10. Each pair was representing different conditions of injury giving a set of controlled burn depths. Additionally, each of measuring points was controlled by full-thickness skin biopsies and followed by histopathologic analysis of burn depth. The results of skin burn degree are in good accordance to the data given by Singer. The pigs were maintained under a surgical plane of anesthesia in conditioned environment of 24°C. The back skin was clipped before creation of burns. The total-body surface area of the burns in each pig was approximately 4%. The animals were observed and treated against pain or discomfort.

After the "burns" experiment was completed, the same pigs were anesthetized, intubated, and used for experimental myocardial infarction by closing the left descending artery. This experiment was lasting 5 h to evidence the nonreciprocal changes of the heart muscle. Full histopathology investigation was performed for objective evaluation of necrosis process.

Basic measurement set-up used in our experiments is composed of instruments shown in Figure 7.4. It consists of the Agema 900 thermographic camera system; IR incandescent lamps of different power or a set of halogen lamps (up to 1000 W) with mechanical shutter or an air-cooling fan as thermal excitation sources; a control unit for driving the system. Additionally a specially designed meter of tissue thermal properties is used to determine the contact reference data necessary for reliable model of the skin or other tissue.

The initial assumptions applied for reconstruction of a tested structure using ADT and TT procedures:

- Tested tissue is a 2- or 3-D medium with two- or three-layer composite structure (e.g., for skin—epidermis, dermis, subcutaneous tissue); (but for simplicity and having limited performance instrumentation even the one-layer model may be useful for diagnostic use in tissue evaluation).
- Tissue properties are independent on its initial temperature.

For skin and under-skin tissues an equivalent three-layer thermal model was developed [11]. For the assumed model its effective parameters have been reconstructed based on the results of transient thermal processes. For known thermal diffusivity and conductivity of specific tissues the local thickness of a two- or three-layer structure may be calculated. In the structural model, each layer should correspond to the anatomy structure of the skin (see Figure 7.11). But in the case of a well-developed burn a two-layer model is sufficient as the skin is totally changed and the new structure is determined by the necrotic layer of burn and modified by injury internal structure of thermal properties drastically different comparing the pre-injury state.

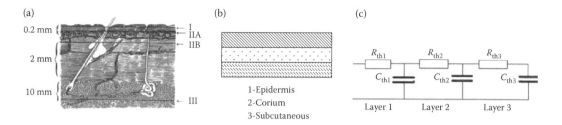

FIGURE 7.11 (a) Anatomy of the skin [59], (b) three-layer structural model, and (c) equivalent thermo-electric model.

7.7.2 Results and Discussion

The related results are divided into groups covering clinical measurements of skin burns and heart surgery and equivalent measurements on pigs. The animal experiments are especially valuable to show importance of the discussed methods. Such experiments assure conditions which may be regarded as reference because fully controlled interventions and objective histopathologic investigations have been performed in this case.

7.7.3 Skin Burns

7.7.3.1 *In Vivo* Experiments on Pigs

The measurements were performed approximately 0.5 h after the burn creation and were repeated after 2 and 5 h and every 24 h during 5 consecutive days after the injury. As the ADT and TT experiments are new in medicine there are several notices concerning methodology of experiments. Especially some biological factors are of high importance. The pigs are growing in extremely fast; therefore, the conditions of measurements are constantly changed. Also the hairs grow anew changing the measurement conditions and cannot be clipped again as the area is affected by injury. This was especially important for direct contact measurements but was not influencing IR measurements. The position of animals in consecutive days was not always the same causing some problems with data interpretation. The results of biopsy and histopathologic observations are used as the reference data of the skin burn thickness. We indicated the same set of parameters as given by Singer, is shown in Figure 7.12 for exemplary burns of one of the pigs. The affected thickness of skin is dependent on the temperature and time of the aluminum and copper bars application. The plots are showing relative thickness of a burn with respect to the thickness of the skin. Such normalization may be important as the skin is of different thickness along the body. All affected points have full histopathologic description. Here only one example is shown for points of the aluminum bar applications of different temperature but constant 30 s time of application.

Some thermographic data of one of pigs taken 5 h after the burn injury are shown in Figure 7.13a [11]. A thermogram while heating is switched off with indicated six measurement areas is shown. Two bottom fields are uninjured reference points, four others are burns made using the aluminum bar—from the left: 100°C/30 s; 100°C/10 s, and 90°C/30 s; 80°C/30 s. The set of halogen lamps was applied for PT and ADT experiment. The result of NDPTI pictures show temperature distribution 100 and 200 ms after the excitation Figure 7.13b,c. The burns are clearly visible and the relative temperature rise is very strongly correlated to the condition of the injury. For the indicated areas the mean value of temperature changes are calculated. This is shown for the first phase of cooling in Figure 7.14. The data might be fitted to thermal models giving ADT specific synthetic pictures of thermal time constants. The quality of the fitting procedure of measurement data to the equivalent model parameters is shown as an example in Figure 7.6a,b for one- and two-layer models.

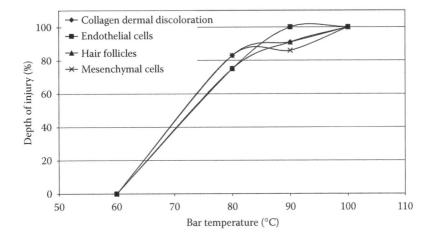

FIGURE 7.12 Relative thickness of the burned skin for several aluminum bar temperatures for time of application equal 30 s—different indicators describing burns are listed (biopsy results).

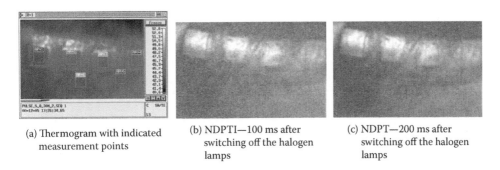

(a) Thermogram with indicated measurement points

(b) NDPTI—100 ms after switching off the halogen lamps

(c) NDPT—200 ms after switching off the halogen lamps

FIGURE 7.13 Classical thermogram (a) and PT normalized differential thermography index pictures of burns (b) and (c).

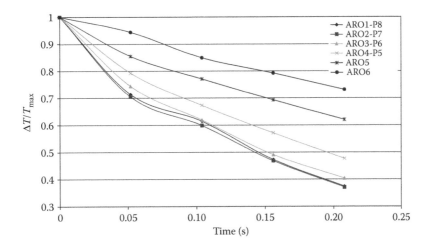

FIGURE 7.14 The normalized averaged temperature changes at the first phase after the excitation is switched off for indicated in Figure 7.13a points.

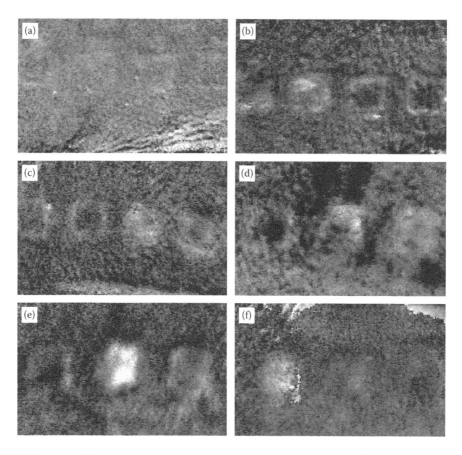

FIGURE 7.15 Thermal time constant representation for the same case as showed in Figures 7.13 and 7.14 in consecutive days, starting from the moment of injury (upper left) and ending on the 5th day (bottom right). The placement of injures seems to be shifted due to different positions of the animal from day-to-day. Specific burn areas are easy recognizable. Additional studies are necessary for making full diagnostic medical recognition of injury, as there are two factors influencing each other—healing process and fast growth of the animal.

There is a possibility to transform each of the pixels into the equivalent model descriptor. In Figure 7.15a–f thermograms taken in consecutive days are transformed according to the formula $A \exp(-R_0 t)$. Distribution of the parameter—the thermal time constant—is shown.

Diagnostic importance of thermal transients is especially clearly visible in Figure 7.16 [18], where different injuries are responding differently to thermal excitation. The line mode of the IR-camera is here applied for fast data registration (>3000 scans/s), allowing reduction of noise.

Based on the same material, the TT procedure and reconstruction of thermal conductivity is also possible [18,54]. Assuming that we know the initial state of the object, the value of excitation flux, thermal response on the surface and spatial distribution of volumetric specific heat within the object it is possible to find the distribution of thermal conductivity:

$$
\left.\begin{array}{l}
T_m^1, \quad m = 1, \ldots, M \\
\quad \Phi_{exc} \\
T_1^p, \quad p = 1, \ldots, P \\
\rho c_m, m = 1, \ldots, M
\end{array}\right\} \Rightarrow k_m, \quad m = 1, \ldots, M \tag{7.23}
$$

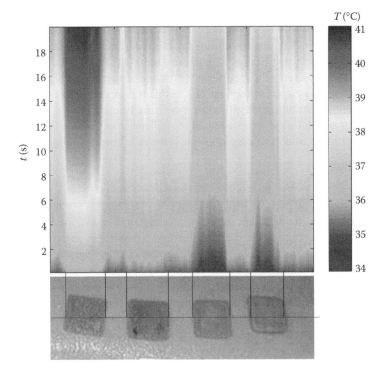

FIGURE 7.16 Thermal transients of burn fields of degree: I, IIa, IIb, IIb.

The inverse problem defined by Equation 7.23 is highly justified due to practical reasons: the volumetric specific heat of biological tissues is much less dependent on water and fat content and other physiological factors than thermal conductivity is (especially effective thermal conductivity, which includes blood perfusion) [45]. Hence, the distribution of $\rho c(x)$ may be regarded with good approximation as known, and irregularities in thermal conductivity distribution should be searched for.

Figure 7.17 presents the results of minimization based on measurements taken during experiments with controlled degree burn fields as it was shown in Figure 7.16. The distribution of specific heat has been assumed identical with the healthy tissue, because of its low dependency on burn [18]. On the reconstructed

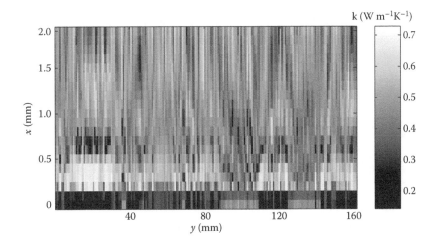

FIGURE 7.17 Results of the reconstruction of thermal conductivity distribution in burn fields.

thermal tomogram, a distinct layered structure of the skin can be seen, as well as the changes in its thermal structure caused by burns. The outermost two slices (0–0.2 mm) represent the epidermis and have thermal conductivity close to 0.3 W m^{-1} K^{-1}, which is almost not affected by burns. Fundamental changes in thermal properties induced by burns may be noticed in the next four slices (0.2–0.5 mm). From the anatomical point of view, this region represents the superficial plexus and adjacent layer of the dermis. As one can expect, for superficial burns we observe increase of k in this region, and decrease of k for severe burns. Based on this observation, quantitative classification criteria can be proposed [18].

For examination of skin burns it seems optimal to implement expression 7.23. It allows obtaining thermal tomography images, which reflect pathologic changes in thermal conductivity distribution up to 2–3 mm.

7.7.3.2 Clinics

We have examined around 100 patients with different skin injuries. Burn injuries were formed in different accidents as a result of flame action directly on skin, hands and faces, and on burning clothes, thorax. The investigations were performed during the first 5 days following an accident. To illustrate the clinical importance of the discussed methods an example of burns caused by an accident is shown in Figure 7.5. The well-determined area of the second-degree burn is evident. The next step—after collecting more clinical data—will be quantitative classification based on data of the burn thickness. Long-lasting effects of treatment and scars formation are still studied and not related here, yet. Already we claim that the value of the thermal time constant taken on the second day after the accident allows quantitative, objective discrimination between IIa and IIb burn [60,61]. More results of skin burns classification using ADT and TT procedures with application of CO_2 external cooling as thermal excitation are in reference [62].

7.7.4 Cardiac Surgery

7.7.4.1 *In Vivo* Experiments on Pigs

Understanding the thermal processes existing during the open chest heart operations is essential for proper surgical interventions, performed for saving the life. Experiments on pigs may be giving answers to several important questions impossible to be responded in clinical situations. This especially may concern sudden heart temperature changes and proper interpretation of the causes.

Clamping the left descending artery and evoking a heart infarct performs the study. Temperature changes have been correlated with the histopathologic observations. The macroscopic thickness of the necrosis was evaluated. This was confirmed by the microscopic data. In Figure 7.18a anatomy of the heart with the clamp is shown. The cross-section line is visible. In the Figure 7.18b the macroscopic cross-section is shown and the evidence of tissue necrosis is by the microscopic histopathology is shown in Figure 7.18c. The widths of the left and right ventricle as well as septum were also measured. The mass

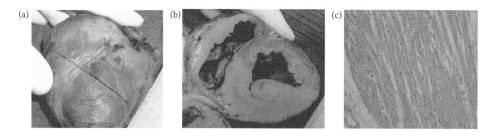

FIGURE 7.18   Pig's heart after heart infarct evoked by the indicated clamp (a), the cross section of the heart with visible area of the stroke (the necrosis of the left ventricle wall—under the finger—is evidenced by darker shadow) (b) and the micro-histopathologic picture showing the cell necrosis (c).

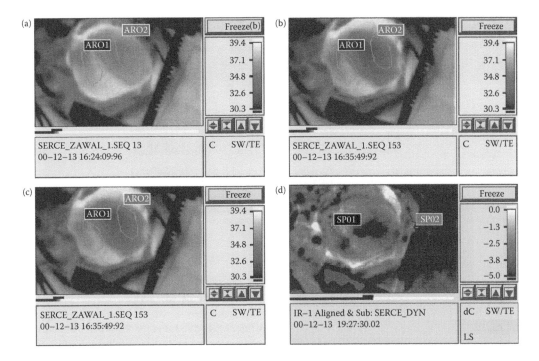

FIGURE 7.19 Thermograms of the heart: before (a) and after the evoked infarct (b). The change of the PT/ADT pictures in time (indicated in the subscript of the thermogram) 0.5 h after the LAD clamping (the tissue is still alive) (c), and 3 h later (the necrosis is evidenced by the change of thermal properties of the tissue) (d).

of the necrosis tissue was calculated. The evident correlation between the thickness of the left and right ventricle walls and the thermographic data are found. Thermographic views of this heart are shown in Figure 7.19a,b respectively before and after application of the clamp. The left ventricle wall is cooling faster than the right one as the left ventricle wall thickness is bigger and the heating effect caused by the flowing blood (of the same temperature in both ventricles) is here weaker. The cooling process of the heart caused by stopping the blood flow in LAD is shown in Figure 7.20—the mean temperature changes are indicated in Figure 7.18a, regions AR01 and AR02 are plotted.

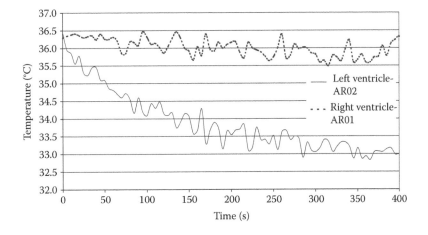

FIGURE 7.20 Temperature changes of the left and right ventricle after the LAD clamping related to Figure 7.19a,b.

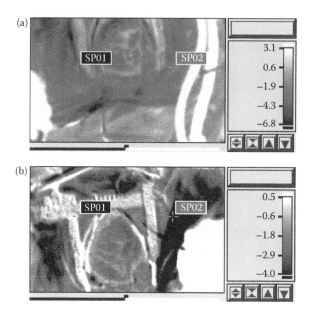

FIGURE 7.21 ADT picture of clinical intervention—before the CABG (a) and after the operation (b). Important is more uniform temperature distribution after the operation proving that intervention was fully successful.

The PT/ADT picture 0.5 h after the clamp was applied (Figure 7.19c) is still showing that thermal properties of the walls are unchanged even the temperature of the left ventricle wall is evidently decreased (Figure 7.19b). Progressing in time the necrosis process is changing thermal tissue properties what is clearly visible after 3 h in Figure 7.19d.

7.7.4.2 Clinics

Operation under extracorporeal circulation is normally a typical clinical situation. Two cases before and after CABG—coronary artery by-pass grafting—are shown for ADT in Figure 7.21a,b respectively. Areas of necrosis and of ischemia as well as volume of the healthy heart muscle were evaluated with respect to coronaroangiographic and radioisotope data showing high correlation. The observations performed before and after CABG show evident recovery of heart functioning. It is evidenced by smaller temperature rise after optical excitation of the heart in the same conditions. ADT shows the level and efficiency of revascularization. The region of heart muscle dysfunction due to heart stroke was possible to be differentiated using both—normal thermography as well as ADT method—what leads to the conclusion that both modalities should be taken into account for diagnostics. The application of thermography for instant evaluation of the quality of the CABG intervention is prompt—the patency of the LIMA—LAD graft is unbeatable by any other method [63], but thermal tissue properties reflecting real state of the tissue is given only by ADT.

Application of all discussed modalities PT, ADT, PPT, and TT during the cardiosurgical interventions gives important indications how to improve surgical procedures.

7.8 Conclusions

The study shows that ADT as well as TT have been successfully used in medical applications giving not only qualitative improvement of diagnostics but in several applications allowing also quantitative evaluation of affected structures.

The importance of active dynamic thermal imaging and thermal tomography applied to burn evaluation and in the inspection of the open chest cardiology interventions was verified. Most probably there are also other attractive fields of medical applications of ADT and TT as, for example, cancer diagnostics [62]. ADT can be applied in medical procedures for monitoring the state of the skin and subdermal tissue structure during burns treatment giving objective, measurable ratings of the treatment procedure. The moment of thermal investigation after an accident is of the highest significance. The most valuable results are obtained during the first and the second days following an accident. The clinical valuable features of the method in skin burns evaluation are the early objective determination of the burn depth—up to two days after an accident and possibility of quantitative evaluation of burn surface and depth as well as objective classification of burn degree. Also, the monitoring role of the method in interoperation of cardiac protection seems to be unbeatable. Changes of tissue vascularization are important for prediction of treatment progress. In all applications, the method is noncontact, noninvasive, clean and nonstressed, allows wide area of investigation and clear and objective documentation of diagnoses and treatment process.

The use of different excitation sources should be limited to those of the best measurement properties. Based on the described experiments most handy are optical sources of irradiation but probably cooling may give even better results in terms of higher signal-to-noise ratio. The operation of such sources is safe, harmless, and easy. The distribution of light may be relatively uniform; the control of irradiated power is also easy.

The work in the field of ADT and TT is still under fast development. Important is extensive IR instrumentation progress giving radical imaging improvement. The problem to be solved is development of a method of noncontact automatic determination of basic properties of thermal parameters of affected tissue what is one of main tasks for the future. Special software is under development for objective and automatic generation of affected tissue depth maps. Important will be to find not only qualitative information but also application of quantitative measures; to describe ratings of the burn wounds or calculate the thickness of affected heart tissues. Still more experiments giving statistically significant knowledge are necessary.

One of important goals is combination of different modalities and application of automatic classification procedures for improved diagnostics. This requires some progress in standardization of IR imaging, which still is not offered in the DICOM standard. For sure, distant consultation of images and automatic data retrieval will be pushing proper work in this field.

Acknowledgments

The author thanks his coworkers and PhD students, who participated in the development of ADT & TT in medicine and who produced the presented data. Most of the reported *in vivo* experiments have been performed with the help of Dr. M. Kaczmarek, Dr. M. Hryciuk, A. Galikowski, and others from the Department of Biomedical Engineering, Gdansk University of Technology. Participation of medical staff: Dr. A. Renkielska and Dr. J. Grudziński (from the Department of Plastic Surgery and Burns), Professor J. Siebert (from the Department of Cardiac Surgery and Cardiology, Medical University of Gdansk) and others are highly appreciated. *In vivo* animal experiments have been performed at the Department of Animal Physiology, Gdansk University with the main assistance of Dr. W. Stojek. Others are listed in the attached bibliography as coauthors of common publications. The work was supported by several grants from KBN (Polish Ministry of Science and Information).

References

1. *Proc. Quantitative InfraRed Thermography:* Chatenay-Malabary-1992, Naples-94, Stuttgart-1996, Lodz-1998, Venice-2000, Reims-2002, Brussels-2004, Dubroynik-2006, Krakow-2008, and Quebec-2010. See also *QIRT Journal*, 2004.
2. Maldague X.P.V. *Theory and Practice of Infrared Technology for Nondestructive Testing*, John Wiley & Sons, Inc., New York, 2001.

3. Van Denhaute E., Ranson W., Cornelis J., Barel A., and Steenhaut O., Contrast enhancement of IR thermographic images by microwave heating in biomedical applications, *Application mikro-und-optolelktronischer Systemelemente*, pp. 71–75, 1985.

4. Steenhaut O., Van Denhaute E., and Cornelis J., Contrast enhancement in IR-thermography by application of microwave irradiation applied to tumor detection, *MECOMBE'86*, pp. 485–488, 1986.

5. Nowakowski A., and Kaczmarek M., Dynamic thermography as a quantitative medical diagnostic tool, *Med. Biol. Eng. Comput. Incorporate Cell. Eng,* 37(Suppl. 1), Part 1, 244–245, 1999.

6. Kaczmarek M., Rumiński J., and Nowakowski A., Measurement of thermal properties of biological tissues—Comparison of different thermal NDT techniques, *Proceedings of the Advanced Infrared Technology and Application*, Venice, 1999, pp. 322–329, 2001.

7. Rumiński J., Kaczmarek M., and Nowakowski A., Data visualization in dynamic thermography, *J. Med. Inform. Technol*, 5, IT29–IT36, 2000.

8. Rumiński J., Nowakowski A., Kaczmarek M., and Hryciuk M., Model-based parametric images in dynamic thermography, *Polish J. Med. Phys. Eng*, 6, 159–164, 2000.

9. Rumiński J., Kaczmarek M., and Nowakowski A., Medical active thermography—A new image reconstruction method, *Lecture Notes in Computer Science*, LNCS 2124, Springer, Berlin, pp. 274–281, 2001.

10. Nowakowski A., Kaczmarek M., and Rumiński J., Synthetic pictures in thermographic diagnostics, *Proceedings of the EMBS-BMES Conference*, CD, Houston, pp. 1131–1132, 2002.

11. Nowakowski A., Kaczmarek M., Rumiński J., Hryciuk M., Renkielska A., Grudziński J., Siebert J. et al., Medical applications of model based dynamic thermography, Thermosense XIII, Orlando, *Proc. SPIE*, 4360, 492–503, 2001.

12. Sakagami T., Kubo S., Naganuma T., Inoue T., Matsuyama K., and Kaneko K., Development of a new diagnosis method for caries in human teeth based on thermal images under pulse heating, Thermosense XIII, Orlando, *Proc. SPIE*, 4360, 511–515, 2001.

13. Vavilov V., and Shirayev V., Thermal Tomograph—USSR Patent no. 1.266.308, 1985.

14. Vavilov V.P, Kourtenkov D., Grinzato E., Bison P., Marinetti S., and Bressan C., Inversion of experimental data and thermal tomography using "Thermo Heat" and "Termidge" Software, *Proc. QIRT '94*, 273–278, 1994.

15. Vavilov V.P., 1D–2D–3D transition conditions in transient IR thermographic NDE, *Seminar 64—Quantitative Infra-Red Thermography—QIRT'2000, Reims*, 74, 2000. http://qirt.gel.ulaval.ca/dynamique/index.php?idD=45.

16. Hryciuk M., Nowakowski A., and Renkielska A., Multi-layer thermal model of healthy and burned skin, *Proc. 2nd European Medical and Biological Engineering Conference*, EMBEC'02, Vol. 3, Pt. 2., pp. 1614–1617, Vienna, 2002.

17. Nowakowski A., Kaczmarek M., and Hryciuk M., Tomografia Termiczna, pp. 615–696, in Chmielewski L., Kulikowski J.L., and Nowakowski A., *Obrazowanie Biomedyczne* (Biomedical Imaging—in Polish) Biocybernetyka i Inżynieria Biomedyczna 2000, 8, Akademicka Oficyna Wydawnicza EXIT, Warszawa, 2003.

18. Hryciuk M., Badanie struktury biologicznych obiektów warstwowych z wykorzystaniem pobudzenia cieplnego (PhD Dissertation—Investigation of layered biological object structure using thermal excitation, *in Polish*), *Politechnika Gdanska*, 2003.

19. Kaczmarek M., Nowakowski A., and Renkielska A., Rating burn wounds by dynamic thermography, in D. Balageas, J. Beaudoin, G. Busse, and G. Carlomagno, eds., *Quantitative InfraRed Thermography* 5, pp. 376–381, Reims, 2000. http://qirt.gel.ulaval.ca/dynamique/index.php?idD=45.

20. Kaczmarek M., Nowakowski A., Renkielska A., Grudziński J., and Stojek W., Investigation of skin burns based on active thermography, *Proc. 23rd Annual International Conference IEEE EMBS*, CD-ROM, Istanbul, 2001.

21. Nowakowski A, Kaczmarek M., Wtorek J., Siebert J., Jagielak D., Roszak K., and Topolewicz J., Thermographic and electrical measurements for cardiac surgery inspection, *Proc. of 23rd Annual International Conference IEEE EMBS*, CD-ROM, Istanbul, 2001.

22. Nowakowski A., Kaczmarek M., Hryciuk M., and Rumiński J., *Postępy termografii–aplikacje medyczne*, Wyd. Gdańskie (Advances of Thermography–Medical Applications, *in Polish*), Gdańsk, 2001.

23. Hryciuk M. and Nowakowski A., Multi-layer thermal model of healthy and burned skin, *Proc. 2nd European Medical and Biological Engineering Conference*, EMBEC'02, Vol. 3, Pt. 2., pp. 1614–1617, Vienna, 2002.

24. Hryciuk M., and Nowakowski A., Evaluation of thermal diffusivity variations in multi-layered structures, *Proc. 6 QIRT*, Zagreb, pp. 267–274, 2003.

25. Kaczmarek M. and Nowakowski A., Analysis of transient thermal processes for improved visualization of breast cancer using IR imaging, *Proc. IEEE EMBC*, Cancun, pp. 1113–1116, 2003.

26. Kaczmarek M., Modelowanie właściwości tkanek ż ywych dla potrzeb termografii dynamicznej, PhD dissertation—Modeling of living tissue properties for dynamic thermography, *in Polish*, Politechnika Gdańska, 2003.

27. Chato J.C., A method for the measurement of the thermal properties of biological materials, thermal problems in biotechnology, *ASME Symp*, Philadelphia, Pennsylvania, pp. 16–25, 1968.

28. Bowman H.F., Cravalho E.G., and Woods M., Theory, measurement and application of thermal properties of biomaterials, *Ann. Rev. Biophys. Bioeng*, 4, 43–80, 1975.

29. Chen M.M., Holmes K.R., and Rupinskas V., Pulse-decay method for measuring the thermal conductivity of living tissues, *J. Biomech. Eng*, 103, 253–260, 1981.

30. Balasubramaniam T.A. and Bowman H.F., Thermal conductivity and thermal diffusivity of biomaterials: A simultaneous measurement technique, *J. Biomech. Eng*, 99, 148–154, 1977.

31. Shitzer A. and Eberhart R.C., *Heat Transfer in Medicine and Biology*, Vols. 1, 2, Plenum Press, New York, London, 1985.

32. Balageas D.L., Characterization of living tissues from the measurement of thermal effusivity, *Innov. Tech. Biol. Med*, 12, 145–153, 1991.

33. Valvano J.W., Cochran J.R., and Diller K.R., Thermal conductivity and diffusivity of biomaterials measured with self-heating thermistors, *Int. J. Thermophys*, 6, 301–311, 1985.

34. Marks R.M., and Barton S.P., *The Physical Nature of the Skin*, Boston, MTP Press, 1998.

35. Janna W.S., *Engineering Heat Transfer*, CRC Press, Washington, DC, 2000.

36. Pennes H.H., Analysis of tissue and arterial blood temperatures in the resting human forearm, *J. Appl. Physiol*, 1, 93–122, 1948.

37. IDEAS operating manual.

38. NASTRAN operating manual.

39. MATLAB operating manual.

40. Mathematica operating manual.

41. Salerno A., Dillenz A., Wu D., Rantala J., and Busse G., Progress in ultrasound lock-in thermography, *Proc. QIRT '98*, pp. 154–160, Lodz, 1998.

42. Maldague X., and Marinetti S., Pulse phase thermography, *J. Appl. Phys*, 79(5), 2694–2698, 1996.

43. Nowakowski A., Kaczmarek M., and Dębicki P., Active thermography with microwave excitation, D. Balageas, J., Beaudoin, G., Busse, G., and Carlomagno, eds., *Quantitative InfraRed Thermography*, Vol. 5, pp. 387–392, 2000. http://qirt.gel.ulaval.ca/dynamique/index.php?idD=45.

44. Nowakowski A., Kaczmarek M., Renkielska A., Grudziński J., and Stojek J., Heating or cooling to increase contrast in thermographic diagnostics, *Proc. EMBS-BMES Conference*, CD, Houston, pp. 1137–1138, 2002.

45. Duck A.F., *Physical Properties of Tissue*, Academic Press, London, 1990.

46. Ring E.F.J., Standardization of thermal imaging technique, *Thermology Oesterrich*, 3, 11–13, 1993.

47. Rumiński J., Kaczmarek M., Nowakowski A., Hryciuk M., and Werra W., Differential analysis of medical images in dynamic thermography, *Proc. V National Conference on Application of Mathematics in Biology and Medicine*, Zawoja, pp. 126–131, 1999.

48. Nowakowski A., Kaczmarek M., and Rumiński J., Synthetic pictures in thermographic diagnostics, *Proc. EMBS-BMES Conf*, 2002, CD, pp. 1131–1132, Houston, 2002.

49. Kaczmarek M., Nowakowski A., and Renkielska A., Rating burn wounds by dynamic thermography, D. Balageas, J. Beaudoin, G. Busse, and G. Carlomagno, eds., *Quantitative InfraRed Thermography*, Vol. 5, pp. 376–381, 2000. http://qirt.gel.ulaval.ca/dynamique/index.php?idD=45.

50. Kaczmarek M., Rumiñski J., Nowakowski A., Renkielska A., Grudziñski J., and Stojek W., In-vivo experiments for evaluation of new diagnostic procedures in medical thermography, *Proceedings of 6th International Conference on Quantitative Infrared thermography*, Proc. *Quantitative Infrared Thermography 6-QIRT'02*, pp. 260–266, Zagreb, 2003.

51. Ibarra-Castanedo C., and Maldague X., Pulsed phase thermography reviewed, *QIRT J*, 1(1), 47–70, 2004.

52. Hryciuk M., and Nowakowski A., Multilayer thermal model of healthy and burned skin, *Proc. 2nd European Medical and Biological Engineering Conference*, EMBEC'02, Vol. 3, Pt. 2, pp. 1614–1617, Vienna, 2002.

53. Hryciuk M. and Nowakowski A., Evaluation of thermal diffusivity variations in multi-layered structures, *Proceedings of 6th International Conference on Quantitative Infrared thermography*, Proc. *Quantitative Infrared Thermography 6-QIRT'02*, pp. 267–274, Zagreb, 2003.

54. Hryciuk M. and Nowakowski A., Formulation of inverse problem in thermal tomography for burns diagnostics, *Proc. SPIE*, 5505, 11–18, 2004.

55. Kaczmarek M., Rumiñski J., and Nowakowski A., Data processing methods for dynamic medical thermography, *Proceedings of International Federation for Medical and Biological Engineering*, EMBEC'02, pp. 1098–1099, Vienna, 2002.

56. Nowakowski A., Kaczmarek M., Siebert J., Rogowski J., Jagielak D., Roszak K., Topolewicz J., and Stojek W., Role of thermographic inspection in cardiosurgery, *Proc. Int. Federation for Medical and Biological Engineering*, EMBEC'02, pp. 1626–1627, Vienna, 2002.

57. Kaczmarek M. and Nowakowski A., Analysis of transient thermal processes for improved visualization of breast cancer using IR imaging, *25th Annual International Conference of the IEEE Engineering in Medicine and Biology Society* "A New Beginning for Human Health," Cancun, Mexico, CD, 2003.

58. Singer A.J., Berruti L., Thode HC J.R., and McClain S.A., Standardized burn model using a multiparametric histologic analysis of burn depth. *Academic Emergency Medicine*, 7(1), 1–6, 2000.

59. Eberhart C. *Heat Transfer in Medicine and Biology, Analysis and Applications*, Vol. 1. Plenum Press, London, 1985.

60. Renkielska A., Nowakowski A., Kaczmarek M., Ruminski J., Burn depths evaluation based on active dynamic IR thermal imaging—A preliminary study, *Burns*, 32(7), 867–875, 2006.

61. Rumiñski J., Kaczmarek M., Renkielska A., Nowakowski A., Thermal parametric imaging in the evaluation of skin burn depth, *IEEE Transactions on Biomedical Engineering*, 54(2), 303–312, 2007.

62. Nowakowski A. et al., Rozwój diagnostyki termicznej metodami detekcji podczerwieni (ilościowa diagnostyka ran oparzeniowych i inne aplikacje), in *Polish*, EXIT, Warszawa, 2009.

63. Kaczmarek M., Nowakowski A., Suchowirski M., Siebert J., Stojek W., Active dynamic thermography in cardiosurgery, *QIRT J.*, 4(1), 107–123, 2007.

8

Dynamic Thermal Assessment

University at Buffalo, State University of New York

8.1 Introduction

The purpose of this chapter is not just to summarize studies on dynamic thermal assessment (DTA) [aka *dynamic infrared imaging* (DIRI) or *dynamic thermal imaging* (DTI)] published since 1987, but to present this topic from a general, critical perspective, focusing on studies published since 2000, which have significantly extended the scope of this promising biomedical technique. The latter most interesting achievements have not been reviewed up-to-date in the context of DTI. This chapter is not a tutorial on DTI or DTA, replicating previously published reviews. Interested readers are, therefore, encouraged to consult appropriate review papers pointed out below. Further, since this chapter is likely to be the last first-hand account of the history of DTI, written by the person who conceptualized and initiated this field of endeavor, it will include historical details that were not published before. The nature of this chapter also allows the author to include personal views on some of the topics discussed.

To bring readers up to speed, they must understand the concept of DTA as compared with classical *static thermal imaging*. Whereas the former methodology monitors or quantitatively measures temporal *changes* in temperature over areas of interest, the latter technique involves observation of temperature distribution over such areas. Thermal changes can be *monotonic*—warming or cooling of areas of interest or parts thereof—or they can be *periodic*, manifested as temperature *modulation*. Monitoring of either kind of temperature change over an area can be done just by one kind of technology—using infrared cameras, preferably in the 8–12 μm range, to monitor the blackbody radiation of the areas of interest. Consequently, one can describe this technique as dynamic *infrared* monitoring or *DIRI*. In brief, whenever we use in this review the terms DTI or DTA, it is implied that the measurement of temperature was done remotely using infrared detection.

Certain clinical applications of DTI have been monitoring monotonic temperature changes—warming of organs following a cold challenge, following surgical reperfusion, or cooling of organs following perfusion with a cold fluid (used occasionally in cardiac surgery). Such slow monotonic warming or cooling processes can be visually monitored in real time. This is one kind of DTI—display of sequential thermal images over a period of time to be observed visually. These sequential observations can also be represented by single images of the *rates* of gradual temperature change, which may be different for different subareas of the region of interest (ROI). The spatial distribution of rates over the ROI can be displayed on color-coded bitmaps.

Physiological processes that affect the temperature of live surfaces (e.g., skin, cornea, or the surface of the brain) are generally *periodic*, resulting in *temperature modulation*. Observing gradual warming of biological surfaces ignores the minute underlying modulations and averages them out. However, those minute modulations, which generally require higher precision of temperature measurement, convey important physiological information. Moreover, those modulations can be monitored when no gross changes in average surface temperature take place. In fact, the use of DTI in this mode, with no prior thermal challenge, is generally far more informative, as it monitors physiology noninvasively under *undisturbed* conditions.

Quantitative assessment of temperature modulation requires accumulation of hundreds or even thousands of sequential thermal images acquired at a significantly higher rate than the frequencies of the physiological or pathophysiological temperature modulation monitored. Presented generally to the end user is the spatial distribution of amplitudes of temperature modulation at discrete frequencies or range of frequencies. This modulation amplitude versus frequency presentation is derived from fast Fourier transformation (FFT) of the observed periodic temperature changes. This spatial distribution can be displayed as color bitmaps of *amplitudes* distribution or of the *microhomogeneity of amplitude* distribution [1], at a given frequency or range of frequencies. These can then be visually spatially correlated with anatomical features over the ROI.

With the exception of real-time continuous DTI, generally used in surgical intraoperative applications, DTI "images" do not represent spatial distribution of infrared radiation flux or of temperature, but are a computerized presentation of periodic temperature *changes*. Furthermore, certain applications of "DIRI" that monitor well-defined anatomical features, requiring no anatomic spatial correlation, do not require display of any images altogether. Thus the term "imaging" in these cases is an utter misnomer.

Notwithstanding semantic finesse, while discussing DTA, we will often use the term "DTI" interchangeably with DTA, because "*thermal* imaging," that is, imaging of thermal phenomena, which are generally *dynamic*, is distinct from classical *static* thermographic "*temperature* imaging."

8.2 Biomedical Dynamic Thermal Assessment

8.2.1 Early History

Dynamic thermal assessment (DTA) was conceptualized in the late 1980s by Michael Anbar [2–5]. Anbar discovered that rapid *changes* in human skin temperature convey valuable physiological and pathophysiological information, information that cannot be derived from static temperature mapping (i.e., measurement of temperature distribution over areas of skin). These ideas crystallized further on the basis of rudimentary experimental studies in the early 1990s [1,6–18].

Since under physiological conditions the temperature of skin depends on blood supply to cutaneous and subcutaneous tissues, blood being the heat exchange fluid, skin temperature manifests a variety of hemodynamic processes. These processes are modulated by pulsatile cardiogenic changes in blood flow as well as by neuronal control of blood flow in the vasculature and microvasculature. Since blood perfusion affects skin temperature differently at different areas of skin, depending on the underlying vascular anatomy, assessment of these processes requires repeated temperature imaging of ROI at frequencies higher than the highest frequency of measurable blood supply modulation. In brief, DTI of ROI can provide new information on the anatomy of the vasculature, on systemic changes in blood flow due to both heart function and neuronal systemic and local control of vascular blood flow and perfusion of the capillary bed.

In addition to useful anatomic and physiologic information, DTI can also be clinically useful. Pathologies that affect any anatomic or physiologic parameters of blood supply can be diagnosed by this technique. For these discoveries Anbar was awarded the President's Award of the American Academy of Thermology for 1990.

The only thermometric technique that can attain the spatial resolution, thermal sensitivity, and speed of data acquisition of skin temperature is infrared imaging with a large ($\Box$256 × 256) solid-state detector array in the 8–12 μm range (temperature can be measured also by infrared sensors in the 3–5 μm range—however, at a significantly lower sensitivity and subject to substantial reflectivity artifacts). In the early 1990s, infrared detection technology was still in its infancy (multidetector two-dimensional arrays were still on the drawing board) and thus it could not meet the requirement of a practical dynamic diagnostic technique. Anbar's early experiments were, therefore, rudimentary, far from offering a practical solution to any clinical problem. Medical technology had to wait, therefore, for additional developments in infrared detection and image-processing technologies.

8.2.2 "Quantitative Dynamic Area Telethermometry in Medical Diagnosis and Management": An Imaginative Speculative Treatise

In 1994, Anbar summarized his ideas and preliminary experimental research in this new field in a monograph titled *Quantitative Dynamic Area Telethermometry in Medical Diagnosis and Management* [19]. The term "area telethermometry" was used to distinguish remote, noncontact, infrared thermal imaging from the, now obsolete, contact thermometric imaging methods (e.g., liquid crystals, thermistor arrays). Today, the term "thermal imaging" implies *infrared* imaging. That monograph described the scientific background of this imaging technique, which is uniquely suited for dynamic applications.

The book projected a large variety of clinical applications that would be feasible using DTI.

It describes potential clinical uses of DTA in a variety of neurological disorders that affect circulatory behavior and/or cutaneous perfusion. These include carpal tunnel syndrome and Raynaud's syndrome, complex regional pain syndrome (earlier named "reflex sympathetic dystrophy"), back pain and leg pain due to spinal stenosis or herniation, as well as migraines and other headaches involving vasospasms.

Among other neurological effects that can be monitored by DTI the book describes sympathetic nervous response to mental stress. The latter entails numerous potential clinical applications in psychiatry both as a diagnostic as well as therapeutic tool (using biofeedback). DTI could as well effectively monitor mental stress induced by deception ("lie detection"). Since sympathetic vasoconstriction or vasodilatation are much more sensitive manifestations of mental stress than perspiration, Anbar suggested that remote monitoring of subjects by DTI (which also readily measures heart rate) is likely to be a far more effective "lie detector" than the best of polygraphs.

Vascular disorders are the next group of clinical applications of DTI discussed in the book. These include occlusions (including peripheral vascular occlusive disease) and aneurysms, as well as vascular effects of diabetes. Anbar suggested that DTI could distinguish between diabetic neuropathies and diabetic vascular occlusive disease, helping optimize treatment of diabetes mellitus.

Then Anbar's book discusses how DTI can be useful in the differential diagnosis and monitoring of treatment of inflammatory processes, including rheumatic disease, local infections, and burns.

Potential surgical applications of DTI are the next topic discussed in the 1994 book. These include monitoring of reperfusion of transplanted organs (e.g., kidneys) and of the heart following cardioplegia in open heart surgery, reperfusion of skin grafts and skin flaps in plastic surgery, and monitoring reperfusion following reconnection of blood vessels and of intestines or in colostomy. It also suggests the use of DTI in monitoring anesthesia.

Quantitative Dynamic Area Telethermometry suggests the use of DTI to monitor metabolic processes also at the cellular level. Metabolic processes are either endothermic or exothermic. At the cellular level they are periodical, that is, oscillatory, owing to thermal and nonthermal autocatalysis, as well as because of the complex interactive kinetics of diffusion of nutrients, oxygen, and metabolites through cellular membranes and cytoplasm. Anbar suggested to monitor the metabolism of mammalian cells in tissue cultures and thus rapidly detect, using DTI, changes in cellular metabolism (a physiological approach) due to the effect of antimetabolites or toxins, instead of waiting for the effects manifested in replication (essentially an anatomic approach).

Clinical applications of the cellular DTI methodology include rapid evaluation of the effect of chemotherapeutic agents on cells extracted by biopsy from malignant tumors, rapid evaluation of the effect of antigens on such cells, and rapid evaluation of their radiation sensitivity. Another potential clinical application of DTI discussed in the book was rapid determination of the sensitivity of bacteria or fungi to antibiotics, without the needed follow-up *in vitro* of proliferation of such pathogens.

A preliminary study done at the University at Buffalo by the author in collaboration with Dr. Malcolm Slaughter indicates the feasibility of cellular DTI. In a study on the thermal behavior of live retinal neurons *in vitro*, retinas of newts were isolated and their thermal behavior was dynamically observed with the aid of a 1:1 germanium macro lens (AIM, Infrarot Module, Heilbrunn, Germany). The study tried to find out if temperature modulation of the neurons (four neurons per pixel), due to cellular metabolism, observed in the 2–8 Hz range, was altered by pulsed light stimulation. The findings were positive though not sufficiently reproducible because of experimental limitations (mainly spatial resolution at 256×256 pixels). However, this has been the first attempt to monitor *in vitro* cellular metabolism by DTI. Commercially available equipment with 16 times higher spatial resolution could be used today to verify DTI's potential in this field.

Cellular DTI could be extended to detection of viral infection of cells in tissue cultures because the metabolism of affected cells changes dramatically owing to the effect of the virus on their transcriptive apparatus. Cellular DTI could be much faster than radiotracer uptake studies. This might allow rapid detection of viruses and testing the effectiveness of antiviral drugs. Although Anbar discussed in his book the use of DTI for detecting viral infections *in vivo*, due to their effect on macrophages, which locally generate nitric oxide (NO) very much like cancerous cells (see below), he overlooked, at the time, this potentially highly important *in vitro* application of cellular DTI in clinical virology. This is, therefore, a new idea whose time will hopefully come soon.

Among the most exciting of the potential clinical applications of DTI discussed by Anbar were those related to oncology. Neoplastic lesions are generally associated with angiogenesis and subsequent local hypervascularization. This by itself must produce aberration in the local anatomy of perfusion. Moreover, newly formed neoplastic blood vessels are likely to be sparsely innervated, if at all, and therefore their response to neuronal vascular control is likely to be abnormal and therefore detectable by DTI, which monitors neuronal modulation of vascular flow. However, even a more specific feature of neoplastic lesions from the standpoint of potential detection and management of cancer by DTI is their generation of NO.

Back in 1978, Anbar has discovered that humans and other mammals produce and carry measurable amounts of endogenic NO in their blood [20]. The physiologic function of this highly reactive substance was unknown at the time and remained unknown until the late 1980s and early 1990s, when it was shown to be a neurotransmitter that induces vasodilatation [21–23]. Since it had been well documented that neoplastic disease is associated with local hyperthermia of the overlying skin, Anbar showed in 1994 by quantitative analysis that this phenomenon must be associated with local vasodilatation [24]. He then advanced a hypothesis that this manifestation of neoplasms is due to generation of NO by cancerous tissues.

He showed that this abnormality of neoplastic cells gives them a substantial advantage over normal cells by enhancing blood supply to the neoplastic tissue (before it induces angiogenesis) and by enhancing potential metastasis [25]. This hypothesis has been experimentally corroborated independently already in 1994 in different types of cancerous cell lines [26–28].

Since it was known by 1994 that NO is a chemical messenger in regulation of vascular tone, Anbar speculated that NO produced by cancerous lesions will interfere with vasomotor regulation of perfusion. Thus Anbar's hypothesis predicted that not only will cancerous lesions enhance regional vascular perfusion but also that perfusion in the surrounding tissues will not be normally modulated. The corollary of this hypothesis is that DTI might detect the effect of cancer-produced NO on the neuronal modulation of perfusion, and thereby detect cancerous lesions more effectively than by just monitoring local hyperthermia. A recent thermographic study confirmed Anbar's conclusion that breast-cancer-induced hyperthermia is not due to local hypervascularity [29], and it must, therefore, be associated with local vasodilatation.

An important parameter conceptualized by Anbar and described in his monograph is the micro-homogeneity of temperature modulation or spatial thermal homogeneity (STH = 1/SD (standard deviation) of temperature values of spots in a given subarea) [1]. Quantitative DTI can also measure the modulation of local homogeneity of subareas at any given frequency or range of frequencies. The latter parameter measures the level of perfusion of a given subcutaneous region: that is, when the region's perfusion reaches a maximum so does the microhomogeneity.

Finally, another novel concept highlighted by Anbar already in 1994 was the direct *objective* use of digital data in clinical diagnosis. DTI is essentially not an anatomic imaging technique but a physiological one. It produces its findings on the oscillatory behavior of the temperature of biological surfaces in terms of frequencies and their relative amplitudes (following FFT, as demonstrated by Anbar in 1991). DTI expresses pathology or biological abnormalities as frequency–amplitude aberrations, analogously to electrocardiogram (ECG) or electroencephalogram (EEG). When these aberrations are monitored over large areas of skin their spatial distribution can be bitmapped, color coded, and visually evaluated by the human eye. However, these are not images of temperature but images of the distribution of temperature *changes*. This is the only meeting point between DTI and anatomy. Now, in many clinical applications the anatomical information is absolutely unnecessary, because the physiological data contain all the diagnostics information. As these data are digital to begin with, they can be processed further to provide *objective* diagnostic measures of pathology, or of the probability of pathology. DTI can therefore become a forerunner of computerized medicine in the twenty-first century.

In brief, at the risk of repetition, it must be concluded that in spite of its common name, DTI is essentially not an imaging technique. Clinical imaging techniques require always evaluation of images by experts to achieve a diagnosis. DTA is, therefore, more than an imaging technique.

Back in 1994, all these scores of novel-computerized clinical applications were essentially just a glimpse in the eye of the author. At that time, this well-documented, scientific book with hundreds of references was still not far from being good science fiction, because none of the applications envisioned there had been reduced to practice. Like Jules Verne in his time or Arthur Clarke, Anbar recognized the potential of a new technology in an entirely different field. In this case it was infrared imaging, which was likely to rapidly develop so to meet emerging needs of military surveillance and targeting. Anbar foresaw that this technology will meet the sensitivity and speed requirements of clinically useful DTI. Also computational speed and computer memory available in 1994 were insufficient to process DTI data in any practical manner. This did not deter Anbar from projecting that these will be available when infrared technology will meet DTI requirements. Many of these "dreams" were realized within less than a decade.

After publicizing his new ideas on DTA [30,31], and extending the ideas on the mechanism of breast cancer hyperthermia [25,32], Anbar advanced a hypothesis on the role of NO in pathophysiological pain, explaining the mechanism of hyperthermia associated with pain [33,34]. Yet all these plausible speculations awaited experimental verification.

8.2.2.1 Hypothesis on the Effect of Cancer on Perfusion Confirmed

Anbar's hypothesis that NO generated in neoplastic cells affects vascular behavior in the vicinity of solid tumor has been confirmed already in the following 3 years [35–39]. The role of NO in cancer biology has been demonstrated in breast cancer [40–42], cancer of the colon [43–45], squamous cell carcinomas of the head and neck [46–48], brain malignancies [49–52], melanoma [53–55], lung cancer [56–58], cancer of the prostate [59], ovarian cancer [60], cancer of the pancreas [61], chondrosarcoma [62], cancer of the bladder [63,64; see, however, 65], and gastric cancer [66,67]. In brief, the production of NO in malignant tissues has been well established, corroborating Anbar's prediction [24,25], as has been its vasodilatatory effect. (For a more detailed discussion see the introduction in Reference 68.)

What remained to be demonstrated was to what extent can the specific effect of solid tumors on the modulation of tissue perfusion be detected at skin level by DTI, refining the information on the tumor-associated local skin hyperthermia (an effect that has been well known for decades—although its mechanism remained obscure before the NO mechanism has been elucidated) [24].

8.2.3 Considering the Realities

The 1994 publication of Anbar's monograph was a landmark in the history of DTI, but this technique would have been forgotten if not for his research efforts to make his dream come through. In making DTI a practical technique one must consider its scope and limitations.

After discussing the wide scope of DTI's potential applications, we must also consider the limitations of this technique when applied to the live human body. It must be realized that because of the very high absorption coefficient of skin tissue to photons in the 8–10 μm range, blackbody radiation emitted from the skin and detectable by infrared cameras represents the temperature of a layer less than 0.1 mm thick of skin. Since no temperature changes can occur intrinsically within that thin skin layer, any modulation in local skin temperature of the upper layer of skin represents temperature modulation of subcutaneous tissues, including blood vessels, conducted to the skin's surface. The modulation of skin temperature is due to modulation of perfusion of the subcutaneous capillary bed as well as modulation of blood flow through vessels in proximity of the skin. These modulations are driven by two processes— cardiogenic pulses and neurological modulation of the vascular tone (i.e., vasodilatation and vasoconstriction). Locally elevated or lower temperature due to hypermetabolism or hypometabolism of tissues, respectively, or to environmental factors, are not expected to manifest any modulation below 0.001 Hz and will not be detectable by appropriately programmed computerized DTI.

Further, modulation of perfusion of tissues or vasculature situated deep (>10 mm) below the skin is unlikely to be detectable at skin level because of heat dissipation, which results in impedance of the modulated thermal signal. The latter effect will result in blurred modulated images of deep vessels detectable at skin level; the deeper the vessel the more blurred will be its modulated thermal image until it fades into the background. Since the impedance of temperature modulation increases with frequency, only low-frequency modulation of heat sources situated deep under the skin will be detectable at skin level. It can be expected that temperature modulation of heat sources below, say, 10 mm, will not be detectable at all at skin level even if unmodulated skin temperature might be higher in a given area due to a low-lying heat source. The level of detectable modulation depends obviously on the precision of temperature measurement. The higher the precision the more sensitive will be the detection of modulation due to perfusion of deeper structures.

In brief, the higher the frequency of the modulation the more pronounced will be the attenuation of the heat-dissipated signals. Consequently, we expect to detect high-frequency modulation (>1 Hz) only of modulated cutaneous microvasculature or over superficially situated large vessels. However, temperature modulation of large vessels will be predominantly driven by cardiogenic pulses with only minor effects of vasodilatation or vasoconstriction that occur at other frequencies. In any case, temperature modulation at the skin surface is necessarily a biased representation of the modulation of subcutaneous tissue perfusion. While modulation frequencies >2 Hz can occasionally be detectable at skin surface (excluding FFT artifactual harmonics of cardiac pulsation) most useful DTI information is expected in the below the 2 Hz frequency region. These may include low-frequency waves due to reflection and interference of cardiogenic waves and to beats of interference between higher frequency neurogenic vasoconstriction or vasodilatation modulations.

It must be concluded, therefore, that assessment of perfusion dynamics by DTI is rather limited in spite of the relative simplicity of measurement, and more precise information on perfusion dynamics of deeper tissues might be obtained by other technologies, such as laser Doppler flowmetry, Doppler ultrasound, or MRI. Yet all the basic predictions developed by Anbar regarding the diagnostic usefulness of monitoring *modulation* of blood perfusion (described above) are valid irrespective of the monitoring technique. It must also be realized that the limitations on DTI of skin do not apply to DTI of cells in tissue cultures or of microorganisms.

8.3 DTA Experimentation (1997–2001)

8.3.1 Early DTA Experimentation

Modulation of skin temperature due to hemodynamic effects is of the order of 10 mK [69], which requires a precision of temperature measurement better than 2 mK. If the hemodynamic modulation frequencies of interest are 2–10 Hz, a data acquisition rate of at least 100 Hz is required as is a stability of 1 mK over >30 min—the duration of a complete multi-image DTI study [70]. Anbar's attempts from 1992 till mid-1997 to experimentally use different commercial infrared camera systems available at that time to verify basic concepts of quantitative DTA on human subjects have ended with ambiguous results because of the inadequate sensitivity, reproducibility, and speed of those systems. None of those failed experiments were published.

Finally, by the end of 1996 Anbar became aware of the new type of fast and sensitive Ga/As 256×256 array quantum-well infrared photodetectors (QWIP) developed by Gunapala *et al.* [71,72] at the Jet Propulsion Laboratory (JPL) in Pasadena. The availability of a sensitive 256×256 focal plane detector array has been a minimal prerequisite for meaningful DTA biomedical applications. At the time, that camera could be transferred to and used in a DOD laboratory only. Consequently, Anbar joined Dr. Kaveh Zamani at the Walter Reed Army Institute of Research (WRAIR) in Washington DC and explored a variety of potential biomedical uses of DTA with this unique camera.

These studies included demonstration of monitoring cardiac pulsatile hemodynamics and measurement of blood flow rate in peripheral vasculature [69,73], and demonstration of DTA use in assessment of cutaneous lesions and neuropathies caused by chemicals and of observation of significant changes in facial perfusion under mental stress [74]. Shortly later, DTI was also applied, though with less advanced equipment, to study joint inflammation and pain, presumably mediated by NO [75–77].

Following the promising preliminary findings at WRAIR, Dr. Gunapala loaned his camera to facilitate preliminary DTI clinical studies at Buffalo. At the Erie County Hospital, in collaboration with Dr. William Flynn of the Department of Surgery, Anbar demonstrated the effective use of DTI in assessment of microsurgical attachment of a severed penis. Then, in collaboration with Dr. Kenneth Eckert, he demonstrated, at the Windsong Clinic of Buffalo, the use of DTA in assessment of hemodynamic behavior of cancerous breasts [78]. These brief preliminary demonstration studies with borrowed equipment confirmed the potential usefulness of DTA in the clinic. The conclusions of these preliminary findings were then summarized in review papers that pointed out the instrumental and software requirements of this technique [70,79–81].

8.3.2 Research at the Millard Fillmore Hospital at Buffalo

By 1999 Anbar received from a company (OmniCorder Technologies Inc., now Advanced Biophotonics Inc., East Setauket, New York) a state-of-the-art commercial fast digital infrared camera with a QWIP 8–9 μm 256×256 detector array, operating at a rate of 100 frames/s (AIM, Infrarot Module, Heilbrunn, Germany). That camera incorporated a highly reliable and stable Stirling helium cooler, which is essential for meaningful DTA studies (because of the high temperature dependence of the sensitivity of QWIP detectors). Anbar received this camera, placed at the Millard Fillmore University Hospital in Buffalo, for experimental DTA studies on breast cancer.

The goal of the research project at Buffalo was twofold:

1. Develop algorithms that will *objectively* differentiate between cancerous and cancer-free breasts without the need for human expertise of "reading" bitmaps of DTI modulation amplitudes. In other words, alleviate *imaging* from DTI and make it a genuine DTA-computerized technique. DTI, which is essentially a wholly computerized technique, which produces FFT-generated modulation amplitude spectra for each pixel of the image, lends itself uniquely to the latter end when the whole organ in question, or well-defined regions of it, are treated as an anatomically undifferentiated ROI [68,70,82].

2. Establish the feasibility of using the algorithms developed meeting the first goal to *objectively* detect breast cancer in patients who have had suspicious x-ray mammograms before undergoing exploratory biopsy, that is, to distinguish between true- and false-positive x-ray mammograms; the DTA findings were to be compared with the pathology of calcified biopsied tissue as the "gold standard."

Using algorithms developed by Dr. Lorin Milescu [83,84], Anbar's group studied DTI data of a total of 100 breasts, 64 free of cancer and 36 with biopsy-confirmed breast cancers (three DTI views were taken for each breast), of patients examined at the Millard Fillmore Hospital at Buffalo and at the Department of Radiology, University Hospital and Medical Center at Stony Brook, New York [68,85]. That was the first study to demonstrate the potential use of DTA as an *objective* quantitative diagnostic technique. Because of this and since this may be the last time this study will be reviewed in the general context of DTA, we shall include here certain clarifying details and somewhat newer and more effective computational analysis procedures used in the last phase of this 3-year research effort.

To summarize, the objective of that study was to demonstrate that DTA can effectively differentiate between cancerous breasts (irrespective of the type of the cancer) and breasts free of cancer. Once this objective has been achieved, a follow-on study would have had to use the criteria developed in the first study to test the effectiveness of the new methodology under clinical field conditions. Although the objective of the first study has been fully achieved [68,85], unfortunately, no follow-up clinical study under similar experimental and computational conditions was implemented up-to-date.

8.3.2.1 Methodology of Assessment

The working hypothesis of the Buffalo study was that the skin of cancerous breasts will manifest significantly lower temperature modulation because of higher perfusion, primarily owing to the vasodilatatory effect of NO.

The following DTA methodology was developed to quantify the temperature modulation of the surface of breasts and identify breasts with low modulation.

Following the acquisition of 1200 sequential thermal images at a rate of 100 frames/s, the projected area of the thermal image of the breast was subdivided into square subareas (spots) of 4×4 pixels each, corresponding to approximately 16 mm² of skin. The total area of interest delineated manually on the primary thermal image, comprised of about 1700± spots, depending on the size of the breast. The average temperature of each of those spots was then calculated, and a time series of 1024 temperature values was obtained for each spot. After linear regression eliminated slow (<0.1 mHz) temperature changes, and after correction for spurious modulation of the camera or of the environment (common interference removal), each time series underwent FFT analysis [69,83]. Using the FFT data of 160 intervals, 50 mHz each, in the frequency range 2–10 Hz, the modulation amplitude of each spot at each of these 160 discrete frequencies was obtained. The cardiogenic modulation frequencies (0.5–2 Hz) were excluded, primarily because their assessment would require longer observation times, exacerbating motion artifacts (primarily patient breathing).

For each discrete frequency a subarea (or subareas) of spots with lowest modulation amplitudes (☐24 spots in the published study) was (were) identified. Such a subarea was named a *cluster* of low-amplitude spots. These clusters were identified and demarcated as follows: all the spots that represented the area of interest (the whole demarcated breast) were rank ordered by their amplitudes (the spot with the lowest amplitude first, the spot with the second lowest second, up to the spot with the highest amplitude). Spots with low amplitude, owing to one or more "dead" pixels, caused by corresponding inactive detector elements, were excluded from the ranking. The position of the spots with the lowest amplitudes was then registered. If these spots were adjacent to other such spots, a cluster of two or more low-amplitude spots was identified. Then spots with the next to lowest amplitude were identified and those in the proximity of a spot with the lowest amplitude were counted as members of a recognized cluster.

This computerized process was continued until a cluster of low-amplitude spots with a specified size was identified (24 spots in this study, corresponding approximately to 4 cm² (that number could be larger than 24 because the last step could have added more than one sequential spot to the low-amplitude cluster, or more than a single cluster with more than 24 spots was formed in the last step of clustering). The amplitude of the spot with the highest amplitude in the identified cluster was then registered. Then the ratio between the number of spots in the identified low-amplitude cluster (or clusters with $\square$24 spots) and all the other spots (nonclustered or less clustered) that, at that frequency, had amplitudes smaller or equal to the spot with the highest amplitude value in the identified low-amplitude cluster was determined. That ratio was defined as "first cluster ratio" (FCR).

Another differentiating parameter used was the "first cluster amplitude ratio" (FCA) = the ratio between the average amplitude of the first cluster and the average of the amplitudes of all the spots over the whole demarcated area of interest (the whole breast). The working hypothesis was that both FCR and FCA in a cancerous breast will be significantly lower than in a cancer-free breast.

As discussed earlier, computerized DTA produced information on modulation of both temperature and STH [1,68]; thus we computed in parallel also FCR and FCA values for STH of the spots.

For each case studied and for each of its three views, the computerized output was a spectrum of FCR and FCA values of both temperature and homogeneity for each of the 160 frequencies analyzed.

This large data matrix was statistically analyzed to find out if these parameters can differentiate between cancerous and cancer-free breasts.

The working hypothesis predicted cancerous breasts to have higher FCR and FCA values, at least at certain frequencies, because local attenuation of modulation by extravascular, cancer-produced NO is expected to result in larger clusters with attenuated temperature amplitudes and greater homogeneity compared with breasts free of cancerous lesions. This hypothesis could be tested by a statistical analysis of the FCR and FCA values obtained for the 100 breasts studied.

8.3.2.2 Statistical Analysis and Findings

Using a macroprocedure, a computer sequentially processed each of the image series (cases) in our database, that is, each of the views of each case. For each view, the researchers obtained as the output four ASCII files that listed the FCR(temp), FCR(STH), FCA(temp), and FCA(STH) values for each of the 160 discrete frequencies analyzed. This output could be displayed as a spreadsheet with 160 columns (frequencies), with each row corresponding to a single case (a single view of a specified breast). The same spreadsheet also contained demographic and clinical data. These spreadsheets were then fed to a statistical program, developed at Buffalo by Dr. Aleksey Naumov, for further analysis. The statistical analysis, described before for other DTA-derived parameters [85], reaches the same conclusion—DTA can highly effectively differentiate between cancerous and cancer-free breasts; sensitivities and diagnostic powers >0.95 were obtained, with STH data providing a somewhat higher significance. It was also found that frontal views of cancerous breasts yield significantly higher sensitivity and specificity values than lateral views of the same breast. A description of the statistical procedures and findings are, however, outside the purview of this review chapter.

8.3.2.3 Limitations

The Buffalo study was limited to assessing only high-frequency modulation >2 Hz because the motion artifacts limited the acquisition time to 12 s. While the discrete structures of the average spectra and the highly significant difference between the averages of the two groups of subjects indicates manifestation of hemodynamic process in the 2–8 Hz region [68,85], it is regrettable that no reliable information could be produced at lower frequencies. (This limitation has been recently removed by sophisticated motion correction algorithms to be published soon, and those algorithms have already been used to process data in a very recent breast cancer study [86]; see below.)

Removal of motion artifacts will evidently revolutionize many of DTI's clinical applications, especially its use in breast cancer detection. It is now possible to study the hemodynamic behavior of the breast or other tissues down to 0.01 Hz. This could open up a new era in the application of DTA to breast cancer detection and evaluation.

8.3.2.4 Physiological Considerations and Their Implications

From the physiological mechanistic standpoint, the positive findings of the Buffalo study pose interesting questions. While it is expected that a hyperperfused breast will appear warmer, as found in classical thermal imaging, how could changes in perfusion modulation kinetics of a tumor situated, say 20 mm below the skin, affect temperature modulation and the homogeneity of temperature distribution at the skin level. It must be concluded, therefore, that the presence of a cancerous lesion inside the breast affects the behavior of the cutaneous capillary bed. Were it only due to the higher heat dissipation of cancerous breasts, DTI would not have had a higher sensitivity, and especially specificity, of breast cancer detection than classical thermography. Moreover, ductal carcinoma *in situ* (DCIS) that is unlikely to affect the heat dissipation of the whole breast would not have been detectable as it was by DTA [68,85]. We must advance the hypothesis that the NO produced in a tumor inside the breast affects *cutaneous* perfusion dynamics measurable by DTI.

Then one must ask whether it is NO that diffuses from the cancerous lesion to the cutaneous capillary bed and, if so, how does it diffuse? Is it carried in the arterial blood supply? This is rather unlikely in view of its short lifetime in the presence of hemoglobin. Does it diffuse through the lymphatic system or in the interstitial space? This again is not very plausible in view of the rate of oxidation of NO in aqueous media.

We venture here the hypothesis that cancerous breasts build up a significant level of NO in their fat wherein the half-life of NO is likely to be quite long. Then the NO diffuses slowly from the fat into the cutaneous capillary bed, affecting its hemodynamic behavior. This hypothesis awaits, evidently, experimental verification. The fact that cancerous breasts were identified irrespective of the size of the tumors by examining temperature modulation over the whole breast, irrespective of the site of the tumor, supports this new hypothesis, presented here for the first time.

The NO–fat hypothesis implies that subcutaneous fat of cancerous breasts retrieved by needle liposuction will have a significant NO content, while fat of cancer-free breasts would be virtually free of NO. This suggests a quasi-invasive preliminary test of breasts that were found suspicious by x-ray mammography. Such a test could be an attractive alternative to exploratory biopsy, which has a significant level of false negatives, in addition to its invasiveness and costs. It is noteworthy that the presence of NO in subcutaneous fat is less dependent on the locale and nature of the cancerous lesion and might be, therefore, a highly sensitive and specific test, actually alleviating the need for DTI for this clinical problem. Furthermore, if this test proved reliable, it could be streamlined to become the first-line diagnostic test to be followed by surgery. If this hypothesis was verified and such a diagnostic test was found effective, it can be considered a conceptual offshoot of the Buffalo DTI study, justifying inclusion of this suggestion in this review paper.

8.3.3 Preliminary DTA Findings on Melanoma and Diabetes

Although the main trust of DTA studies at Buffalo were on breast cancer patients, two other exploratory studies done there warrant being mentioned. Preliminary studies of patients with osteosarcoma and melanoma undergoing chemotherapy gave encouraging results. In the first study, carried out in collaboration with Dr. C. Karakousis of the Department of Surgery, it was found that cancer induced characteristic DTI "signatures" of melanoma disappeared following chemotherapy, suggesting that inoperative metastatic tumors lost their capacity of NO production as a result of chemotherapy, as a result of their metabolic arrest. In other patients DTI was also able to pick up indication of the presence of residual malignant tissue following excision of the primary lesion.

The other preliminary study was done by Dr. C. Carthy in collaboration with Dr. P. Dandona on diabetic patients. These investigators explored the use of DTI in staging of diabetes mellitus by examining the perfusion of the extremities. The preliminary findings, though promising, however, were not conclusive.

8.4 Recent DTA Studies (2000–2005)

8.4.1 Detection of and Treatment of Malignancy

Following the preliminary findings in Buffalo there have been a number of attempts in different clinical research centers to use DTI in oncology. This included detection of malignant lesions as well as follow-up of treatment of cancer.

Janicek et al. [87] studied the response of soft tissue sarcomas to chemotherapy with DTI, in parallel with CT and PET. DTA detected the malignancy in superficially located sarcomas, and the findings were reported to correlate well with assessments by the two other imaging modalities. Janicek et al. [88–90] showed enhanced temperature modulation, probably due to hypervascularization since they monitored in the 0.8–2 Hz range, also in the case of metastatic gastrointestinal stromal tumors; the finding correlated well with those of Doppler ultrasound. Similar results were observed also in the case of malignant lymphomas. The malignant lesions studied by Janicek et al. [91] might have been too deep to exhibit frequencies higher than 2 Hz; thus only the cardiogenic pulsatile modulations were observable. The conclusion of these preliminary clinical studies and suggestions to use DTI in other than breast cancer studies were summarized in 2003.

Lately, a research project was undertaken by Dr. Johan Nilsson at the Karolinska Institute in Stockholm to explore the use of DTI in the detection of metastatic melanoma. In a preliminary series of tests on two patients, melanoma metastases were detectable as spots of enhanced modulation on DTI temperature amplitude bitmaps, both in the cardiogenic frequency range of 0.9–1.7 Hz and in the low vasomotor frequency range of 0.05–0.2 Hz. The DTI detected melanoma metastases, some of which were visible on the skin and palpable, and confirmed *prior* to the DTI study by thin needle aspiration biopsy. Unfortunately, this preliminary study was not extended to include detection of lesions that were not known beforehand, like in the case of the prebiopsy breast cancer study in Buffalo. These preliminary measurements in the <2 Hz modulation frequency range complement Anbar's earlier observations (see above) on the effect of melanoma metastases on localized temperature modulation in the 2–8 Hz range. This far an optimal frequency range for detection of melanoma by DTI has yet to be established.

Following the Buffalo study, there have been just two more DTI studies aimed at breast cancer detection. A major multicenter study by Parisky et al. [92] involved close to a thousand prebiopsy patients. These investigators tried to identify emerging characteristic thermal patterns on thermal images of breasts, in response to external cooling by a stream of cold air. In this study Parisky used hundreds of sequential thermal images. This subset of what has been done in Buffalo 4 years earlier is based on Anbar's hypothesis that vasculature surrounding cancerous lesions is less likely to respond to sympathetic vasoconstriction induced by cold stimuli. The findings corroborated this hypothesis. Unfortunately this extensive study was limited by inferior IR equipment that lacked the sensitivity and speed necessary to monitor temperature modulation. The other limitation of this study was the use of human "experts" in evaluating the computerized images. As was pointed out by Moskowitz [93], the usefulness of this subjective DTI protocol is rather limited.

Another study on breast cancer detection was done by Button et al. [94] at Stony Brook University Hospital. Button used up-to-date equipment and computational procedures in his study of temperature modulation in the cardiogenic frequency range. His findings on a limited number of prebiopsy patients were indicative but nonconclusive, partially because he, like Parisky, used trained human "experts" to evaluate the DTI images.

The latest use of DTI in breast oncology has been in a study by Fanning et al. [86] at the Cleveland Clinic. In their preliminary experiments they assessed the effect of chemotherapeutic agents on established cancerous lesions, on the basis of the assumption that as the viability of the lesion diminishes so will the effect of the NO on temperature modulation. Their preliminary findings were encouraging. However, this use of DTI has limited sensitivity because NO is also produced by phagocytes during the necrosis of the shrinking tumors [33].

In summary, in spite of several encouraging attempts by several groups, the use of DTA in breast oncology still awaits extensive, well-designed clinical trials. A major obstacle to acquisition of reliable DTA data over a wide range of frequencies has been recently removed by motion corrections algorithms. However, this new promising approach to detection of breast cancer has yet to be tested in large-scale clinical studies. Such studies are warranted in view of independent experimental findings that confirmed the vascular effect of cancer-produced NO.

Button et al. [94] have shown in an animal study that the generation of NO by a neoplastic lesion can be detected by DTI. Mice were implanted subcutaneously with human epithelial caner cells. Once these cells established a viable neoplastic lesion 7 days after implantation, the lesion was unambiguously manifested on a DTI bitmap. This manifestation was totally abolished by administration of a conventional NO synthesis inhibitor (LNAME) but reappeared 3 days later, when the effect of the inhibitor abated. This study also demonstrated detection by DTI of the vasodilatatory effect of NO on the vasculature of mice following topical treatment with nitroglycerin.

8.4.2 Use of DTI in the Study of Brain Function and Brain Malignancies

Up to this point we discussed only DTI studies of skin, the huge exposed organ of the body. However, the exposed parenchyma of any other organ could be studied by DTI if its blackbody radiation can be exposed to an IR camera. During open brain surgery for intracranial vascular repair or removal of malignant or benign space-occupying lesions, the outer surface of the brain is exposed, allowing infrared imaging, including DTI.

Alexander Gorbach conceptualized the idea of using infrared imaging to monitor the brain already in 1993 when he was still in Russia [95]. Gorbach persisted with his studies in this field at NIH and has published since 2002 a series of highly impressive studies, starting with animal experimentation [96] and ending with numerous clinical open brain intraoperative investigations. Gorbach's studies covered two major aspects:

1. Monitoring the physiology of neurological functions in different regions of the brain by manifestations of changes in local perfusion and metabolic activity [97,98].
2. Clinical uses of DTI, including real-time follow-up of reperfusion in the course of vascular neurosurgery, and localization of benign or malignant space-occupying lesions by their abnormal vascular structure and perfusion dynamics, allowing then the following up of their surgical removal [96,99,100].

Gorbach has shown that local metabolism even in hypermetabolic brain tissues is just a minor contributor of temperature change following mental stimulation compared with enhanced perfusion of the ROI [97,98]; this confirmed Anbar's conclusion regarding the metabolism versus perfusion in cancerous breasts [24]. Gorbach's findings on regional perfusional changes in response to somatosensory stimulation have been most recently independently confirmed [101].

Unlike Anbar, Gorbach used a 3–5 μm camera with a lower thermal sensitivity and stability, and lower acquisition rate, as well as lower accuracy because of the need for a much greater correction for the reflectivity of the monitored surface. These studies were limited, therefore, to monitoring of changes in brain surface temperature following stimulation or surgical manipulations rather than changes in modulation amplitudes of perfusion. In view of these limitations Gorbach's beautifully designed experiments and their unambiguous conclusions are even more praiseworthy. Moreover, Gorbach did not limit his studies to visual observations of temperature change or of the rate of temperature change (which are generally sufficient for neurosurgeons), but he used computerized data processing, similar to that used by Anbar 4 years earlier [83], to receive off-line *objective* information [97]. Since we believe that this is going to be a major advantage of DTA, the independent use of similar algorithms in processing DTI data is quite gratifying.

In parallel with Gorbach's studies of the brain at NIH, Ecker et al. [102] at the Mayo Clinic in Rochester, MN, carried out a DTI study using up-to-date IR imaging equipment and superior data processing software. The superior equipment used at the Mayo clinic allowed demonstration of the same clinical uses of DTI demonstrated at NIH, that is, visualization of brain reperfusion following vascular operations and localization of brain tumors by their abnormal perfusion patterns and vascular dynamics. The equipment used by Ecker et al. [102] (200 frames/s) and its FFT algorithms allowed detection of pathology with a much higher sensitivity and resolution. This demonstrated again the pronounced advantages of computerized analysis of temporal temperature modulation at high spatial resolution.

Ecker's study, independently corroborated by Gorbach's later findings, unambiguously indicates the use of DTI in neurosurgery. What is preventing today the routine use of this modality are not the cost of the equipment (<$300,000 for a surgical procedure costing >$20,000 each) or its sophistication (it can be fully automated). What is needed is a multicenter clinical study that will demonstrate the added value, in terms of outcomes of open brain surgical procedures, of using DTI as a tool. From the commercial standpoint, such a multimillion dollar study is not attractive because of the relatively limited commercial market for such equipment (brain surgery is not a common medical procedure). Also from the standpoint of public health, problems that call for brain surgery are not very common compared to other public health care problems. Therefore, expenditure of millions on this problem might be of low priority. Neurosurgeons will have to wait, therefore, for manufacturers of DTI equipment to build up a sufficiently large market for their products in other clinical fields to be in position to sponsor an outcome-driven clinical study in cranial neurosurgery, so as to open up another, though limited, market for their equipment.

Amazingly, Ecker's paper does not refer at all to Gorbach's closely related work, whereas Gorbach refers only once to Ecker's study. In the closely knit community of neurosurgeons of the top institutions in the United States, including NIH and the Mayo Clinic, people must have been aware of the parallel efforts. It seems that science calls for more collegial information exchange and collaboration. It is perhaps equally surprising that Ecker, who unlike Gorbach was not a pioneer in this field, did not refer also to Anbar's monograph, which summarized in detail the history of uses of IR imaging in neurosurgery up to 1993 and strongly recommends the use of DTI in this clinical field. Chances are that Ecker got the idea of using DTI in his clinic from that book, as he used Anbar's algorithms in his study (which he does refer to). Moreover, Ecker's paper's title starts with the pretentious phrase: "Vision of the Future": Whose vision has this been to begin with? It is quite conceivable that Ecker would never have undertaken his study without having been made aware of Anbar's monograph by the distributor of the IR equipment he used.

This teaches us something about the sociology of science, studied by Anbar many years earlier [103]. Academicians are often more ready to give due credit to developers of new hardware or software, or to producers of new data, rather than to people who conceptualized new ideas. However, new ideas are those that have driven science through the ages. Scientists ought to be recognized and rewarded for their ideas first of all.

To summarize, the clinical use of DTI in neurosurgery is one of the first to have verified Anbar's vision on the potential clinical value of this science-based approach.

8.4.3 Use of DTI Intraoperatively in Other Surgical Procedures

There are numerous potential uses of DTI used in surgery. We discussed in Section 8.3.2 neurosurgical applications separately, because open cranial surgery has yielded substantial new physiological neurological information, in addition to is potential clinical uses. Other surgical applications to be discussed next are primarily of clinical value only.

Different surgical applications were suggested in Anbar's monograph, based predominantly on previous static infrared imaging studies. All surgical applications of DTI are based on intraoperative monitoring of perfusion. Two groups of surgical applications of DTI have been recently reduced to practice in well-documented clinical studies—uses in reconstructive and vascular surgery (which have been first demonstrated by Anbar in 1998; see above) and uses in cardiac and transplant surgery. These uses will be described below.

8.4.3.1 Reconstructive and Vascular Surgery

Intraoperative DTI is potentially an ideal tool for reconstructive and vascular surgery, as it allows to assess perfusion and reperfusion in real time and thus provide the surgeon immediate feedback. Surgeons can thus take corrective action when it is most effective. Zeroing in on and displaying bitmaps of the spatial distribution of cardiogenic frequency modulation in real time enhances significantly the sensitivity of the technique. DTI can also be a highly useful preoperative and postoperative evaluation tool.

In addition, DTI at cardiogenic frequency allows the precise localization of perforator vessels, which perfuse subcutaneous tissue from below, preoperatively, and thus significantly improving the outcome of grafting. The currently used laser Doppler flowmetry has inferior capacity to provide information on the vasculature over large areas of skin. The application of DTI to this important surgical problem, which has been first demonstrated by Binzoni et al. [104] at the University of Geneva, is likely to be adopted by plastic surgeons as a routine preoperative procedure once a broad scale multicenter clinical study demonstrates its usefulness in a controlled outcome study.

While the use of DTI in reconstructive and vascular surgery is self-evident, it is surprising that only two other clinical studies have been published in this field in recent years. In addition to the pioneering study in Geneva, a study on the intraoperative use of DTI in vascular surgery was carried out in Dundee University in Scotland [105]. A study at Tohoku University demonstrated the advantage of long-wave IR cameras in DTI-assisted vascular surgery [106]. Those three demonstration studies are expected to be just a beginning to a highly beneficial array of routine uses of DTI in these surgical specialties.

8.4.3.2 Cardiac and Transplant Surgery

The use of DTI in open heart surgery is an obvious clinical application, as it involves perfusion and reperfusion of a highly vascular organ. It also involves grafting of vessels (discussed in Section 8.4.3.1). Mohr et al. [107] demonstrated this application on a large number of patients and documented improved outcomes when this technique was used [108,109]. Mohr actually extended the use of DTI in cardiac surgery he pioneered in 1989, using later much better equipment. The former study was cited in Anbar's 1994 monograph as a model for surgical uses of DTI. Surprisingly, in spite of these well-documented positive clinical results, no other studies on the use of DRI in cardiac surgery in other clinical centers were published to date.

Another surgical application that depends on assessment of reperfusion is transplant surgery. Gorbach et al. [110] have demonstrated the effectiveness of using DTI in kidney transplant surgery. They showed the rate of reperfusion to be an important predictive parameter of the success of transplantation, especially in the case of allografts from cadavers that have been subjected to extended periods of ischemic cooling. The history of this application of thermal imaging, which dates back to the early 1970s, was reviewed in Anbar's 1994 monograph. Yet only modern DTI equipment enabled a reproducible quantitative noninvasive assessment of perfusion rates, making this an attractive procedure for routine kidney transplant surgery, as well as for other transplant situations. Like in the case of cardiac surgery, there has been so far no other follow-up studies of these pioneering efforts.

It seems that introduction of a new technology into surgical suits requires more than just good clinical demonstration studies. Surgery is a relatively conservative discipline, and changes in protocol take a considerable effort, especially when they involve the installation of substantial instrumentation in the operating room. It is often easier to find an enthusiastic research clinician to test a new technological modality and get academic credit for it, than to convince a hospital administrator to acquire a new instrument for a clinic with the sole benefit of better clinical outcomes. As long as there is no competition that forces a hospital to modernize, modernization is less likely. Moreover, the use of any new technique in the clinic requires appropriate compensation by medical insurance, Medicare in particular. Insurance companies are, obviously, reluctant to approve new technology unless it cuts the cost of hospitalization. In the case of cardiac surgery it can be readily claimed that early detection of failure in vascular grafting or of rejection in the case of transplant surgery will cut hospitalization costs. Yet we see how slow has been the acceptance of routine use of DTI even in these surgical procedures.

8.4.4 Clinical Applications of DTI in Other Fields

The use of infrared imaging to monitor reperfusion of tissues, of the extremities in particular, is as old as classic infrared imaging [19]. DTI in its most rudimentary mode—quantitative assessment of warming rates at different subareas of the ROI—makes this methodology of clinical value. After our review of its uses in surgery (see above), we shall discuss a few recent studies that have used this approach in nonsurgical clinical situation.

Gold et al. [111] have recently used DTI to diagnose upper extremity musculoskeletal disorders, including carpal tunnel syndrome, by following the perfusion dynamics of hands during and after a standardized typing task. The findings were consistent with earlier ones [19], but the more precise DTI assessment allowed meaningful statistical evaluation. It remains to be seen if this technique, which has been known for decades among neurologists, will become more accepted with more precise computerized equipment available today. As was pointed out by Anbar already in 1994, it is conceivable that more precise differential diagnosis could be achieved if temperature modulation was monitored in addition to gross cooling or warm-up rates.

Rewarming of extremities following brief immersion (10 s) in cold water has been used as a diagnostic test also for neurovascular damage resulting from type-2 diabetes mellitus [112]. Since the test performed used classic thermal imaging at 5 min intervals, it can hardly be classified as a DTI procedure. A similar study, though more sophisticated and precise, was carried out more recently on type-1 diabetes patients by Zotter et al. [113] at the University of Graz. In this study differential diagnosis was based on the relative rewarming rates of different toes, adding an anatomic dimension to the test.

Regretfully, up-to-date, no studies have confirmed the value of DTI in differential diagnosis between diabetes-induced vascular occlusions and vascular neuropathies, which should be possible by comparing thermal modulation of the skin of extremities in the cardiogenic and vasomotor frequency ranges, respectively [19]. This could be done without thermal stress, although response to a transient thermal stress could add diagnostic information. Also no attempts have been made so far to use DTI as a management tool to stage diabetes and monitor the effectiveness of treatment.

The same limitations on the acceptance of DTI routinely in surgery, discussed above, apply also to its use in diabetic clinics. Feasibility studies are first needed, and then large-scale clinical tests must prove the added value of the diagnostic technique, before a new technology reaches routine clinical use.

Finally, in the field of peripheral neuronal diseases we will cite a recent quantitative DTI study on complex regional pain syndrome [114]. Huygen et al. [114] at Rotterdam, studied the dynamics of temperature change in the hands of patients following systemic major thermal challenge by a thermosuit. While the DTI temperature imaging was done at 10 min intervals, their computerized diagnostic test allowed *objective* evaluation of patients, as in Anbar's diagnostic test on breast cancer [83]. Moreover, this application of DTA has been among those suggested by Anbar 10 years earlier [19].

Other early DTI studies have had recent follow-ups. Fifteen years after Anbar had first demonstrated the use of DTI in monitoring the temperature of the cornea [11,115], a 2005 review recommends the use of dynamic ocular thermography in ophthalmology [116]. The early use of DTI in pain assessment [33,76] have been recently followed in patients with postoperative neuralgia [117]. The results in this group of patient have shown, however, no correlation between cutaneous blood flow and tactile induced pain, unlike pain in rheumatic inflammation [76], suggesting that the former pain is not due to a local vasoactive process. The latter suggestion has been corroborated by another independent thermographic study [118].

Completing the review of thermal imaging warm-up studies since 1999, one has to mention a study on the warm up of teeth and gingival tissue following a brief cooling period with a stream of cold air, to assess the perfusion and thereby the vitality of teeth [119,120]. This study essentially establishes that all normal teeth in a patient's mouth have a similar warm-up pattern, implying that an affected tooth will show a lower rate of thermal recovery. The competing technology is, however, laser Doppler flowmetry (LDF), which can provide the same information at a lower cost. This study is an example of an unjustified use of DTI. Since the area of interest is quite small and DTI equipment is bulkier and more expensive than LDF, there is no justification for the use of DTI for this specific dental problem. It is not surprising, therefore, that there has been no follow-up to this study. However, DTI has a potential use in dental surgery (e.g., in the management of TMJ pain [75,76]).

8.5 Conclusion

It has been both a gratifying experience and a disappointment for me writing this chapter. I found that the ideas I advanced in the early 1990s, summarized in my 1994 monograph, on the potential advantages of physiology-based DTI as a clinical tool have merit. Many investigators, most of them new users of thermal imaging, have accepted and adopted these notions. At the same time, static, anatomic thermal imaging, which has been around since the early 1970s, has failed to gain acceptance by the medical community. By now it has been universally recognized that thermal imaging manifests physiology and it must, therefore, measure the *dynamics* of physiological processes. A fair number of clinical applications of DTI suggested in my 1994 monograph have been successfully demonstrated since 2000, when appropriate computerized infrared monitoring equipment became commercially available. By now DTI is a recognized domain in medical technology. This has been truly gratifying.

On the other hand, this far none of those applications has reached the stage of routine use in a clinical setting. This may not be as disappointing as it looks, compared with the long time it took to get ultrasonic imaging or even MRI into routine clinical use. In either of these two cases it took time until major medical equipment manufacturers realized the potential of those technologies and marketed them on a large scale. It all boils down to marketing. DTI has not yet reached this stage.

However, it has become more difficult with time to market new medical technologies because of economic constraints. One must prove that the benefits of using a new technology (in terms of better clinical outcomes—patients' survival and quality of life, days of hospitalization, patients' pain and suffering, and so on) outweigh its costs (including the equipment, space, training of personnel, risks, etc.). Moreover, the benefit/cost of the given technology must exceed those of competing technologies. For reasons I pointed out already in 1994 [19], classic clinical thermal imaging has not met these criteria over 30+ years of its existence, in spite of a few successful preliminary attempts. It was left behind, while ultrasound imaging, CT, and MRI have flourished.

Will DTI be more successful? For one, it is based on firmer biomedical grounds (users can understand what they are measuring in physiological terms), it is quantitative and is potentially the most *objective* among all medical-imaging techniques (as described above, computerized DTA can be used without *any*

imaging output), and its hardware costs a fraction of that of CT, PET, or MRI. It also exceeds competing technologies—laser Doppler flowmetry and ultrasound Doppler imaging—in its ability to acquire
information extremely rapidly on large anatomical areas. Moreover, while DTI can become an invaluable tool in surgery by providing real-time information on facets unavailable to the clinician in any
other way, DTA allows completely automated objective diagnosis, as I have demonstrated. DTA heralds
therefore a new paradigm in clinical medicine alleviating the need for imaging experts, thus substantially reducing heath care costs. It must be realized that DTA poses a threat to radiologists, the experts
who interpret clinical images, and it might be fought by them for obvious reasons. In spite of this, DTI
and especially DTA are, with all likelihood, going to become someday a mainstream biomedical technology. This may take more time, possibly not during the lifetime of this reviewer.

The author is comforted by another personal experience. In the late 1950s Anbar pioneered the use of
F^{18} in nuclear medicine [121,122], and demonstrated its use in brain tumor localization by coincidence
detection of positrons [123–125]. This was long before computers allowed CT to develop. It took more
than 30+ years revolutionary developments in computer technology before PET scanners became clinically useful. Hopefully, DTA will have a similar fate.

The author wishes to conclude this review with a personal note.

Note: In his desire to make this review of DTI as inclusive as possible, the author covered all relevant papers that have appeared up to January 2006 in the National Library of Medicine database, in
the European *EMbase* (Excerpta Medica). Also covered were *Thermology* (discontinued), *Thermology
International* (and its discontinued predecessors *Thermologie Osterreich* and *European Journal of
Thermology*), and *Biomedical Thermology* (Japan), which are not covered in full in the international
medical databases. The author also consulted Professor K. Ammer and Professor K. Mabuchi, the editors of the two latter current journals. Consequently, publications on DTI outside these resources could
have been left out.

References

Note: Review papers are demarcated by an asterisk (*).

1. Anbar M. and Haverly R.F. Local "micro" variance in temperature distribution evaluated by digital
   thermography. *Biomedical Thermology*, 13, 173–187, 1994. Also in *Advanced Techniques and Clinical
   Applications in Biomedical Thermology*. Mabuchi K., Mizushina S. and Harrison B. (eds.) Chur,
   Harwood Academic Publishers, 173–187, 1994.
2. Anbar M. Computerized thermography. The emergence of a new diagnostic imaging modality.
   International Journal of Technology Assessment in Health Care, 3, 613–621, 1987.
3. Montoro J. and Anbar M. New modes of data handling in computerized thermography. *Proceedings
   of the 10th Annual International Conference of the IEEE Engineering in Medicine and Biology Society*,
   Harris G. and Walker C. (eds.), Vol. 10, 845–847, 1988.
4. Montoro J., Hershey L.A., and Anbar M. Enhancement of interpretation of thermograms through
   on-line software. *Thermology*, 3, 121–124, 1989.
5. Montoro J., Lee K.-H., and Anbar M. Study of regulation of skin temperature using dynamic digital
   thermal imaging. *Proceedings of the 11th Annual International Conference of the IEEE Engineering in
   Medicine and Biology Society*. Seattle, WA, 1158–1159, 1989.
6. Montoro J., Lee K.-H., Spangler R.A., and Anbar M. Assessment of skin temperature regulation by
   dynamic digital thermal imaging. *Proceedings of the 34th Annual Meeting of the Biophysical Society.
   Biophysical Journal*, 57, 280a, 1990.
7. Montoro J., Lee K.-H., and Anbar M. Skin temperature regulation parameters derived from temporal
   studies of infrared images. *The FASEB Journal*, Abstracts Part I, 4, A698, 1990.
8. Montoro J. and Anbar M. Visualization and analysis of dynamic thermographic changes. *Proceedings
   of the First Conference on Visualization in Biomedical Computing*, Atlanta, GA, 486–489, 1990.

9. Montoro J., D'Arcy S., and Anbar M. Temporal analysis of thermal images. *Proceedings of the 12th Annual International Conference of the IEEE Engineering in Medicine and Biology Society*, 12, 1578–1579, 1990.

10. Anbar M. Objective assessment of clinical computerized thermal images. *SPIE Proceedings* 1445, 479–484, 1991.

11. Montoro J.C., Haverly R.F., D'Arcy S.J., Gyimesi I.M., Coles W.H., Spangler R.A., and Anbar M. Use of digital infrared imaging to objectively assess thermal abnormalities in the human eye. *Thermology* 3, 242–248, 1991.

12. Anbar M., Montoro J.C., Lee K.-H., and D'Arcy S.J. Manifestation of neurological abnormalities through frequency analysis of skin temperature regulation. *Thermology*, 3, 234–241, 1991.

13. Anbar M., D'Arcy S., and Montoro J. Characteristic frequencies of human skin temperature regulation derived from temporal analysis of infrared images. *The FASEB Journal*, 5, 4303, 1991.

14. Anbar M. and D'Arcy S. Localized regulatory frequencies of human skin temperature derived from the analysis of series of infrared images. *Proceedings of the 4th Annual IEEE Symposium on Computer Based Medical Systems (CBMS '91)*, 184–191, 1991.

15. Anbar M. Recent technological developments in thermology and their impact on clinical applications. *Biomedical Thermology*, 10, 270–276, 1992.

16. Anbar M. Thermoregulatory processes affecting skin temperature derived from time series of infrared images. *Thermologie Osterreich*, 2S, 18, 1992.

17. *Anbar M. Dynamic area telethermometry—a new field in clinical thermology Part I. *Medical Electronics*, 146, 62–73, 1994.

18. *Anbar M. Dynamic area telethermometry—a new field in clinical thermology Part II. *Medical Electronics*, 147, 73–85, 1994.

19. *Anbar M. *Quantitative Dynamic Telethermometry in Medical Diagnosis and Management*. Boca Raton, FL, CRC Press Inc., 1994.

20. Freeman G., Dyer R.L., Juhos L., St. John G.A., and Anbar M. Identification of nitric oxide (NO) in human blood. *Archives of Environmental Health*, 33, 19–23, 1978.

21. Moncada S. The first Robert Furchgott lecture: From endothelium-dependent relaxation to the 811L-arginine: NO pathway. *Blood Vessels*, 27, 208–217, 1990.

22. Luscher T.F. Endothelium-derived nitric oxide: The endogenous nitrovasodilator in the human cardiovascular system. *European Heart Journal*, 12 (Suppl E), 2–11, 1991.

23. Moncada S. and Higgs A. The 811L-arginine-nitric oxide pathway. *New England Journal of Medicine*, 329, 2002–2012, 1993.

24. Anbar M. Hyperthermia of the cancerous breast—analysis of mechanism. *Cancer Letters*, 84, 23–29, 1994.

25. Anbar M. Mechanism of hyperthermia of the cancerous breast. *Biomedical Thermology*, 15, 135–139, 1995.

26. Thomsen L.L., Lawton F.G., Knowles R.G., Beesley J.E., Riveros-Moreno V., and Moncada S. Nitric oxide synthase activity in human gynecological cancer. *Cancer Research*, 54, 1352–1354, 1994.

27. Fujisawa H., Ogura T., Kurashima Y., Yokoyama T., Yamashita J., and Esumi H. Expression of two types of nitric oxide synthase mRNA in human neuroblastoma cell lines. *Journal of Neurochemistry*, 63, 140–145, 1994.

28. Jenkins D.C., Charles I.G., Baylis S.A., Lelchuk R., Radomski M.W., and Moncada S. Human colon cancer cell lines show a diverse pattern of nitric oxide synthase gene expression and nitric oxide generation. *British Journal of Cancer*, 70, 847–849, 1994.

29. Xie W., McCahon P., Jakobsen K., and Parish C. Evaluation of the ability of digital infrared imaging to detect vascular changes in experimental animal tumours. *International Journal of Cancer*, 108, 790–794, 2004.

30. *Anbar M. Quantitative and dynamic telethermometry—a fresh look at clinical thermology. *IEEE Engineering in Medicine and Biology Magazine*, 14, 15–16, 1995.

31. *Anbar M. Dynamic area telethermometry and its clinical applications. *SPIE Proceedings*, 2473, 312–322, 1995.

32. Anbar M. The role of nitric oxide as a synchronizing chemical messenger in the hyperperfusion of the cancerous breast. In *The Biology of Nitric Oxide Part 5*, Moncada S. et al. (eds.), London, Portland Press, pp. 288a–288d, 1996.

33. Anbar M. and Gratt B.M. Role of nitric oxide in the physiopathology of pain. *Journal of Pain and Symptom Management*, 14, 225–254, 1997.

34. Anbar M. Mechanism of the association between local hyperthermia and local pain in joint disorders. *European Journal of Thermology*, 7, 173–188, 1997.

35. Jenkins D.C., Charles I.G., Thomsen L.L., Moss D.W., Holmes L.S., Baylis S.A., Rhodes P., Westmore K., Emson P.C., and Moncada S. Roles of nitric oxide in tumor growth. *Proceedings of the Natural Academy of Science USA*, 92, 4392–4396, 1995.

36. Maeda H., Noguchi Y., Sato K., and Akaike T. Enhanced vascular permeability in solid tumor is mediated by nitric oxide and inhibited by both new nitric oxide scavenger and nitric oxide synthase inhibitor. *Japan Journal of Cancer Research*, 85, 331–334, 1994.

37. Tozer G.M., Prise V.E., and Bell K.M. The influence of nitric oxide on tumour vascular tone. *Acta Oncology*, 34, 373–377, 1995.

38. Tozer G.M., Prise V.E., and Chaplin D.J. Inhibition of nitric oxide synthase induces a selective reduction in tumor blood flow that is reversible with 811L-arginine. *Cancer Research*, 57, 948–955, 1997.

39. Fukumura D. and Jain R.K. Role of nitric oxide in angiogenesis and microcirculation in tumors. *Cancer Metastasis Review*, 17, 77–89, 1998.

40. Thomsen L.L., Miles D.W., Happerfield L., Bobrow L.G., Knowles R.G., and Moncada S. Nitric oxide synthase activity in human breast cancer. *British Journal Cancer*, 72, 41–44, 1995.

41. Zeillinger R., Tantscher E., Schneeberger C., Tschugguel W., Eder S., Sliutz G., and Huber J.C. Simultaneous expression of nitric oxide synthase and estrogen receptor in human breast cancer cell lines. *Breast Cancer Research Treatment*, 40, 205–207, 1996.

42. Duenas-Gonzalez A., Isales C.M., del Mar Abad-Hernandez M., Gonzalez-Sarmiento R., Sangueza O., and Rodriguez-Commes J. Expression of inducible nitric oxide synthase in breast cancer correlates with metastatic disease. *Modern Pathology*, 10, 645–649, 1997.

43. Blachier F., Selamnia M., Robert V., M'Rabet-Touil H., and Duee P.H. Metabolism of 811L-arginine through polyamine and nitric oxide synthase pathways in proliferative or differentiated human colon carcinoma cells. *Biochimica et Biophysica Acta*, 1268, 255–262, 1995.

44. Ambs S., Merriam W.G., Bennett W.P., Felley-Bosco E., Ogunfusika M.O, Oser S.M., Klein S., Shields P.G., Billiar T.R., and Harris C.C. Frequent nitric oxide synthase-2 expression in human colon adenomas: Implication for tumor angiogenesis and colon cancer progression. *Cancer Research*, 58, 334–341, 1998.

45. Kojima M., Morisaki T., Tsukahara Y., Uchiyama A., Matsunari Y., Mibu R., and Tanaka M. Nitric oxide synthase expression and nitric oxide production in human colon carcinoma tissue. *Journal of Surgical Oncology*, 70, 222–229, 1999.

46. Rosbe K.W., Prazma J., Petrusz P., Mims W., Ball S.S., and Weissler M.C. Immunohistochemical characterization of nitric oxide synthase activity in squamous cell carcinoma of the head and neck. *Otolaryngology Head and Neck Surgery*, 113, 541–549, 1995.

47. Gallo O., Masini E., Morbidelli L., Franchi A., Fini-Storchi I., Vergari W.A., and Ziche M. Role of nitric oxide in angiogenesis and tumor progression in head and neck cancer. *Journal of Natural Cancer Institute*, 90, 587–596, 1998.

48. Gavilanes J., Moro M.A., Lizasoain I., Lorenzo P., Perez A., Leza J.C., and Alvarez-Vicent J.J. Nitric oxide synthase activity in human squamous cell carcinoma of the head and neck. *Laryngoscope*, 109, 148–152, 1999.

49. Cobbs C.S., Brenman J.E., Aldape K.D., Bredt D.S., and Israel M.A. Expression of nitric oxide synthase in human central nervous system tumors. *Cancer Research*, 55, 727–730, 1995 and *British Journal of Cancer*, 73, 189–196, 1996.

50. Ellie E., Loiseau H., Lafond F., Arsaut J., and Demotes-Mainard J. Differential expression of inducible nitric oxide synthase mRNA in human brain tumours. *Neuro Report*, 7, 294–296, 1995.

51. Whittle I.R., Collins F., Kelly P.A., Ritchie I., and Ironside J.W. Nitric oxide synthase is expressed in experimental malignant glioma and influences tumour blood flow. *Acta Neurochir (Wien)*, 138, 870–875, 1996.

52. Bakshi A., Nag T.C., Wadhwa S., Mahapatra A.K., and Sarkar C. The expression of nitric oxide synthases in human brain tumours and peritumoral areas. *Journal of Neurology Science*, 155, 196–203, 1998.

53. Joshi M., Strandhoy J., and White W.L. Nitric oxide synthase activity is up-regulated in melanoma cell lines: A potential mechanism for metastases formation. *Melanoma Research*, 6, 121–126, 1996.

54. Joshi M. The importance of 811L-arginine metabolism in melanoma: An hypothesis for the role of nitric oxide and polyamines in tumor angiogenesis. *Free Radical Biology and Medicine*, 22, 573–578, 1997.

55. Ahmad N., Srivastava R.C., Agarwal R., and Mukhtar H. Nitric oxide synthase and skin tumor promotion. *Biochemical and Biophysical Research Communication*, 232, 328–331, 1997.

56. Edwards P., Cendan J.C., Topping D.B., Moldawer L.L., MacKay S., Copeland E.M., and Lind D.S. Tumor cell nitric oxide inhibits cell growth *in vitro*, but stimulates tumorigenesis and experimental lung metastasis *in vivo*. *Journal of Surgical Research*, 63, 49–52, 1996.

57. Liu C.Y., Wang C.H., Chen T.C., Lin H.C., Yu C.T., and Kuo H.P. Increased level of exhaled nitric oxide and up-regulation of inducible nitric oxide synthase in patients with primary lung cancer. *British Journal of Cancer*, 78, 534–541, 1998.

58. Fujimoto H., Ando Y., Yamashita T., Terazaki H., Tanaka Y., Sasaki J., Matsumoto M., Suga M., and Ando M. Nitric oxide synthase activity in human lung cancer. *Japan Journal of Cancer Research*, 88, 1190–1198, 1997.

59. Klotz T., Bloch W., Volberg C., Engelmann U., and Addicks K. Selective expression of inducible nitric oxide synthase in human prostate carcinoma. *Cancer*, 82, 1897–1903, 1998.

60. Thomsen L.L., Sargent J.M., Williamson C.J., and Elgie A.W. Nitric oxide synthase activity in fresh cells from ovarian tumour tissue: Relationship of enzyme activity with clinical parameters of patients with ovarian cancer. *Biochemical Pharmacology*, 56, 1365–1370, 1998.

61. Hajri A., Metzger E., Vallat F., Coffy S., Flatter E., Evrard S., Marescaux J., and Aprahamian M. Role of nitric oxide in pancreatic tumour growth: *In vivo* and *in vitro* studies. *British Journal of Cancer*, 78, 841–849, 1998.

62. Di Cesare P.E., Carlson C.S., Attur M., Kale A.A., Abramson S.B., Della Valle C., Steiner G., and Amin A.R. Up-regulation of inducible nitric oxide synthase and production of nitric oxide by the Swarm rat and human chondrosarcoma. *Journal of Orthopaedic Research*, 16, 667–674, 1998.

63. Swana H.S., Smith S.D., Perrotta P.L., Saito N., Wheeler M.A., and Weiss R.M. Inducible nitric oxide synthase with transitional cell carcinoma of the bladder. *Journal of Urology*, 161, 630–634, 1999.

64. Morcos E., Jansson O.T., Adolfsson J., Kratz G., and Wiklund N.P. Endogenously formed nitric oxide modulates cell growth in bladder cancer cell lines. *Urology*, 53, 1252–1257, 1999.

65. Jansson O.T., Morcos E., Brundin L., Bergerheim U.S., Adolfsson J., and Wiklund N.P. Nitric oxide synthase activity in human renal cell carcinoma. *Journal of Urology*, 160, 556–560, 1998.

66. Eroglu A., Demirci S., Ayyildiz A., Kocaoglu H., Akbulut H., Akgul H., and Elhan H.A. Serum concentrations of vascular endothelial growth factor and nitrite as an estimate of *in vivo* nitric oxide in patients with gastric cancer. *British Journal of Cancer*, 80, 1630–1634, 1999.

67. Goto T., Haruma K., Kitadai Y., Ito M., Yoshihara M., Sumii K., Hayakawa N., and Kajiyama G. Enhanced expression of inducible nitric oxide synthase and nitrotyrosine in gastric mucosa of gastric cancer patients. *Clinical Cancer Research*, 5, 1411–1415, 1999.

68. Anbar M., Naumov A., Milescu L., and Brown C. Objective detection of breast cancer by DAT—an update. *Thermology International*, 11, 11–18, 2001.

69. Anbar M., Grenn M.W., Marino M.T., Milescu L., and Zamani K. Fast dynamic area telethermometry (DAT) of the human forearm with a Ga/As quantum well infrared focal plane array camera. *European Journal of Thermology*, 7, 105–118, 1997.

70. *Anbar M. and Milescu L. Hardware and software requirements of clinical DAT. *SPIE Proceedings*, 3698, 63–74, 1999.
71. Gunapala S.D., Liu J.K., Sundaram M., Bandara S.V., Shott C.A., Hoelter T.R., Maker P.D., and Muller R.E. Long-wavelength 256 × 256 QWIP hand-held camera. *SPIE Proceedings*, 2746, 124–133, 1996.
72. Gunapala S.D., Liu J.K., Park J.S., Sundaram M., Shott C.A., Hoelter T., Lin T.-L., Massie S.T., Maker P.D., Muller R.E., and Sarusi G. 9 µm cutoff 256 × 256 GaAs/AlxGal-xAs quantum well infrared photodetector hand-held camera. *IEEE Transactions of Electron Devices*, 44, 51–57, 1997.
73. Anbar M., Milescu L., Grenn M.W., Zamani K., and Marino M.T. Study of skin hemodynamics with fast dynamic area telethermometry (DAT). *Proceedings of the 19th IEEE EMBS International Conference*, 644–648, 1997.
74. Zamani K., Marino M.T., Bonner M., and Anbar M. Assessment of mental stress as well as neurological effects of chemical warfare agents by dynamic area telethermometry. *Proceedings of the 21th Army Science Conference*, 125–130, 1998.
75. Gratt B.M. and Anbar M. Thermology and facial thermography: Part II. Current and future clinical applications in dentistry. *Journal of Dentomaxillofacial Radiology*, 27, 68–74, 1998.
76. Anbar M. and Gratt B.M. The possible role of nitric oxide in the physiopathology of pain associated with temporomandibular disorders. *Journal of Oral and Maxillofacial Surgery*, 56, 872–882, 1998.
77. Anbar M. and Zamani K. A DAT study of a painful knee—comparison with MRI findings. *Proceedings of the 4th Congress of the International College of Thermology* and *the Annual Meeting of the American Academy of Thermology*, Fort Lauderdale, FL, 25, 1998.
78. Anbar M., Eckhert K.H. Jr., and Milescu L. Preliminary study of women's breasts with dynamic area telethermometry (DAT). *Proceedings of the 4th Congress of the International College of Thermology* and *the Annual Meeting of the American Academy of Thermology*, Fort Lauderdale, FL, 27–28, 1998.
79. *Anbar M. Clinical thermal imaging today—shifting from phenomenological thermography to pathophysiologically based thermal imaging. *IEEE EMBS Magazine*, 17, 25–33, 1998.
80. Anbar M. and Milescu L. Scope and limitations of dynamic area telethermometry (DAT). *Proceedings of the 20th IEEE EMBS International Conference*, Hong Kong, HK, 928–931, 1998.
81. Anbar M., Brown C.A., Milescu L., and Babalola J.A. Clinical applications of DAT using a QWIP FPA camera. *SPIE Proceedings*, Orlando, FL, 3698, 93–104, 1999.
82. Anbar M., Brown C.A., Milescu L., Babalola J.A., and Gentner L. Potential applications of dynamic area telethermometry (DAT) in assessment of cancer in the breasts. *IEEE EMBS Magazine*, 19, 58–62, 2000.
83. Anbar M., Brown C.A., and Milescu L. Objective identification of cancerous breasts by dynamic area telethermometry (DAT). *Thermology International*, 9, 137–143, 1999.
84. Anbar M., Brown C.A., and Milescu L. Objective detection of beast cancer by dynamic area telethermometry. *Proceedings of the 21st Annual International Conference of EMBS*, 1115–1116, 1999.
85. Anbar M., Milescu L., Naumov A., Brown C.A., Button T., Carty C., and AlDulaimi K. Detection of cancerous breasts by dynamic area telethermometry (DAT). *IEEE EMBS Magazine*, 20, 80–91, 2001.
86. Fanning S.R., Short S., Coleman K., Andresen S., Budd G.T., Moore H., Rim A., Crowe J., and Weng D.E. Correlation of dynamic infrared imaging with radiologic and pathologic response for patients treated with primary systemic therapy for locally advanced breast cancer. *American Society of Clinical Oncology (ASCO) Annual Meeting*, Atlanta, GA, 2006.
87. Janicek M.J., Waxman A., Janicek M.R., and Demetri G.D. Assessment of early response to therapy in sarcomas: Dynamic infrared imaging (DIRI), computerized tomography (CT) and F-18 FDG positron emission tomography (PET). *American Society of Clinical Oncology (ASCO) Annual Meeting*, 2001, Abstract #1390.
88. Janicek M.J., van Sonnenberg E., and Demetri G.D. Monitoring of early response to treatment in metastatic gastrointestinal stromal tumor (gist): Model of multimodality approach including f-18 fdg pet, infrared imaging and Doppler ultrasound. *SGR (Society of Gastrointestinal radiologists) Abdominal radiology Conference*, 2001.

89. Janicek M.J., Janicek M.R., Merriam P., Potter A., Silberman S., Dimitrijevic S., Fauci M., and Demetri G.D. Imaging responses to Imatinib mesylate (Gleevec, STI571) in gastrointestinal stromal tumors (GIST): Vascular perfusion patterns with Doppler ultrasound (DUS) and dynamic infrared imaging (DIRI). *American Society of Clinical Oncology (ASCO) Annual Meeting*, 2002, Abstract #333.

90. Janicek M.J., Demetri G.D., Janicek M.R., Shaffer K., and Fauci M.A. Dynamic infrared imaging of newly diagnosed malignant lymphoma compared with Gallium-67 and Fluorine-18 fluorodeoxyglucose (FDG) positron emission tomography. *Cancer Research and Treatment*, 2, 571–578, 2003.

91. Janicek M.J., Demetri G., Janicek M.R., Shaffer K., and Fauci M.A. Dynamic infrared imaging of newly diagnosed malignant lymphoma compared with gallium-67 and fluorine-18 fluorodeoxyglucose (FDG) positron emission tomography. *Technology in Cancer Research and Treatment*, 2, 71–77, 2003.

92. Parisky Y.R., Sardi A., Hamm R., Hughes K., Esserman L., Rust S., and Callahan K. Efficacy of computerized infrared imaging analysis to evaluate mammographically suspicious lesions. *American Journal of Roentgenology*, 180, 263–269, 2003.

93. Moskowitz M. Efficacy of computerized infrared imaging. *American Journal of Roentgenology*, 180, 596, 2003.

94. Button T.M., Li H., Fisher P., Rosenblatt R., Dulaimy K., Li S., O'Hea B., Salvitti M., Geronimo V., Geronimo C., Jambawalikar S., Carvelli P., and Weiss R. Dynamic infrared imaging for the detection of malignancy. *Physics in Medicine and Biology*, 49, 3105–3116, 2004.

95. Gorbach A.M. Infrared imaging of brain function. *Advances in Experimental Medicine and Biology*, 333, 95–123, 1993.

96. Watson J.C., Gorbach A.M., Pluta R.M., Rak R., Heiss J.D., and Oldfield E.H. Real-time detection of vascular occlusion and reperfusion of the brain during surgery by using infrared imaging. *Journal of Neurosurgery*, 96, 918–923, 2002.

97. Gorbach A.M., Heiss J., Kufta C., Sato S., Fedio P., Kammerer W.A., Solomon J., and Oldfield E.H. Intraoperative infrared functional imaging of human brain. *Annals of Neurology*, 54, 297–309, 2003.

98. Gorbach A.M. Local alternated temperature gradients as footprints of cortical functional activation. *Journal of Thermal Biology*, 29, 589–598, 2004.

99. Gorbach A.M., Heiss J.D., Kopylev L., and Oldfield E.H. Intraoperative infrared imaging of brain tumors. *Journal of Neurosurgery*, 101, 960–969, 2004.

100. Gorbach A.M., Ntziachristos V., and Perelman L. Advances in optical imaging of cancer. In *New Techniques in Oncologic Imaging*, Padhani A.M. and Choyke P.L. (eds.), New York, Taylor and Francis, Chapter 18, pp. 351–370, 2005.

101. Trubel H.K., Sacolik L.I., and Hyder F. Regional temperature changes in the brain during somatosomatic stimulation. *Journal Cerebral Blood Flow and Metabolism*, 26, 68–78, 2006.

102. Ecker R.D., Goerss S.J., Meyer F.B., Cohen-Gadol A.A., Britton J.W., and Levine J.A. Vision of the future: Initial experience with intraoperative real-time high-resolution dynamic infrared imaging. *Journal of Neurosurgery*, 97, 1460–1471, 2002.

103. Cohen B.P., Kruse R.J., and Anbar M. The social structure of scientific research teams. *Pacific Sociological Review*, 25, 205–245, 1982.

104. Binzoni T., Leung T., Delpy D.T., Fauci M.A., and Rufenacht D. Mapping human skeletal muscle perforator vessels using a quantum well infrared photodetector (QWIP) might explain the variability of NIRS and LDF measurements. *Physics in Medicine and Biology*, 49, N165–N173, 2004.

105. Campbell P.A., Cresswell A.B., Frank T.G., and Cuschieri A. Real-time thermography during energized vessel sealing and dissection. *Surgical Endoscopy*, 17, 1640–1645, 2003.

106. Nakagawa A., Hirano T., Uenohara H., Utsunomiya H., Suzuki S., Takayama K., Shirane R., and Tominaga T. Use of intraoperative dynamic infrared imaging with detection wavelength of $7–14\ \mu m$ in the surgical obliteration of spinal arteriovenous fistula: Case report and technical considerations. *Minimally Invasive Neurosurgery*, 47, 136–139, 2004.

107. Mohr F.W. and Falk V. Thermal coronary angiography: A method for assessing graft patency and coronary anatomy in coronary bypass surgery. *Annals of Thoracic Surgery*, 47, 441–449, 1989. Updated in 1997 *Annals of Thoracic Surgery*, 63, 1506–1507.

108. Falk V., Walther T., Philippi A., Autschbach R., Krieger H., Dalichau H., and Mohr F.W. Thermal coronary angiography for intraoperative patency control of arterial and saphenous vein coronary artery bypass grafts: Results in 370 patients. *Journal of Cardiac Surgery*, 10, 147–160, 1995.

109. van Son J.A., Falk V., Walther T., Diegeler A., and Mohr F.W. Thermal coronary angiography for intraoperative testing of coronary patency in congenital heart defects. *Annals of Thoracic Surgery*, 64, 1499–1500, 1997.

110. Gorbach A., Simonton D., Hale D.A., Swanson S.J., and Kirk A.D. Objective, real-time, intraoperative assessment of renal perfusion using infrared imaging. *American Journal of Transplantation*, 3, 988–993, 2003.

111. Gold J.E., Cherniack M., and Buchholz B. Infrared thermography for examination of skin temperature in the dorsal hand of office workers. *European Journal of Applied Physiology*, 93, 245–251, 2004.

112. Fujiwara Y., Inukai T., Aso Y., and Takemura Y. Thermographic measurement of skin temperature recovery time of extremities in patients with type 2 diabetes mellitus. *Experimental and Clinical Endocrinology and Diabetes*, 108, 463–469, 2000.

113. Zotter H., Kerbl R., Gallistl S., Nitsche H., and Borkenstein M. Rewarming index of the lower leg assessed by infrared thermography in adolescents with type 1 diabetes mellitus. *Journal of Pediatric Endocrinology Metabolism*, 16, 1257–1262, 2003.

114. Huygen F.J., Niehof S., Klein J., and Zijlstra F.J. Computer-assisted skin videothermography is a highly sensitive quality tool in the diagnosis and monitoring of complex regional pain syndrome type I. *European Journal of Applied Physiology*, 91, 516–524, 2004.

115. Haverly R.F. and Anbar M. A telethermometric study of the age dependence of corneal temperature. *Biomedical Thermology*, 14, 10–26, 1994.

116. Purslow C. and Wolffsohn J.S. Ocular surface temperature: A review. *Eye Contact Lens*, 31, 117–123, 2005.

117. Besson M., Brook P., Chizh B.A., and Pickering A.E. Tactile allodynia in patients with postherpetic neuralgia: Lack of change in skin blood flow upon dynamic stimulation. *Pain*, 117, 154–191, 2005.

118. Kemppainen P., Forster C., and Handwerker H.O. The importance of stimulus site and intensity in differences of pain-induced vascular reflexes in human orofacial regions. *Pain*, 91, 331–338, 2001.

119. Kells B.E., Kennedy J.G., Biagioni P.A., and Lamey P.J. Computerized infrared thermographic imaging and pulpal blood flow: Part 1. A protocol for thermal imaging of human teeth. *International Endodontic Journal*, 33, 442–447, 2000.

120. Kells B.E., Kennedy J.G., Biagioni P.A., and Lamey P.J. Computerized infrared thermographic imaging and pulpal blood flow: Part 2. Rewarming of healthy human teeth following a controlled cold stimulus. *International Endodontic Journal*, 33, 448–462, 2000.

121. Anbar M., Guttmann S., and Lewitus Z. The accumulation of fluoroborate ions in thyroid glands of rats. *Endocrinology*, 66, 888–890, 1960.

122. Anbar M. Application of fluorine 18 in biological studies with special reference to bone physiology. In *Production and Use of Short Lived Radioisotopes*, Vienna, IAEA, Vol. II, pp. 227–245, 1963.

123. Anbar M., Kosary I., Laor Y., Guttmann S., Lewitus Z., and Askenasy H. The localization of intracranial tumors by positron emitting F18 labelled fluoborate. *Proceedings of the Beilinson Hospital*, 10, 50–52, 1961.

124. Askenasy H., Kosary I., Toledo E., Lewitus Z., and Anbar M. The use of radioisotopes in the detection of brain tumors. *Proceedings of the Beilinson Hospital*, 10, 53–58, 1961.

125. Ashkenazy H.M., Anbar M., Laor Y., Kosary I.Z., and Guttmann S. The localization of intracranial space-occupying lesions by fluoroborate ions labelled with fluorine 18. *American Journal of Roentgenology and Nuclear Medicine*, 88, 350–354, 1962.

9

Thermal Texture Mapping: Whole-Body Infrared Imaging and Its Holistic Interpretation

H. Helen Liu
Institute of Holistic Health and Science

Zhong Qi Liu
Academy of TTM Technologies and Bioyear Medical Instrument

9.1 Introduction

A live human body emits heat, part of which is radiated as infrared (IR) photons. Heat is one of a few basic features that distinguish a live body from a dead one, because heat is continuously being generated through metabolic, blood flow, and other important biological processes such as thermogenesis processes. Thus, although IR photons represent only a narrow band of the electromagnetic spectrum, information contained in such images is fundamentally associated with the functional status and movement in our body.

In radiological science, functional imaging is a fast evolving field that focuses on imaging and assessing biological functions of the body. Based on the physical and physiological characteristics of IR imaging, it can fit perfectly into this category and attract tremendous medical interests [1,2]. However, despite many appealing features of IR as compared to other imaging modalities, IR has not entered the field of radiology in mainstream medicine so far. This is an unfortunate fact, and many historical, social–political as well as scientific factors can be attributed to it.

In our opinion, there are two major misconceptions about IR that have hindered its development in the past. The first misconception is related to the depth of IR photons that can penetrate through the skin. At present, IR is generally perceived to be useful for imaging skin, blood flow and conditions near superficial depths only, for example, breast imaging, pain management, skin perfusion, and so on. This perception came into dominance because IR photons can only go through a minuscule distance inside tissues. Thus, it has been conventionally believed that heat from sources deep within the body cannot be easily seen from IR images, and thus, IR has very limited role to play in studying internal organs and in the field of radiology.

The second misconception about IR is that it merely reflects the thermal activity of the body. This misconception came into existence because physics has taught us that heat gives off IR photons, and there is a relationship between temperature and intensity of IR photons. Thus, we tend to believe that IR imaging is only about seeing the thermal activity of the body. This has significantly limited the use of the IR, because heat only represents a narrow range of biological functions in the body.

Quite contrary to these conventionally held beliefs, through continuous research and development efforts over the past 15 years, we have found that first of all, surface thermal patterns are strongly associated with internal organ functions and biological processes. More importantly, IR images represent essentially a type of energy maps of the human body within the IR spectral window, in other words, they are intrinsically holographic or reflexiological maps of the human energy structure.

These important findings came into light when we carried out earlier experiments and theoretical analyses, followed by a large number of clinical observations of various diseases and health conditions. On the basis of continuous availability of clinical data and accumulated experience over a decade of work, a new system of acquiring and interpreting the images was developed. In differentiating traditional thermography from this new approach, it received a new name of "thermal texture mapping" or TTM which refers to the way how the images are acquired, analyzed, and interpreted [3–10]. TTM has evolved to be essentially a new tool to evaluate the functional balance and health environment of the body. IR-TTM is in fact a bridge between mainstream medicine and alternative medicine, the merging of two will lead to a new era of holistic medicine.

In parallel to the research efforts, commercial development was undertaken in the 1990s for TTM. Complete hardware and software systems were developed, facilitating the use of TTM in hospital settings as well as in public health. Applications in mainstream medicine and holistic medicine, and other emerging fields are being actively investigated at present. Meanwhile, we would like to point out that the methodology of TTM is still in its developmental stage. We anticipate its continuous and dynamic growth in the coming years as more basic research and clinical investigations are added to the field.

9.2 Early Experiments and Analyses

A live human body must maintain thermal homeostasis within a few degree Celsius to ensure proper functions of its biological integrity. In the meantime, heat exchange is constantly ongoing to reach a steady state of thermal equilibrium within the body and with the environment. It is then apparent that our body, through millions if not billions of years of natural evolution, must be equipped with extraordinary means of maintaining the thermal homeostasis and equilibrium of itself. Various thermal pathways must exist to conduct the heat in an extremely efficient and harmonious manner for our basic survival. In classic context, it is believed that heat transfer in the body is assisted by blood flow and perfusion, respiration and perspiration, direct heat conduction, convection, and radiation. There are likely other means of transferring the heat efficiently, for example, through meridian and other subtle energy systems [11] of which presently science has little knowledge about.

It is then not difficult to realize that the thermal patterns observed on the skin surface must be affected and related to the internal body function and the entire heat exchange process. In fact, this concept has been used in many industrial and military applications in which underground targets are detected through surface thermal patterns. Even in considering the adipose tissue (fat) which does not contain a large amount of blood vessels and is well insulated to protect the body, heat exchange is constantly ongoing, and the thermal patterns on the skin surface must be related intrinsically to the thermal characteristics underneath it. Earlier works have explored this topic and showed promising results that warrant further investigations [12–17].

This work was inspired by earlier observations of diseases known to have thermogenesis properties, such as inflammation and cancer growth that are highly metabolically active [3,7–10]. From conventional thermographs of human subjects with these health conditions, we observed that certain thermal patterns seen on the skin surface could be closely related to local lesions that are thermal active. Our body seems to be rather thermal transparent, allowing internal heat to be transferred efficiently to the skin surface as a means to maintain thermal homeostasis and equilibrium in the body.

9.2.1 *Ex Vivo* Measurements and Modeling

In an attempt to address the question on how surface heat is related to its internal source, we have conducted experiments and analyses using *ex vivo* animal tissues [6]. In the simulation, heat sources with a temperature close to the human body temperature were imbedded inside freshly dissected animal tissue samples. A typical piece of such tissue sample includes shaved skin, subcutaneous fat, connected tissues, and muscular tissues. The experimental setup is shown in Figure 9.1. In this particular case, a piece of adipose tissue was used, with the size being approximately 15 cm × 15 cm × 2 cm. A brass ball was buried in the tissue as a heat source. The diameter of the ball could be changed from 5 mm to a few centimeters. The surface temperature of the tissue was scanned using a commercial IR camera (spectral response in 8–14 µm) at a distance of approximately 1 m. The brass ball was heated for about 2 h to reach steady state of the surface temperature. Figure 9.2 shows the heat distribution measured from the tissue surface.

We found that the heat going through the tissue sample and emitted from the surface showed a bell-shaped thermal distribution. The center of this bell-shaped function is where the imbedded heat source was the closest to the surface. We found that there is a unique relationship between the depth of the heat source and the width of the bell-shaped function. In other words, we found that the area covered by certain isothermal contours was related to the depth of the heat source itself.

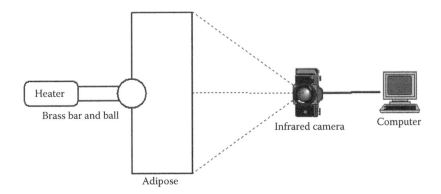

FIGURE 9.1 Illustration of the experimental setup where an internal ball heat source (e.g., brass ball) was imbedded in a piece of adipose tissue. For color figures, see www.holistichealthscience.org.

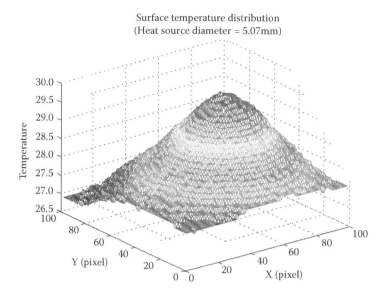

Surface temperature distribution
(Heat source diameter = 5.07mm)

FIGURE 9.2 Illustration of the surface temperature distribution of the internal source of Figure 9.1. For color figures, see www.holistichealthscience.org.

To explore the theoretical basis of this observation, we used the commonly applied bio-heat transfer equation [18],

$$bc\frac{\partial T}{\partial t} = \nabla \cdot (k\nabla T) + Q_b + Q_m \tag{9.1}$$

which describes the energy conservation in a volume of tissue with a constant blood flow rate Q_b and metabolic rate Q_m, and c, b, and k denote the specific heat, the density, and the effective thermal conductivity of the tissue, respectively. If the *ex vivo* tissue sample is in a steady state, Q_b and Q_m equal to zero and the bio-heat transfer equation is reduced to,

$$\nabla \cdot (k\nabla T) = 0 \tag{9.2}$$

To arrive at an analytical solution to Equation 9.2, we consider that the temperature distribution T can be approximated by concentric spheres in a large isotropic phantom, where the center of the spheres coincides with the center of the ball heat source. In spherical coordinates with radius ρ. Equation 9.2 can be expressed as,

$$\frac{d}{d\rho}\left(\rho^2\frac{dT}{d\rho}\right) = 0 \tag{9.3}$$

If the total heat emanating from the ball heat source is Q, the heat density q of the spot at a distance of ρ from the ball center is

$$q = \frac{Q}{4\pi\rho^2} \tag{9.4}$$

The heat density also can be obtained through the Fourier's Law

$$q = -\frac{dT}{d\rho} \tag{9.5}$$

From Equations 9.4 and 9.5, the temperature field function $T(\rho)$ can be found after taking the integral,

$$T(\rho) = T_1 - \frac{Q}{4\pi}\frac{1}{k}\left(\frac{1}{R} - \frac{1}{\rho}\right) \tag{9.6}$$

where T_1 is the temperature of the heat source and R is the radius of the ball. Next, we assume that at any spot on the boundary, heat density transferring from internal heat source is fully exchanged with the environment, that is,

$$q_{in} = q_{out} \tag{9.7}$$

where

$$q_{in} = \frac{Q}{4\pi\rho^2} = k\frac{T_1 - T_D(r)}{(\rho/R)(\rho - R)} \tag{9.8}$$

$$q_{out} = h_r(T_D(r) - T_0) + h_c(T_D(r) - T_0) = \alpha(T_D(r) - T_0) \tag{9.9}$$

in which h_r, h_c, and α are the radiative, convection, and total heat exchange coefficients used to approximate the boundary heat exchange using the first-order approximation. Finally, the surface temperature distribution can be obtained with an analytical solution,

$$T_D(r) = T_0 + \frac{(T_1 - T_0)}{(\alpha/k)\cdot\left(\sqrt{D^2 + r^2}/R\right)\cdot\left(\sqrt{D^2 + r^2} - R\right) + 1} \tag{9.10}$$

where D is the distance from the surface to the center of the heat source.

The calculated thermal distribution $T_D(r)$ was found to behave in a bell-shaped function, similar to those observed in the *ex vivo* tissue testing.

The effort in deriving the analytical solution to Equation 9.2 helped us understand the characteristics of the surface heat distribution versus the heat source itself in an extremely simplified setting. The results were very encouraging because they showed that in *ex vivo* conditions, surface heat is related to the depth of the source D and its spherical radius R. If the source radius R is relatively small compared to the depth of D, then the surface heat is primarily influenced by the depth of the source. This is an important finding, which identifies that the primary influence of the heat distribution is the depth of the heat source, and it is relatively independent of many other factors associated with the heat-exchange process in the tissue. There were earlier experimental work [14,15] which drew essentially the same conclusion.

9.2.2 Phantom Simulations

To further explore the surface heat distribution under more realistic phantom conditions, we simulated a cylindrical phantom where a spherical heat source was imbedded along the central axis of the phantom at different depths. The phantom consists of layers of tissues including skin, fat, and visceral tissues. We relied on solving Equation 9.1 using a finite-element approach (see more details in Reference 19) to obtain the surface temperature distribution. The effects of various parameters related to the heat source, physiological and environmental factors were studied extensively. Similar works have been done by other investigators using different approaches [20,21].

We found that the relative temperature distribution was essentially affected by radius R and depth D of the heat source only, which is consistent to both the analytical results and the one obtained from the *ex vivo* experiment, even though more layers of tissues and blood flow were included in the simulation. Furthermore, the range of the relative temperature distribution was mainly influenced by the depth D of the source only, relatively independent of other parameters used in the simulation (Figure 9.3).

What these results suggested is that when a source is deeply situated in the tissue, the width of the bell-shaped function is larger, resulting in a more gradual change of the surface heat; the shallower the source, the sharper the change of surface heat, resulting in a greater temperature drop away from the central axis of the cylinder (see Figure 9.3). Therefore, the temperature gradient of the surface heat can indicate the depth of the source. This makes obvious sense as we can see that the heat loss from the source center to any point on the surface has to be governed by the distance between these two points. The depth of the source determines how this distance is increased from the source center to peripheral areas. And the shallower the source, the sharper this distance will be increased away from the center, thereby, the greater the temperature will be reduced. Therefore, by evaluating the temperature gradient of the source surface temperature distribution, one could estimate whether the source is close to the skin surface or deeply situated away from the surface.

In practical terms, the temperature gradient can be displayed by isothermal contours for a given heat source, similar to the idea of using topology maps [6,12,13,19]. For a deeply situated source, its surface isothermal contours appear as densely packed, meaning the distance between the isothermal contours is closer because of a more gradual decrease of temperature from the source center, in contrast to a shallow source; whereas for a source situated near the skin surface, the isothermal contours near the source center will spread out with a greater distance in between.

Figure 9.4 illustrates this concept comparing deep and shallow heat sources. Another technique that we have developed to display the topology of the isothermal contours is called "dynamic slicing technique" shown in Figure 9.5. Here we reduce the displayed temperature level one step at a time,

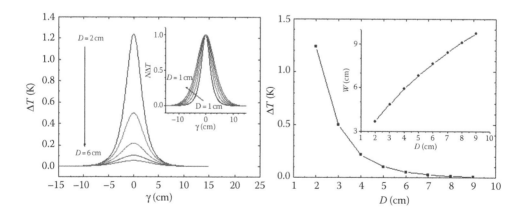

FIGURE 9.3 Left: temperature distributions on the phantom surface of the heat source. Baseline temperature was subtracted to obtain the relative temperature DT(r) and its normalized term NDT(r); r is the radius on the phantom surface away from the center of the source. Right: change of the DT(r=0) and its width W versus the source depth D. For color figures, see www.holistichealthscience.org.

FIGURE 9.4 Illustration of the isothermal contours of the surface temperature distributions of heat sources at different depths. Left: a skin surface source with a greater distance between the isothermal contours. Right: a deep tissue source with closer distance between the isothermal contours. For color figures, see www.holistichealth-science.org.

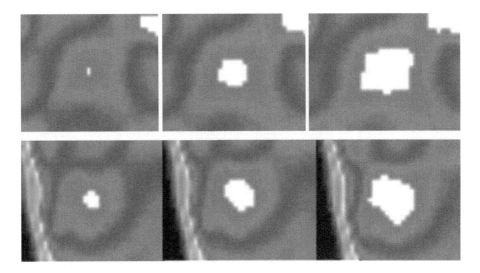

FIGURE 9.5 Illustration of the dynamic slicing technique in displaying the isothermal contours. The white area is the isothermal contour above certain temperature level corresponding to those contours shown in Figure 9.4. The temperature level is dynamically reduced every 0.05°C in a movie mode (three frames of the movie are shown here). Top: a skin surface source with greater changes between the isothermal contours. Bottom: a deep tissue source with gradual changes between the isothermal contours. For color figures, see www.holistichealthscience.org.

for example, every 0.05°C, allowing the surface isothermal contour to appear one level at a time corresponding to the chosen displayed temperature. In visualizing the changes of the isothermal contours dynamically with the displayed temperature, we can compare depths of different heat sources, in a way to differentiate a deep internal source versus skin surface heat source.

9.2.3 Summary of Early Findings

Results of our *ex vivo* experiments and simplified theoretical simulations were quite encouraging, because they pointed to the possibility of identifying deep situated heat sources based on the gradient of their surface temperature distribution. This conclusion was derived based on the condition that the tissue serving as the heat-exchange medium between the source and the surface is relatively homogeneous and isotropic in terms of their thermal properties.

Let us consider a realistic situation where a heat source is imbedded in live tissues, for example, a small tumor growing inside the muscle. The tumor-generated excessive metabolic heat will undergo heat exchange to reach thermal equilibrium with its surrounding tissues and the environment. Under steady-state conditions, heat from the tumor will be eventually distributed to the skin surface, though the surface temperature distribution will also depend on heat transfer processes involving different tissues, for example, blood flow and other thermal pathways. Therefore, whether we could detect deep-tissue heat sources will depend on the temperature contrast between the source and its surrounding tissues, sensitivity of the IR imager itself, and influence of many other physical and physiological factors associated with the heat exchange.

In the next phase of our investigation, we performed a large number of clinical studies and developed corresponding hardware system and software tools in acquiring and processing the image data. Before we present results of these clinical studies, the components of the TTM system and other relevant details are explained here for the purpose of clarification.

9.3 The TTM System

9.3.1 Hardware

To facilitate image acquisition of human subjects in clinical settings, we have developed an IR imaging system in the 1990s. This system consists of a scanner gantry, patient platform, and computer workstations (Figure 9.6). The scanner gantry has an IR camera mounted and a mechanical platform which allows the camera to perform pan, tilt, and vertical movement operations. The patient platform has a base unit where a patient can stand straight on the top and be rotated and scanned from different angles. This hardware system was designed to acquire whole-body images of the patient from any angle and height. The distance between the scanning gantry and the patient platform is about 1.5–2 m depending on the optical property of the camera.

At present, a typical TTM image series consists of 10–20 images with patients performing a sequence of poses. See Figure 9.7 for examples of these poses.

9.3.2 Software

Upon image acquisition, the images are sent to the computer workstation to be reviewed and analyzed. A commercial TTM software package was developed to manage patient database and image processing.

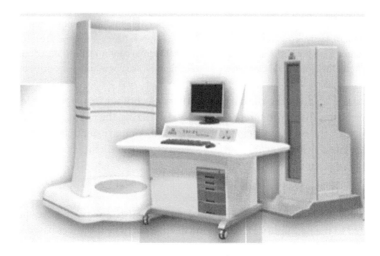

FIGURE 9.6   Illustration of the hardware equipment of a TTM system. For color figures, see www.holistichealth-science.org.

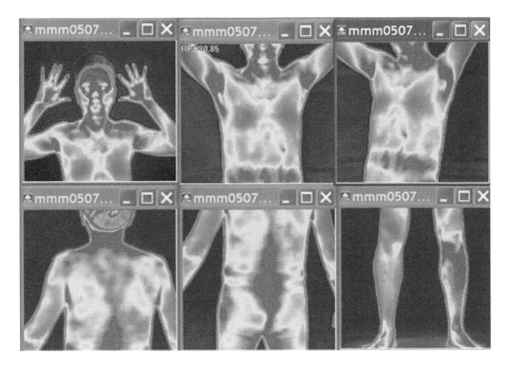

FIGURE 9.7 Typical poses used in a whole-body scan using the TTM system. For color figures, see www.holisti-chealthscience.org.

The image processing tools allow adjustment of color maps, displayed window and level of the images (brightness and contrast), image fusion and comparison, image measurements, and other associated functions. We also implemented tools to display the isothermal contours of the images along with the dynamic slicing technique. These tools help us visualize and analyze the topology, shape, texture, and anatomical features of the images.

The name "thermal texture mapping" or TTM came from the way that these images are displayed and analyzed. The terminology is still being used today for the purpose of continuity only, though the way we use the images today has evolved with a great degree of difference from the earlier versions. TTM differs from traditional thermography on several aspects. First of all, relative temperature and isothermal contour maps are used in TTM rather than the absolute temperature maps in traditional thermography. It is well known that absolute temperature is influenced by many physical and physiological factors and can vary significantly depending on the processes involved. By using the relative temperature maps and isothermal contours, effects of many of these factors, for example, how the IR signal is converted into temperature, environmental conditions, and camera conditions, and so on, can be minimized.

Second, we emphasize on assessing the topology information of the images in TTM in addition to the conventional temperature maps. The topological and morphological properties of the isothermal contours including shapes, textures, connection between different spots and other anatomical features are used to analyze the thermal distributions and their associated physiological implications.

Furthermore, in TTM software, image tools were implemented allowing image registration and fusion of TTM images with standard anatomical template as well as radiological images (x-ray/CT, MRI, PET, etc.). This feature was particularly effective in combining and comparing information obtained from other radiological modalities. An example of such fused TTM image is given in Figure 9.8. There are other unique features developed in interpreting the images in the TTM system. Details will be given in the next sections.

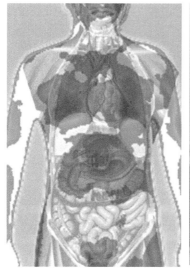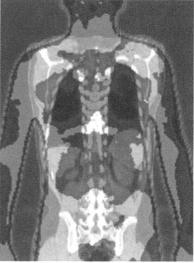

FIGURE 9.8 TTM images are fused with anatomical template (left) and reconstructed CT images of a coronal plane of the studied subject. For color figures, see www.holistichealthscience.org.

9.4 Clinical Observations of Internal Body Heat

9.4.1 Study Procedures

It is known that excessive cell growth, blood flow, or aggregated blood vessels such as those seen in cancer development or inflammation could be associated with a higher metabolic rate and a thermogenesis rate. These types of abnormalities can then serve as internal thermal sources. To investigate surface thermal patterns and how they are affected by the internal heat, we performed multiple studies of commonly seen solitary tumors, hyperplasia, and other tumor types. In addition, diseases associated with local infection and inflammation were observed as well. Interesting studies involved subjects of acquired immune deficiency syndrome (AIDS) and severe acute respiratory syndrome (SARS) which involved infection/inflammation in the lungs [7]. These studies typically included the following procedures.

Patients with one particular type of abnormality, for example, prostate cancer, were recruited with certain predetermined selection criteria (details varied depending on the type of disease and the design of that study). Typically, these patients had clinically confirmed abnormality based on combined radiological studies, clinical tests, and manifested symptoms. Before any surgical procedures and therapeutic intervention, patients were scanned with a TTM system to acquire images of the whole body of the study subject.

When acquiring the images, factors that could possibly influence thermal patterns of the body were minimized, similar to the requirements of conventional thermography procedures. For example, patients were asked to fast for at least 2 h prior to the image scan. Alcohol, use of other stimulants and medications were stopped for at least 8 h before the scan. Skin conditioning such as use of cosmetics and perfume was minimized. Patients were asked to take all clothes off and expose the body in a temperature-controlled room for 10–15 min prior to the scan. Images were then taken for typically about 10–15 min depending on the specific protocol requirements.

We also collected radiological images and other medical reports whenever available, including the lesion's depth, size, clinical stage, and other relevant demographic and clinical data. We used these data in combination with the TTM images to study whether certain thermal signatures were observed near the physical location of the lesion underneath the skin surface. If reconstructed radiological images were available, we also fused such images with TTM images, allowing further examination of the thermal patterns corresponding to the physical location of the lesion identified on the radiological images. These radiological images could come from x-ray/CT, MRI, or PET/CT images available for the studied subjects. Results from representing cases are discussed below.

9.4.2 Case Studies

Figure 9.9 shows a case of liver hemangioma which forms a cluster of entangled blood vessels in the liver. The tumor was approximately 4 cm in size and 5 cm deep inside the liver, diagnosed through an ultrasound exam. This type of tumor serves as an ideal heat source in internal organs because of their simple physiological and anatomical structures. On thermal images, hemangiomas in the liver often appear as distinct hot spots in close proximity to their anatomical locations. Results confirmed our initial finding that under steady-state condition, local heat generated by thermal sources such as a hemangioma will be eventually transferred to skin surface to reach thermal equilibrium. Using the dynamic slicing technique, we can analyze the features of the isothermal contours of the tumor in comparison with other known surface hot spots on the same image. This tumor that is situated inside the liver had isothermal contours that were closely packed in contrast to those of a shallow skin spot. A comparison of the isothermal contours is given in Figure 9.5.

Figure 9.10 shows a stage III stomach cancer case involving the gastrointestinal junction of the stomach and the end of the esophagus. On the thermal map, thermal elevation was seen corresponding to the anatomical location of the lesion and in an agreement with findings from the PET/CT scan. The fact that heat from this deep lesion could be seen on the skin surface was both intriguing and encouraging because of the complexity of the anatomy and in homogeneity of the tissues involved. The heat from the lesion had to go through sections of the stomach, liver, diaphragm, bones of the rib cages, and various connective tissue layers in between to reach the skin surface. However, this complex matrix of tissues surrounding the lesion did not seem to prevent the internal heat from being detected on the skin surface, indicating the capability of heat transfer in the body to reach thermal equilibrium conditions.

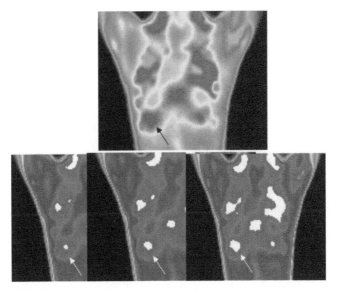

FIGURE 9.9 TTM images of a studied subject with hemangioma of the liver, shown by the arrow. Top: temperature map; bottom: dynamic slicing used in TTM and isothermal contours shown by the white areas. For color figures, see www.holistichealthscience.org.

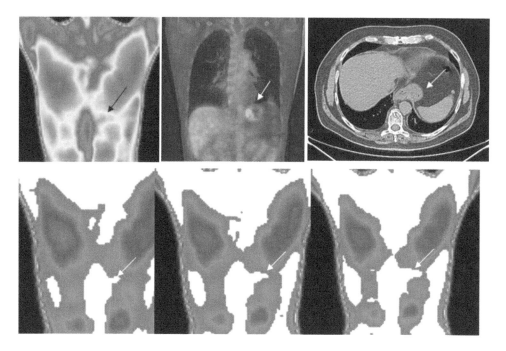

FIGURE 9.10 TTM images of a studied subject with stage III cancer of the stomach, shown by the arrow. Top: temperature map, PET/CT, and a CT axial cut; bottom: dynamic slicing used in TTM and isothermal contours shown by the white areas of the lesion.

Figure 9.11 shows another case of liver cancer, however, with more complex presentations and multiple lesions (stave IV multifocal hepatocellular carcinoma). A cluster of lesions were found from the MRI study near the anterior portion of the liver by the junction of the left and right lobes. TTM images showed that the skin surface near the anatomical location of the liver displayed abnormal hot spots seen from the anterior, right side, and posterior surface of the studied subject. These hot spots were quite irregular, naturally, because of the cancer lesions themselves were at an advanced stage and far more complex than a simple hemoangioma.

9.4.3 Further Discussions

It is known that the angiogenesis process involved in cancer development also helps cancer lesions develop a matrix of blood vessels that supplies metabolic need of the cancer growth. Likely, these vessels are also involved in dissipating the excessive heat from the cancer lesions to their surrounding tissues. At an advanced stage of cancer development, tissue necrosis also occurs because of hypoxia in the core of the lesions. All these factors make the heat-exchange process far more complex than that of an infancy cancer or a simple solid tumor. For such advanced stage cancer cases, we observed several important phenomena that are worth discussing here.

First of all, the topology of the isothermal contours of this type of cancer lesions become far more irregular and complex. The results came as no surprise, because, in addition to the factors of angiogenesis and necrosis, multiple heat sources could be superimposed together, each may have irregular shape and extend to different depths and layers of tissues. Second, the hot-spot locations appearing on the surface thermal maps could deviate from those seen from radiological images, likely caused by the necrosis process and blood flow-related factors as well. Naturally as the cancer lesions grow larger, heat produced by the lesions may be more intense at the edge of the lesions where metabolic rate is greater than the core of the lesions where necrosis could occur. Blood vessels could also carry the heat to surrounding

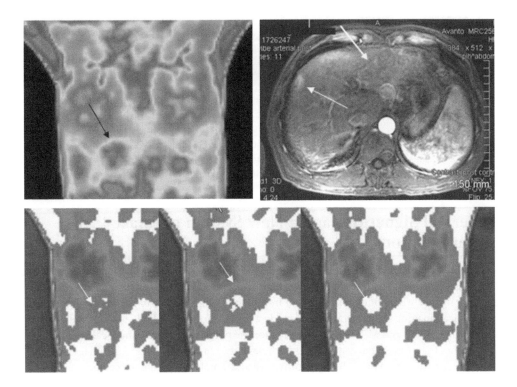

FIGURE 9.11 TTM images of a studied subject with stage IV cancer of the liver, shown by the arrows. Top: temperature map, an MRI axial cut; bottom: dynamic slicing used in TTM and isothermal contours shown by the white areas of the lesions. For color figures, see www.holistichealthscience.org.

tissues and skin surface at locations away from the cancer lesions themselves, making their thermal patterns appear differently than those observed on anatomical images. Third, because of reduced metabolic and thermogenesis rate near the core of the lesions, advanced cancer lesions may not show significant thermal contrast, making it more difficult to distinguish near their corresponding anatomical locations.

Another more interesting and important observation we found is that heat generated by cancer lesions could also be carried elsewhere far away from the source itself through thermal pathways where thermal exchange happens at a greater rate, in other words, pathways that are least thermal resistance. For example, heat transfer through muscular tissue is greater than adipose tissue and bones, and is greater along the muscular fiber orientation than transverse from it. Examples will be given in the next section. There are likely other thermal pathways, for example, involving blood or lymphatic circulation, meridian systems and network of subtle energy [22–24], that can carry the heat away from the source. For stomach and esophageal cancers, for example, we found hot spots that we refer to as "thermal acupoints" that are located near the throat, abdomen and back far away from the stomach itself (Figure 9.12). These acupoints appear as distinct circular or other regularly shaped structures, with locations consistent across different studied subjects. The exact mechanism and physiological basis of these acupoints are unknown, but they seem to indicate a template of such structures in correspondence to the local stress such as cancer growth of the area.

To account for these complex processes and their effects on isothermal contours, we had to extend our original method used for simple solid heat sources in tissues. By reviewing images from many clinical cases on various diseases and health conditions, we found interesting features of thermal patterns or thermal signatures that are associated with disease development and the ways of heat exchange in our body, for example, thermal acupoints as just mentioned.

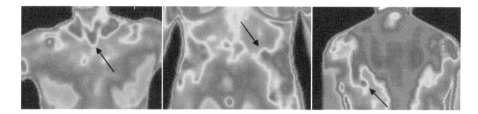

FIGURE 9.12 Thermal acupoints that are associated with stomach disorders. Left: acupoint near the superclavicular notch of the same subject of Figure 9.10. Mid and right: acupoints near the left middle abdomen and left back of another subject. For color figures, see www.holistichealthscience.org.

9.5 Thermal Signatures and a New System of Image Interpretation

It has been long recognized that cancer, autoimmune disorders, and many other serious conditions are systematic diseases that affect the whole body. Thus, disease development not only affects the local area, but may also involve whole-body response and impacts the functionality of the entire system. At present, radiological imaging is often confined to study a limited area of the body because these modalities such as ultrasound, CT and MRI are mainly used for the purpose of examining local anatomy and tissues only rather than functional aspects of the whole body. On the other hand, IR images intrinsically reflect function status of the body. Being noninvasive and simple to operate, IR can be used to scan the whole body. This feature allows us to investigate thermal signatures of diseases, their developmental processes, the overall health environment and body response to such local conditions. Some of the representative cases are presented here to illustrate our findings.

9.5.1 Case Studies

Figure 9.13 shows a case of prostate hyperplasia involving enlargement of the prostate. The thermal image of the front pelvic area indicates the increase of heat dissipated near the local area of the prostate. The back view shows hot spots by both sides of the buttock area, which is a typical feature observed from thermal patterns of prostate hyperplasia and prostate cancer. Likely, sacrum and coccyx of the pelvic girdle are quite thermal insulating, preventing the heat of the prostate from coming to the posterior skin surface directly. Thermal pathways through blood flow and muscular conduction could carry the heat more efficiently farther away from the prostate to the outer edges of the pelvic area.

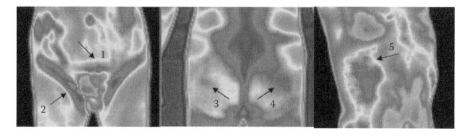

FIGURE 9.13 TTM images of a studied subject with prostate hyperplasia shown by arrow #1. Abnormal thermal activities were also observed at the pelvic lymph node area (arrow #2, back of the buttock (arrow #3, #4), and in the liver area (arrow #5). For color figures, see www.holistichealthscience.org.

In addition to heat associated with local prostate, we also observed the elevation of thermal activities accompanied by thermal asymmetry of the pelvic lymph node area. Thermal asymmetry refers to the temperature difference between the left and right side of the pelvic nodal region seen from the images. This thermal signature may be associated with increased metabolic activity and blood flow of the pelvic nodes which are involved in the process of cancer development.

Furthermore, we also observed that subjects with prostate disorder had accompanying symptoms of abnormal thermal signatures of the liver area as well, indicating certain association between the liver and prostate functions. It is unknown whether such an association is a simple correlation (both occur simultaneously) or it has a causal effect (one causes the other to occur). But this finding has suggested the importance of observing liver function and its effect on the development of prostate cancer or vice versa. To the best of our knowledge, this phenomenon has not been documented in the current clinical research, but it indicates that using whole-body IR can reveal important relationship and interconnection among functions of different organs. One interesting clinical question to be explored is whether treatment of the prostate cancer by local therapy will impact the liver function, and how these two organs are related.

Figure 9.14 shows another clinical case involving cancer of the apex of the right lung. Although PET and CT images showed highlighted regions near this anatomical location only, abnormal thermal activities were seen on the mid right chest and upper left chest, along with right supraclavicular lymph node and right mediastinum lymph node areas. As mentioned earlier, at an advanced stage of disease development, thermal patterns of local diseases can become rather complex, and thermal elevation is no longer constrained to local areas. Local–regional lymph nodes and other abnormal thermal patterns may be involved, indicating the involvement and response of the entire region or even the whole-body under the stress condition. Figure 9.15 shows a case with

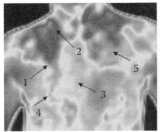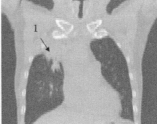

FIGURE 9.14 Images of a studied subject with cancer of the lung shown by arrow #1 on TTM image (left) and chest CT (right). Abnormal thermal activities were also observed near the right supraclavicular and mediastinum lymph node areas (arrow #2, #3, respectively), and mid right chest and upper left chest areas (arrow #4, #5, respectively). For color figures, see www.holistichealthscience.org.

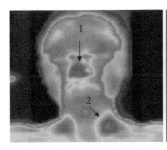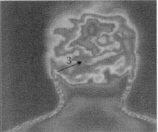

FIGURE 9.15 Images of a studied subject with nasopharynx cancer near the nose cavities (arrow #1). Abnormal thermal activities were also observed near the left neck area and supraclavicular lymph node (arrow #2) and from the back of the head (arrow #3). For color figures, see www.holistichealthscience.org.

nasopharynx cancer. Apparent change of thermal pattern in the nasal cavity area can be observed. In addition, thermal elevation was seen by the left neck area and supraclavicular node. Abnormal thermal activity was also observed on the back of head corresponding to the posterior side of the cancer lesion.

9.5.2 Features of Whole-Body Thermal Signatures

Results from imaging the commonly occurring cancer types and systematic diseases showed that these cases are associated with several important thermal signatures accompanying the thermal activities of local conditions. First of all, local and regional lymph nodes near the cancer lesions have elevated and abnormal thermal activities, shown by the cases above. This observation is in good agreement with standard clinical findings on the role of lymphatic-activated immuno-response in disease development. For example, in addition to thermal patterns observed in conventional thermography studies, breast cancer cases may also have abnormal thermal activities near the axillary nodes, maxillary nodes, and thermal asymmetry of both underarm areas.

Second, in cancer, autoimmune, and other systematic disorders, thermal highlight was seen from the endocrine organs that possibly involve thyroids, pancreas, ovaries, or prostate. Again, in breast cancer cases, for example, abnormal thermal patterns, either hot spots or cold spots of the thyroids, and highlighted area of the ovaries and uterus were often seen. This finding is also consistent with current clinical observation that breast cancer itself may be associated with abnormal hormonal status, and functions of the endocrine and reproductive organs.

Next, certain thermal signatures were seen along the spine areas of the back and other "feature spots" corresponding to the diseases. The mechanism of these thermal features is unknown and may be related to other subtle thermal pathways discussed earlier. The spine is where the central nerve system passes through. This is also an area rich with sympathetic and parasympathetic nerves, as well as acupoints that are associated with important functions of the internal organs in the entire torso. For example, patients with cardiac diseases were found to have the back of the spine highlighted near vertebrae of C7 to T2, in an agreement with the sympathetic nerves and acupoints associated with the heart function.

Furthermore, we have found that irregularity of the thermal patterns is associated with a high mental stress level and in patients with cancer and other systematic diseases. This thermal irregularity has a unique feature that we name as a "leopard spot" pattern of the whole body, which is highly distinguishable on thermal images (Figure 9.16). This pattern was identified initially from subjects who reported themselves to experience a high level of mental stress from their work environment or family relationship. This thermal pattern can be mitigated or smoothed out after studied subjects performed relaxation exercises (e.g., musical therapy or meditation exercises) to reduce mental stress.

Thermal signatures for certain psychological burden and illness are currently under study, for example, depression and bipolar disorder. Abnormal thermal patterns have been observed on the back of the

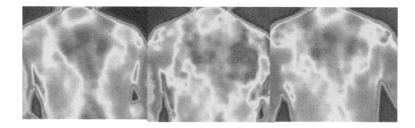

FIGURE 9.16 The effect of stress and its mitigation shown by TTM images. Left: morning condition with low mental stress; middle: evening condition with high mental stress; right: 1 h after the evening scan of the middle panel with 30 min of relaxation meditation. For color figures, see www.holistichealthscience.org.

head, suggesting disorders of brain activities and associated thermogenesis from the area. The results are encouraging because they suggested that IR can serve as an imaging tool to study emotional and mental disorders and their important effects on disease development and physical health. This topic has been quickly recognized in mainstream medicine recently and will continue to attract a surge of scientific interests in medical research [25].

9.5.3 A New System of Image Interpretation

In summary, clinical observations have shown that systematic diseases such as cancer have unique thermal signatures that are not only confined to the local area, but are also presented in the whole body. Using these thermal signatures may aid in the screening and diagnosis processes of these diseases and studying functional relationship among different organs and physiological systems. Based on the characteristics of the thermal signatures, we have developed a new system of image interpretation using the whole-body images in combination with the TTM technique discussed above.

In general, this new system of holistic image interpretation involves the following steps:

1. Assess overall topological features of the whole body, patterns and locations of areas with either thermal elevation or depression;
2. Assess thermal symmetry of the whole body from three axes—front vs. back, left vs. right, and head to feet; (Thermal asymmetry and thermal depression will be discussed further in the next section.)
3. Assess the presence of mental stress and other psychological burden by their thermal signatures;
4. Assess thermal activities of important endocrine organs including the thyroids, pancreas, ovary, and prostate;
5. Assess thermal activities of the regional lymph nodes, and further compare thermal symmetry if the nodal areas are located on either sides of the body (e.g., head-neck, supraclavicular, axillary, and pelvic nodes on both sides of the body);
6. Assess thermal patterns of the central spine area of the back and appearance of specific thermal acupoints associated with certain diseases;
7. Assess local hot spots and surrounding areas by using the TTM technique, measuring the relative temperatures of the suspected hot spots or cold areas, examining thermal topological features and their relationship with other parts of the body;
8. Perform additional stress tests if necessary to examine thermal pattern changes of suspected areas before and after the stress, for example, in taking certain drugs or contrast enhancing agents for suspected diseases (more details will be given in the next section).

We can appreciate that in this new system of holistic image interpretation, eight metrics are used to assess thermal patterns of the whole body and its overall health conditions that we refer to as the "health environment" of an individual. The first six steps have to do with thermal signatures of the functional balance and harmony of the body, and the last two steps are associated with the local suspected areas only.

We believe that the strength of whole-body IR with the use of this set of metrics would enable a comprehensive assessment of the health environment of a person and the status of one's overall wellbeing. In the current medical practice, this information is missing yet extremely important in health evaluation, and none of other radiological imaging test or lab tests can provide such critical information at this point. This is the unique advantage of the whole-body IR and the TTM system, which can serve as an effective tool for wellness focused medicine.

We also want to emphasize that the health environment is a key concept in maintaining the wellbeing of a person and in risk assessment of diseases and health conditions. If the overall health environment is out of order, systematic diseases such as cancer, cardiovascular diseases, autoimmune diseases, and many other types of disorders could arise as a result of such imbalance of the body functions. For example, if the first six metrics are problematic, it is simply a matter of what specific types of diseases will arise subsequently and where the disorder will appear first or to be diagnosed. The focus of the TTM-based health evaluation is always on the overall conditions and environment of the whole-body system rather than appearance of specific diseases.

In other words, if we ignore the overall health environment of a person, and simply focus all attention on diagnosing and treating local disease by itself, then we would lose the overall big picture of the whole body and getting lost in specific details. This is precisely the difference between wellness focused medicine or holistic medicine, versus disease focused medicine or allopathic medicine. In the later case, one may be working hard in removing specific types of symptoms and disorders without appreciating the overall balance of the system. Often such an approach will result in recurrence of the same issues or trigger other more serious conditions as a result of aggressive local treatment if the overall health environment is not restored. We believe that tools such as TTM will help us gain a deeper appreciation of this principle and adopt more holistic approaches in medical practices.

9.6 Other Important Applications of the TTM System

So far, we have been mainly focused on discussing the applications of TTM in studying internal heat sources and their thermal signatures. However, the concept of TTM is not limited to this scope, it can also be applied to study thermal patterns of many other types of diseases and conditions. Clinical studies are being conducted, for example, in the areas of cardiovascular diseases, autoimmune disorders, metabolic disorders, epidemic and infection diseases, and even in mental health. Some of the important cases are shown here to illustrate the use of the TTM concept in these fields.

9.6.1 Thermal Imbalance and Its Associated Disorders

Figure 9.17 shows representative images of subjects who had cardiac ischemia caused by coronary artery diseases and those who had chronic fatty liver, high cholesterol, and high blood pressure syndromes. These examples are shown here to illustrate that certain health conditions can also be manifested as thermal deficiency from excessive coldness of an area on the body. In other words, thermal imbalance in the form of either excessive heat or coldness likely indicates certain disorders or health conditions of the area. In subjects with coronary artery diseases, lack of blood flow to the heart causes reduced perfusion, thus compromised metabolic activity of the cardiac tissues. Asymmetry thermal patterns with coldness in the left chest area seen either from the front or back

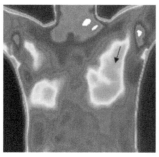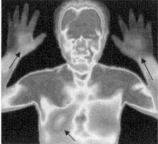

FIGURE 9.17 Left: thermal deficiency in the left chest area (shown by the arrow) associated with a subject with coronary artery disease and cardiac ischemia. Right: thermal deficiency in the right chest area and of the palms and the smaller fingers (shown by the arrows) in a subject who had chronic fatty liver and high cholesterol. High blood pressure experienced by this subject shown as thermal elevation in the head and upper shoulder area. For color figures, see www.holistichealthscience.org.

body surface have been found in such subjects, who have a risk of heart attack, stroke, and other cardiovascular diseases.

Subjects with fatty liver and liver cirrhosis were found to have coldness in the area slightly above the liver on the right side of the chest. These subjects who had accompanying chronic history of high cholesterol also manifest a cold finger symptom (shown in Figure 9.17) in which temperature drops remarkably across from the five fingers likely associated with reduced blood flow and disorder of microcirculation in the smaller fingers. High blood pressure could be experienced by such subjects as well which shows a thermal signature of excessive heat in the entire head or in the forehead area with heat radiating from the neck and upper shoulder. Seen from the whole-body image, thermal imbalance shown as temperature drop from the head to feet direction has been found in these subjects with high blood pressure, consistent with the feelings of head being heavy or hot in the manifestation of the clinical symptoms. Thermal asymmetry has also been observed in patients with recent history of transient ischemia attack or stroke in the brain. The left and right sides of the body showed thermal difference consistent with the loss of function in one side of the body typically experienced by such subjects. In general, healthy normal subjects showed thermal symmetry along three axes of the body: front to back, left to right, and head to feet. Asymmetric temperature distributions could indicate imbalance of energy and functions along either axes of the body.

As emphasized earlier, the advantage of whole-body IR is that it helps us study and explore the inter-connection among different health conditions and underlying relationship weaving through different symptoms that appear to be isolated and uncorrelated. In reality, many of these events are correlated or have causal effects that need to be addressed as a whole rather than being treated as isolated occurrence. For example, we mentioned about the findings in our prostate cancer study that liver and prostate functions were found to be closely related. In studying cardiovascular diseases and metabolic disorders, we also found similar inter-connection among liver conditions, high cholesterol and high blood pressure, and their combined effects on the health of the heart. Kidney function also plays an important role in this group of conditions. Abnormal thermal signatures in the kidney, either as thermal deficiency shown by excessive coldness or local hot spots have been seen in such patient population as well.

9.6.2 Temporal and Functional Changes of the Body

IR imaging and information contained in such images have unique temporal or frequency properties that are very different from those of anatomical images. In the later case, images of the body essentially capture specific physical properties and structures of the body. For example, x-ray CT reflects the medium density, and MRI is about proton density and other related factors. Such anatomical images are static and will not change quickly over time (except in the case that imaging contrast agent is injected in the blood or in the tissue artificially). IR images on the other hand, reflect the metabolic, blood flow, and other activities which occur in a much faster fashion. Such changes could happen in a matter of few minutes to a few hours. Therefore, IR images are more dynamic and fluid compared to anatomical images; this is an intrinsic property of IR that one has to bear in mind in using this modality. In a way, we can observe the body being live and moving in an interactive way in responding to its internal undulations and environmental stimulants.

Results from our clinical observations showed that within the scope of the IR itself, changes of image intensity follow different temporal patterns as well. First of all, fast changes can come from the need to respond to environment (room temperature, air flow, clothing, etc.), body movement, and even mental changes such as mood and emotional swings. Such changes can happen in a matter of a few minutes assisted by hormonal, blood circulation, and metabolic responses of the body. For example, a person's blood pressure could rise in a few minutes in responding to a burst of anger, and heat from the head will increase correspondingly. The next category is normal physiological functions and cycles that are constantly ongoing in the body, for example, digestion, resting, sleeping, menstrual cycle, and so on. Such changes happen in the scale of hours, days, and weeks depending on the specific physiological functions. More significant changes accompanying disease progression or other physiological or psychological factors can last for months or even years.

In general, we found that the body tends to preserve a steady state of equilibrium and thermal pattern of itself. Thus, although small scales of changes are constantly ongoing in the body, there is a stable whole-body thermal pattern that is being maintained for individual people. Each person carries a set of unique thermal signature that can last for months to years. In a good analogy, temperature of different locations on the Earth is constantly changing, but over a scale of decades to centuries, there is an overall stable pattern that is being preserved for the Earth's thermal stability.

Understanding the temporal properties of IR is extremely important in clinical studies, because this feature governs how the images should be acquired, analyzed, and interpreted. For example, precise quantitative measurement of certain temperatures associated with a disease will be both difficult and meaningless when we intend to use IR for disease diagnosis because of the constant changes of the body and person-to-person variations. Rather, we have to look for and capture the essential topographic features of the thermal patterns or thermal signatures qualitatively of a disease. This is what we want to emphasize in TTM image interpretation.

IR is also an effective tool in observing body changes in response to certain diseases, stress, or stimulant conditions [3,4,26–28]. In our experiments, we found that drugs that are taken orally can start to change the body as fast as 10–20 min. The effects are most apparent in the window of about 30–40 min, and stabilize in about 45–60 min. Details of this time sequence depend greatly on the pharmacokinetics of individual drugs and sensitivity of the person. After 1 h, the changes can be brought by other biophysical factors and normal physiological cycle of the body. Therefore, it is important to minimize other environmental, physiological, and psychological factors when IR is used in studying the drug effect and therapy response.

Figure 9.18 shows the changes of the body of a studied subject who underwent cancer chemotherapy and upon oral intake of a dose of herbal supplement to reduce side effects of the chemotherapy. The changes of the body over 2 months before and after the chemotherapy are also shown, illustrating the apparent difference in body functions consistent with lab test results and clinical symptoms of the studied subject. Obviously, the ability of the whole-body IR in observing body response will be very important in drug development and comparing efficacy of different therapies. The principle of the dynamic responding test has also been used in identifying properties of suspected abnormalities under certain stress conditions. For example, cold stress test has been used in thermography to differentiate benign versus malignant lesions in breast cancer [26,27]. It is also possible to use contrast-enhancing agents or drugs to stimulate response of the lesions and observe its temporal course, from which certain physiological characteristics can be derived [29]. This approach has been commonly used in radiological imaging and is included in our metrics of system interpretation of IR.

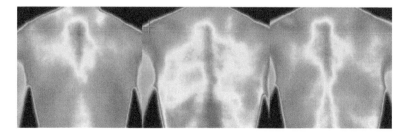

FIGURE 9.18 Thermal pattern changes of a studied subject undergoing cancer chemotherapy. Left: prior to chemotherapy; middle: post chemotherapy 2 months after the first scan; right: 1 h after the scan of the middle panel and with oral intake of a dose of herbal supplement to mitigate side effects of chemotherapy. For color figures, see www.holistichealthscience.org.

9.6.3 Energy Mapping and Subtle Energy Pathways

In fact, whole-body IR is not limited to only study drug or therapy response. We can also use this tool to understand the effects of many other types of stimulants, stress, or change of body energy level on the body thermal patterns. For example, Figure 9.16 shows changes of the body throughout one day, in the morning when the subject reported to be relatively relaxed after a full-night resting, in the evening when the subject reported to experience a high level of fatigue and mental stress after a full-day work, and the effect of taking 30 min relaxation meditation of the subject. We can appreciate the dynamic changes of the body and how it responds to different environmental and mental inputs. In other words, what we can see from the whole-body IR is a dynamic and live body system at work, which constantly undergoes changes and interactions within itself and with its environment.

What forces are driving the life and movement of this dynamic and live system at work? This seems to be an eternal question for us since the beginning of human life, and it is the same question that inspires us in the research of IR and TTM. We can find answers of this question in many branches of science, philosophy, religions and spiritual studies. Indeed, current medical science has started to embrace the holistic view of the body, mind, and spirit connection, and the power of emotional and mental health on our physical body.

Conventionally, IR images have been believed to represent "thermal" activities of the body. This is because in the history of physics, scientists found that heat sources radiate electromagnetic spectrum in the window of the IR photons, and a relationship was thus developed in converting IR energy to temperature according to the theory of black-body radiation. Thus in classic scientific context, we tend to believe that intensity in the IR images is related to heat of the body, and such heat is mainly caused by metabolic activities and blood flow related phenomena.

Although we have been using the word of "thermal" in compliance with conventional terminology so far, our recent work revealed that the IR images of the body reflect essentially the energy system of the human body within the IR window. This is a more accurate and broader description of intrinsic nature of the IR images, in contrast to the conventional belief and a narrow definition that IR images merely reflect "thermal" activities of the body.

Even though IR images occupies only a narrow window on the electromagnetic spectrum, we want to clarify that IR image itself is a "holographic" or "reflexiologic" map of the energy structure of the body. This IR energy map of the body is intrinsically associated with other types of energy that the current science has little knowledge about. This source of energy has been called "subtle energy" in recent years in describing its invisibility and subtleness that are different from other more familiar forms of energy such as electromagnetic energy [11]. In ancient texts, this energy has also been referred to as "Qi" in Chinese medicine and "Prana" in Indian ayurvedic medicine.

Because all energy has continuous spectra and can change freely from one form to the other as a fundamental nature of the energy, it is then not difficult to realize that IR images have to be connected with this subtle form of energy as well. This insight have opened up exciting new areas of research in IR because imaging studies on subtle energy has been extremely limited so far, and they could bring fundamental breakthroughs in medical research and in understanding other subtle forces at work for human health. Thus, in clinical studies of TTM, we have extended its application in holistic medicine, in particular, energy medicine that primarily deals with this form of subtle energy.

Figure 9.19 illustrates examples of a study on meridian system and heat conduction in the body. The images were obtained by heating an acupoint on the shoulder (GB 21 of the gall bladder meridian) with a traditional moxibustion technique and observing the passage of heat along this meridian. Previous publications have shown conflicting results of thermal conduction along the meridian system, which motivated us to pursue further studies on this topic [22–24,30–33].

Results of our experiments showed that not all meridian channels and acupoints are thermal sensitive and conductive. Rather, there are specific acupoints that are highly active and involved in the heat exchange process. When moxibustion is applied to these acupoints, studied subjects experienced

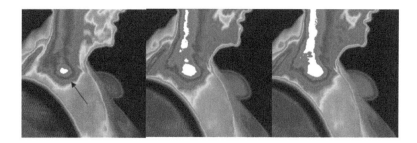

FIGURE 9.19 Thermal pattern changes of a studied subject after applying moxibustion treatment at an acupoint on the shoulder (GB21 of the gall bladder meridian, shown by the arrow). Thermal increase was seen along the side of the neck along this meridian after 5, 10, and 15 min (left, middle, and right panels, respectively) after the treatment and was in agreement with the same feeling of warming up of the neck experienced by the subject. For color figures, see www.holistichealthscience.org.

warming up of certain areas, either along the same meridian channel or at other locations. This finding is consistent with clinical observations that certain diseases are associated with thermal acupoints at distinct locations. Perhaps one of the functions of these acupoints is to provide additional thermal pathways allowing internal heat to be transferred to the body surface.

Therefore, in addition to the regular heat transfer mechanisms including blood flow and conduction, there may be other subtle channels that help the body maintain thermal equilibrium and homeostasis. IR-TTM is a new scientific tool for alternative medicine in studying subtle energy and how it influences our body in intricate ways. Thus, TTM can serve essentially as a bridge between the mainstream and alternative medicine. The merging of the two medical fields will lead to quantum leaps in medicine, revolutionizing the way we look at human health and its improvement.

9.7 Summary

In this chapter, we presented the basic idea of how internal heat source is related to its surface thermal distribution under extremely simplified *ex vivo* conditions. This idea was explored in live human bodies through clinical observations of different types of diseases and conditions. Though the depth of tissue that IR photons can go through is very limited, heat from internal organs can be expressed on the body surface through different mechanisms. The technique of thermal texture mapping or TTM was developed initially to visualize and analyze thermal gradient of heat sources and whether such sources are surface based or deeply situated inside tissues. For live bodies, surface thermal distribution becomes very complex, and the technique of thermal gradient analysis has to be extended in real situations. In this case, topology of the thermal distributions, including shapes, textures, anatomical locations and other morphological features have to be used in understanding the nature of the heat sources.

Results of clinical observations suggested that systematic diseases and health conditions such as cancer and autoimmune disorders are associated with certain whole-body thermal signatures that can be used to assess the presence and risk of such disorders. These thermal signatures were obtained from the atlas of whole-body IR images, and they suggested the association or inter-connection of different parts of the body in responding to certain stress conditions. Development of serious conditions such as cancer not only involves abnormal thermal patterns of the local region, but also presents as unique thermal signatures of the whole body including thermal asymmetry, deficiency, and irregularity. These findings motivated us to develop a new system for image interpretation, which emphasizes on combining information of the whole-body images in a holistic and integrative way. This is an essential extension of the TTM technique in image interpretation and evaluation of the health environment of the whole body.

The concept and methodology of TTM has also been tested in other applications including the study of drug or therapy response, holistic medicine, and epidemic diseases in public health. Findings of these studies showed that IR with TTM is an effective tool in observing functions of the body, its dynamic changes, overall balance and associations of different subcomponents. This modality also helps us explore advanced topics such as meridian channels, subtle energy, and their manifestation on the IR images.

On the basis of our experience from available clinical observations, we want to emphasize in the end that IR images are essentially energy maps of the body within the IR spectral window. Our body energy system is live and dynamic, as our mind and body are constantly undergoing changes and movement inside and outside. In addition, there are other sources or channels of subtle energy that are continuously influencing the intensity of such energy maps. It is important to realize that the information contained in IR images goes far beyond the heat emitted from the metabolic activities and blood flow in the body, which is what we have believed so far in the context of traditional thermography.

The beauty of IR is that it provides us a new angle of seeing the body as a whole, and inter-connected system, live and moving. And the secret of maintaining its health is about the intricate balance inside this system and harmonious interactions with its environment. This new perspective is drastically different from the traditional reductionist view of treating the body as made up of separated mechanical parts that are static and isolated, as how the present anatomical and disease-driven paradigms are dominating in medicine and radiological imaging. The adoption of the new holistic view will fundamentally revolutionize medicine and healthcare, allowing setting up new scientific directions and medical practices in congruence with the operating principles of the nature. We expect that IR will serve an even more important role in the new era of integrative and holistic medicine in the coming years.

References

1. Diakides, N.A. and J.D. Bronzino, Editors, *Medical Infrared Imaging*. CRC Press, Taylor & Francis Group, USA, 2008.
2. Fujimasa, I., T. Chinzei, and I. Saito, Converting far infrared image information to other physiological data. *IEEE Engineering in Medicine and Biology Magazine*, 2000. **19**(3): 71–6.
3. Li, Q., Y.E. Yuan, and W.D. Han, Experimental animal model in breast cancer development using thermal texture maps. *Chinese Journal of Medical Imaging and Technologies*, 2006. **22**(6): 819–20.
4. Li, X., X. Huang, and X. Li, Effects and mechanisms of *Herba dendrobii* on rats with stomach-heat syndrome. *Zhongguo Zhong Yao Za Zhi*, 2010. **35**(6): 750–4.
5. Liu, Z.Q. and C. Wang, *Method and Apparatus for Thermal Radiation Imaging*. Technical Report 6,023,637, United States Patent., 2000.
6. Shang, Z. and G. Jiang, Fundamental theoretic research of thermal texture maps I—Simulation and analysis of the relation between the depth of inner heat source and surface temperature distribution in isotropy tissue. *Conf Proc IEEE Eng Med Biol Soc*, 2004. **7**: 5271–3.
7. Wang, W. et al., Clinical study on using thermal texture maps in SARS diagnosis. *Conf Proc IEEE Eng Med Biol Soc*, 2004. **7**: 5258–64.
8. Yuan, C., C. Wang, and S.T. Song, Thermal texture mapping in breast cancer. *Chinese Journay of Medical Imaging and Technologies*, 2006. **16**(1): 7–10.
9. Yuan, Y. et al., Image analysis of breast tumors using thermal texture mapping (TTM). *Conf Proc IEEE Eng Med Biol Soc*, 2005. **1**: 697–9.
10. Yuan, Y.E., C. Wang, and R. Yuan, Image analysis of thermal texture maps on breast cancer. *Chinese Journal of Medical Imaging and Technologies*, 2004. **20**(12): 1803–05.
11. Dale, C., *The Subtle Body: An Encyclopedia of Your Energetic Anatomy*. Sounds True, Incorporated; 2009.
12. Chunfang, G., L. Kaiyang, and Z. Shaoping, A novel approach of analyzing the relation between the inner heat source and the surface temperature distribution in thermal texture maps. *Conf Proc IEEE Eng Med Biol Soc*, 2005. **1**: 623–6.
13. Deng, Z.S. and J. Liu, Mathematical modeling of temperature mapping over skin surface and its implementation in thermal disease diagnostics. *Comput Biol Med*, 2004. **34**(6): 495–21.

14. Draper, J.W. and J.W. Boag, Skin temperature distributions over veins and tumours. *Phys Med Biol*, 1971. **16**(4): 645–54.

15. Draper, J.W. and J.W. Boag, The calculation of skin temperature distributions in thermography. *Phys Med Biol*, 1971. **16**(2): 201–11.

16. Gustafsson, S.E., S.K. Nilsson, and L.M. Torell, Analytical calculation of the skin temperature distribution due to subcutaneous heat production in a spherical heat source. *Phys Med Biol*, 1975. **20**(2): 219–24.

17. Weinbaum, S., L.M. Jiji, and D.E. Lemons, Theory and experiment for the effect of vascular microstructure on surface tissue heat transfer—Part I: Anatomical foundation and model conceptualization. *J Biomech Eng*, 1984. **106**(4): 321–30.

18. Pennes, H.H., Analysis of tissue and arterial blood temperatures in the resting human forearm. *J Appl Physiol*, 1948. **1**(2): 93–122.

19. Wu, Z. et al., A basic step toward understanding skin surface temperature distributions caused by internal heat sources. *Phys Med Biol*, 2007. **52**(17): 5379–92.

20. Chan, C.L., Boundary element method analysis for the bioheat transfer equation. *J Biomech Eng*, 1992. **114**(3): 358–65.

21. Mital, M. and E.P. Scott, Thermal detection of embedded tumors using infrared imaging. *J Biomech Eng*, 2007. **129**(1): 33–9.

22. Wang, P., X. Hu, and B. Wu, Displaying of the infrared radiant track along meridians on the back of human body. *Zhen Ci Yan Jiu*, 1993. **18**(2): 90–3, 89.

23. Xu, J.S. et al. Comparison of the thermal conductivity of the related tissues along the meridian and the non-meridian. *Zhongguo Zhen Jiu*, 2005. **25**(7): 477–82.

24. Yang, H.Q. et al., Appearance of human meridian-like structure and acupoints and its time correlation by infrared thermal imaging. *Am J Chin Med*, 2007. **35**(2): 231–40.

25. *Advances in Mind – Body Medicine*. InnoVision Communications, USA.

26. Amalu, W.C., Nondestructive testing of the human breast: The validity of dynamic stress testing in medical infrared breast imaging. *Conf Proc IEEE Eng Med Biol Soc*, 2004. **2**: 1174–7.

27. Hu, L. et al., Effect of forced convection on the skin thermal expression of breast cancer. *J Biomech Eng*, 2004. **126**(2): 204–11.

28. Hassan, M., V. Chernomordik, and A. Vogel, *Infrared Imaging for Functional Monitoring of Disease Processes*. Medical Infrared Imaging; Edited by Diakides and Bronzino; CRC Press, USA, 2008.

29. Gescheit, I.M. et al., Minimal-invasive thermal imaging of a malignant tumor: A simple model and algorithm. *Med Phys*, 2010. **37**(1): 211–16.

30. Hu, X., B. Wu, and P. Wang, Displaying of meridian courses travelling over human body surface under natural conditions. *Zhen Ci Yan Jiu*, 1993. **18**(2): 83–9.

31. Litscher, G., Infrared thermography fails to visualize stimulation-induced meridian-like structures. *Biomed Eng Online*, 2005. **4**(1): 38.

32. Lo, S.Y., Meridians in acupuncture and infrared imaging. *Med Hypotheses*, 2002. **58**(1): 72–6.

33. Zhang, D. et al., Research on the acupuncture principles and meridian phenomena by means of infrared thermography. *Zhen Ci Yan Jiu*, 1990. **15**(4): 319–23.

10

Infrared Imaging of the Breast: A Review

William C. Amalu
Pacific Chiropractic and Research Center

William B. Hobbins
Women's Breast Health Center

Jonathan F. Head
Elliot-Elliot-Head Breast Cancer Research and Treatment Center

Robert L. Elliot
Elliot-Elliot-Head Breast Cancer Research and Treatment Center

10.1 Introduction

The use of infrared imaging in health care is not a recent phenomenon. However, its utilization in breast cancer screening is seeing renewed interest. This attention is fueled by research that clearly demonstrates the value of this procedure and the tremendous impact it has on the mortality of breast cancer.

Since the late 1950s, extensive research has been performed on the use of infrared imaging in breast cancer screening. Over 800 papers can be found in the indexed medical literature. In this database, well over 300,000 women have participated as study subjects. The numbers of participants in many studies are very large and range from 10,000 to 85,000 women. Some of these studies have followed patients for up to 12 years in order to investigate and establish the technology's unique ability as a risk marker.

With strict standardized interpretation protocols having been established for over 15 years, infrared imaging of the breast has obtained an average sensitivity and specificity of 90%. As a future risk indicator for breast cancer, a persistent abnormal thermogram carries a 22 times higher risk and is 10 times more significant than a first-order family history of the disease. Studies clearly show that an abnormal infrared image is the single most important risk marker for the existence of or future development of breast cancer.

The first recorded use of thermobiological diagnostics can be found in the writings of Hippocrates around 480 BC [1]. A mud slurry spread over the patient was observed for areas that would dry first and was thought to indicate underlying organ pathology. Since this time, continued research and clinical observations proved that certain temperatures related to the human body were indeed indicative of normal and abnormal physiologic processes.

In the 1950s, military research into infrared monitoring systems for nighttime troop movements ushered in a new era in thermal diagnostics. Once declassified in the mid-1950s, infrared imaging technology was made available for medical purposes. The first diagnostic use of infrared imaging came in 1956 when Lawson discovered that the skin temperature over a cancer in the breast was higher than that of normal tissue [2–4]. He also showed that the venous blood draining the cancer is often warmer than its arterial supply.

The Department of Health Education and Welfare released a position paper in 1972 in which the director, Thomas Tiernery, wrote, "The medical consultants indicate that thermography, in its present state of development, is beyond the experimental state as a diagnostic procedure in the following 4 areas: (1) Pathology of the female breast ..." (1972 HEW position paper). On January 29, 1982, the Food and Drug Administration published its approval and classification of thermography as an adjunctive diagnostic screening procedure for the detection of breast cancer. Since the late 1970s, numerous medical centers and independent clinics have used thermography for a variety of diagnostic purposes.

Since Lawson's groundbreaking research, infrared imaging has been used for over 40 years as an adjunctive screening procedure in the evaluation of the breast. In this time, significant advances have been made in infrared detection systems and the application of sophisticated computerized image processing.

10.2 Fundamentals and Standards in Infrared Breast Imaging

Clinical infrared imaging is a procedure that detects, records, and produces an image of a patient's skin surface temperatures or thermal patterns. The image produced resembles the likeness of the anatomic area under study. The procedure uses equipment that can provide both qualitative and quantitative representations of these temperature patterns.

Infrared imaging does not entail the use of ionizing radiation, venous access, or other invasive procedures; therefore, the examination poses no harm to the patient. Classified as a functional imaging technology, infrared imaging of the breast provides information on the normal and abnormal physiologic functioning of the sensory and sympathetic nervous systems, vascular system, and local inflammatory processes.

10.2.1 Physics

All objects with a temperature above absolute zero (–273 K) emit infrared radiation from their surface. The Stefan–Boltzmann Law defines the relation between radiated energy and temperature by stating that the total radiation emitted by an object is directly proportional to the object's area and emissivity and the fourth power of its absolute temperature. Since the emissivity of human skin is extremely high (within 1% of that of a blackbody), measurements of infrared radiation emitted by the skin can be converted directly into accurate temperature values. This makes infrared imaging an ideal procedure to evaluate surface temperatures of the body.

10.2.2 Equipment Considerations

Infrared rays are found in the electromagnetic spectrum within the wavelengths of 0.75 µm to 1 mm. Human skin emits infrared radiation mainly in the 2–20 µm wavelength range, with an average peak at 9–10 µm [5]. With the application of Planck's equation and Wein's Law, it is found that approximately 90% of the emitted infrared radiation in humans is in the longer wavelengths (6–14 µm).

There are many important technical aspects to consider when choosing an appropriate clinical infrared imaging system. (The majority of which is outside the scope of this chapter.) However, minimum equipment standards have been established from research studies, applied infrared physics, and human anatomic and physiologic parameters [6,7]. Absolute, spatial, and temperature resolution along with thermal stability and adequate computerized image processing are just a few of the critical specifications to be taken into account. Real-time image capture is a basic requirement in modern clinical imaging systems. Absolute resolution, or the number of distinct infrared detectors dedicated to forming the image (resolvable elements per line in a scanning system), must be adequate enough to produce a quality image. The ability of the system to resolve discrete areas on the surface of the skin is defined as spatial resolution. Standards for clinical systems require that the spatial resolution of the camera should be able to resolve 1 mm^2 at the surface of the skin when the patient is placed at 40 cm from the detector (2.5 mrad). The temperature resolution, or thermal sensitivity, of the camera should be 0.1°C. This level of temperature discernment is necessary to maintain adequate objective temperature measurements when performing quantitative analyses. Although these specifications are extremely important, the most fundamental consideration in the selection of clinical infrared imaging equipment is the wavelength sensitivity of the infrared detector. The decision on which area in the infrared spectrum to select a detector from depends on the object one wants to investigate and the environmental conditions in which the detection is taking place. Considering that the object in question is the human body, Planck's equation leads us to select a detector in the 6–14 μm region. Assessment of skin temperature by infrared measurement in the 3–5 μm region is less reliable due to the emissivity of human skin being farther from that of a blackbody in that region [8,9]. The environment under which the examination takes place is well controlled, but not free from possible sources of detection errors. Imaging room environmental artifacts such as reflectance can cause errors when shorter wavelength detectors (under 7 μm) are used [10]. Consequently, the optimum infrared detector to use in imaging the breast, and the body as a whole, would have sensitivity in the longer wavelengths spanning the 9–10 μm range [7–14].

The problems encountered with first-generation infrared camera systems, such as incorrect detector sensitivity (shorter wavelengths), thermal drift, calibration, analog interface, and so forth have been solved for almost two decades. Modern clinical infrared imaging systems exceed the minimum specification standards mentioned earlier. However, no studies have been published demonstrating the need for a change in the standards. Modern computerized infrared imaging systems have the ability to discern minute variations in thermal emissions while producing extremely high-resolution images that can undergo digital manipulation by sophisticated computerized analytical processing.

10.2.3 Laboratory and Patient Preparation Protocols

In order to produce diagnostic quality infrared images, certain laboratory and patient preparation protocols must be strictly adhered to. Infrared imaging must be performed in a controlled environment. The primary reason for this is the nature of human physiology. Changes from a different external (noncontrolled room) environment, clothing, and so forth produce thermal artifacts. In order to properly prepare the patient for imaging, the patient should be instructed to refrain from sun exposure, stimulation or treatment of the breasts, cosmetics, lotions, antiperspirants, deodorants, exercise, and bathing before the examination.

The imaging room must be temperature- and humidity controlled and maintained between 18°C and 23°C, and kept to within 1°C of change during the examination. This temperature range ensures that the patient is not placed in an environment in which their physiology is stressed into a state of shivering or perspiring. The room should also be free from drafts and infrared sources of heat (i.e., sunlight and incandescent lighting). In keeping with a physiologically neutral temperature environment, the floor should be carpeted or the patient must wear shoes in order to prevent increased physiologic stress.

Lastly, the patient must undergo 15 min of waist-up nude acclimation in order to reach a condition in which the body is at thermal equilibrium with the environment. At this point, further changes in the surface temperatures of the body occur very slowly and uniformly; thus, not affecting changes in homologous anatomic regions. Thermal artifacts from clothing or the outside environment are also removed at this time. The last 5 min of this acclimation period is usually spent with the patient placing their hands on top of their head in order to facilitate an improved anatomic presentation of the breasts for imaging. Depending on the patient's individual anatomy, certain positioning maneuvers may need to be implemented such that all of the pertinent surfaces of the breasts may be imaged. In summary, adherence to proper patient and laboratory protocols is absolutely necessary to produce a physiologically neutral image free from artifact and ready for interpretation.

10.2.4 Imaging

The actual process of imaging is undertaken with the intent to adequately detect the infrared emissions from the pertinent surface areas of the breasts. As with mammography, a minimum series of images is needed in order to facilitate adequate coverage. The series includes the bilateral frontal breast along with the right and left oblique views (a right and left single breast close-up view may also be included). The bilateral frontal view acts as a scout image to give a baseline impression of both breasts. The oblique views (~45° to the detector) expose the lateral and medial quadrants of the breasts for analysis. The optional close-up views maximize the use of the detector allowing for the highest thermal and spatial resolution image of each breast. This series of images takes into consideration the infrared analyses of curved surfaces and adequately provides for an accurate analysis of all the pertinent surface areas of the breasts (see Figures 10.1 through 10.5).

Positioning of the patient prior to imaging facilitates acclimation of the surface areas and ease of imaging. Placing the patient in a seated or standing posture during the acclimation period is ideal to facilitate these needs. In the seated position, the patient places their arms on the arm rests away from the body to allow for proper acclimation. When positioning the patient in front of the camera, the use of a rotating chair or having the patient stand makes for uncomplicated positioning for the necessary views.

Because of differing anatomy from patient to patient, special views may be necessary to adequately detect the infrared emissions from the pertinent surface areas of the breasts. The most common problem encountered is inadequate viewing of the inferior quadrants due to nipple ptosis. This is easily remedied by adding "lift views." Once the baseline images are taken, the patient is asked to gently "lift" each breast from the superior most aspect of the Tail of Spence exposing the inferior quadrants for detection. Additional images are then taken in this position in order to maintain the surface areas covered in the standard views.

10.2.5 Special Tests

In the past, an optional set of views may have been added to the baseline images. Additional views would be taken after the patient placed their hands in ice cold water as a thermoregulatory challenge. It was hoped that this dynamic methodology would increase the sensitivity and specificity of the thermographic procedure.

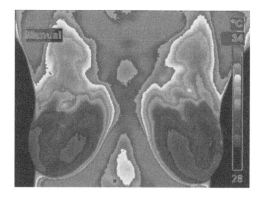

FIGURE 10.1 Bilateral frontal.

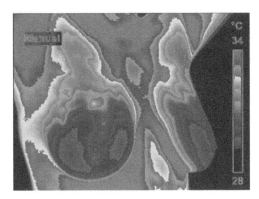

FIGURE 10.2   Right oblique.

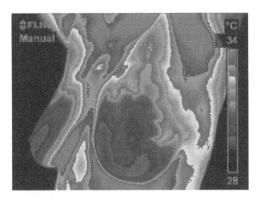

FIGURE 10.3   Left oblique.

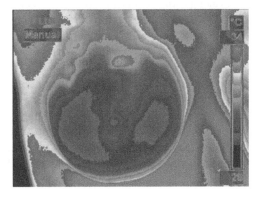

FIGURE 10.4   Right close-up.

In order to understand the hopes placed on this test, one needs to understand the underlying physiologic mechanisms of the procedure. The most common and accepted method of applied thermoregulatory challenge involves ice water immersion of the hands or feet (previous studies investigating the use of fans or alcohol spray noted concerns over the creation of thermal artifacts along with the methods causing a limited superficial effect). The mechanism is purely neurovascular and involves a primitive survival reflex initiated from peripheral neural receptors and conveyed to the central nervous system.

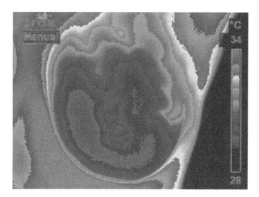

FIGURE 10.5   Left close-up.

To protect the body from hypothermia, the reflex invokes a sympathetically mediated blood vessel constriction in the periphery in an attempt to maintain the normal core temperature set point. This stress test is intended to increase the sensitivity of the thermogram by attempting to identify nonresponding blood vessels such as those involved in angiogenesis associated with neoplasm. Blood vessels produced by cancerous tumors are simple endothelial tubes devoid of a muscular layer and the neural regulation afforded to embryologic vessels. As such, these new vessels would fail to constrict in response to a sympathetic stimulus. In the normal breast, test results would produce an image of relative cooling with attenuation of vascular diameter. A breast harboring a malignancy would theoretically remain unchanged in temperature or demonstrate hyperthermia with vascular dilatation. However, to date it has not been found that the stress test offers any advantage over the baseline images [15].

For well over a decade, the largest infrared breast imaging centers worldwide, along with the leading experts and researchers in the field of infrared breast imaging, have discontinued the use of the cold challenge. Yet, in a 2004 detailed review of the literature combined with an investigational study, Amalu explored the validity of the thermoregulatory challenge test [15]. Results from 23 patients with histologically confirmed breast cancers along with 500 noncancerous patients were presented demonstrating positive and negative responses to the challenge. From the combined literature review and study analysis it was found that the test did not alter the clinical decision-making process for following up suspicious thermograms, nor did it enhance the detection of occult cancers found in normal thermograms. In summary, it was found that there was no evidence to support the use of the cold challenge. The study noted insufficient evidence to warrant its use as a mandated test with all women undergoing infrared breast imaging. It also warned that it would be incorrect to consider a breast thermogram substandard, inaccurate, or incomplete if a thermoregulatory challenge was not included. In conclusion, Amalu stated that "Until further studies are performed and ample evidence can be presented to the contrary, a review of the available data indicates that infrared imaging of the breast can be performed excluding the cold challenge without any known loss of sensitivity or specificity in the detection of breast cancers."

10.2.6 Image Interpretation

Early methods of interpretation of infrared breast images was based solely on qualitative (subjective) criteria. The images were read for variations in vascular patterning with no regard to temperature variations between the breasts (Tricore method) [16]. This led to wide variations in the outcomes of studies preformed with inexperienced interpreters. Research throughout the 1970s proved that when both qualitative and quantitative data were incorporated in the interpretations, an increase in sensitivity and specificity was realized. In the early 1980s, a standardized method of thermovascular analysis was proposed. The interpretation was composed of 20 discrete vascular and temperature attributes

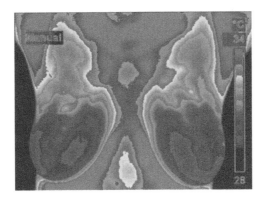

FIGURE 10.6 TH1 (normal uniform nonvascular).

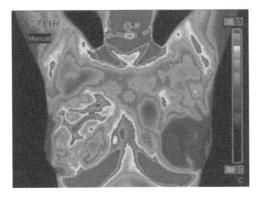

FIGURE 10.7 Right TH5 (severely abnormal).

in the breast [17,18]. This method of analysis was based on previous research and large-scale studies comprising tens of thousands of patients. Using this methodology, thermograms would be graded into one of 5 TH (thermobiological) classifications. Based on the combined vascular patterning and temperatures across the two breasts, the images would be graded as TH1 (normal uniform nonvascular), TH2 (normal uniform vascular), TH3 (equivocal), TH4 (abnormal), or TH5 (severely abnormal) (see Figures 10.6 and 10.7). The use of this standardized interpretation method significantly increased infrared imaging's sensitivity, specificity, positive and negative predictive value, and inter-/intraexaminer interpretation reliability. Continued patient observations and research over the past two decades has caused changes in some of the thermovascular values; thus keeping the interpretation system up-to-date. Variations in this methodology have also been adopted with great success. However, it is recognized that, as with any other imaging procedure, specialized training and experience produces the highest level of screening success.

10.2.7 Correlation between Pathology and Infrared Imaging

The empirical evidence that an underlying breast cancer alters regional skin surface temperatures was investigated early on. In 1963, Lawson and Chughtai, two McGill University surgeons, published an elegant intraoperative study demonstrating that the increase in regional skin surface temperature associated with breast cancer was related to venous convection [19]. This early quantitative experiment added credence to previous research suggesting that infrared findings were linked to increased vascularity.

Infrared imaging of the breast may also have critical prognostic significance since it may correlate with a variety of pathologic prognostic features such as tumor size, tumor grade, lymph node status, and markers of tumor growth [20]. Continued research is underway investigating the pathologic basis for these infrared findings. One possibility is increased blood flow due to vascular proliferation (assessed by quantifying the microvascular density [MVD]) as a result of tumor-associated angiogenesis. Although in one study [21], the MVD did not correlate with abnormal infrared findings. However, the imaging method used in that study consisted of contact plate technology (liquid crystal thermography [LCT]), which is not capable of modern computerized analysis. Consequently, LCT does not possess the discrimination and digital processing necessary to begin to correlate histological and discrete vascular changes [22].

In 1993, Head and Elliott reported that improved images from second-generation infrared systems allowed more objective and quantitative analysis [20], and indicated that growth-rate-related prognostic indicators were strongly associated with the infrared image interpretation.

In a 1994 detailed review of the potential of infrared imaging [23], Anbar suggested that the previous empirical observation that small tumors were capable of producing notable infrared changes could be due to enhanced perfusion over a substantial area of the breast surface via regional tumor induced nitric oxide (NO) vasodilatation. NO is synthesized by nitric oxide synthase (NOS), found both as a constitutive form of NOS, especially in endothelial cells, and as an inducible form of NOS, especially in macrophages [24]. NOS has been demonstrated in breast carcinoma [25] using tissue immunohistochemistry, and is associated with a high tumor grade.

Nitric oxide is a molecule with potent vasodilating properties. It is a simple highly reactive free radical that readily oxidizes to form nitrite or nitrate ions. It diffuses easily through both hydrophilic and hydrophobic media. Thus, once produced, NO diffuses throughout the surrounding tissues, inside and outside the vascular system, and induces a variety of biochemical changes depending on the specific receptors involved. NO exerts its influence by binding to receptor sites in the endothelium of arteries or arterioles. This causes inhibition of sympathetic vasoconstriction. The end result is NO induced vasodilatation, which in turn may produce an asymmetrical thermovascular infrared image.

The largest body of evidence surrounding the physiologic mechanism by which infrared imaging detects precancerous and malignant states of the breast lies in the recruitment of existing blood vessels and the formation of new ones (angiogenesis). The process of angiogenesis begins with the release of angiogenesis factors (AF) from precancerous or cancerous cells. In the early stages of tumor growth, the majority of neoplasms exhibit a lower cellular metabolic demand. As such, the release of AF causes the existing vessels to resist constriction in order to maintain a steady supply of nutrients to the growing mass. As the tumor increases in size, the need for nutrients becomes greater. AF begins to exert its influence by opening the dormant vessels in the breast. Once this blood supply becomes too little to maintain the growth of the neoplasm, AF causes the formation of new blood vessels. These new vessels are simple endothelial tubes connecting the tumor to existing nearby arteries and arterioles. This augmented blood supply produces the increase in heat and vascular asymmetry seen in infrared images.

The concept of angiogenesis, as an integral part of early breast cancer, was emphasized in 1996 by Guidi and Schnitt. Their observations suggested that it is an early event in the development of breast cancer and may occur before tumor cells acquire the ability to invade the surrounding stroma and even before there is morphologic evidence of an *in situ* carcinoma [26]. In 1996, in his highly reviewed textbook entitled *Atlas of Mammography—New Early Signs in Breast Cancer*, Gamagami studied angiogenesis by infrared imaging and reported that hypervascularity and hyperthermia could be shown in 86% of nonpalpable breast cancers. He also noted that in 15% of these cases infrared imaging helped to detect cancers that were not visible on mammography [27].

The greatest evidence supporting the underlying principle by which infrared imaging detects precancerous growths and cancerous tumors surrounds the well-documented recruitment of existing vascularity and angiogenesis that is necessary to maintain the increased metabolism of malignant cellular growth and multiplication. The biomedical engineering evidence of infrared imaging's value, both in model *in vitro* and clinically *in vivo* studies of various tissue growths, normal and neoplastic, has been established [28–34].

10.3 Role of Infrared Imaging in the Detection of Cancer

To determine the value of infrared imaging, two viewpoints must be considered: first, the sensitivity of thermograms taken preoperatively in patients with known breast carcinoma; and second, the incidence of normal and abnormal thermograms in asymptomatic populations (specificity) and the presence or absence of malignancy in each of these groups.

In 1965, Gershon-Cohen, a radiologist and researcher from the Albert Einstein Medical Center, introduced infrared imaging to the United States [35]. Using a Barnes thermograph, he reported on 4000 cases with a sensitivity of 94% and a false-positive rate of 6%. This data was included in a review of the then current status of infrared imaging published in 1968 in *California—A Cancer Journal for Physicians* [36].

In prospective studies, Hoffman first reported on thermography in a gynecologic practice. He detected 23 carcinomas in 1924 patients (a detection rate of 12.5 per 1000), with an 8.4% false-negative (91.6% sensitivity) and a 7.4% false-positive (92.6% specificity) rate [37].

Stark and Way screened 4621 asymptomatic women, 35% of whom were under 35 years of age, and detected 24 cancers (detection rate of 7.6 per 1000), with a sensitivity and specificity of 98.3% and 93.5%, respectively [38].

In a study comprising 25,000 patients screened and 1878 histologically proven breast cancers, Amalric and Spitalier reported on their results with infrared imaging. From this group, a false-negative and false-positive rate of 9% (91% sensitivity and specificity) was found [39].

In a mobile unit examination of rural Wisconsin, Hobbins screened 37,506 women using thermography. He reported the detection of 5.7 cancers per 1000 women screened with a 12% false-negative and 14% false-positive rate. His findings also corroborated with others that thermography is the sole early initial signal in 10% of breast cancers [17,40].

Reporting his Radiology division's experience with 10,000 thermographic studies done concomitantly with mammography over a 3-year period, Isard reiterated a number of important concepts including the remarkable thermal and vascular stability of the infrared image from year to year in the otherwise healthy patient and the importance of recognizing any significant change [41]. In his experience, combining these modalities increased the sensitivity rate of detection by approximately 10%; thus, underlining the complementarity of these procedures since each one did not always suspect the same lesion. It was Isard's conclusion that, had there been a preliminary selection of his group of 4393 asymptomatic patients by infrared imaging, mammographic examination would have been restricted to the 1028 patients with abnormal infrared imaging, or 23% of this cohort. This would have resulted in a cancer detection rate of 24.1 per 1000 combined infrared and mammographic examinations as contrasted to the expected 7 per 1000 by mammographic screening alone. He concluded that since infrared imaging is an innocuous examination, it could be utilized to focus attention upon asymptomatic women who should be examined more intensely. Isard emphasized that, like mammography and other breast imaging techniques, infrared imaging does not diagnose cancer, but merely indicates the presence of an abnormality.

Spitalier and associates screened 61,000 women using thermography over a 10-year period. The false-negative and false-positive rate was found to be 11% (89% sensitivity and specificity). Thermography also detected 91% of the nonpalpable cancers (Grade T0: tumors less than 1 cm in size). The authors noted that of all the patients with cancer, thermography alone was the first alarm in 60% of the cases [42].

Two small-scale studies by Moskowitz (150 patients) [43] and Treatt (515 patients) [44] reported on the sensitivity and reliability of infrared imaging. Both used unknown experts to review the images of breast cancer patients. While Moskowitz excluded unreadable images, data from Threatt's study indicated that less than 30% of the images produced were considered good, the rest being substandard. Both of these studies produced poor results; however, this could be expected considering the lack of adherence to accepted imaging methods and protocols. The greatest error in these studies is found in the methods used to analyze the images. The type of image analysis consisted of the sole use of abnormal vascular pattern recognition. At the time these studies were performed, the accepted method of infrared image interpretation consisted of a combined vascular pattern and quantitative analysis of temperature variations across the breasts. Consequently, the data obtained from these studies is highly questionable. Their findings were also inconsistent with numerous previous large-scale multicenter trials. The authors suggested that for infrared imaging to be truly effective as a screening tool, there needed to be a more objective means of interpretation and proposed that this would be facilitated by computerized evaluation. This statement is interesting considering that recognized quantitative and qualitative reading protocols (including computer analysis) were being used at the time.

In a unique study comprising 39,802 women screened over a 3-year period, Haberman and associates used thermography and physical examination to determine if mammography was recommended. They reported an 85% sensitivity and 70% specificity for thermography. Haberman cautioned that the findings of thermographic specificity could not be extrapolated from this study as it was well-documented that long-term observation (8–10 years or more) is necessary to determine a true false-positive rate. The authors noted that 30% of the cancers found would not have been detected if it were not for thermography [45].

Gros and Gautherie reported on a large-scale study comprising 85,000 patients screened. Culmination of the data resulted in a 90% sensitivity and 88% specificity for thermography [46–49].

In a large-scale multicenter review of nearly 70,000 women screened, Jones reported a false-negative and false-positive rate of 13% (87% sensitivity) and 15% (85% sensitivity), respectively for thermography [50].

In a study performed in 1986, Usuki reported on the relation of thermographic findings in breast cancer diagnosis. He noted an 88% sensitivity for thermography in the detection of breast cancers [51].

Parisky and associates published a study from a multicenter 4-year clinical trial using infrared imaging to evaluate mammographically suspicious lesions. Data from a blinded subject set was obtained in 769 women with 875 biopsied lesions resulting in 187 malignant and 688 benign findings. The index of suspicion resulted in a 97% sensitivity in the detection of breast cancers [52].

In a study comparing clinical examination, mammography, and thermography in the diagnosis of breast cancer, three groups of patients were used: 4716 patients with confirmed carcinoma, 3305 patients with histologically diagnosed benign breast disease, and 8757 general patients (16,778 total participants). This paper also compared clinical examination and mammography to other well-known studies in the literature including the National Cancer Institute (NCI) sponsored Breast Cancer Detection and Demonstration Projects (BCDDPs). In this study, clinical examination had an average sensitivity of 75% in detecting all tumors and 50% in cancers less than 2 cm in size. This rate is exceptionally good when compared to many other studies at between 35% and 66% sensitivity. Mammography was found to have an average of 80% sensitivity and 73% specificity. Thermography had an average sensitivity of 88% (85% in tumors less than 1 cm in size) and a specificity of 85%. An abnormal thermogram was found to have a 94% predictive value. From the findings in this study, the authors suggested that "none of the techniques available for screening for breast carcinoma and evaluating patients with breast-related symptoms is sufficiently accurate to be used alone. For the best results, a multimodal approach should be used" [53].

In a series of 4000 confirmed breast cancers, Thomassin and associates observed 130 subclinical carcinomas ranging in diameter of 3–5 mm. Both mammography and thermography were used alone and in combination. Of the 130 cancers, 10% were detected by mammography, 50% by thermography, and 40% by both techniques. Thus, there was a thermal alarm in 90% of the patients and the only sign in 50% of the cases [54].

In a study performed at Cornell University, a prospective clinical trial comprising 92 patients investigated digital infrared imaging for breast cancer screening. Based on prior mammograms or ultrasound, all 92 patients were recommended for biopsy. Of the 94 biopsies performed 60 were malignant and 34 were benign. Digital infrared imaging identified 58 of 60 malignancies for a sensitivity of 97%. The study summary noted that "Digital infrared thermal imaging is a valuable adjunct to mammography and ultrasound, especially in women with dense breast parenchyma" [55].

Wang and associates conducted a study on 276 women with a comparison made between findings on infrared imaging and mammographic BI-RADS categories. In all, 174 malignant lesions (22 DCIS and

152 invasive carcinomas) were discovered among BI-RADS categories 3 through 5. When compared to the BI-RADS scale, the sensitivities for infrared imaging were highest in the 4–5 categories. For BI-RADS category 3 lesions, infrared imaging correctly identified the only one cancerous lesion. An 82% overall sensitivity was found for infrared imaging across the BI-RADS categories 3–5. A sensitivity of 92% was noted for infrared imaging when compared to the BI-RADS 4 and 5 categories [56].

In a study utilizing a new computerized analysis methodology, 100 women scheduled for biopsy underwent infrared imaging. In total, 106 biopsies were performed with 65 malignant and 41 benign lesions noted. With the use of the computer analysis alone a sensitivity of 78% and specificity of 75% was realized. When combined with mammography, the sensitivity was increased to 89% in women under the age of 50. The study concluded that the combined sensitivity of infrared imaging and mammography in women under 50 was encouraging, suggesting a potential way forward for a dual imaging approach in this younger age group [57].

In a simple review of over 15 large-scale studies from 1967 to 1998, infrared imaging of the breast has showed an average sensitivity and specificity of 90%. With continued technological advances in infrared imaging in the past decade, some studies are showing even higher sensitivity and specificity values. However, until further large-scale studies are performed, these findings remain in question.

10.4 Infrared Imaging as a Risk Indicator

As early as 1976, at the *Third International Symposium on Detection and Prevention of Cancer* held in New York, thermal imaging was established by consensus as the highest risk marker for the possibility of the presence of an undetected breast cancer. It had also been shown to predict such a subsequent occurrence [58–60]. The Wisconsin Breast Cancer Detection Foundation presented a summary of its findings in this area, which has remained undisputed [60]. This, combined with other reports, has confirmed that an abnormal infrared image is the highest risk indicator for the future development of breast cancer and is 10 times as significant as a first-order family history of the disease [61].

In a study of 10,000 women screened, Gautherie found that, when applied to asymptomatic women, thermography was very useful in assessing the risk of cancer by dividing patients into low- and high-risk categories. This was based on an objective evaluation of each patient's thermograms using an improved reading protocol that incorporated 20 thermopathological factors [62].

A screening of 61,000 women using thermography was performed by Spitalier over a 10-year period. The authors concluded that "in patients having no clinical or radiographic suspicion of malignancy, a persistently abnormal breast thermogram represents the highest known risk factor for the future development of breast cancer" [42].

From a patient base of 58,000 women screened with thermography, Gros and associates followed 1527 patients with initially healthy breasts and abnormal thermograms for 12 years. Of this group, 44% developed malignancies within 5 years. The study concluded that "an abnormal thermogram is the single most important marker of high-risk category for the future development of breast cancer" [49].

Spitalier and associates followed 1416 patients with isolated abnormal breast thermograms. It was found that a persistently abnormal thermogram, as an isolated phenomenon, is associated with an actuarial breast cancer, a risk of 26% at 5 years. Within this study, 165 patients with nonpalpable cancers were observed. In 53% of these patients, thermography was the only test which was positive at the time of initial evaluation. It was concluded that:

1. A persistently abnormal thermogram, even in the absence of any other sign of malignancy, is associated with a high risk of developing cancer.
2. This isolated abnormality also carries with it a high risk of developing interval cancer, and as such the patient should be examined more frequently than the customary 12 months.
3. Most patients diagnosed as having minimal breast cancer have abnormal thermograms as the first warning sign [63,64].

In a study by Gautherie and associates, the effectiveness of thermography in terms of survival benefit was discussed. The authors analyzed the survival rates of 106 patients in whom the diagnosis of breast cancer was established as a result of the follow-up of thermographic abnormalities found on the initial examination when the breasts were apparently healthy (negative physical and mammographic findings). The control group consisted of 372 breast cancer patients. The patients in both groups were subjected to identical treatment and followed for 5 years. A 61% increase in survival was noted in the patients who were followed-up due to initial thermographic abnormalities. The authors summarized the study by stating that "the findings clearly establish that the early identification of women at high risk of breast cancer based on the objective thermal assessment of breast health results in a dramatic survival benefit" [65,66].

Infrared imaging provides a reflection of functional tumor-induced angiogenesis and metabolic activity rather than structurally based parameters (i.e., tumor size, architectural distortion, and microcalcifications). Recent advances in cancer research have determined that the biological activity of a neoplasm is far more significant an indicator of aggressiveness than the size of the tumor. As a direct reflection of the biological activity in the breast, infrared imaging has been found to provide a significant biological risk marker for cancer.

10.5 Infrared Imaging as a Prognostic Indicator

Studies exploring the biology of cancers have shown that the amount of thermovascular activity in the breast is directly proportional to the aggressiveness of the tumor. As such, infrared imaging provides the clinician with an invaluable tool in prognosis and treatment monitoring.

In a study of 209 breast cancers, Dilhuydy and associates found a positive correlation between the degree of infrared abnormalities and the existence of positive axillary lymph nodes. It was reported that the amount of thermovascular activity seen in the breast was directly related to the prognosis. The study concluded that infrared imaging provides a highly significant factor in prognosis and that it should be included in the pretherapeutic assessment of a breast cancer [67].

Amalric and Spitalier reported on 25,000 patients screened and 1878 histologically proven breast cancers investigated with infrared imaging. The study noted that the amount of infrared activity in the breast was directly proportional to the survival of the patient. The "hot" cancers showed a significantly poorer prognosis with a 24% survival rate at 3 years. A much better prognosis with an 80% survival rate at 3 years was seen in the more biologically inactive or "cooler" cancers. The study also noted a positive association between the amount of thermal activity in the breast and the presence of positive axillary nodes [68].

Reporting on a study of breast cancer doubling times and infrared imaging, Fournier noted significant changes in the thermovascular appearance of the images. The shorter the tumor doubling time, the more thermographic pathological signs were evident. It was concluded that infrared imaging served as a warning signal for the faster-growing breast cancers [69].

A retrospective analysis of 100 normal patients, 100 living cancer patients, and 126 deceased cancer patients was published by Head. Infrared imaging was found to be abnormal in 28% of the normal patients, compared to 65% of the living cancer patients and 88% of the deceased cancer patients. Known prognostic indicators related to tumor growth rate were compared to the results of the infrared images. The concentration of tumor ferritin, the proportion of cells in DNA synthesis and proliferating, and the expression of the proliferation-associated tumor antigen Ki-67 were all found to be associated with an abnormal infrared image. It was concluded that "the strong relationships of thermographic results with these three growth rate-related prognostic indicators suggest that breast cancer patients with abnormal thermograms have faster-growing tumors that are more likely to have metastasized and to recur with a shorter disease-free interval" [20].

In a paper by Gros and Gautherie, the use of infrared imaging in the prognosis of treated breast cancers was investigated. The authors considered infrared imaging to be absolutely necessary for assessing pretherapeutic prognosis or carrying out the follow-up of breast cancers treated by exclusive radio-therapy. They noted that before treatment, infrared imaging yields evidence of the cancer growth rate (aggressiveness) and contributes to the therapeutic choice. It also indicates the success of radio-sterilization or the suspicion of a possible recurrence or radio-resistance. The authors also noted a weaker 5-year survival with infrared images that showed an increase in thermal signals [70].

In a study by Keyserlingk, 20 women with core-biopsy proven locally advanced breast cancer underwent infrared imaging before and after chemohormonotherapy. All 20 patients were found to have abnormal thermovascular signs prior to treatment. Upon completion of the final round of chemotherapy, each patient underwent curative-intent surgery. Prior to surgery, all 20 patients showed improvement in their initial infrared scores. The amount of improvement in the infrared images was found to be directly related to the decrease in tumor size. A complete normalization of prechemotherapy infrared scores was seen in five patients. In these patients, there was no histological evidence of cancer remaining in the breast. In summary, the authors stated that "Further research will determine whether lingering infrared detected angiogenesis following treatment reflects tumor aggressiveness and ultimately prognosis, as well as early tumor resistance, thus providing an additional early signal for the need of a therapeutic adjustment" [71].

10.6 Breast Cancer Detection and Demonstration Project

The BCDDP is the most frequently quoted reason for the decreased interest in infrared imaging. The BCDDP was a large-scale study performed from 1973 through 1979, which collected data from many centers around the United States. Three methods of breast cancer detection were studied: physical examination, mammography, and infrared imaging.

Just before the onset of the BCDDP, two important papers appeared in the literature. In 1972, Gerald D. Dodd of the University of Texas Department of Diagnostic Radiology presented an update on infrared imaging in breast cancer diagnosis at the *Seventh National Cancer Conference* sponsored by the National Cancer Society and the NCI [72]. In his presentation, he suggested that infrared imaging would be best employed as a screening agent for mammography. He proposed that in any general survey of the female population aged 40 and over, 15–20% of these subjects would have positive infrared imaging and would require mammograms. Of these, approximately 5% would be recommended for biopsy. He concluded that infrared imaging would serve to eliminate 80–85% of the potential mammograms. Dodd also reiterated that the procedure was not competitive with mammography and, reporting the Texas Medical School's experience with infrared imaging, noted that it was capable of detecting approximately 85% of all breast cancers. Dodd's ideas would later help to fuel the premise and attitudes incorporated into the BCDDP.

Three years later, J.D. Wallace presented to another Cancer Conference, sponsored by the American College of Radiology, the American Cancer Society and the Cancer Control Program of the NCI, an update on infrared imaging of the breast [73]. The author's analysis suggested that the incidence of breast cancer detection per 1000 patients screened could increase from 2.72 when using mammography to 19 when using infrared imaging. He then underlined that infrared imaging poses no radiation burden on the patient, requires no physical contact, and, being an innocuous technique, could concentrate the sought population by a significant factor selecting those patients that required further investigation. He concluded that "the resulting infrared image contains only a small amount of information as compared to the mammogram, so that the reading of the infrared image is a substantially simpler task."

Unfortunately, this rather simplistic and cavalier attitude toward the generation and interpretation of infrared images was prevalent when it was hastily added and then prematurely dismissed from the BCDDP, which was just getting underway. Exaggerated expectations led to the ill-founded premise that infrared imaging might replace mammography rather than complement it. A detailed review of the Report of the Working Group of the BCDDP, published in 1979, is essential to understand the subsequent evolution of infrared imaging [74].

The work scope of this project was issued by the NCI on March 26, 1973 with six objectives, the second being to determine if a negative infrared image was sufficient to preclude the use of clinical examination and mammography in the detection of breast cancer. The Working Group, reporting on results of the first 4 years of this project, gave a short history regarding infrared imaging in breast cancer detection. They wrote that, as of the 1960s, there was intense interest in determining the suitability of infrared imaging for large-scale applications, and mass screening was one possibility. The need for technological improvement was recognized and the authors stated that efforts had been made to refine the technique. One of the important objectives behind these efforts had been to achieve a sufficiently high sensitivity and specificity for infrared imaging in order to make it useful as a prescreening device in selecting patients for referral for mammographic examination. It was thought that, if successful, the incorporation of this technology would result in a relatively small proportion of women having mammography (a technique that had caused concern at that time because of the carcinogenic effects of radiation). The Working Group indicated that the sensitivity and specificity of infrared imaging readings, with clinical data emanating from interinstitutional studies, were close to the corresponding results for physical examination and mammography. They noted that these three modalities selected different subgroups of breast cancers, and for this reason further evaluation of infrared imaging as a screening device in a controlled clinical trial was recommended.

10.6.1 Poor Study Design

While the Working Group describes in detail the importance of quality control of mammography, the entire protocol for infrared imaging was summarized in one paragraph and simply indicated that infrared imaging was conducted by a BCDDP-trained technician. The detailed extensive results from this report, consisting of over 50 tables, included only one that referred to infrared imaging showing that it had detected only 41% of the breast cancers during the first screening while the residual were either normal or unknown. There is no breakdown as far as these two latter groups were concerned. Since 28% of the first screening and 32% of the second screening were picked up by mammography alone, infrared imaging was dropped from any further evaluation and consideration. The report stated that it was impossible to determine whether abnormal infrared images could be predictive of interval cancers (cancers developing between screenings) since they did not collect these data.

By the same token, the Working Group was unable to conclude, with their limited experience, whether the findings were related to the then available technology of infrared imaging or with its application. They did, however, conclude that the decision to dismiss infrared imaging should not be taken as a determination of the future of this technique, rather that the procedure continued to be of interest because it does not entail the risk of radiation exposure. In the Working Group's final recommendation, they state that "infrared imaging does not appear to be suitable as a substitute for mammography for routine screening in the BCDDP." The report admitted that several individual programs of the BCDDP had results that were more favorable than what was reported for the BCDDP as a whole. They encouraged investment in the development and testing of infrared imaging under carefully controlled study conditions and suggested that high priority be given to these studies. They noted that a few suitable sites appeared to be available within the BCDDP participants and proposed that developmental studies should be solicited from sites with sufficient experience.

10.6.2 Untrained Personnel and Protocol Violations

JoAnn Haberman, who was a participant in this project [75], provided further insight into the relatively simplistic regard assigned to infrared imaging during this program. The author reiterated that expertise in mammography was an absolute requirement for the awarding of a contract to establish a screening center. However, the situation was just the opposite with regard to infrared imaging—no experience was required at all. When the 27 demonstration project centers opened their doors, only five had any preexisting expertise in infrared imaging. Of the remaining screening centers, there was no experience at all in this technology. Finally, more than 18 months after the project had begun, the NCI-established centers where radiologists and their technicians could obtain sufficient training in infrared imaging. Unfortunately, only 11 of the demonstration project directors considered this training of sufficient importance to send their technologists to learn proper infrared technique. The imaging sites also disregarded environmental controls. Many of the project sites were mobile imaging vans that had poor heating and cooling capabilities, and often kept their doors open in the front and rear to permit an easy flow of patients. This, combined with a lack of adherence to protocols and preimaging patient acclimation, lead to unreadable images.

In summary, with regard to infrared imaging, the BCDDP was plagued with problems and seriously flawed in five critical areas:

1. The study was initiated with an incorrect premise that infrared imaging might replace mammography. A functional imaging procedure that detects metabolic thermovascular aberrations cannot replace a test that looks for specific areas of structural changes in the breast.
2. Completely untrained technicians were used to perform the scans.
3. The study used radiologists who had no experience or knowledge in reading infrared images.
4. Proper laboratory environmental controls were completely ignored. In fact, many of the research sites were mobile trailers with extreme variations in internal temperatures.
5. No standardized reading protocol had yet been established for infrared imaging. It was not until the early 1980s that established and standardized reading protocols were adopted.

Considering these facts, the BCDDP could not have properly evaluated infrared imaging. Since the termination of the BCDDP, a considerable amount of published research has demonstrated the true value of this technology.

10.7 Mammography and Infrared Imaging

From a scientific standpoint, mammography and infrared imaging are completely different screening tests. As a structural imaging procedure, mammography cannot be compared to a functional imaging technology such as infrared imaging. While mammography attempts to detect architectural tissue shadows, infrared imaging observes for changes in the subtle metabolic milieu of the breast. Even though mammography and infrared imaging examine completely different aspects of the breast, research has been performed that allows for a statistical comparison of the two technologies. Since a review of the research on infrared imaging has been covered, data on the current state of mammography is presented.

In a study by Rosenberg, 183,134 screening mammograms were reviewed for changes in sensitivity due to age, breast density, ethnicity, and estrogen replacement therapy. Out of these screening mammograms 807 cancers were discovered at screening. The results showed that the sensitivity for mammography was 54% in women younger than 40 years, 77% in women aged 40–49, 78% in women aged 50–64, and 81% in women older than 64 years. Sensitivity was 68% in women with dense breasts and 74% in estrogen replacement therapy users [76].

Investigating the cumulative risk of a false-positive result in mammographic screening, Elmore and associates performed a 10-year retrospective study of 2400 women aged 40–69 years of age. A total of 9762 mammograms were investigated. It was found that a woman had an estimated 49.1% cumulative risk of having a false-positive result after ten mammograms. Even though no breast cancer was present, over one-third of the women screened were required to have additional evaluations [77].

In a review of the literature, Head investigated the sensitivity, specificity, positive predictive value, and negative predictive values for mammography and infrared imaging. The average reported performance for mammography was 86% sensitivity, 79% specificity, 28% positive predictive value, and 92% negative predictive value. For infrared imaging the averaged performance was: 86% sensitivity, 89% specificity, 23% positive predictive value, and 99.4% negative predictive value [78].

Pisano, along with a large investigational group, provided a detailed report on the Digital Mammographic Imaging Screening Trial (DMIST). The study investigated the diagnostic performance of digital versus film mammography in breast cancer screening. Both digital and film mammograms were taken on 42,760 asymptomatic women presenting for screening mammography. Data were gathered from 33 sites in the United States and Canada. Digital mammography was found to be more accurate in women under age 50 and in women whose breasts were radiographically dense. The sensitivity for both film and digital mammography was found to be 69% [79].

Keyserlingk and associates published a retrospective study reviewing the relative ability of clinical examinations, mammography, and infrared imaging to detect 100 new cases of ductal carcinoma *in situ*, stage I and II breast cancers. Results from the study found that the sensitivity for clinical examination alone was 61%, mammography alone was 66%, and infrared imaging alone was 83%. When suspicious and equivocal mammograms were combined the sensitivity was increased to 85%. A sensitivity of 95% was found when suspicious and equivocal mammograms were combined with abnormal infrared images. However, when clinical examination, mammography, and infrared images were combined a sensitivity of 98% was reached [80].

From a review of the cumulative literature database, it can be found that the average sensitivity and specificity for mammography is, at best, 80% and 79%, respectively, for women over the age of 50 [81–83]. A significant decrease in sensitivity and specificity is seen in women below this age. This same research also shows that mammography routinely misses interval cancers (cancers that form between screening exams) [80] that may be detected by infrared imaging. Taking into consideration all the available data, mammography leaves much to be desired as the current gold standard for breast cancer screening. As a stand alone screening procedure, it is suggested that mammography may not be the best choice. In the same light, infrared imaging should also not be used alone as a screening test. The two technologies are of a complementary nature. Neither used alone are sufficient, but when combined each builds on the deficiencies of the other. In reviewing the literature it seems evident that a multimodal approach to breast cancer screening would serve women best. A combination of clinical examination, mammography, and infrared imaging would provide the greatest potential for breast conservation and survival.

10.8 Current Status of Breast Cancer Detection

Current first-line breast cancer detection strategy still depends essentially on clinical examination and mammography. The limitations of the former, with its reported sensitivity rate often below 65% [80,84] is well-recognized, and even the proposed value of self-breast examination is being contested [85]. While mammography is accepted as the most cost-effective imaging modality, its contribution continues to be challenged with persistent false-negative rates ranging up to 30% [76,79,86,87]; with decreasing sensitivity in younger patients and those on estrogen replacement therapy [76,87,88]. In addition, there is recent data suggesting that denser and lesser informative mammography images are precisely those associated with an increased cancer risk [89,90]. Echoing some of the shortcomings of the BCDDP concerning their study design and infrared imaging, Moskowitz indicated that mammography is also not a procedure to be performed by the inexperienced technician or radiologist [91].

With the current emphasis on earlier detection, there is now renewed interest in the parallel development of complementary imaging techniques that can also exploit the precocious metabolic, immunological, and vascular changes associated with early tumor growth. While promising, techniques such as scintimammography [92], Doppler ultrasound [93], and magnetic resonance imaging (MRI) [94], are associated with a number of disadvantages that include exam duration, limited accessibility, need of intravenous access, patient discomfort, restricted imaging area, difficult interpretation, and limited availability of the technology. Like ultrasound, they are more suited to be used as second-line options to pursue the already abnormal screening evaluations. While practical, this stepwise approach currently results in the nonrecognition, and thus delayed utilization of second-line technology in approximately 10% of established breast cancers [91]. This is consistent with a study published by Keyserlingk et al. [80].

As an addition to the breast health screening process, infrared imaging has a significant role to play. Owing to infrared imaging's unique ability to image the metabolic aspects of the breast, extremely early warning signals (8–10 years before any other detection method) have been observed in long-term studies. It is for this reason that an abnormal infrared image is the single most important marker of high risk for the existence of or future development of breast cancer. This, combined with the proven sensitivity, specificity, and prognostic value of the technology, places infrared imaging as one of the major front-line methods of breast cancer screening.

10.9 Future Advancements in Infrared Imaging

Modern high-resolution uncooled focal plane array cameras coupled with high-speed computers running sophisticated image analysis software are commonplace in today's top infrared imaging centers. However, research in this field continues to produce technological advancements in image acquisition and digital processing.

Research is currently under way investigating the possible subtle alterations in the blood supply to the breast during the earliest stages of malignancy. Evidence suggests that there may be a normal vasomotor oscillation frequency in the arterial structures of the human body. It is theorized that there may be disturbances in this normal oscillatory rate when a malignancy is forming. Research using infrared detection systems capturing 200 frames per second with a sensitivity of 0.009 of a degree centigrade may be able to monitor alterations in this vasomotor frequency band.

Another unique methodology is investigating the possibility of using infrared emission data to extrapolate depth and location of a metabolic heat source within the body. In the case of cancer, the increased tissue metabolism resulting from rapid cellular multiplication and growth generates heat. With this new approach in infrared detection, it is theorized that an analysis based on an analogy to electrical circuit theory—termed the thermal-electric analog—may possibly be used to determine the depth and location of the heat source.

The most promising of all the advances in medical infrared imaging are the adaptations being used from military developments in the technology. Hardware advances in narrow-band filtering hold promise in providing multispectral and hyperspectral images. One of the most intriguing applications of multispectral/hyperspectral imaging may include real-time intraoperative cytology. Investigations are also underway utilizing smart processing, also known as artificial intelligence. This comes in the form of postimage processing of the raw data from the infrared sensor array. Some of the leading-edge processing currently under study includes automated target recognition (ATR), artificial neural networks (ANN), and a host of other proprietary algorithms. The uses of ATR and similar algorithms are dependent on a reliable normative database. The images are processed based on what the system has learned as normal and compares the new image to that database. Unlike ATR, ANN uses data summation to produce pattern recognition. This is extremely important when it comes to the complex thermovascular patterns seen in infrared breast imaging. Ultimately, these advancements should lead to a substantial increase in both objectivity and accuracy as an analytical aid to the human interface.

In another and possibly critical role, infrared imaging of the breast may hold a significant potential in breast cancer prevention. Due to the ability of infrared imaging's detection of changes in the dermal circulation, any exogenous pharmacological intervention or endogenous release of biochemicals that have the propensity to alter vascular profusion may be detected. This is especially true of chemicals that are target specific for the tissues of the breast. Being a primary target tissue for the hormone estrogen, the hormone's effect in the breast is anabolic to the ductal cells. As such, the outcome is one of increased cellular metabolism. This increase in cellular activity necessitates the need for nutrients above and beyond the norm. In order to facilitate this need, an increase in blood supply must occur. This translates to an infrared image demonstrating a uniform increase in vascular patterning. The importance of this observation lies in one of the primary risk factors for breast cancer—lifetime exposure to estrogen. If infrared imaging has the ability to warn of increased thermovascular activity secondary to levels of estrogen in the breast, action can be taken to lower this activity and ultimately the patient's risk for future breast cancer. Treatments can be monitored for positive effects by incorporating infrared imaging as a method of observing these effects. Studies have shown this effect and the positive outcome of pharmacological intervention. Many patients with this condition also demonstrate signs and symptoms that include breast pain, tenderness, cysts, and benign lumps. In many patients, a reversal or reduction in these signs and symptoms are also noted when treatment is initiated [95–97]. Infrared imaging's ability to detect increased thermovascular activity secondary to levels of estrogen in the breast, and to monitor the effects of treatment targeted at the breast, may play a significant role in breast cancer prevention.

New breast cancer treatments are also exploring methods of targeting the angiogenic process. Owing to a tumor's dependence on a constant blood supply to maintain growth, antiangiogenesis therapy is becoming one of the most promising therapeutic strategies and has been found to be pivotal in the new paradigm for consideration of breast cancer development and treatment [98]. The future may see infrared imaging and antiangiogenesis therapy combined as the state-of-the-art in the biological assessment and treatment of breast cancer.

These and other new methodologies in medical infrared imaging are being investigated and may prove to be significant advancements.

10.10 Conclusion

The large patient populations and long survey periods in many of the above clinical studies yields a high significance to the various statistical data obtained. This is especially true for the contribution of infrared imaging to early cancer diagnosis, as an invaluable marker of high-risk populations, and in therapeutic decision making.

Currently available high-resolution digital infrared imaging technology benefits greatly from enhanced image production, computerized image processing and analysis, and standardized image interpretation protocols. Over 40 years of research and 800 indexed papers encompassing well over 300,000 women participants has demonstrated infrared imaging's abilities in the early detection of breast cancer. Infrared imaging has distinguished itself as the earliest detection technology for breast cancer. It may be able to signal an alarm that a cancer may be forming 8–10 years before any other procedure can detect it. In seven out of ten cases, infrared imaging will detect signs of a cancer before it is seen on a mammogram. Clinical trials have also shown that infrared imaging significantly augments the long-term survival rates of its recipients by as much as 61%. And when used as part of a multimodal approach (clinical examination, mammography, and infrared imaging), 95% of all early stage cancers will be detected. Ongoing research into the thermal characteristics of breast pathologies will continue to investigate the relationships between neoangiogenesis, chemical mediators, and the neoplastic process.

It is unfortunate, but many clinicians still hesitate to consider infrared imaging as a useful tool in spite of the considerable research database, steady improvements in both infrared technology and image analysis, and continued efforts on the part of the infrared imaging societies. This attitude may be due in part to the average clinician's unfamiliarity with the physical and biological basis of infrared imaging. The other methods of cancer investigations refer directly to topics of medical teaching. For instance, radiography and ultrasonography refer to structural anatomy. Infrared imaging, however, is based on thermodynamics and thermokinetics, which are unfamiliar to most clinicians; though man is experiencing heat production and exchange in every situation he undergoes or creates.

Considering the contribution that infrared imaging has demonstrated thus far in the field of early breast cancer detection, all possibilities should be considered for promoting further technical, biological, and clinical research along with the incorporation of the technology into common clinical use.

References

1. Adams, F., *The Genuine Works of Hippocrates*. Williams & Wilkins, Baltimore, 1939.
2. Lawson, R. N., Implications of surface temperatures in the diagnosis of breast cancer. *Can. Med. Assoc. J.*, 75, 309, 1956.
3. Lawson, R. N., Thermography—A new tool in the investigation of breast lesions. *Can. Serv. Med.*, 13, 517, 1957.
4. Lawson, R. N., A new infrared imaging device. *Can. Med. Assoc. J.* 79, 402, 1958.
5. Archer, F., Gros, C., Classification thermographique des cancers mammaries. *Bull. Cancer*, 58, 351, 1971.
6. Amalu, W. et al., Standards and protocols in clinical thermographic imaging. International Academy of Clinical Thermology, September 2002.
7. Kurbitz, G., Design criteria for radiometric thermal-imaging devices, in *Thermological Methods*, VCH mbH, 94–100, 1985.
8. Houdas, Y., Ring E.F.J., Models of thermoregulation, in *Human Temperature: Its Measurement and Regulation*, Plenum Press, New York, 136–141.
9. Flesch, U., Physics of skin-surface temperature, in *Thermological Methods*, VCH mbH, 21–33, 1985.
10. Anbar M., *Quantitative Dynamic Telethermometry in Medical Diagnosis and Management*, CRC Press, FL, 106, 1994.
11. Anbar, M., Potential artifacts in infrared thermographic measurements. *Thermology*, 3, 273, 1991.
12. Friedrich, K. (Optic research laboratory, Carl Zeiss—West Germany), Assessment criteria for infrared thermography systems. *Acta Thermographica*, 5, 68–72.
13. Engel, J. M., Thermography in locomotor diseases. *Acta Thermographica*, 5, 11–13.
14. Cuthbertson, G. M., The development of IR imaging in the United Kingdom, in *The Thermal Image in Medicine and Biology*. Uhlen-Verlag, Wien, 21–32, 1995.
15. Amalu, W., Nondestructive testing of the human breast: The validity of dynamic stress testing in medical infrared breast imaging. *Proceedings of the 26th Annual International Conference of the IEEE EMBS*, 1174–1177, 2004.
16. Gautherie, M., Kotewicz, A., Gueblez, P., Accurate and objective evaluation of breast thermograms: Basic principles and new advances with special reference to an improved computer-assisted scoring system: in *Thermal assessment of Breast Health*, MTP Press Limited, 72–97, 1983.
17. Hobbins, W. B., Abnormal thermogram—Significance in breast cancer. *Interamer. J. Rad.*, 12, 337, 1987.
18. Gautherie, M., New protocol for the evaluation of breast thermograms, in *Thermological Methods*, VCH mbH, 227–235, 1985.
19. Lawson, R. N., Chughtai, M. S., Breast cancer and body temperatures. *Can. Med. Assoc. J*, 88, 68, 1963.
20. Head, J. F., Wang, F., Elliott, R. L., Breast thermography is a noninvasive prognostic procedure that predicts tumor growth rate in breast cancer patients. *Ann N Y Acad. Sci*, 698, 153, 1993.
21. Sterns, E. E., Zee, B., Sen Gupta, J., Saunders, F. W., Thermography: Its relation to pathologic characteristics, vascularity, proliferative rate and survival of patients with invasive ductal carcinoma of the breast. *Cancer*, 77, 1324, 1996.
22. Head, J. F., Elliott, R. L., Breast thermography. *Cancer*, 79, 186, 1995.
23. Anbar M., in *Quantitative Dynamic Telethermometry in Medical Diagnosis and Management*. CRC Press, Boca Raton, FL, pp. 84–94, 1994.
24. Rodenberg, D. A., Chaet, M. S., Bass, R. C. et al. Nitric oxide: An overview. *Am. J Surg.* 170, 292, 1995.
25. Thomsen, L. L., Miles, D. W., Happerfield, L. et al. Nitric oxide synthase activity in human breast cancer. *Br. J. Cancer*, 72(1), 41, 1995.
26. Guidi, A. J., Schnitt, S. J., Angiogenesis in pre-invasive lesions of the breast. *The Breast J.*, 2, 364, 1996.
27. Gamagami, P., Indirect signs of breast cancer: Angiogenesis study, in *Atlas of Mammography*, Blackwell Science, Cambridge, MA, 231–226, 1996.

28. Love, T., Thermography as an indicator of blood perfusion. *Proc NY Acad Sci J*, 335, 429, 1980.

29. Chato, J., Measurement of thermal properties of growing tumors. *Proc NY Acad Sci*, 335, 67, 1980.

30. Draper, J., Skin temperature distribution over veins and tumors. *Phys. Med. Biol.,* 16(4), 645, 1971.

31. Jain, R., Gullino, P., Thermal characteristics of tumors: Applications in detection and treatment. *Ann. NY Acad. Sci.*, 335, 1, 1980.

32. Gautherie, M., Thermopathology of breast cancer; measurement and analysis of *in-vivo* temperature and blood flow. *Ann. NY Acad. Sci.*, 365, 383, 1980.

33. Gautherie, M., Thermobiological assessment of benign and malignant breast diseases. *Am. J Obstet. Gynecol.,* 147(8), 861, 1983.

34. Gamigami, P., *Atlas of Mammography: New Early Signs in Breast Cancer*. Blackwell Science, 1996.

35. Gershen-Cohen, J., Haberman, J., Brueschke, E., Medical thermography: A summary of current status. *Radiol. Clin. North Am.,* 3, 403, 1965.

36. Haberman, J., The present status of mammary thermography. *Ca—A Cancer J. Clin.,* 18, 314, 1968.

37. Hoffman, R., Thermography in the detection of breast malignancy. *Am. J Obstet. Gynecol.*, 98, 681, 1967.

38. Stark, A., Way, S., The screening of well women for the early detection of breast cancer using clinical examination with thermography and mammography. *Cancer*, 33, 1671, 1974.

39. Amalric, D. et al., Value and interest of dynamic telethermography in detection of breast cancer. *Acta Thermogr.*, 1, 89–96.

40. Hobbins, W., Mass breast cancer screening. *Proceedings, Third International Symposium on Detection and Prevention of Breast Cancer*, New York City, NY, 637, 1976.

41. Isard, H. J., Becker, W., Shilo, R. et al., Breast thermography after four years and 10,000 studies. *Am. J Roentgenol.*, 115, 811, 1972.

42. Spitalier, H., Giraud, D. et al., Does infrared thermography truly have a role in present-day breast cancer management? *Biomedical Thermology*, Alan R. Liss, New York, NY, 269–278, 1982.

43. Moskowitz, M., Milbrath, J., Gartside, P. et al., Lack of efficacy of thermography as a screening tool for minimal and stage I breast cancer. *N Engl. J Med.*, 295, 249, 1976.

44. Threatt, B., Norbeck, J.M., Ullman, N.S. et al., Thermography and breast cancer: An analysis of a blind reading. *Ann. N Y Acad Sci,* 335, 501, 1980.

45. Haberman, J., Francis, J., Love, T., Screening a rural population for breast cancer using thermography and physical examination techniques. *Ann. NY Acad. Sci.*, 335, 492, 1980.

46. Sciarra, J., Breast cancer: Strategies for early detection, in *Thermal Assessment of Breast Health (Proceedings of the International Conference on Thermal Assessment of Breast Health)*. MTP Press LTD, 117–129, 1983.

47. Gautherie, M., Thermobiological assessment of benign and malignant breast diseases. *Am. J Obstet. Gynecol.*, 147(8), 861, 1983.

48. Louis, K., Walter, J., Gautherie, M., Long-term assessment of breast cancer risk by thermal imaging, in *Biomedical Thermology*. Alan R. Liss Inc., 279–301, 1982.

49. Gros, C., Gautherie, M., Breast thermography and cancer risk prediction. *Cancer*, 45, 51, 1980.

50. Jones, C. H., Thermography of the female breast, in *Diagnosis of Breast Disease*, C.A. Parsons (Ed.), University Park Press, Baltimore, 214–234, 1983.

51. Useki, H., Evaluation of the thermographic diagnosis of breast disease: Relation of thermographic findings and pathologic findings of cancer growth. *Nippon Gan Chiryo Gakkai Shi*, 23, 2687, 1988.

52. Parisky, Y. R., Sardi, A. et al., Efficacy of computerized infrared imaging analysis to evaluate mammographically suspicious lesions. *AJR*, 180, 263, 2003.

53. Nyirjesy, I., Ayme, Y. et al., Clinical evaluation, mammography, and thermography in the diagnosis of breast carcinoma. *Thermology*, 1, 170, 1986.

54. Thomassin, L., Giraud, D. et al., Detection of subclinical breast cancers by infrared thermography, in *Recent Advances in Medical Thermology (Proceedings of the Third International Congress of Thermology)*, Plenum Press, New York, NY., 575–579, 1984.

55. Arora, N., Martins, D., Ruggerio, D. et al., Effectiveness of a noninvasive digital infrared thermal imaging system in the detection of breast cancer. *Am. J Surg.*, 196(4), 523–526, 2008.

56. Wang et al., Evaluation of the diagnostic performance of infrared imaging of the breast: A preliminary study. *Biomedical Engineering*, 9, 3, 2010.

57. Wishart, G. C., Lampiri, M., Boswell, M. et al., The accuracy of digital infrared imaging for breast cancer detection in women undergoing breast biopsy. *Eur. J Surg. Oncol.*, 36(6), 535–540, 2010.

58. Amalric, R., Gautherie, M., Hobbins, W., Stark, A., The future of women with an isolated abnormal infrared thermogram. *La Nouvelle Presse Med*, 10(38), 3153, 1981.

59. Gautherie, M., Gros, C., *Contribution of Infrared Thermography to Early Diagnosis, Pretherapeutic Prognosis, and Post-Irradiation Follow-up of Breast Carcinomas*. Laboratory of Electroradiology, Faculty of Medicine, Louis Pasteur University, Strasbourg, France, 1976.

60. Hobbins, W., Significance of an "isolated" abnormal thermogram. *La Nouvelle Presse Medicale*, 10, 3155, 1981.

61. Hobbins, W., Thermography, highest risk marker in breast cancer. *Proceedings of the Gynecological Society for the Study of Breast Disease*, 267–282, 1977.

62. Gauthrie, M., Improved system for the objective evaluation of breast thermograms, in *Biomedical Thermology*. Alan R. Liss, Inc., New York, NY, 897–905, 1982.

63. Amalric, R., Giraud, D. et al., Combined diagnosis of small breast cancer. *Acta Thermographica*, 1984.

64. Spitalier, J., Amalric, D. et al., The importance of infrared thermography in the early suspicion and detection of minimal breast cancer, in *Thermal Assessment of Breast Health*, MTP Press Ltd., 173–179, 1983.

65. Gautherie, M. et al., Thermobiological assessment of benign and malignant breast diseases. *Am. J Obstet. Gynecol.*, 147(8), 861, 1983.

66. Jay, E., Karpman, H., Computerized breast thermography, in *Thermal Assessment of Breast Health*, MTP Press Ltd., 98–109, 1983.

67. Dilhuydy, M. H. et al., The importance of thermography in the prognostic evaluation of breast cancers. *Acta Thermogr.*, 130–136.

68. Amalric, D. et al., Value and interest of dynamic telethermography in detection of breast cancer. *Acta Thermogr.*, 89–96.

69. Fournier, V. D., Kubli, F. et al., Infrared thermography and breast cancer doubling time. *Acta Thermogr.*, 107–111.

70. Gros, D., Gautherie, M., Warter, F., Thermographic prognosis of treated breast cancers. *Acta Thermogr.*, 11–14.

71. Keyserlingk, J. R., Ahlgren P. D. et al., Preliminary evaluation of high resolution functional infrared imaging to monitor pre-operative chemohormonotherapy-induced changes in neo-angiogenesis in patients with locally advanced breast cancer. Ville Marie Oncology Center/St. Mary's Hospital, Montreal, Canada. In submission for publication, 2003.

72. Dodd, G. D., Thermography in breast cancer diagnosis, in *Abstracts for the Seventh National Cancer Conference Proceedings*. Lippincott, Philadelphia, Los Angeles, CA, 267, 1972.

73. Wallace, J. D., Thermographic examination of the breast: An assessment of its present capabilities, in *Early Breast Cancer: Detection and Treatment*, Gallagher, H.S. (Ed.). American College of Radiology, Wiley, New York, 13–19, 1975.

74. Report of the Working Group to Review the National Cancer Institute Breast Cancer Detection Demonstration Projects. *J. Natl. Cancer Inst.*, 62, 641, 1979.

75. Haberman, J., An overview of breast thermography in the United States, in *Medical Thermography*, M. Abernathy, S. Uematsu (Eds), American Academy of Thermology, Washington, 218–223, 1986.

76. Rosenberg, R. D., Hunt, W. C. et al., Effects of age, breast density, ethnicity, and estrogen replacement therapy on screening mammographic sensitivity and cancer stage at diagnosis: Review of 183,134 screening mammograms in Albuquerque, New Mexico. *Radiology*, 209(2), 511, 1998.

77. Elmore, J. et al, Ten-year risk of false positive screening mammograms and clinical breast examinations. *N. Engl. J Med.*, 338, 1089, 1998.

78. Head, J. F., Lipari, C. A., Elliot, R. L., Comparison of mammography, and breast infrared imaging: Sensitivity, specificity, false negatives, false positives, positive predictive value and negative predictive value. *IEEE*, 1999.

79. Pisano, E. D., Gatsonis, C. et al., Diagnostic performance of digital versus film mammography for breast-cancer screening. *N. Engl. J Med.*, 353, 2005.

80. Keyserlingk, J. R., Ahlgren, P. D. et al., Infrared imaging of the breast; initial reappraisal using high-resolution digital technology in 100 successive cases of stage 1 and 2 breast cancer. *Breast J.*, 4, #4, 1998.

81. Schell, M. J., Bird, R. D., Desrochers, D. A., Reassessment of breast cancers missed during routine screening mammography. *Am. J Roentgenol.*, 177, 535, 2001.

82. Poplack, S. P., Tosteson, A. N., Grove, M. et al., The practice of mammography in 53,803 women from the New Hampshire mammography network. *Radiology*, 217, 832, 2000.

83. Pullon, S., McLeod, D., The early detection and diagnosis of breast cancer: A literature review. General Practice Department, Wellington School of Medicine, December 1996.

84. Sickles, E. A., Mammographic features of "early" breast cancer. *Am. J Roentgenol.*, 143, 461, 1984.

85. Thomas, D. B., Gao, D. L., Self, S. G. et al., Randomized trial of breast self-examination in Shanghai: Methodology and preliminary results. *J. Natl. Cancer Inst.*, 5, 355, 1997.

86. Moskowitz, M., Screening for breast cancer. How effective are our tests? *CA Cancer J. Clin.*, 33, 26, 1983.

87. Elmore, J. G., Wells, C. F., Carol, M. P. et al., Variability in radiologists interpretation of mammograms. *NEJM*, 331(22), 1493, 1994.

88. Gilliland, F. D., Joste, N., Stauber, P. M. et al., Biologic characteristics of interval and screen-detected breast cancers. *J. Natl. Cancer Inst.*, 92, 743, 2000.

89. Laya, M. B., Effect on estrogen replacement therapy on the specificity and sensitivity of screening mammography. *J. Natl. Cancer Inst.*, 88, 643, 1996.

90. Boyd, N. F., Byng, J. W., Jong, R. A. et al., Quantitative classification of mammographic densities and breast cancer risk. *J. Natl. Cancer Inst.*, 87, 670, 1995.

91. Moskowitz, M., Breast imaging, in *Cancer of the Breast*, Donegan, W. L., Spratt, J.S. (Eds), Saunders, New York, 206–239, 1995.

92. Khalkhali, I., Cutrone, J. A. et al., Scintimammography: The complementary role of Tc-99m sestamibi prone breast imaging for the diagnosis of breast carcinoma. *Radiology*, 196, 421, 1995.

93. Kedar, R. P., Cosgrove, D. O. et al., Breast carcinoma: Measurement of tumor response in primary medical therapy with color doppler flow imaging. *Radiology*, 190, 825, 1994.

94. Weinreb, J. C., Newstead, G., MR imaging of the breast. *Radiology*, 196, 593, 1995.

95. Verzini, L., Romani, L., Talia, B., (Radiology department university of Modena (Italy)). Thermographic variations in the breast during the menstrual cycle. *Acta Thermographica,* 143–149, 1980s.

96. Huber, C., Pons, J., Pateau, A., (Gynecology and department of radiology Cretei hospital (Paris) France). Breast fibrocystic disease and thermography. *Acta Thermographica,* 48–50, 1980s.

97. Borten, M., Ransil, B. et al. (Department of obstetrics and gynecology, Beth Israel Hospital, Harvard Medical School). Regional differences in breast surface temperature by liquid crystal thermography. *Thermology,* 1, 216–220, 1986.

98. Love, S. M., Barsky, S. H., Breast cancer: An interactive paradigm. *Breast J.*, 3, 171, 1996.

11

Functional Infrared Imaging of the Breast: Historical Perspectives, Current Application, and Future Considerations

John R. Keyserlingk
Ville Marie Medical and Women's Health Center

P.D. Ahlgren
Ville Marie Medical and Women's Health Center

E. Yu
Ville Marie Medical and Women's Health Center

Normand Belliveau
Ville Marie Medical and Women's Health Center

Mariam Yassa
Ville Marie Medical and Women's Health Center

There is a general consensus that earlier detection of breast cancer should result in improved survival. For the last two decades, first-line breast imaging has relied primarily on mammography. Despite better equipment and regulation, particularly with the recent introduction of digital mammography, variability in interpretation and tissue density can affect mammography accuracy. To promote earlier diagnosis, a number of adjuvant functional imaging techniques have recently been introduced, including Doppler ultrasound and gadolinium-enhanced magnetic resonance imaging (MRI) that can detect cancer-induced regional neovascularity. While valuable modalities, problems relating to complexity, accessibility, cost, and in most cases the need for intravenous access make them unsuitable as components of a first-line imaging strategy.

In order to reassess the potential contribution of infrared (IR) imaging as a first-line component of a multi-imaging strategy, using currently available technology, we will first review the history of its introduction and early clinical application, including the results of the Breast Cancer Demonstration Projects (BCDDP). We will then review the Ville Marie Multidisciplinary Breast Center's more recent experience with currently available high-resolution computerized IR technology to assess IR imaging both as a complement to clinical exam and mammography in the early detection of breast cancer and also as a tool to monitor the effects of preoperative chemohormonotherapy in advanced breast cancer. Our goal is to show that high-resolution IR imaging provides additional safe, practical, cost-effective, and objective information in both of these instances when produced by strict protocol and interpreted by sufficiently trained breast physicians. Finally, we will comment on its further evolution.

11.1 Historical Perspectives

11.1.1 Pre-Breast Cancer Detection Demonstration Projects Era

In 1961, in the *Lancet*, Williams and Handley [1] using a rudimentary handheld thermopile, reported that 54 of 57 of their breast cancer patients were detectable by IR imaging, and "among these were cases in which the clinical diagnosis was in much doubt." The authors reported that the majority of these cancers had a temperature increase of 1–2°C, and that the IR imaging permitted excellent discrimination between benign and malignant processes. Their protocol at the Middlesex Hospital consisted of having the patient strip to the waist and be exposed to the ambient temperature for 15 min.

The authors demonstrated a precocious understanding of the significance of IR imaging by introducing the concept that increased cooling to 18°C further enhanced the temperature discrepancy between cancer and the normal breast. In a follow-up article the subsequent year, Handley [2] demonstrated a close correlation between the increased thermal pattern and increased recurrence rate. While only 4 of 35 cancer patients with a 1–2°C discrepancy recurred, five of the six patients with over 3°C rise developed recurrent cancer, suggesting already that the prognosis could be gauged by the amount of rise of temperature in the overlying skin.

In 1963, Lawson and Chughtai [3], two McGill University surgeons, published an elegant intraoperative study demonstrating that the increase in regional temperature associated with breast cancer was related to venous convection. This quantitative experiment added credence to Handley's suggestion that IR findings were related to both increased venous flow and increased metabolism.

In 1965, Gershon-Cohen [4], a radiologist and researcher from the Albert Einstein Medical Center, introduced IR imaging to the United States. Using a Barnes thermograph that required 15 min to produce a single IR image, he reported 4000 cases with a remarkable true-positive rate of 94% and a false-positive rate of 6%. These data were included in a review of the then current status of infrared imaging published in 1968 in *CA—A Cancer Journal for Physicians* [5]. The author, JoAnn Haberman, a radiologist from Temple University School of Medicine, reported the local experience with IR imaging, which produced a true-positive rate of 84% compared with a concomitant true-positive rate of 80% for mammography. In addition, she compiled 16,409 IR imaging cases from the literature between 1964 and 1968 revealing an overall true-positive rate of 87% and a false-positive rate of 13%.

A similar contemporary review compiled by Jones, consisting of nearly 70,000 cases, revealed an identical true-positive rate of 85% and an identical false-positive rate of 13%. Furthermore, Jones [6] reported on over 20,000 IR imagings from the Royal Marsden Hospital between 1967 and 1972, and noted that approximately 70% of Stage I and Stage II cancers and up to 90% of Stage III and Stage IV cancers had abnormal IR features. These reports resulted in an unbridled enthusiasm for IR imaging as a front-line detection modality for breast cancer.

Sensing a potential misuse of this promising but unregulated imaging modality, Isard made some sobering comments in 1972 [7] in a publication of the *American Journal of Roentengology*, where he emphasized that, like other imaging techniques, IR imaging does not diagnose cancer but merely indicates the presence of an abnormality. Reporting his Radiology division's experience with 10,000 IR studies done concomitantly with mammography between 1967 and 1970, he reiterated a number of important concepts, including the remarkable stability of the IR image from year to year in the otherwise healthy patient, and the importance of recognizing any significant change. Infrared imaging detected 60% of occult cancers in his experience, versus 80% with mammography. The combination of both these modalities increased the sensitivity by approximately 10%, thus underlining the complementarity of both of these processes, since each one did not always suspect the same lesion.

It was Isard's conclusion that, had there been a preliminary selection of his group of 4393 asymptomatic patients by IR imaging, mammography examination would have been restricted to the 1028 patients with abnormal IR imaging (23% of this cohort). This would have resulted in a cancer detection rate of 24.1 per 1000 mammographic examinations, as contrasted to the expected 7 per 1000 by mammographic screening. He concluded that since IR imaging is an innocuous examination, it could be utilized to focus attention upon asymptomatic women who should be examined more intensely.

In 1972, Gerald D. Dodd [8] of the Department of Diagnostic Radiology of the University of Texas presented an update on IR imaging in breast cancer diagnosis at the Seventh National Cancer Conference sponsored by the National Cancer Society and the National Cancer Institute (NCI). He also suggested that IR imaging would be best employed as a screening agent for mammography and proposed that in any general survey of the female population age 40 and over, 15–20% would have positive IR imaging and would require mammograms. Of these, approximately 5% would be recommended for biopsy. He concluded that IR imaging would serve to eliminate 80–85% of the potential mammograms. Reporting the Texas Medical School's experience with IR imaging, he reiterated that IR was capable of detecting approximately 85% of all breast cancers. The false-positive rate of 15–20% did not concern the author, who stated that these were false-positives only in the sense that there was no corroborative evidence of breast cancer at the time of the examination and that they could serve to identify a high-risk population.

Feig et al. [9] reported the respective abilities of clinical exam, mammography, and IR imaging to detect breast cancer in 16,000 self-selected women. While only 39% of the initial series of overall established cancer patients had an abnormal IR imaging, this increased to 75% in his later cohort, reflecting an improved methodology. Of particular interest was the ability of IR imaging to detect 54% of the smallest tumors, four times that of clinical examination. This potential important finding was not elaborated, but it could reflect IR's ability to detect vascular changes that are sometimes more apparent at the initiation of tumor genesis. The authors suggested that the potential of IR imaging to select high-risk groups for follow-up screening merited further investigation.

Wallace [10] presented an update on IR imaging of the breast to another contemporary Cancer Conference sponsored by the American College of Radiology, the American Cancer Society, and the Cancer Control Programme of the NCI. The analysis suggested that the incidence of breast cancer detection per 1000 screenees could increase from 2.72 when using mammography to 19 when using IR imaging. He then underlined that IR imaging poses no radiation burden on the patient, requires no physical contact, and, being an innocuous technique, could concentrate the sought population by a significant factor, selecting those patients that required further investigation. He concluded that "the resulting IR image contains only a small amount of information as compared to the mammogram, so that the reading of the IR image is a substantially simpler task."

Unfortunately, this rather simplistic and cavalier attitude toward the acquisition and interpretation of IR imaging was widely prevalent when it was hastily added to the BCDDP, which was just getting underway. Rather than assess, in a controlled manner, its potential as a complementary first-line detection modality, it was hastily introduced into the BCDDP as a potential replacement for mammography and clinical exam.

11.1.2 Breast Cancer Detection Demonstration Projects Era

A detailed review of the Report of the Working Group of the BCDDP is essential to understand the subsequent evolution of IR imaging [11]. The scope of this project was issued by the NCI on March 26, 1973, with six objectives, the second being to determine if a negative IR imaging was sufficient to preclude the use of clinical examination and mammography in the detection of breast cancer. The Working Group, reporting on results of the first 4 years of this project, gave a short history regarding IR imaging in breast cancer detection. They reported that as of the 1960s, there was intense interest in determining the suitability of IR imaging for large-scale applications, and mass screening was one possibility. The need for technological improvement was recognized and the authors stated that efforts had been made to refine the technique. One of the important objectives behind these efforts had been to achieve a sufficiently high sensitivity and specificity for IR imaging under screening conditions to make it useful as a prescreening device in selecting patients who would then be referred for mammographic examination. It was thought that if successful, this technology would result in a relatively small proportion of women having mammography, a technique that caused concern because of the carcinogenic effects of radiation. The Working Group indicated that the sensitivity and specificity of IR imaging readings from clinical data emanating from interinstitutional studies were close to the corresponding results for physical examination and for mammography. While they noted that these three modalities selected different subgroups of breast cancers, further evaluation of IR imaging as a potential stand-alone screening device in a controlled clinical trial was recommended.

The authors of the BCDDP Working Group generated a detailed review of mammography and efforts to improve its quality control in image quality and reduction in radiation. They recalled that in the 1960s, the Cancer Control Board of the U.S. Public Health Service had financed a national mammography training program for radiologists and their technologists. Weekly courses in mammography were taught at approximately 10 institutions throughout the country with material supplied by the American College of Radiology. In 1975, shortly after the beginning of this project, the NCI had already funded seven institutions in the United States in a 3-year effort aimed at reorienting radiologists and their technologists in more advanced mammographic techniques and interpretation for the detection of early breast cancer.

In the interim, the American College of Radiology and many interested mammographers and technologists had presented local postgraduate refresher courses and workshops on mammography. Every year for the previous 16 years, the American College of Radiology had supported, planned, and coordinated week-long conferences and workshops aimed at optimizing mammography to promote the earlier detection and treatment of breast cancer. It was recognized that the well-known primary and secondary mammographic signs of a malignant condition, such as ill-defined mass, skin thickening, skin retraction, marked fibrosis and architectural distortion, obliteration of the retromammary space, and enlarged visible axillary lymph nodes, could detect an established breast cancer. However, the authors emphasized that more subtle radiographic signs that occur in minimal, clinically occult, and early cancers, such as localized punctate calcifications, focal areas of duct prominence, and minor architectural distortion, could lead to an earlier diagnosis even when the carcinoma was not infiltrating, which was a rare finding when previous mammographic techniques were used.

The authors reiterated that the reproduction of early mammography signs required a constant high-quality technique for fine image detail, careful comparison of the two breasts during interpretation, and the search for areas of bilateral, parenchymal asymmetry that could reflect underlying cancer. The BCDDP Working Group report stated that mammographies were conducted by trained technicians and that, while some projects utilized radiological technicians for the initial interpretation, most used either a radiologist or a combination of technician and radiologist. Quality control for mammography consisted of reviews by the project radiologists and site visits by consultants to identify problems in procedures and the quality of the films.

On the other hand, the entire protocol for IR imaging within this study was summarized in one paragraph, and it indicated that IR imaging was conducted by a BCDDP-trained technician. Initial interpretation was made mostly by technicians; some projects used technicians plus radiologists and a few used radiologists and/or physicians with other specialties for all readings. Quality control relied on review of procedures and interpretations by the project physicians. Positive physical exams and mammographies were reported in various degrees of certainty about malignancy or as suspicious-benign; IR imaging was reported simply as normal or abnormal. While the protocol for the BCDDP required that the three clinical components of this study (physical examination, IR imaging, and mammography) be conducted separately, and initial findings and recommendations be reported independently, it was not possible for the Working Group to assess the extent to which this protocol was adhered to or to evaluate the quality of the examinations.

The detailed extensive results from this Working Group report consisted of over 50 tables. There was, however, only one table that referred to IR imaging, showing that it had detected 41% of the breast cancers during the first screening, while the residuals were either normal or unknown. There is no breakdown as far as these two latter groups were concerned. Since 28% of the first screening and 32% of the second screening were picked up by mammography alone, IR imaging was dropped from any further evaluation and consideration. The report stated that it was impossible to determine whether abnormal IR imaging could be predictive of interval (developing between screenings) cancers, since these data were not collected.

By the same token, the Working Group was unable to conclude, with their limited experience, whether the findings were related to the then existing technology of IR imaging or with its application. They did, however, indicate that the decision to dismiss IR imaging should not be taken as a determination of the future of this technique, rather that the procedure continued to be of interest because it does not entail the risk of radiation exposure. In the Working Group's final recommendation, they state that "infrared imaging does not appear to be suitable as a substitute for mammography for routine screening in the BCDDP" but could not comment on its role as a complementary modality. The report admitted that several individual programs of the BCDDP had results that were more favorable than for the BCDDP as a whole. They also insisted that high priority be given to development and testing of IR imaging under carefully controlled study conditions. They noted that a few suitable sites appeared to be available among the BCDDP and proposed that developmental studies should be solicited from the sites with sufficient experience.

Further insight into the inadequate quality control assigned to IR imaging during this program was provided by Haberman, a participant in that project [12]. The author reiterated that, while proven expertise in mammography was an absolute requirement for the awarding of a contract to establish a Screening Center, the situation was just the opposite as regards IR imaging. As no experience was required, when the 27 demonstration projects opened their doors, only five of the centers had preexisting expertise in IR imaging. Of the remaining screening centers, there was no experience at all in this technology. Finally, more than 18 months after the BCDDP project had begun operating, the NCI, recognizing this problem, established centers where radiologists and their technicians could obtain further training in IR imaging. Unfortunately, only 11 of the demonstration project directors considered this training of sufficient importance to send their technologists. In some centers, it was reported that there was no effort to cool the patient prior to examination. In other centers, there was complete lack of standardization, and a casual attitude prevailed with reference to interpretation of results. While quality control of this imaging technology could be considered lacking, it was nevertheless subjected to the same stringent statistical analysis as was mammography and clinical breast examination.

11.1.3 Post-Breast Cancer Detection Demonstration Projects Era

Two small-scale studies carried out in the 1970s by Moskowitz [13] and Threatt [14] reported on the sensitivity and reliability of IR imaging. Both used "experts" to review the images of breast cancer patients. While Moskowitz excluded unreadable images, data from Threatt's study indicated that less than 30% of the images produced were considered good, with the rest being substandard. Both these studies produced poor results, inconsistent with numerous previous multicenter trials, particularly that of Stark [15] who, 16 years earlier, reported an 81% detection rate for preclinical cancers.

Threatt noted that IR imaging demonstrated an increasing accuracy as cancers increased in size or aggressiveness, as did the other testing modalities (i.e., physical examination and mammography). The author also suggested that computerized pattern recognition would help solve the reproducibility problems sometimes associated with this technology and that further investigation was warranted. Moskowitz also suggested that for IR imaging to be truly effective as a screening tool, there needed to be more objective means of interpretation. He proposed that this would be much facilitated by computerized evaluation. In a frequently cited article, Sterns and Cardis [16] reviewed their group's limited experience with IR in 1982. While there were only 11 cancer cases in this trial, they were concerned about a sensitivity of 60% and a false-positive rate of IR of 12%. While there was no mention of training, they concluded, based on a surprisingly low false-positive rate of clinical exam of only 1.4%, that IR could not be recommended for breast diagnosis or as an indication for mammography.

Thirteen years later, reviewing the then status of breast imaging, Moskowitz [17] challenged the findings of the recent Canadian National Breast Screening Study (NBSS) that questioned the value of mammography, much in the same way that the Working Group of the BCDDP questioned IR imaging 20 years previously. Using arguments that could have qualified the disappointing results of the IR imaging used in the BCDDP study, the author explained the poor results of mammography in the NBSS on the basis of inadequate technical quality. He concluded that only 68% of the women received satisfactory breast imaging.

In addition to the usual causes of poor technical quality, failure to use the medial lateral oblique view resulted in exclusion of the tail of Spence and of much of the upper outer quadrant in many of the subjects screened. There was also a low interobserver agreement in the reading of mammographies, which resulted in a potential diagnostic delay. His review stated that of all noncontrast, nondigital radiological procedures, mammography required the greatest attention to meticulous detail for the training of technologists, selection of the film, contrast monitoring of processing, choosing of equipment, and positioning of the patient. For mammography to be of value, it required dedicated equipment, a dedicated room, dedicated film, and the need to be performed and interpreted by dedicated people. Echoing some of the criticisms that could be pertinent to the BCDDP's use of IR imaging, he indicated that mammography is not a procedure to be performed by the untutored. In rejecting any lack of quality control of IR imaging during the BCDDP studies by stating that "most of the investigators in the BCDDP did undergo a period of training," the author once again suggested that the potential of IR imaging would only increase if there was better standardization of technology and better-designed clinical trials.

Despite its initial promise, this challenge was not taken up by the medical community, who systematically lost interest in this technique, primarily due to the nebulous BCDDP experience. Nevertheless, during the 1980s, a number of isolated reports continued to appear, most emphasizing the risk factors associated with abnormal IR imaging. In *Cancer* in 1980, Gautherie and Gros [18] reported their experience with a group of 1245 women who had a mildly abnormal infrared image along with either normal or benign disease by conventional means, including physical exam, mammography, ultrasonography, and fine needle aspiration or biopsy. They noted that within 5 years, more than one-third of this group had histologically confirmed cancers. They concluded that IR imaging is useful not only as a predictor of breast cancer risk but also to identify the more rapidly growing neoplasms.

The following year, Amalric et al. [19], expanded on this concept by reporting that 10–15% of patients undergoing IR imaging will be found to be mildly abnormal when the remainder of the examination is essentially unremarkable. They noted that among these "false-positive" cases, up to 38% will eventually develop breast cancer when followed closely. In 1981, Mariel [20] carried out a study in France on 655 patients and noted an 82% sensitivity. Two years later, Isard [21] discussed the unique characteristics and respective roles of IR imaging and ultrasonography and concluded that, when used in conjunction with mammography in a multi-imaging strategy, their potential advantages included enhanced diagnostic accuracy, reduction of unnecessary surgery, and improved prognostic ability. The author emphasized that neither of these techniques should be used as a sole screening modality for breast cancer in asymptomatic women, but rather as a complementary modality to mammography.

In 1984, Nyirjesy [22] reported in *Obstetrics and Gynecology* a 76% sensitivity for IR imaging of 8767 patients. The same year, Bothmann [23] reported a sensitivity of 68% from a study carried out in Germany on 2702 patients. In 1986, Useki [24] published the results of a Japanese study indicating an 88% sensitivity.

Despite newly available IR technology, due in large part to military research and development, as well as compelling statistics of over 70,000 documented cases showing the contribution of functional IR imaging in a hitherto structurally based strategy to detect breast cancer, few North American centers have shown an interest, and fewer still have published their experience. This is surprising in view of the current consensus regarding the importance of vascular-related events associated with tumor initiation and growth that finally provide a plausible explanation for the IR findings associated with the early development of smaller tumors. The questionable results of the BCDDP and a few small-scale studies are still being referred to by a dwindling authorship that even mention the existence of this imaging modality. This has resulted in a generation of imagers that have neither knowledge of nor training in IR imaging. However, there are a few isolated centers that have continued to develop an expertise in this modality and have published their results.

In 1993, Head et al. [25] reported that improved images of the second generation of IR systems allowed more objective and quantitative visual analysis. They also reported that growth-rate-related prognostic indicators were strongly associated with the IR results [26]. In 1996, Gamagami [27] studied angiogenesis by IR imaging and reported that hypervascularity and hyperthermia could be shown in 86% of nonpalpable breast cancers. He also noted that in 15% of these cases, this technique helped to detect cancers that were not visible on mammography.

The concept of angiogenesis, suggested by Gamagami as an integral part of early breast cancer, was reiterated in 1996 by Guidi and Schnitt [28], whose observations suggested that angiogenesis is an early event in the development of breast cancer. They noted that it may occur before tumor cells acquire the ability to invade the surrounding stroma and even before there is morphologic evidence of an *in situ* carcinoma.

In contemporary publications, Anbar [29,30], using an elegant biochemical and immunological cascade, suggested that the empirical observation that small tumors capable of producing notable IR changes could be due to enhanced perfusion over a substantial area of breast surface via tumor-induced nitric oxide vasodilatation. He introduced the importance of dynamic area telethermometry to validate IRs full potential.

Parisky and his colleagues working out of six radiology centers [31] published an interesting report in *Radiology* in 2003 relating to the efficacy of computerized infrared imaging analysis of mammographically suspicious lesions. They reported a 97% sensitivity, a 14% specificity, and a 95% negative predictive value when IR was used to help determine the risk factors relating to mammographically noted abnormal lesions in 875 patients undergoing biopsy. They concluded that infrared imaging offers a safe, noninvasive procedure that would be valuable as an adjunct to mammography, determining whether a lesion is benign or malignant.

11.2 Ville Marie Multidisciplinary Breast Center Experience with High-Resolution Digital Infrared Imaging to Detect and Monitor Breast Cancer

11.2.1 Infrared Imaging as Part of a First-Line Multi-Imaging Strategy to Promote Early Detection of Breast Cancer

There is still a general consensus that the crucial strategy for the first-line detection of breast cancer depends essentially on clinical examination and mammography. Limitation of the former, with its reported sensitivity rate below 65% is well recognized [32], and even the proposed value of breast self-examination is now being contested [33]. With the current emphasis on earlier detection, there is an increasing reliance on better imaging. Mammography is still considered our most reliable and cost-effective imaging modality [17]. However, variable interreader interpretation [34] and tissue density, now proposed as a risk factor itself [35] and seen in both younger patients and those on hormonal replacement

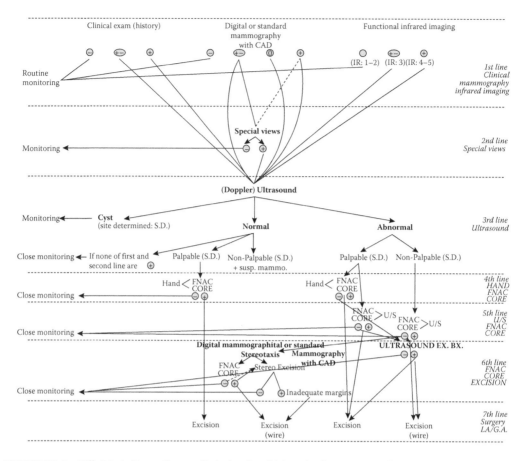

FIGURE 11.1   Ville Marie Breast Center: clinical and multi-imaging breast cancer detection strategy.

[36], prompted us to reassess currently available IR technology spearheaded by military research and development, as a first-line component of a multi-imaging breast cancer detection strategy (Figure 11.1).

This modality is capable of quantifying minute temperature variations and qualifying abnormal vascular patterns, probably associated with regional angiogenesis, neovascularization (Figure 11.2), and nitric oxide-induced regional vasodilatation [29], frequently associated with tumor initiation and progression, and potentially an early predictor of tumor growth rate [26,28]. We evaluated a new fully integrated high-resolution computerized IR imaging station to complement mammography units. To validate its reported ability to help detect early tumor-related regional metabolic and vascular changes [27], we limited our initial review to a series of 100 successive cases of breast cancer who filled the following three criteria: (a) minimal evaluation included a clinical exam, mammography, and IR imaging; (b) definitive surgical management constituted the preliminary therapeutic modality carried out at one of our affiliated institutions; and (c) the final staging consisted of noninvasive cancer ($n = 4$), Stage I ($n = 42$), or Stage II ($n = 54$) invasive breast cancer.

While 94% of these patients were referred to our Multidisciplinary Breast Center for the first time, 65% from family physicians and 29% from specialists, the remaining 6% had their diagnosis of breast cancer at a follow-up visit. Age at diagnosis ranged from 31 to 84 years, with a mean age of 53. The mean histologic tumor size was 2.5 cm. Lymphatic, vascular, or neural invasion was noted in 18% of patients, and concomitant noninvasive cancer was present, along with the invasive component, in 64%. One-third of the 89 patients had axillary lymph node dissection, one-third had involved nodes, and 38% of the tumors were histologic Grade III.

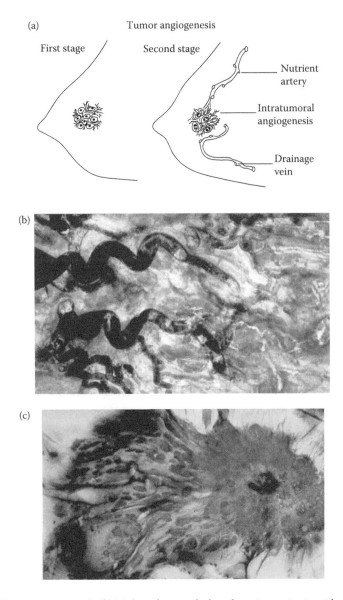

FIGURE 11.2 (a) Tumor angiogenesis. (b) High nuclear-grade ductal carcinoma *in situ* with numerous blood vessels (angiogenesis) (Tabar). (c) Invasive ductal carcinoma with central necrosis and angiogenesis (Tabar).

While most of these patients underwent standard four-view mammography, with additional views when indicated, using a GE DMR apparatus at our center, in 17 cases we relied on recent and adequate quality outside films. Mammograms were interpreted by our examining physician and radiologist, both having access to the clinical findings. Lesions were considered suspicious if either of them noted findings indicative of carcinoma. The remainder were considered either contributory but equivocal or nonspecific. A nonspecific mammography required concordance with our examining physician, radiologist, and the authors.

Our integrated IR station at that time consisted of a scanning-mirror optical system containing a mercury–cadmium–telleride detector (Bales Scientific, CA) with a spatial resolution of 600 optical lines, a central processing unit providing multitasking capabilities, and a high-resolution color monitor

TABLE 11.1   Ville Marie Infrared (IR) Grading Scale

Abnormal signs

1. Significant vascular asymmetry[a]
2. Vascular anarchy consisting of unusual tortuous or serpiginous vessels that form clusters, loops, abnormal arborization, or aberrant patterns
3. A 1°C focal increase in temperature (ΔT) when compared to the contralateral site and when associated with the area of clinical abnormality[a]
4. A 2°C focal ΔT versus the contralateral site[a]
5. A 3°C focal ΔT versus the rest of the ipsilateral breast when not present on the contralateral site[a]
6. Global breast ΔT of 1.5°C versus the contralateral breast[a]

Infrared scale

IR1 = Absence of any vascular pattern to mild vascular symmetry
IR2 = Significant but symmetrical vascular pattern to moderate vascular asymmetry, particularly if stable
IR3 = One abnormal sign
IR4 = Two abnormal signs
IR5 = Three abnormal signs

[a] Unless stable on serial imaging or due to known noncancer causes (e.g., abscess or recent benign surgery).
  Infrared imaging takes place in a draft-free, thermally controlled room maintained between 18°C and 20°C, after a 5 min equilibration period during which the patient is disrobed with her hands locked over her head. Patients are asked to refrain from alcohol, coffee, smoking, exercise, deodorant, and lotions 3 h prior to testing.

capable of displaying 1024×768 resolution points and up to 110 colors or shades of gray per image. IR imaging took place in a draft-free thermally controlled room, maintained at between 18°C and 20°C, after a 5 min equilibration period during which the patient sat disrobed with her hands locked over her head. We requested that the patients refrain from alcohol, coffee, smoking, exercise, deodorant, and lotions 3 h prior to testing.

Four images (an anterior, an undersurface, and two lateral views) were generated simultaneously on the video screen. The examining physician would digitally adjust them to minimize noise and enhance the detection of more subtle abnormalities prior to exact on-screen computerized temperature reading and IR grading. Images were then electronically stored on retrievable laser discs. Our original Ville Marie grading scale relies on pertinent clinical information, comparing IR images of both breasts with previous images. An abnormal IR image required the presence of at least one abnormal sign (Table 11.1). To assess the false-positive rate, we reviewed, using similar criteria, our last 100 consecutive patients who underwent an open breast biopsy that produced a benign histology. We used the Carefile Data Analysis Program to evaluate the detection rate of variable combinations of clinical exam, mammography, and IR imaging.

Of this series, 61% presented with a suspicious palpable abnormality, while the remainder had either an equivocal (34%) or a nonspecific clinical exam (5%). Similarly, mammography was considered suspicious for cancer in 66%, while 19% were contributory but equivocal, and 15% were considered nonspecific. Infrared imaging revealed minor variations (IR-1 or IR-2) in 17% of our patients while the remaining 83% had at least one (34%), two (37%), or three (12%) abnormal IR signs. Of the 39 patients with either a nonspecific or equivocal clinical exam, 31 had at least one abnormal IR sign, with this modality providing pertinent indication of a potential abnormality in 14 of these patients who, in addition, had an equivocal or nonspecific mammography.

Among the 15 patients with a nonspecific mammography, there were 10 patients (mean age of 48; 5 years younger than the full sample) who had an abnormal IR image. This abnormal finding constituted a particularly important indicator in six of these patients who also had only equivocal clinical findings. While 61% of our series presented with a suspicious clinical exam, the additional information provided by the 66 suspicious mammographies resulted in an 83% detection rate. The combination of only suspicious mammograms and abnormal IR imaging increased the sensitivity to 93%, with a further increase to 98% when suspicious clinical exams were also considered (Figure 11.3).

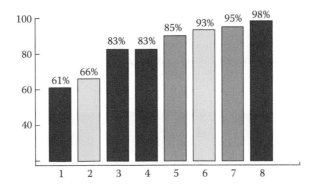

FIGURE 11.3   Relative contribution of clinical exam, mammography, and IR imaging to detect breast cancer in the Ville Marie Breast Center series. In this series, adding infrared imaging to both suspicious and equivocal mammographies increased the detection rate. (1) Suspicious clinical exam; (2) suspicious mammography; (3) suspicious clinical exam or suspicious mammography; (4) abnormal infrared imaging; (5) suspicious or equivocal mammography; (6) abnormal infrared imaging or suspicious mammography; (7) abnormal infrared imaging or equivocal or suspicious mammography; and (8) abnormal infrared imaging or suspicious mammography or suspicious clinical exam.

The mean histologically measured tumor size for those cases undetected by mammography was 1.66 cm while those undetected by IR imaging averaged 1.28 cm. In a concurrent series of 100 consecutive eligible patients who had an open biopsy that produced benign histology, 19% had an abnormal IR image while 30% had an abnormal preoperative mammography that was the only indication for surgery in 16 cases.

The 83% sensitivity of IR imaging in this series is higher than the 70% rate for similar Stage I and II patients tested from the Royal Marsden Hospital two decades earlier [6]. Although our results might reflect an increased index of suspicion associated with a referred population, this factor should apply equally to both clinical exam and mammography, maintaining the validity of our evaluation. Additional factors could include our standard protocol, our physicians' prior experience with IR imaging, their involvement in both image production and interpretation, as well as their access to much improved image acquisition and precision (Figure 11.4).

While most previous IR cameras had 8-bit (one part in 256) resolution, current cameras are capable of one part in 4096 resolution, providing enough dynamic range to capture all images with 0.05°C discrimination without the need for range switching. With the advancement of video display and enhanced gray and colors, multiple high-resolution views can be compared simultaneously on the same monitor. Faster computers now allow processing functions such as image subtraction and digital filtering techniques for image enhancement. New algorithms provide soft tissue imaging by characterizing dynamic heat-flow patterns. These and other innovations have made vast improvements in the medical IR technology available today.

The detection rate in a series where half the tumors were under 2 cm, would suggest that tumor-induced thermal patterns detected by currently available IR technology are more dependent on early vascular and metabolic changes. These changes possibly are induced by regional nitric oxide diffusion and ferritin interaction, rather than strictly on tumor size [29]. This hypothesis agrees with the concept that angiogenesis may precede any morphological changes [28]. Although both initial clinical exam and mammography are crucial in signaling the need for further investigation, equivocal and nonspecific findings can still result in a combined delayed detection rate of 10% [17].

When eliminating the dubious contribution of our 34 equivocal clinical exams and 19 equivocal mammograms, which is disconcerting to both physician and patient, the alternative information provided by IR imaging increased the index of concern of the remaining suspicious mammograms by 27% and the combination of suspicious clinical exams or suspicious mammograms by 15% (Figure 11.3).

An imaging-only strategy, consisting of both suspicious and equivocal mammography and abnormal IR imaging, also detected 95% of these tumors, even without the input of the clinical exam. Infrared imaging's most tangible contribution in this series was to signal an abnormality in a younger cohort of breast cancer patients who had noncontributory mammograms and also nonspecific clinical exams who conceivably would not have been passed on for second-line evaluation.

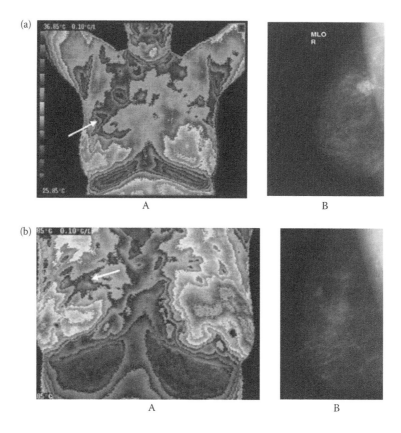

FIGURE 11.4 (a) A 46-year-old patient: Lump in the upper central area of the right breast. _Infrared imaging_ (A): significant vascular asymmetry (SVA) in the upper central area of the right breast (IR-3). _Mammography_ (B): corresponding speculated opacity. _Surgical histology_: 2 cm infiltrating ductal carcinoma with negative sentinel nodes. (b) A 44-year-old patient. _Infrared imaging_ (A): revealed a significant vascular asymmetry in the upper central and inner quadrants of the right breast with a ΔT of 1.8°C (IR-4.8). _Corresponding mammography_ (B): reveals a spiculated lesion in the upper central portion of the right breast. _Surgical histology_: 0.7 cm infiltrating ductal carcinoma. Patient underwent adjuvant brachytherapy. (c) A 52-year-old patient presented with a mild fullness in the lower outer quadrant of the right breast. _Infrared imaging_ (A): left breast (B): right breast reveals extensive significant vascular asymmetry with a ΔT of 1.35°C (IR-5.3). A 2 cm cancer was found in the lower outer area of the right breast. (d) A 37-year-old patient. _Infrared image_ (A): significant vascular asymmetry in the upper inner quadrant of the left breast with a ΔT of 1.75°C (IR-4) (mod IR-4.75). _Mammography_ (B): corresponding spiculated lesion. _Ultrasound_ (C): 6 mm lesion. _Surgical histology_: 0.7 cm infiltrating ductal carcinoma. (e) An 82-year-old patient. _Infrared imaging anterior view_: significant vascular asymmetry and a ΔT of 1.90°C (IR-4) in the left subareolar area. _Corresponding mammography_: (not included) asymmetrical density in the left areolar area. _Surgical histology_: left subareolar 1 cm infiltrating ductal carcinoma. (f) A 34-year-old patient with a palpable fullness in the supra-areolar area of the right breast. _Infrared imaging_ (A): extensive significant vascular asymmetry and a ΔT of 1.3°C in the right supra-areolar area (IR-4) (mod IR-5.3). _Mammography_ (B): scattered and clustered central microcalcifications. _Surgical histology_: after total right mastectomy and TRAM flap: multifocal DCIS and infiltrating ductal CA centered over a 3 cm area of the supra-areolar area.

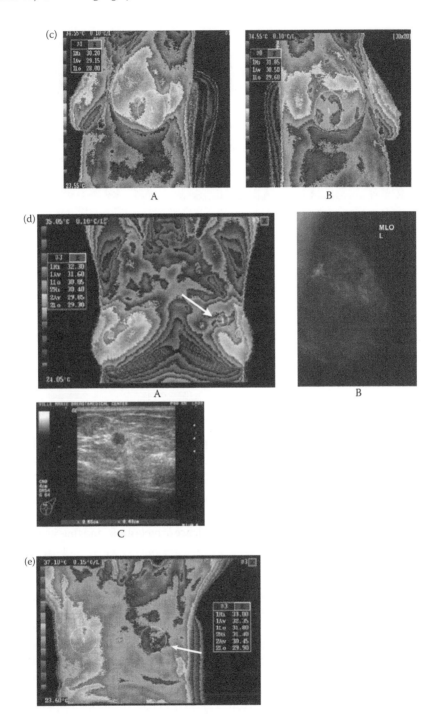

FIGURE 11.4 (Continued).

While 17% of these tumors were undetected by IR imaging, due to either insufficient production or detection of metabolic or vascular changes, the 19% false-positive rate in histologically proven benign conditions, in part a reflection of our current grading system, suggests sufficient specificity for this modality to be used in an adjuvant setting.

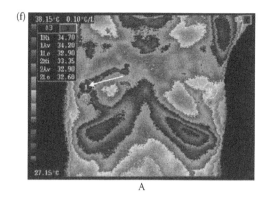

A   B

FIGURE 11.4   (Continued).

Our validation of prior data would also suggest that IR imaging, based more on process than structural changes and not requiring contact, compression, radiation, or venous access, can provide pertinent and practical complementary information to both clinical exam and mammography, our current first-line detection modalities. Quality-controlled abnormal IR imaging heightened our index of suspicion in cases where clinical or mammographic findings were equivocal or nonspecific, thus signaling further investigation rather than observation or close monitoring (Figure 11.5) and to minimize the possibility of further delay (Figure 11.6).

11.2.2 Preliminary Evaluation of Digital High-Resolution Functional Infrared Imaging to Monitor Preoperative Chemohormonotherapy-Induced Changes in Neoangiogenesis in Patients with Advanced Breast Cancer

Approximately 10% of our current breast cancer patients present with sufficient tumor load to be classified as having locally advanced breast cancer (LABC). This heterogeneous subset of patients, usually diagnosed with either stage T3 or T4 lesions without evidence of metastasis, and thus judged as potential surgical candidates, constitutes a formidable therapeutic challenge. Preoperative or neoadjuvant chemotherapy (PCT), hormonotherapy, or both, preferably delivered within a clinical trial, is the current favored treatment strategy.

Preoperative chemotherapy offers a number of advantages, including ensuring improved drug access to the primary tumor site by avoiding surgical scarring, the possibility of complete or partial tumor reduction that could downsize the extent of surgery and also the ability to better plan breast reconstruction when the initial tumor load suggests the need for a possible total mastectomy. In addition, there is sufficient data to suggest that the absence of any residual tumor cells in the surgical pathology specimen following PTC confers the best prognosis, while those patients achieving at least a partial clinical response as measured by at least a 50% reduction in the tumor's largest diameter often can aspire to a better survival than nonresponders [37]. While current clinical parameters do not always reflect actual PCT-induced tumor changes, there is sufficient correlation to make measuring the early clinical response to PTC an important element in assessing the efficacy of any chosen regimen. Unfortunately, the currently available conventional monitoring tools, such as palpation combined with structural/anatomic imaging such as mammography and ultrasound, have a limited ability to precisely measure the initial tumor load, and even less to reflect the effect of PCT [34].

These relatively rudimentary tools are dependent on often-imprecise anatomical and structural parameters. A more effective selection of often quite aggressive therapeutic agents and ideal duration of

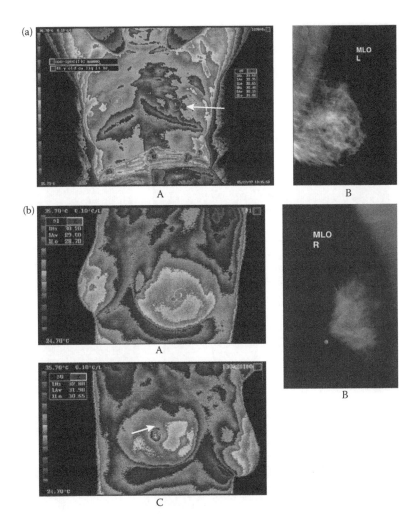

FIGURE 11.5 (a) A 48-year-old patient. *Infrared imaging* (A): significant vascular asymmetry and a Δ*T* of 0.8°C (IR-3) in the lower inner quadrant of the left breast. *Corresponding mammography* (B): a nonspecific density. *Surgical histology*: 1.6 cm left lower inner quadrant infiltrating ductal carcinoma. (b) A 40-year-old patient. *Infrared imaging* (A): left breast, (B) right breast: focal hot spot in the right subareolar area on a background of increased vascular activity with a Δ*T* of 1.1°C (IR-4). *Corresponding mammography* (C): reveals dense tissue bilaterally. *Surgical histology*: reveals a 1 cm right infiltrating ductal carcinoma and positive lymph nodes. (c) A 51-year-old patient. *Infrared imaging* (A): significant vascular asymmetry and a Δ*T* of 2.2°C (IR-5) in the upper central area of the left breast. *Corresponding mammography* (B): mild scattered densities. *Surgical histology*: 2.5 cm infiltrating ductal carcinoma in the upper central area of the left breast. (d) A 44-year-old patient. *Infrared imaging* (A): significant vascular asymmetry and a Δ*T* of 1.58°C (IR-4) in the upper inner quadrant of the left breast. *Corresponding mammography* (B): a nonspecific corresponding density. *Surgical histology*: a 0.9 cm left upper inner quadrant infiltrating ductal carcinoma. (e) A 45-year-old patient with a nodule in central area of left breast. *Infrared imaging* (A): extensive significant vascular asymmetry (SVA) in the central inner area of the left breast with a Δ*T* of 0.75°C (IR-3) (mod IR-4.75). *Mammography* (B): noncontributory. *Surgical histology*: 1.5 cm infiltrating ductal carcinoma with necrosis in the central inner area and 3+ axillary nodes. (f) A 51-year-old patient. *Infrared imaging* (A): extensive significant vascular asymmetry and a Δ*T* of 2.2° (IR-5) (mod IR-6.2) in the upper central area of the left breast. *Corresponding mammography* (B): scattered densities. *Surgical histology*: 2.5 cm infiltrating ductal carcinoma in the upper central area of the left breast. (g) A 74-year-old patient. *Infrared imaging* (A): significant vascular asymmetry in the upper central portion of the right breast with a Δ*T* of 2.8°C (IR-5) (Mod VM IR: 6.8). *Corresponding mammography* (B): bilateral extensive density. *Surgical histology*: 1 cm right central quadrant infiltrating ductal carcinoma.

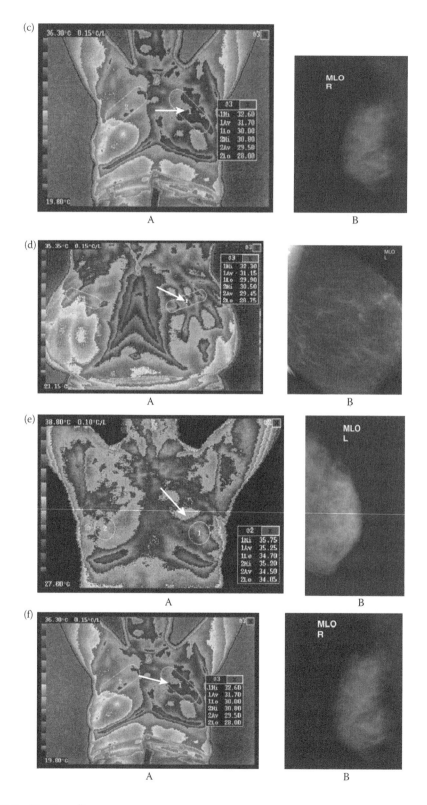

FIGURE 11.5   (Continued).

(g)

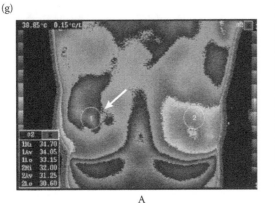
A

B

FIGURE 11.5   (Continued).

their use could be enhanced by the introduction of a convenient, safe, and accessible modality that could provide an alternative and serial measurement of their therapeutic efficacy.

There is thus currently a flurry of interest to assess the potential of different functional imaging modalities that could possibly monitor tumor changes looking at metabolic and vascular features to fill the void. Detecting early metabolic changes associated early tumor initiation and growth using positron emission tomography [38,39], MRI, and Sestamibi scanning are all potential candidates to help monitor PCT-related effects. Unfortunately, they are all hampered by limited access for serial use, duration of the exam, costs, and the need of intravenous access. High-resolution digital infrared imaging, on the other hand, a convenient functional imaging modality free of these inconveniences and not requiring radiation, nuclear material, contact, or compression, can be used repeatedly without safety issues. There are ample data indicating its ability to effectively and reliably detect, in a multi-imaging strategy, neoangiogenesis related to early tumor growth [40]. The premise of our review is that this same phenomenon should even be more obvious when using IR as a monitoring tool in patients with tumors associated with extensive vascular activity as seen in LABC.

To evaluate the ability of our high-resolution digital IR imaging station and a modified Ville Marie scoring system to monitor the functional impact of PCT, 20 successive patients with LABC underwent prospective IR imaging, both prior to and after completion of PCT, usually lasting between 3 and 6 months, which was then followed by curative-intent surgery [41]. Ages ranged between 32 and 82 with a mean of 55. Half of the patients were under 50. Patients presented with T2, T3, or inflammatory carcinoma were all free of any distant disease, thus remaining post-PCT surgical candidates. IR was done at the initial clinical evaluation and prior to core biopsy, often ultrasound guided to ensure optimal specimen harvesting, which was used to document invasive carcinoma. Both sets of IR images were acquired according to our published protocol [38] using the same integrated infrared station described in our previous section.

We used a modification of the original Ville Marie IR scale (Table 11.2) where IR-1 reflects the absence of any significant vascular pattern to minimal vascular symmetry; IR-2 encompasses symmetrical to moderately asymmetrical vascular patterns, including focal clinically related significant vascular asymmetry (SVA); IR-3 implies a regional SVA while an extensive SVA, occupying more than a third of the involved breast, constitutes an IR-4. Mean temperature difference (ΔT) in degrees centigrade between the area of focal, regional, or extensive SVA and the corresponding area of the noninvolved breast is then added, resulting in the final IR score.

Conventional clinical response to PCT was done by measuring the maximum tumor diameter in centimeters, both before beginning and after completion of PCT.

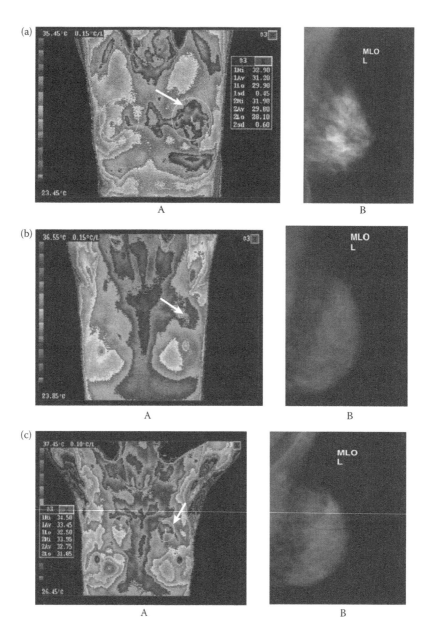

FIGURE 11.6   (a) A 66-year-old patient. Initially seen 2 years prior to diagnosis of left breast cancer for probable fibrocystic disorder (FCD) and scattered cysts, mostly on the left side. *Initial infrared imaging* (A): extensive significant vascular asymmetry left breast and a near global ΔT of 1.4°C (IR-4) (Mod IR: 5.4). *Initial mammography* (B): scattered opacities and ultrasound and cytologies were consistent with FCD. Two years later a 1.7 cm cancer was found in the central portion of the left breast. (b) A 49-year-old patient. *Infrared imaging* 1997 (A): significant vascular asymmetry in the upper central aspect of the left breast (IR-3) *Corresponding mammography* 1997 (B): dense tissue (contd. 6B2). (c) Patient presents 2 years later with a lump in the upper central portion of the left breast. *Corresponding infrared imaging* (A): still reveals significant vascular asymmetry and a ΔT of 0.7°C (IR-3) (Mod VM IR: 3.7). *Corresponding mammography* (B): now reveals an opacity in the upper aspect of the left breast. *Surgical histology*: 1 cm infiltrating ductal carcinoma.

TABLE 11.2   Modified Ville Marie Infrared Scoring Scale

IR-1	Absence of any vascular pattern to mild vascular symmetry
IR-2	Symmetrical vascular pattern to moderate vascular asymmetry, particularly if stable or due to known noncancer causes (e.g., infection, abscess, recent or prior surgery, or anatomical asymmetry). Local clinically related vascular assymetry
IR-3	Regional significant vascular asymmetry (SVA)
IR-4	Extensive SVA, involving at least one-third of the breast area

Add the temperature difference (ΔT) in degrees centigrade between the involved area and the corresponding area of the noninvolved contralateral breast to calculate the final IR score.

Induction PCT in 10 patients, 8 on a clinical trial (NSABP B-27 or B-57) consisted of four cycles of adriamycin (A) 60 mg/m^2 and cyclophosphamide (C) 600 mg/m^2, with or without additional taxotere (T) 100 mg/m^2, or six cycles of AT every 21 days. Eight other patients received AC with additional 5 FU (1000 mg/m^2) and methotrexate (300 mg/m^2). Tamoxifen, given to select patients along with the chemotherapy, was used as sole induction therapy in two elderly patients.

Values in both clinical size and IR scoring both before and after chemotherapy were compared using a paired *t*-test.

11.2.3 Results

All 20 patients in this series with LABC presented with an abnormal IR image (IR $\geq$ 3). The preinduction PCT mean IR score was 5.5 (range: 4.4–6.9). Infrared imaging revealed a decrease in the IR score in all 20 patients following PCT, ranging from 1 to 5 with a mean of 3.1 ($p < .05$). This decrease following PCT reflected the change in the clinical maximum tumor dimension, which decreased from a mean of 5.2 cm prior to PCT to 2.2 cm ($p < .05$) following PCT in the two-thirds of our series who presented with a measurable tumor. Four of the complete pathological responders in this series saw their IR score revert to normal (<3) following PCT (Figure 11.7) while a fifth had a post-PCT IR score of 3.6. An additional seven patients had a final post-PCT IR score that reflected the final tumor size as measured at surgery (Figure 11.8).

LABC is considered an aggressive process that is typically associated with extensive neoangiogenesis required to sustain rapid and continued tumor growth [27]. Functional IR imaging provided a vivid real-time visual reflection of this invasive process in all our patients. The dramatic IR findings associated with LABC, often occupying more than a third of the breast, are further emphasized by the comparative absence of any significant vascular findings in the uninvolved breast. These images thus provided a new parameter and baseline to complement the traditional structurally based imaging, particularly for the seven patients with clinically nonmeasurable LABC.

The significant reduction in the mean IR score following PCT is primarily an indication of its effect on neoangiogenesis. While this reduction can sometimes correspond to tumor size, as it did in half of this series, IR's main contribution concerns functional parameters that can both precede and linger after structural tumor-induced changes occur. Because IR-detected regional angiogenesis responded slightly slower to PCT than did the anatomical parameters in nine patients, it underscores the fundamental difference between functional imaging such as IR and structural-dependent parameters such as clinical tumor dimensions currently used to assess PCT response. IR has the advantage of not being dependent on a minimal tumor size but rather on the tumor's very early necessity to develop an extensive network to survive and proliferate. This would be the basis of IR's ability to sometimes detect tumor growth earlier than can structurally based modalities. The slight discrepancy between the resolving IR score and the anatomical findings in nine of our patients could suggest that this extensive vascular network, most evident in LABC, requires more time to dismantle in some patients. It could reflect the variable volume of angiogenesis, the inability of PCT to affect

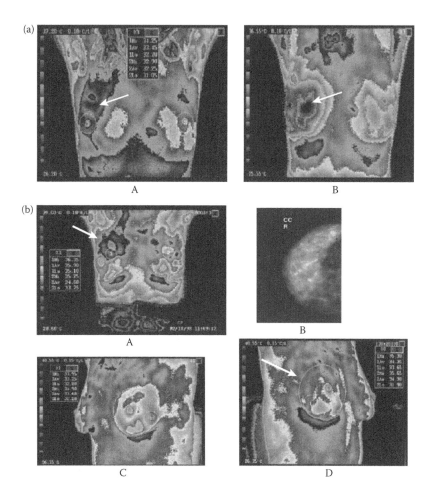

FIGURE 11.7 (a) A 52-year-old patient. *Infrared imaging* (A): extensive vascular asymmetry (SVA) in the right breast with a ΔT of 1.2°C (IR-5.2). The patient was started on preoperative chemotherapy (PCT). *Post-PCT infrared imaging* (B): resolution of previously noted SVA (IR-2). *Surgical histology*: no residual carcinoma in the resected right breast specimen. (b) A 32-year-old patient. *Infrared imaging* (A): extensive significant vascular asymmetry and tortuous vascular pattern and a ΔT of 1.3°C (IR-5.3) in the right breast. *Corresponding mammography* (B): scattered densities. Patient was started on preoperative chemotherapy (PCT). *Post-PCT infrared image* (C—left breast; D—right breast): Notable resolution of SVA, with a whole breast ΔT of 0.7°C (IR-2). *Surgical histology*: no residual right carcinoma and chemotherapy induced changes. (c) A 47-year-old patient. *Infrared imaging* (A): extensive significant vascular asymmetry in the inner half of the left breast with a ΔT of 2°C (IR-6.0). The patient was started on preoperative chemotherapy (PCT). *Post-PCT infrared imaging* (B): no residual asymmetry (IR-1). *Surgical histology*: no viable residual carcinoma, with chemotherapy-induced dense fibrosis surrounding nonviable tumor cells. (d) A 56-year-old patient. *Prechemotherapy infrared imaging* (A): significant vascular asymmetry overlying the central portion of the left breast with a ΔT of 1.35°C (IR-4.35). The patient was given preoperative chemotherapy (PCT). *Post-PCT infrared* (B): mild bilateral vascular symmetry (IR-). *Surgical histology*: no residual tumor. (e) A 54-year-old patient. *Infrared image* (A): extensive significant vascular asymmetry right breast with an rT of 2.8°C (IR-6.8). Received preoperative chemotherapy (PCT). *Post-PCT infrared* (B): mild local residual vascular asymmetry with an rT of 1.65°C (IR-3.65). *Surgical pathology*: no residual viable carcinoma.

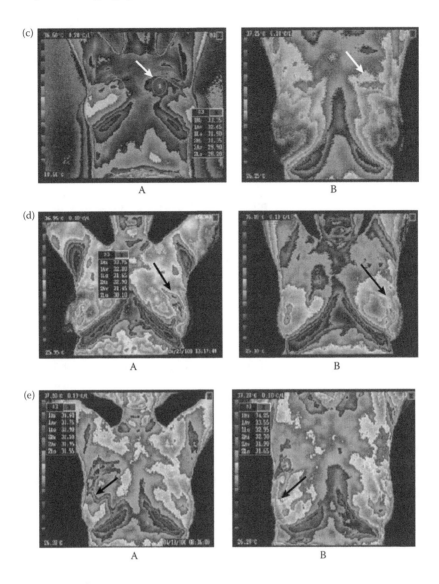

FIGURE 11.7 (Continued).

it, and thus possibly constitute a prognostic factor. This feature could also result from a deficiency in our proposed scoring scale.

Further study and follow-up are needed to better evaluate whether the sequential utilization of this practical imaging modality can provide additional pertinent information regarding the efficacy of our current and new therapeutic strategies, particularly in view of the increasing number with antiangiogenesis properties, and whether lingering IR-reflected neoangiogenesis following PCT ultimately reflects on prognosis.

11.3 Future Considerations Concerning Infrared Imaging

Mammography, our current standard first-line imaging modality only reflects an abnormality that could then prompt the alert clinician to intervene rather than to observe. This decision is crucial since

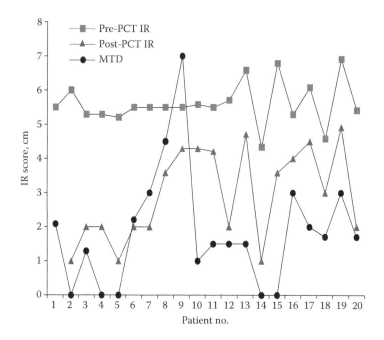

FIGURE 11.8 Pre- and post-PCT IR score and histological maximum tumor dimension (MTD).

it is at this first level that sensitivity and specificity are most vulnerable. There is a clear consensus that we have not yet developed the ideal breast cancer imaging technique and this is reflected in the flurry of new modalities that have recently appeared. While progress in imaging and better training have resulted in the gradual decrease in the average size of breast tumors over the previous decade, the continued search for improved imaging must continue to further reduce the false-negative rate and promote earlier detection.

Digital mammography is already recognized as a major imaging facilitator with the capability to do tomosynthesis and subtraction, along with state-of-the-art 3D and 4D ultrasound. However, there is now new emphasis on developing functional imaging that can exploit early vascular and metabolic changes associated with tumor initiation that often predate morphological changes that most of our current structural imaging modalities still depend on; thus, the enthusiasm in the development of Sestamibi scanning, Doppler ultrasound, and MRI of the breast [42,43]. Unfortunately, as promising as these modalities are, they are often too cumbersome, costly, inaccessible, or require intravenous access to be used as first-line detection modalities alongside clinical exam and mammography.

On the other hand, integrating IR imaging, a safe and practical modality into the first-line strategy, can increase the sensitivity at this crucial stage by also providing an early warning signal of an abnormality that in some cases is not evident in the other components. Combining IR imaging and mammography in an "IR-Assisted Mammography Strategy" is particularly appealing in the current era of increased emphasis on detection by imaging with less reliance on palpation as tumor size further decreases.

Intercenter standardization of a protocol concerning patient preparation, temperature-controlled environment, digital image production, enhanced grading, and archiving, as well as data collection and sharing are all important factors that are beginning to be addressed. New technology could permit real-time image acquisition that could be submitted to computerized assisted image reading which will further enhance the physician's ability to detect subtle abnormalities.

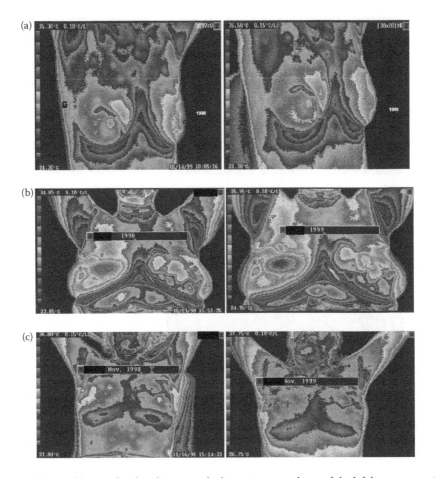

FIGURE 11.9 (a) A stable vascular distribution in the lower inner quadrant of the left breast over a 12-month period. *Infrared imaging* of the breast usually remains remarkably stable in the absence of any ongoing developing significant pathology. It is important in grading these images to determine if the findings are to be considered evolving and important or stable and possibly relating to the patient's vascular anatomy. (b) An identical vascular distribution of the left breast over a 12-month period. (c) A stable vascular distribution in both breasts over a 12-month period.

Physician training is an essential quality control component for this imaging modality to contribute its full potential. A thorough knowledge of all aspects of benign and malignant breast pathology and concomitant access to prior IR imaging that should normally remain stable (Figure 11.9) are all contributory features. This modality needs to benefit from the same quality control previously applied to mammography and should not, at this time, be used as stand-alone. It should benefit from the same interaction between clinical knowledge and other imaging tools as is the case in current breast cancer detection in the clinical environment. This is especially important since there are no current IR regulations, as it poses no health threat and does not use radiation, and could thus fall victim to untrained personnel who could misuse it on unsuspecting patients as was previously the case.

Its future promise, however, resides primarily in its ability to qualify and quantify vascular and metabolic changes related to early tumor genesis. The proposals that a higher temperature difference (ΔT) and increased vascular asymmetry are prognostic factors of tumor aggressivity need to be validated

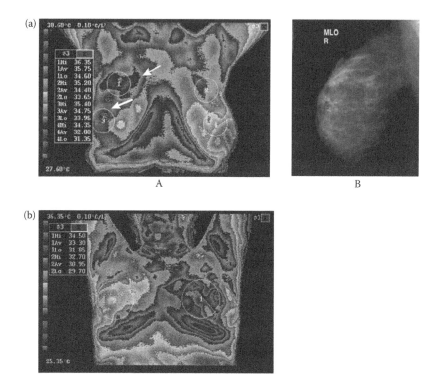

FIGURE 11.10 (a) A 50-year-old patient. *Infrared imaging* (A): significant vascular asymmetry and a Δ*T* of 3.4°C (IR-5) in the right breast. *Corresponding mammography* (B): increased density and microcalcifications in the right breast. Surgical histology: extensive right multifocal infiltrating ductal carcinoma requiring a total mastectomy. (b) A 50-year-old patient. *Infrared imaging*: significant vascular asymmetry with a near global Δ*T* of 2.35°C in the left breast (IR-5). *Mammography* (not available) diffuse density. *Surgical histology*: left multifocal infiltrating ductal carcinoma requiring a total mastectomy.

by further research (Figure 11.10). The same applies to the probability that the reduction of IR changes seen with preoperative chemohormonotherapy reflect reduction in neoangiogenesis and thus treatment efficacy. These remain, at the very least, extremely interesting and promising areas for future research, particularly in view of the current interest in new angiogenesis-related therapeutic strategies. Its contribution to monitoring postoperative patients (Figure 11.11) which is problematic with both mammography and ultrasound [44] and its ability to recognize recurrent cancer (Figure 11.12) are other areas for further clinical trials. Adequate physician training and strict attention to image acquisition are essential to avoid false-negative interpretation (Figure 11.13).

As is the case for all current imaging modalities, the fact that this modality does not detect all tumors should not detract from its contribution as a functional adjuvant addition to our current first-line imaging strategy that is still based essentially on mammography alone, a structural modality that has reached its full potential. There are already sufficient data regarding IR's sensitivity and specificity that has been more recently validated and ongoing data collection will be important to justify its continued use in the ever-evolving breast imaging field. Infrared imaging may become an alternative to a more costly and invasive imaging modality, and possibly integrated as an informative first-line tool. In the meantime, to ignore its contribution to the very complex process of promoting earlier breast cancer detection could be questioned. A good first-line imaging modality must be safe, convenient, and able to help detect primarily the more aggressive tumors where early intervention can have a greater impact on survival.

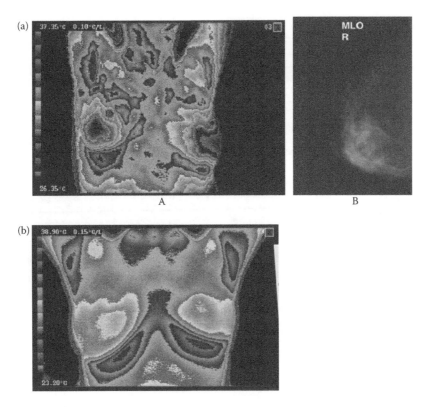

FIGURE 11.11 (a) Four years post-right partial mastectomy, the patient is recurrence free. Current *infrared imaging* (A): shows no activity and *mammography* (B): shows scar tissue. (b) *Infrared imaging*: 5 years following a left partial mastectomy and radiotherapy for carcinoma revealing slight asymmetry of volume, but no abnormal vascular abnormality and resolution of radiation-induced changes. The patient is disease free.

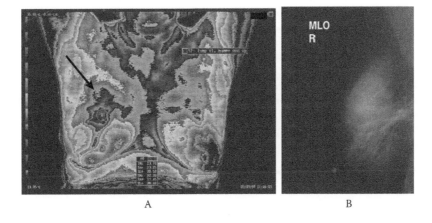

FIGURE 11.12 A 52-year-old patient, 5 years following right partial mastectomy, radiation, and chemotherapy for right breast cancer. Recent follow-up *infrared imaging* (A) now reveals significant vascular asymmetry (SVA) in the right breast with a ΔT of 1.5°C in the area of the previous surgery (IR-4). *Corresponding mammography* (B): density and scar tissue versus possible recurrence. *Surgical histology*: 4 cm, recurrent right carcinoma in the area of the previous resection.

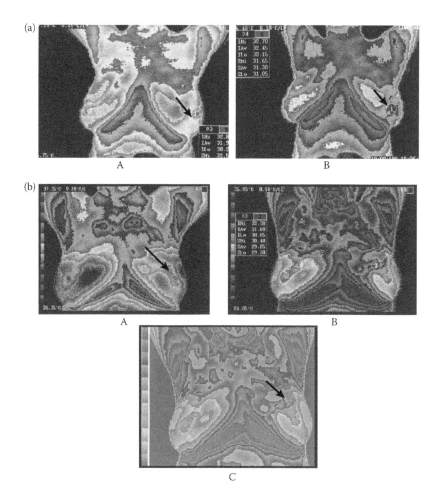

FIGURE 11.13 (a) A 45-year-old patient with small nodular process just inferior and medial to the left nipple areolar complex. *Infrared imaging* (A), without digital enhancement, was carried out and circles were placed on the nipple area rather than just below and medial to it. The nonenhanced IR image was thus initially misinterpreted as normal with a ΔT of 0.25°C. The same image was recalled and *repeated* (B), now with appropriate digital enhancement and once again, documented the presence of increased vascular activity just inferior and medial to the left nipple areolar complex with a ΔT of 1.15°C (IR-4.15). A 1.5 cm cancer was found just below the left nipple. (b) A 37-year-old patient with lump in the upper inner quadrant of the left breast. Mammography confirms spiculated mass. *Infrared imaging* (A): reported as "normal." Infrared imaging was repeated after appropriate *digital adjustment* (B): now reveals obvious significant vascular asymmetry in the same area and a ΔT of 1.75°C (Ir-4) (mod IR-4.75). Using a different IR camera on the *same patient* (C) confirms same finding. *Surgical histology*: 1.5 cm infiltrating ductal CA.

References

1. Lloyd-Williams K. and Handley R.S. Infra-red thermometry in the diagnosis of breast disease. *Lancet* 2, 1371–1381, 1961.
2. Handley R.S. The temperature of breast tumors as a possible guide to prognosis. *Acta Unio. Int. Contra. Cancrum.* 18, 822, 1962.
3. Lawson R.N. and Chughtai M.S. Breast cancer and body temperatures. *Can. Med. Assoc. J.* 88, 68–70, 1963.

4. Gershen-Cohen J., Haberman J., and Brueschke E.E. Medical thermography: A summary of current status. *Radiol. Clin. North. Am.* 3, 403–431, 1965.

5. Haberman J. The present status of mammary thermography. *Ca—Can. J. Clin.* 18, 314–321, 1968.

6. Jones C.H. Thermography of the female breast. In Parsons C.A. (ed.), *Diagnosis of Breast Disease*, University Park Press, Baltimore, pp. 214–234, 1983.

7. Isard H.J., Becker W. et al. Breast thermography after four years and 10,000 studies. *Am. J. Roentgenol.* 115, 811–821, 1972.

8. Dodd G.D. Thermography in breast cancer diagnosis. In *Proceedings of the 7th National Cancer Conference*. Los Angeles, CA, September 27–29, Lippincott, Philadelphia, Toronto, p. 267, 1972.

9. Feig S.A., Shaber G.S. et al. Thermography, mammography, and clinical examination in breast cancer screening. *Radiology* 122, 123–127, 1977.

10. Wallace J.D. Thermographic examination of the breast: An assessment of its present capabilities. In Gallagher H.S. (ed.), *Early Breast Cancer: Detection and Treatment*. American College of Radiology, Wiley, New York, pp. 13–19, 1975.

11. Report of the Working Group to Review the National Cancer Institute Breast Cancer Detection Demonstration Projects. *J. Natl. Cancer Inst.* 62, 641–709, 1979.

12. Haberman J. An overview of breast thermography in the United States. In Abernathy M. and Uematsu S. (eds), *Medical Thermography*. American Academy of Thermology, Washington, pp. 218–223, 1986.

13. Moskowitz M., Milbrath J. et al. Lack of efficacy of thermography as a screening tool for minimal and stage I breast cancer. *N. Engl. J. Med.* 295, 249–252, 1976.

14. Threatt B., Norbeck J.M. et al. Thermography and breast cancer: An analysis of a blind reading. *Ann. N.Y. Acad. Sci.* 335, 501–519, 1980.

15. Stark A. The use of thermovision in the detection of early breast cancer. *Cancer* 33, 1664–1670, 1964.

16. Sterns E. and Cardis C. Thermography in breast diagnosis. *Cancer* 50, 323–325, 1982.

17. Moskowitz M. Breast imaging. In Donegan W.L. and Spratt J.S. (eds), *Cancer of the Breast*. Saunders, New York, pp. 206–239, 1995.

18. Gautherie M. and Gros C.M. Breast thermography and cancer risk prediction. *Cancer* 45, 51–56, 1980.

19. Amalric R., Gautherie M. et al. Avenir des femmes à thermogramme infra-rouge mammaire anormal isolé. *La. Nouvelle. Presse Médicale* 38, 3153–3155, 1981.

20. Mariel L., Sarrazin D. et al. The value of mammary thermography. A report on 655 cases. *Sem. Hop.* 57, 699–701, 1981.

21. Isard H.J. Other imaging techniques. *Cancer* 53, 658–664, 1984.

22. Nyirjesy I. and Billingsley F.S. Detection of breast carcinoma in a gynecological practice. *Obstet. Gynecol.* 64, 747–751, 1984.

23. Bothmann G.A. and Kubli F. Plate thermography in the assessment of changes in the female breast. 2. Clinical and thermographic results. *Fortschr. Med.* 102, 390–393, 1984.

24. Useki H. Evaluation of the thermographic diagnosis of breast disease: Relation of thermographic findings and pathologic findings of cancer growth. *Nippon. Gan. Chiryo. Gakkai. Shi.* 23, 2687–2695, 1988.

25. Head J.F., Wang F., and Elliott R.L. Breast thermography is a noninvasive prognostic procedure that predicts tumor growth rate in breast cancer patients. *Ann. N.Y. Acad. Sci.* 698, 153–158, 1993.

26. Head J.F. and Elliott R.L. Breast thermography. *Cancer* 79, 186–188, 1995.

27. Gamagami P. Indirect signs of breast cancer: Angiogenesis study. In *Atlas of Mammography*. Blackwell Science, Cambridge, MA, pp. 231–26, 1996.

28. Guidi A.J. and Schnitt S.J. Angiogenesis in preinvasive lesions of the breast. *Breast J.* 2, 364–369, 1996.

29. Anbar M. Hyperthermia of the cancerous breast: Analysis of mechanism. *Cancer Lett.* 84, 23–29, 1994.

30. Anbar M. Breast cancer. In *Quantitative Dynamic Telethermometry in Medical Diagnosis and Management.* CRC Press, Ann Arbor, MI, pp. 84–94, 1994.

31. Parisky H.R., Sard A. et al. Efficacy of computerized infrared imaging analysis to evaluate mammographically suspicious lesions. *Am. J. Radiol.* 180, 263–272, 2003.

32. Sickles E.A. Mammographic features of "early" breast cancer. *Am. J. Roentgenol.* 143, 461, 1984.

33. Thomas D.B., Gao D.L. et al. Randomized trial of breast self-examination in Shanghai: Methodology and preliminary results. *J. Natl. Cancer Inst.* 5, 355–365, 1997.

34. Elmore J.G., Wells C.F. et al. Variability in radiologists interpretation of mammograms. *N. Engl. J. Med.* 331, 99–104, 1993.

35. Boyd N.F., Byng J.W. et al. Quantitative classification of mammographic densities and breast cancer risk. *J. Natl. Cancer Inst.* 87, 670–675, 1995.

36. Laya M.B. Effect on estrogen replacement therapy on the specificity and sensibility of screening mammography. *J. Natl. Cancer Inst.* 88, 643–649, 1996.

37. Singletary S.E., McNeese M.D., and Hortobagyi G.N. Feasibility of breast-conservation surgery after induction chemotherapy for locally advanced breast carcinoma. *Cancer* 69, 2849–2852, 1992.

38. Jansson T., Westlin J.E. et al. Position emission tomography studies in patients with locally advanced and/or metastatic breast cancer: A method for early therapy evaluation? *J. Clin. Oncol.* 13, 1470–1477, 1995.

39. Hendry J. Combined positron emission tomography and computerized tomography. Whole body imaging superior to MRI in most tumor staging. *JAMA* 290, 3199–3206, 2003.

40. Keyserlingk J.R., Ahlgren P.D., Yu E., and Belliveau N. Infrared imaging of the breast: Initial reappraisal using high-resolution digital technology in 100 successive cases of stage I and II breast cancer. *Breast J.* 4, 245–251, 1998.

41. Keyserlingk J.R., Yassa, M., Ahlgren, P., and Belliveau N. Tozzi. Ville Marie Oncology Center and St. Mary's Hospital, Montreal, Canada D. *Preliminary Evaluation of Digital Functional Infrared Imaging to Reflect Preoperative Chemohormonotherapy-Induced Changes in Neoangiogenesis in Patients with Locally Advanced Breast Cancer.* European Oncology Society, Milan, Italy, September 2001.

42. Berg W. Tumor type and breast profile determine value of mammography, ultrasound and MR. *Radiology* 233, 830–849, 2004.

43. Oestreicher N. Breast exam and mammography. *Am. J. Radiol.* 151, 87–96, 2004.

44. Mendelson, Berg et al. Ultrasound in the operated breast. Presented at the *RSNA.* Chicago, November 2005.

12

MammoVision (Infrared Breast Thermography) Compared to X-Ray Mammography and Ultrasonography: 114 Cases Evaluated

Reinhold Berz
German Society of Thermography and Regulation Medicine (DGTR)

Claus Schulte-Uebbing
German Society of Thermography and Regulation Medicine (DGTR)

12.1 Introduction

Breast cancer is the leading deadly cancer in women at least in the developed countries. X-ray mammography (MG) is called the gold standard for detecting breast cancer, but the method has some limitations: For many women MG may be painful, and the ionizing radiation could cause malignancy. More importantly: Cancerous lumps must have a diameter of at least 5 mm (often much more) to be detected by MG. Many breast cancers at this stage are 5–10 years old. Earlier detection is requested, but seems to be not easy to do by the established methods. MRI could be a better way, but there is a lack of experience, and it is costly. Ultrasonography (USG) can be helpful, too, but is recognized as a complementary examination.

12.1.1 Breast Metabolism and Heat Signs

There is a well known relationship between breast cancer and breast heat signs. Especially aggressive and fast growing breast cancers have an exaggerated metabolism causing a high blood supply. Intraoperative studies have demonstrated that breast cancer leads to an increased venous flow and to heat convection [1]. Usually the healthy breast is, depending on its size, colder than the surrounding chest and abdominal areas without thermographically visible signs of vessels or hot spots. Both breasts should have an average temperature differing not more than 0.5°C (thermal symmetry, Figure 12.1).

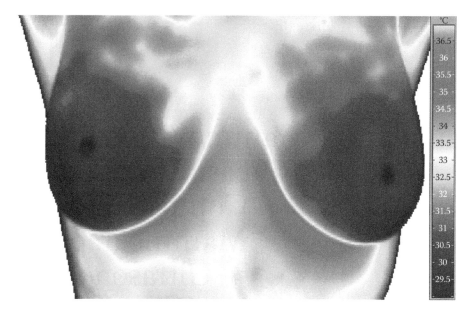

FIGURE 12.1   Normal MammoVision result of a 29-year-old healthy woman: symmetrically cool, no visible vessels, no hot spots, nipple and areola cold, contour without bulge or edge sign.

12.1.2 Infrared Thermography

The heat patterns of the breasts can be recorded with infrared detecting devices (infrared cameras), that have been in use for medical purposes since 1956 [2–5]. Since the late 1990s, due to the technological progress and sensor development infrared imaging devices since the late 1990s are much more suitable for medical measurements [6]. According to PLANCK's law, the infrared radiation of the human skin of 30°C (303 K) peaks at a wavelength of about 10 μm, and so called long wave infrared cameras (LWIR) are suitable to record and measure the heat pattern of the breast (Figure 12.2). Comparable to the US FDA approval, in Europe medical temperature measurement devices have to comply with the Medical Devices Directive, Council Directive 93/42/EEC [7]. This approval regards the whole

FIGURE 12.2   Medically approved infrared camera Jenoptik (formerly Carl Zeiss Jena) VarioCam Head.

equipment: IR camera, cable connections, power supply, computer and other hardware, software and examination rack.

12.2 MammoVision Methodology of Regulation Thermography

There are different approaches to conduct infrared thermography of the female breast. To solve the problem of lacking standardization, in Europe the method of Regulation Thermography, with decades of applications, was chosen. This approach is based on the idea that the response to a stimulus (like in other medical examinations, for example, ECG or EEG) enhances the diagnostic value and acuteness. Regulation Thermography or IRI (Infrared Regulation Imaging) applies a cool air stress exposure to the patient (disrobed at room temperature of 19–21°C) and measures the thermal pattern immediately after disrobing (comfort temperature) and a second time after adapting to the cold ambient (down regulated skin temperature after 10 min) [8–15]. This method has shown to be appropriate for female breast imaging [16]. For other applications different cold stress tests are common [17–20]. The thermal down regulation leads to a skin temperature decrease of about 1.0°C within the breast areas (Figure 12.3).

Very important is the appropriate preparation of the patient, the adaptation to the examination room (patient dressed) of 30 min, the ambient conditions of the examination room (19–21°C, humidity between 30 and 50%, shielding of the patient from sources of heat or cold (Figure 12.4).

12.2.1 MammoVision Equipment

MammoVision requires a complete medically approved temperature measurement system: infrared camera, cable connections, power supply, computer and other hardware, software and examination rack. The IR camera used for this study was a German made Jenoptik (formerly Carl Zeiss Jena) VarioCam

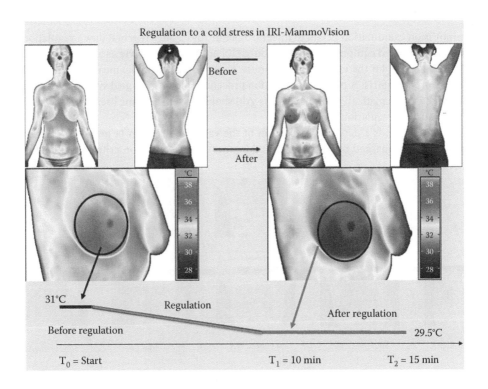

FIGURE 12.3 Principles of IRI (Infrared Regulation Imaging) and MammoVision.

FIGURE 12.4 Early prototype of MammoVision (2000, left) and recent MammoVision device (bioaging Stuttgart, Germany, right).

Head with a spatial resolution of 320×240 sensors, detector pitch 35×35 μm, sensitivity 60 mK and better, accuracy ±0.4°C, stability over time 20 min after onset better than ±0.5°C, lens aperture $f = 1/1.0$, focal length $f = 25$ mm. An additional digital camera (webcam) takes a colour picture in parallel for the documentation of scarves or other visible breast abnormalities.

Core of the MammoVision system is the medically approved software EXAM for administration, examination, measurement and imaging control, data storage and, most importantly, data evaluation, and generating a medical report [21].

12.2.2 MammoVision Evaluation Guideline

The MammoVision examination needs 10 images: 5 before and 5 after thermal down regulation due to the cooling. The patented equipment and procedure [22] ensures that the views are more or less identical, which is important for the evaluation (Figure 12.5). After recording the measurement and images, the evaluation process starts: A patented evaluation procedure including a grid system marks the breast areas that will be mathematically evaluated. This grid should cover the same breast areas of the images taken before and after regulation (Figure 12.6).

The next step demands a detailed description of the vascular situation (a grayscale image is more suitable than a colour masked one, Figure 12.7). MammoVision evaluation criteria include the lateral

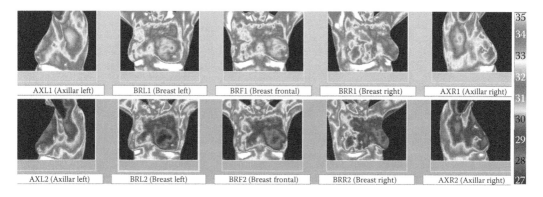

FIGURE 12.5 Measurement views of MammoVision: upper row before cold stimulus, lower row after cold stimulus and cooling down.

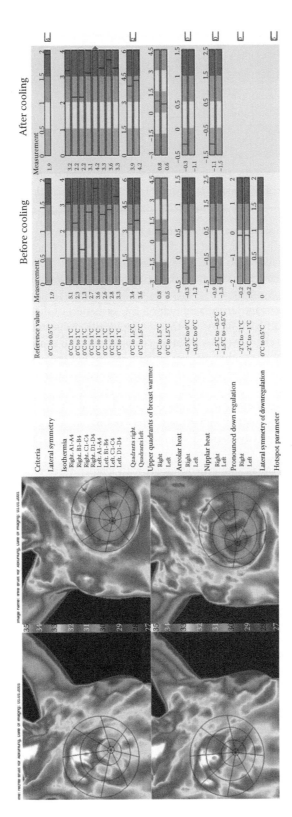

Criteria	Reference value	Before cooling (Measurement)	After cooling (Measurement)
Lateral symmetry	0°C to 0.5°C	1.9	1.9
Isothermia			
Right. A1-A4	0°C to 1°C	3.1	3.2
Right. B1-B4	0°C to 1°C	2.3	2.2
Right. C1-C4	0°C to 1°C	1.3	2.2
Right. D1-D4	0°C to 1°C	2.7	3.1
Left. A1-A4	0°C to 1°C	3.6	4.2
Left. B1-B4	0°C to 1°C	2.6	3.3
Left. C1-C4	0°C to 1°C	2.8	3.6
Left. D1-D4	0°C to 1°C	3.3	3.3
Quadrants right	0°C to 1.5°C	3.4	3.9
Quadrants left	0°C to 1.5°C	3.6	4.2
Upper quadrants of breast warmer			
Right	0°C to 1.5°C	0.8	0.8
Left	0°C to 1.5°C	0.5	0.6
Areolar heat			
Right	-0.5°C to 0°C	-0.5	-0.3
Left	-0.5°C to 0°C	-1.2	-1.1
Nipplar heat			
Right	-1.5°C to -0.5°C	-1.5	-1.1
Left	-1.5°C to -0.5°C	-0.9	-1.5
Pronounced down regulation			
Right	-2°C to -1°C	-1.3	-1.1
Left	-2°C to -1°C	-2	-1.3
Lateral symmetry of downregulation		-0.2	-0.2
Hotspot parameter	0°C to 0.5°C	0	0

FIGURE 12.6 Positioning of the evaluation grid in MammoVision (left): upper row before, lower row after cold stimulus and cooling (woman with hot cancer in the right breast); MammoVision results in an evaluation graph (right).

FIGURE 12.7 Grayscale masked MammoVision image for vascular description (left); MammoVision vascular description form (right).

symmetry of the breasts, the isothermia (homogeneous tempered breast areas with a small span of temperatures), areolar or nipplar heat, the extent of down regulation and the hotspot parameter [23]. The shapes of the breasts and vascular signs have to be assessed.

According to the assessment criteria [24,25] the results have been classified into 5 BIRAS groups (**B**reast **I**nfra**R**ed **A**ssessment **S**ystem):

BIRAS I: Inconspicuous; average lateral symmetry 0.5°C or better; isothermia/homogeneous areas with temperatures in a range of 1.5°C or less; no thermographically visible vascularity or few small vessels with homogeneous vessel characteristics including comparison with contra lateral breast; vascularity thermographically disappearing after cooling; no hot areas or hot spots, nipples and areola colder than the average of the breast, no bulge or edge signs, down regulation after cooling −0.5°C or more;

BIRAS II: Slightly conspicuous; fairly lateral symmetry between 0.5 and 1.0°C; fairly isothermia/homogeneous areas with temperatures in a range between 1.5 and 3.0°C; more impressive, but still physiological vessel characteristics, that can remain after cooling; nipples and areola colder than the average of the breast, no bulge or edge signs, down regulation after cooling −0.5°C or more;

BIRAS III: Conspicuous; often lateral asymmetry of more than 1.0°C; thermal pattern of the breasts more inhomogeneous, isothermia/homogeneous areas with temperatures more than 3.0°C differing; intensive and/or slightly abnormal vascular signs; slight edge or bulk signs; warm areas or warm spots; warm areola; most signs decreasing after cooling;

BIRAS IV: Very conspicuous; obvious lateral asymmetry of more than 1.5°C; pronounced unmodified hot spots and hot areas; very inhomogeneous, isothermia/homogeneous areas with temperatures more than 4.5°C differing; hot spikes; clear edge or bulk signs; areola and/or nipple hot; often one breast unilaterally affected, the other without or with less symptoms and signs; most signs resistive to cooling.

BIRAS V: Significantly conspicuous; like BIRAS IV, impressive lateral asymmetry of more than 2.0°C; cancer-related vascular signs (hot spiders, circular vessels, vascular chaos; very abnormal vascularity); paradoxical heating of spots or areas instead of cooling down regulation.

12.3 Patients Examined

114 women were enrolled in the study at the breast center Prof. Schulte-Uebbing, Munich, who were asked and agreed to participate. All women have had an x-ray mammogram (MG) in a radiological institute within the same time period as the MammoVision had been performed, from January 2006 to June 2007. The reasons for the MG examinations differed: either for screening purposes, or a lump had been detected, or other symptoms were recorded. The age of the patients ranged between 29 and 69 years, with an average of 48.5 years. Based on the MG BI-RADS classification the patients formed 5 groups, because MG is called the gold standard for breast cancer diagnosis. Group BI-RADS I $n = 41$; group BI-RADS II $n = 45$; group BI-RADS III $n = 15$; group BI-RADS IV $n = 8$; group BI-RADS V $n = 5$; totally 114 women.

Most of the patients had additional ultrasonography (USG), some had MRI, in some women biopsies were examined.

12.4 Results

12.4.1 Illustrations of BIRAS Classified Results

The illustrations show women with typical signs and classifications before and after cooling, focused on the symptomatic breast. Figure 12.8 shows a woman with a typical BIRAS I result, completely inconspicuous, symmetrical and homogeneous thermal breast pattern without thermographically visible vascularity.

Figure 12.9 represents a typical BIRAS II finding: still symmetrical, but with a broader range of temperatures within each breast (isothermia/homogeneity slightly affected), nipple cold, areola slightly warm, but cooling down after down regulation, warm vascular pattern above the breast that is sufficiently cooling down.

Figure 12.10 is a good example for a BIRAS III result: especially the very warm periareolar areas on both breasts are suspicious and should give reason for further examinations; even after down regulation, the warm areolar region is visible.

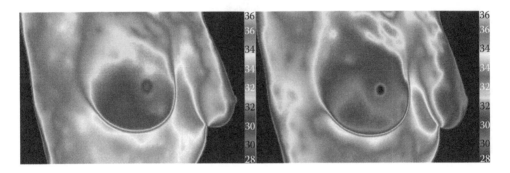

FIGURE 12.8 BIRAS I example; 29-year-old woman, left picture before cold stimulus, right picture after cold stimulus and cooling down; no clinical signs of breast disorder; in MammoVision very homogeneous, symmetrical thermal patterns, no vessels visible.

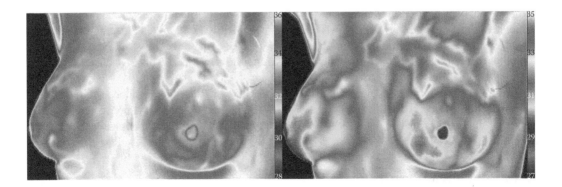

FIGURE 12.9 BIRAS II example; 42-year-old woman, left picture before cold stimulus, right picture after cold stimulus and cooling down; mastopathia, no other breast disease; MammoVision: symmetrical thermal pattern, areola and nipple cool, sufficient thermal down cold stimulus, small vessels that mostly disappear to be visible after cooling.

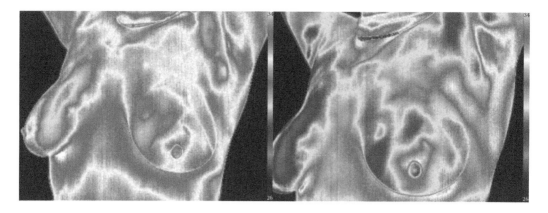

FIGURE 12.10 BIRAS III example; 43-year-old woman, left picture before cold stimulus, right picture after cold stimulus and cooling down; mastopathia, palpable nodules. MammoVision: suspicious result, areolar heat both sides, intensive down cold stimulus.

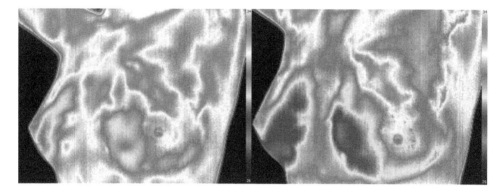

FIGURE 12.11 BIRAS IV example; 55 year old woman, left picture before cold stimulus, right picture after cold stimulus and cooling down; palpable lump in the left breast, questionable palpable axillary lymph nodes; MammoVision: very suspicious, highly asymmetrical, insufficient down cold stimulus, inflammatorious area above left areola; persisting heat even after cooling; biopsy: invasive ductal carcinoma pT2 pN1, G3, ER +, PR +, HER –.

The woman presented in Figure 12.11 shows a highly suspicious result. The thermal symmetry is lost, and the left breast seems clearly affected with heat signs. Impressively the areolar heat increases instead of cooling after thermal down regulation. In the upper quadrants there are classified hot spot/hot areas signs, like an inflammation. This woman had a biopsy, and the result was an invasive ductal carcinoma (pT2 pN1, G3, ER +, PR +, HER –).

Figure 12.12 shows a rare sign of malignancy in the right breast. The whole right breast is over heated compared to the left side; especially the vascular pattern is highly suspicious due to the circular vessel with aspects of a thermal vascular spider. Vascular abnormalities like this are highly conspicuous that there is already breast cancer. In this case it was an invasive ductal carcinoma; pT2 pN1, G3, ER +, PR +, HER +.

12.4.2 Statistical Results

All patients of the BI-RADS IV and V groups (highly suspicious to have breast cancer) have had clearly pathological signs in the MammoVision examination. The same holds true, with one exception, for the women of the BI-RADS III group. But in the BI-RADS I and II groups (not suspicious to

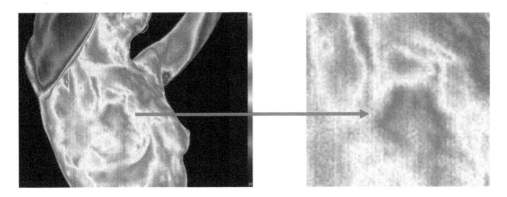

FIGURE 12.12 BIRAS V example; MammoVision very suspicious: extreme asymmetry, right breast heated intensively, absolute pathological vessel (circular structure with vascular spider aspects); biopsy: invasive ductal carcinoma; pT2 pN1, G3, ER +, PR +, HER +.

have breast cancer) there were remarkable differences between the MG classification and many of the MammoVision results. Especially in group BI-RADS II, half of the patients had slightly to very conspicuous thermal signs.

For a more detailed evaluation each BI-RADS group (I to V) was separately compared with the differentiated MammoVision results (BIRAS I to V) as presented in Figure 12.13. It can be stated that in the x-ray MG BI-RADS I group the MammoVision findings mostly are classified as non conspicuous. In 9 of 41 women there were BIRAS II results to be found.

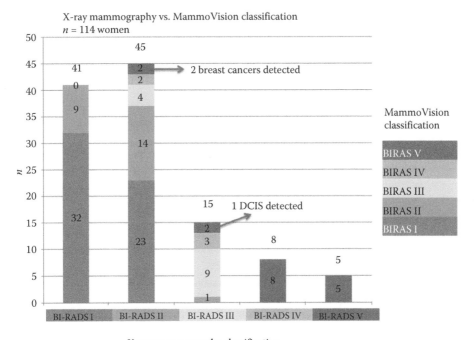

FIGURE 12.13 Comparison of x-ray mammography classification (BI-RADS I to V) versus MammoVision classification (BIRAS I to V).

The MG BI/RADS III group looked very inhomogeneous regarding the MammoVision results. About 50% of the women hat inconspicuous MammoVision findings, 30% slightly suspicious results, but 8 of 45 women showed more conspicuous results, two of them classified as BIRAS IV and two as BIRAS V. They were therefore consequently examined by USG and MRI. Three of them had a biopsy, and in two of these women cancer was found.

Also in the BI-RADS III group, MammoVision gave the reason for additional USG, MG and MRI examinations. Two women underwent biopsy, one of them had DCIS.

12.5 Conclusion

The results of this study should be regarded as preliminary. The number of women included is too small to obtain statistically valid conclusions.

In the past only few studies have stressed the issue of comparing x-ray MG to infrared thermography [26–29]. One problem of comparing infrared thermography to x-ray MG is the amount of false positive and false negative results of MG. Even in this limited 114 women study, x-ray MG failed to detect 2 cases of invasive breast cancer and one case of DCIS, whereas the MammoVision results gave reason for further examinations and at last for confirmation of these three cases of malignancy. Infrared thermography of the breast should therefore be compared to MRI (which actually is proving to be more accurate than MG) or to biopsy results. But MRI is expensive and rarely applied, and biopsy is invasive and only done in highly suspicious cases.

Regarding the future applications of breast thermography, this study wanted to contribute and to pinpoint the increased possibilities given by last generation high accuracy and reliable thermal measurement equipment. Further comparison and outcome research for infrared thermography of the breast is urgently needed.

References

1. Lawson, R.N. and Chughtai, M.S., Breast cancer and body temperatures. *Can Med Assoc J* 88, 68–70, 1963.
2. Lawson, R.N., Implications of surface temperature in the diagnosis of breast lesions. *Canad Med Assoc J* 75, 309, 1956.
3. Lloyd Williams, K., Thermography in breast cancer. *Br J Radiol* 5, 75, 1969.
4. Lloyd Williams, K., Lloyd Williams, F., and Handley, R.S., Infrared thermometry in the diagnosis of breast diseases. *Lancet* 2, 1378, 1971.
5. Lewis, J.D., Milbrath, J.R., Shaffer, K.A., and Das Gupta, T.K., Implications of suspicious findings in breast cancer screening. *Arch Surg* 110, 903–907, 1975.
6. Berz, R. and Sauer, H., The medical use of infrared-thermography history and recent applications. In: Deutsche Gesellschaft für Zerstörungsfreie Prüfung e.V. (Ed): Thermografie-Kolloquium 2007. DGZfP-Berichtsband BB 107-CD. Berlin 2007.
7. Council Directive 92/42/EEC of 14 January 1993 on Medical Devices, 1993.
8. Schwamm, E., Thermoregulation und Thermodiagnostik. In Rost, A. (Ed), Thermographie und Thermoregulationsdiagnostik. Uelzen, 96–107, 1980.
9. Heim, G., Thermographie im Zeitablauf unter verschiedenen Belastungsformen. In Rost, A. (Ed), Thermographie und Thermoregulationsdiagnostik. MLV, Uelzen 64–73, 1980.
10. Berz, R., Das Wärmebild und die Reaktion auf Abkühlung bei jungen gesunden Probanden. *Ärztezeitschrift f Naturheilverfahren* 26, 237–243, 1985.
11. Berz, R., Regulation thermography—a survey. In: Baleagas, D., Busse, G., Carlomagno, C.M., Wiecek, B. (Eds), *Medical InfraRed Thermography MIRT'98*. Technical University of Lodz, Poland, 1998.

12. Berz, R., *About Regulation Thermography—A Sensitive Tool for Early Diagnose and Therapy Control.* Proceedings of the Focus Symposium on Health, Healing and Medicine, V 01 V. University of Windsor, Canada 1999.

13. Berz, R., Infrared Regulation Imaging (IRI)—a different approach to health, wellness, and to prevention. ThermoMed 16, 49–58, 2000.

14. Berz, R., Introducing Regulation into Infrared Imaging: ReguVision and MammoVision. In: Institute of Electronics, Technical University of Lodz (Ed): Proceedings of the 4th National Conference "Termografia i Termometria w Podczerwieni" TTP 2000. Lodz, Poland, 206–212, 2000.

15. Berz, R., MammoVision—a new approach to diagnosis and prevention of breast cancer. In: Benkö, I., Kovaczicz, I., and Lovak, I. (Eds.): 12th Inter-national Conference on Thermal Engineering and Thermogrammetry (THERMO), June 2001, Budapest. Mate, Hungary, 265–272, 2001.

16. Berz, R. and Sauer, H., Comparing effects of thermal regulation tests (cool air stimulus vs. cold water stress test) on infrared imaging of the female breast. In: *Institute of Physics and Engineering in Medicine* Ed), Clinical temperature measurement and thermography. York, UK, 36–41, 2007.

17. Ring, E.F.J., Aarts, N.J.M., Black, C.M., and Boesiger, P., Raynaud's phenomenon assessment by thermography. *Thermology* 3, 69–73, 1988.

18. Ring, E.F.J., Cold stress test for the hands. In Ammer, K. and Ring, E.F.J. (Eds), *The Thermal Image in Medicine and Biology*. Uhlen, Vienna, 237–240, 1995.

19. Pascoe, D.D., Purohit, R.C., Shanley, L.A., and Herrick, R.T., Pre- and post-operative evaluation of carpal tunnel syndrome. In Ammer, K. and Ring, E.F.J. (Eds), *The Thermal Image in Medicine and Biology*. Uhlen, Vienna, 188–190, 1995.

20. Berz R. and Bucher W., Procedure for the evaluation of thermograhically images of a female or male breast (Verfahren zur Auswertung von Wärmebildern einer weiblichen oder männlichen Brust), German Patent Nr. 101 50 918, Munich, 2003.

21. Medical Device Certification GmbH Stuttgart (Germany), CE-0483, 2007.

22. Berz R., Equipment for measuring the temperature pattern at the body surface of a person (Vorrichtung zur Bestimmung der Temperaturverteilung an der Körperoberfläche einer Person), German Patent Nr. 199 56 346, Munich, 2004.

23. Boesiger, P. and Stucki, D., Quantitative auswertung und interpretation von infrarotthermogrammen der weiblichen Brust. In Lauth, G. and Eulenburg, R. (Eds), *Thermographie der weiblichen Brust.* VCH, Weinheim, 185–219, 1986.

24. Schulte-Uebbing, C., Breast Cancer and Heavy Metals, New Aspects for Diagnosis and Therapy, Oncology Meeting, Bad Aibling, Germany, February 2008.

25. Schulte-Uebbing, C., Mammovision—ein komplementäres Mamma—Diagnostik—Verfahren, zaenmagazin 2, 13–19, 2010.

26. Parisky, Y.R., Sardi, A., Hamm, R., Hughes, K., Esserman, L., Rust, S., and Callahan, K., Efficacy of computerized infrared imaging analysis to evaluate mammographically suspicious lesions. *Am J Roentgenol* 180, 263–269, 2003.

27. Ng, E.Y.K., Ung, L.N., Ng, F.C., and Sim, L.S.J., Statistical analysis of healthy and malignant breast thermography. *J Med Eng Tech*, 25, 253–263, 2001.

28. Qi, H., Liu, Z.Q., and Wang, C., Breast cancer identification through shape analysis in thermal texture maps. Annual International Conference of the IEEE Engineering in Medicine and Biology Proceedings, 2, 1129–1130, 2002.

29. Qi, H., Kuruganti, P.T., and Snyder, W.E., Detecting breast cancer from thermal infrared images by asymmetry analysis. In: Diakides, N.A., Bronzino, J.D. (Eds), *Medical Infrared Imaging*. CRC Press, Boca Raton, 11-1–11-13, 2008.

13

Detecting Breast Cancer from Thermal Infrared Images by Asymmetry Analysis

Hairong Qi
University of Tennessee

Phani Teja Kuruganti
Oak Ridge National Laboratory

Wesley E. Snyder
North Carolina State University

One of the popular methods for breast cancer detection is to make comparisons between contralateral images. When the images are relatively symmetrical, small asymmetries may indicate a suspicious region. In thermal infrared (IR) imaging, asymmetry analysis normally needs human intervention because of the difficulties in automatic segmentation. In order to provide a more objective diagnostic result, we describe an automatic approach to asymmetry analysis in thermograms. It includes automatic segmentation and supervised pattern classification. Experiments have been conducted based on images provided by Elliott Mastology Center (Inframetrics 600M camera) and Bioyear, Inc. (Microbolometer uncooled camera).

13.1 Introduction

The application of IR imaging in breast cancer study starts as early as 1961 when Williams and Handley first published their results in the *Lancet* [1]. However, the premature use of the technology and its poorly controlled introduction into breast cancer detection in the 1970s have led to its early demise [2]. IR-based diagnosis was criticized as generating a higher false-positive rate than mammography, and thus was not recommended as a standard modality for breast cancer detection. Therefore, despite its deployment in many areas of industry and military, IR usage in medicine has declined [3]. Three decades later, several papers and studies have been published to reappraise the use of IR in medicine [2,3] for the following three reasons: (1) we have greatly improved IR technology. New generations of IR cameras have been developed with much enhanced accuracy; (2) we have much better capabilities in image processing. Advanced techniques including image enhancement, restoration, and segmentation have been effectively used in processing IR images; and (3) we have a deeper understanding of the patho-physiology of heat generation.

The main objective of this work is to evaluate the viability of IR imaging as a noninvasive imaging modality for early detection of breast cancer so that it can be performed both on the symptomatic and the asymptomatic patient and can thus be used as a complement to traditional mammography. This report summarizes how the identification of the asymmetry can be automated using image segmentation, feature extraction, and pattern recognition techniques. We investigate different features that contribute the most toward the detection of asymmetry. This kind of approach helps reduce the false-positive rate of the diagnosis and increase chances of disease cure and survival.

13.1.1 Measuring the Temperature of Human Body

Temperature is a long-established indicator of health. The Greek physician, Hippocrates, wrote in 400 B.C. "In whatever part of the body excess of heat or cold is felt, the disease is there to be discovered" [4]. The ancient Greeks immersed the body in wet mud and the area that dried more quickly, indicating a warmer region, was considered the diseased tissue. The use of hands to measure the heat emanating from the body remained well into the sixteenth and the seventeenth centuries. It was not until Galileo, who made a thermoscope from a glass tube, that some form of temperature-sensing device was developed, but it did not have a scale. It is Fahrenheit and later Celsius who have fixed the temperature scale and proposed the present-day clinical thermometer. The use of liquid crystals is another method of displaying skin temperature. Cholesteric esters can have the property of changing colors with temperature and this was established by Lehmann in 1877. It was involved in use of elaborative panels that encapsulated the crystals and were applied to the surface of the skin, but due to large area of contact, they affected the temperature of the skin. All the methods discussed above are contact based.

Major advances over the past 30 years have been with IR thermal imaging. The astronomer, Sir William Herschel, in Bath, England discovered the existence of IR radiation by trying to measure the heat of the separate colors of the rainbow spectrum cast on a table in the darkened room. He found that the highest temperature was found beyond the red end of the spectrum. He reported this to the Royal society as Dark Heat in 1800, which eventually has been turned the IR portion of the spectrum. IR radiation occupies the region between visible and microwaves. All objects in the universe emit radiations in the IR region of the spectra as a function of their temperature. As an object gets hotter, it gives off more intense IR radiation, and it radiates at a shorter wavelength [3]. At moderate temperatures (above 200°F), the intensity of the radiation gets high enough that the human body can detect that radiation as heat. At sufficiently high temperatures (above 1200°F), the intensity gets high enough and the wavelength gets short enough that the radiation crosses over the threshold to the red end of the visible light spectrum. The human eye cannot detect IR rays, but they can be detected by using the thermal IR cameras and detectors.

13.1.2 Metabolic Activity of Human Body and Cancer Cells

Metabolic process in a cell can be briefly defined as the sum total of all the enzymatic reactions occurring in the cell. It can be further elaborated as a highly coordinated, purposeful activity in which many sets of interrelated multienzyme systems participate, exchanging both matter and energy between the cell and its environment. Metabolism has four specific functions: (1) to obtain chemical energy from the fuel molecules; (2) to convert exogenous nutrients into the building blocks or precursor of macromolecular cell components; (3) to assemble such building blocks into proteins, nucleic acids, lipids, and other cell components; and (4) to form and degrade biomolecules required in specialized functions of the cell.

Metabolism can be divided into two major phases, catabolism and anabolism. Catabolism is the degradative phase of metabolism in which relatively large and complex nutrient molecules (carbohydrates, lipids, and proteins) are degraded to yield smaller, simpler molecules such as lactic acid, acetic acid, CO_2, ammonia, or urea. Catabolism is accompanied by conservation of some of the energy of the nutrient in the form of phosphate bond energy of adenosine triphosphate (ATP). Conversely, anabolism is the building up phase of metabolism, the enzymatic biosynthesis of such molecular components of cells as the nucleic acids, proteins, lipids, and carbohydrates from their simple building block precursors. Biosynthesis of organic molecules from simple precursors requires input of chemical energy, which is furnished by ATP generated during catabolism. Each of these pathways is promoted by a sequence of specific enzymes catalyzing consecutive reactions. The energy produced by the metabolic pathways is utilized by the cell for its division. Cells undergo mitotic cell division, a process in which a single cell divides into many cells and forms tissues, leading further into the development and growth of the multicellular organs. When cells divide, each resultant part is a complete relatively small cell. Immediately after division the newly formed cells grow rapidly soon reaching the size of the original cell. In humans, growth occurs through mitotic cell division with subsequent enlargement and differentiation of the reproduced cells into organs. Cancer cells also grow similarly but lose the ability to differentiate into organs. So, a cancer may be defined as an actively dividing undifferentiated mass of cells called the "tumor."

Cancer cells result from permanent genetic change in a normal cell triggered by some external physical agents such as chemical agents, x-rays, UV rays, and so on. They tend to grow aggressively and do not obey normal pattern of tissue formation. Cancer cells have a distinctive type of metabolism. Although they possess all the enzymes required for most of the central pathways of metabolism, cancer cells of nearly all types show an anomaly in the glucose degradation pathway (namely, glycolysis). The rate of oxygen consumption is somewhat below the values given by normal cells. However, the malignant cells tend to utilize anywhere from 5 to 10 times as much glucose as normal tissue and convert most of it into lactate instead of pyruvate (lactate is a low energy compound whereas pyruvate is a high-energy compound). The net effect is that in addition to the generation of ATP in mitochondria from respiration, there is a very large formation of ATP in extra mitochondrial compartment from glycolysis. The most important effect of this metabolic imbalance in cancer cells is the utilization of a large amount of blood glucose and release of large amounts of lactate into blood. The lactate so formed is recycled in the liver to produce blood glucose again. Since the formation of blood glucose by the liver requires six molecules of ATP whereas breakdown of glucose to lactate produces only two ATP molecules, the cancer cells are looked upon as metabolic parasites dependent on the liver for a substantial part of their energy. Large masses of cancer cells thus can be a considerable metabolic drain on the host organism. In addition to this, the high metabolic rate of cancer cells causes an increase in local temperature as compared to normal cells. Local metabolic activity ceases when blood supply is stopped since glycolysis is an oxygen-dependent pathway and oxygen is transported to the tissues by the hemoglobin present in the blood; thus, blood supply to these cells is important for them to proliferate. The growth of a solid tumor is limited by the blood supply. If it were not invaded by capillaries a tumor would be dependent on the diffusion of nutrients from its surroundings and could not enlarge beyond a diameter of a few millimeters. Thus, in order to grow further the tumor cells stimulate the blood vessels to form a capillary network that invades the tumor mass. This phenomenon is popularly called "angiogenesis," which is a process of vascularization of a tissue involving the development of new capillary blood vessels.

Vascularization is a growth of blood vessels into a tissue with the result that the oxygen and nutrient supply is improved. Vascularization of tumors is usually a prelude to more rapid growth and often to metastasis (advanced stage of cancer). Vascularization seems to be triggered by angiogenesis factors that stimulate endothelial cell proliferation and migration. In the context of this chapter the high metabolic rate in the cancer cells and the high density of packaging makes them a key source of heat concentration (since the heat dissipation is low) thus enabling thermal IR imaging as a viable technique to visualize the abnormality.

13.1.3 Early Detection of Breast Cancer

There is a crucial need for early breast cancer detection. Research has shown that if detected earlier (tumor size <10 mm), the breast cancer patient has an 85% chance of cure as opposed to 10% if the cancer is detected late [5].

Different kinds of diagnostic imaging techniques exist in the field of breast cancer detection. The most popularly used method presently is x-ray mammography. The drawback of this technique is that it is invasive and experts believe that electromagnetic radiation can also be a triggering factor for cancerous growth. Because of this, periodic inspection might have a negative effect on the patient's health. Research shows that the mammogram sensitivity is higher for older women (age group 60–69 years) at 85% compared with younger women (<50 years) at 64% [5]. A new study in a British medical journal (the *Lancet* [6]) has asserted that there is no reliable evidence that screening with mammography for breast cancer reduces mortality. They show that screening actually leads to more aggressive treatment, increasing the number of mastectomies by about 20%, and the number of mastectomies and tumorectomies by about 30%.

In contrast to this, IR imaging uses a noninvasive imaging technique as the diagnostic tool. The main source of IR rays is heat emitted from different bodies whose temperature is above absolute zero. Thus a thermogram of a patient provides the heat distribution in the body. The cancerous cells, due to high metabolic rates and angiogenesis, are at a higher temperature than the normal cells around it. Thus the cancer cells can be imaged as hotspots in the IR images. The thermogram provides more dynamic information of the tumor since the tumor can be small in size but can be fast growing making it appear as a high-temperature spot in the thermogram [7,8]. However, this is not the case in mammography, in which unless the tumor is beyond certain size, it cannot be imaged as x-rays essentially pass through it unaffected. This qualifies IR imaging as an effective diagnostic tool for early detection of breast cancer. Keyserlingk et al. [2] reported that the average tumor size undetected by thermal imaging is 1.28 and 1.66 cm by mammography. It is also reported that thermography can provide results that can be correct even 8–10 years before mammography can detect a mass in the patient's body [9,10].

13.2 Asymmetric Analysis in Breast Cancer Detection

Radiologists routinely make comparisons between contralateral images. When the images are relatively symmetrical, small asymmetries may indicate a suspicious region. This is the underlying philosophy in the use of asymmetry analysis for mass detection in breast cancer study [11]. Unfortunately, due to various reasons such as fatigue, carelessness, or simply because of the limitation of human visual system, these small asymmetries might not be easy to detect. Therefore, it is important to design an automatic approach to eliminate human factors.

There have been a few papers addressing techniques for asymmetry analysis of mammograms [11–16]. Head et al. [17,18] recently analyzed the asymmetric abnormalities in IR images. In their approach, the thermograms are segmented first by an operator. Then breast quadrants are derived automatically based on unique point of reference, that is, the chin, the lowest, rightmost, and leftmost points of the breast.

This chapter describes an automatic approach to asymmetry analysis in thermograms. It includes automatic segmentation and pattern classification. Hough transform is used to extract the four feature curves that can uniquely segment the left and right breasts. The feature curves include the left and the right body boundary curves, and the two parabolic curves indicating the lower boundaries of the breasts. Upon segmentation, different pattern recognition techniques are applied to identify the asymmetry.

Both segmentation and classification results are shown on images provided by Elliott Mastology Center (Inframetrics 600M camera) and Bioyear, Inc. (Microbolometer uncooled camera).

13.2.1 Automatic Segmentation

There are several ways to perform segmentation, including *threshold-based* techniques, *edge-based* techniques, and *region-based* techniques. Threshold-based techniques assign pixels above (or below) a specified threshold to be in the same region. Edge-based techniques assume that the boundary formed by adjacent regions (or segments) has high edge strength. Through edge detection, we can identify the boundary that surrounds a region. Region-based methods start with elemental regions and split or merge them [19].

13.2.1.1 Edge Image Generation

In this research work, we choose to use the edge-based segmentation technique as we have identified four dominant curves in the edge image, which we call "feature curves," including the left and right body boundaries, and the two lower boundaries of the breasts (Figure 13.3) or the two shadow boundaries that are below the breasts (Figure 13.2) whichever is stronger.

Edges are areas in the image where the brightness changes suddenly. One way to find edges is to calculate the amplitude of the first derivative (Equation 13.1) that generates the so-called gradient magnitude image where $\partial f/\partial x$ is the derivative of the image f along the x direction, and $\partial f/\partial y$ is the derivative along the y direction (the x and y directions are orthogonal to each other) and to threshold the amplitude image.

$$\text{Gradient magnitude} = \sqrt{\left(\frac{\partial f}{\partial x}\right)^2 + \left(\frac{\partial f}{\partial y}\right)^2} \tag{13.1}$$

Another way is to calculate the second derivative and to locate the zero-crossing points, as the second derivative at edge pixels would appear as a point where the derivatives of its left and right pixels change signs.

Although effective, both derivative images are very sensitive to noise. In order to eliminate or reduce the effect of noise, a Gaussian smoothing low-pass filter is applied before taking the derivatives. The smoothing process, on the other hand, also results in thicker edges. The Canny edge detector [20] solves this problem by taking two extra steps, the *nonmaximum suppression* (NMS) step and the *hysteresis thresholding* step, in order to locate the true edge pixels (only one pixel wide along the gradient direction).

For each edge pixel in the gradient magnitude image, the NMS step looks one pixel in the direction of the gradient, and another pixel in the reverse direction. If the magnitude of the current pixel is not the maximum of the three, then this pixel is not a true edge pixel. NMS helps locate the true edge pixels properly. However, the new edges still suffer from extra edge points problem due to noise and missing edge points.

The hysteresis thresholding improves the quality of edge image by using a dual-threshold approach, in which two thresholds, τ_1 and τ_2 (τ_2 is significantly larger than τ_1), are applied on the NMS-processed edge image to produce two binary edge images, denoted as T_1 and T_2, respectively. Since T_1 was created using a lower threshold, it will contain more extra edge pixels than T_2. Edge pixels in T_2 are therefore considered to be parts of true edges. When some discontinuities in edges occur in the T_2 image, the pixels at the same location of the T_1 image is looked up that could be continuations of the edge. If such pixels are found, they are included in the true edge image. This process continues until the continuation of pixels in T_1 connects with an edge pixel in T_2 or no connected T_1 points are found. (See Figures 13.2 and 13.3 for examples of the edge image derived from Canny edge detector.)

13.2.1.2 Hough Transform for Edge Linking

From Figures 13.2 and 13.3, we find that the edge images still cannot be used directly for segmentation as the edge detector picks up all the intensity changes, which would complicate the segmentation. On the other hand, many edges show gaps among edge pixels that would result in segments that are not closed. We have identified four feature boundaries that could enclose the breast. The body boundaries are easy to detect. Difficulties lie in the detection of the lower boundaries of the breasts or the shadow. We observe that the breast boundaries are generally parabolic in shape. Therefore, the Hough transform [21] is used to detect the parabola.

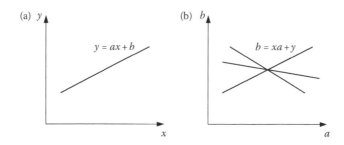

FIGURE 13.1 Illustration of the Hough transform. (a) The original image domain and (b) the parametric domain.

The Hough transform is one type of parametric transforms, in which the object in an image is expressed as a mathematical function of some parameters and the object can be represented in another transformation domain by these parameters. Take a straight edge in an image as an example, in the original image representation domain (or the x–y domain), we can use $y = ax + b$ to describe this edge with a slope of a and an intercept of b, as shown in Figure 13.1a. In order to derive the two parameters, a and b, we can convert the problem to a parametric domain (or the a–b domain) from the original x–y space, and treat x and y as parameters. We find that for each point (x, y) on the edge, there are infinite number of possible corresponding (a, b)s and they form a line in the a–b space, $b = xa + y$. We can imagine that for all the points on the edge, if each of which corresponds to a straight line in the a–b space, then in theory, these lines must intersect at one and only one point (Figure 13.1b), which indicates the true slope and intercept of the edge.

Similarly, the problem of deriving the parameters that describe the parabola can be formulated in a three-dimensional (3D) parametric space of x_0–y_0–p as there are three unknown parameters:

$$y - y_0 = p(x - x_0)^2$$

(13.2)

Each point on the parabola in the x–y space corresponds to a parabola in the x_0–y_0–p space. All the points on the parabola in the x–y space intersect at one point in the x_0–y_0–p space. In order to locate this intersection point in the parametric space, the idea of *accumulator array* is used in which the number of times that a certain pixel in the parametric space is "hit" by a transformed curve (line, parabola, etc.) is treated as the intensity of the pixel. Therefore, the value of the parameters can be derived based on the coordinates of the brightest pixel in the parametric space. The readers are referred to Reference [19] for details on how to implement this accumulator array.

The coordinates of the two brightest spot in the parametric space are used to describe the parabolic functions that form the lower boundaries of the breasts, as shown in Figures 13.2 and 13.3.

13.2.1.3 Segmentation Based on Feature Boundaries

Segmentation is based on three key points, the two armpits (P_L, P_R) derived from the left and right body boundaries by picking up the point where the largest curvature occurs and the intersection (O) of the two parabolic curves derived from the lower boundaries of the breasts/shadow of the breasts. The vertical line that goes through point O and is perpendicular to line $P_L P_R$ is the one used to separate the left and right breasts.

13.2.1.4 Experimental Results

The first set of testing images are obtained using the Inframetrics 600M camera, with a thermal sensitivity of 0.05 K. The images are collected at Elliott Mastology Center. Results from two testing images (*lr*, *nb*) are shown in Figure 13.2 that includes the intermediate results from edge detection, feature curve extraction, and segmentation. From the figure, we can see that Hough transform can derive the parabola at the accurate location.

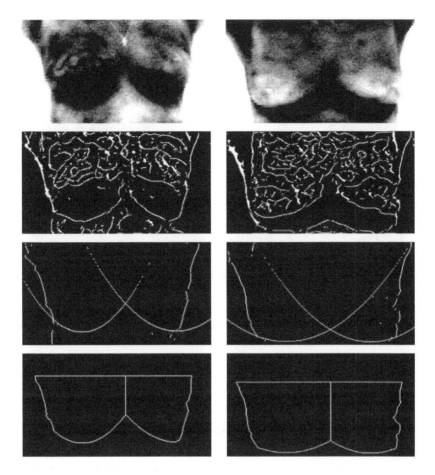

FIGURE 13.2 Segmentation results of two images. Left: results from *lr*. Right: results from *nb*. From top to bottom: original image, edge image, four feature curves, and segments. (Copyright 2005 IEEE.)

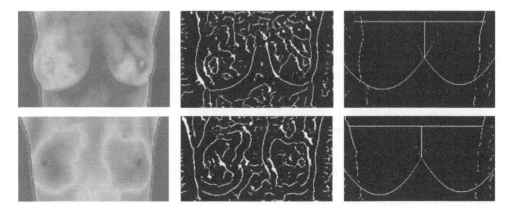

FIGURE 13.3 Hough transform-based image segmentation. Top: image of a patient with cancer. Bottom: image of a patient without cancer. From left to right: the original TIR image, the edge image using Canny edge detector, and the segmentation based on Hough transform. (Copyright 2005 IEEE.)

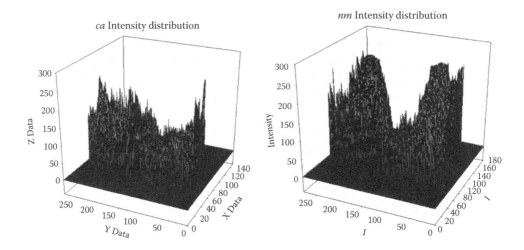

FIGURE 13.4   The left figure show the intensity distribution of a cancerous image and the right figure shows the same for a noncancerous image. The cancerous image is more asymmetrical than the noncancerous one.

Another set of images are obtained using Microbolometer uncooled camera, with a thermal sensitivity of 0.05 K. Some examples of the segmented images are shown in Figure 13.3.

Figure 13.4 shows the 3D histogram of the thermal distribution described in the intensity component of the cancerous (*ca*) and noncancerous (*nm*) images. From the graphs, we observe that the *ca* image is more asymmetrical than the *nm* image.

13.2.2 Asymmetry Identification by Unsupervised Learning

Pixel values in a thermogram represent the thermal radiation resulting from the heat emanating from the human body. Different tissues, organs, and vessels have different amount of radiation. Therefore, by observing the heat pattern, or in other words, the pattern of the pixel value, we should be able to discover the abnormalities if there are any.

Usually, in pattern classification algorithms, a set of training data are given to derive the decision rule. All the samples in the training set have been correctly classified. The decision rule is then applied to the testing data set where samples have not been classified yet. This classification technique is also called supervised learning. In unsupervised learning, however, data sets are not divided into training sets or testing sets. No *a priori* knowledge is known about which class each sample belongs to.

In asymmetry analysis, none of the pixels in the segment knows its class in advance, thus there will be no training set or testing set. Therefore, this is an unsupervised learning problem. We use *k*-means algorithm to do the initial clustering. *k*-Means algorithm is described as follows:

1. Begin with an arbitrary set of cluster centers and assign samples to nearest clusters
2. Compute the sample mean of each cluster
3. Reassign each sample to the cluster with the nearest mean
4. If the classification of all samples has not changed, then stop, else go to step 2

After each sample is relabeled to a certain cluster, the cluster mean can then be calculated. The segmented image can also be displayed in labeled format. From the difference of mean distribution, we can tell if there is any asymmetric abnormalities.

Figure 13.5 provides the histogram of pixel value from each segment that generated in Figure 13.2 with 10-bin setup. We can tell just from the shape of the histogram that *lr* is more asymmetric than *nb*. However, the histogram only reveals global information. Figure 13.6 displays the classification results for

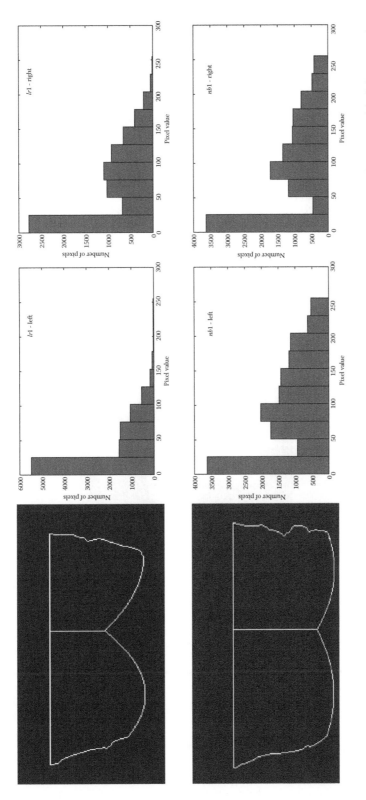

FIGURE 13.5   Histogram of the left and right segments. Top: results from *lr*. Bottom: results from *nb*. From left to right: the segments, histogram of the left segment, histogram of the right segment. (Copyright 2005 IEEE.)

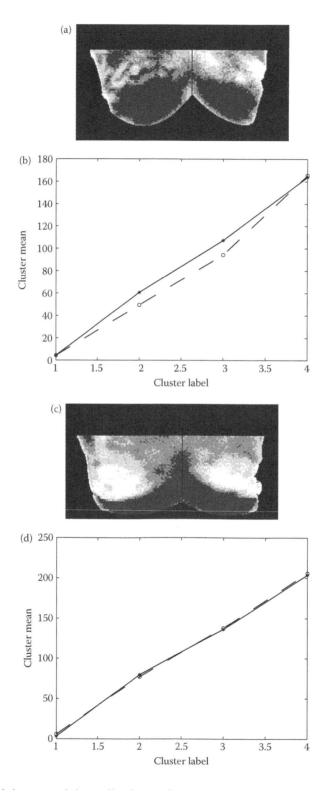

FIGURE 13.6 Labeled image and the profile of mean for each cluster. (a) Results from *lr*, (b) labeled image, (c) results from *nb*, and (d) average pixel value profile of each cluster. (Copyright 2005 IEEE.)

each segment in its labeled format. Here, we choose to use four clusters. The figure also shows the mean difference of each cluster in each segmented image. From Figure 13.6, we can clearly see the much bigger difference shown in the mean distribution or image *lr*, which can also be observed from the labeled image.

13.2.3 Asymmetry Identification Using Supervised Learning Based on Feature Extraction

Feature extraction is performed on the segmented images. The aim of this research is to identify the effectiveness of the features in contributing toward the asymmetry analysis.

As discussed earlier, TIR imaging is a functional imaging technique representing thermal information as a function of intensity. The TIR image is a pseudo-colored image with different colors assigned to different temperature ranges.

The distribution of different intensities can now be quantified by calculating some high-order statistics as feature elements. We design the following features to form the feature space:

- *Moments of the intensity image*: The intensity component of the image directly corresponds to the thermal energy distribution in the respective areas. The histogram describing the intensity distributions essentially describes the texture of the image. The moments of the histogram give statistical information about the texture of the image. Figure 13.5 shows the intensity distribution of images of a cancerous patient and noncancerous patient. The four moments, mean, variance, skewness, and kurtosis are taken as

$$\text{Mean } \mu = \frac{1}{N} \sum_{j=1}^{N} p_j \tag{13.3}$$

$$\text{Variance } \sigma^2 = \frac{1}{N-1} \sum_{j=1}^{N} (p_j - \mu)^2 \tag{13.4}$$

$$\text{Skewness} = \frac{1}{N} \sum_{j=1}^{N} \left[\frac{p_j - \mu}{\sigma} \right]^3 \tag{13.5}$$

$$\text{Kurtosis} = \frac{1}{N} \sum_{j=1}^{N} \left(\frac{p_j - \mu}{\sigma} \right)^4 \tag{13.6}$$

where p_j is the probability density of the *j*th bin in the histogram, and *N* is the total number of bins.

- *The peak pixel intensity of the correlated image*: The correlated image between the left and right (reflected) breasts is also a good indication of asymmetry. We use the peak intensity of the correlated image as a feature element since the higher the correlation value, the more symmetric the two breast segments.
- *Entropy*: Entropy measures the uncertainty of the information contained in the segmented images. The more equal the intensity distribution, the less information. Therefore, the segment with hot spots should have a lower Entropy.

$$\text{Entropy } H(X) = -\sum_{j=1}^{N} p_j \log p_j \tag{13.7}$$

TABLE 13.1   Moments of the Histogram

	Cancerous		Noncancerous	
Moments	Left	Right	Left	Right
Mean	0.0010	0.0008	0.0012	0.0010
Variance (10^{-6})	2.0808	1.1487	3.3771	2.7640
Skewness (10^{-6})	2.6821	1.1507	4.8489	4.5321
Kurtosis (10^{-8})	1.0481	0.3459	2.1655	2.3641

- *Joint Entropy*: The higher the joint entropy between the left and right breast segments, the more symmetric they are supposedly to be, and the less possible of the existence of tumor.

$$\text{Joint Entropy } H(X,Y) = \sum_{i=1}^{N_X} \sum_{j=1}^{N_Y} p_{ij} \log(p_{ij}) \tag{13.8}$$

where p_{ij} is the joint probability density, N_X and N_Y are the number of bins of the intensity histogram of images X and Y, respectively.

From the set of features derived from the testing images, the existence of asymmetry is decided by calculating the ratio of the feature from the left segment to the feature from the right segment. The closer the value to 1, the more correlated the features or the less asymmetric the segments. Classic pattern classification techniques like the maximum posterior probability and the kNN classification [22] can be used for the automatic classification of the images. Table 13.1 describes the typical moments for the cancerous and noncancerous images.

The typical values of the cancerous images and noncancerous images are tabulated in Table 13.2. The asymmetry can be clearly stated with a close observation of the given feature values. We used six normal patient thermograms and 18 cancer patient thermograms. With a larger database, a training feature set can be derived and supervised learning algorithms such as discriminant function or kNN classification can be implemented for a fast, effective, and automated classification.

Figure 13.7 evaluates the effectiveness of the features used. The first data point along the *x*-axis indicates *entropy*, the second to the fifth points indicate the four statistical moments (*means, variance, skewness,* and *kurtosis*). The *y*-axis shows the *closeness* metric we defined as

$$\text{Bilateral ratio closeness to } 1 = \left| \frac{\text{feature value from left segment}}{\text{feature value from right segment}} - 1 \right| \tag{13.9}$$

From the figure, we observe that the high-order statistics are the most effective features to measure the asymmetry, while low-order statistics (*mean*) and *entropy* do not assist asymmetry detection.

TABLE 13.2   Entropy and Correlation Values

Feature	Cancerous	Noncancerous
Correlation	$\times 10^8$	2.35719×10^8
Joint entropy	9.0100	17.5136
Entropy		
Left	1.52956	1.70684
Right	1.3033	1.4428

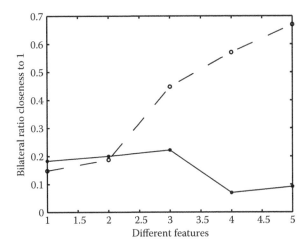

FIGURE 13.7 Performance evaluation of different feature elements. Solid line: noncancerous image; dashed line: cancerous image. The five data points along the *x*-axis indicate (from left to right): entropy, mean, variance, skewness, and kurtosis. (Copyright 2005 IEEE.)

13.3 Conclusions

This chapter develops a computer-aided approach for automating asymmetry analysis of thermograms. This kind of approach will help the diagnostics as a useful second opinion. The use of TIR images for breast cancer detection and the advantages of thermograms over traditional mammograms are studied. From the experimental results, it can be observed that the Hough transform can be effectively used for breast segmentation. We propose two pattern classification algorithms, the unsupervised learning using k-means and the supervised learning using kNN based on feature extraction. Experimental results show that feature extraction is a valuable approach to extract the signatures of asymmetry, especially the joint entropy. With a larger database, supervised pattern classification techniques can be used to attain more accurate classification. These kind of diagnostic aids, especially in a disease such as breast cancer where the reason for the occurrence is not totally known, will reduce the false-positive diagnosis rate and increase the survival rate among the patients since the early diagnosis of the disease is more curable than in a later stage.

References

1. Llyod-Williams, K. and Handley, R.S., Infrared thermometry in the diagnosis of breast disease. *Lancet*, 2, 1378–1381, 1961.
2. Keyserlingk, J.R., Ahlgren, P.D., Yu, E., Belliveau, N., and Yassa, M., Functional infrared imaging of the breast. *IEEE Engineering in Medicine and Biology*, 19, 30–41, 2000.
3. Jones B.F., A reappraisal of the use of infrared thermal image analysis in medicine. *IEEE Transactions on Medical Imaging*, 17, 1019–1027, 1998.
4. Thermology, http://www.thermology.com/history.htm
5. Ng, E.Y.K. and Sudarshan, N.M., Numerical computation as a tool to aid thermographic interpretation. *Journal of Medical Engineering and Technology*, 25, 53–60, 2001.
6. Oslen, O. and Gotzsche, P.C., Cochrane review on screening for breast cancer with mammography. *Lancet*, 9290, 1340–1342, 2001.
7. Hay, G.A., *Medical Image: Formation, Perception and Measurement*. The Institute of Physics, John Wiley & Sons, New York, NY, 1976.

8. Watmough, D.J., The role of thermographic imaging in breast screening, discussion by C.R. Hill. In *Medical Images: Formation, Perception and Measurement, 7th L H Gray Conference: Medical Images*, pp. 142–158, 1976.

9. Gautheire, M., *Atlas of breast thermography with specific guidelines for examination and interpretation*. Milan, Italy, PAPUSA, 1989.

10. Ng, E.Y.K., Ung, L.N., Ng, F.C., and Sim, L.S.J., Statistical analysis of healthy and malignant breast thermography. *Journal of Medical Engineering and Technology*, 25, 253–263, 2001.

11. Good, W.F., Zheng, B., Chang, Y. et al., Generalized procrustean image deformation for subtraction of mammograms. In *Proceedings of SPIE Medical Imaging—Image Processing*, Vol. 3661, pp. 1562–1573, San Diego, CA, SPIE, 1999.

12. Shen, L., Shen, Y.P., Rangayyan, R.M., and Desautels J., Measures of asymmetry in mammograms based upon shape spectrum. In *Proceedings of the Annual Conference on EMB*, Vol. 15, pp. 48–49, San Diego, CA, 1993.

13. Yin, F.F., Giger, M.L., Doi, K. et al., Computerized detection of masses in digital mammograms: Analysis of bilateral subtraction images. *Medical Physics*, 18, 955–963, 1991.

14. Yin, F.F., Giger, M.L., Doi, K. et al., Computerized detection of masses in digital mammograms: Automated alignment of breast images and its effect on bilateral-subtraction technique. *Medical Physics*, 21, 445–452, 1994.

15. Yin, F.F., Giger, Vyborny C.J. et al., Comparison of bilateral-subtraction and single-image processing techniques in the computerized detection of mammographic masses. *Investigative Radiology*, 6, 473–481, 1993.

16. Zheng, B., Chang, Y.H., and Gur, D., Computerized detection of masses from digitized mammograms: Comparison of single-image segmentation and bilateral image subtraction. *Academic Radiology*, 2, 1056–1061, 1995.

17. Head J.F., Lipari, C.C., and Elliott R.L., Computerized image analysis of digitized infrared images of the breasts from a scanning infrared imaging system. In *Proceedings of the 1998 Conference on Infrared Technology and Applications XXIV, Part I*, Vol. 3436, pp. 290–294, San Diego, CA, SPIE, 1998.

18. Lipari C.A. and Head J.F., Advanced infrared image processing for breast cancer risk assessment. In *Proceedings for 19th International Conference of IEEE/EMBS*, pp. 673–676, Chicago, IL, Oct. 30–Nov. 2. IEEE, 1997.

19. Snyder, W.E. and Qi, H., *Machine Vision*, Cambridge University Press, New York, 2004.

20. Canny, J., A computational approach to edge detection. *IEEE Transactions on Pattern Analysis and Machine Intelligence*, 6, 679–698, 1995.

21. Jafri M.Z. and Deravi, F., Efficient algorithm for the detection of parabolic curves. In *Vision Geometry III*, Vol. 2356, pp. 53–62, SPIE, 1998.

22. Duda, R.O., Hart, P.E., and Strok, D.G., *Pattern Classification*, 2nd ed. John Wiley & Sons, New York, NY, 2001.

14

Application of Nonparametric Windows in Estimating the Mutual Information between Bilateral Breasts in Thermograms

M. Etehadtavakol
Isfahan University of Medical Science

E.Y.K. Ng
School of Mechanical and Aerospace Engineering

Nanyang Technological University

Caro Lucas
University of Tehran

S. Sadri
Isfahan University of Technology

Isfahan University of Medical Science

N. Gheissari
Isfahan University of Technology

Isfahan University of Medical Science

14.1 Preamble

All things in the universe radiate infrared radiation as a function of their temperature [1]. The surface temperature of the human body is influenced by the level of blood perfusion which directs the infrared radiation from the skin. The variations in temperature can be captured by a sensitive infrared (IR) camera [2]. Cancerous cells send signals to the surrounding normal host tissue by which certain genes in the host tissue can make some proteins that can develop new blood vessels [3]. The cells are shown as hot spots in the thermal images.

Since symmetry is usually a sign of healthy subjects [4], an asymmetrical temperature distribution between the right breast and the left breast could be a sign of abnormality [5–10]. Studies have shown that some factors such as environmental stresses and genetic mutations would affect individuals and consequently the homeostatic mechanisms that preserve the symmetry of paired structures such as the breast tend to demote [11]. There are also studies concerning the relationships between asymmetry and hormonal concentrations. Estrogen is extremely essential in the development and growth of the breast. An individual's ability to tolerate disruptive hormonal variations can affect the symmetrical breast development [12]. Consequently, an increase in the unstable asymmetry of paired structures could be a sign of poor health.

The mutual information (MI) is a similarity measure that can be used to indicate the temperature distribution similarity between the two breast IR images. In order to estimate it, the joint PDF of the two IR breast images and the marginal PDF of each breast IR image are required.

Nonparametric (NP) windows are signal densities and distribution estimators that are based on the Shannon–Whittaker–Nyquist theory of sampled signals. It estimates signal statistics by directly calculating each piecewise section of the signal. Interpolation and smoothing are performed in the signal domain rather than in the probability domain. The advantages of NP windows make the technique attractive to estimate MI. NP windows are applied to estimate the joint distribution and consequently the joint intensity histogram of image pairs.

14.2 Mutual Information

Information has a broad concept. When any probability distribution entropy is defined, it has many properties that agree with the notion of measuring information [13].

The MI can be defined as the relative entropy between the joint distribution and the product distribution.

$$MI(x, y) = \sum_{x \in X} \sum_{y \in Y} p(x, y) \log \frac{p(x, y)}{p(x)p(y)} \tag{14.1}$$

MI is a nonlinear measure and it can be used to measure both linear and nonlinear correlations [14]. Independent component analysis [15,16], registration of remote sensing [17], and medical images [18] are some of its applications.

In this chapter, the similarity between the two breast images is measured by calculating the MI which is a similarity measure of intensity between the two regions.

When a tumor is initiated, its growth can be described by a separate mathematical function or a stochastic process model. The tumor growth function starting from a single cell may be described by a simple logistic cell kinetic function predicting the volume of cells viable in the tumor as a logistic function. For example, a gamma-distributed mixture of exponential tumor growth functions [19]. As a result, the variations of the right and the left breasts' IR images are due to the cancerous cells that are comparatively nonlinear. Consequently, the MI technique for the detection of the asymmetry between the two breasts becomes very useful.

Estimation of MI can be difficult which has been a common problem in many MI applications and this estimation sometimes is unreliable, noisy, and even biased.

Parametric and NP approaches are the two basic approaches for the estimation of MI. In NP approaches, the form of the function is not given and the estimation techniques include nearest neighbor, histogram-based, adaptive partitioning, spline, and kernel density [20].

In parametric estimation, it is assumed that the data belong to a given density function and the form of the density function is given. In this study, the accuracy of the estimated MI is important and it is estimated from the joint intensity histogram of the image pair.

The joint intensity histogram can be described as a two-dimensional (2D) histogram of combinational intensity pairs, I1, I2, where I1 and I2 denote the intensity level (0–255) of image 1 and image 2, respectively. Traditionally, histograms are built-up by using each intensity sample to populate the histogram bin. The limitation of the quantization of intensity and the number of intensity samples available to populate the histogram is a disadvantage of this method. For example, the number of bins are limited with too few samples and consequently an under populated histogram is resulted, so that usable MI values are not achievable from such a histogram.

To avoid under-populated histograms, several algorithms have been proposed [21–25]. For this achievement, the number of bins in the histogram and also an appropriate kernel size are needed. Practically, there are some limitations for estimating the joint histogram.

By proposing NP windows, Kadir and Brady solved this problem [26,27]. NP is a technique which is established by the Shannon–Whittaker–Nyquist theory of the sampled signal. For a given interpolation model, signal density and distribution are estimated by directly calculating the distribution of each piecewise section of a signal. In this technique, interpolation or smoothing is accomplished in the signal domain and not in the probability domain which is contrary to the previous methods. The benefits of NP windows make it appropriate for the estimation of MI [28].

By using the joint PDF of the two breast images with their respective intensities i_1 and i_2, the MI between the two breast images can be obtained as Equation 14.2

$$MI = \sum_{i_1}\sum_{i_2} p_{f_1 f_2}(i_1,i_2)\log\frac{p_{f_1 f_2}(i_1,i_2)}{p_{f_1}(i_1)p_{f_2}(i_2)} \tag{14.2}$$

$p_{f_1 f_2}$ denotes the joint PDF of f_1 and f_2. The marginal PDFs are $p_{f_1} = \sum_{i_2} p_{f_1 f_2}$ and $p_{f_2} = \sum_{i_1} p_{f_1 f_2}$. The finite bin size of the histogram from which $p_{f_1 f_2}$ is calculated can be indicated by discrete variables i_1 and i_2. Bin size is inversely proportional to the number of bins in the histogram.

14.3 Nonparametric Windows to Estimate the Joint Histogram for a Pair of Images

In this study, two interpolation methods are investigated; bilinear interpolation and half-bilinear interpolation.

14.3.1 Bilinear Interpolation

Bilinear interpolation can be regarded as the intuitive generalization of the one-dimensional linearly interpolated signal to the two-dimensional form. The linear interpolation is accomplished in one direction first and then again in the other direction. Suppose, the value of the unknown function f at the point $m = (x,y)$ is derived, the values of f at the four points are assumed to be f_1, f_2, f_3, and f_4.

The following equations are obtained by doing the linear interpolation in the X direction as shown in Figure 14.1.

$$f(R_1) = \frac{x_1 - x}{x_1 - x_2} f_1 + \frac{x_2 - x}{x_2 - x_1} f_2 \tag{14.3}$$

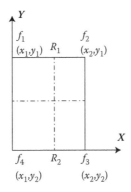

FIGURE 14.1 f at the four points (x_1,y_1), (x_2,y_2), (x_3,y_3), and (x_4,y_4) are f_1, f_2, f_3, and f_4, respectively and are known values. The desired point for interpolation is $m = (x,y)$.

$$f(R_2) = \frac{x_1 - x}{x_1 - x_2} f_4 + \frac{x_2 - x}{x_2 - x_1} f_3 \tag{14.4}$$

Then, interpolating in the Y direction we obtain Equation 14.5

$$f(m) = \frac{y_1 - y}{y_1 - y_2} f(R_1) + \frac{y_2 - y}{y_1 - y_2} f(R_2) \tag{14.5}$$

$$f(m) = \frac{y_1 - y}{y_1 - y_2}\left(\frac{x_1 - x}{x_1 - x_2} f_1 + \frac{x_2 - x}{x_2 - x_1} f_2\right)$$
$$+ \frac{y_2 - y}{y_1 - y_2}\left(\frac{x_1 - x}{x_1 - x_2} f_4 + \frac{x_2 - x}{x_2 - x_1} f_3\right) \tag{14.6}$$

A coordinate system in which the four points are (0,0), (0,1), (1,0), and (1,1) may be chosen, then the interpolant can be obtained as

$$f(p) = axy + bx + cy + d \tag{14.7}$$

For a pair of two-dimensional images and by applying the bilinear interpolation, we obtain Equations 14.8 and 14.9.

$$f_1 = a_1 x_1 x_2 + b_1 x_1 + c_1 x_2 + d_1 \tag{14.8}$$

$$f_2 = a_2 x_1 x_2 + b_2 x_1 + c_2 x_2 + d_2 \tag{14.9}$$

The probability of f can be developed by using the transformation formula. It depends on the absolute value of the gradient x with respect to f (Jacobian) [29]

$$p_f(f) = \left|\frac{\partial x}{\partial f}\right| p_x(x(f)) \tag{14.10}$$

The joint probability is derived as Equation 14.11 for this case.

$$p_f(f_1, f_2) = \left|\det(J_{x\circ f})\right| p_x(x(X)) \tag{14.11}$$

The Jacobian is also acquired as Equation 14.12

$$\left|\det J_{xof}\right| = [(b_2 c_1 - b_1 c_2 + a_2(d_1 - f_1) - a_1(d_2 - f_2))^2$$
$$- 4(a_2 c_1 - a_1 c_2)(b_2(d_1 - f_1) - b_1(d_2 - f_2))]^{-\frac{1}{2}} \tag{14.12}$$

A polygon bounded by four edges is the valid region for this PDF. The total probability over a particular bin covered by the polygon is achieved by integrating Equation 14.12 directly to determine the histogram.

Integration of $|\det(J_{x \circ f})|$ is almost unworkable as it is shown in Equation 14.13. However, if the half-bilinear interpolation which is illustrated in the following section is applied, an estimation of the histogram would be possible.

$$\rho = k_1 \sqrt{\phi(f_1,2)} + \phi(f_1)(k_2 + \log[\phi(f_1) + k_3 \sqrt{\phi(f_1,2)}]) + k_4 \log[\phi(f_1)]$$
$$+ k_5 \sqrt{\phi(f_1,2)}] + k_6 \log \left[\frac{\phi(f_1) + k_7 \sqrt{\phi(f_1,2)}}{\phi(f_1)} \right]$$
(14.13)

14.3.2 Half-Bilinear Interpolation

Suppose the value of the unknown function f at the point $m = (x,y)$ is derived assuming that $f_1, f_2,$ and f_3 are known and are the values of f at the three points $(x_1,y_1), (x_2,y_2,),$ and $(x_3,y_3),$ respectively are indicated in Figure 14.2.

$A_1, A_2,$ and A_3 are areas of triangle f_2pf_3, triangle f_1pf_3, and triangle f_1pf_2, respectively, are defined in Equations 14.14 through 14.16.

$$A_1 = \frac{1}{2} \det \begin{bmatrix} x & x_2 & x_3 \\ y & y_2 & y_3 \\ 1 & 1 & 1 \end{bmatrix}$$
(14.14)

$$A_2 = \frac{1}{2} \det \begin{bmatrix} x & x_1 & x_3 \\ y & y_1 & y_3 \\ 1 & 1 & 1 \end{bmatrix}$$
(14.15)

$$A_3 = \frac{1}{2} \det \begin{bmatrix} x & x_1 & x_2 \\ y & y_1 & y_2 \\ 1 & 1 & 1 \end{bmatrix}$$
(14.16)

$$A = \frac{1}{2} \det \begin{bmatrix} x_1 & x_2 & x_3 \\ y_1 & y_2 & y_3 \\ 1 & 1 & 1 \end{bmatrix}$$
(14.17)

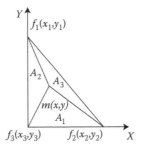

FIGURE 14.2 f at the three points (x_1,y_1), (x_2,y_2), and (x_3,y_3) are $f_1, f_2,$ and f_3, respectively which are known values. The desired point for interpolation is $m = (x,y)$.

$$f(m) = \frac{A_1}{A} f_1 + \frac{A_2}{A} f_2 + \frac{A_3}{A} f_3 \tag{14.18}$$

$$f(m) = ax + by + c \tag{14.19}$$

If the cross-term of Equation 14.7 is removed from the bilinear interpolant, Equation 14.19 would be obtained. Only three basis samples (α, β, γ) are required because there are only three coefficients in Equation 14.20 which must be computed. Their relation yields

$$\begin{pmatrix} a \\ b \\ c \end{pmatrix} = \begin{pmatrix} -1 & 1 & 0 \\ 0 & -1 & 1 \\ 1 & 0 & 0 \end{pmatrix} \begin{pmatrix} \alpha \\ \beta \\ \gamma \end{pmatrix} \tag{14.20}$$

The interpolated intensity $f(x_1, x_2)$ for an image is depicted in Figure 14.3, where dark indicates low intensity and light indicates high intensity.

Suppose we have two images instead of one image. For the images, the image lattice may be separated to neighbors of regular 45° right-angled nonoverlapping triangles.

Assuming that the basis samples lay on the triangle $(\alpha \, \beta \, \gamma)$ where β is 90° angle. If for a pair of images the half-bilinear interpolation is applied, the following functions would be derived:

$$f_1 = a_1 x_1 + b_1 x_2 + c_1 \tag{14.21}$$

$$f_2 = a_2 x_1 + b_2 x_2 + c_2 \tag{14.22}$$

The inverse functions are obtained as Equations 14.23 and 14.24

$$x = \frac{c_1 b_2 - b_1 c_2 - b_2 f_1 + b_1 f_2}{b_1 a_2 - a_1 b_2} \tag{14.23}$$

$$y = \frac{c_1 a_2 - a_1 c_2 - a_2 f_1 + a_1 f_2}{-b_1 a_2 + a_1 b_2} \tag{14.24}$$

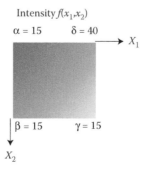

Intensity $f(x_1, x_2)$

$\alpha = 15$ $\delta = 40$ X_1

$\beta = 15$ $\gamma = 15$

X_2

FIGURE 14.3 An example of interpolated image $(f(x_1, x_2))$ light shade indicates high intensity and dark shade indicates low intensity.

and then the Jacobian is given as Equation 14.25

$$\det(J_{x \circ f}) = \begin{vmatrix} \dfrac{\partial x_1}{\partial f_1} & \dfrac{\partial x_1}{\partial f_2} \\ \dfrac{\partial x_2}{\partial f_1} & \dfrac{\partial x_2}{\partial f_2} \end{vmatrix} = \frac{1}{|a_1 b_2 - b_1 a_2|} \tag{14.25}$$

According to the standard probability theory, the joint probability distribution of a pair of two-dimensional signals with transformation yields

$$p_f(f_1, f_2) = \left| \det(J_{x \circ f}) \right| p_x(x(X)) \tag{14.26}$$

and it demonstrates that the absolute determinant of the Jacobian for a pair of basis sample triplets is constant. In addition, the area of the triangle given by (α_1, α_2), (β_1, β_2), and (γ_1, γ_2) is described as Equation 14.27.

$$\frac{1}{2} \left[(\beta_1 - \alpha_1)(\gamma_2 - \beta_2) - (\gamma_1 - \beta_1)(\beta_2 - \alpha_2) \right] = \frac{1}{2}(a_1 b_2 - b_1 a_2) \tag{14.27}$$

Multiplying this area by the probability within the triangle accompanies to 1/2. It is an unsurprised value, because only half a pixel has been calculated.

Suppose we have a cell of four basis samples $(\alpha_1, \beta_1, \gamma_1, \delta_1)$ and $(\alpha_2, \beta_2, \gamma_2, \delta_2)$. By each neighborhood (a single pair of triangles), the PDF contribution is constant and bounded within a triangle. Accordingly, the vertex coordinating in the joint PDF are the intensity values in the two images.

First, the individual contribution of each neighborhood is weighted to normalize it and then the results are aggregated to determine the overall PDF. The smaller triangle possesses a higher weighting and the probability over each triangle aggregates to 1/2. They are depicted in Figure 14.4.

Two special cases may occur geometrically: the triangle yields a line and the triangle yields a point. The three possible cases are demonstrated in Figure 14.5.

Since the probability value within the regions of integration is always constant. Hence, the integral corresponds to calculating three cases: an area for triangles and a length for lines, and a value for points. It takes the value of one for points. Assuming that the vertices of one of the triangles are shown in Figure 14.6a.

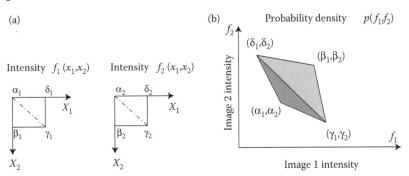

FIGURE 14.4 (a) Two images with $f_1(x_1, x_2)$ and $f_2(x_1, x_2)$ intensities. (b) Triangles of uniform probability in the joint PDF of intensity, for each triangle, the probability must integrate to one, the smaller triangle possesses a higher weight, and the vertices of the two triangles are matching the intensity values in part (a).

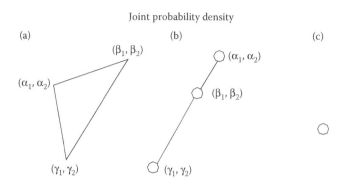

Joint probability density

FIGURE 14.5 Three possible cases occur for joint half-bilinear interpolation image. (a) A triangle, (b) a line and (c) a point.

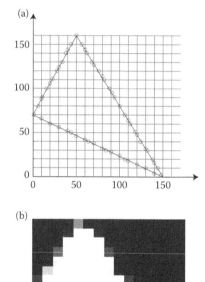

FIGURE 14.6 (a) Example of a triangle with circles indicating the intersection of the triangle sides and bin boundaries. (b) The proportional coverage of the triangle over each histogram bin is obtained and cumulatively added.

We suggest an algorithm in order to calculate the proportional coverage of the triangle over each histogram bin. It consists of the following steps:

  Step 1. The intersections of the sides of the triangles with bin boundaries (they are depicted by circles) are found.
  Step 2. The inside region from the outside region of the triangle is separated (zero code for inside region and one for outside region).
  Step 3. Regions with code one come into the polygon shape in different bins and the vertices for each bin are found.

Step 4. Vertices are ordered in counter clockwise.

Step 5. For each bin the area of the polygon is calculated.

Step 6. The area of the polygon for each bin is divided by the area of the triangle.

In the above example, the result is shown in Figure 14.6b.

The histogram is an image whose bins are individual pixels and the probability in each bin is the image intensity.

14.4 Experimental Results

In the following experiments, we applied the MI algorithm on 60 breast-simulated images which are in six groups. For instance, image a is the real image of a patient's right breast which is accessible by the Sunstate Thermal Imaging Center in Australia [30]. Simulated images b through f are retained as follows: mirroring image a, image b is achieved. Images c through f are retained by sequential color changes to the preceding images. By pseudo-color algorithms, the main output images of IR cameras which are gray-scale images are transformed to RGB images. The map of the heat profile is experimentally evident by inspecting the real color images, so that it is possible to detect low-temperature parts from the high-temperature parts. Fittingly, in our simulated images, we have tried to simulate the color changes in one breast in comparison with the other one just for the hotter parts. In consonance with natural phenomena, the changes between the sequential images are kept small. In each step, the MI between the pairs of real image a and image b through f is calculated. In this way we study the power of MI technique for monitoring sequential images of a patient. One image from each group is depicted in Figure 14.7 and their matching, calculated, and normalized MI values are presented in Table 14.1.

As it is shown in Table 14.1, the more similar the IR image of the right breast to the IR image of the left breast, the closer will be the normalized MI value which is equal to 1. Means and standard deviations for each group of simulated images are shown in Table 14.2. The standard deviation and the mean of the MI between the pairs of real image a and image b through f are provided bold in Table 14.2. The mean of abnormal and also normal cases are provided bold in Table 14.3.

Estimated normalized MI value of an investigated patient, as an additional feature, may be then fed into a trained classification system such as support vector machine to achieve more reliable and more authentic results to identify malignant anomalies in breast thermograms. Furthermore, the regions of interest can be different matching quadrants of each breast.

14.5 The Algorithm for Real Case Issue

In the recommended protocols for the measurement of the human body temperature [31–34], all the instrumental setup, environmental control, patients' preparation, and data-collecting procedures performed should follow the recommended guidelines [31–33], relevant standards, and protocols in clinical thermographic imaging [34]. This practice is to minimize as much as possible those potential background uncertainties which may affect the false-positive prediction.

In practice, most of the real IR breast images which are accommodated under the standard protocol of IR screening contrasting our simulated images do not have symmetry boundaries. This phenomenon can be studied in real images that are used in this study. These real images are provided by Sunstate Thermal Imaging Center in Australia [30].

Therefore, a registration for two breasts is required for real images. Applying shape contexts, an approach was proposed by Belongie and Malik [35]. In this study, the algorithm follows the below-mentioned steps:

1. Correspondences between the points on the boundaries on the two breasts are solved.
2. The correspondences are used to estimate an aligning transform (mapping function). The procedures have been shown in Figures 14.8 and 14.9.

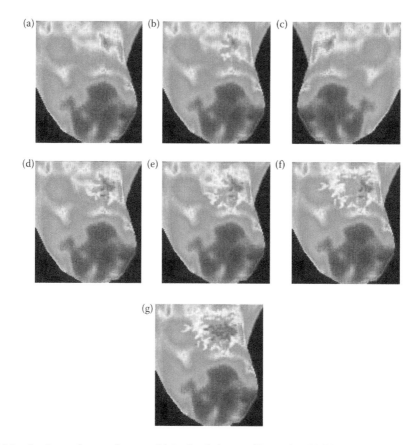

FIGURE 14.7 One image from each group. (a) Real right breast, (b) simulated left breast, (c) simulated left breast with small changes with respect to b, (d) simulated left breast with small changes with respect to c, (e) simulated left breast with small changes with respect to d, (f) simulated left breast with small changes with respect to e, (g) simulated left breast with small changes with respect to f.

TABLE 14.1 Normalized Mutual Information Values Corresponding to the Images of Figure 14.7

A,f	A,e	A,d	A,c	A,b	A,g	Images
0.9475	0.9585	0.9615	0.9833	1	0.9189	Normalized mutual information

TABLE 14.2 Means and Standard Deviations for Six Groups of Simulated Images

A,f		A,e		A,d		A,c		A,b		A,g		Images
SD	Mean	SD	Mean	SD	SD	SD	SD	SD	Mean	SD	Mean	Normalized mutual information
0.0838	0.9351	0.0701	0.9531	0.0681	0.0539	0.0574	0.0539	0.0539	1	0.0539	0.9183	

TABLE 14.3 Means and Standard Deviations for Normal and Abnormal Simulated Images along with P-Value

Normal Mean ± SD	Abnormal Mean ± SD	P-Value
0.9829 ± 0.0139	**0.9355** ± 0.0153	0.0294

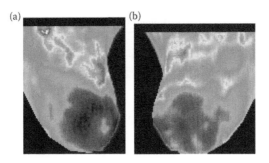

FIGURE 14.8 Original images (a) left breast, (b) right breast.

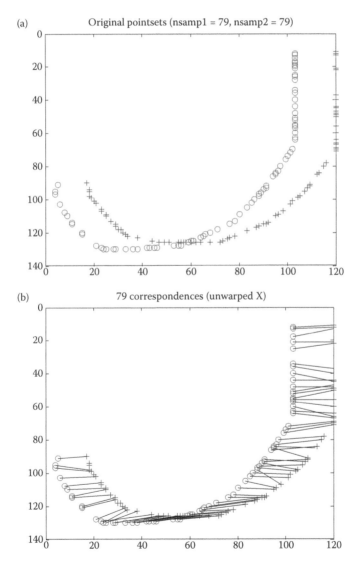

FIGURE 14.9 (a) Boundary samples of right breast (+), boundary samples of left breast (o), (b) unwarped boundaries, (c) warped boundaries, (d) boundaries of the two breasts after employing the mapping function.

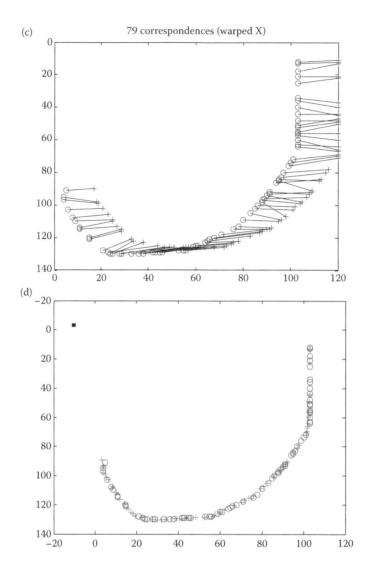

FIGURE 14.9 (Continued).

3. Calculated mapping function of boundary points is used to map the interior points of the breasts. In digital images, the discrete picture elements, or pixels, are restricted to lie on a sampling grid, taken to be in an integer lattice. The output grid does not generally coincide with the integer lattice. Rather, the positions of the grid points may take on any of the continuous values that are allocated by the mapping function.

   Consequently, an interpolation must be presented to fit a continuous surface through the input data samples. The continuous surface may then be sampled at noninteger positions of the output grid. The accuracy of the interpolation has a significant influence on the quality of the output image.

4. Cubic spline for interpolation is used to determine the R, G, B amounts of the output grid appropriately.

The algorithm is implemented for 15 normal and nine abnormal cases. Results are shown in Table 14.4.

TABLE 14.4   Means and Standard Deviations for Normal and Abnormal Real Cases

Normalized Mutual Information	Normal	Abnormal
Mean	0.9601	0.9249
Standard deviation	0.0167	0.0354

14.6 Conclusion

Symmetry of the two breasts is usually a sign of a healthy breast. Collation between contra-lateral breast images is one of the impressive methods in breast cancer detection. Unhealthy breasts have asymmetric temperature distribution.

The MI is a measure that grasps linear and nonlinear dependencies, without demanding the specification of any kind of model of dependence. Hence, it is appropriate for our investigation. We applied NP windows to estimate the MI.

Although NP windows is a computationally expensive technique, it does not demand parameters such as prior selection of a kernel. It obtains accurate histograms since it coordinates an interpolation model which enhances the resolution to a highly oversampled image.

For our determination, we used 60 simulated breast thermal images. Experimental results indicate that the more similar the thermal image of the left breast to the thermal image of the right breast, the closer will be the normalized MI value equal to 1. Hence, we are capable of capturing very small dissimilarities between the two breasts. Future work could be focusing on the different regions of interest such as the different matching quadrants of each breast.

References

1. R. Siegel, J. R. Howell, *Thermal Radiation Heat Transfer*, Hemisphere, Washington, DC, 1992.
2. J. M. Barreiro, F. M. Sanchez, V. Maojo, F. Sanz, *Biological and Medical Data Analysis, 5th International Symposium*, Barcelona, Spain, Springer, November 18–19, 2004.
3. Y. Singh, Tumor angiogenesis: Clinical implications, *Nepal Journal of Neuroscience*, 1(1): 61–63, 2004.
4. S. Uematsu, Symmetry of skin temperature comparing one side of the body to the other, *International Journal of Thermology*, 1(1): 4–7, 1985.
5. H. Qi, P. T. Kuruganti, W. E. Snyder, Detecting breast cancer from thermal infrared images by asymmetry analysis, *Biomedical Engineering Handbook*, CRC Press, Boca Raton, FL, Chapter 27, pp. 1–14, 2006.
6. N. Diakides, J. D. Bronzino, *Medical Infrared Imaging*, CRC Press, Taylor & Francis Group, Boca Raton, FL, 2008.
7. T. Z. Tan, C. Quek, G. S. Ng, E. Y. K. Ng, A novel cognitive interpretation of breast cancer thermography with complementary learning fuzzy neural memory structure, *Expert Syst. Appl.*, 33(3): 652–666, 2007.
8. M. Frize, C. H. Herry, R. Roberge, Processing of thermal images to detect breast cancer: Comparison with previous work, *IEEE EMBS/BMES Conference*, Houston, Texas, Vol. 2, pp. 1159–1160, 2002.
9. T. Jakubowska, B. Wiecek, M. Wysocki, C. Drews Peszynski, Thermal signatures for breast cancer screening comparative study, In *Proceedings of the 25th Annual International Conference of the IEEE EMBS Conference*, Cancun, Mexico, Vol. 2, pp.1117–1120, 2003.
10. G. Schaefer, M. Zavisek, T. Nakashima, Thermography based breast cancer analysis using statistical features and fuzzy classification, *Pattern Recognition*, 42(6): 1133–1137, 2009.
11. N. H. Eltonsy, A. S. Elmaghraby, G. D. Tourassi, Bilateral breast volume asymmetry in screening mammograms as a potential marker of breast cancer: Preliminary experience, *Image Processing, IEEE International Conference in Image Processing*, San Antonio, Texas, Vol. 5, pp. 5–8, 2007.
12. D. Scutt, G. A. Lancaster, J. T. Manning, Breast asymmetry and predisposition to breast cancer, *Breast Cancer Research*, doi: 10.1186/bcr1388, 8 R14, 2006.

13. T. M. Cover, J. A. Thomas, *Elements of Information Theory*, John Wiley & Sons, Inc., New York, 1991.

14. F. Rossi, A. Lendasse, D. François, V. Wertz, M. Verleysen, Mutual information for the selection of relevant variables in spectrometric nonlinear modeling, *Chemometrics and Intelligent Laboratory Systems*, 80, 215–226, 2006.

15. K. C. Chiu, Z. Y. Liu, L. Xu, A statistical approach to testing mutual independence of ICA recovered sources, *4th International Symposium on Independent Component Analysis and Blind Signal Separation*, Nara, Japan, April 2003.

16. G. A. Darbellay, The mutual information as a measure of statistical dependence, *IEEE International Symposium on Information Theory*, Ulm, Germany, 405pp., 1997.

17. H. M. Chen, P. Varshney, M. K. Arora, A study of joint histogram estimation methods to register multi sensor remote sensing images using mutual information, *IEEE Geoscience and Remote Sensing Symposium*, Toulouse, France, Vol. 6, pp. 4035–4037, July 2003.

18. P. W. Josien, J. B. Pluim, A. Maintz, M. A. Viergever, Mutual information based registration of medical images: A survey, *IEEE Transactions on Medical Imaging*, 22(8): 986–1004, 2003.

19. K. G. Manton, I. Akushevick, J. Kravchenko, *Cancer Mortality and Morbidity Patterns in the US Population*, Statistics for Biology and Health, Springer, New York, 2009.

20. J. W. Williams, Y. Li, *Estimation of Mutual Information: A Survey*, Springer-Verlag, Berlin Vol. 5589/2009, pp. 389–396, 2009.

21. E. Parzen, On estimation of a probability density function and mode, *Annals of Mathematical Statistics*, 33(3): 1065–1076, 1962.

22. P. Thevenaz, M. Unser, Optimization of mutual information for multi-resolution image registration, *IEEE Transactions on Image Processing*, 9(12): 2083–2099, 2000.

23. F. Maes, A. Collignon, D. Vandermeulen, G. Marchal, P. Suetens, Multimodality image registration by maximization of mutual information, *IEEE Transactions on Medical Imaging*, 16(2): 187–198, 1997.

24. H. Chen, P. Varshney, Mutual information-based CT-MR brain image registration using generalised partial volume joint histogram estimation, *IEEE Transactions on Medical Imaging*, 22(9): 1111–1119, 2003.

25. E. D'Agostino, F. Maes, D. Vandermeulen, P. Suetens, An information theoretic approach for nonrigid image registration using voxel class probabilities, *Medical Image Analysis*, 10, 413–430, 2006.

26. T. Kadir, M. Brady, Estimating statistics in arbitrary regions of interest, *Proceedings of the 16th British Machine Vision Conference*, Oxford, Vol. 2, pp. 589–598, Sept. 2005.

27. T. Kadir, M. Brady, Nonparametric estimation of probability distributions from sampled signals, *Technical Report OUEL No: 2283/05*, Robotics Research Laboratory, Oxford University, July 2005.

28. N. Dowson, T. Kadir, R. Bowden, Estimating the joint statistics of images using nonparametric windows with application to registration using mutual information, *IEEE Transactions on Pattern Analysis and Machine Intelligence*, 30(10): 1841–1857, 2008.

29. A. Papoulis, S. U. Pillai, *Probability, Random Variables and Stochastic Processes*, McGraw-Hill, New York, 4th Edition, 2002.

30. STImaging: *http://www.stimaging.com.au/page2.html* (last accessed March 2010).

31. E. F. J. Ring, Progress in the measurement of human body temperature, *Engineering in Medicine & Biology Magazine*, 17: 19–24, 1998.

32. E. Y. K. Ng, A review of thermography as promising non-invasive detection modality for breast tumour. *International Journal of Thermal Sciences*, 48(5): 849–855, 2009.

33. E. F. J. Ring, H. McEvoy, A. Jung, J. Zuber, G. Machin, New standards for devices used for the measurement of human body temperature. *Journal of Medical Engineering & Technology*, 34(4): 249–253, 2010.

34. *Thermography Guidelines: Standards and Protocols in Clinical Thermographic Imaging*, Sept. 2002, http://www.iact-org.org/professionals/thermog-guidelines.html, also: http://www.blatmanpainclinic.com/blat_cancerscreening.htm

35. S. Belongie, J. Malik, J. Puzicha, Shape matching and object recognition using shape contexts, *IEEE Transactions on Pattern Analysis and Machine Intelligence*, 24(24): 509–522, 2002.

15

Boguslaw Wiecek
Technical University of Lodz

Maria Wiecek
Technical University of Lodz

Robert Strakowski
Technical University of Lodz

M. Strzelecki
Technical University of Lodz

T. Jakubowska
Technical University of Lodz

M. Wysocki
Technical University of Lodz

C. Drews-Peszynski
Technical University of Lodz

Breast Cancer Screening Based on Thermal Image Classification

Thermal imaging can be a useful technique for early detection of many diseases, for example, skin lesions, breast cancers, benign and malignant skin tumors, and so on. Thermal image features (indexes, signatures) and image classification can be one of the possible approaches for fast, noninvasive, contactless screening. There are many different methods that describe image features. A large group of methods is based on statistical parameters [4,9,10]. Parameters such as mean value, standard deviation, skewness, kurtosis, and so on can be used to compare and differentiate thermal images. We consider both the first and the second-order statistical parameters [3,4,7,10,15–17]. The first-order ones use histograms of the thermal image to compute signatures (see Figure 15.1), while the second-order statistical parameters are defined for the so-called co-occurrence matrix of the image [4,7,10].

In the medical applications, one of the basic features of a thermal image is its axis symmetry of temperature patterns. The human body and its thermal pattern are symmetric in most of the physiological cases, while they are usually asymmetric in the pathological conditions. However, sometimes a significant asymmetry can indicate the physiologic abnormality as well. Thermal asymmetry may correspond to pathology, including cancer, fibrocystic diseases, an infection (see Figure 15.2) or a vascular disease or it might indicate an anatomical variant [1,2,11,12,16]. This makes the difficulty to achieve the high efficiency of the image classifications and screening and it is the main reason that the researchers are still looking for the new effective methods of image processing for medical applications.

The next group of methods is based on image transformations, such as linear and nonlinear filtering, Fourier or wavelet analysis. All these methods allow regenerating an image (data) in another domain, and in consequence, the thermal signatures are defined based on new kind of data corresponding to frequency, time delay, scale in wavelet transform, and so on. Well-known Karhunen–Loeve transform can be implemented in the form of principle component analysis (PCA). PCA is a technique that is usually used for reducing the dimensionality and decorrelating of multivariate data preserving most of the variance [3,6,9,14].

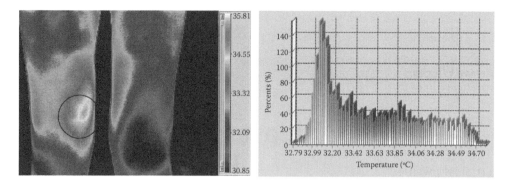

FIGURE 15.1   Thermal image and its histogram.

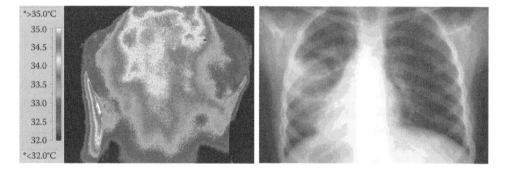

FIGURE 15.2   Thermal and x-ray images of pneumonia with asymmetric distribution of temperature.

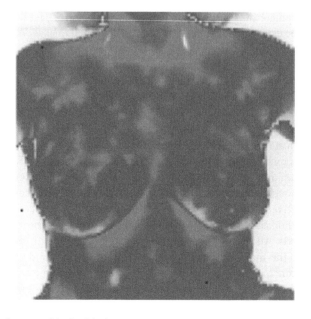

FIGURE 15.3   Thermal image of the healthy breast.

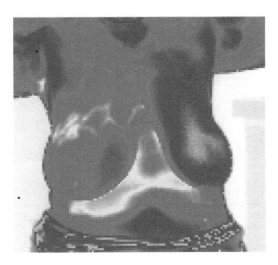

FIGURE 15.4 Thermal image of the breast with malignant tumor (left side of the image).

Thermal image classification is a powerful tool for many medical diagnostic protocols, including breast cancer screening [15–17]. Figures 15.3 and 15.4 show thermal images of a healthy breast and that with malignant tumor, respectively. Among the variety of different image features, statistical thermal signatures (first and second order) have been already effectively used for classification of images represented by raw data [15–17]. In another approach, the features obtained from the wavelet transformation can also be used for successful classification [8,16,17].

It is possible to define many features for an image, and obviously, the selection and reduction of them are needed. Two approaches may be applied, based on Fischer coefficient as well as by using minimization of classification error probability (POE) and average correlation coefficients (ACC) between chosen features [2,10,17]. It can reduce the number of features to a few ones. Features preprocessing which generates new parameters after linear or nonlinear transformations can be the next step in the procedure. It allows getting less correlated and lower order data. Two approaches are used, that is, PCA and linear discriminant analysis (LDA) [6,10]. Finally, classification can be performed using different methods, for example, artificial neural network (ANN) or nearest-neighbor classification (NNC) [16,17].

15.1 Histogram-Based First-Order Statistical Thermal Signatures

An image is assumed to be a rectangular matrix of discretized data (pixels) pix[m,n], where $m = 0, 1, …, M$, $n = 0,1, … N$. Each pixel takes a value from the range $i \in <0,L-1>$. The histogram describes the frequency of existence of pixels of the same intensity in whole image or in the region of interest (ROI). Formally, the histogram represents the distribution of the probability function of existence of the given image intensity and it is expressed using Kronecker delta function as

$$H(i) = \sum_{n}^{N-1} \sum_{m}^{M-1} \delta(p[m,x],i) \quad \text{for } i = 0,1,...,L-1, \tag{15.1}$$

where

$$\delta(p[m,n],i) = \begin{cases} 1 & \text{for } p[m,n] = i \\ 0 & \text{for } p[m,n] \neq i \end{cases} \tag{15.2}$$

Histogram is used for defining first-order statistical features of the image, such as mean value μ_H, variance σ_H, skewness, kurtosis, energy, and entropy. The definitions of these parameters are given below.

$$\mu_H = \sum_{i=0}^{L-1} ip(i)$$

$$\sigma_H = \sum_{i=0}^{L-1} (i - \mu_H)^2 p(i)$$

$$\text{skewness} = \frac{\sum_{i=0}^{L-1} (i - \mu_H)^3 p(i)}{\sigma_H^{3}}$$

$$\text{kurtosis} = \frac{\sum_{i=0}^{L-1} (i - \mu_H)^4 p(i)}{\sigma_H^{4}} - 3 \qquad (15.3)$$

$$\text{energy} = \sum_{i=0}^{L-1} p^2(i)$$

$$\text{entropy} = -\sum_{i=0}^{L-1} \log_2(p(i))p(i)$$

Histogram describes the global information of an image (see Figure 15.1). By processing the histogram, one can obtain very useful image improvement, such as contrast enhancement or intensity adjustment [3,9]. First-order statistical parameters can be employed to separate physiological and pathological cases using infrared (IR) imaging. The preliminary results for breast cancer screening are presented in the following subchapters.

In the first clinical trial, 32 healthy patients and 10 patients with malignant tumors were analyzed using thermography. There were four images registered for every patient that represented each breast in direct (frontal) and lateral direction to the camera. Histograms were created for these images and on the basis of statistical parameters, the following features were calculated: mean temperature, standard deviation, variance, skewness, and kurtosis. Afterward, differences of parameter values for left and right breast were calculated. The degree of symmetry on the basis of these differences was then estimated (see Figures 15.5 and 15.6).

The mean temperature in the healthy group was estimated at the level 30.2 ± 1.8°C in the direct position and 29.7 ± 1.9°C in the lateral one. The mean temperature was higher in eight cases out of 10 in malignant tumor group. Moreover, six cases out of 32 in the healthy group with mean temperatures exceeded normal level. Therefore, we have found that it is necessary to analyze symmetry between left and right breast. Comparison of mean temperature was insufficient to separate physiological and pathological images.

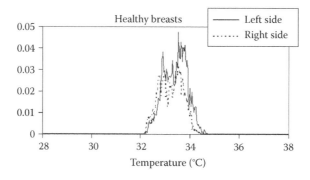

FIGURE 15.5   Histograms of thermal image for healthy breasts, see Figure 15.3.

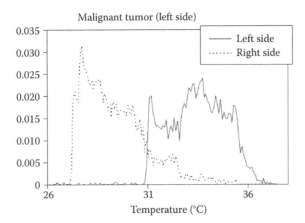

FIGURE 15.6 Histograms of thermal images for breasts presented in Figure 15.4 (malignant tumor on the left side).

Among analyzed parameters, skewness was the most promising for successful classification of thermal images. Absolute differences of skewness for left and right side was equal to $0.4 \pm 0.3°C$ in frontal position and $0.6 \pm 0.4°C$ in lateral one for the healthy group. These differences were higher for images in lateral position for all cases in the pathological group in comparison to the healthy patients' images.

The images in Figures 15.3 and 15.4 confirm the evidence of asymmetry between left and right side for healthy and malignant tumor cases.

Analyzing the first-order statistical parameter, let us conclude that it is quite hard to use them for the image classification and detecting tumors with high efficiency. Only mean temperature and skewness allow separating and classifying thermal images of breasts with and without malignant tumors. Frontal and lateral positions were used during this investigation, but no significant difference of the obtained result was noticed.

It is concluded that the first-order parameters do not give the satisfactory results, and due to some physiological changes of the breast, we could observe that these parameters do not allow separating patients with and without tumors. That was the main reason, that the second-order statistical parameters were used for further investigations.

15.2 Second-Order Statistical Parameters

More advanced statistical information on thermal images can be derived from the second-order parameters. They are defined using the so-called co-occurrence matrix [2,10,16,17]. Such a matrix represents the joint probability of two pixels having ith and jth intensity (temperature) at the different distances d, in the different directions. Co-occurrence matrix gives more information on intensity distribution over the whole image, and in this sense, it can effectively be used to separate and classify thermal images.

Let us assume that each pixel has eight neighbors lying in four directions: horizontal, vertical, diagonal, and antidiagonal. One can consider only the nearest neighbors, so the distance $d = 1$ (see Figure 15.7). As an example let us take an image 4×4 with four intensity levels given as (see Figure 15.8).

For horizontal direction, the co-occurrence matrix for the data presented in Figure 15.8 takes a form:

$$m_{\text{horizontal}} = \begin{bmatrix} 2 & 1 & 0 & 0 \\ 1 & 4 & 2 & 2 \\ 0 & 2 & 4 & 2 \\ 0 & 2 & 2 & 0 \end{bmatrix} \tag{15.4}$$

The co-occurrence matrix is always square and diagonal with the dimension equal to the number of intensity levels L in the image. After normalization we get the matrix of the probabilities $p(i,j)$.

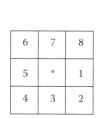 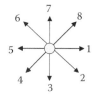

FIGURE 15.7 Eight neighboring pixels in four direct horizontal, vertical, diagonal, and antidiagonal.

FIGURE 15.8 Example of 4 × 4 image with four discrete intensity levels.

Normalization is done by dividing all elements by the number of possible couple pixels for a given direction of analysis. For horizontal, vertical, and diagonal directions this numbers are equal to $2N(M-1)$, $2M(N-1)$, and $2(M-1)(N-1)$, respectively.

Second-order parameters are presented by Equations 15.5 and 15.6. As an example of comparing thermal images, mean values and standard deviations (μ_x, μ_y, σ_x, σ_y) for the elements of co-occurrence matrixes were calculated for horizontal and vertical directions, respectively. The quantitative results are presented in Figures 15.9 and 15.10.

Second-order statistical parameters can effectively be used to discriminate the physiological and pathological cases, for example, breast cancers. Most of them successfully discriminate healthy and malignant tumor cases. The protocol of the investigation assumes the symmetry analysis in the following way. At first, a polygon-shape ROI were chosen for analysis of the part of the skin corresponding to the breast. Then, the co-occurrence matrixes were calculated for left and right breasts to evaluate

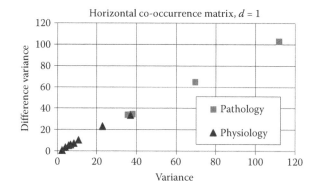

FIGURE 15.9 Difference variance versus variance obtained from co-occurrence matrix for horizontal direction.

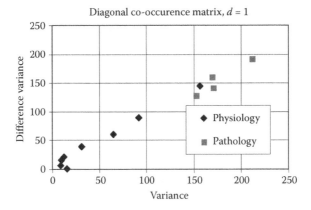

FIGURE 15.10 Difference variance versus variance obtained from co-occurrence matrix for diagonal direction.

second-order statistical parameters for different directions. Only the neighbor pixels are considered in these investigations ($d = 1$). Finally, the differences of the values of these parameters for left and right sides were used for the image classification. Figures 15.9 and 15.10 illustrate that the differences of second-order parameters for left and right sides are typically greater for pathological cases. Taking only two second-order parameters, that is, difference variance and variance allow separating successfully almost all healthy and malignant tumor cases.

$$\text{Variance} = \sum_{i=0}^{L-1} \sum_{j=0}^{L-1} (i - j)^2 p(i,j)$$

$$\text{Angular second moment} = \sum_{i=0}^{L-1} \sum_{j=0}^{L-1} p(i,j)^2$$

$$\text{Contrast} = \sum_{n=0}^{L-1} n^2 \sum_{\substack{i=0 \\ |i-j|=n}}^{L-1} \sum_{j=0}^{L-1} p(i,j)$$

$$\text{Correlation} = \frac{\sum_{i=0}^{L-1} \sum_{j=0}^{L-1} (ij)p(i,j) - \mu_x \mu_y}{\sigma_x \sigma_y}$$

$$\text{Sum of squares} = \sum_{i=0}^{L-1} \sum_{j=0}^{L-1} (i-\mu_x)^2 p(i,j)$$

$$\text{Inverse difference moment} = \sum_{i=0}^{L-1} \sum_{j=0}^{L-1} \frac{p(i,j)}{1 + (i-j)^2}$$

$$\text{Sum average} = \sum_{i=0}^{2(L-1)} i p_{x+y}(i)$$

$$\text{where } p_{x+y}(l) = \sum_{i=0}^{L-1} \sum_{j=0}^{L-1} p(i,j), \text{ for } |i+j| = l, l = 0,1,...,2(L-1)$$

$$\text{Sum variance} = \sum_{i=0}^{2(L-1)} (i - \text{Sum average})^2 p_{x+y}(i)$$

$$\text{Sum entropy} = -\sum_{i=0}^{2(L-1)} p_{x+y}(i)\log_2\left[p_{x+y}(i)\right]$$

$$\text{Entropy} = -\sum_{i=0}^{L-1}\sum_{j=0}^{L-1} p(i,j)\log_2\left[p(i,j)\right]$$

$$\text{Difference variance} = Var(p_{x-y})$$

$$\text{where } p_{x-y}(l) = \sum_{i=0}^{L-1}\sum_{j=0}^{L-1} p(i,j), \text{ for } |i-j| = l, l = 0,1,...,L-1$$

$$\text{Difference entropy} = -\sum_{i=0}^{L-1} p_{x-y}(i)\log_2\left[p_{x-y}(i)\right]$$

$$\text{Energy} = \sum_{i=0}^{L-1}\sum_{j=0}^{L-1} p^2(i,j) \tag{15.5}$$

where

$$p_{x-y}(l) = \sum_{i=0}^{L-1}\sum_{j=0}^{L-1} p(i,j), \text{ for } |i-j| = l, l = 0,1,...,L-1 \tag{15.6}$$

15.3 D-Wavelet Transformation

Among many known 2-dimensional (2D) transformations, wavelet method becomes more and more useful tool for image processing [8,15–17]. Wavelets allow performing time–frequency analysis, similar to the so-called short-time Fourier and more general Gabor transforms. Practically, it works as 1D low (L) and high (H) pass filtering for an image represented by its rows and columns alternatively (see Figure 15.11). After each step of filtering, decimation is applied to reduce the size of the result image. Four images (LL_1, LH_1, HL_1, HH_1) are the results of such a processing, as shown in Figure 15.12. Next, LL_1 image becomes an input data for the next level of processing, where exactly the same procedure is repeated. Theoretically, it can be performed many times until the result images have 1×1 dimensions. In practice, the algorithm stops maximally after 3–4 steps (levels).

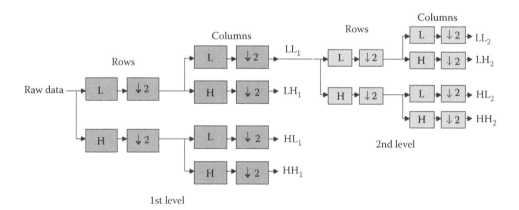

FIGURE 15.11   2D wavelet transform concept.

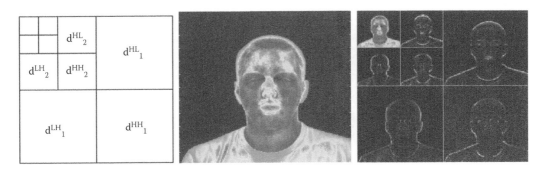

FIGURE 15.12 Typical result of 2D wavelet filtering, images' map, original and filtered images.

On each level, four images are available for further processing. These images are now used to calculate image features of the IR image. The technique of representing an image by the set of features is well known and frequently used in medical imaging. The features are then used in the classification process.

Tumors typically have better perfusion. They should have higher temperature, but in many cases, the network of blood vessels are much better developed and it has a special structural pattern. In such cases, the second-order parameters seem to be more effective to separate and classify healthy and pathological thermal images.

There are tens of thermal signatures that can be easily calculated. One of the problems in choosing the right thermal parameters is their dependency on the noise and size of the image taken for the investigations. Noise strongly depends on the type of the camera used in measurements [15]. This issue seems to be important as we have today two main thermovision camera types on the market—microbolometer, uncooled and photon, cooled ones. Obviously, uncooled, cost-effective cameras have the thermal resolution significantly lower, reaching the level 40–50 mK in contrast to 15–20 mK for cooled ones. The practical question can be posed, if there is a way to compensate the lower performance of the equipment by choosing the thermal signature less dependent on noise. In addition, because of the need of standardization of thermal images [7,15], it is necessary to estimate the right distance between the camera and the subject. Definitely, the closer the patient, the more accurate is the thermal imaging, but due to practical reason, it is very difficult to keep the constant distance in time, during massive screening. For this reason the sensitivity of the parameters as a function of distance and size of the subject should be as low as possible.

In order to find thermal parameters less dependent on the level of additive noise and the size of ROI, the preliminary research has been performed. The first- and second-order statistical parameters were considered. In addition, 2D wavelet transform was applied to calculate both histogram and co-occurrence matrix-based thermal signatures. A hypothesis was assumed that wavelet transform based on filtering (see Figures 15.11 and 15.12), should reduce the noise influence. Additionally, the considered parameter should allow differentiating the pathological and physiological cases in the acceptable range of noise and size variation.

As an example, the investigations of 10 patients with recognized breast tumors, as well as 30 healthy patients are presented. All healthy and pathological cases were confirmed by other diagnostic methods, such as mammography, USG, and/or biopsy. We have used thermographic camera to take two images for each breast: frontal and lateral ones. Each patient was relaxing before the experiment for 15 min in a room where temperature was stabilized at 20°C. Figure 15.13 presents the pathological case. Left breast has evidently higher temperature and the temperature distribution is very asymmetric.

In order to verify the noise and image size influence on the chosen features, the numerical procedure using 2D wavelet transformation was implemented. The wavelet transformation with biorthogonal filters was used. In order to differentiate healthy and pathological cases in breast investigation,

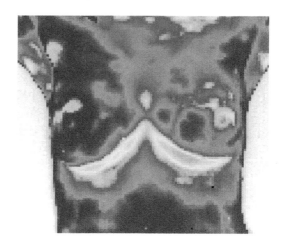

FIGURE 15.13 Example of thermal image showing the tumor (left breast).

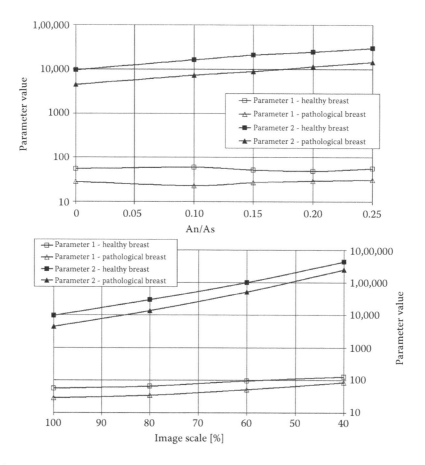

FIGURE 15.14 Second-order contrast (parameter 1) and correlation (parameter 2) chosen for breast investigation, parameter values versus noise level (top), and image scale (bottom).

second-order contrast and correlation were selected (see Figure 15.14). The chosen parameters depend on noise and size, but can still be used for classification in the wide range of noise and size variation. For example, decreasing the size of the thermal image to 40% of the original one and adding 25% noise to the image, change the parameter values both for healthy and pathological cases, but preserves possibility of the tumor detection (see Figure 15.14).

The investigations confirmed the expectation that among many different thermal signatures, the parameters calculated after 2D wavelet transformation using second-order statistical parameters are very promising for the screening purposes. Definitely, this investigation proved the necessity of selecting the best parameters. Different criteria have to be considered for such a selection. The proper discriminating healthy and pathological images have to be the main criterion. This investigation proved that such an approach can be a powerful tool for the medical diagnosis, and can be implemented as an automatic or semiautomatic software procedure.

15.4 ANN Classification

Thermal image classification can be a powerful tool for many medical diagnostic protocols such as breast cancer screening [15–17]. The classification process begins with image features calculation (see Figure 15.15). Because it is possible to define hundreds of different features for an image, obviously selection and reduction are needed. Many approaches are known for feature selection. One can use Fischer coefficient or POE together with ACC. These methods allow selecting the most effective and discriminative features [2,9,10]. Features' preprocessing which generates new parameters after linear or nonlinear transformation can be the next step in the overall procedure. It allows getting less correlated data and data of the lower order. Two approaches are used, PCA (Principal Component Analysis) and LDA [2,10]. Finally, classification can be performed using different ANNs, with or without additional hidden layers, and with different number of neurons. Alternatively, NNC can also be effective for thermal image classification.

ANN is typically used for classification. The selected image features are used as inputs. It means that the number of input nodes is equal to number of features. Number of neurons in the first hidden layer can be equal, lower or higher than the number of features in the classification, as illustrated in Figure 15.16. ANN can have user-defined next hidden layers which allow additional nonlinear processing of the input features. As ANN is the nonlinear system, such technique allows the additional decorrelating and data reduction.

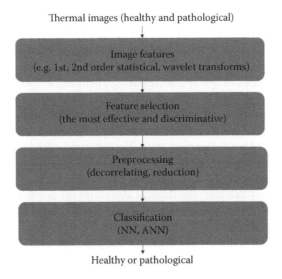

FIGURE 15.15 A typical image classification procedure.

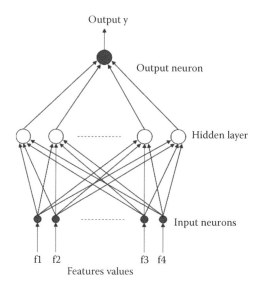

FIGURE 15.16   A typical one-directional ANN structure for thermal image classification.

Decorrelation of data, if possible, is always recommended for classification. Features of thermal images are more or less correlated with each other. It is because the change of certain thermal image content simultaneously changes the values of different features. Decorrelated set of features, describing different properties of thermal images are the best for classification. It can be achieved by preprocessing using PCA or Nonlinear Discriminant Analysis (NDA) [5,6,14].

It is well known that a proper training of ANN is the very important step in the entire protocol. It is a multivariable optimization problem, typically based on back-propagation technique. In the general case it can lead to wrong solutions if there is no single minimum of an error function. Therefore, one needs sufficient data size during training phase, and sometimes it is better or necessary to repeat training of ANN with different initial values of the neuron weight coefficients.

15.5 Software for Medical Diagnosis and Screening

To verify the research assumptions, novel software was created in MATLAB® environment (see Figures 15.18 and 15.19). In Laser Diagnostic and Therapy Center, at Technical University of Lodz, the laboratory for parallel diagnosis of breast diseases using mammography, ultrasonography, and thermography was created a few years ago and now is continuously using for medical treatment. In the same place and at the same time, a patient can be diagnosed using thermography, digital mammography, and ultrasonography. In our laboratory, a screening program has just been started, so we hope to collect enough images for ANN learning process. The software which is discussed in this chapter is suitable for features' calculations and image classification either for x-ray, acoustic or thermal images.

As shown in Figure 15.8, thermal camera operator can capture thermal image, define ROI, generate the histogram, and finally calculate the thermal signatures for raw data (first and second order). Having the left and right breast images, it is possible to evaluate the asymmetry of temperature patterns in the chosen ROI. In addition, the group of methods based on image transformations is implemented in the software. It is linear and nonlinear filtering, and Fourier or wavelet analysis (see Figure 15.18). Finally, classification can be performed using different ANN, with or without additional hidden layers, and with different number of neurons. NN method can also be used, as an alternative images' classification.

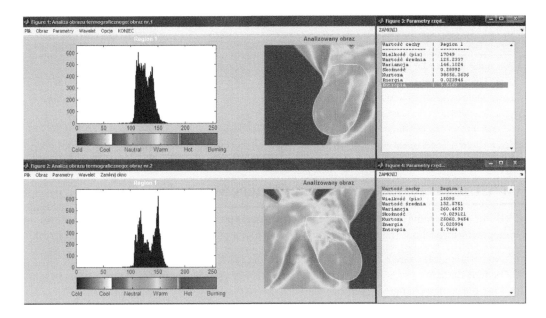

FIGURE 15.17   Application window with selected region and displayed the first-order features values.

In this research, both first- and second-order thermal features for left and right breasts were calculated (see Figures 15.17 and 15.18). Table 15.1 shows the values of chosen exemplary parameters for two breasts of the same patient, one is healthy and the second one is from the risky group. They can easily differentiate the thermal patterns on both images. These features were selected manually, just to confirm their usefulness in the thermal image classification. In practice, the automatic objective method can be used for features selection.

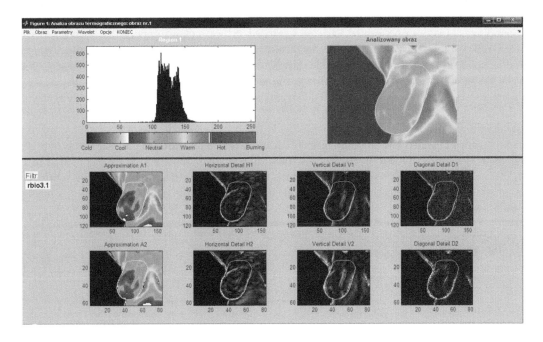

FIGURE 15.18   Results after wavelet transform for first and second filtering steps.

TABLE 15.1   Exemplary Parameter Values Differentiating Left and Right Breasts Shown in Figure 15.17

Parameters	Left Breast	Right Breast
First order, skewness for raw image	−0.029	0.29
Second order in vertical direction, correlation for raw image	107	1.24
First-order skewness after wavelets, LL1 image	0.0255	0.252
First-order variance after wavelets, LH2 image	452	223
Second order in diagonal direction, variance after wavelets, LH2 image	3060	599
Second order in antidiagonal direction, variance after wavelets, LH2 image	223	79.9

15.6 Clinical Results of Breast Cancer Screening

The preliminary screening investigations using thermal image processing has been made during the routine mammography inspection. The mammography investigations gave the results divided in five-stage scale.

1. No pathological change
2. Benign change
3. Possible benign change
4. Suspicious pathological change
5. Pathological change

The investigations have been carried out in the Laser Diagnostics and Therapy Centre at the Technical University of Lodz. During the medical diagnostics, 240 images were taken, two frontal and two lateral ones for 60 patients. Based on the mammography investigation, each breast image was classified into one of five groups. The same classification was made for the patients. There were patients with two healthy breasts without any pathological change classified to group no. 1. In addition, we had found patients with one breast which was selected to one of four groups other than the groups mentioned above. In this way, the small database of thermal images and patients was created (see Table 15.2). The images of patient breasts from group nos. 3 and 4 were recognized as suspicious. No patients with pathological change were found during this investigation.

Four different tests were performed. Each test was based on the simple assumption that having N images in each trail, $N-1$ images were used for training the neural network and one separated for testing. Then the one directional neural network with the hidden layer was trained using all images except the one separated. Finally, the separated image was used for the classification test. In order to minimize the statistical error, this process was repeated several times with different images for testing, and the average result was taken.

Test no. 1

The first test was carried out for 15 images of breasts classified into the 3rd and 4th risk group. The next 15 images were selected to the group without any change. The training data consisted of 29 sets of selected parameters. Frontal thermal images were taken into account only. The results are presented in Table 15.3. Selection using Fisher coefficient reduces the set of features into six ones as below.

- Contrast—Second-order statistical parameter after Haar wavelet, matrix HL-2, horizontal direction with distance $d = 1$.

TABLE 15.2   Number of Patients for Five Risk Groups during the Screening Investigations

No. of group	1	2	3	4	5
No. of patients	25	20	10	5	0
No. of images	170	40	20	10	0

TABLE 15.3 Classification Results for Frontal Thermal Images

No.	Risk Group	% of Correctly Classified	No. of Correctly Classified	No. of Classification Trials
1	3	27.3	3	11
2	3	81.8	9	11
3	3	72.7	8	11
4	3	72.7	8	11
5	3	63.6	7	11
6	3	90.9	10	11
7	3	81.8	9	11
8	3	63.6	7	11
9	3	54.5	6	11
10	3	72.7	8	11
11	4	72.7	8	11
12	4	27.2	3	11
13	4	27.2	3	11
14	4	18.2	2	11
15	4	72.7	8	11
16	1	18.2	2	11
17	1	0	0	11
18	1	18.2	2	11
19	1	100	11	11
20	1	36.4	4	11
21	1	36.4	4	11
22	1	18.2	2	11
23	1	72.7	8	11
24	1	63.6	7	11
25	1	72.7	8	11
26	1	54.5	6	11
27	1	81.8	9	11
28	1	72.7	8	11
29	1	100	11	11
30	1	63.6	7	11

- Sum variance—Second-order statistical parameter after Haar wavelet, matrix HL-2, antidiagonal direction (135°) with distance $d = 1$.
- Difference entropy—Second-order statistical parameter after Haar wavelet, matrix HL-1, diagonal direction (45°) with distance $d = 2$.
- Skewness—First-order statistical parameter after Haar wavelet, matrix LH-1.
- Kurtosis skewness—First-order statistical parameter after reverse biorthogonal 3.1 wavelet, matrix LL-2.
- Difference variance—First-order statistical parameter, diagonal direction (45°) with distance $d = 2$.

Test no. 2

Test no. 2 was made for the same group of patients as test no. 1. The lateral thermal images were analyzed (see Table 15.4 and Figure 15.19). Another six features were chosen using the Fisher criterion.

- Sum entropy—Second-order statistical parameter after Haar wavelet, matrix HL-1, diagonal direction (45°) with distance $d = 3$.
- Difference entropy—Second-order statistical parameter after Haar wavelet, matrix HL-1, antidiagonal direction (135°) with distance $d = 3$.

TABLE 15.4   Classification Results for Lateral Thermal Images

No.	Risk Group	% of Correctly Classified	No. of Correctly Classified	No. of Classification Trials
1	3	63.6	7	11
2	3	45.5	5	11
3	3	36.4	4	11
4	3	81.8	9	11
5	3	81.8	9	11
6	3	54.5	6	11
7	3	63.6	7	11
8	3	81.8	9	11
9	3	27.2	3	11
10	3	9.1	1	11
11	4	81.8	9	11
12	4	54.5	6	11
13	4	45.4	5	11
14	4	81.8	9	11
15	4	90.9	10	11
16	1	18.1	2	11
17	1	9.1	1	11
18	1	45.4	5	11
19	1	0	0	11
20	1	45.4	5	11
21	1	9.1	1	11
22	1	63.6	7	11
23	1	54.5	6	11
24	1	45.4	5	11
25	1	27.2	3	11
26	1	18.1	2	11
27	1	54.5	6	11
28	1	45.4	5	11
29	1	36.3	4	11
30	1	27.2	3	11

- Entropy—First-order statistical parameter after Haar wavelet, matrix HL-1.
- Difference entropy—First-order statistical parameter, horizontal direction with distance $d = 1$.
- Short run emphasis inverse moments—Run length matrix-based parameters [10,17].
- Kurtosis for gradient-based parameters [10,17].

Test no. 3

Test no. 3 was carried out for 40 patients. Twenty-five women were healthy (with both breasts belonging to group no. 1), and 15 with changes from 3rd and 4th risk group (only one breast belonged to group no. 3 or 4). Simulation in this case took into account the difference of chosen parameters for both patients' breasts seen in front (see Table 15.5). The following six features were used in the calculations.

- Contrast—Second-order statistical parameter after Haar wavelet, matrix LH-1, diagonal direction (45°) with distance $d = 1$.
- Difference variance—Second-order statistical parameter after Haar wavelet, matrix HL-2, horizontal direction with distance $d = 1$.
- Sum of squares—Second-order statistical parameter after Haar wavelet, matrix LH-2, diagonal direction (45°) with distance $d = 1$.

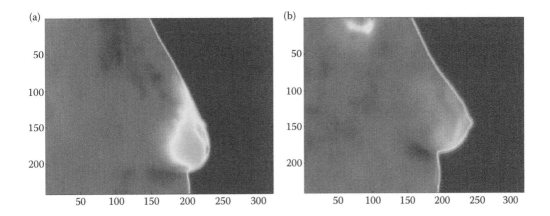

FIGURE 15.19

- Difference entropy—Second-order statistical parameter after reverse biorthogonal 3.1 wavelet, matrix LH-1, vertical direction with distance $d = 2$.
- Sum average—Second-order statistical parameter after reverse biorthogonal 3.1 wavelet, matrix LH-1, antidiagonal direction (135°) with distance $d = 1$.
- Variance—First-order statistical parameter after reverse biorthogonal 3.1 wavelet, matrix LH-2.

Test no. 4

The test no. 4 was made for the same group of patients as test no. 3. Simulation in this case took into account the difference of chosen parameters for both breasts seen from the lateral side (see Table 15.6). The same (as in test no. 3) six features were used in the calculations.

- Contrast—Second-order statistical parameter after Haar wavelet, matrix LH-1, diagonal direction (45°) with distance $d = 1$.
- Difference variance—Second-order statistical parameter after Haar wavelet, matrix HL-2, horizontal direction with distance $d = 1$.
- Sum of squares—Second-order statistical parameter after Haar wavelet, matrix LH-2, diagonal direction (45°) with distance $d = 1$.
- Difference entropy—Second-order statistical parameter after reverse biorthogonal 3.1 wavelet, matrix LH-1, vertical direction with distance $d = 2$.
- Sum average—Second-order statistical parameter after reverse biorthogonal 3.1 wavelet, matrix LH-1, anti-diagonal direction (135°) with distance $d = 1$.
- Variance—First-order statistical parameter after reverse biorthogonal 3.1 wavelet, matrix LH-2.

Table 15.7 presents the summarized results of the classification. The efficiency of detection is higher for pathological changes than for healthy cases. The lateral thermal imaging of breasts allows getting the better results. The best detection of pathological changes was for test no. 4, that is, for features calculated as the difference of chosen parameters for healthy and suspicious breasts seen from the lateral direction. Detectivity for pathological changes was at 70%, while 57% for healthy cases. One can notice that it is a good result for low number of images available for training neural network and classification.

15.7 Conclusions

This chapter presents the preliminary results of the feature analysis for thermal images for different medical applications, mainly used in breast oncology. Thermography as the additional and adjacent method can be very helpful for early screening it helps to detect and even recognize the tumors. At first, we consider first- and second-order statistical parameters. Although we do not have many pathological

TABLE 15.5 Classification Results for Frontal Thermal Images for the Parameter Difference between Pathological and Healthy Breast

No.	Risk Group	% of Correctly Classified	No. of Correctly Classified	No. of Classification Trials
1	3	57.1	4	7
2	3	28.6	2	7
3	3	85.7	6	7
4	3	57.1	4	7
5	3	42.9	3	7
6	3	85.7	6	7
7	3	85.7	6	7
8	3	42.9	3	7
9	3	71.4	5	7
10	3	100	7	7
11	4	100	7	7
12	4	85.7	6	7
13	4	85.7	6	7
14	4	28.6	2	7
15	4	0	0	7
16	1	85.7	6	7
17	1	85.7	6	7
18	1	71.4	5	7
19	1	57.1	4	7
20	1	71.4	5	7
21	1	71.4	5	7
22	1	42.9	3	7
23	1	57.1	4	7
24	1	42.9	3	7
25	1	71.4	5	7
26	1	42.9	3	7
27	1	28.6	2	7
28	1	57.1	4	7
29	1	28.6	2	7
30	1	0	0	7
31	1	71.4	5	7
32	1	0	0	7
33	1	85.7	6	7
34	1	57.1	4	7
35	1	85.7	6	7
36	1	57.1	4	7
37	1	71.4	5	7
38	1	14.3	1	7
39	1	71.4	5	7
40	1	0	0	7

cases for investigations yet, the first results are very promising. The second-order parameters are more sensitive to the overall structure of an image. Lately, the study has been extended by choosing second-order parameters for multivariate data classification. The presented approach includes the PCA analysis to reduce the dimensionality of the problem and by selecting the eigen vectors it is possible to generate data which represents the tumors more evidently.

Breast cancer screening is a challenge today for medical engineering. Breast temperature depends not only because of some pathological changes, but it also varies in normal physiological situations, even

TABLE 15.6 Classification Results for Lateral Thermal Images for the Parameter Difference between Pathological and Healthy Breast

No.	Risk Group	% of Correctly Classified	No. of Correctly Classified	No. of Classification Trials
1	3	57.14	4	7
2	3	28.57	2	7
3	3	85.71	6	7
4	3	57.14	4	7
5	3	42.86	3	7
6	3	85.71	6	7
7	3	85.71	6	7
8	3	42.86	3	7
9	3	71.43	5	7
10	3	100	7	7
11	4	100	7	7
12	4	85.71	6	7
13	4	85.71	6	7
14	4	28.57	2	7
15	4	100	7	7
16	1	85.71	6	7
17	1	85.71	6	7
18	1	71.43	5	7
19	1	57.14	4	7
20	1	71.43	5	7
21	1	71.43	5	7
22	1	42.86	3	7
23	1	57.14	4	7
24	1	42.86	3	7
25	1	71.43	5	7
26	1	42.86	3	7
27	1	28.57	2	7
28	1	57.14	4	7
29	1	28.57	2	7
30	1	0	0	7
31	1	71.43	5	7
32	1	0	0	7
33	1	85.71	6	7
34	1	57.14	4	7
35	1	85.71	6	7
36	1	57.14	4	7
37	1	71.43	5	7
38	1	14.29	1	7
39	1	71.43	5	7
40	1	100	7	7

it is a consequence of emotional state of a patient. It was a main reason that we are looking for more advanced methods of thermal image processing that could give satisfactory results.

One of the possible alternatives for such processing in ANN Classification based on multidimensional feature domain, with use of image transformations, for example, wavelet analysis. The preliminary investigations were quite successful, and can be improved by increasing the number of samples taken for processing.

Future research will concentrate around selection of features and adjusting wavelet transformation parameters to get the best classification. We assume that the more satisfactory results can be obtained

TABLE 15.7   Summary of the Classification Tests

	True Positive (%)	False Negative (%)	True Negative (%)	False Positive (%)
Test 1	60	40	54	46
Test 2	60	40	33	67
Test 3	64	36	53	47
Test 4	70	30	57	43

by using features based on asymmetry between left and right side of a patient. It could help for one-side cancerous lesion classification, what is the most typical pathological case and frequently happens today.

References

1. W.C. Amalu, W.B. Hobbins, F.J. Head Elliot, Infrared imaging of the breast—An overview, *Medical Devices and Systems*, Ed. J.D. Bronzino, Boca Raton, FL, CRC Press, 2006, Section III, Chapters 25-1 to 25-21.
2. M. Bennett Breast cancer screening using high-resolution digital thermography, *Total Health*, 22(6) 44, 1985.
3. R. Causton, *A Biologist's Advanced Mathematics*, London, Allen and Unwin, 1987.
4. P. Cichy, Texture analysis of digital images–doctoral thesis, *Technical University of Lodz, Institute of Electronics*, Lodz, 2000, in Polish.
5. P. Debiec, M. Strzelecki, A. Materka, Evaluation of texture generation methods based on CNN and GMRF image texture models, *Proceedings of International Conference on Signals and Electronic Systems*, October 17–20, 2000, Ustron, pp. 187–192.
6. I. T. Jolliffe, *Principal Component Analysis*, New York, Springer-Verlag, 1986.
7. T. Jakubowska, B. Wiecek, M. Wysocki, C. Drews-Peszynski, Thermal signatures for breast cancer screening comparative study, *Proc. IEEE EMBS Conf.*, Cancun, Mexico, Sept. 17–21, 2003.
8. M. Kociolek, A. Materka, M. Strzelecki, P. Szczypinski, Discrete wavelet transform-derived features for digital image texture analysis, *Proceedings of International Conference on Signals and Electronic Systems ICSES'2001*, Lodz, 18–21 Sept. 2001, pp. 111–116.
9. B.F.J. Manly, *Multivariate Statistical Method: A Primer*. London, Chapman & Hall, 1994.
10. A. Materka, M. Strzelecki, R. Lerski, L. Schad, Evaluation of texture features of test objects for magnetic resonance imaging, *Infotech Oulu Workshop on Texture Analysis in Machine Vision*, Oulu, Finland, June 1999.
11. E.Y.K. Ng, L.N. Ung, F.C. Ng, L.S.J. Sim, Statistical analysis of healthy and malignant breast thermography, *Journal of Medical Engineering and Technology*, 25(6), 253–263, 2001.
12. E.Y-K. Ng, A review of thermography as promising non-invasive detection modality for breast tumour, *International Journal of Thermal Sciences*, 2008, DOI: 10.1016/j.ijthermalsci.2008.06.015.
13. E.Y-K. Ng, and N.M. Sudharsan, Numerical modelling in conjunction with thermography as an adjunct tool for breast tumour detection, *BMC Cancer, Medline Journal* 4(17), 1–26, 2004.
14. J. Schürman, *Pattern Classification*, John Wiley & Sons, 1996.
15. B. Wiecek, S. Zwolenik, Thermal wave method—Limits and potentialities of active thermography in biology and medicine, *2nd Joint EMBS-BMES Conference, 24th Annual International Conference of the IEEE Engineering in Medicine and Biology Society*, BMES-EMS 2002, Houston, Oct. 23–26, 2002.
16. B. Wiecek, M. Strzelecki, T. Jakubowska, M. Wysocki, C. Drews-Peszynski, Advanced thermal image processing, *Handbook of Medical Devices and Systems*, Ed. Joseph D. Bronzino, Boca Raton, FL, CRC Press, 2006, Chapters 28-1–28-13.
17. M. Wiecek, R. Strakowski, T. Jakubowska, B. Wiecek, Software for classification of thermal imaging for medical applications, *9th International Conference on Quantitative InfraRed Thermography QIRT2008, Inżynieria Biomedyczna*, Vol. 14, nr 2/2008, str. 143.

16

Fuzzy C Means Segmentation and Fractal Analysis of the Benign and Malignant Breast Thermograms

M. Etehadtavakol
Isfahan University of Medical Science

E.Y.K. Ng
School of Mechanical and Aerospace Engineering

Nanyang Technological University

Caro Lucas
University of Tehran

S. Sadri
Isfahan University of Technology

Isfahan University of Medical Science

16.1 Preamble

Breast cancer is one of the main problems in women's health today. Early detection of breast cancer plays a significant role in lowering the mortality rate. Cancer threats could be halted by identifying and removing malignant tumors in the early stages before they metastasize and spread to adjacent regions. Breast thermography [1] is a potential early detection method which is fast, nonradiating, noninvasive, low cost, passive, painless, and risk free with no contact with the body [2–4]. It is effective for women with all sizes of breast as well as with all ages, for women with breast with dense tissue, and for nursing or pregnant women [5,6].

In 1956, Lawson declared that the skin temperature of breast with cancer was hotter than the normal one. Hence, he proposed that in the infrared (IR) images, the cancer tissues can be differentiated as hot spots [7]. The cancerous tissue emits with angiogenesis and it has an inflammation temperature pattern different from the healthy one. In IR pseudocolor images, different colors indicate different rates of temperature. Color segmentation of IR thermal images can be very useful in detecting the tumor regions.

Cancer is often designated as a chaotic, poorly regulated growth [8]. Cancerous cells and tumors have irregular shapes, and traditional Euclidean geometry based on smooth shapes such as line, plane, cylinder, and sphere are not able to delineate them. When the focus is on irregularities of tumor growth, fractal geometry is thus useful. Fractal geometry can be a more powerful means of quantifying the spatial complexity of real objects [9].

16.2 Color Segmentation

16.2.1 K Means Clustering

The K means algorithm that has been used in many pattern recognition problems was suggested in the 1960s [10]. It is one of the most straightforward unsupervised learning techniques for clustering.

It divides N data points to K disjoint subsets S_j where $j = 1,2,...,K$.

It works with the minimization of the objective function which is expressed as

$$J = \sum_{j=1}^{K} \sum_{n \in S_j} \left\| x_n - \mu_j \right\|^2 \tag{16.1}$$

where x_n is a vector for the nth data point in S_j and μ_j is the geometric centroid of data points. The algorithm consists of two steps: the number of cluster k must be chosen, and then the K means clustering to the image must be implied.

16.2.2 Mean Shift Clustering

Mean shift clustering (MS) was proposed by Fukunaga and Hosteler in 1975 [11]. The mean shift algorithm is a nonparametric clustering technique which does not require prior knowledge of the number of clusters, and does not constrain the shape of the clusters [12–15]. Assuming x_i ($i = 1,...,n$) be a set of feature vectors in a d-dimensional feature space. The density at any point x can be estimated by the Parzen window kernel density estimator $K(x)$ with window size h as described below.

$$f_{h,K}(x) = \frac{c_{k,d}}{nh^d} \sum_{i=1}^{n} k\left(\left\| \frac{x - x_i}{h} \right\|^2 \right) \tag{16.2}$$

The normalization constant $c_{k,d}$ assures that $K(x)$ integrates to one. Three different kernel estimators are Epanechnikov, Uniform, and Normal. They are introduced in Equations 16.3, 16.4, and 16.5, respectively.

$$K(x) = \begin{bmatrix} c(1 - \|x\|^2) & \text{for} \|x\| \leq 1 \\ 0 & \text{otherwise} \end{bmatrix} \tag{16.3}$$

$$K(x) = \begin{bmatrix} c & \text{for} \|x\| \leq 1 \\ 0 & \text{otherwise} \end{bmatrix} \tag{16.4}$$

$$K(x) = c.\exp\left(-\frac{1}{2} \|x\|^2 \right) \tag{16.5}$$

Taking the gradient of Equation 16.2 leads to

$$0 = \nabla f_{h,K}(x) = \frac{2c_{k,d}}{nh^{(d+2)}} \sum_{i=1}^{n} (x - x_i) g\left(\left\|\frac{x - x_i}{h}\right\|^2\right) \tag{16.6}$$

$$= \frac{2c_{k,d}}{nh^{(d+2)}} \left[\sum_{i=1}^{n} g\left(\left\|\frac{x - x_i}{h}\right\|^2\right)\right] \left(\frac{\sum_{i=1}^{n} x_i g\left(\left\|\frac{x - x_i}{h}\right\|^2\right)}{\sum_{i=1}^{n} g\left(\left\|\frac{x - x_i}{h}\right\|^2\right)} - x\right) \tag{16.7}$$

where $g(s) = -K'(s)$. Equation 16.7 is a product of two terms. The first term is proportional to the density estimate at x, and the second term is the mean shift vector as expressed below as

$$m_h(x) = \frac{\sum_{i=1}^{n} x_i g\left(\left\|\frac{x - x_i}{h}\right\|^2\right)}{\sum_{i=1}^{n} g\left(\left\|\frac{x - x_i}{h}\right\|^2\right)} - x \tag{16.8}$$

It is observed that the mean shift vector always points toward the direction of the maximum increase in the density. If we start from a point X^t in feature space, we move with the mean shift vector to the point X^t+1. The mean shift procedure is obtained by successive

1. Computation of the mean shift vector $m_h(X^t)$
2. Translation of the window $X^t+1 = X^t + m_h(X^t)$

It is important to note that the number of clusters is not a parameter, but rather an output of the clustering algorithm. The only parameter of the mean shift clustering is the window size parameter h and its influence on the obtained results is significant.

16.2.3 Fuzzy C Means Clustering

Fuzzy C means (FCM) was first suggested by Bezdek et al. [16]. A value between 0 and 1, called the partition matrix, is given by the algorithm. It means the degree of membership between each data and centers of clusters is determined by minimizing the objective function as shown below

$$J_m(U,C) = \sum_{i=1}^{c} \sum_{k=1}^{n} u_{ik}^{m} \|X_k - C_i\|^2 \tag{16.9}$$

where $X_1, X_2, \ldots, X_n$ are n data sample vectors and m is any real number greater than one. $U \equiv [u_{ik}]$ is a $c \times n$ matrix, where u_{ik} is the ith membership value of the kth input sample X_k such that $\sum_{i=1}^{c} u_{ik} = 1$ and $C \equiv \{C_1, C_2, \ldots, C_c\}$ are cluster centers.

The fuzziness of the membership function is controlled by $m \in [1,\infty)$ which is an exponent weight factor. The similarity between any input sample and its corresponding cluster center is expressed by $\|*\|$.

The objective function of Equation 16.10 is optimized with updating of the membership u_{ik}, and cluster center C_i

$$u_{ik} = \frac{1}{\sum_{j=1}^{C} \left\{ \frac{\|X_k - C_i\|}{\|X_k - C_j\|} \right\}^{\frac{2}{m-1}}}$$

(16.10)

$$C_i = \frac{\sum_{k=1}^{n} u_{ik}^m . X_k}{\sum_{k=1}^{n} u_{ik}^m}$$

(16.11)

The standard FCM involves the following steps:

1. The number of clusters C is chosen.
2. The exponent weight m is chosen.
3. The membership u_{ik} is initialized.
4. The cluster center C_i (Equation 16.11) is calculated.
5. The membership u_{ik} (Equation 16.10) is updated for $i = 1,2,\ldots,c$ and $k = 1,2,\ldots,n$.
6. Steps 4 and 5 are repeated until the distortion is less than a specified value.

In color segmentation of IR images by FCM clustering, the colors are compared in a relative sense and grouped in clusters which are not with crisp boundaries. Moreover, data points can belong to more than one cluster. That is, pixels can belong to many clusters with different degrees of membership. The constraint of $\sum_{i=1}^{c} u_{ik} = 1$ for $1 \le k \le n$ points out that the sum of the membership values for all clusters of a data vector equals one. Hence, the trivial solution $U = 0$ is prevented and reasonable results with no empty cluster are provided.

It is worth noting that the data points in this implementation for segmentation of the IR image are pixels of the corresponding color space CIELAB (L*a*b*).

16.3 Results for Color Segmentation

We studied 15 breast thermograms available from the Ann Arbor thermography center [17], thermal imaging lab in the San Fransisco Bay Area [18], American College of Clinical Thermology [19], Thermography of Iowa [20], and Sunstate Thermal Imaging Center in Australia [21].

All cases studied are implemented using K means, mean shift, and C means algorithms. In some trials, empty clusters appeared when applying the K means algorithm, and if the algorithm is repeated several times, the results for different trials may be different showing that K means clustering is not stable. Figure 16.1b shows the trial in which two clusters are empty for a cancer case of Figure 16.1a. Figure 16.2b shows the trial in which three clusters are empty for a normal case of Figure 16.2a.

Mean shift clustering is very sensitive to the window size parameter h. Results of implementation of the mean shift clustering with $h = 12$ and $h = 15$ are presented in Figure 16.2c and d, respectively. In this implementation, 11 empty clusters and four empty clusters are found with $h = 12$ and $h = 15$, respectively.

As we expected, by implementing the FCM algorithm for both cases, as shown in Figures 16.1d and 16.2e, however, no empty cluster appeared. Hence, by using this segmentation technique for breast IR images, we are able to find the first and the second hottest regions for each case where some useful features could be extracted.

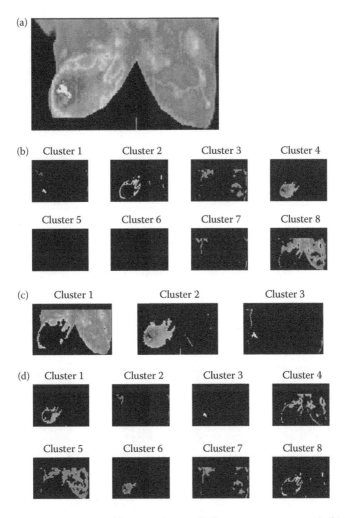

FIGURE 16.1 Inflammatory cancer case. (a) Original image (inflammatory cancer case). (b) Color segmentation by K means; two empty clusters (inflammatory cancer case). (c) Color segmentation by mean shift with $h = 35$. (d) Color segmentation by fuzzy C means.

In our work, despite the fact that different images are taken from different thermography with different color palettes and different numbers of clusters, the FCM approach is however capable of identifying the first and the second hottest regions by comparing their colors with the color palette spectrum which has been applied. With segmentation of thermal images, the hottest cluster can be recognized. We are capable of extracting some useful information from the suspicious regions by comparing the hottest region color with the spectrum of color palette used.

Next, we will examine and have a better understanding of those factors that are affecting the breast surface temperature.

16.4 Factors That Determine Breast Temperature

16.4.1 Angiogenesis

There have been a number of studies on the angiogenesis of tumors. The vascular construction of tumors is experienced to be remarkably different from that of normal tissues [22].

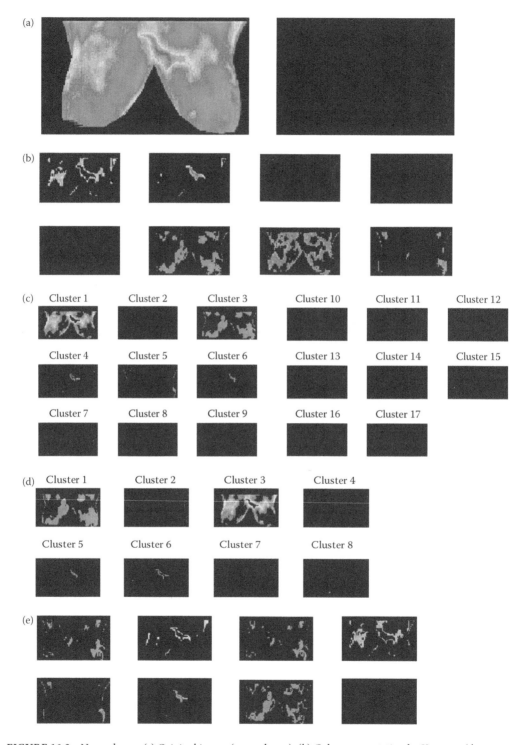

FIGURE 16.2 Normal case. (a) Original image (normal case). (b) Color segmentation by K means (three empty clusters). (c) Color segmentation by mean shift with $h = 12$ (11 empty clusters). (d) Color segmentation by mean shift with $h = 15$ (4 empty clusters). (e) Color segmentation by fuzzy C means.

In normal vascular networks, tree-like branching patterns or nearly constant intravascular distances are persistent features, while no such features exist in disorganized tumor vascular networks. One normal situation where angiogenesis occurs is wound repairing where the fundamental process of the development and growth of new blood vessels from the preexisting vasculature is extremely regulated. A developing child in the mother's womb where the vast network of arteries, veins, and capillaries are created is another example of normal angiogenesis [23]. Some studies have been directed with regard to angiogenesis in breast thermography. Gamagami discovered hypervascularity and hyperthermia in 1996 in 86% of nonpalpable breast cancers with thermograms [24]. He also observed that in 15% of those cases, thermography helped to catch cancers that were not visible through mammography. Guido and Schnitt [25] described that angiogenesis is an early event in the development of breast cancer that might occur before tumor cells achieved the ability to invade the surrounding stoma and even before there was any morphologic evidence of a ductal carcinoma *in situ* (DCIS). Correspondingly, angiogenesis has been discovered as a critical event in tumor growth. Alarcon et al. investigated the influences of blood flow and red blood cell heterogeneity on tumor growth and angiogenesis with a mathematical model [26]. Another model of tumor angiogenesis taking into consideration the biochemical processes was proposed by Smiley and Levine [27].

16.4.2 High-Metabolic-Rate Angiogenesis

Tumor angiogenesis relates the proliferation of a network of blood vessels which supply nutrients and oxygen into tumor cells and remove waste products from them [28]. Molecules that send signals to surrounding normal host tissue are released by cancerous tumor cells. Certain genes (oncogenes) in the host tissue could be activated and make the proteins raise the growth of new blood vessels by these signals.

The local interaction between cells and the vascular system is modeled by Scalerandi et al. [29].

16.4.3 Local Vasodilatation

Anbar observed that notable IR changes in small tumor were produced by enhanced perfusion over a substantial area of breast surface via tumor-induced nitric oxide vasodilatation [30,31].

Nitroxide distribution is different between the tumor and normal tissues. This fact may reflect the differences in vasculatures of microenvironment associated with tumors. Also, in tumors of different sizes, the nitroxide distribution differs substantially [32].

16.5 Applications of Fractal Analysis in Biomedical Images and Fractal Dimension

In biomedical areas, fractal analysis has been widely used. Lee et al. compared several shape factors, including fractal dimension (FD) on the irregularity of the borders of melanocytic lesions [33]. Zheng and Chan proposed a model to employ fractals to detect abnormal regions in mammograms [34]. Guo et al. [35] computed FD in mammograms to identify the abnormality of breast masses by employing a support vector machine [35]. FD was also computed by Caldwell et al. [36] and Byng et al. [37] using the box counting method (BCM) to characterize the breast tissue. Gazit et al. used the fractal concept to analyze the vessel networks that surround a tumor and the hemodynamics within these vessel structures [38]. Grizzi et al. introduced the surface FD that was able to explain the geometric complexity of cancerous vascular networks [39]. They indicated that the number of vessels and their patterns of distribution have a significant impact on the surface FD. Investigations were further carried out by applying FD for classification of tumors in magnetic resonance images of brain [40,41], images related to colonic cancer [42], and ultrasonic images of liver [43]. Rangayyan and Nguyen applied fractal analysis for the classification of breast masses by using only their contours [44].

Fractal analysis plays an important role in discriminating malignant tumors from benign tumors in mammograms [45]. The edge sharpness of malignant mass as well as the benign mass has been studied in several investigations. It has been shown that malignant masses are rough and have complicated boundaries while benign masses are usually round and smooth with well-defined boundaries [46,47]. We can evaluate geometrical complexity by quantifying the irregularities of the boundary.

16.6 Fractal Dimension

A fractal is a nonregular geometric shape that can be split into parts which possess self-similarity or have the same degree of irregularity on all scales [48–53]. FD is a statistical quantity that denotes how completely a fractal would fill the space in different scales or magnification in a fractal geometry. The concept of fractal was introduced by Mandelbrot to denote an object whose Hausdorff dimension is greater than its topological dimension [48].

The relation between D, the self-similarity dimension, and a, the number of self-similar pieces at reduction factor (1/S) is defined by the following power law:

$$a \propto \frac{1}{S^D} \tag{16.12}$$

D is expressed as

$$D = \frac{\log(a)}{\log\left(\frac{1}{S}\right)} \tag{16.13}$$

That is, D is estimated by the slope of the straight line approximation for a plot of log (a) versus log (1/S). Analytical and box counting methods are among several approaches for estimating FD. By using an analytical rule which is based upon a recursive mathematical relation, we can generate fractals. To explain BCM, we may partition the image into square boxes of equal sizes and then count the number of boxes which contain a part of the image. The process is repeated with partitioning the images into smaller and smaller size of boxes. To compute the FD of pattern, we may use the plot of log of the number of boxes counted versus the log of the magnification index for each stage of partitioning. The slope of the best-fitting line to the aforementioned plot is the FD of pattern.

Supposing a grid of boxes is laid over the curve as shown in Figure 16.3. The number of grid boxes that contain a part of the curve (viz. the boxes having intersections with the curve, shaded in gray) is calculated. In Figure 16.3, the number of these boxes is 10 out of 25 boxes, hence, $a = 10$, $1/S = 25$. This is kept on for increased numbers of squares, and the FD is estimated by the gradient of the logarithm of the number of squares occupied by the edge contour, viz. log(a), over the logarithm of the number of squares, log(1/S).

Ten contour shapes of different irregularities are shown in Figure 16.4. Their FDs are determined by BCM and results are included in Table 16.1. We may consider that the more complicated the shape boundary is, the larger the value of FD would be.

FD can be a potentially valuable tool for explaining the pathological architecture of breast tumors and providing insights into the mechanisms of tumor growth as it is demonstrated above. In this study, we examine whether the vascular networks in thermal images own a fractal nature and if so, what would be the FD values in different stages of abnormality.

16.7 Processing Steps and Results for Fractal Analysis

If we share the IR images with others or use them for research, it is more convenient by using a calibrated standard IR camera; then, we can ensure that relevant IR images and accurate temperatures can be viewed since each color is associated with a specific temperature. Although only mapping the local temperature differences are enough to accomplish the imaging for tumors, an accurate and stable calibration

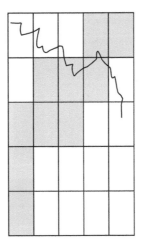

FIGURE 16.3 Box counting method.

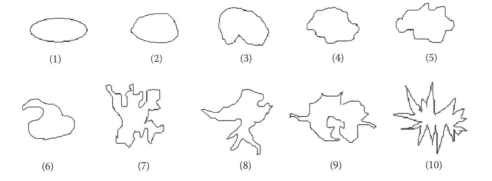

FIGURE 16.4 Sample contours of different irregularities.

TABLE 16.1 Calculated FDs for 10 Different Contour Shapes

Contours	1	2	3	4	5	6	7	8	9	10
FD	1.0711	1.1573	1.2162	1.2204	1.2750	1.2962	1.4076	1.4194	1.4370	1.4464

in order to satisfy the temperature sensitivity of the camera is significant. For the breast cancer case, the calibration method must be carried out such that the temperature difference is about one tenth of 37°C [3]. There are also other crucial factors so as to make thermal images in a standard form. The environmental conditions of thermal imaging such as humidity, ambient temperature, and illumination are essential to be controlled. Also, preparation of the patient in certain conditions is necessary.

Although the images we used are from sources which are varied in their resolutions and generally did not follow a unified protocol, our fractal analysis could demonstrate significant difference between the benign and malignant cases.

In this work, the right breast was separated from the left breast automatically. Canny edge detector and the morphological bridging operations for obtaining closed contour regions were used to extract left and right body boundaries. The two lower boundaries of the breasts were extracted as follows: for a data set, nine landmark points for two breasts were localized in a training procedure. The points with maximum curvature of two breasts correspond to the first and the last points. Then, for a new case, the

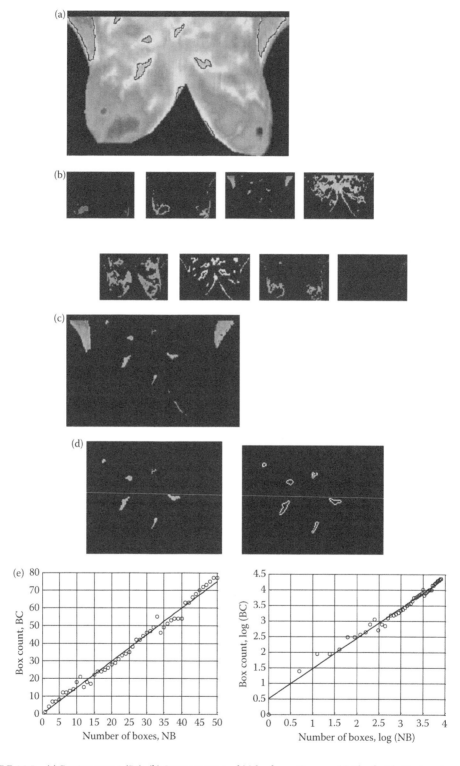

FIGURE 16.5 (a) Benign case 1 (B1). (b) Segmentation of (a) by fuzzy C mean. (c) The first hottest regions of B1. (d) B1: (1) The first hottest regions after removing the axilla and close sternal boundaries. (2) Boundaries of part (1). (e) B1: (1) Box count (BC) versus number of boxes (NB). (2) log(BC) versus log (NB).

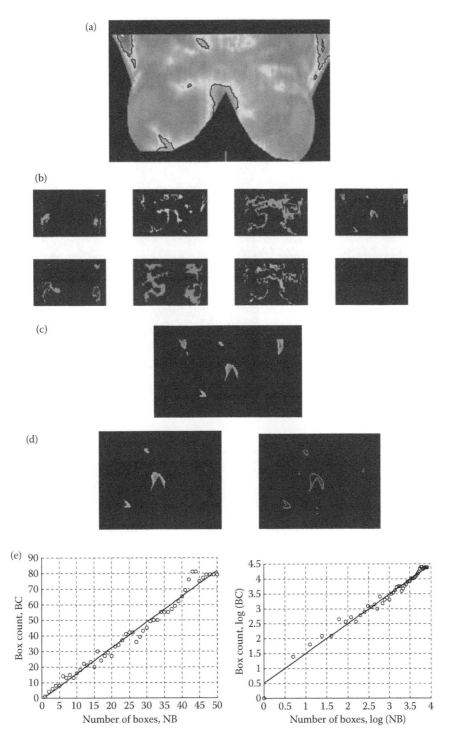

FIGURE 16.6 (a) Benign case 2 (B2). (b) Segmentation of (a) by fuzzy C mean. (c) The first hottest regions of B2. (d) B2: (1) The first hottest regions after removing the axilla boundaries. (2) Boundaries of part (1). (e) B2: (1) Box count (BC) versus number of boxes (NB). (2) log(BC) versus log (NB).

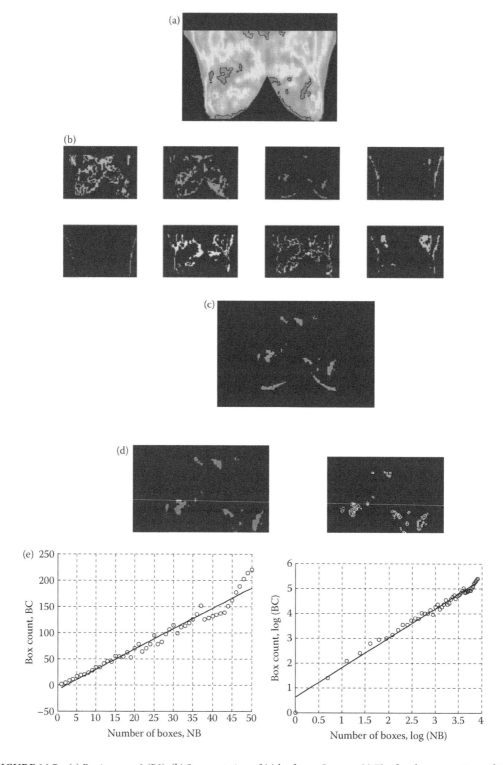

FIGURE 16.7 (a) Benign case 3 (B3). (b) Segmentation of (a) by fuzzy C mean. (c) The first hottest regions of B3. (d) B3: (1) The first hottest regions after removing close sternal boundaries. (2) Boundaries of part (1). (e) B3: (1) Box count (BC) versus number of boxes (NB). (2) log(BC) versus log (NB).

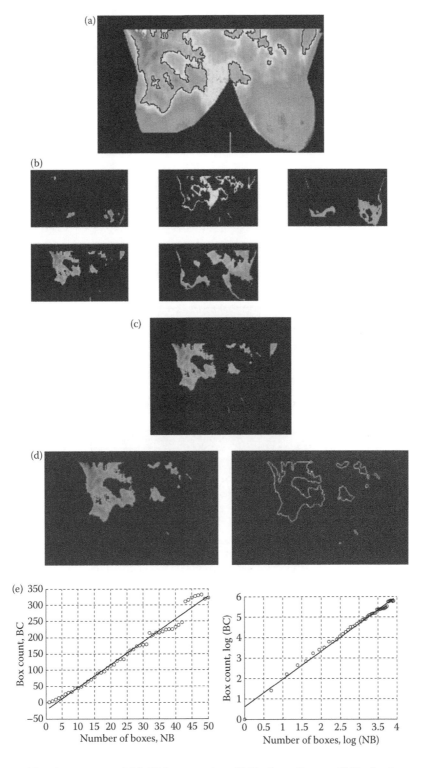

FIGURE 16.8 (a) Malignant case 1 (M1). (b) Segmentation of (a) by fuzzy C mean. (c) The first hottest regions of M1. (d) M1: (1) The first hottest regions after removing the axilla boundaries. (2) Boundaries of part (1). (e) M1: (1) Box count (BC) versus number of boxes (NB). (2) log(BC) versus log (NB).

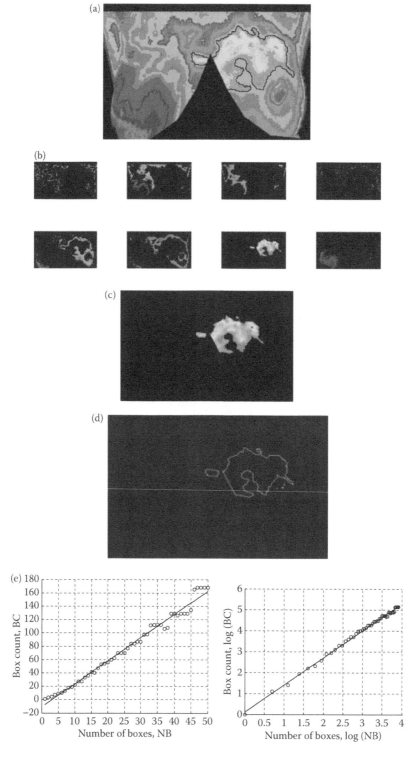

FIGURE 16.9 (a) Malignant case 2 (M2). (b) Segmentation of (a) by fuzzy C mean. (c) The first hottest regions of M2. (d) Boundaries of the first hottest regions of M2. (e) M2: (1) Box count (BC) versus number of boxes (NB). (2) log(BC) versus log (NB).

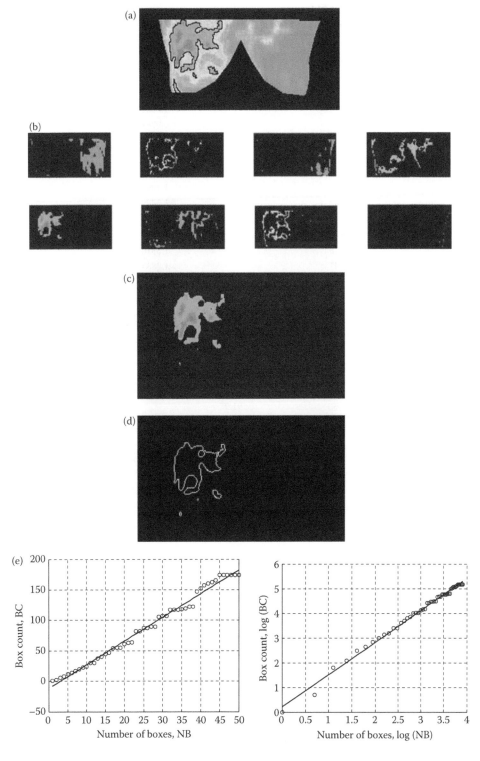

FIGURE 16.10 (a) Malignant case 3 (M3). (b) Segmentation of (a) by fuzzy C mean. (c) The first hottest regions of M3. (d) Boundaries of the first hottest regions of M3. (e) M3: (1) Box count (BC) versus number of boxes (NB). (2) log(BC) versus log (NB).

TABLE 16.2 Calculated FDs for Eight
Benign and Seven Malignant Cases by BCM

B1	0.9675
B2	0.9911
B3	1.1321
B4	1.1060
B5	1.0382
B6	0.9781
B7	1.1218
B8	1.0081
M1	1.3538
M2	1.2992
M3	1.3479
M4	1.2787
M5	1.4001
M6	1.3310
M7	1.2884

B: benign, M: malignant.

TABLE 16.3 Mean and Standard Deviation of Calculated FDs for
Eight Benign and Seven Malignant Cases

	Benign	Malignant
Mean	1.0429	1.3284
Standard deviation	0.2600	0.2074
Accuracy rate (%)	91.9	93.1

two points with maximum curvature of breasts were determined and geometrically transformed to the first and the last points of the averaged set points of the training results. Next, to fit two curves between the points with maximum curvature of each breast and the fifth point of it, a linear interpolation was employed. It performed faultlessly for 90% of all the cases at hand.

The proposed algorithm for detecting malignancy or benignity of breast tumors is then applied to both the left and the right breasts as follows:

1. The breast IR images are segmented by FCM algorithm. The number of clusters depends on the number of the camera palette colors.
2. The first hottest regions which have the color matching to the maximal temperature are diagnosed. The axilla and close sternal boundaries are removed and are not considered.
3. The FD of step (2) is estimated by BCMs as follows:
   - The image of the first hottest regions is made in binary.
   - The edges are detected.
   - A grid of boxes is set up.
   - The number of occupied boxes is counted.
   - The boxes sizes are changed and the previous step is repeated.
   - The slope of the best-fitting line to the plot of the log of the number of box count versus the log of the number of boxes is calculated.

It is notable that the breast IR images are normalized before using the BCM.

Three typical benign cases of breast thermograms are shown in Figures 16.5a, 16.6a, and 16.7a, respectively. Their images are color segmented by using the FCM algorithm and are illustrated respectively in Figures 16.5b, 16.6b, and 16.7b. The first hottest regions of thermal images are shown in Figures 16.5c, 16.6c, and 16.7c, with the axilla and close sternal boundaries removed in Figures 16.5d1, 16.6d1, and 16.7d1, and their corresponding boundaries are presented in Figures 16.5d2, 16.6d2, and 16.7d2. Plot of the number of box count (BC) versus the number of boxes (NB) and plot of the log(BC) versus log(NB) are depicted in Figures 16.5e, 16.6e, and 16.7e, respectively. For the three typical malignant cases, similar information is presented in Figures 16.8, 16.9, and 16.10, respectively.

The FDs for eight benign and seven malignant cases were estimated as tabulated in Table 16.2, and the mean and standard deviation of estimated FDs are presented in Table 16.3. The FD results for benign cases are not significantly different from 1, the topological dimension of a line, while those for malignant cases are however significantly greater.

16.8 Conclusion

Although there is no conclusive evidence that pseudocolor coding of gray images provides more information to the observer, the fact shows that carefully designed pseudocolor image coding does give lesion detection performance that equals gray scale and improves the performance of other tasks such as recognition and interpretation of a lesion. For color segmentation of IR breast images, we used the following three techniques.

Mean shift algorithm is very sensitive to the window size parameter h. Hence, in this investigation, we are frequently confronted with empty clusters, and choosing an appropriate h was not an easy task. K means algorithm minimizes the sum of within-cluster results mathematically and when clusters are compact and well separated, it provides accurate results. However, in many situations, K means clustering is not suitable where clusters are not disjoint and a sample pattern or a pixel in an image may belong to different clusters. It is very important for K means to have good initial cluster centers. Poor initial centers may cause empty clusters. There are some algorithms developed recently to improve the initial centers. In FCM clustering, pixels may belong to many clusters with different degrees of membership. In this study, due to the fuzzy nature of the thermal images, FCM technique gives more accurate results to segmentation of the IR images.

We also analyzed thermal images of breast using FD to identify the possible difference between malignant and benign patterns. The present numerical experimental results verify the theoretical concepts and show a significant difference in FD between malignant and benign cases, with the FDs for benign cases close to 1, while those for malignant cases significantly greater. This suggests that fractal analysis may potentially improve the reliability of thermography in breast tumor detection. Fractal dimension is very sensitive to the algorithm that segments images. The fuzzy nature of IR breast images helps the FCM segmentation and provides more accurate results than the others with no empty cluster.

For future work with a vast library of IR images at hand, validation of FCM to extract pathologically relevant structures can be demonstrated through comparing the regions segmented by FCM with the regions found by other modalities such as mammograms. Likewise, other popular descriptors, including principal components, area/perimeter, and so on, can be applied to characterize the regularity of the shape so as to compare with the FD. Lastly, one may try a better clustering algorithm such as Kohonen map and compare with K means.

References

1. Jones, B.F., A reappraisal of the use of infrared thermal image analysis in medicine. *IEEE Transactions on Medical Imaging.* 17:6,1019–1027, 1998, doi:10.1109/42.746635.
2. Ng, E.Y.K., A review of thermography as promising non-invasive detection modality for breast tumour. *International Journal of Thermal Sciences.* 48:5,849–855, 2009, doi:10.1016/j.ijthermalsci.2008.06.015.

3. Diakides, N., and Bronzino, J.D., *Medical Infrared Imaging*. New York: CRC, Taylor & Francis, 3rd edition, 2008.

4. Ng, E.Y.K., and Kee, E.C., Integrative computer-aided diagnostic with breast thermogram, *Journal of Mechanics in Medicine and Biology*, 7(1):1–10, 2007.

5. www.earlycancerdetection.com/breast_thermo.html (last accessed August 2010).

6. Foster, K.R., Thermographic detection of breast cancer, *IEEE Engineering in Medicine and Biology Magazine*, 17:610–614, 1998, doi:10.1109/51.734241.

7. Lawson, R.N., Implications of surface temperature in the diagnosis of breast cancer, *Canadian Medical Association Journal*, 75:4309–4310, 1956.

8. Ahmed, E., Fractals and chaos in cancer models, *International Journal of Theoretical Physics*, 32(2):353–355, 2004.

9. Baish, J.W., and Jain, R.K., *Fractals and Cancer, American Association for Cancer Research*, *Cancer Research*, 60, 3683–3688, 2000.

10. Zhou, X., Zhang, C., and Li, S. A perceptive uniform pseudo-color coding method of SAR images, Radar, CIE. *468 International Conference*, Shanghai, China, October 2006, IEEE, pp. 1–4.

11. Li, H., and Burgess, A.E., Evaluation of signal detection performance with pseudo-color display and lumpy backgrounds. In: Kundel H.L. (Ed.), *SPIE, Medical Imaging: Image Perception*, Vol 3036, Newport Beach, CA, USA, pp. 143–149, 1997.

12. Connolly, C., and Fliess, T., A study of efficiency and accuracy in the transformation from RGB to CIELAB color space, *IEEE Transactions on Image Processing*, 6(7):1046–1048, 1997.

13. Cheng, Y., Mean shift, mode seeking, and clustering, *IEEE Transactions on Pattern Analysis and Machine Intelligence*, 17(8):790–799, 1995.

14. Kybic, J., Mean shift segmentation, Winter Semester Course 2007, http://cmp.felk.cvut.cz/cmp/courses/33DZOzima2007/slidy/meanShiftSeg.pdf.

15. Mayer, A. and Greenspan, H., Segmentation of brain MRI by adaptive mean shift, *3rd IEEE International Symposium on Biomedical Imaging: Macro to Nano*, Arlington, VA, pp. 319–322, 2006.

16. Bezdek, J.C., Keller, J., Krisnapuram, R., and Pal, N.R. *Fuzzy Models and Algorithms for Pattern Recognition and Image Processing*, Norwell, MA: Kluwer, 1999.

17. AAT: http://aathermography.com (last accessed July 2010).

18. MII: http://www.breastthermography.com/case_studies.htm (last accessed July 2010).

19. ACCT: www.thermologyonline.org/Breast/breast_thermography_what.htm (last accessed July 2010).

20. http://www.thermographyofiowa.com/casestudies.htm (last accessed July 2010).

21. STImaging: http://www.stimaging.com.au/page2.html (last accessed July 2010).

22. Gazit, Y., Baish J.W., Safabakhsh, N., Leuning M., Baxter, L.T., and Jaim, R.K., Fractal characteristics of tumor vascular architecture during tumor growth and regression, *Microcirculation*, 4(4):395–402, 1997.

23. NCI: www.cancer.gov/cancertopics/UnderstandingCancer/angiogenesis (last accessed July 2009).

24. Gamagami, P. (Ed.), Indirect signs of breast cancer: Angiogenesis study, In: *Atlas of Mammography*, Cambridge, MA: Blackwell Science, pp. 231–236, 1996.

25. Guidi, A.J., and Schnitt, S.J., Angiogenesis in preinvasive lesions the breast, *The Breast Journal*, 2(4):364–369, 1996.

26. Alarcon, T., Byrne, H.M., and Maini, P.K., A cellular automaton model for tumour growth in inhomogeneous environment, *Journal of Theoretical Biology*, 225(2):257–274, 2003.

27. Smiley, M.W., and Levine, H.A., Numerical simulation of capillary formation during the onset of tumor angiogenesis, *Proceedings of the 4th International Conference on Dynamical System and Differential Equations*, Wilmington, NC, USA, 817–826, 2000.

28. Singh, Y., Tumor angiogenesis: Clinical implications, *Nepal Journal of Neuroscience*, 1(2):61–63, 2004.

29. Pisano, E.D., *Breast Imaging, Vol. 13 Breast Disease*, 130 pp, ISBN: 978-1-58603-168-8, VA, USA: IOS Press, 2002.

30. Scalerandi, M., Pescarmona, G.P., Delsanto, P.P., and Capogrosso Sansone B., Local interaction simulation approach for the response of the vascular system to metabolic changes of cell behavior, *Physical Review. E, Statistical, Nonlinear and Soft Matter Physics*, 63(1 Pt 1):011901, 2001.

31. Anbar, M., Hyperthermia of the cancerous breast: Analysis of mechanism, *Cancer Letters*, 84(1):23–29, 1994.

32. Anbar, M. (Ed.), Breast cancer. In: *Quantitative Dynamic Telethermometry in Medical Diagnosis and Management*, Ann Arbor, MI: CRC Press, pp. 84–94, 1994.

33. Lee, T.K., McLean, D.I., and Atkins, M.S., Irregularity index, a new border irregularity measure for cutaneous melanocytic lesions, *Medical Image Analysis*, 7(1):47–64, 2003.

34. Zheng, L., and Chan A.K., An artificial intelligent algorithm for tumor detection in screening mammogram, *IEEE Transactions on Medical Imaging*, 20(7):559–567, 2001.

35. Guo, Q., Ruiz, V., Shao, J., and Guo, F., A novel approach to mass abnormality detection in mammographic images, *Proceedings of the IASTED International Conference on Biomedical Engineering*, Innsbruck, Austria, pp. 180–185, 2005.

36. Caldwell, C.B., Stapleton, S.J., Holdsworth, D.W., Jong, R.A., Weiser, W.J., Cooke, G., and Yaffe, M.J., Characterization of mammographic parenchymal pattern by fractal dimension, *Physics in Medicine and Biology*, 35(2):235–247, 1990.

37. Byng, J.W., Boyd, N.F., Fishell, E., Jong, R.A., and Yaffe, M.J., Automated analysis of mammographic densities, *Physics in Medicine and Biology*, 41:909–923, 1996.

38. Gazit, Y., Berk, D.A., Leunig, M., Baxter, L.T., and Jain, R.K., Scale-invariant behavior and vascular network formation in normal and tumor tissue, *Physical Review Letters*, 75(12):2428–2431, 1995.

39. Grizzi, F., Russo, C., Colombo, P., Franceschini, B., Frezza, E., Cobos, E., and Chiriva-Internati, M., Quantitative evaluation and modeling of two-dimensional neovascular network complexity: The surface fractal dimension, *BMC Cancer*, 5(14), DOI: 10.1186/1471-2407-5-14, 2005.

40. Liu, J.Z., Zhang, L.D., and Yue, G.H., Fractal dimension in human cerebellum measured by magnetic resonance imaging, *Biophysical Journal*, 85(6):4041–4046, 2003.

41. Kuczynski, K., and Mikotajczak, P., *Magnetic Resonance Image Classification Using Fractal Analysis, Information Technologies in Biomedicine*, Berlin: Springer, 2008.

42. Esgiar, A.N., and Chakravorty, P.K., Fractal based classification of colon cancer tissue images, *IEEE, 9th International Symposium on Signal Processing and Its Application*, Sharjah, United Arab Emirates, February 1–4, 2007.

43. Lee, W.L., Chen, Y.C., and Chen, Y. Ch., Unsupervised segmentation of ultrasonic liver images by multiresolution fractal feature vector, *Information Science*, 175(3):177–195, 2005.

44. Rangayyan, R.M., and Nguyen, T.M., Fractal analysis of contours of breast masses in mammograms, *Journal of Digital Imaging*, 20(3):223–237, 2007.

45. Mastsubara, T., Fujita, H., Kasai, S., Goto, M., Tani, Y., Hara, T., and Endo, T., Development of new schemes for detection and analysis of mammographic masses, *Proceedings of the 1997 IASTED International Conference on Intelligent Information Systems (IIS97)*, Grand Bahamas Island, Bahamas, December 1997, pp. 63–66.

46. Homer, M.J., *Mammographic Interpretation, A Practical Approach*, Boston, MA: McGraw-Hill, 2nd edition, 1997.

47. Reston, V.A., *American College of Radiology, Illustrated Breast Imaging Reporting and Data System (BI-RADSTM)*, 3rd edition, Reston, VA: American College of Radiology, 1998.

48. Peitgen, H.O. Jurgens, H., and Saupe, D., *Chaos and Fractal, New Frontiers of Science*, New York, NY: Springer, 2004.

49. Liu, S.H., Formation and anomalous properties of fractals, *IEEE Engineering in Medicine and Biology Magazine*, 11(2):28–39, 1992.

50. Deering, W., and West, B.J., Fractal physiology, *IEEE Engineering in Medicine and Biology Magazine*, 11(2):40–46, 1992.

51. Schepers, H.E., Van Beek, J.H.G.M., and Bassingthwaighte, J.B., Four methods to estimate the fractal dimension from self affine signals, *IEEE Engineering in Medicine and Biology Magazine*, 11(2):57–64, 1922.
52. Fortin, C.S., Kumaresan, R., Ohley, W.J., and Hoefer, S., Fractal dimension in the analysis of medical images, *IEEE Engineering in Medicine and Biology Magazine*, 11(2):65–71, 1992.
53. Goldberger, A.L., Rigney, D.R., and West, B.J., Chaos and fractals in human physiology, *Scientific American*, 262(2):42–49, 1990.

17

The Role of Thermal Monitoring in Cardiosurgery Interventions

Antoni
Nowakowski
Gdansk University of Technology

Mariusz Kaczmarek
Gdansk University of Technology

Jan Rogowski
Gdansk University of Medicine

17.1 Introduction

Applications of nondestructive testing based on IR-thermal imaging (NDT TI) are of a very high importance in industry. Classical quantitative IR-thermal imaging (QIRT) and active dynamic thermography (ADT) belong to this field of technology. During the last 25 years, one may observe a very rapid development of solutions such as lock-in, pulsed phase (PPT), ADT, and thermal tomography (TT) in a variety of different applications and just recently also in medical diagnostics [1]. This is due to increased availability of mature technology based on uncooled FPA (focal plane array) IR-detectors, significant decrease of equipment prices, as well as implementation of advanced digital image analysis methods. Therefore, it is not surprising that IR-thermal imaging was also implemented as a useful tool in surgery and cardiosurgery inspection. Probably the first published case of such an application was the use of a thermal camera in the Department of Cardiac Surgery, Heartcenter, University of Leipzig [2–4]. We are involved in the practical use of IR-thermal imaging in cardiosurgery for more than 12 years [5,6], also implementing ADT and TT procedures unknown in medical applications.

In this chapter, the results of a research project devoted to the analysis of the value of IR-thermal imaging in monitoring the quality of cardiosurgery interventions are presented. The research was performed at the Department of Biomedical Engineering, Gdansk University of Technology in cooperation with the Department of Cardiosurgery, Gdansk University of Medicine. *In vivo* experiments on animals have been performed at the Department of Animal Physiology, Gdansk University. The aim of the project was to show which procedures may be directly applied in clinics and which one in specific fields of research. To better understand physiology, apart from thermal studies, active and passive electrical properties of the tested tissues have also been studied. In the following text, basic problems related to practical applications of QIRT in the evaluation of cardiosurgery procedures quality are discussed. The results of the research have already been published, for example, Reference 7.

It should be underlined that medical applications of QIRT in comparison to technical NDT TI applications are much more difficult as heat transfers in living tissues are far more complicated compared to technical applications of nondestructive testing based on IR-thermal imaging. Additionally, in cardiosurgery, the basic problem is nonlinear and the strongly stochastic nature of the beating heart mechanical action which is changing all the dimensions in the tested field. Also, the natural biofeedback reactions of living organisms and tissues should be avoided to get a clear interpretation of measurement results.

One important feature of the application of ADT in medicine should be underlined—analysis of a given problem requires high skills in the modeling of a tested case as all processes of diagnostics use objective model-based evaluation of measurement data. There are also other problems, which have to be taken into account to perform reliable analysis of measurement data, such as elimination of displacement errors, choice of proper measurement conditions, and so on. The following text contains a discussion of problems present in QIRT and ADT IR-thermal analysis of living objects—here, the beating heart.

17.2 Measurement Set-Up

The general concept (block diagram of the measurement set) of measurements performed in QIRT and ADT adapted to cardiosurgery is shown in Figure 17.1. The basic components applied as individual instruments are exchangeable. IR-thermal camera is applied for measurements of temperature distribution at the surface of the heart muscle. To avoid problems with the interpretation of series of thermal images of the beating heart (problems of the heart motion), the registration of images may be synchronized with the heart action using the QRS detector. Versatile acquisition of electrical and thermal data may be set using automatic procedures run in the applied computer-driven system. The electrical instruments allow measurements of cardiac potential and collection of electrical impedance spectroscopy data. Additionally, the thermal excitation source may be applied to perform ADT experiments.

First, the steady-state temperature distribution on the tested surface is recorded using an IR camera. Next, in ADT, an external thermal excitation is applied to force thermal transients at the tested surface. Recording of the temperature allows the calculation of thermal parameters, such as thermal

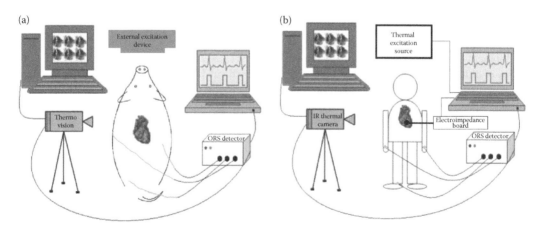

FIGURE 17.1 Measurement set-up for QIRT, ADT, and electrical experiments in cardiosurgery applications: (a) animal experiment, (b) in clinics.

time constants, dependent on the state of the tested structure. We apply external pulse excitation lasting several seconds, up to 1 min, using several solutions, as an apparatus for cryotherapy for cooling or optical excitation for heating. The IR camera is used for capturing series of thermal images, allowing measurements of temperature transients on the tested surface.

The IR camera synchronized with an excitation source (in the experiments described here, this was a set of halogen lamps or an air cooling system, but we are now mainly using CO_2 cryotherapy instrumentation for cooling) allows the surface temperature of the object under examination to be recorded at a speed of 30 frames/s for the AGEMA 900 SW 0.1°C resolution system; similar conditions are used for the uncooled Flir A320G system and 60 frames/s for the SC 3000 QWIP FPA LW 0.02°C system. Changes in tested surface temperature are caused by an external source and are dependent on the internal state of the heart muscle. More detailed description of the instrumentation and procedures applied in thermal monitoring are discussed in Reference 1, where the concept of nondestructive testing using thermal instrumentation is generally described. Our approaches are also presented in several former publications, for example, References 6–8.

To show that the application of the IR camera and an external excitation source is not disturbing the work of doctors, a view of the ADT experiment with halogen lamp excitation is presented in Figure 17.2a, animal experiments, and Figure 17.2b, at the clinic. There is plenty of space for a surgeon to perform interventions as the optical path must be free only at the moment of ADT measurements, while for the rest of the time the image may be disturbed as even a glimpse of the surgeon is sufficient for control of the situation. Additionally, a typical high-resolution RGB camera may be applied for continuous observation of surgical interventions. A prototype system developed in our project with all elements allowing QIRT and ADT experiments is shown in Figure 17.3.

A typical procedure of thermal image capture after cooling is shown in Figure 17.4. Moments of measurements are dotted. Usually, image registration should be synchronized with the heart action to obtain a stable position of the heart in thermal images.

Visual and IR images may be matched, as it is shown in Figure 17.5. Such a procedure may be important for the precise localization of the heart muscle position in thermal images. Also, external markers may be applied for the stabilization of images in time.

Single excitation and registration of transient temperatures at the position x–y allows the identification of the structure "in depth" on the basis of the equivalent thermal model. For all pixels, reconstruction of shallow 3D images of the tested object is possible. In the ADT case, the temperature of each pixel

(a)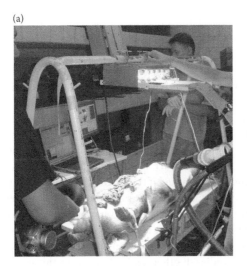
(b)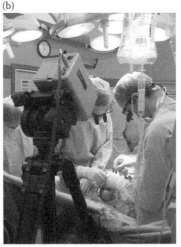

FIGURE 17.2 Placement of an IR-camera during cardiosurgery. (a) Agema THV 900, (b) FLIR SC3000.

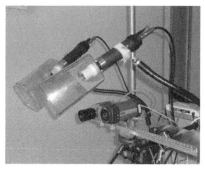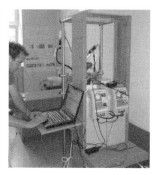

FIGURE 17.3 The experimental set—view from three perspectives. IR and RGB cameras are visible as well as cooling heads of the CO_2 cryotherapy system.

is recorded to calculate thermal time constants specific at each pixel. The parametric image of the time constants then enables the tested surface to be visualized, owing the high degree of correlation with the depth of the affected tissue.

17.3 Instrumentation: Software Problems

There are at least three levels of the software necessary to perform data acquisition and analysis of collected data, leading to diagnostic conclusions in QIRT and ADT. The first one combines specific software for the execution of all necessary actions of the system, as starting measurements, synchronization of excitation sources and recording cameras, and so on. If several imaging modalities are applied, visible and IR, then it is important that synchronization is performed between different cameras.

The second level of the software combines all elements of communication interfaces, including user-friendly manipulation of the system (GUI). In a typical solution, such a software is written in C++ and is specific for the technology applied.

Still, and within the third level of the software, there are different practical problems requiring additional software tools, for example, to eliminate natural movements of a living object or synchronization of images during cardiosurgery interventions, when the beating heart is changing its volume, position, and shape in a nonlinear way. This leads to the question how to analyze thermal transients and apply procedures of automatic understanding of image content, when in the following moments, each detector of the FPA sees not the same region of a tested object and additionally, the temperature at each pixel is also changing in time. Two factors are influencing the measurements based on series of images: first, displacement of a patient, and second, change of the thermal content in the following images. To solve the problems, some actions are needed to eliminate unwanted factors responsible for errors induced in sequences of thermal images collected in time. The first obvious action must be matching (stabilization) of following images to see the same area of a tested object/region, to allow further thermal analysis of series of thermal images. This task is not always easy, especially when the object is moving its position and shape in time. This is a typical case of the heart during open-chest cardiosurgical interventions. Dynamic changes of the heart in time combine not only change of the position but also the angle of observation, volume, and temperature at the surface of the heart, limiting the possibility of using ADT for proper diagnostics of the state of the heart muscle. We asked if there are any possibilities to monitor cardiosurgery interventions to assure a high quality of such operations based on quantitative evaluation of the content of the following images. Unfortunately, the heart during one cycle of action is strongly changing not only its position but also its volume. Practically, all visible elements at its surface are changing positions and are stressed and expanded in a nonlinear way. In effect, the application of simple affine transformations, typically sufficient in studies of other applications, is in this case not effective. What algorithms would be sufficient to make such transformations that will force the structure of

(a)

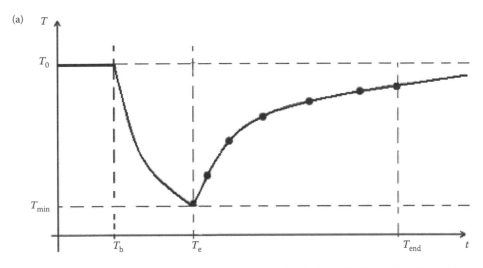

(b)

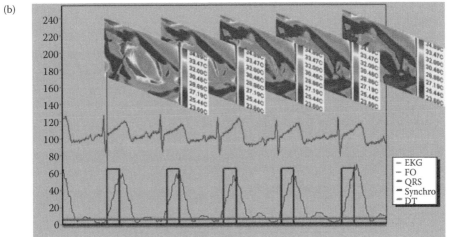

FIGURE 17.4 Measurement procedure requires registration of temperature transients at any pixel x,y in time. The example (a) shows the trace of temperature at x,y—cooling excitation; $t_{b/e}$—moments at which the cooling source is switched *on* (beginning) and *off* (end of excitation), respectively, t_{end}—time of termination of the recording; dots are indicating moments of thermal image capture; (b) shows that thermal images are measured synchronically with the heart rate, according to QRS.

images to unchanged positions? This means the elimination of all artifacts of position, angle of observation, and dimensions of the heart muscle in such a way that those pixels representing specific regions of the heart surface always stay in the same position, independently of the evolution of the tested structure in time. Unfortunately, solving such problems is not trivial and there are no software packages available that support this task. We tried several algorithms of image stabilization and matching but the problem is still generally unsolved. For objective evaluation of different correction algorithms, we prepared software "phantoms" of easy modified distortions, based on real images of the beating heart [9]. For the objective comparison of images after transformations, the following figures of merit are applied: correlation of images (CORR) (Equation 17.1)—which is the best in terms of a clear description of image differences (the value 1 means full identity); root mean square (RMS) (Equation 17.2)—0 means the identity; and the normalized mean square error (NMSE) (Equation 17.3)—the smaller the value the better the fitting of images:

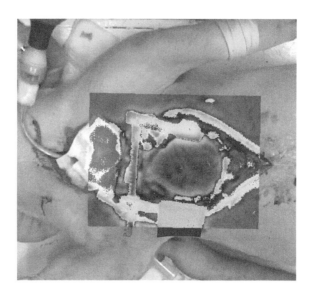

FIGURE 17.5 Matching of visible and IR images may help in recognition of the region of interest.

$$CORR = \frac{\sum_{x=0}^{X}\sum_{y=0}^{Y}[F(x,y)-\overline{F}][G(x,y)-\overline{G}]}{\sqrt{\sum_{x=0}^{X}\sum_{y=0}^{Y}[F(x,y)-\overline{F}]^2[G(x,y)-\overline{G}]^2}} \tag{17.1}$$

$$RMS = \frac{1}{N}\sum_{x=0}^{X}\sum_{y=0}^{Y}\sqrt{[F(x,y)-G(x,y)]^2} \tag{17.2}$$

$$NMSE = \frac{\sum_{x=0}^{X}\sum_{y=0}^{Y}[F(x,y)-G(x,y)]^2}{\sum_{x=0}^{X}\sum_{y=0}^{Y}[F(x,y)]^2} \tag{17.3}$$

where $F(x,y)$ is the pixel value at the point (x,y) in the unchanged image, $G(x,y)$ is the pixel value at the point (x,y) in the deformed image, N is the number of the gray-scale levels, and $\overline{F}$ is the mean value of the image intensity.

To show the simulation methodology, a thermal image of the heart during open-chest cardiosurgery animal experiment is shown in Figure 17.6. The region of interest (ROI), here showing the position of the tested heart, is calculated and indicated as the bright polygon (Figure 17.6a). This area is cut off and a new polygon of assumed parameters is inserted, representing the required mechanical and thermal changes (Figure 17.6b). In this case, the arrows show the direction of shrinking forces and the mechanical deformation is clearly visible in the deformed grid. The next step is the application of a reconstruction algorithm. One has to apply an algorithm generating the anti-deformed grid, as it is shown in Figure 17.6c. Unfortunately, usually the reconstruction is not perfect due to the loss of some data caused by, for example, numerical errors. Additionally, a specific pattern of thermal fields may be applied. Also, this pattern is suffering transformation errors, resulting in the limitation of visualization accuracy. In this way, a series of images in time may be simulated, giving data for further analysis of reconstruction algorithms or thermal tomography procedures to be applied.

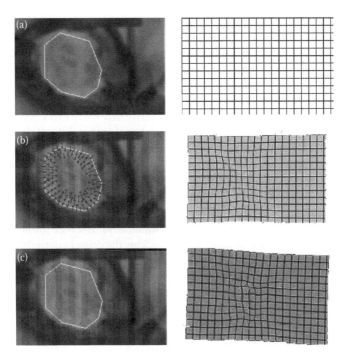

FIGURE 17.6 (a) Original thermal image of a heart and regular geometry transformation grid; (b) the same image after forward transformation (directions of the inserted geometry distortions are indicated by arrows and the grid is respectively deformed); additionally, pattern showing modulation of external temperature is visible; (c) application of the reverse algorithm represented by the geometry of the modified (anti-deformed) grid.

One should note that the segmentation of the heart from thermal images is not an easy task as the differences of temperature between the heart tissues and surrounding tissues may be not high enough to differentiate tissues or organs. Also, during the collection of the series of images, the temperature field in different parts of the ROI is usually changed, which makes the problem even more complicated. To enhance and facilitate the segmentation procedure, an additional RGB camera may be applied and a fusion of the visible and IR images may significantly improve this step of the analysis. Generally, using different approaches, one may regard the segmentation to be successful if the correlation of the indicated ROI and the real ROI exceeds the value 0.96. It means that, for the heart, the probability that a chosen pixel represents the proper part of the heart equals 96%. This work is still continuing.

17.4 Experiments and Measurements

The following text is divided into two parts, the first devoted to animal experiments, to answer several research questions, and the second to clinical applications, to show the practical value of the technology. In both cases, to meet the high accuracy of the experiments, all reference environmental and procedural conditions required for static thermography examination have been secured. The experiments have been conducted in an air-conditioned environment, the temperature and humidity being, respectively, typically 20°C and 60% in the operation theatre and 22–23°C and 45–55% in the animal quarters.

17.4.1 Animal Experiments: Methods and Materials

Pigs were taken as the experimental animals due to their closeness to the human physiological and anatomical structure of the heart muscle and circulatory system. Studies based on *in vivo* animal experiments were performed according to all legal regulations and permission by the Gdansk Ethics Commission for Experimentation on Animals.

Optimized solutions of instrumentation have been analyzed during the experiments, such as evoked heart stroke, coronary artery by-pass grafting CABG, off pump OP CABG procedures, and others. One of the important problems was to compare thermal and electrical measurements, as the response of the heart muscle to electrical stimulation may be nonlinear. We expect that such a measurement may hold important diagnostic information as the level of nonlinearity may carry extra information of the heart tissue properties. The question was—is there a correlation between thermal and electrical tissue properties? Here, the results presented in References 10, 11 are summarized.

The experiments were conducted on several young domestic pigs, each weighing approximately 25 kg. Anesthesia and analgesia were obtained by the administration of ketamine (im), pentobarbital (iv), and fentanyl (iv) at doses of 20, 30–50, and 0.5–0.1 mg/kg, respectively. The animal chest was open by sternotomy; the pericardium was stitched off to have full access to the heart muscle as it is shown in Figure 17.7. The intubation procedure was applied to keep the animal alive. To evoke the heart muscle ischemia, the left descending artery (LAD) was clamped, totally blocking the delivery of the blood (see Figure 17.8).

The short-term effects of ischemia preconditioning and blood arrest in the left ventricle were studied in a swine model of beating heart coronary artery revascularization using ADT instrumentation and image processing. The open-chest experiments with full accessibility to the heart were performed to test typical surgical procedures.

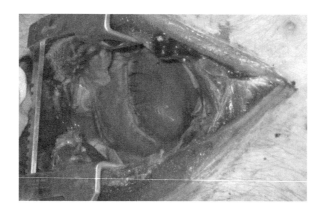

FIGURE 17.7   Exposition of a pig heart before clamping LAD.

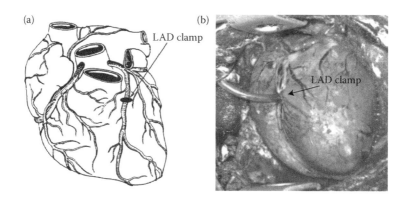

FIGURE 17.8   (a) Diagram of the heart muscle circulatory system [12]; (b) the heart photo with indicated site of clamped LAD.

Permanent clamping of the LAD procedure was damaging the heart muscle in the same way as the heart infarct, resulting in a strong modification of tissue structure and changes of thermal and electrical properties of the heart muscle in time. The heart fibrillation and following death of the animal was observed typically 35–70 min following permanent clamping of LAD. The thermal state of the heart was monitored using an IR thermal camera and simple recording of radiation emitted by the observed surface of the heart. Additionally, the thermal properties of the heart muscle were calculated from ADT experiments based on external thermal excitation.

One of the several protocols applied in the research experiments is the protocol presented in Figure 17.9. The ischemia episodes and thermal investigation are marked on the time chart. The camera indicates the registration of static thermography and ADT—the active dynamic thermography procedure with cooling excitations. The two ischemic preconditioning episodes lasting 2 min each was followed by the permanent clamping of LAD. Seven measuring sessions were performed during the total period of experiment. Several similar experiments have been performed to obtain reliable conclusions.

Ischemic preconditioning (IP), defined as a rapid adaptation response to a brief ischemic episode, was tested as a protecting procedure following the high-risk cardiosurgery. Evidence of ischemic preconditioning has been shown in a large number of previous studies developed in animal and human procedures, for example, References 13–15. IP has been found to increase myocardial tolerance for blood arrest, to reduce the rate of anaerobic glycolysis, and to reduce the area of necrosis during prolonged ischemia. On the other hand, this technique has also been found to fail, and since there is no reliable and clinically suitable method for intrasurgery monitoring of efficacy, the IP is performed "blindly." To our knowledge, presented here (after the paper [10]) are the first ADT parametric images of the myocardium recorded just after removing occlusion from LAD during the preconditioning procedure. One of the goals of this study was to understand ischemic preconditioning in protecting the heart muscle and its role in therapeutic procedures. We asked if one may see the effect of preconditioning on ADT images.

The ADT experiment was performed using a cooling device, generating an air stream with a temperature of 5°C, and the timing was 30 s of the stimulus and 90 s of the recording temperature at the recovery phase (self-rewarming). As an example of measurements performed using the acquisition system, which is shown in Figure 17.1a and according to the experimental protocol shown in Figure 17.9, typical results are presented in Tables 17.1 and 17.2 and in Figures 17.10 through 17.12. The chosen ROIs represent the left (AR01, AR02) and the right ventricle (AR03) surfaces. The AR01 is located above the clamp site and the AR02 below the site of clamping the LAD.

The two exponential thermal model for the calculation of parametric images is applied. The results for τ_1 and τ_2 at ROI for the self-rewarming phase are presented in Tables 17.1 and 17.2.

Strong changes of tissue properties are visible for both time constants. The physical factors responsible for decrease or increase in time of the value of time constants are not fully recognized. Several

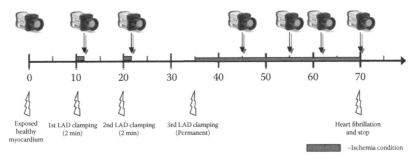

Time of experiment [min]

FIGURE 17.9 Experiment schedule with indicated surgery incidents and measurements points; static thermography and ADT as well as electrical impedance measurements.

TABLE 17.1   Time Constants τ_1 (s) for the Schedule Shown in Figure 17.9

	ROI		
Time (min)	AR01	AR02	AR03
0	2.059 ± 0.226	2.672 ± 0.441	3.709 ± 0.309
12	1.130 ± 0.174	0.797 ± 0.081	0.806 ± 0.145
24	0.426 ± 0.107	0.187 ± 0.057	1.705 ± 0.231
45	2.352 ± 1.526	3.114 ± 0.276	2.007 ± 0.489
55	2.166 ± 0.556	3.441 ± 0.319	3.856 ± 0.427
62	2.035 ± 0.299	2.976 ± 0.325	0.664 ± 0.112
70	4.220 ± 0.229	4.926 ± 0.235	0.618 ± 0.071

TABLE 17.2   Time Constants τ_2 (s) for the Schedule Shown in Figure 17.9

	ROI		
Time (min)	AR01	AR02	AR03
0	23.876 ± 0.969	14.971 ± 1.168	19.660 ± 1.800
12	19.473 ± 0.568	15.276 ± 0.323	10.449 ± 0.331
24	17.449 ± 0.770	11.709 ± 0.388	10.684 ± 0.770
45	26.273 ± 4.090	39.601 ± 2.373	13.552 ± 2.007
55	32.364 ± 4.053	45.636 ± 2.353	14.463 ± 2.637
62	42.052 ± 5.097	47.051 ± 3.956	8.418 ± 0.281
70	35.253 ± 2.017	48.586 ± 3.751	21.122 ± 0.245

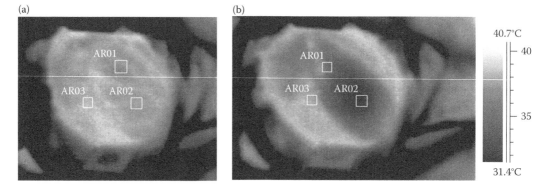

FIGURE 17.10   Thermograms of the myocardium with marked regions of interest AR0x; (a) for normal condition, (b) after clamping of LAD.

factors, mainly the changes in vascularization and changes in cell structures in the affected regions, what can be confirmed by histopathology examination, may be regarded as important. Influence of red blood cells or/and content of water extravasated within the interstitial space, drying of the myocardium surface, and some other phenomena may be taken into account too.

We found that ischemic preconditioning usually is followed by coronary flow increase as the autonomic reaction to blood delivery decrease. It is manifested by decreasing values of the time constants τ_1 and τ_2 for all ROI for measurements collected in the 24th minute, compared to measurements performed at the 0 minute, according to the experiment schedule. This observation is similar to that described in References 16, 17 for ultrasonic imaging.

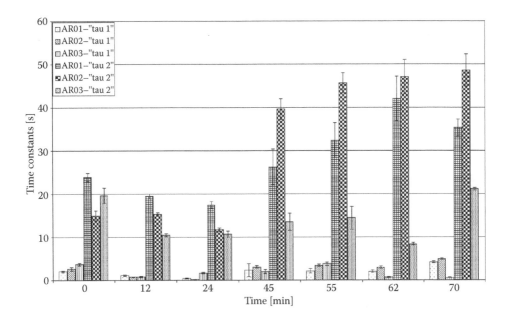

FIGURE 17.11 Estimated time constants (τ_1 and τ_2) for regions of interest AR0x; measurements according to the schedule shown in Figure 17.9.

Measurements acquired after the permanent clamping of LAD show that calculated time constants are getting longer during the blood arrest episode for the left ventricle area—about 4 times longer than for IP condition, while for the right ventricle the time constants stay almost the same or are even decreased. Parametric images of the time constant based on the sequence of thermograms recorded during the self-rewarming transient thermal processes show exactly the area of limited or increased blood perfusion (Figure 17.12).

For the measurements acquired just after the second preconditioning episode, the temperature and the time constant images are relatively uniform with visible big blood arteries (LAD and diagonal branches). In the 20 min following the permanent clamping of LAD, the static temperature image shows that the left ventricle of the heart is of lower temperature than the right ventricle. This temperature is also lower compared to the measurement taken just after the second preconditioning episode. But this lower temperature is almost the same as for the measurements taken in the 10 and the 27 min following the permanent clamping of LAD! We conclude that the static temperature of the left ventricle during blood arrest is significantly lower than for the right ventricle, but after few minutes following clamping of LAD temperature, it reaches the stabilized level. So, the static temperature could not be a discriminating parameter for monitoring the myocardium perfusion quality. The time constant image (Figure 17.12d) shows the area of limited blood flow region characterized by longer time constants. However, there is still a serious lack of information concerning thermal and physiological processes during and after ischemic preconditioning episodes and during progressed heart infarct. Furthermore, the cardioprotective effect (increase of perfusion) of ischemic preconditioning is evident. In our further work, we want to correlate thermal imaging data with bioimpedance measurements of the myocardium.

These preliminary findings show that the proposed methodology is able to determine the changes of blood flow intensity in the myocardium.

Apart from ADT, the thermal behavior of the heart muscle using classical QIRT measurements is very interesting, as illustrated in Figure 17.13. In contrast to ADT, where structural thermal tissue properties are measured, a very important functional information is evidenced here. On the other hand,

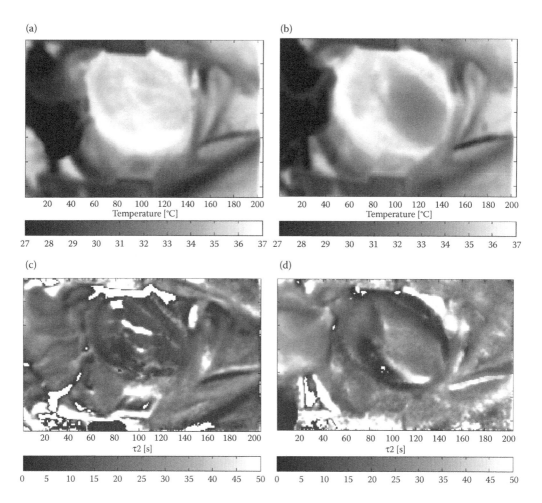

FIGURE 17.12 (a) and (b) Static thermograms; (c) and (d) estimated time constants τ_2 images for measurements taken in 24 min (the first column) and 55 min (the second column) of the experiment schedule presented in Figure 17.9.

there is a strong correlation between both modalities. The analysis of thermal images also allows the understanding of ADT experiment, what is evidenced in Figure 17.14. Series of thermal images show rapid change of vascularization and functionality of the affected tissue after clamping the LAD.

Based on images shown in Figures 17.13 and 17.14, one may calculate ADT parametric images, as shown in Figure 17.15. The content of such images relates to the change of thermal tissue properties in consequence of the time passing after clamping LAD and physiological changes caused by ischemia. There is still a problem of calibration of such images which requires more experience.

Another example of a research problem solved based on thermal investigations is the question of the quality of OP CABG. A special mechanical heart stabilizer (holder) is applied to hold the place where a by-pass graft should be inserted. This region of the heart should be stable in space and time, allowing normal action of the heart during OP CABG. We asked a question that if by using such a mechanical holder, additional control of the tissue electrical properties would be possible by the application of electrodes inserted to the holder. Another question was the quality of such intervention, the time duration of the operation to be safe for a patient, and so on. A prototype of a holder with inserted electrodes for electrical impedance measurements is shown in Figure 17.16a. The four-pole configuration of electrodes is typically applied as shown in Figure 17.16b. Current electrodes are placed at the outer part of the

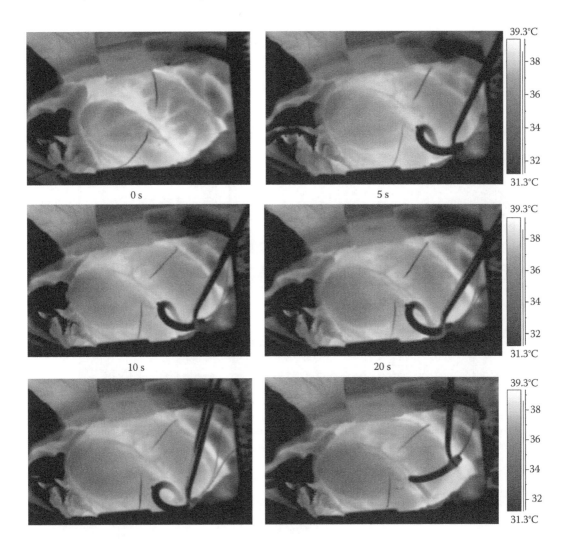

FIGURE 17.13 The static thermograms of the pig heart taken at the indicated times after clamping LAD taken by a camera synchronized with the heart rate. Temperature at the affected region is slowly decreasing; affected vascularization after clamping LAD is evident.

sucking arms. In our experimental arrangement, several voltage electrodes are applied, also allowing measurements of electrical field distribution. The applied Solartron Impedance Analyzer 1260 allows spectral measurements of electrical impedance in a broad range of frequencies. Placing a set of electrodes to the suckers of the stabilizer allows continuous measurements of electrical signals, at all times of the use of the stabilizer, also assuring a very good electrical contact of the electrodes to the heart muscle. Unfortunately, the use of the stabilizer may affect the heart muscle, as illustrated in Figure 17.17. Still, this effect belongs to a procedure claimed to be "minimally invasive."

The important confirmation of the correlation of thermal and electrical tissue properties is illustrated in Figure 17.18, showing change of electrical impedance in time after clamping LAD and showing the devastating influence of ischemia. This experiment is in a perfect agreement with the changes of parametric ADT images in time after forming ischemia and is confirmed by IR-thermal observation. Both functional and structural electrical and thermal properties of the heart muscle are affected by ischemia. This is not only due to limited vascularization but also due to visible changes in cell structures in the affected regions. An example of thermal monitoring of clinical OPCABG is shown in the following section.

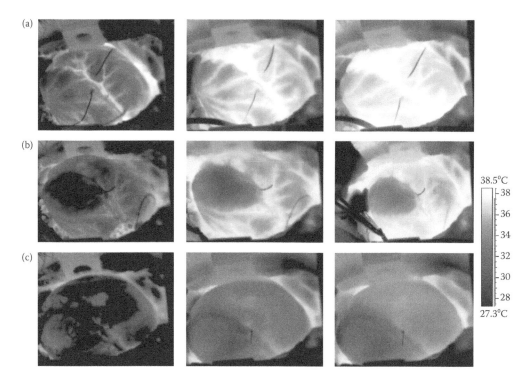

FIGURE 17.14 ADT experiment on the heart during induced heart infarct for different phases of the blood arrest. Thermograms at different moments after stopping cold excitation using CO_2; first column—end of cold excitation, second column—20 s following stopping of cold excitation, third column—90 s following stopping of cold excitation, rows: (a) examination of a healthy heart (before clamping LAD according to Figure 17.9); (b) 40 min after stopping blood flow in LAD; (c) 80 min after blood arrest.

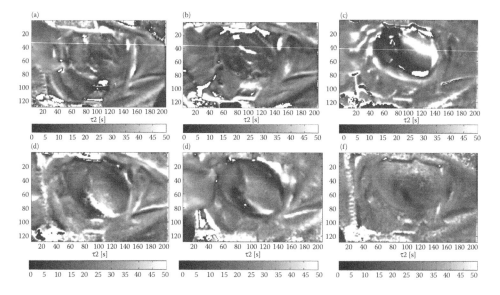

FIGURE 17.15 Images of thermal time constant τ_2 calculated for the phase of rewarming, after excitation using the cryo-instrument generating cold air of temperature 5°C during 30 s; (a) the healthy heart; (b) the heart muscle after the preconditioning lasting 2 min; (c) 10 min after clamping LAD; (d) 20 min after clamping LAD; (e) after 27 min; (f) just after the stop of the heart beating (30 min after clamping LAD).

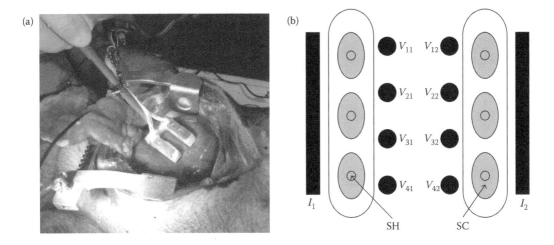

FIGURE 17.16 (a) Mechanical stabilizer with electroimpedance electrodes applied to the heart muscle; (b) the schematic diagram (bottom view) of the probe combined with sucking holder; I and V stand for current and voltage electrodes while SH and SC mark sucking holes and cups, respectively.

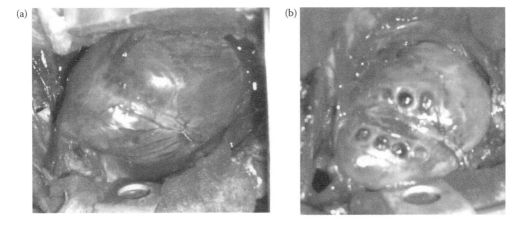

FIGURE 17.17 (a) The pig heart before application of the stabilizer; (b) after intervention—the hematoma visible at the positions of the sucking chambers.

17.4.2 Clinical Applications

Several examples of clinical thermal measurements are already shown in Reference 1. Here, we will summarize the final results of the research based on comparison of the value of applied monitoring tools and also illustrated by a few practical cases in clinics using absolute temperature and temperature courses during surgical interventions, including the ADT method.

17.5 Thermal Monitoring of Surgical Procedures during CABG

Although CABG interventions are today regarded as very invasive and are replaced by OP CABG, the thermal monitoring of several procedures typical in CABG still seems to be important. In all cases, when blood circulation is totally blocked even for a while, such a state is dangerous as there is a deficit of oxygen and metabolites necessary for the survival of the affected tissue. To minimize the negative effects

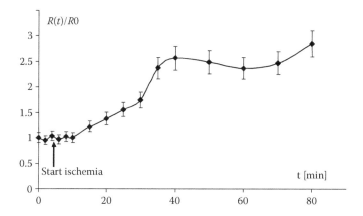

FIGURE 17.18 Resistivity measured at 5 kHz for the set of electrodes in Figure 17.16 (I_1, I_2, V_{21}, and V_{22}) and the mechanical stabilizer placed close to the center of the affected vascularization region of the heart muscle.

of anoxemia, deep hypothermia of the heart muscle is necessary. The heart muscle should be cooled to a temperature of 10–14°C, which allows for a safe surgical intervention lasting up to 120–180 min. Deep cooling is possible by the application of liquid cardioplegia solution at a temperature of 4°C to blood vessels, aorta, and/or descending artery. There are different strategies to perform cardioplegia, called antegrade and retrograde cardioplegia. Examples of effective procedures are shown in Figures 17.19 and 17.20 where images taken in time evidence progressive cooling of the heart muscle.

Thermal monitoring allows precise control of the state of the heart muscle. Typically, during cardio-surgery intervention, the temperature rises slowly, which may be dangerous, causing fibrillation of the heart muscle. Even though the aorta is blocked (clammed), there exists some inflow of blood via arteries, which heats the heart muscle slowly—this is shown in Figure 17.21.

The first thermogram is taken at the moment of stopping cardioplegia. The following images are taken after 3 and 13 s. Control of the temperature allows for a reduction of the fibrillation risk and intraoperation heart stroke.

Additionally, thermal images may be useful in finding coronary arteries to be grafted. Usually, such arteries are covered by a layer of fat which leads to problems in the precise localization using the traditional approach (palpation). After cardioplegia, those arteries are clearly visible at thermograms.

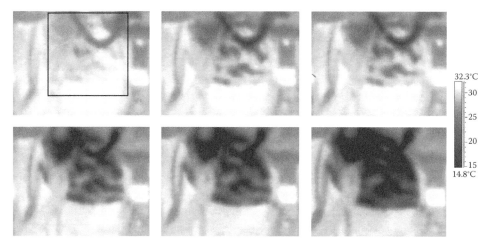

FIGURE 17.19 Series of thermograms during antegrade cardioplegia taken every 10 s; one can see a progressive decrease of the heart muscle temperature; position indicated by the rectangle.

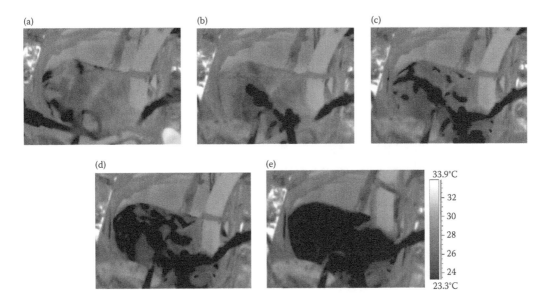

FIGURE 17.20 Series of thermograms during cardioplegia (retro/ante); (a) application of 500 mL of cardioplegia liquid retrograde; (b) continuation using antegrade cardioplegia at 10 s after beginning of antegrade, (c) 25 s, (d) 45 s, (e) 75 s.

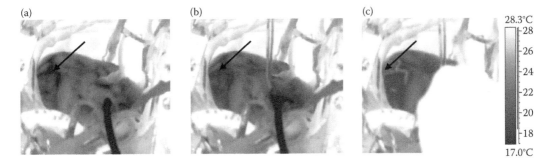

FIGURE 17.21 (a) Thermogram after stopping cardioplegia and respectively (b) after 3 s and (c) after 13 s; the arrow shows LAD.

The next very important surgical procedure is the evaluation of the quality of an inserted graft. Here, one may use thermography just to see the temperature distribution on the myocardium; after a successful CABG operation, it should be a uniform thermal image. On the other hand, all grafts are clearly visible. Especially, the moment of blood circulation recovery is important as one may see an instant rise of temperature on thermal images in all of the important places. Based on thermal images, it is even possible to evaluate the rate of the blood flow, assuming the simplest exponential model of temperature changes:

$$T(t) = T_0 - \Delta T \cdot (1 - e^{-\frac{t}{\tau}}) \tag{17.4}$$

where $T(t)$ is the temperature at the heart surface, ΔT is the increase of temperature due to the inflowing blood, T_0 is the temperature of the myocardium, τ is the time constant—a factor related to the value of free flow-out of blood from LIMA (left internal mammary artery), and t is the time. Two cases of such analysis are presented in Figure 17.22 for correct and failed cases of grafts inserted during CABG

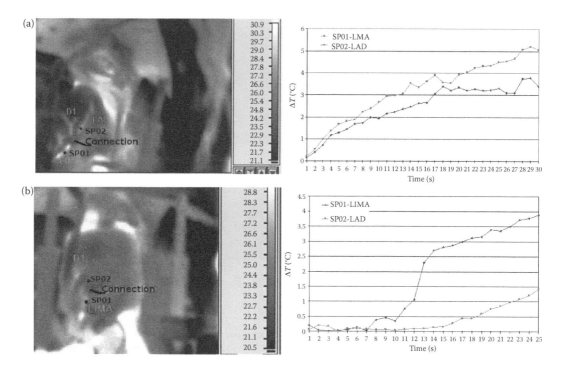

FIGURE 17.22 Quantitative analysis of the graft LIMA-LAD; (a) correct case with temperatures rising in time at the points SP01 and SP02; (b) nonfunctioning graft at SP02—lack or affected rise of temperature in LAD.

interventions. Thermal images and traces of temperature in time after opening the blood circulation in specific points of interest are presented. A noncorrect flow (Figure 17.22b) is manifested by lack of temperature rise below the place of the graft insertion.

Also, ADT parametric images allow for clear evaluation of the graft quality. This is illustrated in Figure 17.23 for two cases: the correct one, the left column, and the graft requiring reoperation, the right column. The time constants at the regions where blood is supplied properly are of the value of 20 s, while in ischemic parts, the time constants are very long. The images show (a) temperature distribution at 20 s after unclamping the graft, (b) temperature difference at this moment, and (c) time constants distribution. At the parametric images, it is clearly visible which fragments of myocardium are characterized by big changes of temperature and those being at almost steady state (the uniform area). For the left column, one may see the branch of LAD with diagonal vessels. For the right case, the only region of increased temperature is in pectoral artery till the LAD joint; the rest is cold. Similarly, in ΔT images, the increase of temperature is around 4°C.

17.6 Thermal Monitoring of Surgical Procedures during OP CABG

OP CABG is regarded as a minimally invasive procedure, even if it is a drastic intervention inside the body of a patient. The practical use of a mechanical stabilizer holding the heart muscle in the place of graft insertion is shown in Figure 17.24a. Practically, it allows maintaining the heart action during the intervention, without applying the procedure of external blood circulation. Especially important is the state of the myocardium, its temperature, in the vicinity of the holder. The decrease of temperature below 28°C may be dangerous, leading to fibrillation. We suggested some changes in the new construction of a heart stabilizer just to decrease the thermal conductivity of the construction, to decrease heat out-flow, and to increase the time of possible surgical intervention.

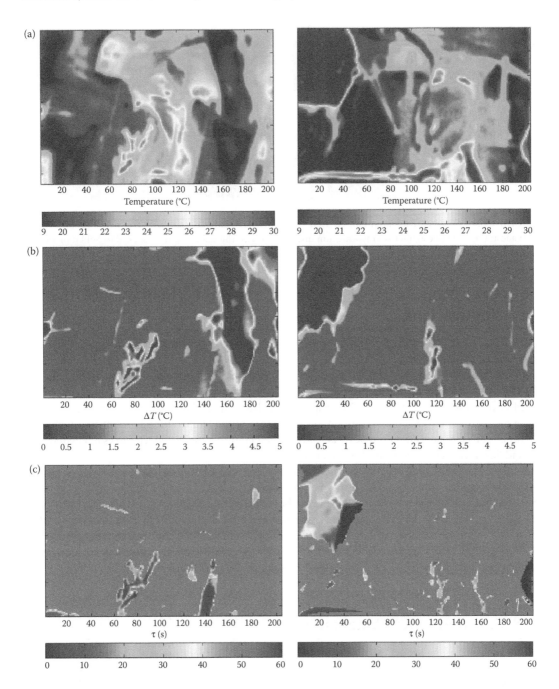

FIGURE 17.23 Correct (left column) and requiring reoperation (right column) LIMA-LAD grafts; (a) thermal images at 20 s after unclamping the graft; (b) recorded rise of temperature; (c) time constants according to Equation 17.4.

Another important problem is finding the optimal position where the graft should be inserted. On thermograms, it is usually clearly visible in the place where the artery is blocked, as is shown in Figure 17.24b (see arrow). The next procedure is the evaluation of the quality of the graft. A properly inserted graft responds to the unclamping procedure by a rapid increase in the temperature of the myocardium below the joint as illustrated in Figure 17.24c.

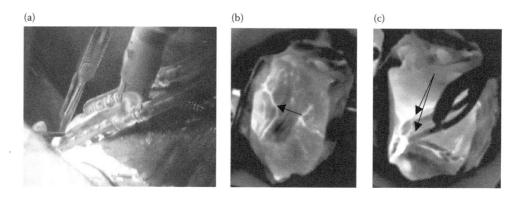

FIGURE 17.24 OP CABG intervention: (a) view of the mechanical heart stabilizer, (b) thermogram showing blockade at the LAD (see arrow), (c) myocardium under LAD with the stabilizer arms sucked to hold it in a stable position (see arrows).

17.7 ADT Monitoring of the Heart Muscle Perfusion and Other Applications

The final effect of different cardiosurgical procedures is properly distributed vascularization of myocardium and effective mechanical heart action. Thermography allows for the evaluation of the global state of the heart muscle. A positive result of any of the interventions may be therefore evidenced by using the described technologies. As an example, the case of CABG intervention is shown in Figure 17.25, based on QIRT measurements. Similarly, ADT images might also be applied, but data treatment is in this case much more advanced. One may see an important change of the temperature distribution which after CABG is much more uniform and the mean value of temperature is higher, which shows that blood supply and flow is regular in all parts of the treated heart.

Another possibility of using classical thermography concerns studies such as the analysis of specific surgery procedures. In Figure 17.26, a thermogram shows the vascularization of an arm after physical exercise. This measurement was performed in studies of extracting the hand radial artery for a by-pass. The same patient, after the extraction and using his artery for a successful by-pass operation within the next 3 months, was under the same examination, and only a slight decrease of temperature, not exceeding 0.5°C, has been noticed. The result is very positive in terms of full recovery of vascularization in the hand. This study on 50 patients showed reasonable use of this artery in cardiosurgery by-pass operations.

17.8 Summary and Conclusions

IR-thermal measurements are noncontact and fully aseptic. Temperature is a functional parameter allowing for nonspecific evaluation of complex physiological processes existing in the tested organ. Observation of thermal fields at the surface of tested organs allows for fast and quantitative evaluation of complex physiological processes. It is similar in ADT measurements but this is a structural parameter showing thermal properties of a tested tissue, thermal conductivity, and thermal capacity. In both cases, thermal visualization allows for an elegant, quantitative illustration of tested processes, as cardiosurgery interventions. ADT allows for precise planning of surgical interventions without the need for big safety margins. Both modalities are complementing each other, allowing better understanding of the state of a tested organ. This also allows for fast and objective monitoring of cardiosurgical interventions, quantitative evidence of the results of such interventions, as well as for objective observation of treatment progress.

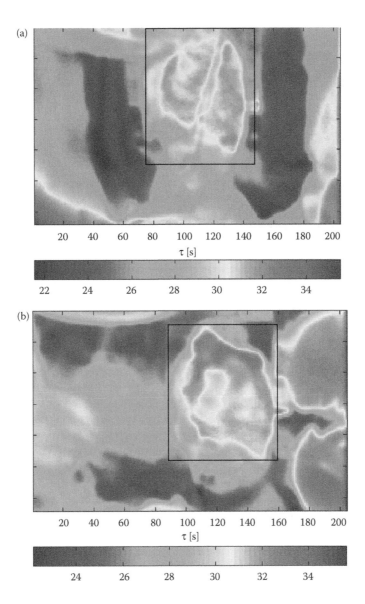

FIGURE 17.25 Perfusion of myocardium (a) before and (b) after CABG; the heart region is indicated by a rectangle.

We have shown an important role of both modalities in research and also in clinics. The boundary conditions are still very important in the interpretation of thermal and ADT parameters; therefore, this should be the main condition for assuring a high quality of such measurements. Especially, proper segmentation of thermal images would be impossible without assuring standard conditions of measurements, which should also be regarded as the highest possible knowledge of boundary conditions.

Absolute temperature may be an important parameter in the monitoring of the state of a patient before, during, and after surgical interventions. Unfortunately, there are no defined standards—clear boundaries of the values of temperature for specific interventions and treatment procedures. This means that further investigations are necessary for filling the existing gap in such knowledge. This notice concerns both experiments on animals and clinical practice.

Temperature allows monitoring of vascularization in each phase of cardiosurgery interventions. This is a perfect method for the evaluation of the quality of the inserted graft, of the efficiency of cardioplegia,

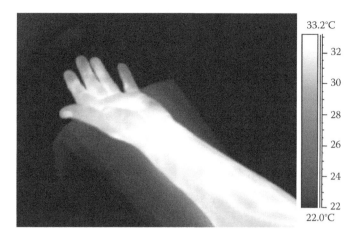

FIGURE 17.26 Vascularization of a hand after mechanical exercise before extracting the radial artery for a by-pass. This is a reference image for further analysis of hand function recovery after reconvalescence lasting typically 3 months.

and of the quality of many surgical procedures in clinical practice. Such monitoring is prompt, easy, and objective, especially if dynamic processes are investigated. It allows for noncontact observations of the blood circulation, especially after the elimination of any blockades.

Temperature measurements allow for easy discrimination of the regions of affected circulation and also of hyperemia or inflammation, and any increased or decreased metabolism.

ADT identification of regions of affected thermal properties may be used for the evaluation of tissues and organs for transplants, and so on. These parameters are also important in the evaluation of the state of blood deficit or progress of tissue necrosis. This is an important factor in cardiosurgery interventions as repairing of the heart is nowadays more important than transplantations. Very important are observations of changes of tissue parameters in time. Necrosis is evidenced within around 30 min after blood arrest!

Visualization of temperature distribution at the surface of the heart muscle is less invasive than electroimpedance measurements, being a fully noncontact procedure. The only condition for measurements in this case is visual accessibility to the tested region, what is not always possible, as thermal inspection should not disturb surgical interventions.

The set of ADT parametric images gives rather structural information of the tested tissue. Properties of tissues are dependent on the blood flow and the physical structure which is devastated after long-lasting processes of ischemia. Blood arrest has a damaging influence on the structure of cells which may be easily evidenced by histopathology. Therefore, the ADT images are differentiating tissues already affected but are not sensitive to short deficits in blood flow. Analysis of the ADT measurement results leads to the conclusion that special care must be taken to secure proper measurement conditions.

Information gathered in both modalities is of different character. Thermography shows regions of increased or decreased temperature, usually caused by a change of vascularization due to different clinical reasons—increased metabolism, necrosis, ischemia, and so on. The minimal requirement for instrumentation is the use of a thermal camera of a resolution of at least 0.1°C and a lens with FOV adequate for the chosen application.

IR thermography instrumentation and application of dedicated data processing for the analysis of thermal transients after external excitations allow to obtain objective, quantitative thermal data of tested tissues; therefore, it has a potential value as an intraoperative monitoring and predictive tool, especially in beating heart myocardial revascularization procedures, where, during the grafting process, the heart muscle is forced to work in ischemic conditions and there is a high possibility that the revascularization procedure can fail or be incomplete. Generally, this approach allows validation of ADT in medical applications such as diagnostics of the heart transplants but also open-heart surgery evaluation.

ADT allows the determination of the thermal properties of a tested region. It employs the external thermal excitation source with the aim of analysis of thermal transients to determine thermal conductivity and thermal capacity, represented by thermal time constants, being the product of both parameters.

The choice of SW or LW IR-camera is of secondary importance, as modern cameras in both ranges of radiation are of similar thermal resolution and speed of operation. We prefer LW spectral region because it is easier to avoid interaction while optical excitation is applied.

Acknowledgments

The authors acknowledge contributions of all coworkers from the Department of Biomedical Engineering, Gdansk University of Technology, and from the Department of Cardiosurgery and other departments of the Gdansk Medical University, especially professor J. Siebert and doctors J. Topolewicz, B. Trzeciak, Ł. Jaworski, K. Roszak, and S. Beta as well as engineering staff: M. Suchowirski, M. Bajorek, A. Galikowski and K. Kudlak. *In vivo* animal experiments have been performed at the Department of Animal Physiology, Gdansk University with the main assistance of Dr. W. Stojek. Coworkers are listed in the attached bibliography as coauthors of common publications. This is to underline that most of the illustrations presented here are taken from the cited own publications. The work was supported by several grants from KBN and recently by the development grant R13 027 01 from the Polish Ministry of Science and Higher Education.

References

1. Nowakowski A., Quantitative active dynamic thermal IR-imaging and thermal tomography in medical diagnostics, *The Biomedical Engineering Handbook, Third Edition, Medical Devices and Systems*, ed. J. B. Bronzino, CRC Press, Taylor & Francis, Boca Raton, FL, III Infrared Imaging, 22, pp. 22-1–22-30, 2006.
2. Mohr F. W. et al., IMA-graft patency control by thermal coronary angiography during coronary bypass surgery, *Eur. J. Cardio-Thorac. Surg.*, 5, 534–541, 1986.
3. Mohr F. W. et al., Intraoperative assessment of internal mammary artery bypass graft patency by thermal coronary angiography, *Cardiovasc-Surg.*, 2(6), 703–710, 1994.
4. Falk V., Walther T., Diegeler A., Rauch T., Kitzinger H., Mohr F.W., Thermal-coronary-angiography (TCA) for intraoperative evaluation of graft patency in coronary artery bypass surgery, *Eutotherm. Seminar 50 Proc. QIRT*, 96, 348–353, 1996.
5. Kaczmarek M., Nowakowski A., Siebert J., Rogowski J., Intraoperative thermal coronary angiography—Correlation between internal mammary artery (IMA) free flow and thermographic measurement during coronary grafting, *Seminar 60—Quantitative InfraRed Thermography—QIRT'98, Proc.*, 1, 250–258, 1998.
6. Kaczmarek M., Nowakowski A., Siebert J., Rogowski J., Infrared thermography—Applications in heart surgery, *Proc. SPIE*, 3730, 184–188, 1999.
7. Nowakowski A., ed., Analiza technik diagnostycznych i terapeutycznych w celu oceny procedur kardiochirurgicznych (in Polish) [Analysis of diagnostic and therapeutic methods in terms of evaluation of cardiosurgery procedures], AOW EXIT, Warsaw, 2008.
8. Nowakowski A, Kaczmarek M., Dynamic thermography as a quantitative medical diagnostic tool, *Medical and Biological Engineering and Computing Incorporate Cellular Engineering*, 37(Suppl. 1, Part 1), 244–245, 1999.
9. Suchowirski M., Nowakowski A., Problems in analysis of thermal images sequences for medical diagnostics, EMBEC 2008, *4th European Conference of the International Federation for Medical and Biological Engineering*, Antwerp, Belgium, 23–27 November, abstracts, 355, 2008.

10. Kaczmarek M., Nowakowski A., Stojek W., Topolewicz J., Siebert J., Rogowski J., Thermal monitoring of the myocardium under blood arrest—Preliminary study, *Proceedings of the 29th IEEE EMBS Annual International Conference*, Lyon, CD, 254–257, 2007.

11. Nowakowski A., Kaczmarek M., Wtorek J., Stojek W., Rogowski J., Siebert J., Electrical and thermal monitoring during cardiosurgery interventions, *IFMBE Proceedings*, 17, 150–153, 2007.

12. Sokołowska-Pituchowa J., Ed., in Polish: Anatomia człowieka [Human anatomy], p. 249, Warszawa, PZWL, 1992.

13. Murry C.E., Jennings R.B., Reimer K.A., Preconditioning with ischemia: A delay of lethal cell injury in ischemic myocardium, *Circulation*, 74(5), 1124–1136, 1986.

14. Flameng W.J., Role of myocardial protection for coronary bypass grafting on the beating heart, *Ann. Thorac. Surg.*, 63, 18–22, 1997.

15. Yellon D.M., Baxter G.F., Garcia-Dorado D., Heusch G., Sumeray M., Ischaemic preconditioning: Present position and future directions, *Cardiovasc. Res.*, 37(1), 21–33, 1998.

16. Hao X., Bruce Ch., Pislaru C., Greenleaf J., Characterization of reperfused infarcted myocardium from high-frequency intracardiac ultrasound imaging using homodyned K-distribution, *IEEE Transactions on Ultrasonics, Ferroelectrics, and Frequency Control*, 49(11), 1530–1542, 2002.

17. Hossaek J., Li Y., Yang Z., French B., Assessment of transient myocardial perfusion defects in intact mice using a microbubble contrast destruction/refill approach, *IEEE International Ultrasonics, Ferroelectrics and Frequency Control, Joint 50th Anniversary Conference*, 9–12, 2004.

18

Physiology-Based Face Recognition in the Thermal Infrared Spectrum

Pradeep Buddharaju
University of Houston

Ioannis Pavlidis
University of Houston

18.1 Introduction

Biometrics has received a lot of attention during the past few years from both the academic and business communities. It has emerged as a preferred alternative to traditional forms of identification, like card IDs, which are not embedded into one's physical characteristics. Research into several biometric modalities, including face, fingerprint, iris, and retina recognition, has produced varying degrees of success [1]. Face recognition stands as the most appealing modality, since it is the natural mode of identification among humans and is totally unobtrusive. At the same time, however, it is one of the most challenging modalities [2]. Research into face recognition has been biased toward the visible spectrum for a variety of reasons. Among those is the availability and low cost of visible band cameras and the undeniable fact that face recognition is one of the primary activities of the human visual system. Machine recognition of human faces, however, has proven more problematic than the seemingly effortless face recognition performed by humans. The major culprit is light variability, which is prevalent in the visible spectrum owing to the reflective nature of incident light in this band. Secondary problems are associated with the difficulty of detecting facial disguises [3].

As a solution to the aforementioned problems, researchers have started investigating the use of thermal infrared (IR) for face recognition purposes [4–6]. However, many of these research efforts in thermal face recognition use the thermal IR band only as a way to see in the dark or reduce the deleterious effect of light variability [7,8]. Methodologically, they do not differ very much from face recognition algorithms in the visible band, which can be classified as appearance- [9,10] and feature-based approaches [11,12]. Recently, attempts have been made to fuse the visible and IR modalities to increase the performance of face recognition [13–16].

In this chapter, we present a novel approach to the problem of thermal facial recognition that realizes the full potential of the thermal IR band. It consists of a statistical face segmentation and a physiological feature extraction algorithm tailored to thermal phenomenology. The use of vein structure for human identification has been studied during the recent years using traits such as hand vein patterns [17,18] and finger vein patterns [19,20]. Prokoski and Riedel [21] anticipated the possibility of extracting the vascular network from thermal facial images and using it as a feature space for face recognition. However, they did not present an algorithmic approach for achieving this. To the best of our knowledge, this is the first attempt to develop a face recognition system using physiological information on the face. An early stage of this research was reported briefly in the *Proceedings of the 2005 IEEE Conference on Advanced Video and Signal Based Surveillance* [22]. Our goal is to promote a different way of thinking in the area of face recognition in thermal IR, which can be approached in a distinct manner when compared with other modalities.

Figure 18.1 shows the architecture of the proposed system. The goal of face recognition is to match a query face image against a database of facial images to establish the identity of an individual. Our system operates in the following two phases to achieve this goal:

1. *Off-line phase*: The thermal facial images are captured by an IR camera. For each subject to be stored in the database, we record five different poses. A two-step segmentation algorithm is applied on each pose image to extract the vascular network from the face. Thermal minutia points (TMPs) are detected on the branching points of the vascular network and are stored in the database (see Figure 18.1).

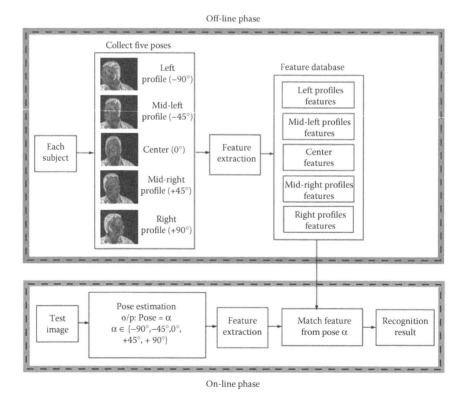

FIGURE 18.1 Architecture of our face recognition system.

2. *On-line phase*: Given a query image, TMPs of its vascular network are extracted and are matched against those of the corresponding pose images stored in the database (see Figure 18.1).

In the following sections, we will describe our face recognition system in detail. In Section 18.2, we present the feature extraction algorithm. In Section 18.3, we discuss our approach for vascular network matching. In Section 18.4, we present the experimental results and attempt a critical evaluation. We conclude this chapter in Section 18.5.

18.2 Feature Extraction

A thermal IR camera with good sensitivity provides the ability to directly image superficial blood vessels on the human face [23]. The pattern of the underlying blood vessels (see Figure 18.2) is characteristic to each individual, and the extraction of this vascular network can provide the basis for a feature vector. Figure 18.3 outlines the architecture of our feature extraction algorithm.

18.2.1 Face Segmentation

Owing to its physiology, a human face consists of "hot" parts that correspond to tissue areas that are rich in vasculature and "cold" parts that correspond to tissue areas with sparse vasculature. This casts the human face as a bimodal temperature distribution entity, which can be modeled using a mixture of two normal distributions. Similarly, the background can be described by a bimodal temperature distribution with walls being the "cold" objects and the upper part of the subject's body dressed in cloths being the "hot" object. Figure 18.4b shows the temperature distributions of the facial skin and the background from a typical IR facial image. We approach the problem of delineating facial tissue from background using a Bayesian framework [22,24] since we have *a priori* knowledge of the bimodal nature of the scene.

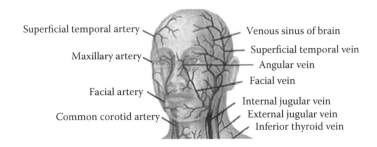

FIGURE 18.2   Superficial blood vessels on the face.

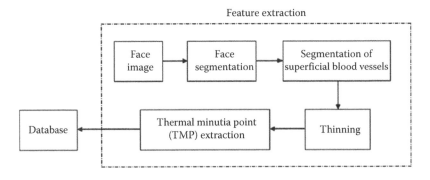

FIGURE 18.3   Architecture of feature extraction algorithm.

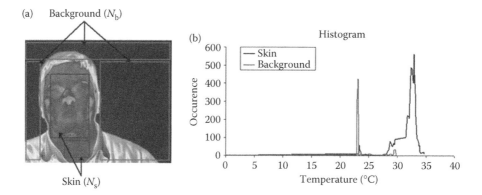

FIGURE 18.4 Skin and background: (a) selection of samples for EM algorithm and (b) corresponding bimodal temperature distributions.

We call θ the parameter of interest, which takes two possible values (skin s or background b) with some probability. For each pixel x in the image at time t, we draw our inference of whether it represents skin (i.e., $\theta = s$) or background (i.e., $\theta = b$) based on the posterior distribution $p^{(t)}(\theta|x_t)$ given by

$$p^{(t)}(\theta \mid x_t) = \begin{cases} p^{(t)}(s \mid x_t), & \text{when } \theta = s \\ p^{(t)}(b \mid x_t) = 1 - p^{(t)}(s \mid x_t), & \text{when } \theta = b \end{cases} \tag{18.1}$$

We develop the statistics only for skin, and then the statistics for the background can easily be inferred from Equation 18.1.

According to the Bayes' theorem

$$p^{(t)}(s|x_t) = \frac{\pi^{(t)}(s)f(x_t|s)}{\pi^{(t)}(s)f(x_t|s) + \pi^{(t)}(b)f(x_t|b)} \tag{18.2}$$

Here, $\pi^{(t)}(s)$ is the prior skin distribution and $f(x_t|s)$ is the likelihood for pixel x representing skin at time t. In the first frame ($t = 1$), the prior distributions for skin and background are considered equiprobable:

$$\pi^{(1)}(s) = \frac{1}{2} = \pi^{(1)}(b) \tag{18.3}$$

For $t > 1$, the prior skin distribution $\pi^{(t)}(s)$ at time t is equal to the posterior skin distribution at time $t - 1$:

$$\pi^{(t)}(s) = p^{(t-1)}(s \mid x_{t-1}) \tag{18.4}$$

The likelihood $f(x_t|s)$ of pixel x representing skin at time $t = 1$ is given by

$$f(x_t \mid s) = \sum_{i=1}^{2} w_{s_i}^{(t)} N(\mu_{s_i}^{(t)}, \sigma_{s_i}^{2(t)}) \tag{18.5}$$

where the mixture parameters w_{s_i} (weight), μ_{s_i} (mean), $\sigma_{s_i}^2$ (variance) $: i = 1, 2$, and $w_{s_2} = 1 - w_{s_1}$ of the bimodal skin distribution can be initialized and updated using the Expectation-Maximization (EM) algorithm. For that, we select N representative facial frames (off-line) from a variety of subjects that we call the training set. Then, we manually segment, for each of the N frames, skin (and background) areas, which yields N_s skin (and N_b background) pixels as shown in Figure 18.4a.

To estimate the mixture parameters for the skin, we initially provide the EM algorithm with some crude estimates of the parameters of interest: $w_{s_0}, \mu_{s_0}, \sigma^2_{s_0}$. Then, we apply the following loop for $k = 0,1,\ldots$:

$$z_{ij}^{(k)} = \frac{w_{s_i}^{(k)} (\sigma_{s_i}^{(k)})^{-1} \exp - \left\{ \frac{1}{2(\sigma_{s_i}^{(k)})^2} (x_j - \mu_{s_i}^{(k)})^2 \right\}}{\sum_{t=1}^{2} w_{s_t}^{(k)} (\sigma_{s_t}^{(k)})^{-1} \exp \left\{ -\frac{1}{2(\sigma_{s_t}^{(k)})^2} (x_j - \mu_{s_t}^{(k)})^2 \right\}}$$

$$w_{s_i}^{(k+1)} = \frac{\sum_{j=1}^{N_s} z_{ij}^{(k)}}{N_s}$$

$$\mu_{s_i}^{(k+1)} = \frac{\sum_{j=1}^{N_s} z_{ij}^{(k)} x_j}{N_s w_{s_i}^{(k+1)}}$$

$$(\sigma_{s_i}^{(k+1)})^2 = \frac{\sum_{j=1}^{N_s} z_{ij}^{(k)} (x_j - \mu_{s_i}^{(k+1)})^2}{N_s w_{s_i}^{(k+1)}}$$

where $i = 1,2$ and $j = 1,\ldots,N_s$. Then, we set $k = k + 1$ and repeat the loop. The condition for terminating the loop is

$$| w_{s_i}^{(k+1)} - w_{s_i}^{(k)} | < \varepsilon, \quad i = 1,2 \tag{18.6}$$

We apply a similar EM process for determining the initial parameters of the background distributions. Once a data point x_t becomes available, we decide that it represents skin if the posterior distribution for the skin, $p^{(t)}(s|x_t) > 0.5$ and that it represents background if the posterior distribution for the background, $p^{(t)}(b|x_t) > 0.5$. Figure 18.5b depicts the visualization of Bayesian segmentation on the subject shown in Figure 18.5a. Part of the subject's nose has been erroneously classified as background and a couple of cloth patches from the subject's shirt have been erroneously marked as facial skin. This is due to occasional overlapping between portions of the skin and background distributions. The isolated nature of these mislabeled patches makes them easily correctable through postprocessing. We apply our three-step postprocessing algorithm on the binary segmented image. Using foreground (and

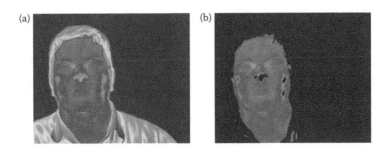

(a) (b)

FIGURE 18.5 Segmentation of facial skin region: (a) original thermal facial image and (b) result of Bayesian segmentation.

background) correction, we find the mislabeled pixels in the foreground (and background) and remove them. Following is the algorithm for achieving this:

1. Label all the regions in the foreground and background using a simple floodfill or connected component labeling algorithm [25]. Let the foreground regions be $R_f(i)$, $i = 1,...,N_f$, where N_f represents the number of foreground regions, and let the background regions be $R_b(j)$, $j = 1,...,N_b$, where N_b represents the number of background regions.
2. Compute the number of pixels in each of the foreground and background regions. Find the maximum foreground (R_f^{max}) and background (R_b^{max}) areas:

$$R_f^{max} = \max\{R_f(i), i = 1,...,N_f\}$$
$$R_b^{max} = \max\{R_b(i), i = 1,...,N_b\}$$

3. Change all foreground regions that satisfy the condition $R_f(i) < R_f^{max}/4$ to background. Similarly, change all background regions that satisfy the condition $R_b(i) < R_b^{max}/4$ to foreground. We found experimentally that outliers tend to have an area smaller than one-fourth of the maximum area, and hence can be corrected with the above conditions. Figure 18.6 shows the result of our post-processing algorithm.

18.2.2 Segmentation of Superficial Blood Vessels

Once a face is delineated from the rest of the scene, the segmentation of superficial blood vessels from the facial tissue is carried out in the following two steps [23,24]:

1. The image is processed to reduce noise and enhance edges.
2. Morphological operations are applied to localize the superficial vasculature.

In thermal imagery of human tissue, the major blood vessels have weak sigmoid edges, which can be handled effectively using anisotropic diffusion. The anisotropic diffusion filter is formulated as a process that enhances object boundaries by performing intraregion as opposed to interregion smoothing. The mathematical equation for the process is

$$\frac{\partial I(\bar{x},t)}{\partial t} = \nabla(c(\bar{x},t)\nabla I(\bar{x},t)) \tag{18.7}$$

In our case, $I(\bar{x},t)$ is the thermal IR image, $\bar{x}$ refers to the spatial dimensions, and t to time. $c(\bar{x},t)$ is called the diffusion function. The discrete version of the anisotropic diffusion filter of Equation 18.7 is as follows:

$$I_{t+1}(x,y) = I_t + \frac{1}{4} * [c_{N,t}(x,y)\nabla I_{N,t}(x,y) + c_{S,t}(x,y)\nabla I_{S,t}(x,y)$$
$$+ c_{E,t}(x,y)\nabla I_{E,t}(x,y) + c_{W,t}(x,y)\nabla I_{W,t}(x,y)] \tag{18.8}$$

The four diffusion coefficients and four gradients in Equation 18.8 correspond to four directions (i.e., north, south, east, and west) with respect to the location (x,y). Each diffusion coefficient and the corresponding gradient are calculated in the same manner. For example, the coefficient along the north direction is calculated as follows:

$$c_{N,t}(x,y) = \exp\left(\frac{-\nabla I_{N,t}^2(x,y)}{k^2}\right) \tag{18.9}$$

where $I_{N,t} = I_t(x,y+1) - I_t(x,y)$.

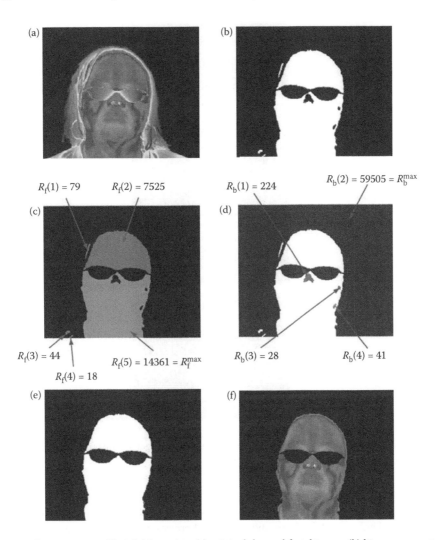

FIGURE 18.6 Segmentation of facial skin region: (a) original thermal facial image; (b) binary segmented image; (c) foreground regions each represented in different color; (d) background regions each represented in different color; (e) binary mask after foreground and background corrections; and (f) final segmentation result after postprocessing.

Image morphology is then applied on the diffused image to extract the blood vessels that are at a relatively low contrast compared with that of the surrounding tissue. We employ for this purpose a top-hat segmentation method, which is a combination of erosion and dilation operations. Top-hat segmentation takes two forms. The first form is the white top-hat segmentation that enhances the bright objects in the image, while the second one is the black top-hat segmentation that enhances dark objects. In our case, we are interested in the white top-hat segmentation because it helps with enhancing the bright ("hot") ridge-like structures corresponding to the blood vessels. In this method the original image is first opened and then this opened image is subtracted from the original image as shown below:

$$I_{open} = (I \ominus S) \oplus S$$
$$I_{top} = I - I_{open}$$

(18.10)

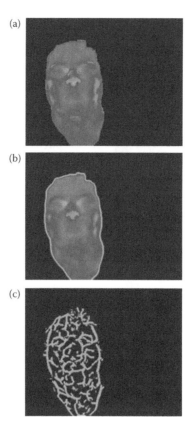

FIGURE 18.7 Vascular network extraction: (a) original segmented image; (b) anisotropically diffused image; and (c) blood vessels extracted using white top-hat segmentation.

where I, I_{open}, and I_{top} are the original, opened, and white top-hat segmented images, respectively, S is the structuring element, and $\ominus$, $\oplus$ are morphological erosion and dilation operations, respectively. Figure 18.7a depicts the result of applying anisotropic diffusion to the segmented facial tissue shown in Figure 18.4b, and Figure 18.7b shows the corresponding blood vessels extracted using white top-hat segmentation.

18.2.3 Extraction of TMPs

The extracted blood vessels exhibit different contour shapes between subjects. We call the branching points of the blood vessels TMPs. TMPs can be extracted from the blood vessel network in ways similar to those used for fingerprint minutia extraction. A number of methods have been proposed [26] for robust and efficient extraction of minutia from fingerprint images. Most of these approaches describe each minutia point by at least three attributes, including its type, its location in the fingerprint image, and the local vessel orientation. We adopt a similar approach for extracting TMPs from vascular networks, which is outlined in the following steps:

1. The local orientation of the vascular network is estimated.
2. The vascular network is skeletonized.
3. The TMPs are extracted from the thinned vascular network.
4. The spurious TMPs are removed.

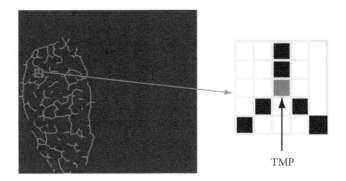

FIGURE 18.8 TMP extracted from the thinned vascular network.

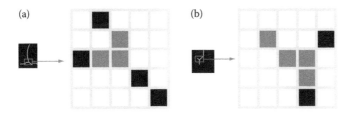

FIGURE 18.9 Spurious TMPs: (a) clustered TMPs and (b) spike formed due to a trivially short branch.

Local orientation $\psi(x,y)$ is the angle formed at (x,y) between the blood vessel and the horizontal axis. Estimating the orientation field at each pixel provides the basis for capturing the overall pattern of the vascular network. We use the approach proposed in Reference 27 for computing the orientation image because it provides pixel-wise accuracy.

Next, the vascular network is thinned to one-pixel thickness [28]. Each pixel in the thinned map contains a value of 1 if it is on the vessel and 0 if it is not. Considering eight-neighborhood $(N_0, N_1, \ldots, N_7)$ around each pixel, a pixel (x,y) represents a TMP if $(\sum_{i=0}^{7} N_i) > 2$ (see Figure 18.8).

It is desirable that the TMP extraction algorithm does not leave any spurious TMPs since this will adversely affect the matching performance. Removal of clustered TMPs (see Figure 18.9a) and spikes (see Figure 18.9b) helps to reduce the number of spurious TMPs in the thinned vascular network.

The vascular network of a typical facial image contains around 50–80 genuine TMPs whose location (x,y) and orientation (ψ) are stored in the database. Figure 18.10 shows the results of each stage of the feature extraction algorithm on a thermal facial image.

18.3 Matching

Each subject's record in the database consists of five different poses to account for pose variation during the testing phase. Since facial images from the same person look quite different across multiple views, it is very important that the search space includes facial images with pose similar to the pose of the test image. Given a test image, we first estimate its pose. Then, the task is simply to match the TMP network extracted from the test image against the TMP database corresponding to the estimated pose.

18.3.1 Estimation of Facial Pose

To the best of our knowledge, it is the first time that the issue of pose estimation in thermal facial imagery is addressed. However, as it is the case with face recognition in general, a number of efforts have been

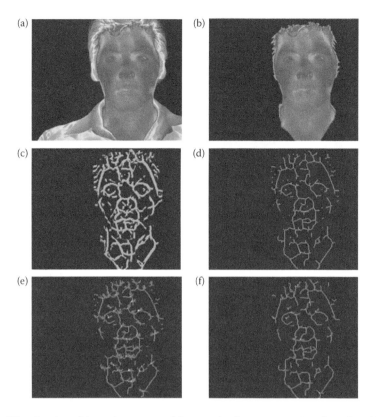

FIGURE 18.10 Visualization of the various stages of the vascular feature extraction algorithm: (a) a typical thermal facial image; (b) facial tissue delineated from the background; (c) network of vascular contours extracted from the thermal facial image; (d) skeletonized vessel map; (e) extracted TMPs from branching points; and (f) cleaned TMP set.

made to address the issue of facial pose estimation in visible band imagery [29,30]. We capitalize upon the algorithm proposed in Reference 29 for estimating head pose across multiple views. We apply principal component analysis (PCA) on the thermal facial images in the training set to reduce the dimensionality of the training examples. Figure 18.11 illustrates sample face images in the database across multiple views. Then, we train the support vector machine (SVM) classifier with the PCA vectors of face samples. Given a test image, SVM can classify it against one of the five poses (center, mid-left profile, left profile, mid-right profile, and right profile) under consideration.

18.3.2 Matching of TMPs

Numerous methods have been proposed for matching fingerprint minutiae, most of which try to simulate the way forensic experts compare fingerprints [26]. Popular techniques are alignment-based point pattern matching, local structure matching, and global structure matching. Local minutiae matching algorithms are fast, simple, and more tolerant to distortions. Global minutiae matching algorithms feature high distinctiveness. A few hybrid approaches [31,32] have been proposed where the advantages of both local and global methods are exploited. We use such a method [31] to perform TMP matching.

For each TMP $M(x_i, y_i, \psi_i)$ that is extracted from the vascular network, we consider its N nearest-neighbor TMPs $M(x_n, y_n, \psi_n)$, $n = 1, \ldots, N$. Then, the TMP $M(x_i, y_i, \psi_i)$ can be defined by a new feature vector:

$$L_M = \{\{d_1, \varphi_1, \vartheta_1\}, \{d_2, \varphi_2, \vartheta_2\}, \ldots, \{d_N, \varphi_N, \vartheta_N\}, \Psi_i\} \tag{18.11}$$

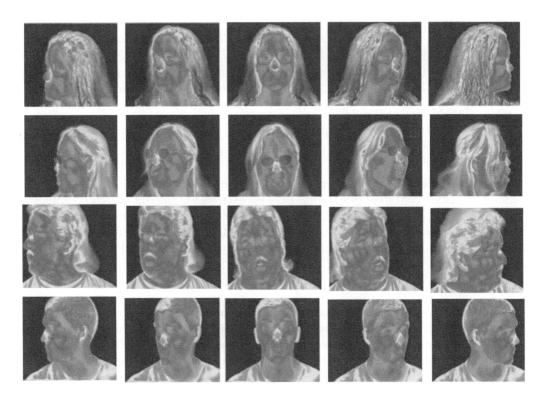

FIGURE 18.11 Samples from our training set featuring five different poses per subject. From left to right the views depicted are left profile, mid-left profile, center, mid-right profile, and right profile.

where

$$d_n = \sqrt{(x_n - x_i)^2 + (y_n - y_i)^2}$$
$$\varphi_n = \text{diff}(\Psi_n, \Psi_i), \quad n = 1, 2, \ldots, N$$
$$\vartheta_n = \text{diff}\left(\arctan\left(\frac{y_n - y_i}{x_n - x_i}\right), \Psi_i\right)$$

(18.12)

The function diff () calculates the difference of two angles and scales the result within the range $[0, 2\pi)$ [32]. Given a test image I_t, the feature vector of each of its TMP is compared with the feature vector of each TMP of a database image. Two TMPs M and M' are marked to be a matched pair if the absolute difference between corresponding features is less than specific threshold values $\{\delta_d, \delta_\varphi, \delta_\vartheta, \delta_\psi\}$. The threshold values should be chosen in such a way that they accommodate linear deformations and translations. The final matching score between the test image and a database image is given by

$$\text{Score} = \frac{\text{NUM}_{\text{match}}}{\max(\text{NUM}_{\text{test}}, \text{NUM}_{\text{database}})}$$

(18.13)

where $\text{NUM}_{\text{match}}$ represents the number of matched TMP pairs, and NUM_{test} and $\text{NUM}_{\text{database}}$ represent the number of TMPs in test and database images, respectively. If the highest matching score between the test and database images is greater than a specific threshold, the corresponding database image is classified as a match. If not, the test image is classified to be not in the database.

18.4 Experimental Results

We collected a large database of thermal facial images in our laboratory from volunteers representing different sex, race, and age groups. The images were captured using a high-quality mid-wave IR Phoenix camera produced by Indigo Systems. The following are the specifications of the camera:

Detector: InSb 640 × 512 element FPA
Spectral range: 3.0–5.0 μm
NETD (sensitivity): 0.01°C
Focal length: 50 mm

We used a subset of the dataset for evaluating the performance of the proposed face recognition algorithm. The dataset consists of 7590 thermal facial images from 138 different subjects (55 images per subject) with varying pose and facial expressions. Five images from each subject (each image representing one of the five training poses) were used for training, TMPs of which were extracted and stored in the database. The remaining 50 images per subject at arbitrary poses were used for testing.

18.4.1 Low Permanence Problem

A major challenge associated with thermal face recognition is the recognition performance over time [33]. Facial thermograms may change depending on the physical condition and environmental conditions. This makes the task of acquiring similar features for the same person over time difficult. A few approaches that use direct temperature data for recognition reported degraded performance over time [10]. However, our approach attempts to solve this problem by using facial anatomical information as feature space, which is unique to each person and at the same time is invariable to physical and environmental conditions as shown in Figure 18.12. The vascular network extracted from the same person with a time gap of about 6 months exhibits a similar pattern.

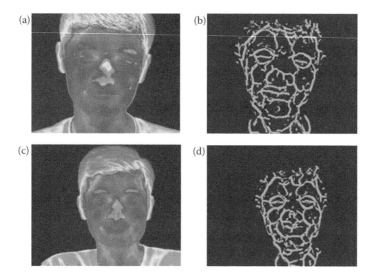

FIGURE 18.12 (a) Thermal facial image of a subject acquired on October 17, 2003; (b) corresponding vascular network; (c) thermal facial image of the same subject acquired on April 29, 2004; and (d) corresponding vascular network.

18.4.2 Frontal Pose and Arbitrary Pose Experiments

Many face recognition algorithms that perform well on the frontal image dataset often have problems when tested on images with arbitrary poses [2]. Our face recognition algorithm overcomes this problem by using multiple pose images for training, which allows pose invariance in the test image. We found experimentally that the five poses we used for training our face recognition algorithm are sufficient to accommodate all yaw rotations (including tilt rotations to a certain extent) without confusing our matching algorithm significantly. As shown in Figure 18.13, when an image that is close to the mid-left profile is queried, pose estimation picked the corresponding mid-left profile image from the training dataset for matching. The small variation in pose that still exists between the query and database images might cause minor position and angle differences in corresponding TMPs extracted from those images. This can be compensated by choosing appropriate values for thresholds $\{\delta_d, \delta_\varphi, \delta_\partial, \delta_\psi\}$, discussed in Section 18.3.2.

We conducted two sets of experiments in order to evaluate the performance of our face recognition system. The first experiment is the frontal pose experiment where the test set contains images with poses between mid-left and mid-right profiles. This test set is matched against only frontal images of the training database. This is the typical experimental procedure used for testing most of the current face recognition algorithms. The second experiment is the arbitrary pose experiment where test set contains all possible poses between left and right profiles. This test set is matched against the entire training set containing five pose images per subject. Figures 18.14 and 18.15 show comparative results of these two experiments. We noticed that the arbitrary pose experiment showed better results when compared to the frontal pose experiment in both cases.

Figure 18.14 shows the cumulative math characteristic (CMC) curves of the two experiments, and Figure 18.15 shows the receiver operating characteristic (ROC) curves based on various threshold values for the matching score discussed in Section 18.3.2. The results demonstrate the promise as well as some problems with our proposed approach. Specifically, CMC shows that rank 1 recognition is over 86% and rank 5 recognition is over 96%. This performance puts a brand new approach very close to the performance of mature visible band recognition methods. In contrast, ROC reveals a weakness of the current method, as it requires false acceptance rate over 20% to reach positive acceptance rate above the 86% range. To address this problem, we believe we need to estimate and eliminate the incorrect TMPs and nonlinear deformations in the extracted vascular network.

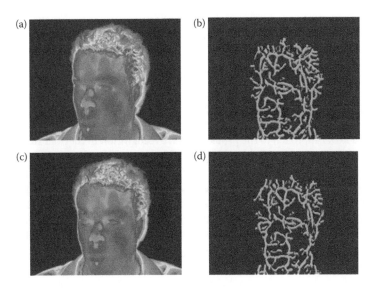

FIGURE 18.13 (a) Test image and (b) corresponding vascular network. (c) Mid-left profile image picked from training database by pose estimation and (d) corresponding vascular network.

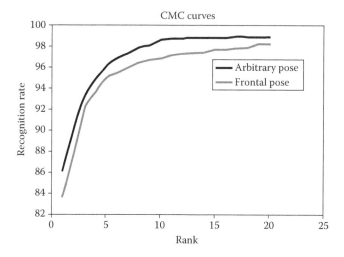

FIGURE 18.14 CMC curves of our method for the frontal and arbitrary pose experiments.

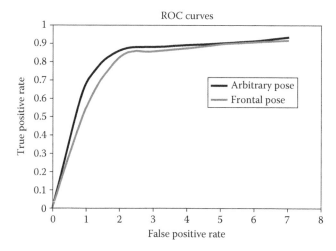

FIGURE 18.15 ROC curves of our method for the frontal and arbitrary pose experiments.

18.5 Conclusions and Future Work

We have outlined a novel approach to the problem of face recognition in thermal IR—one of the fastest-growing biometrics. The cornerstone of the approach is the use of unique and time-invariant physiological information as feature space. We collect five different poses for each subject to be stored in the training database. We have shown that these five poses are capable of accommodating all yaw rotations in the test image. The facial tissue is first separated from the background using a Bayesian segmentation method. The vascular network on the surface of the skin is then extracted based on a white top-hat segmentation preceded by anisotropic diffusion. TMPs are extracted from the vascular network and are used as features for matching the test to database images. The experimental results demonstrate that our approach is very promising.

The method, although young, performed well in a nontrivial database. Our ongoing work is directed toward improving the sophistication of the method regarding nonlinear deformations of the vascular network and testing it comparatively in larger databases.

References

1. Jain, A., Bolle, R., and Pankanti, S., *Biometrics: Personal Identification in Networked Society*, 1st edn., Kluwer Academic Publishers, Norwell, MA, USA, 1999.
2. Zhao, W., Chellappa, R., Phillips, P.J., and Rosenfeld, A., Face recognition: A literature survey, *ACM Computing Surveys (CSUR)*, 35, 399, 2003.
3. Pavlidis, I. and Symosek, P., The imaging issue in an automatic face/disguise detection system. In *Proceedings of IEEE Workshop on Computer Vision Beyond the Visible Spectrum: Methods and Applications*, Hilton Head Island, SC, USA, 2000, p. 15.
4. Prokoski, F., History, current status, and future of infrared identification. In *Proceedings of IEEE Workshop on Computer Vision Beyond the Visible Spectrum: Methods and Applications*, Hilton Head Island, SC, USA, 2000, p. 5.
5. Socolinsky, D.A. and Selinger, A., A comparative analysis of face recognition performance with visible and thermal infrared imagery. In *Proceedings of 16th International Conference on Pattern Recognition*, 4, Quebec, Canada, 2002, p. 217.
6. Wilder, J., Phillips, P.J., Jiang, C., and Wiener, S., Comparison of visible and infrared imagery for face recognition. In *Proceedings of the Second International Conference on Automatic Face and Gesture Recognition*, Killington, VT, 1996, p. 182.
7. Socolinsky, D.A., Wolff, L.B., Neuheisel, J.D., and Eveland, C.K., Illumination invariant face recognition using thermal infrared imagery. In *Proceedings of the IEEE Computer Society Conference on Computer Vision and Pattern Recognition (CVPR 2001)*, 1, Kauai, H1, USA, 2001, p. 527.
8. Selinger, A. and Socolinsky, D.A., Face recognition in the dark. In *Proceedings of the Joint IEEE Workshop on Object Tracking and Classification Beyond the Visible Spectrum*, Washington, DC, 2004.
9. Cutler, R., Face recognition using infrared images and eigenfaces, cs.umd.edu/rgc/face/face.htm, 1996.
10. Chen, X., Flynn, P.J., and Bowyer, K.W., PCA-based face recognition in infrared imagery: Baseline and comparative studies. In *Proceedings of the IEEE International Workshop on Analysis and Modeling of Faces and Gestures*, Nice, France, 2003, p. 127.
11. Srivastava, A. and Liu, X., Statistical hypothesis pruning for recognizing faces from infrared images, *Journal of Image and Vision Computing*, 21, 651, 2003.
12. Buddharaju, P., Pavlidis, I., and Kakadiaris, I., Face recognition in the thermal infrared spectrum. In *Proceedings of the Joint IEEE Workshop on Object Tracking and Classification Beyond the Visible Spectrum*, Washington, DC, 2004.
13. Heo, J., Kong, S.G., Abidi, B.R., and Abidi, M.A., Fusion of visual and thermal signatures with eyeglass removal for robust face recognition. In *Proceedings of the Joint IEEE Workshop on Object Tracking and Classification Beyond the Visible Spectrum*, Washington, DC, 2004.
14. Gyaourova, A., Bebis, G., and Pavlidis, I., Fusion of infrared and visible images for face recognition. In *Proceedings of the eighth European Conference on Computer Vision*, Prague, Czech Republic, 2004.
15. Socolinsky, D.A. and Selinger, A., Thermal face recognition in an operational scenario. In *Proceedings of the IEEE Computer Society Conference on Computer Vision and Pattern Recognition*, 2, Washington DC, 2004, p. 1012.
16. Wang, J.G., Sung, E., and Venkateswarlu, R., Registration of infrared and visible-spectrum imagery for face recognition. In *Proceedings of the Sixth IEEE International Conference on Automatic Face and Gesture Recognition*, Seoul, Korea, 2004, p. 638.
17. Lin, C.L. and Fan, K.C., Biometric verification using thermal images of palm-dorsa vein patterns, *IEEE Transactions on Circuits and Systems for Video Technology*, 14, 199, 2004.
18. Im, S.K., Choi, H.S., and Kim, S.W., A direction-based vascular pattern extraction algorithm for hand vascular pattern verification, *ETRI Journal*, 25, 101, 2003.
19. Shimooka, T. and Shimizu, K., Artificial immune system for personal identification with finger vein pattern. In *Proceedings of the Eighth International Conference on Knowledge-Based Intelligent*

Information and Engineering Systems, Lecture Notes in Computer Science, Wellington, New Zealand, 3214, 2004, p. 511.

20. Miura, N., Nagasaka, A., and Miyatake, T., Feature extraction of finger vein patterns based on iterative line tracking and its application to personal identification, *Systems and Computers in Japan*, 35, 61, 2004.

21. Prokoski, F.J. and Riedel, R., *BIOMETRICS: Personal Identification in Networked Society*, Infrared Identification of Faces and Body Parts, Kluwer Academic Publishers, Norwell, MA, USA, 1998, Chapter 9.

22. Buddharaju, P., Pavlidis, I.T., and Tsiamyrtzis, P., Physiology-based face recognition. In *Proceedings of the IEEE Advanced Video and Signal based Surveillance*, IEEE Conference on Advanced Video and Signal Based Surveillance (AVSS 2005), Como, Italy, 2005.

23. Manohar, C., Extraction of superficial vasculature in thermal imaging, Master's thesis, University of Houston, Houston, TX, 2004.

24. Pavlidis, I., Tsiamyrtzis, P., Manohar, C., and Buddharaju, P., *Biomedical Engineering Handbook*, Biometrics: Face Recognition in Thermal Infrared, CRC Press, Boca Raton, FL, 2006, Chapter 22.

25. Di Stefano, L. and Bulgarelli, A., Simple and efficient connected components labeling algorithm. In *Proceedings of the International Conference on Image Analysis and Processing*, Venice, Italy, 1999, p. 322.

26. Maltoni, D., Maio, D., Jain, A.K., and Prabhakar, S., *Handbook of Fingerprint Recognition*, Springer-Verlag, NY, USA, 2003.

27. Oliveira, M.A. and Leite, N.J., Reconnection of fingerprint ridges based on morphological operators and multiscale directional information, In *The Brazilian Symposium on Computer Graphics and Image Processing*, Curitiba, PR, Brazil, 2004, p. 122.

28. Jang, B.K. and Chin, R.T., One-pass parallel thinning: Analysis, properties, and quantitative evaluation, *IEEE Transactions on Pattern Analysis and Machine Intelligence*, 14, 1129, 1992.

29. Yang, Z., Ai, H., Wu, B., Lao, S., and Cai, L., Face pose estimation and its application in video shot selection. In *Proceedings of the 17th International Conference on Pattern Recognition*, 1, Cambridge, UK, 2004, p. 322.

30. Li, Y., Gong, S., Sherrah, J., and Liddell, H., Support vector machine based multi-view face detection and recognition, *Image and Vision Computing*, 22, 413, 2004.

31. Yang, S. and Verbauwhede, I.M., A secure fingerprint matching technique. In *Proceedings of the 2003 ACM SIGMM Workshop on Biometrics Methods and Applications*, Berkley, CA, 2003, p. 89.

32. Jiang, X. and Yau, W.Y., Fingerprint minutiae matching based on the local and global structures. In *Proceedings of the 15th International Conference on Pattern Recognition*, 2, Barcelona, Catalonia, Spain, 2000, p. 1038.

33. Socolinsky, D.A. and Selinger, A., Thermal face recognition over time. In *Proceedings of the 17th International Conference on Pattern Recognition*, 4, Cambridge, UK, 2004, p. 23.

19

Moinuddin Hassan
U.S. Food and Drug Administration (FDA)

Jana Kainerstorfer
National Institutes of Health

Victor Chernomordik
National Institutes of Health

Abby Vogel
Georgia Institute of Technology

Israel Gannot
Tel Aviv University

Richard F. Little
National Institutes of Health

Robert Yarchoan
National Institutes of Health

Amir H. Gandjbakhche
National Institutes of Health

Noninvasive Infrared Imaging for Functional Monitoring of Disease Processes

Noninvasive imaging techniques are emerging into the forefront of medical diagnostics and treatment monitoring. Both near- and mid-infrared imaging techniques have provided invaluable information in the clinical setting. Most *in vivo* biomedical applications of functional imaging use light in the near-infrared spectrum. The main advantage of the interaction of near-infrared light with tissue is increased penetration: light with wavelengths between 700 and 1100 nm passes through skin and other tissues better than visible light.

In the infrared thermal waveband, information about blood circulation, local metabolism, sweat gland malfunction, inflammation, and healing can be extracted. Originally used to detect breast carcinoma, IR imaging was subsequently reported to have clinical utility in a multitude of neuromusculoskeletal, vascular, and rheomatolgy disorders. There is an especially strong interest in developing optical technologies that have the capability of performing *in situ* tissue diagnosis without the need for sample excision and processing. At present, excisional biopsy followed by histology is considered to be the "gold standard" for the diagnosis of early neoplastic changes and carcinoma. In some cases, cytology, rather than excisional biopsy, is performed. These techniques are powerful diagnostic tools because they provide high-resolution spatial and morphological information of the cellular and subcellular structures of tissues. The use of staining and processing can enhance visual contrast and specificity of histopathology. Both these diagnostic procedures, however, require physical removal of specimens followed by tissue processing in a laboratory. The current status of modern infrared imaging is that of a first-line supplement to both clinical exams and current imaging methods. Using infrared imaging to detect breast pathology is based on the principle that both metabolic and vascular activity in the tissue surrounding a new and developing tumor is usually higher than in normal tissue. Early cancer growth is dependent on increasing blood circulation by creating new blood vessels (angiogenesis). This process results in regional variations that can often be detected by infrared imaging.

The spectroscopic power of light, along with the revolution in molecular characterization of disease processes have given rise to new methods and instrumentation for the early or noninvasive diagnosis of various medical conditions, including arteriosclerosis, heart arrhythmia, cancer, and many other diseases. Near-infrared imaging has been used to functionally monitor diseases processes, including cancer and lymph node detection and optical biopsies. Spectroscopic imaging modalities have been shown to improve the diagnosis of tumors and add new knowledge about the physiological properties of the tumor and surrounding tissues. Particular emphasis should be placed on identifying markers that predict the risk of precancerous lesions progressing to invasive cancers, thereby providing new opportunities for cancer prevention. This might be accomplished through the use of markers as contrast agents for imaging using conventional techniques or through refinements of newer technologies such as magnetic resonance imaging (MRI) or positron emission tomography (PET) scanning.

Infrared imaging techniques have the potential for performing *in vivo* diagnosis on tissue without the need for sample excision and processing. Another advantage of diagnosis through infrared imaging is that the resulting information can be available in real time. In addition, since removal of tissue is not required for diagnosis, a more complete examination of the organ of interest can be achieved than with excisional biopsy or cytology.

Section 19.1 discusses near-infrared imaging and its applications in imaging biological tissues. Mid-infrared thermal imaging techniques, calibration, and a current clinical trial of Kaposi's sarcoma (KS) are described in Section 19.2.

19.1 Near-Infrared Quantitative Imaging of Deep Tissue Structure

In vivo optical imaging has traditionally been limited to superficial tissue surfaces, directly or endoscopically accessible. These methods are based on geometric optics. Most tissues scatter light so strongly, however, that for geometric optics-based equipment to work, special techniques are needed to remove multiply scattered light (such as pinholes in confocal imaging or interferometry in optical coherence microscopies). Even with these special designs, high-resolution optical imaging fails at depths of more than 1 mm below the tissue surface.

Collimated visible or infrared (IR) light impinging upon thick tissue is scattered many times in a distance of ~1 mm, so the analysis of light–tissue interactions requires theories based on the diffusive nature of light propagation. In contrast to x-ray and PET, a complex underlying theoretical picture is needed to describe photon paths as a function of scattering and absorption properties of the tissue.

Approximately two decade ago, a new field called "photon migration" was born that seeks to characterize the statistical physics of photon motion through turbid tissues. The goal is to image macroscopic structures in 3D at greater depths within tissues and to provide reliable pathlength estimation for non-invasive spectral analysis of tissue changes. Although geometrical optics fails to describe light propagation under these conditions, the statistical physics of strong, multiply scattered light provides powerful approaches to macroscopic imaging and subsurface detection and characterization. Techniques using visible and near-infrared light offer a variety of functional imaging modalities, in addition to density imaging, while avoiding ionizing radiation hazards.

In Section 19.1.1, the optical properties of biological tissue are discussed. Section 19.1.2 is devoted to different measurement methods. Theoretical models for spectroscopy and imaging are discussed in Section 19.1.3. In Sections 19.1.4 and 19.1.5, two studies on breast imaging and the use of exogenous fluorescent markers are presented as examples of near-infrared spectroscopy. Finally, the future direction of the field is discussed in Section 19.1.6.

19.1.1 Optical Properties of Biological Tissue

The difficulty of tissue optics is to define optical coefficients of tissue physiology and quantify their changes to differentiate structures and functional status *in vivo*. Light–tissue interactions dictate the way that these parameters are defined. The two main approaches are the wave and particle descriptions of light propagation. Wave propagation uses Maxwell's equations and therefore quantifies the spatially varying permittivity as a measurable quantity. For simplistic and historic reasons, the particle interpretation of light has been used more often (see Section 19.1.3). In photon transport theory, one considers the behavior of discrete photons as they move through the tissue. This motion is characterized by absorption and scattering, and when interfaces (e.g., layers) are involved, refraction. The absorption coefficient, μ_a (mm^{-1}), represents the inverse mean pathlength of a photon before absorption. The distance in a medium where intensity is attenuated by a factor of 1/e (Beer–Lambert law) is considered to be $1/\mu_a$. Absorption in tissue is strongly wavelength dependent and is due to chromophores including water. Among the chromophores in tissue, the dominant component is the hemoglobin in blood. In Figure 19.1, hemoglobin absorption is divided into oxygenated and deoxygenated hemoglobin. As seen in this figure, in the visible range (600–700 nm), the blood absorption is relatively high compared to absorption in the near-infrared. By contrast, water absorption is low in the visible and NIR regions and increases rapidly above approximately 950 nm. Thus, for greatest penetration of light in tissue, wavelengths in the 650–950 nm spectra are used most often. This region of the light spectrum is called the "therapeutic window." One should note that different spectra of chromophores allow one to separate the contribution of varying functional species in tissue (e.g., quantification of oxy- and deoxyhemoglobin to study tissue oxygenation). Similarly, scattering is characterized by a coefficient, μ_s, which is the inverse mean free path of photons between scattering events. The average size of the scattered photons in tissue, proportional to the wavelength of light, places the scattering in the Mie region. In the Mie region, a scattering event does not result in isotropic scattering angles [1,2]. Instead, the scattering in tissue is biased in the forward direction.

For example, by studying the development of neonatal skin, Saidi et al. were able to show that the principal sources of anisotropic scattering in muscle are collagen fibers [3]. The fibers were determined to have a mean diameter of 2.2 μm. In addition to the Mie scattering from the fibers, there is isotropic Rayleigh scattering due to the presence of much smaller scatterers such as organelles in cells.

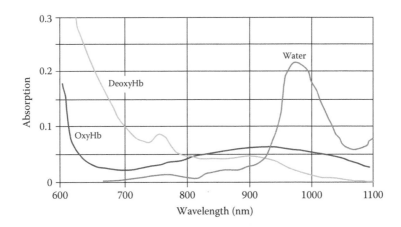

FIGURE 19.1 Absorption spectra of the three major components of tissue in the NIR region: oxy- and deoxyhemoglobin, and water.

Anisotropic scattering is quantified in a coefficient, the mean cosine of the scattering angle g:

$$g \equiv \langle \cos(\theta) \rangle \equiv \frac{\int_0^\pi p(\theta)\cos(\theta)\sin(\theta)d\theta}{\int_0^\pi p(\theta)\sin(\theta)d\theta} \tag{19.1}$$

where, $p(\theta)$ is the probability of a particular scattering angle. For isotropic scattering, $g = 0$. For complete forward scattering, $g = 1$, and for complete back scattering, $g = -1$. In tissue, g is typically 0.7–0.98 [3–5].

Likewise, different tissue types have differing scattering properties that are wavelength dependent. The scattering coefficients of many soft tissues have been measured at a variety of optical wavelengths, and are within the range of 10–100 mm^{-1}. Compared to absorption, however, scattering changes, as a function of wavelength, are more gradual and have smaller extremes. Abnormal tissues such as tumors, fibro-adenomas, and cysts all have scattering properties that are different from normal tissue [6,7]. Thus, the scattering coefficient of an inclusion may also be an important clue to disease diagnosis.

Theories of photon migration are often based on isotropic scattering. Therefore, one must find the appropriate scaling relationships that will allow for an isotropic scattering model. For the case of diffusion-like models (e.g., [8]), it has been shown that one may use an isotropic scattering model with a corrected scattering coefficient, μ_s', and obtain equivalent results, where

$$\mu_s' = \mu_s(1 - g) \tag{19.2}$$

The corrected scattering coefficient is smaller than the actual scattering that corresponds to a greater distance between isotropic scattering events than would occur with anisotropic scattering. For this reason, μ_s' is typically called the transport-corrected scattering coefficient.

There are instances when the spectroscopic signatures will not be sufficient to detect disease. This can occur when the specific disease results in only very small changes to the tissue's scattering and absorption properties, or when the scattering and absorption properties are not unique to the disease. Although it is not clear what the limits of detectability are in relation to diseased tissue properties, it is clear that there will be cases for which optical techniques based on elastic absorption are inadequate. In such cases, another source of optical contrast, such as fluorescence, will be required to detect and localize the disease. The presence of fluorescent molecules in tissue can provide useful contrast mechanisms. Concentration of these endogenous fluorophores in the body can be related to functional and metabolic activities, and therefore to the disease processes. For example, the concentrations of fluorescent molecules such as collagen and NADH have been used to differentiate between normal and abnormal tissue [9].

Advances in the molecular biology of disease processes, new immunohistopathological techniques, and the development of fluorescently labeled cell surface markers have led to a revolution in specific molecular diagnosis of disease by histopathology, as well as in research on molecular origins of disease processes (e.g., using fluorescence microscopy in cell biology). As a result, an exceptional level of specificity is now possible due to the advances in the design of exogenous markers. Molecules can now be tailor-made to bind only to specific receptor sites in the body. These receptor sites may be antibodies or other biologically interesting molecules. Fluorophores may be bound to these engineered molecules and injected into the body, where they will preferentially concentrate at specific sites of interest [10,11].

Furthermore, fluorophore may be used as a probe to measure environmental conditions in a particular locality by capitalizing on changes in fluorophore lifetime [12,13]. Each fluorophore has a characteristic lifetime that quantifies the probability of a specific time delay between fluorophore excitation and emission. In practice, this lifetime may be modified by specific environmental factors such as temperature, pH, and concentrations of substances such as oxygen. In these cases, it is possible to quantify local concentrations of specific substances or specific environmental conditions by measuring the lifetime of fluorophores at the site. Whereas conventional fluorescence imaging is very sensitive to nonuniform fluorophore transport and distribution (e.g., blood does not transport molecules equally to all parts of the body), fluorescence lifetime imaging is insensitive to transport nonuniformity as long as a detectable quantity of fluorophores is present in the site of interest. Throughout the following sections, experimental techniques and differing models used to quantify these sources of optical contrast are presented.

19.1.2 Measurable Quantities and Experimental Techniques

Three classes of measurable quantities prove to be of interest to transform results of remote sensing measurements in tissue into useful physical information. The first is the spatial distribution of light or the intensity profile generated by photons reemitted through a surface and measured as a function of the radial distance from the source and the detector when the medium is continually irradiated by a point source (e.g., a laser). This type of measurement is called continuous wave (CW). The intensity, nominally, does not vary in time. The second class is the temporal response to a very short pulse (~ps) of photons impinging on the surface of the tissue. This technique is called time-resolved and the temporal response is known as the time-of-flight (TOF). The third class is the frequency-domain technique in which an intensity-modulated laser beam illuminates the tissue. In this case, the measured outputs are the AC modulation amplitude and the phase shift of the detected signal. These techniques can be implemented in geometries with different arrangements of source(s) and detector(s): (a) In reflection mode, source(s) and detector(s) are placed at the same side of the tissue. (b) In transmission mode, source(s) and detector(s) are located on opposite sides of the tissue. In the latter, the source(s) and detector(s) can move in tandem while scanning the tissue surface and detectors with lateral offsets can also be used. (c) Tomographic sampling often uses multiple sources and detectors placed around the circumference of the target tissue.

For CW measurements, the instrumentation is simple and requires only a set of light sources and detectors. In this technique, the only measurable quantity is the intensity of light, and, due to multiple scattering, strong pathlength dispersion occurs, which results in a loss of localization and resolution. Hence, this technique is widely used for spectroscopic measurements of bulk tissue properties in which the tissue is considered to be homogeneous [14,15]. However, CW techniques for imaging abnormal targets that use only the coherent portion of light, and thereby reject photons with long pathlengths, have also been investigated. Using the transillumination geometry, collimated detection is used to isolate nonscattered photons [16–18]. Spatial filtering has been proposed which employs a lens to produce the Fourier spectrum of the spatial distribution of light from which the high-order frequencies are removed. The resulting image is formed using only the photons with angles close to normal [19]. Polarization discrimination has been used to select those photons which undergo few scattering events and therefore preserve a fraction of their initial polarization state, as opposed to those photons which experience multiple scattering resulting in complete randomization of their initial polarization state [20]. Several investigators have used heterodyne detection which involves measuring the beat frequency generated by the spatial and temporal combination of a light beam and a frequency modulated reference beam. Constructive interference occurs only for the coherent portion of the light [20–22]. However, the potential of direct imaging using CW techniques in very thick tissue (e.g., breast) has not been established. On the other hand, use of models of photon migration implemented in inverse method based on backprojection techniques has shown promising results. For example, Phillips Medical has used 256 optical fibers placed at the periphery of a white conical-shaped vessel. The area of interest, in this case the breast, is suspended in the vessel, and surrounded by a matching fluid. Three CW laser diodes sequentially illuminate the breast using one fiber. The detection is done simultaneously by 255 fibers.

It is now clear that CW imaging cannot provide direct images with clinically acceptable resolution in thick tissue. Attempts are underway to devise inverse algorithms to separate the effects of scattering and absorption and therefore use this technique for quantitative spectroscopy as proposed by Phillips [23]. However, until now, clinical application of CW techniques in imaging has been limited by the mixture of scattering and absorption of light in the detected signal. To overcome this problem, time-dependent measurement techniques have been investigated.

Time-domain techniques involve the temporal resolution of photons traveling inside the tissue. The basic idea is that photons with smaller pathlengths are those that arrive earlier to the detector. In order to discriminate between unscattered or less scattered light and the majority of the photons, which experience a large number of multiple scattering, sub-nanosecond resolutions are needed. This short time gating of an imaging system requires the use of a variety of techniques involving ultra-fast phenomena and/or fast detection systems. Ultra-fast shuttering is performed using the Kerr effect. The birefringence in the Kerr cell, placed between two crossed polarizers, is induced using very short pulses. Transmitted light through the Kerr cell is recorded, and temporal resolution of a few picoseconds is achieved [19]. When an impulse of light (~picoseconds or hundreds of femtoseconds) is launched at the tissue surface, the whole temporal distribution of photon intensity can be recorded by a streak camera. The streak camera can achieve temporal resolution on the order of few picoseconds up to several nanosececonds detection time. This detection system has been widely used to assess the performance of breast imaging and neonatal brain activity [24,25]. The time of flight recorded by the streak camera is the convolution of the pulsed laser source (in practice with a finite width) and the actual temporal point spread function (TPSF) of the diffuse photons. Instead of using very short pulse lasers (e.g., Ti-Sapphire lasers), the advent of pulse diode lasers with relatively larger pulse width (100–400 ps) have reduce the cost of time-domain imaging much lower. However, deconvolutions of the incoming pulse and the detected TPSF have been a greater issue. Along with diode laser sources, several groups have also used time-correlated single photon counting with photomultipliers for recording the TPSF [26,27]. Fast time gating is also obtained by using stimulated Raman scattering. This phenomenon is a nonlinear Raman interaction in some materials such as hydrogen gas involving the amplification of photons with Stokes shift by a higher energy pump beam. The system operates by amplifying only the earliest-arriving photons [28]. Less widely used techniques such as second-harmonic generation [29], parametric amplification [30], and a variety of others have been proposed for time domain (see an excellent review in [31]).

For frequency-domain measurements, the requirement is to measure the DC amplitude, the AC amplitude, and the phase shift of the photon density wave. For this purpose, a CW light source is modulated with a given frequency (~100 MHz). Lock-in amplifiers and/or phase-sensitive CCD camera have been used to record the amplitude and phase [32,33]. Multiple sources at different wavelengths can be modulated with a single frequency or multiple frequencies [6,34]. In the latter case, a network analyzer is used to produce modulation swept from several hundreds of MHz to up to 1 GHz.

19.1.3 Models of Photon Migration in Tissue

Photon migration theories in a biomedical optics have been borrowed from other fields such as astrophysics, atmospheric science, and specifically from nuclear reactor engineering [35,36]. The common properties of these physical media and biological tissues are their characterization by elements of randomness in both space and time. Because of many difficulties surrounding the development of a theory based on a detailed picture of the microscopic processes involved in the interaction of light and matter, investigations are often based on statistical theories. These can take a variety of forms, ranging from quite detailed multiple-scattering theories [36] to transport theory [37]. However, the most widely used theory is the time-dependent diffusion approximation to the transport equation:

$$\vec{\nabla} \cdot (D\vec{\nabla}\Phi(\vec{r},t)) - \mu_a\Phi(\vec{r},t) = \frac{1}{c}\frac{\partial\Phi(\vec{r},t)}{\partial t} - S(\vec{r},t) \qquad (19.3)$$

where $\vec{r}$ and t are spatial and temporal variables, c is the speed of light in tissue, and D is the diffusion coefficient related to the absorption and scattering coefficients as follows:

$$D = \frac{1}{3\,[\mu_a + \mu_s']} \tag{19.4}$$

The quantity $\Phi(\vec{r},t)$ is called the fluence, defined as the power incident on an infinitesimal volume element divided by its area. Note that the equation does not incorporate any angular dependence, therefore assuming an isotropic scattering. However, for the use of the diffusion theory for anisotropic scattering, the diffusion coefficient is expressed in terms of the transport-corrected scattering coefficient. $S(\vec{r},t)$ is the source term. The gradient of fluence, $J(\vec{r},t)$, at the tissue surface is the measured flux of photons by the detector:

$$J(\vec{r},t) = -D\vec{\nabla}\Phi(\vec{r},t) \tag{19.5}$$

For CW measurements, the time dependence of the flux vanishes, and the source term can be seen as the power impinging in its area. For time-resolved measurement, the source term is a Dirac delta function describing a very short photon impulse. Equation 19.3 has been solved analytically for different types of measurements such as reflection and transmission mode assuming that the optical properties remain invariant through the tissue. To incorporate the finite boundaries, the method of images has been used. In the simplest case, the boundary has been assumed to be perfectly absorbing which does not take into account the difference between indices of refraction at the tissue–air interface. For semi-infinite and transillumination geometries, a set of theoretical expressions has been obtained for time-resolved measurements [38].

The diffusion approximation equation in the frequency domain is the Fourier transformation of the time domain with respect to time. Fourier transformation applied to the time-dependent diffusion equation leads to a new equation:

$$\vec{\nabla} \cdot (D\vec{\nabla}\Phi(\vec{r},\omega)) - \left[\mu_a + \frac{i\omega}{c}\right]\Phi(\vec{r},\omega) + S(\vec{r},\omega) = 0 \tag{19.6}$$

Here, the time variable is replaced by the frequency ω. This frequency is the modulation angular frequency of the source. In this model, the fluence can be seen as a complex number describing the amplitude and phase of the photon density wave, dumped with a DC component:

$$\Phi(\vec{r},\omega) = \Phi_{AC}(\vec{r},\omega) + \Phi_{DC}(\vec{r},0) = I_{AC}\exp(i\theta) + \Phi_{DC}(\vec{r},0) \tag{19.7}$$

In the right-hand side of Equation 19.7, the quantity θ is the phase shift of the diffusing wave. For a nonabsorbing medium, its wavelength is

$$\lambda = 2\pi\sqrt{\frac{2c}{3\mu_s'\omega}} \tag{19.8}$$

Likewise in the time domain, Equation 19.3 has an analytical solution for the case where the tissue is considered homogeneous. The analytical solution permits one to deduce the optical properties in a spectroscopic setting.

For imaging where the goal is to distinguish between structures in tissue, the diffusion coefficient and the absorption coefficient in Equations 19.3 and 19.6 become spatial-dependent and be replaced by $D(r)$ and $\mu_a(r)$. For the cases where an abnormal region is embedded in otherwise homogeneous tissue, perturbation methods based on Born approximation or Rytov approximation have been used (see excellent review in [39]). However, for the cases where the goal is to reconstruct the spectroscopic signatures inside the tissue, no analytical solution exists. For these cases, inverse algorithms are devised to map the spatially varying optical properties. Numerical methods such as finite-element or finite-difference methods have been used to reconstruct images of breast, brain, and muscle [40–42]. Furthermore, in those cases where structural heterogeneity exists, *a priori* information from other image modalities can be useful such as MRI. An example is given in Figure 19.2. Combining MRI and NIR imaging, rat cranium functional imaging during changes in inhaled oxygen concentration was studied [43]. Figure 19.2a and b corresponds to the MRI image and the corresponding constructed finite-element mesh. Figure 19.2c and d corresponds to the oxygen map of the brain with and without incorporation of MRI geometry and constraints.

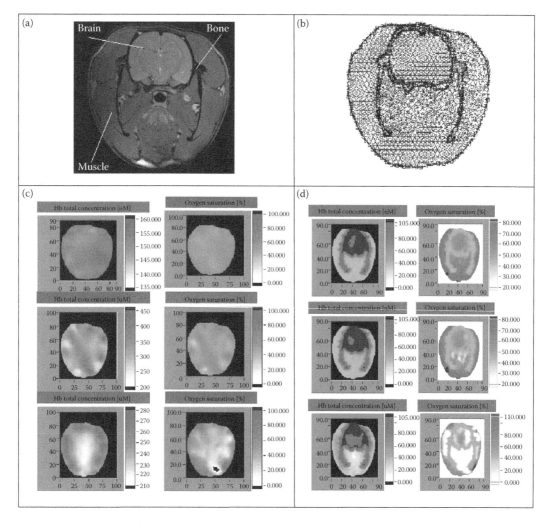

FIGURE 19.2 Functional imaging of rat cranium during changes in inhaled oxygen concentration. (a) MRI image; (b) creation of the mesh to distinguish different compartments in the brain; (c) map of hemoglobin concentration and oxygen saturation of the rat brain without structural constraints from MRI; (d) same as (c) with structural constraints, including tissue heterogeneity. In (c) and (d), the rows from top correspond to 13%, 8%, and 0% (after death) oxygen inhaled. (Courtesy of Dartmouth College.)

The use of MRI images has improved dramatically the resolution of the oxygen map. The use of optical functional imaging in conjunction with other imaging modalities has opened new possibilities in imaging and treating diseases at the bedside.

The second theoretical framework used in tissue optics is the random walk theory (RWT) on a lattice developed at the National Institutes of Health [44,45] and historically precedes the use of the diffusion approximation theory. It has been shown that RWT may be used to derive an analytical solution for the distribution of photon pathlengths in turbid media such as tissue [44]. RWT models the diffusion-like motion of photons in turbid media in a probabilistic manner. Using RWT, an expression may be derived for the probability of a photon arriving at any point and time given a specific starting point and time.

Tissue may be modeled as a 3D cubic lattice containing a finite inclusion, or region of interest, as shown in Figure 19.3. The medium has an absorbing boundary corresponding to the tissue surface, and the lattice spacing is proportional to the mean photon scattering distance, $1/\mu_s'$. The behavior of photons in the RWT model is described by three dimensionless parameters, ρ, n, and μ, which are respectively the radial distance, the number of steps, and the probability of absorption per lattice step. In the RWT model, photons may move to one of the six nearest-neighboring lattice points, each with probability 1/6. If the number of steps, n, taken by a photon traveling between two points on the lattice is known, then the length of the photon's path is also known.

RWT is useful in predicting the probability distribution of photon pathlengths over distances of at least five mean-photon scattering distances. The derivation of these probability distributions is described in review papers [44,45]. For simplicity in this derivation, the tissue–air interface is considered to be perfectly absorbing; a photon arriving at this interface is counted as arriving at a detector on the tissue surface. The derivation uses the central limit theorem and a Gaussian distribution around lattice points to obtain a closed-form solution that is independent of the lattice structure.

The dimensionless RWT parameters, ρ, n, and μ, described above, may be transformed to actual parameters, in part, by using time, t, the speed of light in tissue, c, and distance traveled, r, as follows:

$$\rho \to \frac{r\mu_s'}{\sqrt{2}}, n \to \mu_s'ct, \mu \to \frac{\mu_a}{\mu_s'} \tag{19.9}$$

As stated previously, scattering in tissue is highly anisotropic. Therefore, one must find the appropriate scaling relationships that will allow the use of an isotropic scattering model such as RWT. Like diffusion theory, for RWT [46], it has been shown that one may use an isotropic scattering model with

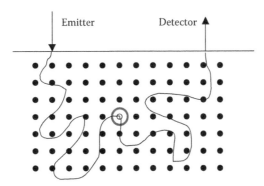

FIGURE 19.3 2D random walk lattice showing representative photon paths from an emitter to a specific site and then to a detector.

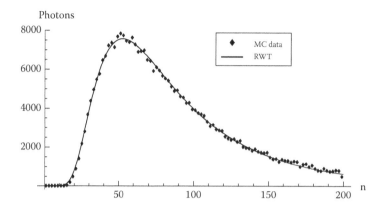

FIGURE 19.4 RWT prediction and Monte Carlo simulation results for transillumination of a 15-mm-thick slab with scattering 1/mm and 109 photons.

a corrected scattering coefficient, μ_s', and obtain equivalent results. The corrected scattering coefficient is smaller than the actual scattering that corresponds to a greater distance between isotropic scattering events than would occur with anisotropic scattering. RWT has been used to show how one would transition from the use of μ_s to μ_s' as the distance under consideration increases [47].

As an example, for a homogeneous slab into which a photon has been inserted, the probability P of a photon arriving at a point ρ after n steps is [48]

$$P(n,\rho) = \frac{\sqrt{3}}{2}\left[\frac{1}{2\pi(n-2)}\right]^{\frac{3}{2}} e^{\frac{-3\rho^2}{2(n-2)}} \sum_{k=-\infty}^{\infty}\left[e^{\frac{-3\left[(2k+1)L-2\right]^2}{2(n-2)}} - e^{\frac{-3\left[(2k+1)L\right]^2}{2(n-2)}}\right]e^{-n\mu} \quad (19.10)$$

where L is the thickness of the slab. The method of images has been used to take into account the two boundaries of the slab. Plotting Equation 19.10 yields a photon arrival curve as shown in Figure 19.4; Monte Carlo simulation data are overlaid. In the next two sections, the use of RWT for imaging will be presented.

19.1.4 RWT Applied to Quantitative Spectroscopy of the Breast

One important and yet extremely challenging area to apply diffuse optical imaging of deep tissues is the human breast (see the review article of Hawrysz and Sevick-Muraca [49]. It is clear that any new imaging or spectroscopic modalities that can improve the diagnosis of breast tumors or can add new knowledge about the physiological properties of the breast and surrounding tissues will have a great significance in medicine.

Conventional transillumination using CW light was used for breast screening several decades ago [50]. However, because of the high scattering properties of tissue, this method resulted in poor resolution. In the late 1980s, time-resolved imaging techniques were proposed to enhance spatial resolution by detecting photons with very short TOF within the tissue. In this technique, a very short pulse, of picosecond duration, impinges upon the tissue. Photons experience dispersion in their pathlengths, resulting in temporal dispersion in their TOF.

To evaluate the performance of time-resolved transillumination techniques, RWT on a lattice was used. The analysis of breast transillumination was based on the calculation of the point spread function (PSF) of time-resolved photons as they visit differing sites at different planes inside a finite slab. The PSF

[51] is defined as the probability that a photon inserted into the tissue visits a given site, is detected at the nth step (i.e., a given time), and has the following rather complicated analytical expression:

$$W_n(\mathbf{s},\mathbf{r},\mathbf{r}_0) = \sum_{l=0}^{n} p_l(\mathbf{r},\mathbf{s}) p_{n-l}(\mathbf{s},\mathbf{r}_0) = \frac{9}{16\pi^{5/2} n^{3/2}} \sum_{k=-\infty}^{\infty} \sum_{m=-\infty}^{\infty} \{F_n[\alpha_+(k),\beta_+(m,\rho)] + \tag{19.11}$$
$$F_n[\alpha_-(k),\beta_-(m,\rho)] - F_n[\alpha_+(k),\beta_-(m,\rho)] - F_n[\alpha_-(k),\beta_+(m,\rho)]\}$$

$$F_n(a,b) = \left(\frac{1}{a} + \frac{1}{b}\right) \exp\left[-\frac{(a+b)^2}{n}\right] \tag{19.12}$$

$$\alpha_\pm(k) = \left\{\frac{3}{2}\left[s_1^2 + (s_3 + 2kN \pm 1)^2\right]\right\}^{1/2} \tag{19.13}$$

$$\beta_\pm(k,\rho) = \left\{\frac{3}{2}\left[(\rho - s_1)^2 + (N - s_3 + 2kN \pm 1)^2\right]\right\}^{1/2} \tag{19.14}$$

where $N = (L\mu_s'/\sqrt{2}) + 1$ is the dimensionless RWT thickness of the slabs and $\bar{s}(s_1,s_2,s_3)$ are the dimensionless coordinates (see Equation 19.9). Evaluation of time-resolved imaging showed that strong scattering properties of tissues prevent direct imaging of abnormalities [52]. Hence, devising theoretical constructs to separate the effects of the scattering from the absorption was proposed, thus allowing one to map the optical coefficients as spectroscopic signatures of an abnormal tissue embedded in thick, otherwise normal tissue. In this method, accurate quantification of the size and optical properties of the target becomes a critical requirement for the use of optical imaging at the bedside. RWT on a lattice has been used to analyze the time-dependent contrast observed in time-resolved transillumination experiments and deduce the size and optical properties of the target and the surrounding tissue from these contrasts. For the theoretical construction of contrast functions, two quantities are needed. First, the set of functions [51] defined previously. Second, the set of functions [53] defined as the probability that a photon is detected at the nth step (i.e., time) in a homogeneous medium (Equation 19.10) [48].

To relate the contrast of the light intensity to the optical properties and location of abnormal targets in the tissue, one can take advantage of some features of the theoretical framework. One feature is that the early time response is most dependent on scattering perturbations, whereas the late time behavior is most dependent on absorptive perturbations, thus allowing one to separate the influence of scattering and absorption perturbations on the observed image contrast. Increased scattering in the abnormal target is modeled as a time delay. Moreover, it was shown that the scattering contrast is proportional to the time-derivative of the PSF, dW_n/dn, divided by P_n [53]. The second interesting feature in RWT methodology assumes that the contrast from scattering inside the inclusion is proportional to the cross section of the target (in the z direction) [51,53], instead of depending on its volume as modeled in the perturbation analysis [54].

Several research groups intend to implement their theoretical expressions into general inverse algorithms for optical tomography, that is, to reconstruct three-dimensional maps of spatial distributions of tissue optical characteristics [49], and thereby quantify optical characteristics, positions, and sizes of abnormalities. Unlike these approaches, the method is a multistep analysis of the collected data. From images observed at differing flight times, construct the time-dependent contrast functions, fit theoretical expressions, and compute the optical properties of the background, and those of the abnormality along with its size. The outline of the data analysis is given in Reference 55.

By utilizing the method for different wavelengths, one can obtain diagnostic information (e.g., estimates of blood oxygenation of the tumor) for corresponding absorption coefficients that no other imaging modality can provide directly. Several research groups have already successfully used multiwavelength measurements using frequency-domain techniques, to calculate physiological parameters (oxygenation, lipid, water) of breast tumors (diagnosed with other modalities) and normal tissue [56].

Researchers at Physikalich.-Techniche-Bundesanstalt (PTB) of Berlin have designed a clinically practical optical imaging system, capable of implementing time-resolved *in vivo* measurements on the human breast [27]. The breast is slightly compressed between two plates. A scan of the whole breast takes but a few minutes and can be done in mediolateral and craniocaudal geometries. The first goal is to quantify the optical parameters at several wavelengths and thereby estimate blood oxygen saturation of the tumor and surrounding tissue under the usual assumption that the chromophores contributing to absorption are oxy- and deoxyhemoglobin and water. As an example, two sets of data, obtained at two wavelengths ($\lambda = 670$ and 785 nm), for a patient (84 year old) with invasive ductal carcinoma, were analyzed. Though the images exhibit poor resolution, the tumor can be easily seen in the optical image shown in Figure 19.5a. In this figure, the image is obtained from reciprocal values of the total integrals of the distributions of TOFs of photons, normalized to a selected "bulk" area. The tumor center is located at $x = -5$, $y = 0.25$ mm.

The best spatial resolution is observed, as expected, for shorter time-delays allowing one to determine the position of the tumor center on the 2-D image (transverse coordinates) with accuracy ~2.5 mm. After preliminary data processing that includes filtering and deconvolution of the raw time-resolved data, we created linear contrast scans passing through the tumor center and analyzed these scans, using the algorithm. It is striking that one observes similar linear dependence of the contrast amplitude on the derivative of PSF ($\lambda = 670$ nm), as expected in the model (see Figure 19.5b). The slope of this linear dependence was used to estimate the amplitude of the scattering perturbation [55].

Dimensions and values of optical characteristics of the tumor and surrounding tissues were then reconstructed for both wavelengths. Results show that the tumor had larger absorption and scattering than the background. Estimated parameters are presented in Table 19.1.

Both absorption and scattering coefficients of the tumor and background all proved to be larger at the red wavelength (670 nm). Comparison of the absorption in the red and near-infrared range is used to estimate blood oxygen saturation of the tumor and background tissue. Preliminary results of the analysis gave evidence that the tumor tissue is in a slightly deoxygenated state with higher blood volume, compared to surrounding tissue.

The spectroscopic power of optical imaging, along with the ability to quantify physiological parameters of human breast have opened a new opportunity for assessing metabolic and physiological activities of human breast during treatment.

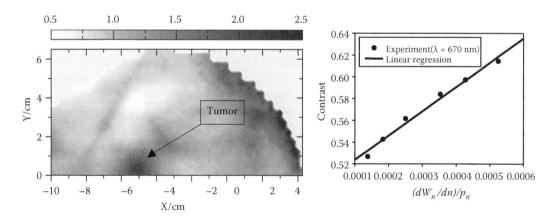

FIGURE 19.5 (a) 2-D optical image of the breast with the tumor. (Courtesy of Physikalich.-Techniche-Bundesanstalt, Berlin.) (b) Contrast obtained from linear scan through the tumor plotted versus the derivative of PSF. From the linear regression, the scattering coefficient of the tumor is deduced.

TABLE 19.1   Optical Parameters of Tumor and Background Breast Tissue

Unknown Coefficients	Reconstructed Values, $\lambda = 670$ nm	Reconstructed Values, $\lambda = 785$ nm
Absorption (background)	0.0029 mm^{-1}	0.0024 mm^{-1}
Scattering (background)	1.20 mm^{-1}	1.10 mm^{-1}
Absorption (Tumor)	0.0071 mm^{-1}	0.0042 mm^{-1}
Scattering (Tumor)	1.76 mm^{-1}	1.6 mm^{-1}

19.1.5 Quantitative Fluorescence Imaging and Spectroscopy

As mentioned in Section 19.1.1, advances in the molecular biology of disease processes, new immuno-histopathological techniques, and the development of specific fluorescently labeled cell surface markers have led a revolution in research on the molecular origins of disease processes. On the other hand, reliable, sensitive, and specific, noninvasive techniques are needed for *in vivo* determinations of abnormalities within the tissue. If successfully developed, noninvasive "optical biopsies" may replace conventional surgical biopsies and provide the advantages of smaller sampling errors, reduction in cost and time for diagnosis, resulting in easier integration of diagnosis and therapy by following the progression of disease or its regression in response to therapy. Clinically practical fluorescence imaging techniques must meet several requirements. First, the pathology under investigation must lie above a depth where the attenuation of the signal results in a poor signal-to-noise ratio and resolvability. Second, the specificity of the marker must be high enough that one can clearly distinguish between normal and abnormal lesions. Finally, one must have a robust image reconstruction algorithm which enables one to quantify the fluorophore concentration at a given depth.

The choices of projects in this area of research are dictated by the importance of the problem, and the impact of the solution on health care. Below, the rationale of two projects is described in Section on Analytical and Functional Biophotonics (SAFB), at the National Institutes of Health is pursuing.

Sjøgren's syndrome (SS) has been chosen as an appropriate test case for developing a noninvasive optical biopsy based on 3D localization of exogenous specific fluorescent labels. SS is an autoimmune disease affecting minor salivary glands which are near (0.5–3.0 mm below) the oral mucosal surface [57]. Therefore, the target pathology is relatively accessible to noninvasive optical imaging. The hydraulic conductivity of the oral mucosa is relatively high, which, along with the relatively superficial location of the minor salivary glands, makes topical application and significant labeling of diseased glands with large fluorescent molecules easy to accomplish. Fluorescent ligands (e.g., fluorescent antibodies specific to CD4$^+$ T cell-activated lymphocytes infiltrating the salivary glands) are expected to bind specifically to the atypical cells in the tissue, providing high contrast and a quantitative relationship to their concentration (and therefore to the stage of the disease process). The major symptoms (dry eyes and dry mouth due to decreased tear and saliva secretion) are the result of progressive immune-mediated dysfunction of the lacrimal and salivary glands. Currently, diagnosis is made by excisional biopsies of the minor salivary glands in the lower lip. This exam, though considered the best criterion for diagnosis, involves a surgical procedure under local anesthesia followed by postoperative discomfort (swelling, pain) and frequently a temporary loss of sensation at the lower lip biopsy site. Additionally, biopsy is inherently subject to sampling errors and the preparation of histopathological slides is time consuming, complicated, expensive, and requires the skills of several professionals (dentist, pathologist, and laboratory technician). Thus, there is a clear need for a noninvasive diagnostic procedure which reflects the underlying gland pathology and has good specificity. A quantitative noninvasive assay would also allow repetition of the test to monitor disease progression and the effect of treatment. However, the quantification of fluorophore concentration within the tissue from surface images requires determining the intensities of different fluorophore sources, as a function of depth and transverse distance and predicting the 3-D distribution of fluorophores within the tissue from a series of images [58].

The second project involves the lymphatic imaging-sentinel node detection. The stage of cancer at initial diagnosis often defines prognosis and determines treatment options. As part of the staging procedure of melanoma and breast cancer, multiple lymph nodes are surgically removed from the primary lymphatic draining site and examined histologically for the presence of malignant cells. Because it is not obvious which nodes to remove at the time of resection of the primary tumor, standard practice involves dissection of as many lymph nodes as feasible. Since such extensive removal of lymphatic tissue frequently results in compromised lymphatic drainage in the examined axilla, alternatives have been sought to define the stage at the time of primary resection. A recent advance in lymph node interrogation has been the localization and removal of the "sentinel" node. Although there are multiple lymphatic channels available for trafficking from the primary tumor, the assumption was made that the anatomic location of the primary tumor in a given individual drains into lymphatic channels in an orderly and reproducible fashion. If that is in fact the case, then there is a pattern by which lymphatic drainage occurs. Thus, it would be expected that malignant cells from a primary tumor site would course from the nearest and possibly most superficial node into deeper and more distant lymphatic channels to ultimately arrive in the thoracic duct, whereupon malignant cells would gain access to venous circulation. The sentinel node is defined as the first drainage node in a network of nodes that drain the primary cancer. Considerable evidence has accrued validating the clinical utility of staging breast cancer by locating and removing the sentinel node at the time of resection of the primary tumor. Currently, the primary tumor is injected with a radionucleotide one day prior to removal of the primary tumor. Then, just before surgery, it is injected with visible dye. The surgeon localizes crudely the location of the sentinel node using a hand-held radionucleotide detector, followed by a search for visible concentrations of the injected dye. The method requires expensive equipment and also presents the patient and hospital personnel with the risk of exposure to ionizing radiation. As an alternative to the radionucleotide, we are investigating the use of IR-dependent fluorescent detection methods to determine the location of sentinel node(s).

For *in vivo* fluorescent imaging, a complicating factor is the strong attenuation of light as it passes through tissue. This attenuation deteriorates the signal-to-noise ratio of detected photons. Fortunately, the development of fluorescent dyes (such as porphyrin and cyanine) that excite and reemit in the "biological window" at near-infrared (NIR) wavelengths, where scattering and absorption coefficients are relatively low, have provided new possibilities for deep fluorescence imaging in tissue. The theoretical complication occurs at depths greater than 1 mm where photons in most tissues enter a diffusion-like state with a large dispersion in their pathlengths. Indeed, the fluorescent intensity of light detected from deep tissue structures depends not only on the location, size, concentration, and intrinsic characteristics (e.g., lifetime, quantum efficiency) of the fluorophores but also on the scattering and absorption coefficients of the tissue at both the excitation and emission wavelengths. Hence, in order to extract intrinsic characteristics of fluorophores within tissue, it is necessary to describe the statistics of photon pathlengths which depend on all these differing parameters.

Obviously, the modeling of fluorescent light propagation depends on the kinds of experiments that one plans to perform. For example, for frequency-domain measurements, Patterson and Pogue [59] used the diffusion approximation of the transport equation to express their results in terms of a product of two Green's function propagators multiplied by a term that describes the probability of emission of a fluorescent photon at the site. One Green's function describes the migration of an incident photon to the fluorophore, and the other describes migration of the emitted photon to the detector. In this representation, the amount of light emitted at the site of the fluorophore is directly proportional to the total amount of light impinging on the fluorophore, with no account for the variability in the number of visits by a photon before an exciting transformation. Since a transformation on an early visit to the site precludes a transformation on all later visits, this results in an overestimation of the number of photons which have a fluorescence transformation at a particular site. This overestimation is important when fluorescent absorption properties are spatially inhomogeneous and largest at later arrival times. RWT has been used to allow for this spatial inhomogeneity by introducing the multiple-passage probabilities concept, thus rendering the model more physically plausible [60]. Another incentive to devise a general theory of diffuse fluorescence photon migration is the capability to quantify local changes in fluorescence lifetime. By selecting fluorophore probes with known lifetime dependence on specific environmental variables, lifetime imaging enables one to localize and quantify such metabolic parameters as temperature and pH, as well as changes in local molecular concentrations *in vivo*.

In the probabilistic RWT model, the description of a photon path may be divided into three parts: the path from the photon source to a localized, fluorescing target; the interaction of the photon with the fluorophore; and finally, the path of the fluorescently emitted photon to a detector. Each part of the photon path may be described by a probability: first, the probability that an incident photon will arrive at the fluorophore site; second, the probability that the photon has a reactive encounter with the fluorophore and the corresponding photon transit delay, which is dependent on the lifetime of the fluorophore and the probability of the fluorophore emitting a photon; and third, the probability that the photon emitted by the fluorophore travels from the reaction site to the detector. Each of these three sequences is governed by a stochastic process. The mathematical description of the three processes is extremely complicated. The complete solution for the probability of fluorescence photon arrival at the detector is [61]

$$\hat{\gamma}(r,s,r_0) = \frac{\eta \; \Phi \hat{p}'_\xi(r|s) \; \hat{p}_\xi(s|r_0)}{\langle \Delta n \rangle (1-\eta)\left[\exp(\xi)-1\right] + \left\{\eta \langle \Delta n \rangle \left[\exp(\xi)-1\right]+1\right\}\left\{1+\left[(1/8)(3/\pi)^{3/2}\sum_{j=1}^{\infty}(\exp(-2j\xi)/j^{3/2})\right]\right\}}$$

(19.15)

where η is the probability of fluorescent absorption of an excitation wavelength photon, Φ is the quantum efficiency of the fluorophore which is the probability that an excited fluorophore will emit a photon at the emission wavelength, $\langle \Delta n \rangle$ is the mean number of steps the photon would have taken had the photon not been exciting the fluorophore (which corresponds to the fluorophore lifetime in random walk parameters), and ξ is a transform variable corresponding to the discrete analog of the Laplace transform and may be considered analogous to frequency. The probability of a photon going from the excitation source to the fluorophore site is $\hat{p}_\xi(s|r_0)$, and the probability of a fluorescent photon going from the fluorophore site to the detector is $\hat{p}'_\xi(r|s)$; the prime indicates that the wavelength of the photon has changed and therefore the optical properties of the tissue may be different. In practice, this solution is difficult to work with, so some simplifying assumptions are desired. With some simplification the result in the frequency domain is

$$\hat{\gamma}(r,s,r_0) = \eta \Phi \left\{ \hat{p}'_\xi(r|s) \; \hat{p}_\xi(s|r_0) - \xi \langle \Delta n \rangle \hat{p}'_\xi(r|s) \hat{p}_\xi(s|r_0) \right\}.$$

(19.16)

The inverse Laplace transform of this equation gives the diffuse fluorescent intensity in the time domain, and the integral of the latter over time leads to CW measurements. The accuracy of such cumbersome equations is tested in well-defined phantoms and fluorophores embedded in *ex vivo* tissue. In Figure 19.6, a line scan of fluorescent intensity collected from 500 μm^3 fluorescent dye (Molecular Probe, far red microspheres: 690 nm excitation; 720 nm emission), embedded in 10.4 mm porcine tissue with a lot of heterogeneity (e.g., fat), are presented. The dashed line is the corresponding RWT fit. The inverse algorithm written in C++ was able to construct the depth of the fluorophore with 100% accuracy. Knowing the heterogeneity of the tissue (seen in the intensity profile), this method presents huge potential to interrogate tissue structures deeply embedded in tissue for which specific fluorescent labeling such as antibodies for cell surfaces exists.

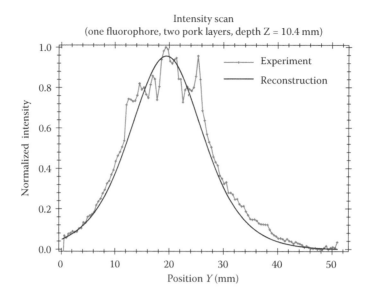

FIGURE 19.6   Intensity scan of a fluorophore 10.4 mm below the tissue surface.

19.1.6 Future Directions

A clinically useful optical imaging device requires multidisciplinary and multistep approaches. At the desk, one devises quantitative theories, and develop methodologies applicable to *in vivo* quantitative tissue spectroscopy and tomographic imaging in different imaging geometries (i.e., transmission or reflection), different types of measurements (e.g., steady-state or time-resolved). Effects of different optical sources of contrast such as endogenous or exogenous fluorescent labels, variations in absorption (e.g., hemoglobin or chromophore concentration) and scattering should be incorporated in the model. At the bench, one designs and conducts experiments on tissue-like phantoms and runs computer simulations to validate the theoretical findings. If successful, one tries to bring the imaging or spectroscopic device to the bedside. For this task, one must foster strong collaborations with physicians who can help to identify physiological sites where optical techniques may be clinically practical and can offer new diagnostic knowledge and/or less morbidity over existing methods. An important intermediate step is the use of animal models for pre-clinical studies. Overall, this is a complicated path. However, the spectroscopic power of light, along with the revolution in molecular characterization of disease processes has created a huge potential for *in vivo* optical imaging and spectroscopy. Maybe the twenty-first century will be the second "*siecle des lumieres*."

19.2 Monitoring of Disease Processes: Clinical Study

19.2.1 Thermal Imaging

The relationship between a change in body temperature and health status has been of interest to physicians since Hippocrates stated "should one part of the body be hotter or colder than the rest, then disease is present in that part." Thermography provides a visual display of the surface temperature of the skin. Skin temperature recorded by an infrared scanner is the resultant balance of thermal transport within the tissues and transport to the environment. In medical applications, thermal images of human skin contain a large amount of clinical information that can help to detect numerous pathological conditions ranging from cancer to emotional disorders. For the clinical assessment of cancer, physicians need to determine the activity of the tumor and location, extent, and response to therapy. All of these factors make it possible for tumors to be examined using thermography. Advantages to using this method are that it is completely nonionizing, safe, and can be repeated as often as required without exposing the patient to risk. Unfortunately, the skin temperature distribution is misinterpreted in many cases, because any high skin temperature does not always indicate a tumor. Therefore, thermography requires thorough investigation to interpret the temperature distribution patterns as well as additional research to clarify various diseases based on skin temperature.

Before applying the thermal technique in the clinical setting, it is important to consider how to avoid misinterpretation of the results. Before the examination, the body should attain thermal equilibrium with its environment. A patient should be unclothed for at least 20 min in a controlled environment at a temperature of approximately 22°C. Under such clinical conditions, thermograms will show only average temperature patterns over an interval of time. The evaluation of surface temperature by infrared techniques requires wavelength and emissive properties of the surface (emissivity) to be examined over the range of wavelengths to which the detector is sensitive. In addition, a thermal camera should be calibrated with a known temperature reference source to standardize clinical data.

An accurate technique for measuring emissivity is presented in Section 19.2.1. In Section 19.2.2, a procedure for temperature calibration of an infrared detector is discussed. The clinical application of thermography with KS is detailed in Section 19.2.3.

19.2.1.1 Emissivity Corrected Temperature

Emissivity is described as a radiative property of the skin. It is a measure of how well a body can radiate energy compared to a black body. Knowledge of emissivity is important when measuring skin temperature with an infrared detector system at different ambient radiation temperatures. Currently, different spectral band infrared detector systems are used in clinical studies such as 3–5 and 8–14 μm. It is well known that the emissivity of the skin varies according to the spectral range. The skin emits infrared radiation mainly between 2 and 20 μm with maximum emission at a wavelength around 10 μm [62]. Jones [63] showed with an InSb detector that only 2% of the radiation emitted from a thermal black body at 30°C was within the 3–5 μm spectral range; the wider spectral response of HgCdTe detector (8–14 μm) corresponded to 40–50% of this black body radiation.

Many investigators have reported on the values for emissivity of skin *in vivo*, measured in different spectral bands with different techniques. Hardy [64] and Stekettee [65] showed that the spectral emissivity of skin was independent of wavelength (λ) when $\lambda > 2$ μm. These results contradicted those obtained by Elam et al. [66]. Watmough and Oliver [67] pointed out that emissivity lies within 0.98 to 1 and was not less than 0.95 for a wavelength range of 2–5 μm. Patil and Williams [68] reported that the average emissivity of normal breast skin was 0.99 ± 0.045, 0.972 ± 0.041, and 0.975 ± 0.043 within the ranges 4–6, 6–18, and 4–18 μm, respectively. Steketee [65] indicated that the average emissivity value of skin was 0.98 ± 0.01 within the range 3–14 μm. It is important to know the precise value of emissivity because an emissivity difference of 0.945 to 0.98 may cause an error of skin temperature of 0.6°C [64].

There is considerable diversity in the reported values of skin emissivity even in the same spectral band. The inconsistencies among reported values could be due to unreliable and inadequate theories and techniques employed for measuring skin emissivity. Togawa [69] proposed a technique in which the emissivity was calculated by measuring the temperature upon a transient stepwise change in ambient radiation temperature [69,70] surrounding an object surface as shown in Figure 19.7.

The average emissivity of skin for the 12 normal subjects measured by a radiometer and infrared camera are presented in Table 19.2. The emissivity values were found to be significantly different between the 3–5 and 8–14 μm spectral bands ($P < 0.001$). An example of a set of images obtained during measurement using an infrared camera (3–5 μm bands) on the forearm of a healthy male subject is shown in Figure 19.8.

An accurate value of emissivity is important because an incorrect value of emissivity can lead to a temperature error in radiometric thermometry especially when the ambient radiation temperature varies widely. The extent to which skin emissivity depends on the spectral range of the infrared

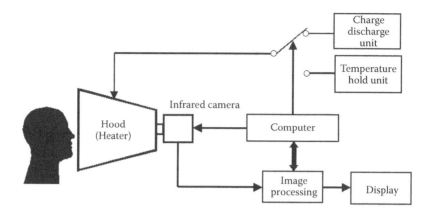

FIGURE 19.7 Schematic diagram of the emissivity measurement system. (From M. Hassan and T. Togawa, *Physiol. Meas.*, 22, 187–200, 2001.)

TABLE 19.2 Average Normal Forearm Skin of 12 Subjects

Emissivity Values	
Infrared camera (3–5 μm)	0.958 ± 0.002
Radiometer (8–14 μm)	0.973 ± 0.0003

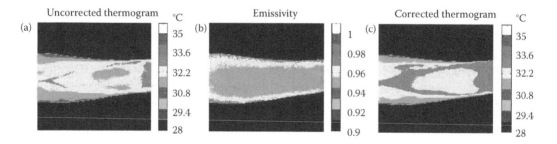

FIGURE 19.8 An example of images obtained from the forearm of a normal healthy male subject. (a) Original thermogram; (b) emissivity image; and (c) thermogram corrected by emissivity.

detectors is demonstrated in Table 19.2, which shows emissivity values measured at 0.958 ± 0.002 and 0.973 ± 0.003 by an infrared detector with spectral bands of 3–5 and 8–14 μm, respectively. These results can give skin temperatures that differ by 0.2°C at a room temperature of 22°C. Therefore, it is necessary to consider the wavelength dependence of emissivity, when high precision temperature measurements are required.

Emissivity not only depends on the wavelength but is also influenced by surface quality, moisture on the skin surface, and so on. In the infrared region of 3–50 μm, the emissivity of most nonmetallic substances is higher for a rough surface than a smooth one [71]. The presence of water also increases the value of emissivity [72]. These influences may account for the variation in results.

19.2.1.2 Temperature Calibration

In infrared thermography, any radiator is suitable as a temperature reference if its emissivity is known and constant within a given range of wavelengths. Currently, many different commercial blackbody calibrators are available to be used as temperature reference sources. A practical and simple blackbody

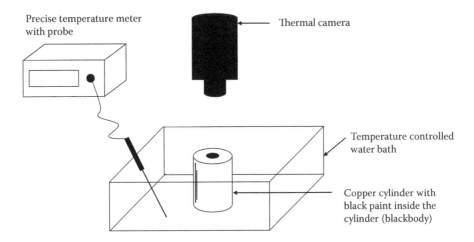

FIGURE 19.9 Schematic diagram of temperature calibration system.

radiator with a known temperature and measurement system is illustrated in Figure 19.9. The system consists of a hollow copper cylinder, a temperature-controlled water bath and a precise temperature meter with probe. The height of the cylinder is 15 cm and the diameter is 7.5 cm. The cylinder is closed except for a hole in the center of the upper end which is 2 cm in diameter. To make the blackbody radiator, the inner surface of the cylinder is coated with black paint (3M Velvet Coating no. 2010) with emissivity of 0.93. Before the calibration, 3/4 of the cylinder is placed vertically in the water and the thermal camera is placed on the top of the cylinder in a vertical direction with a distance of focus length between the surface of the hole and the camera. The water temperature ranges from 18°C to 45°C by 2°C increments. This range was selected since human temperature generally varies from 22°C to 42°C in clinical studies. After setting the water temperature, the thermal camera measures the surface temperature of the hole while the temperature meter with probe measures the water temperature. The temperature of the camera is calibrated according to the temperature reading of the temperature meter.

19.2.2 Near-Infrared Multispectral Imaging

Near-infrared multispectral imaging is most closely related to visual assessment. At National Institutes of Health (NIH) in collaboration with the Lawrence Livermore National Laboratory, a portable spectral imaging system was designed. The system captures images with a high-resolution CCD camera at six near-infrared wavelengths (700, 750, 800, 850, 900, and 1000 nm). Based on differences in absorption coefficients, tissue differences can be assessed. A white light held approximately 45 cm from tissue illuminates the surface uniformly. Using optical filters, images are obtained at the six wavelengths and the intensity images are used in a mathematical optical model of skin containing two layers: an epidermis and much thicker, highly scattering dermis. Each layer contains major chromophores that determine absorption in the corresponding layer and the layers together determine the total reflectance of the skin. Local variations in melanin, oxygenated hemoglobin (HbO_2), and blood volume can be reconstructed through a multivariate analysis.

For the mathematical optical skin model, the effect of the thin epidermis layer on the intensity of the diffusely reflected light is determined by the effective attenuation of light, A_{epi}:

$$A_{epi}(\lambda) = e^{-\mu_{a(epi)}(\lambda)t}$$

where $\mu_{a(epi)}(\lambda)$ is the epidermis absorption coefficient (mm^{-1}), λ is the wavelength (nm), and t is the thickness of the epidermis (mm). The epidermis absorption coefficient is determined by the percentage of melanin, the absorption coefficient of melanin, and the absorption coefficient of normal tissue. Researchers have used different equations to calculate the melanin [73,74] and baseline skin [73–76] absorption coefficients. This model uses the equations of Meglinski and Matcher [77] and Jacques [73] for the melanin and baseline skin absorption coefficients, respectively. The influence of the much thicker, highly scattering dermis layer on the skin reflectance should be estimated by a stochastic model of photon migration, for example, RWT. Fitting the known random walk expression for diffuse reflectivity of the turbid slab [45] yields a formula that depends on the reduced scattering coefficient and dermis absorption coefficient. The dermis absorption coefficient is based on the volume of blood in the tissue and hemoglobin oxygenation, that is, relative fractions of HbO_2 and deoxygenated hemoglobin (Hb). At wavelengths greater than 850 nm, the contribution of water and lipids should be taken into account. The absorption coefficient of blood was calculated by the volume fraction of HbO_2 times the absorption coefficient of HbO_2 plus the volume fraction of Hb times the absorption coefficient of Hb. In the dermis, large cylindrical collagen fibers are responsible for Mie scattering, while smaller-scale collagen fibers and other microstructures are responsible for Rayleigh scattering [73]. The reduced scattering coefficient was calculated by combining Mie and Rayleigh components [78].

Each multispectral image was corrected for the light intensity by calibration of the light source, and camera. Then, each was divided by a weight factor to bring the intensity of the images into the physiologically acceptable range. A best-fit procedure was used to reconstruct for HbO_2 fraction, and blood volume fraction. Melanin concentrations as well as the epidermis thickness were assumed to be constant. The epidermis thickness was assumed to be 60 μm [78] and the melanin content was based on [79].

Thus, near-infrared diffuse multispectral imaging of the skin combined with an analytical, numerical, or stochastic skin model can provide this information by producing spatial maps of skin chromophore concentrations [80–82]. The disadvantage of finding these parameters by fitting the data to an analytical skin model lies in the computationally expensive data postprocessing. This makes immediate conclusions difficult or even impossible if the image size is large, whereas in clinical routines it is often desired to assess the metabolic state of a tumor in real time. An alternative to image reconstruction has been proposed, which can provide real-time blood volume and blood oxygenation maps, which is based on principal component analysis (PCA). PCA [83] linearly transforms the data into an orthogonal coordinate system whose axes correspond to the principal components in the data, that is, the first principal component accounts for as much variance in the data as possible and, successively, further components capture the remaining variance. Through an eigenanalysis, the principal components are determined as eigenvectors of the dataset's covariance matrix and the corresponding eigenvalues refer to the variance that is captured within each eigenvector. It has been shown that, when applied to multispectral images in the near-infrared range (700 nm–850 nm), the first eigenvector corresponds to blood volume and the second to blood oxygenation [84].

19.2.2.1 Validation of Multispectral Images

In order to validate the use of multispectral imaging for assessing fractional blood volume and blood oxygenation, images from healthy volunteers were acquired. A pressure cuff was used to occlude the upper right arm with 180 mm Hg pressure. This amount of pressure was chosen to achieve arterial occlusion and the pressure lasted for 5 min. Multispectral images were taken every 30 s before occlusion, during occlusion, and for 5 min afterwards, resulting in 21 time points in total [80,84,85]. Occlusion experiments were chosen as the behavior of blood volume and blood oxygenation over time is well known. Blood volume remains constant, blood oxygenation undergoes ischemia during, and reactive hyperemia after occlusion [80,86–88].

Figure 19.10 shows 2D reconstruction maps of fractional blood volume and blood oxygenation concentrations over time for one representative healthy volunteer's lower forearm over time. The first row shows the baseline before occlusion, and row 2 and 3 show the results during and after occlusion,

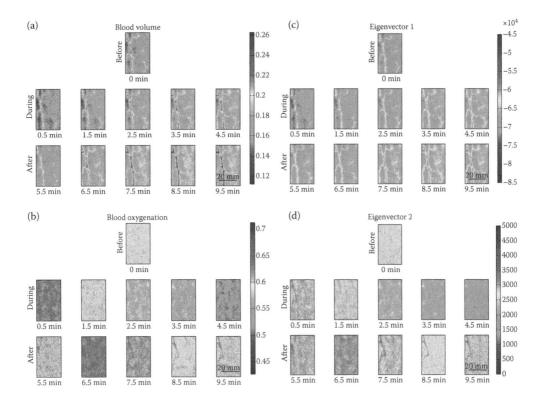

FIGURE 19.10 Typical multispectral results from a healthy volunteer's forearm. Reconstructed fractional blood volume (a) and blood oxygenation (b) over time. Principal component analysis results with eigenvector 1 (c) and eigenvector 2 (d). (From J. Kainerstorfer, M. Ehler, F. Amyot et al., *J. Biomed. Opt.*, 15, 046007, 2010. With permission.)

respectively. Veins contain more blood than the surrounding tissue and are clearly separable in the reconstruction maps by increased blood volume. The overall tissue oxygenation follows the typical expected trend of ischemia during occlusion (drop of oxygenation compared to baseline) and reactive hyperemia after occlusion (overshoot compared to baseline).

The corresponding PCA results show the same spatial distribution, as well as temporal behavior, validating PCA as an alternative tool for chromophore assessment.

19.2.3 Clinical Study: Kaposi's Sarcoma

The oncology community is testing a number of novel targeted approaches such as antiangiogenic, antivascular, immuno-, and gene therapies for use against a variety of cancers. To monitor such therapies, it is desirable to establish techniques to assess tumor vasculature and changes with therapy [89]. Currently, several imaging techniques such as dynamic contrast-enhanced MRI [90–92], PET [93–95], computed tomography (CT) [96–99], color Doppler ultrasound (US) [100,101], and fluorescence imaging [102,103] have been used in angiogenesis-related research. With regard to monitoring vasculature, it is desirable to develop and assess noninvasive and quantitative techniques that can not only monitor structural changes, but can also assess the functional characteristics or the metabolic status of the tumor. There are currently no standard noninvasive techniques to assess parameters of angiogenesis in lesions of interest and to monitor changes in these parameters with therapy. For antiangiogenic therapies, factors associated with blood flow are of particular interest.

KS is a highly vascular tumor that occurs frequently among people infected with acquired immunodeficiency syndrome (AIDS). During the first decade of the AIDS epidemic, 15–20% of AIDS patients developed this type of tumor [104]. Patients with KS often display skin and oral lesions and KS frequently involves lymph nodes and visceral organs [105]. KS is an angio-proliferative disease characterized by angiogenesis, endothelial spindle-cell growth (KS cell growth), inflammatory-cell infiltration, and edema [106]. A gamma herpesvirus called KS-associated herpesvirus (KSHV) or human herpesvirus type 8 (HHV-8) is an essential factor in the pathogenesis of KS [107]. Cutaneous KS lesions are easily accessible for noninvasive techniques that involve imaging of tumor vasculature, and they may thus represent a tumor model in which to assess certain parameters of angiogenesis [108,109].

Two potential noninvasive imaging techniques, infrared thermal imaging (thermography) and laser Doppler imaging (LDI), have been used to monitor patients undergoing an experimental anti-KS therapy [110,111]. Thermography graphically depicts temperature gradients over a given body surface area at a given time. It is used to study biological thermoregulatory abnormalities that directly or indirectly influence skin temperature [112–116]. However, skin temperature is only an indirect measure of skin blood flow, and the superficial thermal signature of skin is also related to local metabolism. Thus, this approach is best used in conjunction with other techniques. LDI can more directly measure the net blood velocity of small blood vessels in tissue, which generally increases as blood supply increases during angiogenesis [117,118]. Thermal patterns were recorded using an infrared camera with a uniform sensitivity in the wavelength range of 8–12 μm and LDI images were acquired by scanning the lesion area of the KS patients at two wavelengths, 690 nm and 780 nm.

An example of the images obtained from a typical KS lesion using different modalities is shown in Figure 19.11 [111]. As can be seen in the thermal image, the temperature of the lesion was approximately 2°C higher than that of the normal tissue adjacent to the lesion. Interestingly, in a number of lesions, the area of increased temperature extended beyond the lesion edges as assessed by visual inspection or palpation. This may reflect relatively deep involvement of the tumor in areas underlying normal skin. However, the thermal signature of the skin not only reflects superficial vascularity but also deep tissue metabolic activity. In the LDI image of the same lesion, there was increased blood flow in the area of the lesion as compared to the surrounding tissue, with a maximum increase of over 600 AU (arbitrary units). Unlike the thermal image, the increased blood velocity extended only slightly beyond the area of this visible lesion, possibly because the tumor process leading to the increased temperature was too deep to be detected by LDI. Both these techniques were used successfully to visualize KS lesions, and although each measure an independent parameter (temperature or blood velocity), there was a strong correlation in a group of 16 patients studied by both techniques (Figure 19.12). However, there were

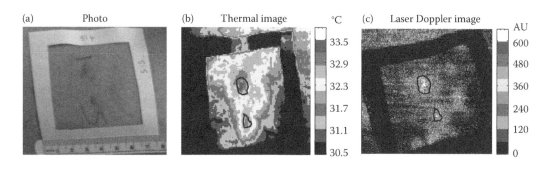

FIGURE 19.11 Typical multimodality images obtained from a patient with KS lesion. The numbers "1" and "5" in the visual image were written on the skin to identify the lesions for tumor measurement. The solid line in the thermal and LDI demarks the border of the visible KS lesion. Shown is a representative patient from the study reported in Reference 111.

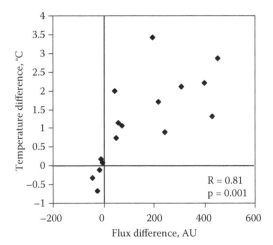

FIGURE 19.12   Relationship between the difference in temperature and flux assessed by LDI of the lesion and surrounding area of the lesion of each subject. A positive correlation was observed between these two methods ($R = 0.8$, $p < 0.001$). (From M. Hassan, R. F. Little, A. Vogel et al., *TCRT*, 3, 451–457, 2004. With permission.)

some differences in individual lesions since LDI measured blood flow distribution in the superficial layer of the skin of the lesion, whereas the thermal signature provided a combined response of superficial vascularity and metabolic activities of deep tissue.

In patients treated with an anti-KS therapy, there was a substantial decrease in temperature and blood velocity during the initial 18-week treatment period as shown in Figure 19.13 [111]. The changes in

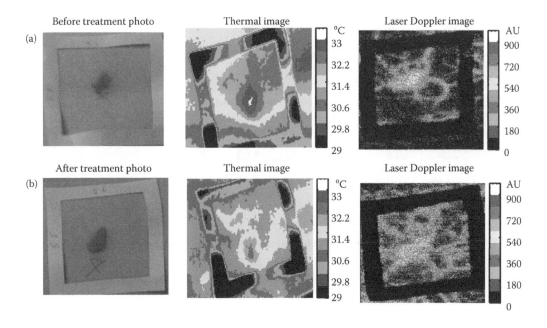

FIGURE 19.13   Typical example of lesion obtained from a subject with KS (a) before and (b) after the treatment. Improvement after the treatment can be assessed by the thermal or LDI images after 18 weeks. Shown is a patient from the clinical trial reported in Reference 111.

these two parameters were generally greater than those assessed by either measurement of tumor size or palpation. In fact, there was no statistically significant decrease in tumor size overall. These results suggest that thermography and LDI may be relatively more sensitive in assessing the response of therapy in KS than conventional approaches. Assessing responses to KS therapy is now generally performed by visual measuring and palpating the numerous lesions and using rather complex response criteria. However, the current tools are rather cumbersome and often subject to observer variation, complicating the assessment of new therapies. The techniques described earlier, possibly combined with other techniques to assess vasculature and vessel function, have the potential of being more quantitative, sensitive, and reproducible than established techniques.

Diffuse multispectral imaging for noncontact and noninvasive monitoring changes in concentrations of blood volume and oxygenated and deoxygenated hemoglobin has also been been used in this clinical study to assess the pathogenesis of the status and changes of KS lesions during therapy. Such an approach can be used to provide early markers for tumor responses and to learn about the pathophysiology of the disease and its changes in response to treatment.

An example of multimodality images obtained from a KS patient before any anti-KS therapy is shown in Figure 19.14. Relatively high contrast of oxygenated hemoglobin and tissue blood volume are observed in the tumor region, which is expected for a metabolically active tumor. The normal tissue blood volume fraction is approximately 5%. This follows previous research that the volume fraction of blood in tissue is 0.2–5% [119]. Another example of multispectral images of a KS patient over the course of treatment is seen in Figure 19.15 in combination of PCA results [120]. Results of oxygenation and blood volume showed an increase and decrease, respectively, inside the lesion over the course of the treatment. Results indicated remission of the disease already at week 14, even though the lesion on the surface did not clinically decrease in size. Pathologic complete remission of disease was confirmed by the clinic in week 48.

The novel imaging modality could potentially be used as predictive tools for the outcome, and therefore also used for individualization of therapeutic strategies.

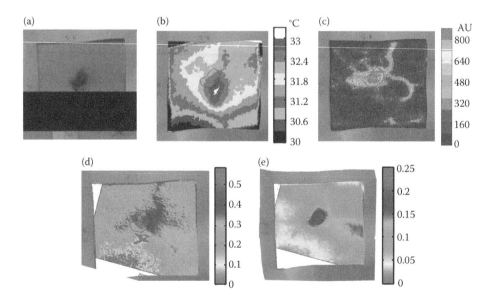

FIGURE 19.14 Set of comparative images of a KS patient. Visual, thermal, laser Doppler, HbO$_2$ fraction, and tissue blood volume fraction images are provided. (From A. Vogel, M. Hassan, F. Amyot et al., *Biomedical Optics 2006, Technical Digest*, p. SG2, 2006.)

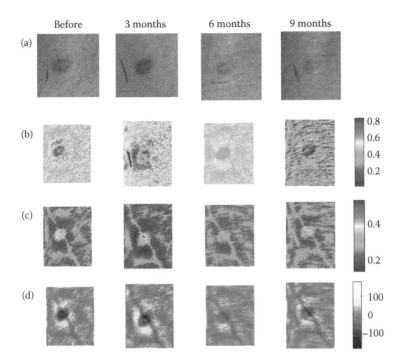

FIGURE 19.15 Digital images of KS lesion over time. Left to right: week 0, 14, 26, and 38 (a); blood oxygenation over time (b); blood volume over time (c); projection along the first principal component (d). (From Kainerstorfer, J. M., F. Amyot, M. Hassan et al., *Biomedical Optics, OSA Technical Digest*, p. BME6, 2010. With permission.)

Acknowledgments

Special thanks go to Dr. Herbert Rinneberg (Physikalich.-Techniche-Bundesanstalt, Berlin) and Dr. Brian Pogue (Dartmouth College) for providing optical images. The authors also wish to express their thanks to Dr. Tatsuo Togawa, a former professor of the Institutes of Biomaterials and Bioengineering, Tokyo Medical and Dental University, Tokyo, Japan for his valuable suggestions and allowing emissivity measurements to be performed in his lab. The authors also thank Stavros Demos at the Lawrence Livermore National Laboratory for helping to design and construct the multispectral imaging system used in the KS clinical trials. This research is supported by the Intramural Research Program of the Eunice Kennedy Shriver National Institute of Child Health and Human Development and the National Cancer Institute, National Institutes of Health.

References

1. M. Born and E. Wolf, *Principles in Optics*, 7th ed. Cambridge: Cambridge University Press, 1999.
2. A. T. Young, Rayleigh scattering, *Phys. Today*, 35, 42–48, 1982.
3. I. S. Saidi, S. L. Jacques, and F. K. Tittel, Mie and Rayleigh modeling of visible-light scattering in neonatal skin, *Appl. Opt.*, 34, 7410, 1995.
4. M. J. C. Van Gemert, S. L. Jacques, H. J. C. M. Sterenberg, and W. M. Star, Skin optics, *IEEE Trans.*, 36, 1146, 1989.
5. R. Marchesini, A. Bertoni, S. Andreola, E. Melloni, and A. Sicherolli, Extinction and absorption coefficients and scattering phase functions of human tissues *in vitro*, *Appl. Opt.*, 28, 2318, 1989.

6. J. Fishkin, O. Coquoz, E. Anderson, M. Brenner, and B. Tromberg, Frequency-domain photon migration measurements of normal and malignant tissue optical properties in a human subject, *Appl. Opt.*, 36, 10, 1997.

7. T. L. Troy, D. L. Page, and E. M. Sevick-Muraca, Optical properties or normal and diseased breast tissues: prognosis for optical mammography, *J. Biomed. Opt.*, 1, 342, 1996.

8. A. H. Gandjbakhche, R. F. Bonner, and R. Nossal, Scaling relationships for anisotropic random walks, *J. Statistical Phys.*, 69, 35, 1992.

9. G. A. Wagnieres, W. M. Star, and W. B. C, *In vivo* fluorescence spectroscopy and imaging for oncological applications, *Photochem. Photobiol.*, 68, 603, 1998.

10. R. Weissleder, A clearer vision for *in vivo* imaging, *Nat. Biotechnol.*, 19, 316, 2001.

11. V. F. Kamalov, I. A. Struganova, and K. Yoshihara, Temperature dependent radiative lifetime of J-aggregates, *J. Phys. Chem.*, 100, 8640, 1996.

12. S. Mordon, J. M. Devoisselle, and V. Maunoury, *In vivo* pH measurement and imaging of a pH-sensitive fluorescent probe (5–6 carboxyfluorescein): Instrumental and experimental studies, *Photochem. Photobiol.*, 60, 274, 1994.

13. C. L. Hutchinson, J. R. Lakowicz, and E. M. Sevick-Muraca, Fluorescence lifetime-based sensing in tissues: A computational study, *Biophys. J.*, 68, 1574, 1995.

14. F. F. Jobsis, Noninvasive infrared monitoring of cerebral and myocardial oxygen sufficiency and circulatory parameters, *Science*, 198, 1264, 1977.

15. T. J. Farrell, M. S. Patterson, and B. Wilson, A diffusion theory model of spatially resolved, steady-state diffuse reflectance for the noninvasive determination of tissue optical properties *in vivo*, *Med. Phys.*, 9, 879, 1992.

16. P. C. Jackson, P. H. Stevens, J. H. Smith, D. Kear, H. Key, and P. N. T. Wells, Imaging mammalian tissues and organs using laser collimated transillumination, *J. Biomed. Eng.*, 6, 70, 1987.

17. G. Jarry, S. Ghesquiere, J. M. Maarek, F. Fraysse, S. Debray, M.-H. Bui, and D. Laurent, Imaging Mammalian tissues and organs using laser collimated transillumination, *J. Biomed. Eng.*, 6, 70, 1984.

18. M. Kaneko, M. Hatakeyama, P. He et al., Construction of a laser transmission photo-scanner: Pre-clinical investigation, *Radiat. Med.*, 7, 129, 1989.

19. L. Wang, P. P. Ho, C. Liu, G. Zhang, and R. R. Alfano, Ballistic 2-D imaging through scattering walls using an ultrafast optical Kerr gate, *Science*, 253, 769, 1991.

20. A. Schmitt, R. Corey, and P. Saulnier, Imaging through random media by use of low-coherence optical heterodyning, *Opt. Lett.*, 20, 404, 1995.

21. H. Inaba, M. Toida, and T. Ichmua, Optical computer-assisted tomography realized by coherent detection imaging incorporating laser heterodyne method for biomedical applications, *SPIE Proc.*, 399, 108, 1990.

22. H. Inaba, Coherent detection imaging for medical laser tomography, *Medical Optical Tomography: Functional Imaging and Monitoring*, ed. Muller, G., SPIE Optical Engineering Press, Bellingham, WA, p. 317, 1993.

23. S. B. Colak, D. G. Papaioannou, G. W. T'Hoooft, M. B. van der Mark, H. Schomberg, J. C. J. Paasschens, J. B. M. Melissen, and N. A. A. J. van Austen, Tomographic image reconstruction from optical projections in light diffusing media, *Appl. Opt.*, 36, 180, 1997.

24. J. C. Hebden, D. J. Hall, M. Firbank, and D. T. Delpry, Time-resolved optical imaging of a solid tissue-equivalent phantom, *Appl. Opt.*, 34, 8038, 1995.

25. J. C. Hebden, Evaluating the spatial resolution performance of a time-resolved optical imaging system, *Med. Phys.*, 19, 1081, 1992.

26. R. Cubeddu, A. Pifferi, P. Taroni, A. Torriceli, and G. Valentini, Time-resolved imaging on a realistic tissue phantom: $\mu_s{}'$ and μ_a images versus time-integrated images, *Appl. Opt.*, 35, 4533, 1996.

27. D. Grosenick, H. Wabnitz, H. Rinneberg, K. T. Moesta, and P. Schleg, Development of a time-domain optical mammograph and first in-vivo application, *Appl. Opt.*, 38, 2927, 1999.

28. M. Bashkansky, C. Adler, and J. Reinties, Coherently amplified Raman polarization gate for imaging through scattering media, *Opt. Lett.*, 19, 350, 1994.

29. K. M. Yoo, Q. Xing, and R. R. Alfano, Imaging objects hidden in highly scattering media using femtosecond second-harmonic-generation cross-correlation time gating, *Opt. Lett.*, 16, 1019, 1991.

30. G. W. Faris and M. Banks, Upconverting time gate for imaging through highly scattering media, *Opt. Lett.*, 19, 1813, 1994.

31. J. C. Hebden, S. R. Arridge, and D. T. Delpry, Optical imaging in medicine I: Experimental techniques, *Phys. Med. Biol.*, 42, 825, 1997.

32. J. R. Lakowitz and K. Brendt, Frequency domain measurements of photon migration in tissues, *Chem. Phys. Lett.*, 166, 246, 1990.

33. M. A. Franceschini, K. T. Moesta, S. Fantini, G. Gaida, E. Gratton, H. Jess, W. W. Mantulin et al., Frequency-domain techniques enhance optical mammography: Initial clinical results, *Proc. Natl. Acad. Sci., Med. Sci.*, 94, 6468, 1997.

34. B. Tromberg, O. Coquoz, J. B. Fishkin, T. Pham, E. Anderson, J. Butler, M. Cahn et al., Non-invasive measurements of breast tissue optical properties using frequency-domain photon migration, *Philos. Trans. R. Soc. London Ser.*, B352, 661, 1997.

35. J. J. Duderstadt and L. J. Hamilton, *Nuclear Reactor Analysis*. New York: Wiley, 1976.

36. K. M. Case and P. F. Zweifel, *Linear Transport Theory*. Reading: Addison Wesley, 1967.

37. A. Ishimaru, *Wave Propagation and Scattering in Random Media*. New York: Academic Press, 1978.

38. M. S. Patterson, B. Chance, and B. Wilson, Time resolved reflectance and transmittance for the non-invasive measurement of tissue optical properties, *Appl. Opt.*, 28, 2331, 1989.

39. S. R. Arridge and J. C. Hebden, Optical imaging in medicine: II. Modelling and reconstruction, *Phys. Med. Biol.*, 42, 841, 1997.

40. S. R. Nioka, M. Miwa, S. Orel, M. Schnall, M. Haida, S. Zhao, and B. Chance, Optical imaging of human breast cancer, *Adv. Exp. Med. Biol.*, 361, 171, 1994.

41. S. Fantini, S. A. Walker, M. A. Franceschini, M. Kaschke, P. M. Schlag, and K. T. Moesta, Assessment of the size, position, and optical properties of breast tumors *in vivo* by noninvasive optical methods, *Appl. Opt.*, 37, 1982, 1998.

42. M. Maris, E. Gratton, J. Maier, W. Mantulin, and B. Chance, Functional near-infrared imaging of deoxygenated haemoglobin during exercise of the finger extensor muscles using the frequency-domain techniques, *Bioimaging*, 2, 174, 1994.

43. B. W. Pogue and K. D. Paulsen, High-resolution near-infrared tomographic imaging simulations of the rat cranium by use of *a priori* magnetic resonance imaging structural information, *Opt. Lett.*, 23, 1716, 1998.

44. R. F. Bonner, R. Nossal, S. Havlin, and G. H. Weiss, Model for photon migration in turbid biological media, *J. Opt. Soc. Am. A*, 4, 423, 1987.

45. A. H. Gandjbakhche and G. H. Weiss, Random walk and diffusion-like models of photon migration in turbid media, *Progress in Optics*, ed. Wolf, E. Elsevier Science B.V., North Holland, Amsterdam, vol. XXXIV, p. 333, 1995.

46. A. H. Gandjbakhche, R. Nossal, and R. F. Bonner, Scaling relationships for theories of anisotropic random walks applied to tissue optics, *Appl. Opt.*, 32, 504, 1993.

47. V. Chernomordik, R. Nossal, and A. H. Gandjbakhche, Point spread functions of photons in time-resolved transillumination experiments using simple scaling arguments, *Med. Phys.*, 23, 1857, 1996.

48. A. H. Gandjbakhche, G. H. Weiss, R. F. Bonner, and R. Nossal, Photon path-length distributions for transmission through optically turbid slabs, *Phys. Rev. E*, 48, 810, 1993.

49. D. J. Hawrysz and E. M. Sevick-Muraca, Developments toward diagnostic breast cancer imaging using near-infrared optical measurements and fluorescent contract agents, *Neoplasia*, 2, 388, 2000.

50. M. Cutler, Transillumination as an aid in the diagnosis of breast lesions, *Surg. Gynecol. Obstet.*, 48, 721, 1929.

51. A. H. Gandjbakhche, V. Chernomordik, J. C. Hebden, and R. Nossal, Time-dependent contract functions for quantitative imaging in time-resolved transillumination experiments, *Appl. Opt.*, 37, 1973, 1998.

52. A. H. Gandjbakhche, R. Nossal, and R. F. Bonner, Resolution limits for optical transillumination of abnormalities deeply embedded in tissues, *Med. Phys.*, 21, 185, 1994.

53. V. Chernomordik, D. Hattery, A. Pifferi, P. Taroni, A. Torricelli, G. Valentini, R. Cubeddu et al., A random walk methodology for quantification of the optical characteristics of abnormalities embedded within tissue-like phantoms, *Opt. Lett.*, 25, 951, 2000.

54. M. Morin, S. Verreault, A. Mailloux, S. Frechette, Y. Chatingny, Y. Painchaud, and P. Beaudry, Inclusion characterization in a scattering slab with time-resolved transmittance measurements: Perturbation analysis, *Appl. Opt.*, 39, 2840–2852, 2000.

55. V. Chernomordik, D. W. Hattery, D. Grosenick, H. Wabnitz, H. Rinneberg, K. T. Moesta, P. M. Schlag et al., Quantification of optical properties of a breast tumor using random walk theory, *J. Biomed. Opt.*, 7, 80–7, 2002.

56. A. P. Gibson, J. C. Hebden, and S. R. Arridge, Recent advances in diffuse optical imaging, *Phys. Med. Biol.*, 50, R1–43, 2005.

57. R. I. e. Fox, Treatment of the patient with Sjogren syndrome, *Rheumatic Dis. Clinic N. Am.*, 18, 699–709, 1992.

58. V. Chernomordik, D. Hattery, I. Gannot, and A. H. Gandjbakhche, Inverse method 3D reconstruction of localized in-vivo fluorescence. Application to Sjogren syndrome, *IEEE J. Sel. Topics Quant. Elec.*, 5, 930, 1999.

59. M. S. Patterson and B. W. Pogue, Mathematical model for time-resolved and frequency-domain fluorescence spectroscopy in biological tissue, *Appl. Opt.*, 33, 1963, 1994.

60. A. H. Gandjbakhche, R. F. Bonner, R. Nossal, and G. H. Weiss, Effects on multiple passage probabilities on fluorescence signals from biological media, *Appl. Opt.*, 36, 4613, 1997.

61. D. Hattery, V. Chernomordik, M. Loew, I. Gannot, and A. H. Gandjbakhche, Analytical solutions for time-resolved fluorescence lifetime imaging in a turbid medium such as tissue, *JOSA(A)*, 18, 1523, 2001.

62. E. Samuel, Thermography—some clinical applications, *Biomed. Eng.*, 4, 15–9, 1969.

63. C. H. Jones, Physical aspects of thermography in relation to clinical techniques, *Bibl. Radiol.*, 1–8, 1975.

64. J. Hardy, The radiation power of human skin in the infrared, *Am. J. Physiol.*, 127, 454–462, 1939.

65. J. Steketee, Spectral emissivity of skin and pericardium, *Phys. Med. Biol.*, 18, 686–94, 1973.

66. R. Elam, D. Goodwin, and K. Willams, Optical properties of human epidermics, *Nature*, 198, 1001–1002, 1963.

67. D. J. Watmough and R. Oliver, Emissivity of human skin in the waveband between 2 micron and 6 micron, *Nature*, 219, 622–624, 1968.

68. K. D. Patil and K. L. William, Spectral study of human radiation. Non-ionizing Radiation, *Non-ionizing Radiat.*, 1, 39–44, 1969.

69. T. Togawa, Non-contact skin emissivity: Measurement from reflectance using step change in ambient radiation temperature, *Clin. Phys. Physiol. Meas.*, 10, 39–48, 1989.

70. M. Hassan and T. Togawa, Observation of skin thermal inertia distribution during reactive hyperaemia using a single-hood measurement system, *Physiol. Meas.*, 22, 187–200, 2001.

71. W. H. McAdams, *Heat Transmission*, New York: McGraw Hill, p. 472, 1954.

72. H. T. Hammel, J. D. Hardy, and D. Murgatroyd, Spectral transmittance and reflectance of excised human skin, *J. Appl. Physiol.*, 9, 257–64, 1956.

73. S. L. Jacques, Skin Optics, http://omlc.ogi.edu/news/jan98/skinoptics.html, 1998.

74. R. Zhang, W. Verkruysse, B. Choi, J. A. Viator, B. Jung, L. O. Svaasand, G. Aguilar et al., Determination of human skin optical properties from spectrophotometric measurements based on optimization by genetic algorithms, *J. Biomed. Opt.*, 10, 024030, 2005.

75. L. F. A. Douven and G. W. Lucassen, Retrieval of optical properties of skin from measurement and modeling the diffuse reflectance, *Proc. SPIE*, 3914, 312–323, 2000.

76. I. O. Svaasand, L. T. Norvang, E. J. Fiskerstrand, E. K. S. Stopps, M. W. Berns, and J. S. Nelson, Tissue parameters determining the visual appearance of normal skin and port-wine stains, *Laser Med. Sci.*, 10, 55–65, 1995.

77. I. V. Meglinski and S. J. Matcher, Quantitative assessment of skin layers absorption and skin reflectance spectra simulation in the visible and near-infrared spectral regions, *Physiol. Meas.*, 23, 741–53, 2002.

78. I. Nishidate, Y. Aizu, and H. Mishina, Estimation of absorbing components in a local layer embedded in the turbid media on the basis of visible and near-infrared (VIS-NIR) reflectance spectra, *Opt. Rev.*, 10, 427–435, 2003.

79. S. L. Jacques, Origins of tissue optical properties in the UVA, visible, and NIR regions, *OSA TOPS on Advances in Optical Imaging and Photon Migration*, eds, R.R. Alfano and James G. Fujimoto, vol. 2, Optical Society of America, Washington, DC, pp. 364–371, 1996.

80. J. Kainerstorfer, F. Amyot, S. G. Demos, M. Hassan, V. Chernomordik, C. K. Hitzenberger, A. H. Gandjbakhche et al., Quantitative assessment of ischemia and reactive hyperemia of the dermal layers using multi—Spectral imaging on the human arm, *Progr. Biomed. Opt. Imaging—Proc. SPIE*, 7369, 2009.

81. A. Vogel, V. V. Chernomordik, J. D. Riley, M. Hassan, F. Amyot, B. Dasgeb, S. G. Demos et al., Using noninvasive multispectral imaging to quantitatively assess tissue vasculature, *J. Biomed. Opt.*, 12, 051604, 2007.

82. A. Vogel, B. Dasgeb, M. Hassan, F. Amyot, V. Chernomordik, Y. Tao, S. G. Demos et al., Using quantitative imaging techniques to assess vascularity in AIDS-related Kaposi's sarcoma, *Conf. Proc. IEEE Eng. Med. Biol. Soc.*, 1, 232–5, 2006.

83. K. Pearson, On lines and planes of closest fit to systems of points in space, *Philos. Mag.*, 2, 559–572, 1901.

84. J. Kainerstorfer, M. Ehler, F. Amyot, M. Hassan, S. G. Demos, V. Chernomordik, C. K. Hitzenberger, A. H. Gandjbakhche, and J. D. Riley, Principal component model of multi spectral data for near real time skin chromophore mapping, *J. Biomed. Opt.*, 15, 046007, 2010.

85. J. Kainerstorfer, F. Amyot, M. Ehler, M. Hassan, S. G. Demos, V. Chernomordik, C. K. Hitzenberger et al., Direct curvature correction for non-contact imaging modalities—applied to multi-spectral imaging, *J. Biomed. Opt.*, 15, 2010.

86. S. H. Tseng, P. Bargo, A. Durkin, and N. Kollias, Chromophore concentrations, absorption and scattering properties of human skin *in-vivo*, *Opt. Express*, 17, 14599–617, 2009.

87. D. J. Cuccia, F. Bevilacqua, A. J. Durkin, F. R. Ayers, and B. J. Tromberg, Quantitation and mapping of tissue optical properties using modulated imaging, *J. Biomed. Opt.*, 14, 024012, 2009.

88. U. Merschbrock, J. Hoffmann, L. Caspary, J. Huber, U. Schmickaly, and D. W. Lubbers, Fast wavelength scanning reflectance spectrophotometer for noninvasive determination of hemoglobin oxygenation in human skin, *Int. J. Microcirc. Clin. Exp.*, 14, 274–81, 1994.

89. D. M. McDonald and P. L. Choyke, Imaging of angiogenesis: From microscope to clinic, *Nat. Med.*, 9, 713–25, 2003.

90. J. S. Taylor, P. S. Tofts, R. Port, J. L. Evelhoch, M. Knopp, W. E. Reddick, V. M. Runge et al., MR imaging of tumor microcirculation: Promise for the new millennium, *J. Magn. Reson. Imaging*, 10, 903–907, 1999.

91. K. L. Verstraete, Y. De Deene, H. Roels, A. Dierick, D. Uyttendaele, and M. Kunnen, Benign and malignant musculoskeletal lesions: Dynamic contrast-enhanced MR imaging—parametric "first-pass" images depict tissue vascularization and perfusion, *Radiology*, 192, 835–43., 1994.

92. L. D. Buadu, J. Murakami, S. Murayama, N. Hashiguchi, S. Sakai, K. Masuda, S. Toyoshima et al., Breast lesions: Correlation of contrast medium enhancement patterns on MR images with histopathologic findings and tumor angiogenesis, *Radiology*, 200, 639–49, 1996.

93. A. Fredriksson and S. Stone-Elander, PET screening of anticancer drugs. A faster route to drug/target evaluations *in vivo*, *Methods Mol. Med.*, 85, 279–94, 2003.

94. G. Jerusalem, R. Hustinx, Y. Beguin, and G. Fillet, The value of positron emission tomography (PET) imaging in disease staging and therapy assessment, *Ann. Oncol.*, 13, 227–34., 2002.

95. H. C. Steinert, M. Hauser, F. Allemann, H. Engel, T. Berthold, G. K. von Schulthess, and W. Weder, Non-small cell lung cancer: Nodal staging with FDG PET versus CT with correlative lymph node mapping and sampling, *Radiology*, 202, 441–446, 1997.

96. S. D. Rockoff, The evolving role of computerized tomography in radiation oncology, *Cancer*, 39, 694–6, 1977.

97. K. D. Hopper, K. Singapuri, and A. Finkel, Body CT and oncologic imaging, *Radiology*, 215, 27–40., 2000.

98. K. A. Miles, M. Hayball, and A. K. Dixon, Colour perfusion imaging: A new application of computed tomography, *Lancet*, 337, 643–645, 1991.

99. K. A. Miles, C. Charnsangavej, F. T. Lee, E. K. Fishman, K. Horton, and T. Y. Lee, Application of CT in the investigation of angiogenesis in oncology, *Acad. Radiol.*, 7, 840–50, 2000.

100. N. Ferrara, Role of vascular endothelial growth factor in physiologic and pathologic angiogenesis: Therapeutic implications, *Semin. Oncol.*, 29, 10–4, 2002.

101. D. E. Goertz, D. A. Christopher, J. L. Yu, R. S. Kerbel, P. N. Burns, and F. S. Foster, High-frequency color flow imaging of the microcirculation, *Ultrasound Med. Biol.*, 26, 63–71, 2000.

102. E. M. Gill, G. M. Palmer, and N. Ramanujam, Steady-state fluorescence imaging of neoplasia, *Methods Enzymol.*, 361, 452–81, 2003.

103. K. Svanberg, I. Wang, S. Colleen, I. Idvall, C. Ingvar, R. Rydell, D. Jocham et al., Clinical multi-colour fluorescence imaging of malignant tumours—initial experience, *Acta Radiol.*, 39, 2–9, 1998.

104. V. Beral, T. A. Peterman, R. L. Berkelman, and H. W. Jaffe, Kaposi's sarcoma among persons with AIDS: A sexually transmitted infection?, *Lancet*, 335, 123–8, 1990.

105. B. A. Biggs, S. M. Crowe, C. R. Lucas, M. Ralston, I. L. Thompson, and K. J. Hardy, AIDS related Kaposi's sarcoma presenting as ulcerative colitis and complicated by toxic megacolon, *Gut*, 28, 1302–1306, 1987.

106. E. Cornali, C. Zietz, R. Benelli, W. Weninger, L. Masiello, G. Breier, E. Tschachler et al., Vascular endothelial growth factor regulates angiogenesis and vascular permeability in Kaposi's sarcoma, *Am. J. Pathol.*, 149, 1851–69, 1996.

107. Y. Chang, E. Cesarman, M. S. Pessin, F. Lee, J. Culpepper, D. M. Knowles, and P. S. Moore, Identification of herpesvirus-like DNA sequences in AIDS-associated Kaposi's sarcoma, *Science*, 266, 1865–1869, 1994.

108. R. Yarchoan, Therapy for Kaposi's sarcoma: Recent advances and experimental approaches, *J. Acquir. Immune Defic. Syndr.*, 21(Suppl 1), S66–S73, 1999.

109. R. F. Little, K. M. Wyvill, J. M. Pluda, L. Welles, V. Marshall, W. D. Figg, F. M. Newcomb et al., Activity of thalidomide in AIDS-related Kaposi's sarcoma, *J. Clin. Oncol.*, 18, 2593–602, 2000.

110. M. Hassan, D. Hattery, V. Chernomordik, K. Aleman, K. Wyvill, F. Merced, R. F. Little et al., Non-invasive multi-modality technique to study angiogenesis associated with Kaposi's sarcoma, *Proc. EMBS BMES*, 1139–1140, 2002.

111. M. Hassan, R. F. Little, A. Vogel, K. Aleman, K. Wyvill, R. Yarchoan, and A. Gandjbakhche, Quantitative assessment of tumor vasculature and response to therapy in Kaposi's sarcoma using functional non-invasive imaging, *TCRT*, 3, 451–457, 2004.

112. C. Maxwell-Cade, Principles and practice of clinical thermography, *Radiography*, 34, 23–34, 1968.

113. J. F. Head and R. L. Elliott, Infrared imaging: making progress in fulfilling its medical promise, *IEEE Eng. Med. Biol. Mag.*, 21, 80–85, 2002.

114. S. Bornmyr and H. Svensson, Thermography and laser-Doppler flowmetry for monitoring changes in finger skin blood flow upon cigarette smoking, *Clin. Physiol.*, 11, 135–41, 1991.

115. K. Usuki, T. Kanekura, K. Aradono, and T. Kanzaki, Effects of nicotine on peripheral cutaneous blood flow and skin temperature, *J. Dermatol. Sci.*, 16, 173–81, 1998.

116. M. Anbar, Clinical thermal imaging today, *IEEE Eng. Med. Biol. Mag.*, 17, 25–33, 1998.
117. J. Sorensen, M. Bengtsson, E. L. Malmqvist, G. Nilsson, and F. Sjoberg, Laser Doppler perfusion imager (LDPI)—for the assessment of skin blood flow changes following sympathetic blocks, *Acta Anaesthesiol. Scand.*, 40, 1145–1148, 1996.
118. A. Rivard, J. E. Fabre, M. Silver, D. Chen, T. Murohara, M. Kearney, M. Magner et al., Age-dependent impairment of angiogenesis, *Circulation*, 99, 111–20, 1999.
119. A. Vogel, M. Hassan, F. Amyot, V. Chernomordik, S. Demos, R. Little, R. Yarchoan et al., Using multi-modality imaging techniques to assess vascularity in AIDS-related Kaposi's sarcoma, *Biomedical Optics 2006, Technical Digest*, p. SG2, 2006.
120. Kainerstorfer, J. M., F. Amyot, M. Hassan, M. Ehler, R. Yarchoan, K. M. Wyvill, T. Uldrick et al., Reconstruction-free imaging of Kaposi's Sarcoma using multi-spectral data, *Biomedical Optics, OSA Technical Digest*, p. BME6, 2010.

20

Biomedical Applications of Functional Infrared Imaging

Arcangelo Merla
University of Chieti-Pescara

Gian Luca Romani
University of Chieti-Pescara

20.1 Introduction

Infrared imaging provides quantitative representation of the surface thermal distribution of the human body. The skin temperature distribution of the human body depends on the complex relationships defining the heat exchange processes between skin tissue, inner tissue, local vasculature, and metabolic activity. All of these processes are mediated and regulated by the sympathetic and parasympathetic activity to maintain the thermal homeostasis. The presence of a disease can affect the heat balance or exchange processes, resulting in an increase or in a decrease of the skin temperature and altered dynamics of the local control of the skin temperature. Therefore, the characteristic parameters modeling the activity of the skin thermoregulatory system can be used as diagnostic parameters. The functional infrared (fIR) imaging is the study for diagnostic purposes, based on the modeling of the bio-heat exchange processes, of the functional properties and alterations of the human thermoregulatory system. In this chapter, we present some of the clinical applications in order to show the potentialities of the technique.

20.2 Diagnosis of Varicocele and Follow-Up of Treatment

Varicocele is a widely spread male disease consisting of a dilatation of the pampiniform venous plexus and of the internal spermatic vein. Consequences of such a dilatation are an increase of the scrotal temperature and a possible impairment of the potential fertility [22,31].

In normal men, testicular temperature is 3–4°C lower than core body temperature [22]. Two thermoregulatory processes maintain this lower temperature: heat exchange with the environment through the scrotal skin and heat clearance by blood flow through the pampiniform plexus. Venous stasis due to the varicocele may increase the temperature of the affected testicle or pampiniform plexus. Thus, an abnormal temperature difference between the two hemiscrota may suggest the presence of varicocele [15,31] (see Figure 20.1).

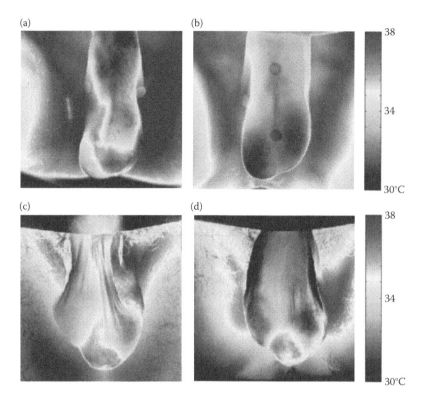

FIGURE 20.1   (a) Second-grade right varicocele. The temperature distribution all over the scrotum clearly highlights significant differences between affected and unaffected testicles. (b) The same scrotum after varicocelectomy. The surgical treatment reduced the increased temperature on the affected hemiscrotum and restored the symmetry in the scrotal temperature distribution. (c) Third-grade left varicocele. (d) The same scrotum after varicocelectomy. The treatment was unsuccessful in repairing the venous reflux, as documented by the persisting asymmetric scrotal distribution.

Telethermography can reveal abnormal temperature differences between the two testicles and altered testicular thermal recovery after an induced cold stress. Affected testicles return to prestress equilibrium temperatures faster than do normal testicles [15].

fIR imaging has been used to determine whether altered scrotal thermoregulation is related to subclinical varicocele [19]. In a study conducted in 2001, Merla and Romani enrolled 60 volunteers, 18–27 years of age (average age, 21 ± 2 years), with no symptoms or clinical history of varicocele. After clinical examination, echo color Doppler imaging (the gold standard) and functional infrared imaging were performed. fIR imaging evaluation consisted of obtaining scrotal images, measuring the basal temperature at the level of the pampiniform plexus (T_p) and the testicles (T_t), and determining thermal recovery of the scrotum after cold thermal stress.

The temperature curve of the hemiscrotum during rewarming showed an exponential pattern and was, therefore, fitted to an exponential curve. The time constant τ of the best exponential fit depends on the thermal properties of the scrotum and its blood perfusion [15,17]. Therefore, τ provides a quantitative parameter assessing how much the scrotal thermoregulation is affected by varicocele. Cooling was achieved by applying a dry patch to the scrotum that was 10°C colder than the basal scrotal temperature. The fIR measurements were performed accordingly with usual standardization procedures [14]. The basal prestress temperature and the recovery time constant τ_p at the level of the pampiniform plexus and of the testicles (τ_t) were evaluated on each hemiscrotum. A basal testicular temperature greater than 32°C and basal pampiniform plexus temperature greater than 34°C were considered warning thresholds. Temperature differences among testicles (ΔT_t) or pampiniform plexus ΔT_p temperature greater than 1.0°C were also considered warning values, as were $\Delta\tau_t$ and $\Delta\tau_t$ values longer than 1.5 min.

The fIR imaging evaluation classified properly the stages of disease, as confirmed by the echo color Doppler imaging and clinical examination in a blinded manner.

In 38 subjects, no warning basal temperatures or differences in rewarming temperatures were observed. These subjects were considered to be normal according to infrared functional imaging. Clinical examination and echo color Doppler imaging confirmed the absence of varicocele ($p < 0.01$, one-way ANOVA test).

In 22 subjects, one or more values were greater than the warning threshold for basal temperatures or differences in rewarming temperatures. Values for ΔT_p and the $\Delta\tau_p$ were higher than the warning thresholds in 8 of the 22 subjects, who were classified as having grade 1 varicocele. Five subjects had ΔT_t and $\Delta\tau_t$ values higher than the threshold. In nine subjects, three or more infrared functional imaging values were greater than the warning threshold values. The fIR imaging classification was grade 3 varicocele. Clinical examination and echo color Doppler imaging closely confirmed the fIR imaging evaluation of the stage of the varicocele. fIR imaging yielded no false-positive or false-negative results. All participants with positive results on infrared functional imaging also had positive results on clinical examination and echo color Doppler imaging. The sensitivity and specificity of fIR test were 100% and 93%, respectively.

An abnormal change in the temperature of the testicles and pampiniform plexus may indicate varicocele, but the study demonstrated that impaired thermoregulation is associated with varicocele-induced alteration of blood flow. Time to recovery of prestress temperature in the testicles and pampiniform plexus appears to assist in the classification of the disease. fIR imaging accurately detected 22 asymptomatic varicocele.

The control of the scrotum temperature should improve after varicocelectomy as a complementary effect of the reduction of the blood reflux. Moreover, follow-up of the changes in scrotum thermoregulation after varicocelectomy may provide early indications on possible relapses of the disease.

To answer the above questions, Merla et al. [13] used fIR imaging to study changes in the scrotum thermoregulation of 20 patients (average age, 27 ± 5 years) that were judged eligible for varicocelectomy on the basis of the combined results of the clinical examination, Echo color Doppler imaging, and spermiogram. No bilateral varicoceles were included in the study.

Patients underwent to clinical examination, echo color Doppler imaging and instrument varicocele grading, and infrared functional evaluation before varicocelectomy and every 2 weeks thereafter, up to the 24th week. Fourteen out of the 20 patients suffered from grade 2 left varicocele. All of them were characterized by basal temperatures and recovery time after cold stress according to Reference 13. Varicoceles were surgically treated via interruption of the internal spermatic vein using modified Palomo's technique. fIR imaging documented the changes in the thermoregulatory control of the scrotum after the treatment as follows: 13 out of the 14 grade 2 varicocele patients exhibited normal basal T_t, T_p on the varicocele side of the scrotum, and normal temperature differences ΔT_t and ΔT_p starting from the 4th week after varicocelectomy. Their $\Delta\tau_t$ and $\Delta\tau_p$ values returned to the normal range from the 4th to the 6th week. Four out of the six grade 3 varicocele patients exhibited normal basal T_t, T_p on the varicocele side of the scrotum, and normal temperature differences ΔT_t and ΔT_p starting from the 6th week after varicocelectomy. Their $\Delta\tau_t$ and $\Delta\tau_p$ values returned to normal range from the 6th to the 8th week. The other three patients did not return to normal values of the above specified parameters. In particular, $\Delta\tau_t$ and $\Delta\tau_p$ remained much longer than the threshold warning values [13] up to the last control (Figure 20.1). Echo color Doppler imaging and clinical examination assessed relapses of the disease.

The study proved that the surgical treatment of the varicocele induces modification in the thermoregulatory properties of the scrotum, reducing the basal temperature of the affected testicle and pampiniform plexus, and slowing down its recovery time after thermal stress. Among the 17 with no relapse, 4 exhibited return to normal T_t, T_p, ΔT_t, and ΔT_p for the latero-anterior side of the scrotum, while the posterior side of the scrotum remained hyperthermal or characterize by ΔT_t and ΔT_p higher than the threshold warning value.

This fact suggested that the surgical treatment via interruption of the internal spermatic vein using Palomo's technique may not be the most suitable method for those varicoceles. The time requested by the scrotum to restore normal temperature distribution and control seems to be positively correlated to the volume and duration of the blood reflux lasting: the greater the blood reflux, the longer the time.

The study demonstrated that IR imaging may provide an early indication on the possible relapsing of the disease and may be used as a suitable complementary follow-up tool.

20.3 Raynaud's Phenomenon and Scleroderma

Raynaud's phenomenon (RP) is defined as a painful vasoconstriction—that may follow cold or emotional stress—of small arteries and arterioles of extremities, like fingers and toes. RP can be primary (PRP) or secondary (SSc) to scleroderma. The latter is usually associated with a connective tissues disease.

RP precedes the systemic autoimmune disorders development, particularly scleroderma, by many years and it can evolve into secondary RP. The evaluation of vascular disease is crucial in order to distinguish between PRP and SSc. In PRP, episodic ischemia in response to cold exposure or to emotional stimuli is usually completely reversible: absence of tissue damage is the typical feature [1], but mild structural changes are also demonstrated [30]. In contrast, scleroderma RP shows irreversible tissue damage and severe structural changes in the finger vascular organization [25].

None of the physiological measurement techniques currently in use, except infrared imaging, are completely satisfactory in focusing primary or secondary RP [10]. The main limit of such techniques (nail fold capillary microscopy, cutaneous laser Doppler flowmetry, and plethysmography) is the fact that they can proceed just into a partial investigation, usually assessing only one finger once. The measurement of skin temperature is an indirect method to estimate change in skin thermal properties and blood flow.

Thermography protocols [4,6,7,9,10,12,23,26,29] usually include cold patch testing to evaluate the capability of the patient hands to rewarm. The pattern of the rewarming curves is usually used to depict the underlying structural diseases. Analysis of rewarming curves has been used in several studies to differentiate healthy subjects from PRP or SSc Raynaud's patients. Parameters usually considered so far are the lag time preceding the onset of rewarming or to reach a preset final temperature; the rate of the rewarming and the maximum temperature of recovery; and the degree of temperature variation between different areas of the hands.

Merla et al. [18] proposed to model the natural response of the fingertips to exposure to a cold environment to get a diagnostic parameter derived by the physiology of such a response. The thermal recovery following a cold stress is driven by thermal exchange with the environment, transport by the incoming blood flow, conduction from adjacent tissue layers, and metabolic processes. The finger temperature is determined by the net balance of the energy input/output. The more significant contributions come from the input power due to blood perfusion and the power lost to the environment [18]:

$$\frac{dQ}{dt} = -\frac{dQ_{env}}{dt} + \frac{dQ_{ctr}}{dt} \tag{20.1}$$

Normal finger recovery after a cold stress is reported in Figure 20.2.

In absence of thermoregulatory control, fingers exchange heat only with the environment: in this case, their temperature T_{exp} follows an exponential pattern with time constant τ given by

$$\tau = \frac{\rho \cdot c \cdot V}{h \cdot A} \tag{20.2}$$

where ρ is the mass density, c the specific heat, V the finger volume, h the combined heat transfer coefficient between the finger and the environment, and A the finger surface area. Thanks to the

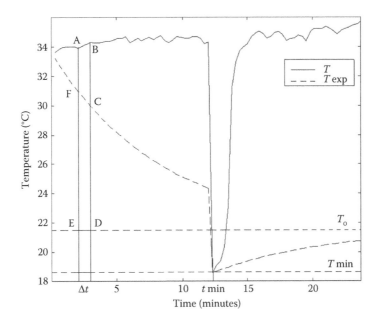

FIGURE 20.2 Experimental rewarming curves after cold stress in normal subjects. The continuous curve represents the recorded temperature finger. The outlined curve represents the exponential temperature pattern exhibited by the finger in the absence of thermoregulatory control. In this case, the only heat source for the finger is the environment. (From Merla, A. et al., Infrared functional imaging applied to Raynaud's phenomenon, *IEEE Eng. Med. Biol. Mag.*, 21, 73, 2002, by permission of the editor.)

thermoregulatory control, the finger maintains its temperature T greater than T_{exp}. For a Δt time, the area of the trapezoid $ABCF$ times hA in Figure 20.2 computes the heat provided by the thermoregulatory system, namely ΔQ_{ctrl}. This amount summed to ΔQ_{env} yields Q, the global amount of heat stored in the finger.

Then, the area of the trapezoid $ABDE$ is proportional to the amount Q of heat stored in the finger during a Δt interval. Therefore, Q can be computed integrating the area surrounded by the temperature curve T and the constant straight line T_o:

$$Q = -h \cdot A \cdot \int_{t_1}^{t_2} \left(T_0 - T(\varsigma) \right) d\varsigma \tag{20.3}$$

where the minus sign takes into account that the heat stored by the finger is counted as positive. Q is intrinsically related to the finger thermal capacity, according to the expression

$$\Delta Q = \rho \cdot c \cdot V \cdot \Delta T \tag{20.4}$$

Under the hypothesis of constant T_o, the numerical integration in (Equation 20.3) can be used to characterize the rewarming exhibited by a healthy or a suffering finger.

The Q parameter has been used in References 16, 18 to discriminate and classify PRP, SSc, and healthy subjects on a set of 40 (20 PRP, 20 SSc) and 18 healthy volunteers. For each subject, the response to a mild cold challenge of hands in water was assessed by fIR imaging. Rewarming curves were recorded for each of the five fingers of both hands; the temperature integral Q was calculated along the 20 min following the cold stress. Ten subjects, randomly selected within the 18 normal ones, repeated two times

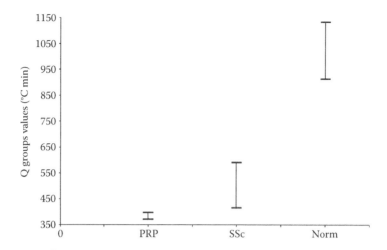

FIGURE 20.3   One-way ANOVA test applied to the Q parameter calculated for each group (PRP, SSc, and healthy). The Q parameter clearly discriminates the three groups. (From Merla, A. et al., Infrared functional imaging applied to Raynaud's phenomenon, *IEEE Eng. Med. Biol. Mag.*, 21, 73, 2002, by permission of the editor.)

and in different days the test to evaluate the repeatability of the fIR imaging findings. The repeatability test confirmed that fIR imaging and Q computation is robust tool to characterize the thermal recovery of the fingers.

The grand average Q values provided by the first measurement was 1060.0 ± 130.5°C min, while for the second assessment it was 1012 ± 135.1°C min ($p > 0.05$, one-way ANOVA test). The grand average Q values for PRP, SSc, and healthy subjects groups are in shown in Figure 20.3, whereas single values obtained for each finger of all of the subjects are reported in Figure 20.4.

The results in References 16, 18 highlight that the PRP group features low intraindividual and interindividual variability whereas the SSc group displays a large variability between healthy and unhealthy fingers. Q values for SSc finger are generally greater than PRP ones.

The temperature integral at different finger regions yields very similar results for all fingers of the PRP group, suggesting common thermal and blood flow (BF) properties. SSc patients showed different thermoregulatory responses in the different segments of the finger. This feature is probably due to the local modification in the tissue induced by the scleroderma. Scleroderma patients also featured a significantly different behavior across the five fingers depending on the disease involvement.

In normal and PRP groups, all fingers show a homogeneous behavior and PRP fingers always exhibit a poorer recovery than normal ones. Additionally, in both groups, the rewarming always starts from the finger distal area differently from what happens in SSc patients. The sensitivity of the method in order to distinguish patients from normal is 100%. The specificity in distinguishing SSc from PRP is 95%.

Q clearly highlights the difference between PRP, SSc, and between and normal subjects. It provides useful information about the abnormalities of their thermoregulatory finger properties.

The PRP patients exhibited common features in terms of rewarming. Such behavior can be explained in terms of an equally low and constant BF in all fingers and to differences in the amount of heat exchanged with the environment [25].

Conversely, no common behavior was found for the SSc patients, since their disease determines—for each finger—very different thermal and blood perfusion properties. Scleroderma seems to increase the tissue thermal capacity with a reduced ability to exchange. As calculated from the rewarming curves, the Q parameter seems to be particularly effective to describe the thermal recovery capabilities of the finger. The method clearly highlighted the difference between PRP and SSc patients and provides useful information about the abnormalities of their thermal and thermoregulatory finger properties.

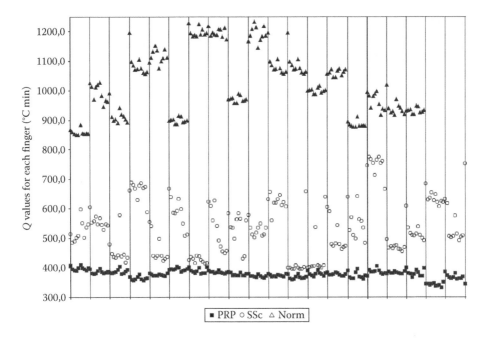

FIGURE 20.4 Q values calculated for each finger of each subject. Vertical grid lines are placed to discriminate the 10 fingers. PRP fingers are characterized by a strong intra- and interindividual homogeneity. Greater mean Q values and greater intra- and interindividual variations characterize the SSc fingers. (From Merla, A. et al., Infrared functional imaging applied to Raynaud's phenomenon, *IEEE Eng. Med. Biol. Mag.*, 21, 73, 2002, by permission of the editor.)

In consideration of the generally accepted theory that the different recovery curves of the patients is a reflection of the slow deterioration of the microcirculation, so that over time in the same patients it is possible to observe changes in the thermal recovery curves, the method described above could be used to monitor the clinical evolution of the disease. In addition, pharmacological treatment effects could be advantageously followed up.

The contribution that functional IR imaging may provide to the clinicians also regards other body regions potentially affected by scleroderma, especially in the systemic sclerosis form [20]. Men with systemic sclerosis (SSc) present an increased risk of developing erectile dysfunction (ED). In a recent study, Merla et al. [20] evaluated the extent of penile vascular damage in sclerodermic patients using Duplex ultrasonography.

The aim of that study was to investigate whether there exist penile thermal differences among sclerodermic patients and healthy controls. For this reason 10 men with systemic sclerosis under current treatment for their disease and 10 healthy controls were enrolled; penile thermal properties were assessed functional IR. Erectile function was evaluated using the sexual health inventory for men (SHIM) questionnaire [27]. The SHIM results confirmed the presence of ED in sclerodermic patients. Baseline penile temperature in patients ($32.1 \pm 1.4°C$) was lower than in controls ($34.1 \pm 0.9°C$). Recovery from cooling test was seen to be faster in healthy controls than in patients, both in terms of recovery amplitude (patients $3.75 \pm 2.09°C$, controls $9.80 \pm 2.77°C$) and amplitude to time constant ratio (patients $1.21 \pm 0.64°C/min$, controls $1.96 \pm 0.48°C/min$).

The results indicated that penile thermal abnormalities occur in almost all sclerodermic patients. Noncontact thermal imaging not only identifies thermal alterations but also clearly distinguishes between SSc patients and healthy controls and therefore could represent a valuable instrument in identifying early ED in systemic sclerosis patients.

20.4 Comparison of Functional IR and Laser Doppler Imaging in Assessment of Cutaneous Perfusion in Healthy Controls and Scleroderma Patients

In recent years, laser Doppler imaging (LDI) has been widely used to quantify microvascular flow and to provide a flux map showing possible microcirculatory defects [2,3,5,11,28]. It is based on the laser Doppler technique that measures blood flow in very small blood vessels of the microvascular network, such as low-speed flows associated with nutritional blood flow in capillaries close to the skin surface and in the underlying arterioles and venules involved in skin temperature regulation. LDI also permits the noninvasive assessment of modifications in cutaneous blood flow (CBF). In fact, although the flow signals are generated by the movement of blood cells (the blood cell flux), it has been demonstrated that relative changes in CBF can be monitored as well [2].

The term commonly used to describe blood flow measured by the laser Doppler technique is "flux": a quantity proportional to the product of the average speed of the blood cells and their numerical concentration (often referred to as blood volume). This is expressed in arbitrary "perfusion units." Standardization of LDI instrument measurements in perfusion units can be achieved by measuring a flux due to the Brownian motion of polystyrene microspheres in water [3].

Temperature measurement is widely used as an indirect assessment of cutaneous circulation. Spatial distribution and time-course of the cutaneous temperature (T_c) can be effectively obtained by means of thermal infrared imaging (IR), that can be then used for the indirect assessment of possible impairments of cutaneous microvasculature or tissues [16,17].

Few studies quantitatively compare CBF and T_c. Bornmyr et al. [2] studied the interrelation between changes in skin temperature (measured by means of contact probes) and changes in blood flow (measured by laser Doppler techniques) in healthy volunteers' feet. The main findings indicated an exponential interrelation between laser Doppler flowmetry and temperature readings, and that relative changes in LDI and temperatures showed a weak linear relationship. Clark et al. [5] compared IR thermography and LDI in the assessment of digital blood flow in subjects with RP. They found a poor correlation between the outcomes obtained with the two techniques and they concluded that one technique cannot substitute the other.

However, CBF is just one of the factors concurring in determining the actual T_c values, as the latter also depend on the tissue metabolism, heat exchange with surrounding tissue and environment, and local thermophysical properties. Therefore, a direct relationship between CBF and T_c may not be observable, especially in the presence of tissue pathologies.

Bio-heat transfer models permit the calculation of CBF from high-resolution IR image series (9). A major advantage for computing CBF from thermal imagery is that CBF images can be obtained at approximately the same frame rate as thermal imaging (up to 100 complete 524×524 pixels images per second using the most advanced commercially available thermal cameras), thus overcoming one of the main limits for LDI, that is low time resolution. In any case, quantitative comparison between LDI-measured CBF values and CBF values obtained from thermal imagery has not been previously evaluated.

In a recent study Merla et al. [21] compared CBF obtained from thermal imagery and CBF measured with LDI in order to verify whether combined LDI and thermal imaging provide useful and effective diagnostic information concerning tissue and/or microvascular impairment. Therefore, the comparison was extended to both healthy subjects and patients suffering from systemic sclerosis.

The model adopted derives from previous works of Fujimasa [8] and Pavlidis [24]. At thermal equilibrium (i.e., stationary state), the heat balance equation for cutaneous tissue can be expressed as

$$Q_r + Q_e + Q_f = Q_c + Q_m + Q_b \qquad (20.5)$$

where Q_r is the heat radiated from the skin to air; Q_e is the basic evaporated heat; Q_f is the heat loss via convection to the neighboring air; Q_c is the heat conducted by subcutaneous tissue; Q_m is the heat generated by tissue metabolism; and Q_b is the heat gain/loss via convection attributable to blood flow of subcutaneous blood vessels.

According to Pavlidis and Levine [24], cutaneous temperature change (ΔT_c) over a short time period (Δt) is expressed by the following equation:

$$C\Delta T_c \cong (\Delta Q_c + \Delta Q_b) \tag{20.6}$$

where C is the heat capacity of the cutaneous tissue and the terms ΔQ_b and ΔQ_c are defined as

$$\Delta Q_b = \alpha C_b \cdot S \cdot [\omega_{c2}(T_b - T_{c2}) - \omega_{c1}(T_b - T_{c1})] \tag{20.7}$$

and

$$\Delta Q_c = (K_c/3d) \cdot [(T_b - T_{c2}) - (T_b - T_{c1})] \tag{20.8}$$

where α is the countercurrent heat exchange coefficient in normal condition; C_b is the heat capacity of blood; S is the thickness of the skin; ω_{ci}, with $i = 1,2$ is the cutaneous blood flow rate at times t_1 and t_2; T_b is the blood temperature in the core; T_{ci}, with $i = 1,2$ is the cutaneous temperature at times t_1 and t_2; K_c is the thermal conductivity of the skin; and d is the depth of the core temperature point from the cutaneous surface.

Equation 20.6 can be then solved using numerical methods to compute the time derivative of the blood flow rate ω_c according to the following equation:

$$d\omega_c/dt = [\psi/(T_b T_c)^2]dT_c/dt \tag{20.9}$$

where Ψ is a constant acting like a scale factor and it is given by

$$\psi = C - (K_c/3d) \tag{20.10}$$

The expression for $d\omega_c/dt$ can be integrated numerically twice over time to obtain an estimate for ω_c (i.e., the blood flow rate) and then the cutaneous blood flow V_c (i.e., the LDI flux equivalent):

$$V_c = V_c(\psi(C,K_c,d),c_1,c_2) \tag{20.11}$$

where c_1 and c_2 are numerical integration constants.

Given a series of thermal digital images of a region of interest, the above-described algorithm can be repeated for each pixel (x,y) of each raw thermal image of the series. Therefore, it is possible to transform raw thermal image series in blood flow image series. CBF-like images from raw thermal data (namely, IR-CBF) can thus be obtained. Such images can then be quantitatively compared to the CBF-LDI images.

Both methods, LDI and thermal imaging based, report CBF in arbitrary units. Therefore, an unknown scale factor between CBF values from LDI and thermal imaging must be introduced.

Each full hand LDI recording lasted 3 min and produced one $n \times n$ (with $n = 64$) pixels CBF image (henceforth named LDI-CBF image). In the same time period, $n + 1$ (=65) 256×256 pixels IR images were recorded. From this set of 65 IR images, we computed $n = 64$ IR-CBF images, as the CBF computation required a frame-by-frame time derivative. This meant that during the time interval required to obtain one full IR-CBF image, the LDI imager scanned just one row out of the $n = 64$ forming the full LDI-CBF image. Therefore, it is necessary to fuse in a single image the series of the n CBF images, opportunely resampled over time, for a meaningful comparison of IR-CBF and LDI-CBF images.

For this reason, each IR-CBF image was resampled in a $n \times n$ grid creating a new "fused" IR-CBF image where the i-th row was copied from the i-th row of the i-th image of the IR-CBF images series,

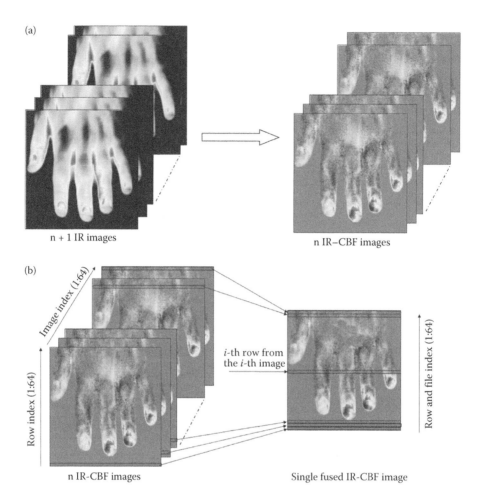

FIGURE 20.5 From the thermal IR images series to the IR-CBF image. (a) The series of $n + 1$ IR images is converted into a series of n IR-CBF images by applying the proposed bio-heat model. (b) The series of n IR-CBF images is then condensed in a single IR-CBF image. Each i-th row of the IR-CBF image is obtained by pasting the i-th row ($i = 1{:}64$) of the i-th image ($i = 1{:}64$) of the IR-CBF image series. In this way, each row out of the 64 rows of the final and single IR-CBF image is synchronous with the corresponding row out of the 64 ones in the LDI-IR image.

with I ranging from 1 to n (Figure 20.5). Therefore, each row of the final IR-CBF image and of the LDI-CBF image resulted synchronized over time and a meaningful quantitative comparison between the same regions of interest could be possible.

In order to quantitatively compare the CBF estimated by the two methods, after having performed skin segmentation, the mean CBF value for each LDI and IR-CBF image, namely V_{LDI} and V_{IR}, respectively, were computed. Correlation between V_{LDI} and V_{IR} values was assessed through Pearson product moment correlation.

Figure 20.5 shows the IR-CBF image obtained from a series of thermal images for a healthy subject. Figure 20.6 shows an example of IR-CBF and LDI-CBF images reporting, in arbitrary units, the map of the cutaneous perfusion for a representative subject. Figure 20.7 shows the scatter plots of the mean values of the cutaneous perfusion obtained with the two methods for the two groups. While a linear correlation between perfusion values obtained through the two methods was found for the healthy group ($V_{IR} = \alpha \cdot V_{LDI} + \beta$, $R = 0.85$, $\alpha = 0.35$, and $\beta = 0.06$), a correlation among V_{LDI} and V_{IR} values for

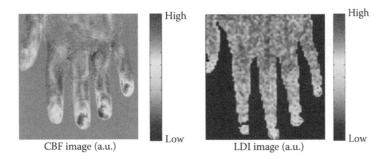

CBF image (a.u.)   LDI image (a.u.)

FIGURE 20.6 CBF images (in arbitrary units) obtained with the proposed method (upper panel, IR-CBF) and recorded with LDI imager (lower panel). Color bar reports false-color visualization of the perfusion distribution. The overall distributions appear to be consistent, both images similarly showing the same high-perfusion and low-perfusion regions.

the SSc group was not observed. The lacking of correlation indicates that the bio-heat transfer exchanges among microvasculature and cutaneous tissues are altered in presence of SSc. Therefore, normal bio-heat transfer models do not apply for SSc.

According to Equation 20.11, actual V_{IR} values depended on local skin thermal conductivity K, depth of core temperature point from cutaneous surface d, and tissue heat capacity C. Such parameters are lumped together in the Ψ value (see Equation 20.10) that played the role of a scale factor with respect to the unknown absolute perfusion value.

Absolute perfusion values from thermal imaging can be obtained only by adopting realistic models of heat exchanges at cutaneous and subcutaneous levels, and taking into account the specific microvasculature geometry. This would also require an *in vivo* estimation of the heat capacity of blood and tissues and the quantification of countercurrent heat exchanges between arterioles, veins, and capillaries.

With reference to the comparison of the two methods, it must to be pointed out that both methods assume that all of the relevant thermophysical features for estimating perfusion values did not change

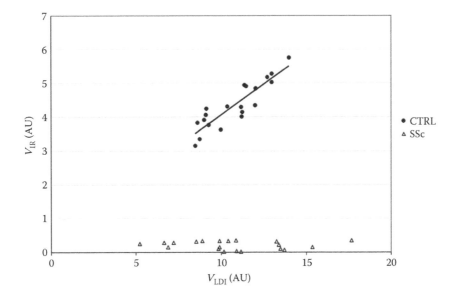

FIGURE 20.7 V_{IR} versus V_{LDI} scatter plot with regression line. Measures are reported in arbitrary units. V_{IR} and V_{LDI} are linearly correlated ($V_{IR} = 0.35V_{LDI} + 0.06$; $R = 0.85$) for the healthy group, but not for the SSc.

among individuals. Such an assumption was correct for the healthy controls group, but it may not have been in the presence of ailments involving cutaneous tissue or cutaneous microvasculature, as is the case of scleroderma. In fact, V_{IR} and V_{LDI} values correlated positively for the healthy controls group. Therefore, in spite of the fact that *in vivo* values of the bio-heat parameters remains unknown, the bio-heat transfer model itself and the assumptions made for computing CBF from thermal IR data appeared to be well-sounded. The proportionality constant α between V_{IR} and V_{LDI} values depends on the normalization factors and the actual choice for the parameters involved in the bio-heat model. Such a constant could be regarded as a conversion factor between the output of the two methods.

The parameter β indicated that zero V_{LDI} readings did not correspond to zero V_{IR} value. Therefore, β could be regarded as a tuning factor of the bio-heat model parameters to achieve the perfect agreement between the readings of the two methods. Also, it could represent the parametric error associated to the bio-heat model.

Interestingly, the bio-heat model failed when applied to the SSc group. This fact suggests that the model parameters, that had been calibrated for normal conditions, does not apply in the presence of a pathology, as the thermophysical properties of the tissue and microvasculature are altered by the morphofunctional changes induced by the disease itself. In fact, SSc determines a global skin tissue rearrangement, which is characterized by decreased capillary density, interstitial edema, especially in the early stage of inflammation, and extracellular matrix deposition. These modifications may lead to local or global thermophysical parameters changes as, for example, an increase of the local tissue thermal capacity due to the local blood flow decrease and to a concomitant increase of water content of the interstitial tissue.

The thermal imaging-based method provided faster and better time-resolved imaging of cutaneous perfusion than standard laser Doppler techniques as thermal cameras can provide up to 100 full 524×524 pixels images per second, thus allowing real-time monitoring of tissue perfusion rate.

The results reported hereby suggest that combined bio-heat models and thermal imaging data may provide accurate and fast assessment of the cutaneous tissue perfusion rate and flow. Also, they suggest that combined LDI and thermal imaging may allow an effective discrimination between healthy versus impaired cutaneous tissue thermal properties and cutaneous vasculature, thus providing a potential effective imaging-based tool for a variety of biomedical and clinical applications, ranging from diagnostics to the follow-up of treatments.

20.5 Discussion and Conclusion

fIR imaging is a biomedical imaging technique that relies on high-resolution infrared imaging and on the modeling of the heat exchange and control processes at the skin layer. fIR imaging is aimed to provide quantitative diagnostic parameters through the functional investigation of the thermoregulatory processes. It is also aimed to provide further information about the studied disease to the physicians, like explanation of the possible physics reasons of some thermal behaviors and their relationships with the physiology of the involved processes. One of the great advantages of fIR imaging is the fact that is not invasive and it is a touchless imaging technique. fIR is not a static imaging investigation technique. Therefore, data for fIR imaging need to be processed adequately for movement. Adequate bio-heat modeling is also required. The medical fields for possible applications of fIR imaging are numerous, ranging from those described in this chapter, to psychometrics, cutaneous blood flow modeling, peripheral nervous system activity, and some angiopathies. The applications described in this chapter show that fIR imaging provides highly effective diagnostic parameters. The method is highly sensitive, but also highly specific in discriminating different conditions of the same disease. For the studies reported hereby, fIR imaging is sensitive and specific as the corresponding golden standard techniques, at least. In some cases, fIR represents a useful follow-up tool (like in varicocelectomy to promptly assess possible relapses) or even an elective diagnostic tool, as in the RP. More advantages from this technique may come from advancement and development in bio-heat transfer processes modeling.

References

1. Allen, E.V. and Brown, G.E., Raynaud's disease: A critical review of minimal requisites for diagnosis, *Am. J. Med. Sci.*, 183,187, 1932.
2. Bornmyr, S., Svensson, H., Lilja, B., Sundkvist, G. Skin temperature changes and changes in skin blood flow monitored with laser Doppler flowmetry and imaging: A methodological study in normal humans. *Clin. Physiol.*, 17, 71–81, 1997.
3. Briers, J.D. Laser Doppler, speckle and related techniques for blood perfusion mapping and imaging. *Physiol. Meas.*, 22, R35–R66, 2001.
4. Clarks, S. et al., The distal-dorsal difference as a possible predictor of secondary Raynaud's phenomenon, *J. Rheumatol.*, 26, 1125, 1999.
5. Clark, S. et al., Laser Doppler imaging—A new technique for quantifying microcirculatory flow in patients with primary Raynaud's phenomenon and systemic sclerosis. *Microvasc. Res.*, 57, 284–291, 1999.
6. Clark S. et al., Comparison of thermography and laser Doppler imaging in the assessment of Raynaud's phenomenon. *Microvasc. Res.*, 66, 73–76, 2003.
7. Del Bianco et al., Raynaud's phenomenon (primary or secondary to systemic sclerosis). The usefulness of laser-Doppler flowmetry in the diagnosis. *Int. Angiol.*, 20, 307–313, 2001.
8. Fujimasa, I., Chinzei, T., Saito, I. Converting far infrared image information to other physiological data. *IEEE Eng. Med. Biol. Mag.*, 19, 71–76, 2000.
9. Guiducci, S., Giacomelli, R., Cerinic, M.M., Vascular complications of scleroderma. *Autoimmun. Rev.*, 6, 520–23, 2007.
10. Herrick, A.L. and Clark, S., Quantifying digital vascular disease in patients with primary Raynaud's phenomenon and systemic sclerosis, *Ann. Rheum. Dis.*, 57, 70, 1998.
11. Herron, G.S., Romero, L. Vascular abnormalities in scleroderma. *Semin. Cutan. Med. Surg.*, 17(1), 12–17, 1998.
12. Javanetti, S. et al., Thermography and nailfold capillaroscopy as noninvasive measures of circulation in children with Raynaud's phenomenon. *J. Rheumatol.*, 25, 997, 1998.
13. Merla, A., Ledda, A., Di Donato, L., Romani, G.L., Assessment of the effects of the varicocelectomy on the thermoregulatory control of the scrotum. *Fertil. Steril.*, 81, 471, 2004.
14. Merla, A. and Romani, G. L. Functional infrared imaging in clinical applications, in *The Biomedical Engineering Handbook*, ed., J.D. Bronzino, CRC Press, Boca Raton, FL, 32.1–32.13, 2006.
15. Merla, A. et al., Dynamic digital telethermography: A novel approach to the diagnosis of varicocele, *Med. Biol. Eng. Comp.*, 37, 1080, 1999.
16. Merla, A. et al., Infrared functional imaging applied to Raynaud's phenomenon. *IEEE Eng. Med. Biol. Mag.*, 21, 73, 2002.
17. Merla, A. et al., Quantifying the relevance and stage of disease with the Tau image technique. *IEEE Eng. Med. Biol. Mag.*, 21, 86, 2002.
18. Merla, A. et al., Raynaud's phenomenon: Infrared functional imaging applied to diagnosis and drugs effects. *Int. J. Immun. Pharm.*, 15, 41, 2002.
19. Merla, A. et al., Use of infrared functional imaging to detect impaired thermoregulatory control in men with asymptomatic varicocele. *Fertil. Steril.*, 78, 199, 2002.
20. Merla, A. et al., Penile cutaneous temperature in systemic sclerosis: A thermal imaging study. *Int. J. Immunopathol. Pharmacol.*, 20(1):139–44, 2007.
21. Merla, A. et al., Comparison of thermal infrared and laser Doppler imaging in the assessment of cutaneous tissue perfusion in scleroderma patients and healthy controls. *Int. J. Immunopathol. Pharmacol.*, 21(3), 679–86, 2008.
22. Mieusset, R. and Bujan, L., Testicular heating and its possible contributions to male infertility: A review. *Int. J. Andr.*, 18,169, 1995.

23. O'Reilly, D. et al., Measurement of cold challenge response in primary Raynaud's phenomenon and Raynaud's phenomenon associated with systemic sclerosis. *Ann. Rheum. Dis.*, 51, 1193, 1992.
24. Pavlidis, I. and Levine, J. Thermal image analysis for polygraph testing. *IEEE Eng. Med. Biol. Mag.*, 21, 56–64, 2002.
25. Prescott et al., Sequential dermal microvascular and perivascular changes in the development of scleroderma, *J. Pathol.*, 166, 255, 1992.
26. Ring, E.F.J., Cold stress test for the hands. in *The Thermal Image in Medicine and Biology*, Ammer, K. and Ring, E. F. G., ed., Uhlen Verlag, Wien, 1995.
27. Rosen, R.C. et al., The International Index of Erectile Function (IIEF): A multidimensional scale for assessment of erectile dysfunction. *Urology*, 49, 822, 1997.
28. Salsano, F. et al., Significant changes of peripheral perfusion and plasma adrenomedullin levels in *N*-acetylcysteine long term treatment of patients with sclerodermic Raynauds phenomenon. *Int. J. Immunopathol. Pharmacol.*, 18, 761–70, 2005.
29. Schuhfried, O. et al., Thermographic parameters in the diagnosis of secondary Raynaud's phenomenon, *Arch. Phys. Med. Rehabil.*, 81, 495, 2000.
30. Subcommittee for Scleroderma Criteria of the American Rheumatism Association Diagnostic and Therapeutic Criteria Committee, Preliminary criteria for the classification of systemic sclerosis (scleroderma). *Arthritis Rheum.*, 23, 581, 1980.
31. Tucker, A., Infrared thermographic assessment of the human scrotum. *Fertil. Steril.*, 74, 802, 2000.

21

Modeling Infrared Imaging Data for the Assessment of Functional Impairment in Thermoregulatory Processes

Alessandro Mariotti
G. d'Annunzio University

Arcangelo Merla
University of Chieti-Pescara

21.1 Introduction

Infrared imaging permits the study of cutaneous thermoregulation through its capability of accurate recording of the cutaneous temperature dynamics. Such a dynamics depends on the complex heat exchange processes between cutaneous tissue, inner tissue, local vasculature, and metabolic activity, all of these processes being mediated and regulated by the sympathetic and parasympathetic activity. The main function of such processes is to preserve the homeostasis. The presence of a disease alters this heat exchange processes, with respect to the corresponding healthy condition, resulting in an increase or in a decrease of the cutaneous temperature and in an alteration of local thermoregulation. The characteristic parameters modeling the activity of the cutaneous thermoregulatory system can be used as diagnostic parameters, thus increasing the diagnostic specificity of the technique.

While fine *in vivo* modeling of the heat exchange processes is indeed extremely complex, the mathematical theory of the control system offers easy-to-manage tools for modeling feedback processes, which are the basis of homeostatic controls. These tools are often used in bioengineering to model biological functions as in the devices for automatic glucose control or pace maker. Very few studies combining thermal imaging data and automatic system control theory are available. In this chapter, we provide examples of such an approach for two important pathologies: Raynaud's phenomenon and varicocele.

21.2 Modeling and Assessment of Thermoregulatory Impairment in Raynaud's Phenomenon and Scleroderma

Raynaud's phenomenon (RP) is a paroxysmal vasospastic disorder of small arteries, precapillary arteries, and cutaneous arteriovenous shunts of the extremities. RP is typically induced by cold exposure and emotional stress [3]. The main clinical signature of RP is cutaneous triphasic color changes: well-demarcated pallor due to a sudden vasospasm, cyanotic phase secondary to local hemoglobin desaturation, and then red flush caused by reactive hyperemia. RP usually involves fingers and toes, even though tongue, nose, ears, and nipples may also be involved. The presence of the initial ischemic phase is mandatory for diagnosis, whereas the reactive hyperemic phase may not occur.

RP can be classified as *primary* (PRP), with no identifiable underlying pathological disorder, and *secondary*, usually associated with a connective tissue disease, use of certain drugs, or exposition to toxic agents [16]. Secondary RP is frequently associated with systemic sclerosis. In this case, RP typically may precede the onset of other symptoms and signs of disease by several years [2]. While PRP is generally characterized by an abnormal vasospastic response in the absence of specific structural abnormalities, RP secondary to systemic sclerosis is characterized by a peculiar rearrangement of the microvascular structures [20,38].

It has been estimated that 12.6% of patients suffering from primary RP develop a secondary disease. In particular, 5–20% of subjects suffering from secondary RP will develop either limited or diffuse systemic sclerosis [45]. On the other hand, all of the systemic sclerosis subjects (SSc) will experience episodes of RP. This indicates the importance of early and proper differential diagnosis.

IR imaging is a noninvasive technique which provides a map of the superficial temperature of a given body by measuring emitted infrared energy [24,28]. Several IR imaging studies have been performed to differentiate primary from secondary RP, often in combination with the monitoring of the finger thermoregulatory response to a controlled cold challenge [6,8,15,17,24,31].

Cutaneous circulation is a major effector of finger thermoregulation [18]. Exposure to cold stress elicits generalized cutaneous vasoconstriction, which may be extremely pronounced at the fingertip surface. Cutaneous vasoconstriction is a response mediated by a sympathetic control process triggered partly by stimulation of the cutaneous cold receptors in the cooled area, and partly by cooled blood returning to the general circulation, which stimulates the temperature-regulating center in the anterior hypothalamus [39,42]. Homeostasis is basically maintained by a negative feedback loop, similar to a thermostat [34], which regulates the energy exchange with the environment at the cutaneous level through metabolic and hemodynamic processes that determine finger temperature at any given time [37]. Employing control system theory, the homeostatic process can be seen as a feedback-controlled system. This kind of system considers a reference signal (i.e., prestress finger temperature) to produce the desired output (i.e., the actual finger temperature). A cold challenge induces a finger temperature (plant-controlled output) change from the basal value (reference value). The difference between the plant-controlled output and the reference value (i.e., output error) prompts the thermoregulatory reaction in order to restore the basal value. In other words, the thermoregulatory reaction steers the output error to zero. The temperature–time evolution can be recorded by means of thermal IR imaging [24]. Examples of temperature versus time curves obtained from experimental recovery data in HCS, SSc, and PRP are reported in Figure 21.1. According to the control system theory, differences in the temperature recovery curves depend on the efficacy of the finger thermoregulatory system, which in turn can be represented by the actual values of a given set of functional modeling parameters.

The finger thermoregulatory response after the cold stress challenge is not instantaneous. The time delay from the onset of the rewarming process and the end of the cold challenge is often referred to as lag time (LT). LT in healthy controls (HCS) is usually around 4–5 min long [43], while it may vary largely in PRP and SSc (Figure 21.1). LT may depend on both physics and physiological factors such as environmental temperature, basal finger temperature, vascular smooth muscle tone,

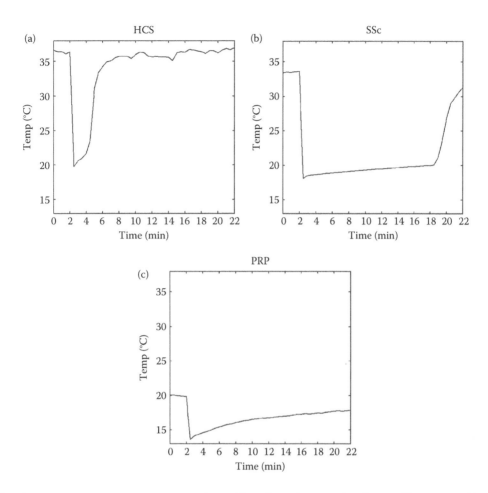

FIGURE 21.1 Temperature versus time curves obtained from thermal imaging data during cold stress test in (a) HCS, (b) SSc, and (c) PRP. (From Mariotti, A., G. Grossi, P. Amerio et al. *Ann. Biomed. Eng.* 37(12):2631–2639, 2009.)

cold-receptor efficacy, and autonomic activity. The cold challenge test activates specific responses of the thermoregulatory system, which operate at both local (i.e., peripheral) and systemic (i.e., central) levels attempting to restore the basal temperature [5]. These two levels of the thermoregulatory system can be modeled through two hierarchical control units: a higher-level unit (supervisor), and a feedback lower-level executor driven by the supervisor [1,11,46]. The supervisor sets the reference signal on the basis of the basal prestress temperature and the onset time. The overall performance of the thermoregulatory system depends on the activity of both the supervisor and the executor. Besides the contribution of the thermoregulatory system, the finger temperature (i.e., system output) is also influenced by the thermal exchange between the finger and the surrounding environment. This thermal exchange depends on the temperature difference, which constitutes the external input to the thermoregulatory system [5,37].

Figure 21.2a shows the overall architecture of the system. The only observable output is the finger cutaneous temperature $y(t)$, obtained through thermal IR imaging. Even though no information about internal variables is available, the theory of automatic system control allows to both quantitatively and qualitatively describe the control action, on the basis of the assumption of an input/output feedback control [32]. The system is characterized by an external input (room temperature) and a

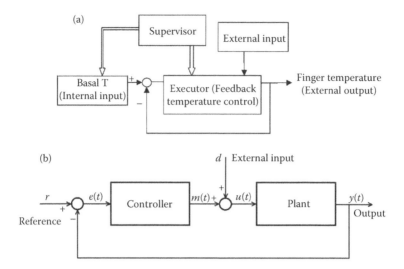

FIGURE 21.2 (a) Logical architecture of the thermoregulatory system. (b) Block diagram of executor unit. (From Mariotti, A., G. Grossi, P. Amerio et al. *Ann. Biomed. Eng.* 37(12):2631–2639, 2009.)

steady-state regime reference signal (basal finger cutaneous temperature). The reference signal can be measured by IR imaging before the exposure to the cold stress, and averaged over time to provide a constant reference value *r*. In particular, the dynamics of the thermal recovery after cold stress (Figure 21.1) classifies the system as a second-order time-invariant feedback system [35]. In particular, the feedback lower-level unit is composed of a controller and a plant block in sequence (Figure 21.2b), both assumed to be time-invariant systems described by first-order transfer functions. The theory provides the differential equation to model the plant output $y(t)$ (i.e., the finger temperature) in the time domain:

$$\dot{y}(t) = -a \cdot y(t) + b \cdot u(t) \tag{21.1}$$

where *u* is the plant input, and *a* and *b* are constant coefficients.

The cold stress stimulation directly affects the system output, lowering the cutaneous temperature. The poststress temperature $y(0)$ (i.e., the temperature measured immediately after the cessation of the cold stress) constitutes the initial condition for the response of the control system. The plant input $u(t)$ is then the sum of the feedback controller output $m(t)$ plus the additional external input *d* as shown in Figure 21.2b:

$$u(t) = m(t) + d \tag{21.2}$$

Input *d* represents passive heat exchange with the environment. Therefore, it depends on room temperature and $y(t)$. In other words, input *d* can be seen as the uncontrolled effect of environmental conditions on the finger temperature.

The feedback controller block generates the signal $m(t)$ stimulated by the difference between the system output and the reference signal *r*, namely, *output error* $e(t)$:

$$e(t) = r - y(t) \tag{21.3}$$

The feedback controller acts on the plant by the signal $m(t)$ to steer the *output error* to zero.

Common approaches for modeling homeostatic processes are based on an integral-type feedback controller system, which nullifies stepwise variation of the error signal [46]. The differential equation that describes the controller behavior in the time domain is

$$\dot{m}(t) = k \cdot e(t) \tag{21.4}$$

where k is a proportionality constant.

The supervisor unit activates this controller by means of logic signals (on/off transition). When the supervisor unit logical output is "on," the feedback is closed on the integral-type controller and then the active temperature recovery can start. Otherwise, when the supervisor unit logical output is "off" (during the LT), the controller is disabled to restore the initial condition, while the external input d is independent of this switching logic.

The evolution of the plant can be described more easily in the Laplace domain. The Laplace transform (L-transform) of Equation 21.1 is given by

$$Y(s) = \frac{1}{(s+a)} \cdot y(0) + \frac{b}{(s+a)} \cdot U(s) \tag{21.5}$$

where s is the Laplace variable, and $Y(s)$ and $U(s)$ are the output and input L-transforms, respectively. The ratio between plant input and output defines the plant transfer function $P(s)$, which is computed assuming null $y(0)$ [22]:

$$P(s) = \frac{Y(s)}{U(s)} = \frac{b}{(s+a)} \tag{21.6}$$

where b is the plant gain coefficient and $s = -a$ is its pole, that is, the negative reciprocal of the plant time constant.

According to Equation 21.4, the transfer function of the integral-type controller $G(s)$ is given by

$$G(s) = \frac{k}{s} \tag{21.7}$$

where k is the controller gain.

As described above and as depicted in Figure 21.3, the overall model works in open loop for $t < \text{LT}$

$$Y(s) = P(s) \cdot d \tag{21.8}$$

and in closed loop for $t > \text{LT}$:

$$Y(s) = \frac{G(s) \cdot P(s)}{1 + G(s) \cdot P(s)} \cdot r + \frac{P(s)}{1 + G(s) \cdot P(s)} \cdot d \tag{21.9}$$

Therefore, the model (Figure 21.3) is unequivocally described by $-a$, k, d, and LT, which can be estimated based on measurements of r and $y(t)$.

In particular, the reciprocal of the plant time constant a represents the speed of the response of the thermal process to external and internal stimuli. A low a value is indicative of a very slow recovery process. The controller gain k refers to the control action and determines the efficiency of the feedback control system in achieving the steady state, restoring the reference basal conditions. This parameter

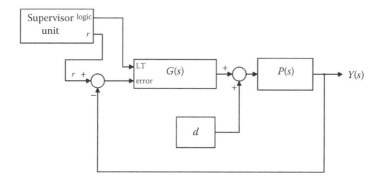

FIGURE 21.3 Thermoregulatory model block diagram for the finger. (From Mariotti, A., G. Grossi, P. Amerio et al. *Ann. Biomed. Eng.* 37(12):2631–2639, 2009.)

quantifies the power of an active and internal vasodilation process. The disturbance input d is related to the response induced by an external input, namely thermal exchange of the system with the environment. d, which is one of the fitted parameters, represents passive heat exchange with the environment, and, therefore, depends on room temperature and $y(t)$. LT is the time interval between the end of the cold stress and the onset of internal rewarming process. During this time, the thermal variations are mostly attributable to the passive heat exchange with the environment. Once LT is finished, there is the onset of the rewarming process and the controller starts to restore the reference basal conditions r.

The model was tested in a previous experimental study by Mariotti et al. [23], which can be referred to for detailed information. Fourteen SSc, 14 patients affected by PRP, and 16 HCS participated in this study. SSc and PRP patients were classified according to the criteria established in 2001 by the American College of Rheumatology [20,21,47].

For each subject, the functional response to a mild cold challenge of hands in water was assessed by functional IR imaging [23,24]. Thermal IR imaging was performed by means of a digital thermal camera (FLIR SC3000, FlirSystems, Sweden), with a Focal Plane Array of 320×240 QWIP detectors, capable of collecting the thermal radiation in the 8–9 μm band, with a 0.02 s time resolution, and 0.02 K temperature sensitivity. Cutaneous emissivity was estimated as $\varepsilon \approx 0.98$. The thermal camera response was blackbody-calibrated to null noise effects related to the sensor drift/shift dynamics and optical artifacts. Thermal images of the dorsum of each subject's hands were recorded. Recording was performed for 25 min, acquiring images every 30 s, and recording five thermal images before the cold stress to obtain the baseline of finger temperature. The thermal camera was placed 1.5 m away from the dorsum. Each image series was corrected for motion artifacts by means of a contour alignment algorithm. The cold stress was achieved by immersing the hands (protected from getting wet by thin, disposable latex gloves) for 3 min in a 3-L water bath maintained at 10°C. After removal of the gloves, the hands were placed in the same position as before the cold stress. Rewarming curves were obtained for each of the five fingertips of both hands, by averaging the temperature of the pixels within the nail-bed region.

For data and graphic analysis, a self-implemented software was used under a MATLAB® platform (www.themathworks.com). The control model was implemented using the Matlab Simulink Graphical User Interface®. The model parameters (a, k, d, and LT) were computed and optimized through the Matlab Simulink Parameter Estimation Toolbox®, by using a nonlinear least square algorithm, while r and $y(t)$ were directly measured by fIRI. Thermoregulation model responses were simulated by the variable-step ODE45 (Dormand-Price) solver. The distributions of the averaged parameters for each class of subjects were tested for normality by visual inspection of the frequency distribution and Shapiro–Wilk test, and then compared through a Student's t-test (STATISTICA, www.statsoft.it). The level of statistical significance was fixed at 0.01.

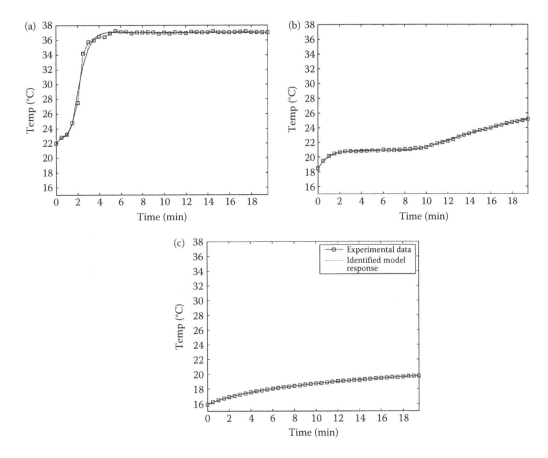

FIGURE 21.4   Comparison between experimental data and identified model response, for (a) HCS, (b) SSc, and (c) PRP. (From Mariotti, A., G. Grossi, P. Amerio et al. *Ann. Biomed. Eng.* 37(12):2631–2639, 2009.)

Figure 21.4 shows an example of the comparison between the identified response and the experimental temperature curves for three representative subjects randomly chosen, from HCS, SSc, and PRP groups. Mean and confidence interval for each parameter are reported in Figure 21.5. The mean values of k and a resulted statistically different in the three groups. The PRP group presented the lowest a and k values, with the lowest interindividual variability. The HCS group was instead characterized by the highest values for both parameters, with the largest interindividual variability for only the k parameter. SSc presented intermediate average values for both parameters and the highest variability with respect to the mean a value. d appeared to clearly discriminate PRP from SSc and HCS. The PRP group presented the highest average for d while the SSc and HCS groups showed lower values, without statistically significant intergroup differences. The LT parameter distinguished SSc from the other two groups, with the highest value. HCS and PRP presented very similar LT values. Mean values of r appeared to replicate the same behavior as the a parameter.

HCS parameters indicate that healthy thermoregulatory systems are fast and efficient in reestablishing the reference basal conditions. High a and k values are suggestive of fast and efficient control over vasomotor and metabolic processes. This observation is in agreement with the fact that the steady-state basal temperature, r, is the highest. HCS d values indicate that the system is driven less by heat exchange with the surrounding environment and dominated more by active temperature control through the controller unit. HCS LT values are the smallest, indicating a prompt reaction of the

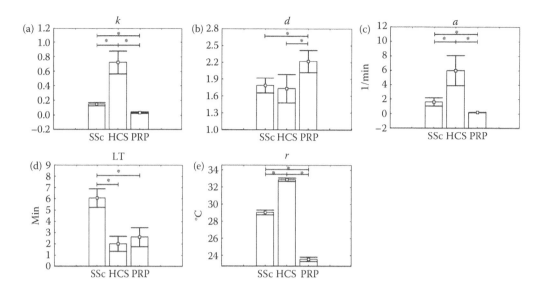

FIGURE 21.5 Comparison of model parameters among groups (Whisker Box plot) Mean and 95% confidence interval are reported for each parameter. The asterisks on the horizontal bars specify the groups with significant statistical difference. (From Mariotti, A., G. Grossi, P. Amerio et al. *Ann. Biomed. Eng.* 37(12):2631–2639, 2009.)

thermoregulatory process to the thermal stimuli. LT values in SSc indicated that the onset of the active control was strongly delayed in comparison to both HCS and PRP. SSc rewarming processes are slower than HCS, with the controllers less effective in achieving the steady state. In fact, for several fingers in several cases, SSc were not able to restore the reference basal conditions within the poststress monitoring time. The variability of a and k values in SSc was high even within the fingers of the same hand, which may be linked to variable structural and functional vascular changes at different stages of scleroderma in each finger. Systemic sclerosis is a generalized connective tissue disease, which is pathologically characterized by microvascular injury and immune activation, resulting in increased synthesis of extracellular matrix components, including collagen, fibronectin, and proteoglycans, cutaneously and viscerally. Interstitial fibrosis and occlusive microvascular damage induce a progressive decrease of capillary density [7]. These structural alterations lead to a reduction in cutaneous blood flow with increased periods of stasis. The estimated LT values for the PRP group were similar to HCS, indicating a proper onset of active recovery. The efficiency of the controller, however, was extremely weak, almost negligible. In fact, PRP fingers had the slowest and weakest recovery, as indicated by a and k values. All PRP fingers exhibited a very similar rewarming pattern, never able to restore the reference basal conditions during the monitoring time. PRP d values were the highest with low controller gain levels indicating that heat exchange with the environment dominated the thermal recovery process. This observation is in agreement with previous fIR imaging studies which reported homogeneous digital thermal behavior for PRP fingers in response to cold stress [24], which may be related to the hypothesis that in PRP subjects an excessive vasoconstrictor tone and a weak systemic vasodilation process, both centrally mediated, do not allow an active thermal recovery [15].

Experimental result from Mariotti et al. [23] and those reported in the literature prove that SSc, HCS, and PRP exhibit different thermoregulatory responses to the standardized external stimuli, administered in strictly controlled environmental conditions. These different responses appear to be due to pathological functional and morphological alterations at the local or central level. The goal of this chapter was to identify a feedback thermoregulatory model to describe and to distinguish the functional differences in the recoveries, at both central and peripheral levels. In fact, the parameters so far identified to define each model describe how the system acts at the two different levels, in terms of process speed, efficiency of control action, and heat exchanges mechanisms, in order to restore the prestress basal conditions. Several advantages may be derived from this approach. The estimated model parameter values for each subject can aid in estimating the level of functional impairment expressed in the different forms of the disease. Thus, it is possible to more effectively monitor both the evolution of the disease and the efficacy of the treatment and its follow-up. The application of this control system theory model to SSc and primary RP diagnosis would also help to treat the causes of different conditions appropriately. In future studies, the proposed approach could be applied to distinguish additional expressions or causes of secondary RP such as limited or diffused scleroderma, lupus, or other prompting factors of the syndrome.

21.3 Scrotal Thermoregulatory Model Assessing Functional Impairment in Varicocele

In healthy men, testicular temperature is 3–4°C lower than the core body temperature. Two main thermoregulatory processes control testicular temperature: heat exchange with the environment through the scrotum skin and heat clearance by blood flow through the pampiniform plexus. Varicocele is defined as the pathological dilatation of the pampiniform plexus and scrotal veins with venous blood reflux. It is present in 15% of the adult male population, in 35% of men with primary infertility, and in 80% of men with secondary infertility [36]. Varicocele usually affects the left hemiscrotum because of valve insufficiency and right-angled mouth of the left testicular vein into the left renal vein. The venous reflux into the testicular vein and the pampiniform plexus results in venous hyperemia leading to hyperthermia and hypoxia of testicular tissue, which may cause infertility [12,29]. The assessment of the vascular impairment is generally performed by means of echo color Doppler, while scrotal hyperthermia is usually evaluated through the measurement of the scrotal cutaneous temperature by means of thermal infrared (IR) imaging [24,26,27,30,41,46]. It has been demonstrated that alteration of the scrotal blood flow secondary to varicocele impairs scrotal thermoregulation. Patients suffering from varicocele present higher scrotal cutaneous temperature and shorter time constant to recover from mild controlled cold stress of the scrotum [26] (Figure 21.6).

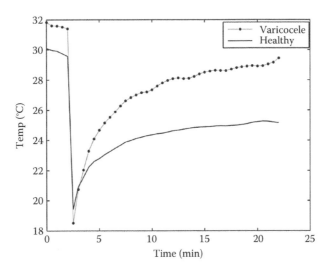

FIGURE 21.6 Cutaneous temperature versus time curves obtained from thermal imaging of the scrotum during cold stress test.

Scrotal thermoregulation serves to liberate the large amount of heat produced during spermatogenesis. A number of supporting mechanisms like thin skin with abundant vascularization, numerous sweat glands, and absence of subcutaneous fat facilitates heat exchange and contributes to maintaining the testicular temperature below body temperature [44].

Venous stasis associated with varicocele increases the cutaneous temperature of the affected testicle or pampiniform plexus [26]. Exposure to cold stress elicits cutaneous vasoconstriction, accompanied by increased skin rugosity to reduce the surface area involved in the heat exchange with the environment [39]. After the cessation of cold exposure, homeostatic processes restore basal prestress conditions, mostly through vasodilatation, favoring heat exchange with deeper layers [24]. In the presence of varicocele, affected testicles return to prestress equilibrium temperature faster than normal testicles [26]. Homeostasis is basically maintained by a negative feedback loop, similar to a thermostat [33] which regulates the energy exchange with the environment at the cutaneous level through metabolic and hemodynamic processes that determine the cutaneous temperature at any given time [37].

Experimental evidence shows that thermoregulatory response after the cold stress challenge is instantaneous. Thus, it can be assumed that the regulatory processes are activated instantaneously and simultaneously just after the cold stress, determining the actual value of the cutaneous temperature through a lumped action. From the control system theory, the basic functional components of homeostasis can be thought as arranged in a feedback loop: a controlled plant (scrotum thermal processes) whose output (cutaneous scrotum temperature) is constrained to follow a given set-point (reference basal value) through an internal feedback loop (homeostatic mechanism). The cutaneous basal temperature can be considered the reference value and assumed to be almost constant. To study the system dynamics and to evoke thermoregulatory processes that produce recovery patterns, the system has to be stimulated by a proper functional input. Namely, a cold challenge induces a scrotum cutaneous temperature (plant-controlled output) change from the basal value (reference value). In other words, the cold stress induces a variation from basal value of the controlled output. In feedback systems, the output signal is compared with the reference value generating an error signal, which stimulates the thermoregulatory control reaction in order to restore the basal value. In other words, the thermoregulatory reaction steers the error to zero.

Time evolution of scrotal temperature can be recorded by means of thermal IR imaging. Examples of temperature versus time curves obtained from experimental recovery data are reported in Figure 21.6. The temperature recovery curves can be interpreted, according to the control system theory, as the feedback system response to a perturbation of operative conditions. Differences in recovery curves depend on the capability of the control system to recover and to calibrate heat generation, thus determining internal and cutaneous temperature up to possible functional impairment associated with the presence of varicocele.

Varicocele can affect just one or both hemiscrota [29] and the only affected hemiscrotum (generally the left) presents marked alteration of thermoregulatory control [26]. In this chapter, we characterize differences in model parameters between healthy and left affected hemiscrota.

As mentioned, experimental evidence showed that scrotum thermoregulatory response after the cold stress challenge is instantaneous. Different from other body regions, the recovery patterns did not show a time delay from the onset of the rewarming process and the end of the cold challenge, often referred to as lag time (LT) [43]. The rewarming pattern after a cold challenge test was preliminarily modeled as an exponential function [25]. The cold challenge test activates specific responses of the thermoregulatory system, which operate at both local (i.e., peripheral) and systemic (i.e., central) levels, attempting to restore basal temperature [10]. These two levels are here modeled through two hierarchical control units: a high-level unit (supervisor) driving a feedback low-level executor [23]. The supervisor sets the reference signal on the basis of the basal prestress temperature. Albeit the overall performance of the thermoregulatory system is also affected by the surrounding environment, the speed of rewarming and the hemiscrotum temperature at the end of recovery allows the neglect of the thermal exchange with the environment with respect to thermoregulatory system contribution.

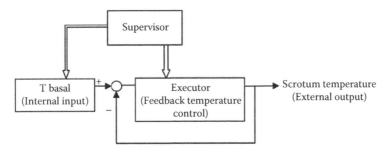

FIGURE 21.7 Logical architecture of the thermoregulatory system for the scrotum.

The model adopted is very similar to the one formulated for the finger thermoregulation and described in the previous section. Figure 21.7 shows the overall architecture of the system. The only measurable output is the scrotum (i.e., hemiscrotum) cutaneous temperature $y(t)$, obtained through thermal IR imaging. The steady-state regime reference signal (basal cutaneous temperature of hemiscrotum) has been measured before the exposure to the cold stress, and time-averaged to provide a constant reference value r. The system is a second-order time-invariant feedback system [35], with the feedback lower-level unit composed of a cascaded controller and plant blocks (Figure 21.8a), both assumed to be time-invariant systems described by first-order transfer functions.

As for the previous model, the plant evolution can be described more easily in the Laplace domain. The transfer function of the controller $G(s)$ is given by

$$G(s) = \frac{k}{s} + h \qquad (21.10)$$

where k is the integral controller gain, h is the gain of the proportional-type controller, and s is the Laplace variable. The model (Figure 21.9) is unequivocally described by the triple a, k, h, which can be estimated by measurements of r and $y(t)$.

The controller gain k refers to the control action and determines the efficiency of the feedback control system in achieving the steady state, restoring the reference basal conditions. This parameter quantifies the power of an active and systemic vasodilation process. h is the gain of the proportional-type controller and represents a measure of the efficiency of the thermal exchange between the cutaneous layer and

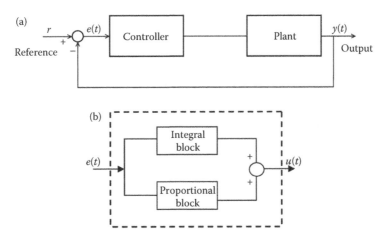

FIGURE 21.8 (a) Block diagram of executor unit. (b) Controller structure.

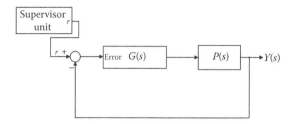

FIGURE 21.9 Thermoregulatory model block diagram for the scrotum.

the inner structures. a, the reciprocal of the plant time constant, represents the speed of the response of the thermal process to external and internal stimuli. A low a value is indicative of a slow recovery process.

The model was tested in 90 young patients (average age: 20 ± 4 years) suffering from left varicocele, and 40 healthy controls (average age: 21 ± 3 years) participated in this study. Only pure left varicoceles were included in this study to avoid possible confounding due to right or bilateral varicocele.

After clinical examination, participants underwent echo color Doppler imaging (ATL 5000 echo color Doppler imaging system, Philips Medical System, Eindhoven, The Netherlands) which is so far considered the gold standard for varicocele diagnosis.

For each subject, the functional response to a mild cold challenge of scrotum was assessed by thermal infrared imaging. All participants were asked to refrain from physical activities and intake of vasoactive substances for 2 h prior to the measurements. Participants were seated comfortably in an environment-controlled room. Before undergoing measurements, the subjects observed a 20-min acclimatization period to the recording room, which was set at standardized temperature (23°C), humidity (50–60%), and without direct ventilation. Thermal infrared imaging was performed by means of a digital thermal camera (FLIR SC3000, FlirSystems, Sweden), with a Focal Plane Array of 320×240 QWIP detectors, capable of collecting the thermal radiation in the 8–9 μm band, with a 0.02 s time resolution, and 0.02 K temperature sensitivity. Cutaneous emissivity was estimated as $\varepsilon \approx 0.98$. The thermal camera response was blackbody-calibrated to null noise effects related to the sensor drift/shift dynamics and optical artifacts. Thermal images of the scrotum of each subject were recorded for 25 min, acquiring images every 30 s. Five thermal images were recorded before the cold stress to obtain the baseline of scrotum temperature. Each image series was corrected for motion artifacts by means of a contour alignment algorithm. The cold stress was achieved by applying a dry patch to the scrotum maintained at 10°C. Rewarming curves were obtained separately for each of the two hemiscrota, by averaging the temperature of the pixels within the cutaneous projection of the testis.

For data and graphic analysis, a self-implemented software was used under MATLAB platform (www.themathworks.com). The control model was implemented using the Matlab Simulink Graphical User Interface. The model parameters (a, k, and h) were computed and optimized through the Matlab Simulink Parameter Estimation Toolbox, by using a nonlinear least square algorithm, while r and $y(t)$ were measured directly by thermal infrared imaging. Thermoregulatory model responses were simulated by the variable-step ODE45 (Dormand-Price) solver. The statistical analysis was performed to compare differences in model parameters and cutaneous temperature between healthy individuals and those affected by varicocele hemiscrota. All parameters for each group were compared through Wilcoxon–Mann–Whitney test [14].

The comparison between the identified response and the experimental temperature curves for two representative cases randomly chosen from varicocele and healthy control groups is shown in Figure 21.10. The mean and confidence interval of each parameter for each hemiscrota are reported in Figure 21.11. None of the parameters were statistically different between hemiscrota in healthy subjects. The mean values of k are not statistically different among hemiscrota, in both healthy and left

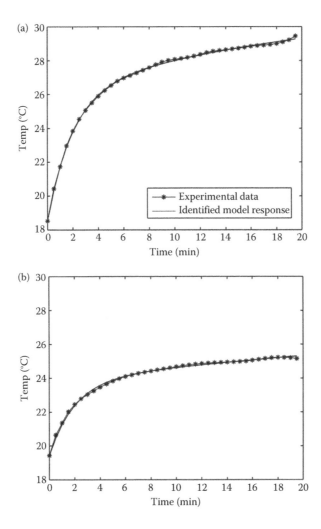

FIGURE 21.10 Comparison between experimental data and identified model response for (a) left varicocele and (b) healthy control.

varicocele groups, and between groups. h and a resulted significantly different between groups for left hemiscrotum ($p < 0.001$; $p < 0.01$, respectively). r resulted significantly different between groups for both left ($p < 0.001$) and right hemiscrotum ($p < 0.01$).

The first important issue that comes from these results is that the presence of left varicocele induced a faster rewarming of the affected hemiscrotum with respect to the homolateral healthy control. In fact the group average a value is lower for the left varicocele group with respect to the healthy side, thus indicating that the presence of this pathology accelerated the return to the basal homeostatic thermoregulatory conditions. This result is concordant with previous published results [26,29].

Basal conditions for left varicocele corresponded to higher cutaneous temperature of the left hemiscrotum with respect to the healthy left hemiscrotum, as proved by higher r values. This result was in agreement with previously reported indications [26,29]. Moreover, the left hemiscrotum affected by varicocele presents higher h values while no difference appears to characterize the average value of the k parameter. Overall, these results suggest that the accelerated recovery of the left hemiscrotum in the presence of varicocele was mostly due to the increased rate, with respect to normal conditions, of heat exchange from inner scrotal structures (veins and testes) with the skin. In fact, the active processes of

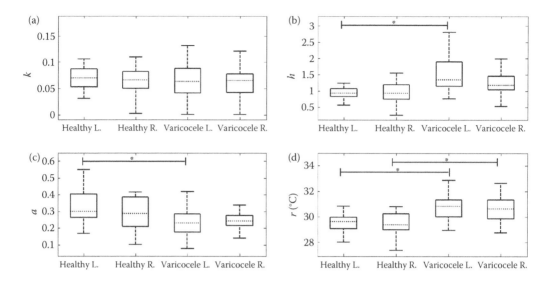

FIGURE 21.11 Comparison of model parameters among groups (Whisker Box plot). Median, first and third quartile, minimum and maximum are reported for each parameter. The asterisks on the horizontal bars specify the groups with statistically significant differences.

thermoregulatory vasodilatation induced by the cold exposure similarly acted in either the presence or the absence of varicocele, as proven by the average values of k, not significantly different between the two groups. This finding suggested that the true altered process in scrotal cutaneous hyperthermia secondary to varicocele did not depend on an increased effectiveness of the active thermoregulatory control, but was attributable to larger thermal exchange, probably by convection, between inner structures and the skin. In other words, it appeared that the increased blood reflux in the spermatic vein may have induced a diffused reflux in the testicular vessel network [19,40].

Average values for modeling parameters in healthy controls did not present statistically significant differences between right and left hemiscrota. This fact indicated and confirmed that in the absence of increased reflux secondary to varicocele, there were no thermoregulatory differences between healthy hemiscrota.

Interestingly, and not previously reported, right hemiscrotum in the presence of left varicocele presented higher basal temperature (i.e., r values) than healthy right hemiscrotum, thus suggesting that left hyperthermic hemiscrota induced a warming effect on the contralateral site. Differences in other functional parameters did not have a statistically significant relevance. Therefore, it seemed that such thermal interaction between hemiscrota in the presence of left varicocele may have been due to the proximity of the hemiscrota or intrascrotal vascular shunts and were not related to systemic vascular control [9,13].

The goal of this study was to identify a thermoregulatory feedback model to describe and to distinguish the functional differences in the recoveries, at both central and peripheral levels. In fact, the parameters so far identified to define each model described how the system acts at the two different levels, in terms of process speed, efficiency of control action, and heat exchanges mechanisms, in order to restore the prestress basal conditions. Several advantages may be derived from this approach. The estimated model parameter values for each subject can aid in estimating the level of functional impairment expressed in the different stages of the disease. Thus, it is possible to more effectively monitor both the evolution of the disease and the efficacy of the treatment and its follow-up. The application of this model to scrotal hyperthermia secondary to varicocele would also help to understand at which specific functional level is the possible spermatogenesis damage determined. The study proved that for scrotal hyperthermia, even pure left varicocele appeared to behave like a bilateral disease, thus providing further elements to understand the functional impairment.

21.4 Discussion and Conclusions

The goal of this chapter was to show that control system theory applied to cutaneous thermal data obtained from thermal imaging may provide a powerful tool to identify a thermoregulatory feedback model. Such an approach is particularly relevant as the cutaneous thermoregulation is based on feedback mechanisms. The model parameters describe and distinguish the functional differences in the recoveries, at both central and peripheral levels. In fact, the parameters so far identified to define each model described how the system acts at the two different levels, in terms of process speed, efficiency of control action, and heat exchanges mechanisms, in order to restore the prestress basal conditions. Several advantages may be derived from this approach. The estimated model parameter values for each subject can aid in estimating the level of functional impairment expressed in the different stages of the disease. Thus, it is possible to more effectively monitor both the evolution of the disease and the efficacy of the treatment and follow-up. The application of this model would also help to understand at which specific functional stage of the regulatory process is the possible functional impairment determined. Functional infrared imaging can take great advantage from this approach, in a variety of clinical applications expressed through abnormalities and deviation from normal thermoregulatory behavior.

References

1. Agresti, A. *Categorical Data Analysis* (2nd edition). Hoboken, NJ: Wiley InterScience, pp. 165–166, 2002.
2. Belch, J. Raynaud's phenomenon. Its relevance to scleroderma. *Ann. Rheum. Dis.* 50:839–845, 2005.
3. Block, J.A., and W. Sequeira. Raynaud's phenomenon. *Lancet.* 357:2042–2048, 2001.
4. Chakravarthy, N., K. Tsakalis, S. Sabesan, and L. Iasemidis. Homeostasis of brain dynamics in epilepsy: A feedback control system perspective seizure. *Ann. Biomed. Eng.* 37(3):565–585, 2009.
5. Charkoudian, N. Skin blood flow in adult human thermoregulation: How it works, when it does not, and why. *Mayo Clin. Proc.* 78:603–612, 2003.
6. Clark, S., F. Campbell, T. Moore, M.I.V. Jayson, T.A. King, and A.L. Herrick. Laser Doppler imaging. A new technique for quantifying microcirculatory flow in patients with primary Raynaud's phenomenon and systemic sclerosis. *Microvasc. Res.* 57:284–291, 1999.
7. Cutolo, M., C. Pizzorni, and A. Sulli. Capillaroscopy. *Best Pract. Res. Clin. Rheum.* 3:437–452, 2005.
8. Di Carlo, A. Thermography and the possibilities for its applications in clinical and experimental dermatology. *Clin. Dermatol.* 13:329–336, 1995.
9. Elbendary, M., and A. Elbadry. Right subclinical varicocele: How to manage in infertile patients with clinical left varicocele? *Fertil. Steril.* 92(6):2050–2053, 2009.
10. Fette, A., and J.M. Austria. Homeostatic control mechanism. *J. Pediatr. Surg.* 35(8):1222–1225, 2002.
11. Forssel, U., and L. Ljung Closed loop identification revisited. *Automatica.* 35:1215–1241, 1999.
12. Gat, Y., G. Bachar, Z. Zukerman, A. Belenky, and M. Gorenish. Physical examination may miss the diagnosis of bilateral varicocele: A comparative study of 4 diagnostic modalities. *J. Urol.* 172(4):1414–1417, 2004.
13. Gat, Y., G. Bachar, Z. Zukerman, A. Belenky, and M. Gornish. Varicocele: A bilateral disease. *Fertil. Steril.* 81(2):424–429, 2004.
14. Glantz, S. *Primer of BioStatistics* (6th edition.). New York: McGraw-Hill, pp. 353–358, 2005.
15. Hahn, M., C. Hahn, M. Jünger, A. Steins, D. Zuder, T. Klyscz, A. Büchtemann, G. Rassner, and V. Blazek. Local cold exposure test with a new arterial photoplethysmography sensor in healthy controls and patients with secondary Raynaud's phenomenon. *Microvasc. Res.* 57:187–198, 1999.
16. Herrick, A.L. Pathogenesis of Raynaud's phenomenon. *Rheumatology.* 44:587–596, 2005.
17. Herrick, A.L., and S. Clark. Quantifying digital vascular disease in patients with primary Raynaud's phenomenon and systemic sclerosis. *Ann. Rheum. Dis.* 57:70–78, 1998.

18. Kellog, D.L. A physiological systems approach to human and mammalian thermoregulation. *J. Appl. Physiol.* 100:1709–1718, 2006.

19. Kessler, A., S. Meirsdorf, M. Graif, P. Gottlieb, and S. Strauss. Intratesticular varicocele: Gray scale and color Doppler sonographic appearance. *J. Ultrasound Med.* 24(12):1711–1716, 2005.

20. Kuryliszin-Moskal, A., P.A. Klimiuk, and S. Sierakowski. Soluble adhesion molecules (sVCAM-1, sE-selectin), vascular endothelial growth factor (VEGF) and endothelin-1 in patient with systemic sclerosis: Relationship to organ systemic involvement. *Clin. Rheumatol.* 24:111–116, 2005.

21. LeRoy E.C., and T.A. Medsger. Criteria for the classification of early systemic sclerosis. *J. Rheumatol.* 28:1573–1576, 2001.

22. Ljung, L. *System Identification.* Upper Saddle River: Prentice Hall PTR, pp. 511–512, 2007.

23. Mariotti, A., G. Grossi, P. Amerio, G. Orlando, P.A. Mattei, A. Tulli, G.L. Romani, and A. Merla. Finger thermoregulatory model assessing functional impairment in Raynaud's phenomenon. *Ann. Biomed. Eng.* 37(12):2631–2639, 2009.

24. Merla, A., L. Di Donato, S. Di Luzio, G. Farina, S. Pisarri, M. Proietti, F. Salsano, and G.L. Romani. Infrared functional imaging applied to Raynaud's phenomenon. *IEEE Eng. Med. Biol. Mag.* 6:73–79, 2002.

25. Merla, A., L. Di Donato, S. Di Luzio, and G.L. Romani. Quantifying the relevance and stage of disease with the Tau image technique. *IEEE Eng. Med. Biol. Mag.* 21(6):86–91, 2002.

26. Merla, A., A. Ledda, L. Di Donato, S. Di Luzio, and G.L. Romani. Use of infrared functional imaging to detect impaired thermoregulatory control in men with asymptomatic varicocele. *Fertil. Steril.* 18(1):199–200, 2002.

27. Merla, A., A. Ledda, L. Di Donato, and G.L. Romani. Assessment of the effects of varicocelectomy on the thermoregulatory control of the scrotum. *Fertil. Steril.* 81(2):471–472, 2004.

28. Merla, A., G.L. Romani, S. Di Luzio, L. Di Donato, G. Farina, M. Proietti, S. Pisarri, and S. Salsano. Raynaud's phenomenon: Infrared functional imaging applied to diagnosis and drug effect. *Int. J. Immunopathol. Pharmacol.* 15:41–52, 2002.

29. Mieusset, R., and L. Bujan. Testicular heating and its possible contributions to male infertility: A review. *Int. J. Androl.* 14(8):169–184, 1995.

30. Nogueira, F., L. Barroso, E. Miranda, J.D. Castro, and F.M. Filho. Infrared digital telethermography: A new method for early detection of varicocele. *Fertil. Steril.* 92(1):361–362, 2009.

31. O'Reilly, D., L. Taylor, K. El-Hadivi, and M.I. Jayson. Measurement of cold challenge response in primary Raynaud's phenomenon and Raynaud's phenomenon associated with systemic sclerosis. *Ann. Rheum. Dis.* 11:1193–1196, 1992.

32. Oussar, Y., and G. Dreyfus. How to be a gray Box: Dynamic semi-physical modelling. *Neural Networks.* 14:1161–1172, 2001.

33. Ren, T., and D. Thieffry. Dynamical behaviour of biological regulatory networks-I. Biological role of feedback loops and practical use of the concept of the loop-characteristic state. *Bull. Math. Biol.* 57:274–276, 1995.

34. Renè, T., and D. Thieffry. Dynamical behaviour of biological regulatory networks- I. Biological role of feedback loops and practical use of the concept of the loop-characteristic state. *Bull. Math. Biol.* 57:247–276, 1995.

35. Rollins, D., N. Bhabdar, and S. Hulting. System identification of the human thermoregulatory system using continuous-time block-oriented predictive modelling. *Chem. Eng. Sci.* 61:1516–1527, 2006.

36. Romeo, C., and G. Santoro. Varicocele and infertility: Why a prevention? *J. Endocrinol. Invest.* 32(6):559–561, 2009.

37. Sanial, D.C., and N.K. Maji. Thermoregulation through cutaneous under variable atmospheric and physiological conditions. *J. Theor. Biol.* 208:451–456, 2001.

38. Sato, S., M. Hasegawa, K. Takehara, and T.F. Tedder. Altered B lymphocyte function induces systemic autoimmunity in systemic sclerosis. *Mol. Immunol.* 42:821–831, 2005.

39. Sawasaki, N., S. Iwase, and T. Mano. Effect of cutaneous sympathetic response to local or systemic cold exposure on thermoregulatory functions in humans. *Auton. Neurosci.* 87:274–281, 2001.
40. Shafik, A. Advances in male contraception. *Arch. Androl.* 45(3):155–167, 2000.
41. Shiraishi, K., K. Naito, and H. Takihara. Indication of varicocelectomy in the era of assisted reproductive technology: Prediction of treatment out come by noninvasive diagnostic methods. *Arch. Androl.* 49(6):475–478, 2003.
42. Shitzer, A., On the thermal efficiency of cold-stressed finger. *Ann. NY Acad. Sci.* 858:74–87, 1998.
43. Shitzer, A., A. Stroscheine, R.R. Gonzalez, and K.B. Pandolf. Lumped parameter tissue temperature-blood perfusion model of a cold stressed finger. *J. Appl. Physiol.* 80:1829–1834, 1996.
44. Skandhan, K., and A. Rajahariprasad. The process of spermatogenesis liberates significant heat and the scrotum has a role in body thermoregulation. *Med. Hypotheses.* 68:303–307, 2007.
45. Suter, L.G., J.M. Murabito, D.T. Felson, and L. Fraenkel. The incidence and natural history of Raynaud's phenomenon in the community. *Arthritis Rheum.* 52(4):1259–1263, 2005.
46. Waterhouse, J. Homeostatic control mechanism. *Anaesth. Intensive Care.* 5:236–240, 2004.
47. Lonzetti, L.S., F. Joyal, J.P. Raynauld, A. Roussin, J.R. Goulet, E. Rich, D. Choquette, Y. Raymond, and J.L. Senécal. Updating the American College of Rheumatology preliminary criteria for systemic sclerosis: Addition of severe nailfold capillaroscopic abnormalities markedly increases sensitivity for limited scleroderma. *Arthritis Rheum.* 44:735–736, 2001.

22

Infrared Thermal Imaging Standards for Human Fever Detection

Francis J. Ring
University of Glamorgan

E.Y.K. Ng
School of Mechanical and Aerospace Engineering

Nanyang Technological University

22.1 Introduction

Infrared (IR) thermal imaging provides an efficient means of recording the surface temperatures of the human body. It is a noncontact radiometric technique that has improved considerably since its early trials in medicine in the late 1950s.[1]

In recent years concerns about the spread and containment of infectious diseases, especially among the travelling public has brought this technology into use as a means of detecting high fever in passengers. The Severe Acute Respiratory Syndrome (SARS) outbreak around 2003 saw the introduction on IR imaging cameras in Chinese airports.[2,3] While there were a small percentage of passengers who were stopped because they had elevated facial temperatures, the policy of locating these cameras high above a crowd of passers by, is now recognized as being not fully efficient. Passengers wearing hats and face masks, for example, may have a large area of their face covered and escape recognition.

The Singapore Standards Organisation SPRING examined the use of these cameras in 2003 and published two excellent guides to the most suitable devices and how they should be calibrated and deployed.[4,5] Subsequently the International Standards Organisation (ISO) was invited to consider these documents and review their potential application in the international community.

As a result an international writing group worked on the contents of these reports, to update and widen the details and references to the supporting information. It also was able to link sections and definitions of the new documents to existing standards, which provided valuable, cross referencing.

Following the form of the SPRING documents, the ISO produced two new documents, the first to describe the essential performance specification of suitable imaging systems for fever detection in humans, and the second on the effective deployment, installation, maintenance, and staff training required to obtain optimum performance in the field. Full details of these documents should be obtained by all who are directly concerned with the specification manufacture and sales of these devices.[6,7] Equally, the detail of the correct deployment of the fever screening installations is found in the full ISO documentation. This chapter aims to merely highlight some of those details, and some of which could have a useful bearing on the clinical application of IR imaging.

22.2 Definitions

The standard documents contain full details of definitions used throughout, and it is not intended to reproduce all in this chapter. Many of the terms and definitions are those that are found across the same technical area, but additional terms have been added for this specific application.

Examples of these additional terms are:

Calibration Source 201.3.202
IR radiation blackbody reference source of known and traceable temperature and emissivity

Detector 201.3.203
IR thermal sensor or array of sensors able to detect IR thermal energy radiating from the surface of the face or other object

External Temperature Reference Source 201.3.205
Part of the screening thermograph that is used to ensure accurate operation between calibration using IR radiation source of known temperature and emissivity. Note: the external temperature reference source is normally imaged in each thermogram or prior to each thermogram.

Face 201.3.206
Anterior cranial face of the patient being measured.

Screening Thermograph 201.3.209 and 201.7.3.101
Medical electrical equipment or ME system (201.3.209) that:
Detects IR radiation emitted from the face from which a thermogram is obtained from the target.
Detects IR radiation emitted from an external temperature radiation source;
Displays a radiometric image;
Obtains a temperature reading from the target and
Compares that temperature reading to the threshold temperature to determine if the patient is febrile
The technical description (201.7.3.101) shall include:

A. A reference to ISO/TR13154 Medical electrical equipment—deployment, implementation, and operational guidelines for identifying febrile humans using a screening thermograph (the guidance document for the application of screening thermographs)
B. A recommendation that the relative humidity in the area of screening should be maintained below 50% and temperature below 24°C to achieve the intended use and an explanation of the effects of elevated humidity and ambient temperature on the temperature reading caused by sweating.

Note 1: The measurements provided by a screening thermograph in intended use can be influenced when the patient is sweating. Sweating thresholds can vary, according to the patient's fitness level, environment of residence, length of adaptation, and the relative humidity.

C. An explanation of the effects due to environmental IR sources such as sunlight, nearby electrical sources and lighting, and instructions on how these should be minimized.

Note 2: The responsible organization needs to be aware of the type of lighting used in the screening area. Lighting such as incandescent, halogen, quart tungsten halogen, and other type of lamps that produce significant interference (heat) should be avoided.
Note 3: The area chosen for screening should have a nonreflective background and minimal reflected IR radiation from the surroundings.

D. An explanation of the effects due to airflow, and instructions that this should be minimized.

Note 4: Drafts from air conditioning ducts can cause forced cooling or heating of the face and should be baffled or diffused to prevent airflow from blowing directly onto the patient.

Skin Temperature 201.3.210

Skin surface temperature as measured from the workable target plane of a screening thermograph, with an appropriate adjustment for skin emissivity.

Note: the emissivity of dry human skin is accepted to be 0.98

Target Plane 201.3.213

Infocus plane perpendicular to the line of sight of a screening thermograph.

22.3 Discussion

From the above it is clear that the standard requires more from the manufacturer than a simple IR camera and temperature reference source. The minimal acceptable performance of the camera is clearly defined, and the accuracy and stability required are based on the expectation that the device will operate continuously throughout a full day. The recommendation is that an un-cooled detector system is used since there is a shortened life expectancy of a cooling system that can be expensive to replace or renew.

The manufacturer needs to be fully aware of the specification details and ensure that the end user (described as the responsible organization) is alerted to the ISO document that specifies optimal conditions for installation and management of the screening set up. There is evidence to indicate that some manufacturers seem to be unaware of this standard, and certainly failing to pass on the kind of information shown in the above extract. However, clinical users of IR thermal imaging will be already familiar with most of these criteria, that play an important role is achieving consistent and reliable thermograms used in medicine. It is interesting that the ISO document chooses to refer to the human target for

FIGURE 22.1 Setup for a typical IR camera system at a medical center entrance.

FIGURE 22.2   Operator's viewpoint (with subject standstill within a designated box for a few seconds).

screening as the patient on the grounds that this technique is designed to indicate a person who is likely to have a fever, and on subsequent medical examination will have that diagnosis accepted or rejected. It is also the reason why the screening thermograph is described as "medical equipment."

In this latter case, this is a new departure from other screening systems used in security screening in airports. Detection of metal objects and drugs are now routine operations for airport passengers.

However, thermal imaging brought into use in a pandemic fever, can be used in other buildings and organizations. Schools, factories, and civil public buildings can equally use this technique. In the Guidance document on deployment, there are two situations described. A high volume transit, where screening can be carried out at maximum speed using automatic software to alert the operator of temperature found in excess of 38°C, from whence the passenger is diverted to a medical check that will include a questionnaire and a thermometry test. However, in a smaller application less-expensive installations can be used where the operator may be required to position the camera, and read the result from the screen image. Figures 22.1 and 22.2 show a typical set-up for the IR system at an entrance to a hospital and an operator's viewpoint.

One critical part of the specification for use in both documents is the positioning and location of the site for temperature measurement.

The forehead, lateral temporal artery locations have both been used, but are less reliable as a site for indicating fever. The inner canthus of the eye, has been shown to be reliable. It is fed from the internal carotid artery, and is less influenced by ambient and physiological stress, than other areas of the face. For this reason the standard specifies that all spectacles must be removed, and hats, and facemasks, if they interfere with a clear reading from the eye region. It is also clear that at a distance from the camera lens it will be impossible to obtain sufficient pixels and temperature readings to read the temperature with certainty. Oblique angles between camera and patient are also inaccurate. It is therefore necessary to stop the passenger for a few seconds (Figures 22.1 and 22.2), to ensure that the camera is in line with the eyes before the image is registered.[8] It is anticipated that only positive findings leading to a thermogram being stored with its appropriate identity including passport details. Only time will reveal if this is adequate given the possible litigation should a febrile passenger be missed in screening, and go on to infect a large number of fellow travellers.

References

1. Ring, E.F.J., Progress in the measurement of human body temperature. *Engineering in Medicine & Biology Magazine*, 17, 1998, 19–24.
2. Bitar, D., Goubar, A., and Desenclos, J.C., International travels and fever screening during epidemics: A literature review on the effectiveness and potential use of non-contact infrared thermometers. *Euro Surveill*, 14(6), 2009, 1–5.
3. Ng, E.Y.K. and Rajendra Acharya, U., A review of remote-sensing infrared thermography for indoor mass blind fever screening in containing an epidemic. *Engineering in Medicine and Biology*, 28(1), 2009, 76–83.
4. Thermal Imagers for human temperature screening Part 1: Requirements and test methods. *Technical Reference TR15:* part 1:2003 SPRING Singapore.
5. Thermal imagers for human temperature screening Part 2: Implementation and Guidelines. *Technical Reference TR15:* part 2:2004 SPRING Singapore.
6. Particular requirements for the basic safety and essential performance of screening thermographs for human febrile temperature screening. *ISO TC121/SC3-IEC SC62D*, 2008.
7. Medical Electrical Equipment—Deployment, implementation and operational guidelines for identifying febrile humans using a screening thermograph. *ISO/TR 13154; ISO/TR 80600*, 2009.
8. Ring, E.F.J., McEvoy, H., Jung, A., Zuber, J., and Machin, G., New standards for devices used for the measurement of human body temperature. *Journal of Medical Engineering & Technology*, 34(4), 2010, 249–253.

23

Francis J. Ring
University of Glamorgan

A. Jung
Military Institute of Medicine

B. Kalicki
Military Institute of Medicine

J. Zuber
Military Institute of Medicine

A. Rustecka
Military Institute of Medicine

R. Vardasca
Polytechnic Institute of Leiria

Infrared Thermal Imaging for Fever Detection in Children

23.1 Introduction

Over recent years, pandemic influenza virus infections have caused concerns about an ever-increasing mobility of populations, especially in air travel. Limiting exposure to fellow travellers who may be suffering from a febrile high temperature has been employed with some success. Pandemic influenza virus is a virulent human form that causes global outbreak, or pandemic, of serious illness. Because there is little natural immunity, the disease can spread easily from person to person. There have been pandemic outbreaks in the United States in 1918, 1958, and 1968 with varying degrees of morbidity and virulence. Children under 18 years have tended to have the highest rates of illness, though not of severe disease and death.[1]

Infrared thermal imaging has in recent years become more accessible and affordable as a means of remote sensing for human body temperature. Historically, a clinical fever has been checked with a clinical thermometer, and these devices have been slowly replaced by infrared radiometers mainly for inner ear temperature measurement. However, the number of scientific reports and papers confirming the reliability of the infrared methods are limited. Furthermore, there is a need for critical technique, which has not always been well defined in these publications.[2]

Some use of infrared imaging for fever screening has been made since 2006 in international airports, most of which have not used optimal technique. For this reason, the International Standards Organisation produced two documents in 2008 and 2009 (ISO TC121/SC3-IEC SC62D)[3,4] (see Chapter 22 by Ring and Ng).

One important issue that is most important to a correct technique is the positioning of both camera and subject. Typical screening installations use cameras mounted at an angle to survey passing groups of passengers. The best site for temperature measurement is the inner corner (or canthus) of the eye.

This requires the camera to be mounted close up and on level with the eyes, thus providing a cluster of image pixels in the region of interest. The study described here has employed a good specification radiometric camera at a correct position, to ensure a minimum of 9 pixels in each eye canthus.

23.2 Method

Cohorts of children at the Pediatric Clinic in Warsaw were checked for fever by three different methods. The routine department procedure was carried out by a nurse using a clinical thermometer under the armpit (axilla), and left there for a full 5 min.

They then entered a room for thermal imaging maintained at 22–23°C where they were seated before the FLIR infrared camera SC640. The image of the face was set to occupy at least 75% of the image, and regions around the inner canthi (i.c.) were measured for temperature. A second area of interest over the forehead was also recorded, as some claims have been made that forehead temperature would be an adequate target for fever. Both inner ear measurements (tympanic membrane) with a clinical radiometer were then recorded and documented, with the child's demographic data.

23.3 Results

A total of 402 children were examined between 2006 and 2011. The majority of children attending the hospital were not febrile, they were compared with those known to have a clinically defined fever, and examined prior to medication. Of this group 350 (85%) were found to be free from fever, and 52 (15%) cases of definite fever were recorded. There were 192 male and 210 female individuals.

All subjects had infrared thermograms and axilla recordings, not all subjects were measured by the ear tympanic membrane thermometer, due to equipment failure.

The age distribution is shown in Table 23.1.

23.3.1 Temperature Data

The data were analyzed for any differences between male and female individuals, no statistical differences were found for sex dependency. Similarly for age dependency, again no statistical differences were found. In the control subjects (afebrile) left eye vs right eye data were compared and found to be not significant. As a result both readings were combined and the mean temperature of both eyes was used for each subject (Table 23.2). While temperatures of 38°C and above are considered to be found in children over 5 years and adults in fever, 37.5°C and above is considered to be a fever level in under 5-year olds. In this study we have categorized all temperatures over 37.6°C as fever. In most cases they are higher than 38°C (Table 23.3).

To examine the possible relationship between the different measurement sites and methods, Pearson's correlation tests were performed on the temperature data. See Tables 23.4a and 23.4b.

This shows the existence of a linear relationship between the measurements among sites in the cases of fever. This is highest when comparing the inner canthus eye measurements by thermography and axilla temperature measured by the clinical contact thermometer. The forehead temperatures and ear measurements correlate less well with statistical significance ($p < 0.1$).

TABLE 23.1   Age Distribution of Subjects

<3 years	50
3–6	128
7–12	136
13–16	82
>16	6

TABLE 23.2   Temperature Data: Nonfever Group

Afebrile	Mean (°C)	Std. Dev.	Number
Eyes i.c.	36.48	0.49	354
Forehead	36.44	0.65	326
Axilla	36.34	0.59	347
Ear	36.12	0.71	178

TABLE 23.3   Temperature Data: Fever Group (49% Male, 51% Female)

Febrile	Mean (°C)	Std. Dev.	Number
Eyes i.c.	38.9	0.84	52
Forehead	34.7	0.86	52
Axilla	38.9	0.68	52
Ear	37.4	1.41	24

TABLE 23.4A   Afebrile

Pair	Pearson's Correlation	p-Value
Eyes i.c./axilla	0.507	0.0006
Forehead/axilla	0.432	0.000
Ear/axilla	0.420	0.014

TABLE 23.4B   Febrile (Fever) Cases

Pair	Pearson's Correlation	p-Value
Eyes i.c./axilla	0.587	0.000
Forehead/axilla	0.432	0.022
Ear/axilla	0.276	0.267

In the analysis of these data it was found that in febrile children the eye temperature measurements and the thermometric axilla temperatures are closer that those measured from the forehead (thermographic) or ear (tympanic membrane radiometer). The eye and axilla measurements also showed more internal consistency than the others (reliability coefficient alpha of 0.724).

23.4 Discussion

This study has been performed on children in Warsaw, Poland according to the recommended technique described in the new ISO standards of 2008/2009. The advantage of performing the study in a hospital environment is that the diagnosis was known prior to the study measurements, but was not conveyed to those undertaking the study until after the data had been examined. There were no cases of clinical fever that were undetected by the study.

It had been questioned if another source of mild infection such as sinusitis could confuse the data and cause a false-positive result. Such cases have been documented by the pediatric department and there was known to be a clear difference between sinusitis and generalized fever. In the case of sinusitis, there is often a bilateral increase in heat affecting the face, and rarely affecting the i.c. temperatures.

The clinical fever cases in this study showed excellent symmetry, so that a close correlation was always found between the left and the right eye, in both the febrile and afebrile patients. Those with sinusitis or dental infection showed a characteristic localized and asymmetric temperature increase.

The examination room was thermally stable, and air conditioning could be used prior to a clinical session if required. However, unlike the normal protocols recommended for clinical tests using infrared imaging, the patients were not required to rest and stabilize prior to the test. This would not be practical in a traveller screening situation, and the promise that the i.c. of the eye should be the most stable site for measurement was borne out by this study. The forehead on the other hand, also measured from the thermogram using a rectangular region of interest occupying some 60% of the exposed area, did not show such consistency. It was sometimes necessary to use a disposable paper hat to contain long hair prior to the imaging procedure.

For the most efficient positioning of the subject before the camera a stool was used, which limited the range of height adjustment required for the camera stand. This was a counterbalanced single pillar stand as used in photographic studios, because, the ISO has specifically warned of the potential for error in reproducible positions, when using a pan and tilt head tripod as the infrared camera mount.

This study has also endorsed the methodology recommended in the ISO document for fever screening where a close-up stationary image is recorded of the subject's face. With the current infrared cameras, even using the best available, the system error could be as high as 0.4°C which is a large proportion of the actual difference in temperatures found between afebrile and febrile cases. It is therefore clear that a camera mounted over a meter from the subject, and possibly also at an oblique angle will fail to have sufficient pixels in the relatively small target area of the i.c. of an eye. Increasing the number of pixels over the measurement site is not only good but essential practice in this application.

Fever screening in a global pandemic has yet to be taken seriously, but it is clear that there can be legal repercussions from false-positive measurements due to bad technique or inadequate technology.

In this study also we decided to include very young children over the age of 6 months. At this age a parent facing the camera held them, and because the distance between subject and camera lens was short, there was no difficulty in filling the screen with the child's face. It was also not a problem if the child was asleep, but only in the case of persistent crying, producing tears, the temperature readings were aborted. In a few cases, they were resumed after the child had been pacified, and a bright colored object moved above the lens of the camera was sufficient to make the child look in the right direction.

In this study, the commercial software proved entirely adequate for the analysis of two regions of the face. However, some recent studies have proposed software that will automatically track the human face and obtain measurements with minimal intervention by the operator. This approach will lead the way to rapid screening in the future.[5,6]

23.5 Conclusion

The use of an infrared radiometric camera can be a reliable tool for the detection of fever in children. The technique recommended by the ISO in document TR 13154:2009 ISO/TR80600 is endorsed by the results of this study. There was a significant difference between the temperatures measured in afebrile patients and those with known fever, with the thermal imaging of the eye region being the most rapid noncontact site for measurement.

Acknowledgments

We are extremely grateful for the complete support of FLIR Infrared Systems, Warsaw, for the regular support of Mr. Pawel Rutkowski, and the loan of the Infrared camera. We also thank the staff and patients of the Military Institute of Medicine, Warsaw for their willing participation in this study.

References

1. *Pandemic Influenza: Preparedness, Response & Recovery. Guide for Critical Infrastructure and Key Resources.* Homeland Security. Sept. 19th, 2006, USA.
2. Dodd S.R., Lancaster G.A., Craig J.V., Smyth R.L., Williamson P.R. In a systematic review, infrared ear thermometry for fever diagnosis in children finds poor sensitivity. *J. Clin. Epidemiol.* 59(4), 354–57, 2006.
3. ISO TC121/SC3-IEC SC62D Particular requirements for the basic safety and essential performance of screening thermographs for human febrile temperature screening, 2008.
4. ISO/TR 13154:2009 *ISO/TR 8-600: Medical Electrical Equipment—Deployment, Implementation and Operational Guidelines for Identifying Febrile Humans Using a Screening Thermograph.*
5. Strakowska M., Strzelecki M., Wiecek B. Automatic measurement of human body temperature in the eye canthi using a thermovison camera. *Thermology International* 21(2), 57–58, 2011.
6. Bajwa U.I., Taj I.A., Bhatti Z.E. A comprehensive comparative performance analysis of Laplacian faces and Eigen faces for face recognition. *The Imaging Science Journal* 59, 32–40, 2011.

24

Thermal Imager as Fever Identification Tool for Infectious Diseases Outbreak

E.Y.K. Ng
*School of Mechanical and
Aerospace Engineering*

*Nanyang Technological
University*

24.1 Introduction

Thermography is a noninvasive diagnostic method that is economic, quick, and does not inflict any pain on the subject. It is a relatively straightforward method of imaging that detects the variation of temperature on the surface of the human skin. Thermography is widely used in the medical arena [1–29]. This includes the detection of an elevated body temperature [6–29], which is the focus of this chapter. Thermograms alone will not be sufficient for the medical practitioner to make a diagnosis. Analytical tools such as bio-statistical methods and artificial neural network (ANN) such as radical basis function network (RBFN) is utilized to analyze the thermograms. Neural network is a pattern recognition program that has the ability to predict the outcome based on the various inputs fed into the program. For an elevated body temperature thermography, the program will predict if the subject is febrile or nonfebrile. Figures 24.1 and 24.2 reveal two typical examples of normal and febrile thermograms.

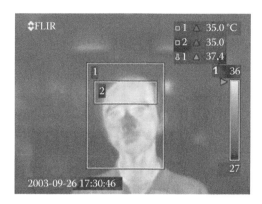

FIGURE 24.1   Nonfebrile thermogram (aural temperature = 36.9°C).

In the late 2009, the World Health Organization (WHO) is urging the world to brace for a second wave of the H1N1 pandemic as the heavily populated Northern Hemisphere edges toward the cooler season when flu thrives. There had been second and third waves in previous pandemics. Thus, one has to be prepared for whatever surprises this capricious new virus would deliver in the coming years. More than 6000 people have died since H1N1 was uncovered in Mexico nearly 7 months ago (April to mid-October 2009). Twenty-two million Americans were infected by the H1N1 during the earlier period of which 4000 were killed after a new counting method yielded an estimate six times higher than the last one. The virus is more infectious than seasonal flu and more durable through the warmer months. It is mystified at the "most worrying" characteristic of this virus. Nearly 40% of the most severe or fatal cases occur in people who are in perfect health.

Fever has been regarded as an important symptom of pandemic influenza though a recent research reveals that one in five of the H1N1 or swine flu pandemic will not be detected, some have mild or no fever, and their general symptoms make them hard to identify and still pass on the virus. It was reported lately that only half had high fever of 37.8°C or more which is one of the symptoms the United States uses to identify H1N1 cases. If we use this cut-off, then 46% of the confirmed cases would not have been picked up. Cough was the most common symptom; affecting four in five patients, although only half had sore throats. For most H1N1, the fever receded a day after they started taking Tamiflu. Thus, the doctors do not know whether the virus is alive or dead and hence they are uncertain if the person is still infectious. However, an improvement in the accuracy of using noncontact infrared (IR) system to detect feverish subjects is useful for other flu pandemics such as SARS and H5N1.

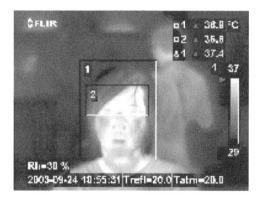

FIGURE 24.2   Febrile thermogram (aural temperature = 38.4°C).

There are many well-documented articles including comprehensive review papers in renowned journals [7,11,12,16,18,21,22,24,26–28] as well as the recent International Standard Organization (ISO) procedures and standards (ISO/TR 13154 and IEC 80601-2-59) as published recently [29–30] that are closely related to the present topic. These standards are related to "Deployment, implementation, and operational guidelines for identifying febrile humans using a screening thermograph" and "Particular requirements for the basic safety and essential performance of screening thermographs for human febrile temperature screening."

24.2 Methodology

24.2.1 Data Acquisition

The blind mass data were collected (502 without duplicate measurements, confirmed later as 86 febrile and 416 healthy cases with ear thermometer) from the designated SARS hospital (A & E Department, Tan Tock Seng Hospital [TTSH]) and the Civil Defense Force Academy [SCDF] in Singapore (in-door screening with ambient temperature of 25 ± 2°C, humidity $\approx$60%) in which thermal imagers are used as a first-line tool for the blind screening of hyperthermia. The subjects are considered febrile if his/her mean ear temperature is $\geq$37.7°C for adults (37.9°C for children) using Braun Thermoscan IRT 3520+. Results are drawn for the two important pieces of information: the best and yet a practical region on the face to screen and guidance on optimal preset threshold temperature for the same handheld radiometric IR ThermaCAM S60 FLIR system [31]. The focal length from the subject to the scanner was fixed at 2 m and the duration of time patients scanned was 3 s. Figure 24.3 illustrates an example of temperature profiles from a temperature operation using thermal imager with temperature reading. Visitors were directed to line up in a single file with the aid of barricades. Stand in position, remove spectacles, and look at the imager so that his/her face fills at least 1/3 vertical height of the display screen (Figures 24.4 and 24.5). The detector was a focal plane array, an uncooled microbolometer of 320×240 pixels with a thermal sensitivity of 0.06°C at 30°C, a spectral range of 7.5–13 µm, and its measurement accuracy at $\pm$2% of the real-time reading [31]. The average temperature of the skin surface was measured from the field-of-view of the thermal imager with an appropriate adjustment for skin emissivity. Human skin emissivity may vary from site to site ranging from 0.94 to 0.99 (0.98 was used here). Figures 24.6 and 24.7 present an example of the processed thermal images of both frontal and side (left) profiles from

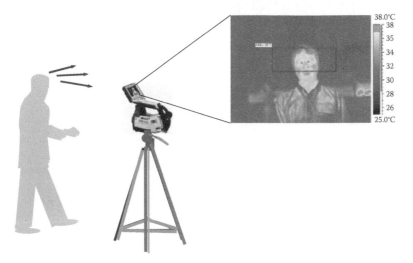

FIGURE 24.3   A typical thermal imager with direct threshold temperature setting. (Adapted from Zugo. www. zugophotonics.com/)

FIGURE 24.4   A yellow square marked on the ground for visitors to stand in.

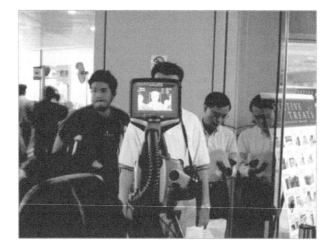

FIGURE 24.5   Face fills at least one-third vertical height for the field of view.

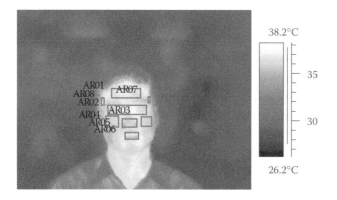

FIGURE 24.6   Processed thermal images of the frontal profile.

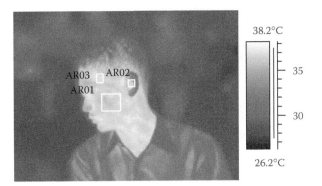

FIGURE 24.7   Processed thermal images of the side profile.

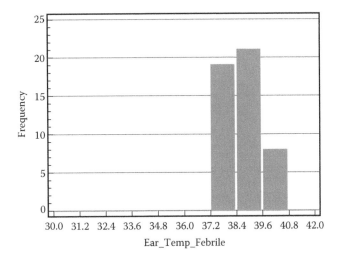

FIGURE 24.8   Distribution of ear temperature for febrile cases.

the same subject. The following spots were logged and analyzed from the subjects with frontal and side profiles: forehead, eye region, average cheeks, nose, mouth (closed), average temple, side face, ears, and side temple (the last three are for the side profile).

Reproducibility of both the instrument and physiological assumptions was established by comparing paired left–right readings of the temples and cheeks [32,33]. Comparing the ear's core temperature of febrile and nonfebrile data (Figures 24.8 through 24.11), it is noted that the mode of the nonfebrile data falls between 36.0°C and 37.2°C. For febrile, it spread over a larger region, from 37.2°C to 39.6°C.

24.2.2 Artificial Neural Networks

ANN are a group of techniques for numerical learning [34]. They are made up of many nonlinear computational elements called neurons. These neurons, also known as network nodes (NN), are linked to one another. Through this weighted interconnection, they formed the main architect of the NN. To draw an analogy, ANN is similar to the neurological system in humans and animals, which are made up of real NN. One important point to note is that ANN is much less complex than the biological NN (BNN). As a result, it is not realistic to expect ANN to emulate BNN, which is responsible for the behavior of humans and animals. However, ANN has the capability to assist us in some tasks. This includes nonlinear estimation, classification, clustering, and content-addressable memory.

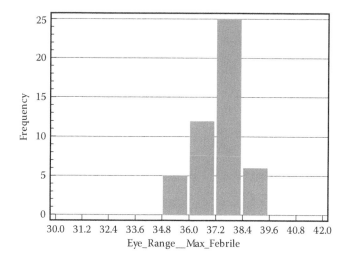

FIGURE 24.9   Distribution of maximum temperature at eye range for febrile cases.

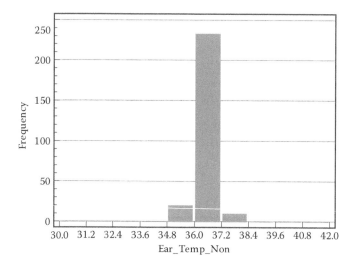

FIGURE 24.10   Distribution of ear temperature for nonfebrile cases.

Two or more inputs are connected to a node in an ANN [34]. Each of them has a weighted linkage attached to it (Figure 24.12). Based on the input values, a node has the ability to perform simple calculation. Both inputs and outputs are real numbers or integers between −1 and 1. All the input data have to be normalized before being fed into the program. The output from one individual node can either be inputted into another node or be a part of the NN's overall output. Each node performs its calculation and function independently from the rest of the nodes. The only association between the nodes is that the output from a node might be the input for another node. This type of architect is also known as a parallel structure, which allows for the exploration of numerous hypotheses. In addition, this parallel architect also permits the NN to make full use of the conventional personal computers.

The main advantage of ANN is that the tolerance of failure of an individual node or neuron is relatively high. This includes the weighted interconnection, because it might be erroneous too. The weights can be obtained by utilizing a trained algorithm, through iteration, and adjustments. The eventual transfer function is obtained with regard to the desired output.

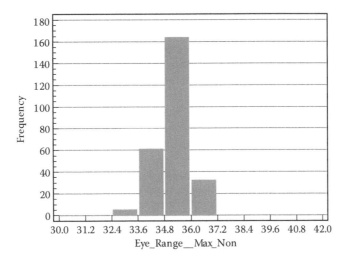

FIGURE 24.11 Distribution of maximum temperature at eye region for nonfebrile cases.

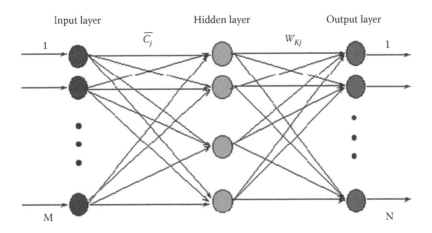

FIGURE 24.12 RBFN architect [35,36].

When a set of inconsistent or incomplete data is given, ANN is able to give an approximate answer rather than a wrong answer. The performance of the NN will undergo a gradual degradation should there be any failures from individual nodes in the network. This is very useful in the medical arena as many a times, it is difficult to run a comprehensive test. The disadvantage of using ANN is that it does not have the capability to predict and forecast accurately beyond the range of previously trained data. In other words, the predicted outcome is based on the available set of data.

24.2.3 Radial Basis Function Network

Radial basis function network is a kind of feed-forward and unsupervised learning paradigm. A simple RBFN consists of three separate layers—input layer, hidden layer, and output layer as shown in Figure 24.12. The first part of the training cycle involves the clustering of input vectors. Mathematically, the clustering is done using Dynamic K-Means algorithm [35]. At the end of the clustering process, the radius of the Gaussian functions at the middle of the clusters will be equivalent to the distance between the two nearest cluster centers [34].

During the training, the RBFN is required to fulfill two tasks. First, it is to determine the middle of each hypersphere (circle in 2-D and hypersphere in n-dimensional pattern space) and second, to obtain its radius. For the first task, it is done by allocating the weights of the processing elements. This can be done by using an unsupervised clustering algorithm. It is important to note that the output neuron in the prototypical layer of a RBFN is in a function of the Euclidean distance. This distance measures from the input vector to the weighted vector. The unsupervised learning phase in the hidden layer of RBFN is followed by another different supervised learning phase. This is the stage where the output neurons are trained to associate each individual cluster with their own distinct shapes and sizes. RBFN is selected for the current work since its training speed is faster than the Back-Propagation network, able to detect data that are not within the norm, and make a better decision during classification problems.

The input and output neurons of RBFN and perceptron are alike [34]. The major difference lies in the hidden neuron. In most cases, it is governed by the Gaussian function. This is different from other processing neurons that produce an output based on the weighted sum of the inputs. On the contrary, the input neurons of the RBFN are not involved in the processing of information. Their sole function is to input the given data to the receiving nodes. Using a linear transfer function, these receiving nodes will decide the weights to be allocated to each subsequent processing element. They are governed by the transfer functions:

$$y_i = f_r(r_i), \quad r_i = \sqrt{\sum_{j=1}^{n}(x_j - w_{ij})^2}$$

where x_j is the input vector, W_{ij} represents the amount of weights allocated to the inputs of the neuron i. f_r is a symmetrical function known as the radial basis function or the Gaussian function, which is the preferred choice of most researchers.

$$f_r(r_i) = \exp\left(\frac{-r_i^2}{2\sigma_i^2}\right)$$

where σ_i is the standard deviation of the Gaussian distribution. Every neuron at each hidden layer will have its own unique σ_i value.

24.2.4 Bio-Statistical Methods

24.2.4.1 Regression Analysis

Regression (least-squares) analysis is a statistical technique used to determine the unique curve or a line that "best fits" all the data points (Figure 24.13). The underlying principle is to minimize the square of the distance of each data point to the line itself. In the regression analysis, there are two variables— namely, dependent and independent. The former is the variable to be estimated or predicted.

The most important result obtained in the analysis is the R-squared, or the coefficient of determination. R-squared is an indication of how tightly or sparsely clustered the data points are and it is a value that lies between 0 and 1. Thus, it is a measure of the correlation between the two variables. Correlation is the predictability of the change in the dependent variable given a change in the independent variable.

Parabolic regression refers to using a parabolic curve to fit the data points. It is a simple yet effective way to obtain the correlation between the two variables. However, a few assumptions are made in using the learn rule (LR). First, a parabolic relationship is assumed between the two variables, which might not always be the case. Second, the dependent variable is assumed to be normally distributed with the same variance with its corresponding value of the independent variable. Mathematically, the parabolic

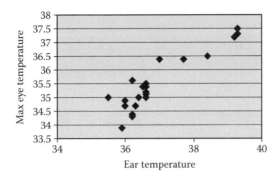

FIGURE 24.13 A typical scattered plot.

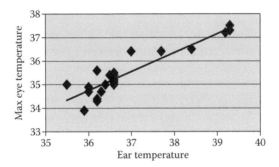

FIGURE 24.14 A regression line being fitted.

regression (PR) model is given by $Y = Ax^2 + Bx + C$ (Figure 24.14). PR usually offers a more realistic and better correlation between the two variables compared with LR.

24.2.4.2 Receiver Operating Characteristics Curve

Receiver operating characteristics (ROC) curves are used to assess the diagnostic performance of a medical test to discriminate unhealthy cases from healthy cases [36]. Very often, in a medical test, the perfect separation between unhealthy and healthy cases is not possible if we were to discriminate them based on a threshold value. To illustrate this phenomenon, let us call the threshold value γ.

Figure 24.15 suggests that at the threshold value γ, the majority of those without the disease will be correctly diagnosed as healthy (TN). Similarly, the majority of those with the disease will be correctly diagnosed as unhealthy (TP). However, there will also be one group of diseased patients wrongly

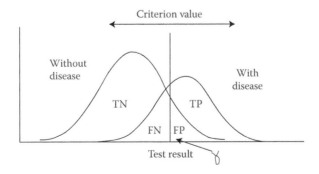

FIGURE 24.15 Discrimination curve.

TABLE 24.1   Basic Mathematical Formulae for ROC Analysis

Test	Disease	Number	Disease	Number	Total
Result	Present	n	Absent	m	
Positive	True positive	a	False positive	c	$a + c$
Negative	False negative	b	True negative	d	$b + d$
Total		$a + b$		$c + d$	

TABLE 24.2   Important Terminology for ROC Analysis

Sensitivity	$a/(a + b)$
Specificity	$d/(c + d)$
Positive predictive value	Sensitivity/(1 − specificity)
Negative predictive value	(1 − sensitivity)/specificity

diagnosed as healthy (FN) and one group of healthy patients wrongly diagnosed as unhealthy (FP). Table 24.1 summarizes all the possibilities—TN, TP, FN, and FP and their respective algebraic representation.

With that, four important criterions can be defined—sensitivity,[*] specificity,[†] positive predictive value (PPV),[‡] and negative predictive value (NPV)[§] and they are commonly used in the ROC analysis to assess the credibility of the test. The mathematical formulas are summarized in Table 24.2.

In the ROC curves analysis result (Figure 24.16), both sensitivity and specificity will be displayed for all criterions. This will allow the user to choose the optimum criterion, which ought to have a high value for both sensitivity and specificity. The value of sensitivity is inversely proportional to that of specificity. This can be easily illustrated by the threshold value γ. A low γ will ensure that those with the disease will be detected. But this will also cause those without the disease to be classified as diseased. On the other hand, a high γ will allow to correctly categorize the healthy group but will miss out on the diseased group.

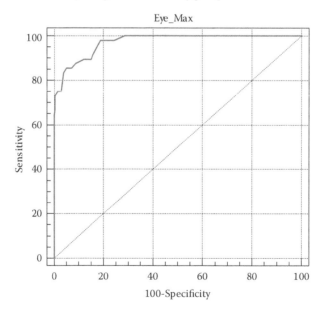

FIGURE 24.16   A typical ROC curve [34].

---

[*] Sensitivity—The probability that test is positive in the unhealthy population.

[†] Specificity—The probability that the test is negative in the healthy population.

[‡] PPV—Given a positive forecast, the probability that it is correct.

[§] NPV—Given a negative forecast, the probability that it is correct.

An example of the ROC curves is presented in Figure 24.16. The vertical axis shows the sensitivity while the horizontal axis shows the (100—specificity). Once again, this reinforces the fact that there is a trade-off between sensitivity and specificity.

The area under the ROC curve is an important information obtained in the analysis. The value lies between 0.5 and 1. A value of 0.5 implies that the test cannot discriminate the unhealthy from the healthy group, whereas a value of 1 implies that the test can distinguish the two groups perfectly.

24.3 Designed-Integrated Approach

24.3.1 Case 1: Advanced-Integrated Technique (Parabolic Regression + Artificial Neural Network Radial Basis Function Network + Receiver Operating Characteristics) for Febrile and Nonfebrile Cases

The proposed approach is a multipronged approach that comprises of PR, RBFN, and ROC analysis. It is a novel, integrative, and powerful technique that can be used to analyze complicated and large numerical data.

24.3.1.1 Step 1: Parabolic Regression

PR reflects the correlation between the variables and the actual health status (febrile or nonfebrile) of the subject, which is decided by the means of a thermometer placed in the ear. The output is either 1 or 0, corresponding to febrile and nonfebrile cases, respectively. The two input variables with the best correlation are chosen. The rational behind using PR over LR is it offers a more accurate and realistic approach in providing the correlation coefficient (LR results are also tabulated here for comparison purposes). Table 24.3 summarizes the temperature data from the thermograms [27].

24.3.1.2 Step 2: ANN Radial Basis Function Network

On the basis of the various inputs fed into the network, RBFN is trained to produce the desired outcome, which are either positive (1) for febrile cases, or negative (0) for nonfebrile cases. When this is done, the RBFN algorithm will possess the ability to predict the outcome when there are new input variables.

24.3.1.3 Step 3: ROC Analysis

Next, ROC is used to evaluate the accuracy, sensitivity, and specificity of the outcome of RBFN test files (i.e., Is RBFN well built or not?).

Table 24.4 and Figure 24.17 reveal the software needed for the entire process, including the steps prior to the advanced integrated technique.

24.3.2 Case 2: Conventional Bio-Statistical Technique (LR + ROC) for Febrile and Nonfebrile Cases

The conventional bio-statistical technique comprises of LR and ROC to analyze the data collected from the thermal imager [22]. Similar to the previous approach, regression is used to select the variable and the strongest correlation with the outcome (health status of the patient). Subsequently, ROC is applied to

TABLE 24.3 Temperature Data of Forehead and Near Eye Regions

Forehead Region	Near Eye Region
Minimum temperature	Minimum temperature
Maximum temperature	Maximum temperature
Average temperature	Average temperature

TABLE 24.4   Software Used for Advanced Integrated Technique for Febrile and Nonfebrile Thermograms

Purpose	Software
Views thermograms from thermal imager and extracts temperature data	Image J
Normalizes raw temperature data and performs statistical analysis (e.g., mean, median, and standard deviation)	MS Excel Statistical Toolbox
Determines the correlation of each variable with the output (health status)	MedCal
Training and testing of data, building an algorithm for the data	NeuralWorks Pro II
To evaluate the effectiveness of the computed method	MedCal

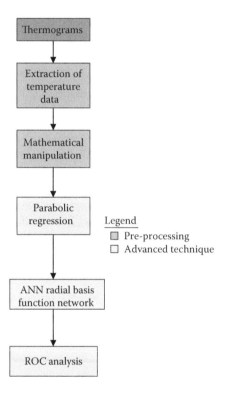

FIGURE 24.17   Flow chart for advanced integrated technique.

obtain the optimal preset temperature based on the chosen variables. The temperature is dependent on the values of sensitivity and specificity from the ROC analysis results. The chosen threshold temperature must have high values of sensitivity and specificity. Also, the area under the ROC curve is to be high to determine whether or not a subject should be considered as febrile (diseased) or nonfebrile (healthy). Table 24.5 and Figure 24.18 show the procedure of the conventional technique used during the SARS-2003 outbreak [22–24,27–28].

24.4 Results and Discussion

24.4.1 Case 1: Advanced Integrated Technique (PR + ANNRBFN + ROC) for Febrile and Nonfebrile Cases

Table 24.6 tabulates the results for PR and LR results are also included for comparison purposes. The PR coefficient of determination is always higher than that of LR. Thus, using simple LR with the present nonlinear data set are not always the best possible ways to "fit the data," as it is frequently used.

TABLE 24.5 Software Used for Conventional Approach

Purpose	Software
Views thermograms from thermal imager and extracts temperature data	Image J
Normalizes raw temperature data	MS Excel statistical toolbox
Performs statistical analysis (e.g., mean, median, and standard deviation)	
Determines correlation of each variable on the output (health status)	MedCal
Determines the optimal preset temperature and evaluates the effectiveness on the basis of sensitivity and specificity	MedCal

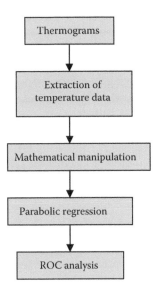

FIGURE 24.18 Flow chart for conventional approach.

When a parabolic curve is used to generate the correlation coefficient, the *maximum temperature of the near eye* and *the maximum temperature of the forehead regions* still remain to be the best correlated spots on the frontal face with regard to the core temperature. Hence, temperature data from these two regions are selected as input variables for the training of ANN. Although, using the parabolic function gives the better correlation coefficient, the outcome of the NN remains the same even if LR is used.

TABLE 24.6 Summarized Results for Step 1 for Advanced Integrated Technique

	Coefficient of Determination	
Independent Variable	Linear	Parabolic
Maximum temperature at eye range	0.5507	0.6315
Minimum temperature at eye range	0.0672	0.1114
Standard deviation at eye range	0.0303	0.0588
Total average at eye range	0.4489	0.5721
Maximum temperature at forehead range	0.4973	0.6362
Minimum temperature at forehead range	0.1169	0.1798
Standard deviation at forehead range	0.0053	0.0053
Total average at forehead range	0.3759	0.5379

TABLE 24.7   Selected Results for RBFN SLP of Advanced Integrated Technique

Learn Rule	Transfer Rule	Score (%)
Delta rule	Sigmoid	96
Delta rule	DNNA	96
Norm-cum-delta	Linear	96
Norm-cum-delta	TanH	96
Norm-cum-delta	Sigmoid	96
Norm-cum-delta	Sine	96
Ext DBD	Linear	96

Table 24.7 gives the selected results for RBFN SLP with a various combination of learn rule and transfer rate. The results for the selected combination of learn rule and transfer rule (with various *options* tested) for RBFN SLP are included in Table 24.8.

Various combinations of learn rule, transfer rule, and *options* were tested. With the inclusion of *options* (e.g., Connect Prior, Connect Bias), more NNs with an accuracy of 96% can be generated. The RBFN is credible and has the ability to differentiate the febrile from the nonfebrile cases to a very large extent. There are always four input data which the model always predicts wrongly and accounts for the 4% error. This is due to the inconsistencies between the patient's facial temperature (deduced from the thermograms) and his core temperature. For example, the Max Temp in the near eye region and Max Temp in the forehead region are very high and it indicates the fact that the person is having fever. Hence, ANN predicts that the person is having fever. However, the core temperature taken by the thermometer suggests that the person is not having fever. Thus, ANN's prediction is wrong. This is certainly not the ANN's fault

TABLE 24.8   Selected Results for RBFN SLP (with *Options* for Advanced Integrated Technique)

Learn Rule	Transfer Rule	Option	Score (%)
Delta rule	Sigmoid	Connect Prior	96
Delta rule	Sigmoid	Connect Bias	96
Delta rule	Sigmoid	MinMax Table	96
Delta rule	DNNA	Connect Prior	96
Delta rule	DNNA	Linear O/P	96
Delta rule	DNNA	Softmax O/P	96
Delta rule	DNNA	Connect Bias	96
Delta rule	DNNA	MinMax Table	96
Norm-cum-delta	TanH	Connect Prior	96
Norm-cum-delta	TanH	Linear O/P	96
Norm-cum-delta	TanH	Connect Bias	96
Norm-cum-delta	TanH	MinMax Table	96
Norm-cum-delta	Sigmoid	Connect Prior	96
Norm-cum-delta	Sigmoid	Linear O/P	96
Norm-cum-delta	Sigmoid	Connect Bias	96
Norm-cum-delta	Sigmoid	MinMax Table	96
Norm-cum-delta	Sine	Connect Prior	96
Norm-cum-delta	Sine	Linear O/P	96
Norm-cum-delta	Sine	Connect Bias	96
Norm-cum-delta	Sine	MinMax Table	96
Ext DBD	Linear	Connect Prior	96
Ext DBD	Linear	Linear O/P	96
Ext DBD	Linear	Connect Bias	96
Ext DBD	Linear	MinMax Table	96

TABLE 24.9   ROC Results for RBFN SLP with Various Combination of Learn Rule and Transfer Rule

No.	Learn Rule	Transfer Rule	Score (%)	Area Under Curve	Sensitivity	Specificity
1	Delta rule	Sigmoid	96	0.972	100	84.1
2	Delta rule	DNNA	96	0.971	91.7	94.3
3	Norm-cum-delta	Sigmoid	96	0.975	100	88.6

TABLE 24.10   ROC Results for RBFN SLP with Selected Combination of Learn Rule and Transfer Rule (with Various *Options* Tested)

No.	Learn Rule	Transfer Rule	Option	Score (%)	Area Under Curve	Sensitivity	Specificity
1	Delta rule	Sigmoid	Connect prior	96	0.972	100	84.1
2	Delta rule	Sigmoid	MinMax table	96	0.974	91.7	94.3
3	Delta rule	DNNA	Connect prior	96	0.971	91.7	94.3
4	Delta rule	DNNA	Linear O/P	96	0.971	91.7	94.3
5	Delta rule	DNNA	Softmax O/P	96	0.971	91.7	94.3
6	Delta rule	DNNA	Connect bias	96	0.970	91.7	94.3
7	Norm-cum-delta	TanH	Connect bias	96	0.973	91.7	94.3
8	Norm-cum-delta	TanH	MinMax table	96	0.978	100	87.5
9	Norm-cum-delta	Sigmoid	Connect prior	96	0.975	100	88.6
10	Norm-cum-delta	Sigmoid	Linear O/P	96	0.975	100	88.6
11	Norm-cum-delta	Sigmoid	Connect bias	96	0.970	100	85.2
12	Norm-cum-delta	Sigmoid	MinMax table	96	0.981	100	94.3
13	Norm-cum-delta	Sine	Connect bias	96	0.975	91.7	94.3
14	Norm-cum-delta	Sine	MinMax table	96	0.975	91.7	94.3
15	Ext DBD	Linear	Connect bias	96	0.980	100	93.2
16	Ext DBD	Linear	MinMax table	96	0.984	100	94.3

because in these cases, the person's facial temperature has a poor correlation with his core temperature. Without these exceptional cases, ANN should achieve a higher accuracy rate. Tables 24.9 and 24.10 (with various *options* tested) summarize the selected results (of ROC area >0.970) for ROC analysis.

The ROC area under the curve for all the RBFNs shown in Tables 24.9 and 24.10 is larger than 0.97. These RBFNs also have high sensitivities (>90%) and high specificities (>80%). This suggests that the RBFN is well-built and the overall diagnostic performance is reliable, and can be used for mass screening of febrile subjects. The best performing RBFN is a single-layered perceptron with Ext DBD as the learn rule, linear function as the transfer rule, and MinMax table as the selected option. The area under the ROC curve is 0.984 and its sensitivity and specificity are 100% and 94.3%, respectively.

24.4.2 Case 2: Conventional Bio-Statistical Technique (LR + ROC) for Febrile and Nonfebrile Cases

The LR analysis shows that the particular area on the skin surface that will produce the most consistent results with regard to the core temperature (taken using ear scanner) is the maximum temperature in the eye region with a coefficient correlation of 0.5507 (Table 24.11). The least correlated independent variable with the core temperature is the standard deviation for the forehead and eye regions (0.0053 and 0.0303), and the minimum temperature in the eye region (0.0672). Figure 24.19 shows the ROC plot in which the false positives are weighted similarly as false negatives. Table 24.12 summarizes the sensitivity and specificity for various preset scanner temperature.

TABLE 24.11 Summarized Results of LR for Conventional Approach

Independent Variable	Coefficient of Determination (linear)
Maximum temperature at eye range	0.5507
Minimum temperature at eye range	0.0672
Standard deviation at eye range	0.0303
Total average at eye range	0.4489
Maximum temperature at forehead range	0.4973
Minimum temperature at forehead range	0.1169
Standard deviation at forehead range	0.0053
Total average at forehead range	0.3759

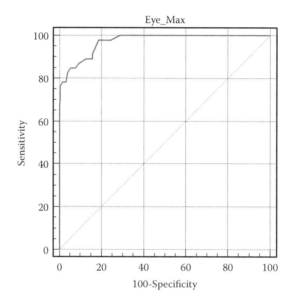

FIGURE 24.19 ROC curve for conventional technique (area of ROC = 0.972) [22–24].

ROC curve analysis for the maximum eye region shows that the optimal preset (cut-off) temperature is 36.3°C. If a subject's maximum temperature in the eye region exceeds 36.3°C, it implies that one is having fever. At this cut-off temperature, the sensitivity and specificity are 85.4% and 95.0%, respectively, with an accuracy rate of 93%.

24.4.3 Comparison between the Advanced Integrated Technique and Conventional Bio-Statistical Technique

The advanced technique achieves 96% of the accuracy rate whereas the conventional bio-statistical technique has 93% accuracy rate. Hence, there is a greater promise in using the advanced integrated technique for the thermogram analysis. In the conventional technique, after the "Max Temp in the Eye region" is found to have the strongest correlation with the output, the rest of the variables (e.g., Max Temp in the forehead region, Min Temp in the forehead region, Min Temp in the eye region, mode, median, etc.) are discarded. This implies that they will no longer be used in ANN. But in the advanced technique, it is possible and a common practice to use more than one input variable (besides Max Temp in the eye region). For this study Max Temp in the forehead region is used as the second input variable for ANN training and testing. These are the two variables with the strongest correlation with the output.

TABLE 24.12 Selected Results of ROC Analysis for Conventional Bio-Statistical Technique [22–24]

Criterion	Sensitivity	Specificity
36.1	85.4	92.7
36.2	85.4	93.9
36.3[a]	85.4	95.0
36.4	83.3	96.2
36.5	75	96.9

[a] Selected criterion for preset temperature: 36.3°C.

Further studies could be carried out to check if third, fourth, fifth, … input variables would further improve the effectiveness of the advanced technique.

24.5 Conclusion and Future Work

The chapter focuses on the numerical analyses of the data, to the detriment of performing a scientifically rigorous test of the hypotheses that febrile patients can be detected using simple thermograms. Through the use of ANN and bio-statistical methods, progress is made in thermography application with regard to achieving a higher level of consistency. This is made possible with the introduction of the novel advanced integrated technique in thermogram analysis.

The advanced technique has a high level of accuracy rate in prediction on the basis of the temperature data extracted from the thermograms. It improves the correlation and may prove more efficacious for mass fever screening. For elevated body temperature thermography, the advanced technique enables us to have in place a reliable system for the mass screening of fever cases. The proposed approach (PR + ANN + ROC) has surpassed the conventional bio-statistical approach (LR + ROC), which was used for analytical purposes during the SARS-2003 pandemic. In other words, the advanced technique enables us to possibly differentiate the febrile from nonfebrile cases in a short time . This is important with regard to the SARS outbreak in 2003 and the potentially lethal Avian flu or malaria. Recently Indonesia has reported the world's first laboratory-confirmed cluster of human-to-human transmission of bird flu, although scientists are as yet unsure of the significance for the multiple mutations in the H5N1 virus. In the event of such a virus being pandemic, we will be better prepared to set up thermography units in public places to mass screen populations for epidemic outbreaks.

In brief, thermography application is like an unpolished gemstone, waiting for us to unlash its full potential. The future development of an integrative fever screening system may incorporate the effectiveness of Laser Doppler Flowmeter (for heart rate), microwave radar (for respiration rate), and thermography (for skin temperature) to eliminate setbacks or noises that are prevalent in each individual device. All the medical images/data obtainable from these screening devices can then be supplied into the self-developed software that is built on the basis of ANN. With sufficient data collected from the affected patients, the software can be trained to carry out the automatic feature definition and image classification objectively.

Acknowledgments

To pay a tribute to the late Dr. Nicholas Diakides, this chapter is a revised version of Chapter 16, *Biomedical Engineering Handbook, Infrared Imaging Spin-off Edition*, ISBN-0-8493-9027-3, CRC Press.

References

1. Ring, E.F.J., Progress in the measurement of human body temperature, *Engineering in Medicine and Biology Magazine*, 17, 1998, 19–24.

2. Merla, A., Donato, L.D., Luzio, S.D., Farina, G., Pisarri, S., Proietti, M., Salsano, F., Romani, G.L., Infrared functional imaging applied to Raynaud's phenomenon, *IEEE Engineering in Medicine and Biology Magazine*, 21(6), 2002, 73–79.
3. Ng, E.Y.-K., Chen, Y., Segmentation of breast thermogram: Improved boundary detection with modified snake algorithm, *Journal of Mechanics in Medicine and Biology*, 6(2), 2006, 123–136.
4. Ng, E.Y.-K., A review of thermography as promising non-invasive detection modality for breast tumour, *International Journal of Thermal Sciences*, 48(5), 2009, 849–855.
5. Tan, J.H., Ng, E.Y.-K., Acharya R.U., Chee, C., Infrared thermography on ocular surface temperature: A review, *Infrared Physics and Technology*, 52(4), 2009, 97–108.
6. Pang, C., Gu, D.L., Some problems about detecting the suspected cases of SARS according to the local skin temperatures on face, *Space Medical and Medical Engineering*, 16(3), 2003, 231–234.
7. Chan, L.S., Cheung, G.T.Y., Lauder, I.J., Kumana, C.R., Screening for fever by remote-sensing infrared thermographic camera, *Journal of Travel Medicine*, 11, 2004, 273–278.
8. Peacock, G.R., Human radiation thermometry and screening for elevated body temperature in humans, *Thermosense XXVI, Proceedings of SPIE*, Vol. 5405, 2004, pp. 48–53, Orlando, USA.
9. Wu, M., Stop outbreak of SARS with infrared cameras, *Thermosense XXVI, Proceedings of SPIE*, Vol. 5405, 2004, pp. 98–105, Orlando, USA.
10. Samaan, G., Patel, M., Spencer, J., Roberts, L., Border screening for SARS in Australia: What has been learnt? *The Medical Journal of Australia*, 180(5), 2004, 220–223.
11. Hay, A.D., Peters, T.J., Wilson, A., Fahey, T., The use of infrared thermometry for the detection of fever, *The British Journal of General Practice*, 54, 2004, 448–450.
12. Liu, C.C., Chang, R.E., Chang, W.C., Limitations of forehead infrared body temperature detection for fever screening for SARS, *Infection Control and Hospital Epidemiology*, 25(12), 2004, 1109–1111.
13. Itoi, M., Yanai, Y., Abe, S., Is the using of thermography useful for the active surveillance of infectious disease at quarantine spot? *Journal of Japanese Quarantine Medicine*, 6, 2004, 125–132.
14. Health Canada. Thermal image scanners to detect fever in airline passengers, Vancouver and Toronto, 2003. *Canada Communicable Disease Report*, 30, (19), 2004, 165–167.
15. Ng, D.K., Chan, C.H., Lee, R.S., Leung, L.C., Non-contact infrared thermometry temperature measurement for screening fever in children, *Annals of Tropical Paediatrics*, 25(4), 2005, 267–275.
16. Shu, P.Y., Chien, L.J., Chang, S.F., Su, C.L., Kuo, Y.C., Liao, T.L., Fever screening at airports and imported dengue, *Emerging Infectious Diseases*, 11(3), 2005, 460–462.
17. Chiu, W.T., Lin, P.W., Chiou, H.Y., Lee, W.S., Lee, C.N., Yang, Y.Y., Lee, H.M., Hsieh, M.S., Hu, C.J., Ho, Y.S., Deng, W.P., Hsu, C.Y., Infrared thermography to mass-screen suspected SARS patients with fever, *Asia-Pacific Journal of Public Health*, 17(1), 2005, 26–28.
18. St John, R.K., King, A., de Jong, D., Bodie-Collins, M., Squires, S.G., Tam, T.W., Border screening for SARS, *Emerging Infectious Diseases*, 11(1), 2005, 6–10.
19. Wong, J.J., Wong, C.Y., Non-contact infrared thermal imagers for mass fever screening: State of the art or myth? *Hong Kong Medical Journal*, 12(3), 2006, 242–244.
20. Matsui, T., Suzuki, S., Ujikawa, K., Usui, T., Gotoh, S., Sugamata, M., Abe, S., The development of a non-contact screening system for rapid medical inspection at a quarantine depot using a laser Doppler blood-flow meter, microwave radar and infrared thermography, *Journal of Medical Engineering and Technology*, 33(6), 2009, 481–487.
21. Bitar, D., Goubar, A., Desenclos, J.C., International travels and fever screening during epidemics: A literature review on the effectiveness and potential use of non-contact infrared thermometers, *Eurosurveillance*, 14(6), 2009, 1–5.
22. Ng, E.Y.-K., Kaw, G.J.L, Chang, W.M., Analysis of IR thermal imager for mass blind fever screening, *Microvascular Research*, 68(2), 2004, 104–109.
23. Ng, E.Y.-K., Chong, C., Kaw, G.J.L, Classification of human facial and aural temperature using neural networks and IR fever scanner: A responsible second look, *Journal of Mechanics in Medicine and Biology*, 5(1), 2005, 165–190.

24. Ng, E.Y.-K., Is thermal scanner losing its bite in mass screening of fever due to SARS?, *Medical Physics*, 32(1), 2005, 93–97.
25. Ng, E.Y.-K., Chong, C., ANN based mapping of febrile subjects in mass thermogram screening: Facts and myths, *International Journal of Medical Engineering and Technology*, 30(5), 2006, 330–337.
26. Ng, E.Y.-K., Rajendra Acharya, U., A review of remote-sensing infrared thermography for indoor mass blind fever screening in containing an epidemic, *IEEE Engineering in Medicine and Biology*, 28(1), 2009, 76–83.
27. Ng, E.Y.-K., Kaw, G.J.L., IR scanners as fever monitoring devices: Physics, physiology and clinical accuracy, in *Biomedical Engineering Handbook*, CRC Press, Boca Raton, FL, N. Diakides, ed., pp. 24-1 to 24-20. (Mar. 2006).
28. Ng, E.Y.-K., Wiryani, M., Wong, B.S., Study of facial skin and aural temperature using IR with and w/o TRS, *IEEE Engineering in Medicine and Biology Magazine*, 25(3), 2006, 68–74.
29. ISO TC121/SC3-IEC SC62D:2008 Particular requirements for the basic safety and essential performance of screening thermographs for human febrile temperature screening.
30. ISO/TR 13154:2009 ISO/TR 80600: Medical electrical equipment:Deployment, implementation and operational guidelines for identifying febrile humans using a screening thermograph.
31. FLIR Systems: http://www.flir.com (accessed 17th July 2010).
32. Togawa, T., Body temperature measurement, *Clinical Physics Physiology Measure*, 6, 1985, 83–108.
33. Kaderavek, F., Clinical thermometry (Czech), *Casopis Lekaru Ceskych*, 111, 1972, 1135–1138.
34. Hopgood, A. (2000). Intelligent systems for engineers and scientists. Library of Congress Cataloging-in-Publication Data.
35. Battelle: http://www.battelle.org/pipetechnology/ (assessed 10 June 2009).
36. Receiver operating characteristics (ROC): http://www.medcalc.be/ (accessed 17 July 2010).

25

Thermal Imaging in Diseases of the Skeletal and Neuromuscular Systems

Francis J. Ring
University of Glamorgan

Kurt Ammer
Ludwig Boltzmann Research Institute for Physical Diagnostics

University of Glamorgan

25.1 Introduction

Clinical medicine has made considerable advances over the last century. The introduction of imaging modalities has widened the ability of physicians to locate and understand the extent and activity of a disease. Conventional radiography has dramatically improved, beyond the mere demonstration of bone and calcified tissue. Computed tomography ultrasound, positron emission tomography, and magnetic resonance imaging are now available for medical diagnostics.

Infrared imaging has also added to this range of imaging procedures. It is often misunderstood, or not been used, due to lack of knowledge of thermal physiology and the relationship between temperature and disease.

In rheumatology, disease assessment remains complex. There are a number of indices used, which testify to the absence of any single parameter for routine investigation. Most indices used are subjective. Objective assessments are of special value, but may be more limited due to their invasive nature. Infrared imaging is noninvasive, and with modern technology has proved to be reliable and useful in rheumatology.

From early times, physicians have used the cardinal signs of inflammation, that is, pain, swelling, heat, redness, and loss of function. When a joint is acutely inflamed, the increase in heat can be readily detected by touch. However, subtle changes in joint surface temperature occur and increase and decrease in temperature can have a direct expression of reduction or exacerbation of inflammation.

25.2 Inflammation

Inflammation is a complex phenomenon, which may be triggered by various forms of tissue injury. A series of cellular and chemical changes take place that are initially destructive to the surrounding tissue. Under normal circumstances, the process terminates when healing takes place, and scar tissue may then be formed.

A classical series of events take place in the affected tissues. First, a brief arteriolar constriction occurs, followed by a prolonged dilatation of arterioles, capillaries, and venules. The initial increased blood flow caused by the blood vessel dilation becomes sluggish and leucocytes gather at the vascular endothelium. Increased permeability to plasma proteins causes exudates to form, which is slowly absorbed by the lymphatic system. Fibrinogen, left from the reabsorption, partly polymerizes to fibrin. The increased permeability in inflammation is attributed to the action of a number of mediators, including histamines, kinins, and prostaglandins. The final process is manifest as swelling caused by the exudates, redness, and increased heat in the affected area resulting from the vasodilation, and increased blood flow. Loss of function and pain accompany these visible signs.

Increase in temperature and local vascularity can be demonstrated by some radionuclide procedures. In most cases, the isotope is administered intravenously and the resulting uptake is imaged or counted with a gamma camera. Superficial increases in blood flow can also be shown by laser Doppler imaging, although the response time may be slow. Thermal imaging, based on infrared emission from the skin is both fast and noninvasive.

This means that it is a technique that is suitable for repeated assessment, and especially useful in clinical trials of treatment whether by drugs, physical therapy, or surgery.

Intra-articular injection, particularly to administer corticosteroids came into use in the middle of the last century. Horvath and Hollander in 1949 [1] used intra-articular thermocouples to monitor the reduction in joint inflammation and synovitis following treatment. This method of assessment while useful to provide objective evidence of anti-inflammatory treatment was not universally used for obvious ethical reasons.

The availability of noncontact temperature measurement for infrared radiometry was a logical progression. Studies in a number of centers were made throughout the 1960s to establish the best analogs of corticosteroids and their effective dose. Work by Collins and Cosh in 1970 [2] and Ring and Collins in 1970 [3] showed that the surface temperature of an arthritic joint was related to the intra-articular joint, and to other biochemical markers of inflammation obtained from the exudates. In a series of experiments with different analogs of prednisolone (all corticosteroids), the temperature measured by thermal imaging in groups of patients can be used to determine the duration and degree of reduction in inflammation [4,5].

At this time, a thermal challenge test for inflamed knees was being used in Bath, UK, based on the application of a standard ice pack to the joint. This form of treatment is still used, and results in a marked decrease of joint temperature, although the effect may be transient.

The speed of temperature recovery after an ice pack of 1 kg of crushed ice to the knee for 10 min was shown to be directly related to the synovial blood flow and inflammatory state of the joint. The mean temperature of the anterior surface of the knee joint could be measured either by infrared radiometry or by quantitative thermal imaging [6].

A number of new nonsteroid anti-inflammatory agents were introduced into rheumatology in the 1970s and 1980s. Infrared imaging was shown to be a powerful tool for the clinical testing of these drugs, using temperature changes in the affected joints as an objective marker. The technique had been successfully used on animal models of inflammation, and effectively showed that optimal dose–response curves could be obtained from temperature changes at the experimental animal joints. The process with human patients suffering from acute rheumatoid arthritis was adapted to include a washout period for previous medication. This should be capable of relieving pain but no direct anti-inflammatory action per se. The compound used by all the pharmaceutical companies was paracetamol. It was shown by

Bacon et al. [7] that small joints such as fingers and metacarpal joints increased in temperature quite rapidly while paracetamol treatment was given, even if pain was still suppressed. Larger joints, such as knees and ankles required more than 1 week of active anti-inflammatory treatment to register the same effect. Nevertheless, the commonly accepted protocol was to switch to the new test anti-inflammatory treatment after 1 week of washout with the analgesic therapy. In every case, if the dose was ineffective, the joint temperature was not reduced. At an effective dose, a fall in temperature was observed, first in the small joints, then later in the larger joints. Statistical studies were able to show an objective decrease in joint temperature by infrared imaging as a result of a new and successful treatment. Not all the new compounds found their way into routine medicine; a few were withdrawn as a result of undesirable side effects. The model of infrared imaging to measure the effects of a new treatment for arthritis was accepted by all the pharmaceutical companies involved and the results were published in the standard peer-reviewed medical journals. More recently, attention has been focused on a range of new biological agents for reducing inflammation. These are also being tested in trials that incorporate quantitative thermal imaging.

To facilitate the use and understanding of joint temperature changes, Ring and Collins [3] and Collins et al. [8] devised a system for quantitation. This was based on the distribution of isotherms from a standard region of interest. The thermal index was calculated as the mean temperature difference from a reference temperature. The latter was determined from a large study of 600 normal subjects where the average temperature threshold for ankles, knees, hands, elbows, and shoulder were calculated. Many of the clinical trials involved the monitoring of hands, elbows, knees, and ankle joints. Normal index figure obtained from controls under the conditions described was from 1 to 2.5 on this scale. In inflammatory arthritis, this figure was increased to 4–5, while in osteoarthritic joints, the increase in temperature was usually less, 3–4. In gout and infection, higher values around 6–7 on this scale were recorded.

However, to determine normal values of finger joints is a very difficult task. This difficulty arises partly from the fact that cold fingers are not necessarily a pathological finding. Tender joints showed higher temperatures than nontender joints, but a wide overlap of readings from nonsymptomatic and symptomatic joints was observed [9]. Evaluation of finger temperatures from the reference database of normal thermograms [10] of the human body might ultimately solve the problem of being able to establish a normal range for finger joint temperatures in the near future.

25.3 Paget's Disease of Bone

The early descriptions of Osteitis Deformans by Czerny [11] and Paget [12] refer to "chronic inflammation of bone." An increased skin temperature over an active site of this disease has been a frequent observation and that the increase may be around 4°C. Others have shown an increase in peripheral blood flow in almost all areas examined. Increased periosteal vascularity has been found during the active stages of the disease. The vascular bed is thought to act as an arterio-venous shunt, which may lead to high output cardiac failure. A number of studies, initially to monitor the effects of calcitonin, and later bisphosphonate therapy, have been made at Bath. As with the clinical trials previously mentioned, a rigorous technique is required to obtain meaningful scientific data. It was shown that the fall in temperature during calcitonin treatment was also indicated more slowly, by a fall in alkaline phosphatase, the common biochemical marker. Relapse and the need for retreatment was clearly indicated by thermal imaging. Changes in the thermal index often preceded the onset of pain and other symptoms by 2–3 weeks. It was also shown that the level of increased temperature over the bone was related to the degree of bone pain. Those patients who had maximal temperatures recorded at the affected bone experienced severe bone pain. Moderate pain was found in those with raised temperature, and no pain in those patients with normal temperatures. The most dramatic temperature changes were observed at the tibia, where the bone is very close to the skin surface. In a mathematical model, Ring and Davies [13] showed that the increased temperature measured over the tibia was primarily derived from osseous blood flow and not from metabolic heat. This disease is often categorized as a metabolic bone disease.

25.4 Soft Tissue Rheumatism

25.4.1 Muscle Spasm and Injury

Muscle work is the most important source for metabolic heat. Therefore, contracting muscles contribute to the temperature distribution at the body's surface of athletes [14,15]. Pathological conditions such as muscle spasms or myofascial trigger points may become visible at regions of increased temperature [16]. An anatomic study from Israel proposes in the case of the levator scapulae muscle that the frequently seen hot spot on thermograms of the tender tendon insertion on the medial angle of the scapula might be caused by an inflamed bursae and not by a taut band of muscle fibers [17].

Acute muscle injuries may also be recognized by areas of increased temperature [18] due to inflammation in the early state of trauma. However, long-lasting injuries and also scars appear at hypothermic areas caused by reduced muscle contraction and therefore reduced heat production. Similar areas of decreased temperature have been found adjacent to peripheral joints with reduced range of motion due to inflammation or pain [19]. Reduced skin temperatures have been related to osteoarthritis of the hip [20] or to frozen shoulders [21,22]. The impact of muscle weakness on hypothermia in patients suffering from paresis was discussed elsewhere [23].

25.4.2 Sprains and Strains

Ligamentous injuries of the ankle [24] and meniscal tears of the knee [25] can be diagnosed by infrared thermal imaging. Stress fractures of bone may become visible in thermal images prior to typical changes in x-rays [26]. Thermography provides the same diagnostic prediction as bone scans in this condition.

25.4.3 Enthesopathies

Muscle overuse or repetitive strain may lead to painful tendon insertions or where tendons are shielded by tendon sheaths or adjacent to bursae, to painful swellings. Tendovaginitis in the hand was successfully diagnosed by skin temperature measurement [27]. The acute bursitis at the tip of the elbow can be detected through an intensive hot spot adjacent to the olecranon [28]. Figure 25.1 shows an acute tendonitis of the Achilles tendon in a patient suffering from inflammatory spondylarthropathy.

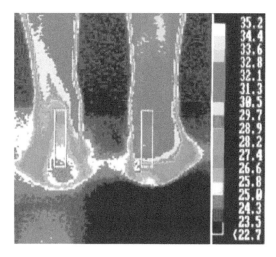

FIGURE 25.1 Acute tendonitis of the right Achilles tendon in a patient suffering from inflammatory spondylarthropathy.

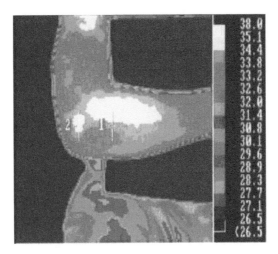

FIGURE 25.2 Tennis elbow with a typical hot spot in the region of tendon insertion.

25.4.3.1 Tennis Elbow

Painful muscle insertion of the extensor muscles at the elbow is associated with hot areas on a thermogram [29]. Thermal imaging can detect persistent tendon insertion problems of the elbow region in a similar way as isotope bone scanning [30]. Hot spots at the elbow have also been described as having a high association with a low threshold for pain on pressure [31]. Such hot areas have been successfully used as outcome measure for monitoring treatment [32,33]. In patients suffering from fibromyalgia, bilateral hot spots at the elbows is a common finding [34]. Figure 25.2 is the image of a patient suffering from tennis elbow with a typical hot spot in the region of tendon insertion.

25.4.3.2 Golfer Elbow

Pain due to altered tendon insertions of flexor muscles on the medial side of the elbow is usually named Golfer elbow. Although nearly identical in pathogenesis as the tennis elbow, temperature symptoms in this condition were rarely found [35].

25.4.3.3 Periarthropathia of Shoulder

The term "periarthropathia" includes a number of combined alterations of the periarticular tissue of the humero-scapular joint. The most frequent problems are pathologies at the insertion of the supraspinous and infraspinous muscles, often combined with impingement symptoms in the subacromial space. Long-lasting insertion alteration can lead to typical changes seen on radiographs or ultrasound images, but unfortunately there are no typical temperature changes caused by the disease [22,36]. However, persistent loss in range of motion will result in hypothermia of the shoulder region [21,22,36,37]. Figure 25.3 gives an example of an area of decreased temperature over the left shoulder region in patient with restricted range of motion.

25.4.4 Fibromyalgia

The terms "tender points" (important for the diagnosis of fibromyalgia) and "trigger points" (main feature of the myofascial pain syndrome) must not be confused. Tender points and trigger points may give a similar image on the thermogram. If this is true, patients suffering from fibromyalgia may present with a high number of hot spots in typical regions of the body. A study from Italy could not find different patterns of heat distribution in patients suffering from fibromyalgia and patients with osteoarthritis of the spine [38]. However, they reported a correspondence of nonspecific hyperthermic patterns with painful

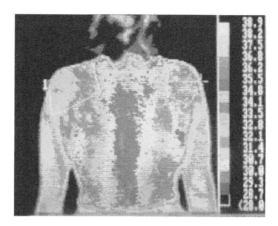

FIGURE 25.3 Decreased temperature in patient with a frozen shoulder on the left-hand side.

muscle areas in both groups of patients. Our thermographic investigations in fibromyalgia revealed a diagnostic accuracy of 60% of hot spots for tender points [34]. The number of hot spot was greatest in fibromyalgia patients and the smallest in healthy subjects. More than 7 hot spots seem to be predictive for tenderness of more than 11 out of 18 specific sites [39]. Based on the count of hot spots, 74.2% of 252 subjects (161 fibromyalgia, 71 with widespread pain but less than 11 tender sites out of 18, and 20 healthy controls) have been correctly diagnosed. However, the intra- and interobserver reproducibility of hot spot count is rather poor [40]. Software-assisted identification of hot or cold spots based on the angular distribution around a thermal irregularity [41] might overcome that problem of poor repeatability.

25.5 Peripheral Nerves

25.5.1 Nerve Entrapment

Nerve entrapment syndromes are compression neuropathies at specific sites in the human body. These sites are narrow anatomic passages where nerves are situated. The nerves are particularly prone to extrinsic or intrinsic pressure. This can result in paresthesias such as tingling or numb feelings, pain, and ultimately in muscular weakness and atrophy.

Uematsu [42] has shown in patients with partial and full lesion of peripheral nerves that both conditions can be differentiated by their temperature reaction to the injury. The innervated area of partially lesioned nerve appears hypothermic, caused by the activation of sympathetic nerve fibers. Fully dissected nerves result in a total loss of sympathetic vascular control and therefore in hyperthermic skin areas.

The spinal nerves, the brachial nerve plexus, and the median nerve at the carpal tunnel are the most frequently affected nerves with compression neuropathy.

25.5.1.1 Radiculopathy

A slipped nucleus of an intervertebral disk may compress the adjacent spinal nerve or better the sensory and motor fibers of the dorsal root of the spinal nerve. This may or must not result in symptoms of compression neuropathy in the body area innervated by these fibers.

The diagnostic value of infrared thermal imaging in radiculopathies is still under debate. A review by Hoffman et al. [43] from 1991 concluded that thermal imaging should be used only for research and not in clinical routine. This statement was based on the evaluation of 28 papers selected from a total of 81 references.

The study of McCulloch et al. [44], planned and conducted at a high level of methodology, found thermography not valid. However, the applied method of recording and interpretation of thermal images was not sufficient. The chosen room temperature of 20–22°C might have been too low for the identification of hypothermic areas. Evaluation of thermal images was based on the criterion that at least 25% of a dermatome present with hypothermia of 1°C compared to the contralateral side. This way of interpretation might be feasible for contact thermography, but does not meet the requirements of quantitative infrared imaging.

The paper of Takahashi et al. [45] showed that the temperature deficit identified by infrared imaging is an additional sign in patients with radiculopathy. Hypothermic areas did not correlate with sensory dermatomes and only slightly with the underlying muscles of the hypothermic area. The diagnostic sensitivity (22.9–36.1%) and the positive predictive value (25.2–37.0%) were low both for muscular symptoms such as tenderness or weakness and for spontaneous pain and sensory loss. In contrast, high specificity (78.8–81.7%), high negative predictive values (68.5–86.2%), and a high diagnostic accuracy were obtained.

Only the papers by Kim and Cho [46] and Zhang et al. [47] found thermography of high value for the diagnosis of both lumbosacral and cervical radiculopathies. However, these studies have several methodological flaws. Although a high number of patients were reported, healthy control subjects were not mentioned in the study on lumbosacral radiculopathy. The clinical symptoms are not described and the reliability of the used thermographic diagnostic criteria remains questionable.

25.5.1.2 Thoracic Outlet Syndrome

Similar to fibromyalgia, the disease entity of the thoracic outlet syndrome (TOS) is under continuous debate [48]. Consensus exists, that various subforms related to the severity of symptoms must be differentiated. Recording thermal images during diagnostic body positions can reproducibly provoke typical temperature asymmetries in the hands of patients with suspected TOS [49,50]. Temperature readings from thermal images from patients passing that test can be reproduced by the same and by different readers with high precision [51]. The original protocol included a maneuver in which the fist was opened and closed 30 times before an image of the hand was recorded. As this test did not increase the temperature difference between index and little finger, the fist maneuver was removed from the protocol [52]. Thermal imaging can be regarded as the only technique that can objectively confirm the subjective symptoms of mild TOS. It was successfully used as outcome measure for the evaluation of treatment for this pain syndrome [53]. However, in a patient with several causes for the symptoms paresthesias and coldness of the ulnar fingers, thermography could show only a marked cooling of the little finger, but could not identify all reasons for that temperature deficit [54]. It was also difficult to differentiate between subjects whether they suffer from TOS or carpal tunnel syndrome (CTS). Only 66.3% of patients were correctly allocated to three diagnostic groups, while none of the CTSs have been identified [55].

25.5.1.3 Carpal Tunnel Syndrome

Entrapment of the median nerve at the carpal tunnel is the most common compression neuropathy. A study conducted in Sweden revealed a prevalence of 14.4%; for pain, numbness, and tingling in the median nerve distribution in the hands. Prevalence of clinically diagnosed CTS was 3.8% and 4.9% for pathological results of nerve conduction of the median nerve. Clinically and electrophysiologically confirmed CTS showed a prevalence of 2.7% [56].

The typical distribution of symptoms leads to the clinical suspect of CTS [57], which must be confirmed by nerve conduction studies. The typical electroneurographic measurements in patients with CTS show a high intra- and interrater reproducibility [58]. The course of nerve conduction measures for a period of 13 years in patients with and without decompression surgery was investigated and it was shown that most of the operated patients presented with less pathological conduction studies within 12 months after operation [59]. Only 2 of 61 patients who underwent a simple nerve decompression by division of the carpal ligament as therapy for CTS had pathological findings in nerve conduction studies 2–3 years after surgery [60].

However, nerve conduction studies are unpleasant for the patient and alternative diagnostic procedures are welcome. Liquid crystal thermography was originally used for the assessment of patients with suspected CTS [61–64]. So et al. [65] used infrared imaging for the evaluation of entrapment syndromes of the median and ulnar nerves. Based on their definition of abnormal temperature difference to the contralateral side, they found thermography without any value for assisting diagnosis and inferior to electrodiagnostic testing. Tchou reported infrared thermography of high diagnostic sensitivity and specificity in patients with unilateral CTS. He has defined various regions of interest representing mainly the innervation area of the median nerve. Abnormality was defined if more than 25% of the measured area displayed a temperature increase of at least 1°C when compared with the asymptomatic hand [66].

Ammer has compared nerve conduction studies with thermal images in patients with suspected CTS. Maximum specificity for both nerve conduction and clinical symptoms was obtained for the temperature difference between the third and fourth finger at a threshold of 1°C. The best sensitivity of 69% was found if the temperature of the tip of the middle finger was by 1.2°C less than the temperature of the metacarpus [67].

Hobbins [68] combined the thermal pattern with the time course of nerve injuries. He suggested the occurrence of a hypothermic dermatome in the early phase of nerve entrapment and hyperthermic dermatomes in the late phase of nerve compression. Ammer et al. [69] investigated how many patients with a distal latency of the median nerve greater than 6 ms present with a hyperthermic pattern. They reported a slight increase of the frequency of hyperthermic patterns in patients with severe CTS, indicating that the entrapment of the median nerve is followed by a loss of the autonomic function in these patients.

Ammer and Melnizky [70] has also correlated the temperature of the index finger with the temperature of the sensory distribution of the median nerve on the dorsum of the hand and found nearly identical readings for both areas. A similar relationship was obtained for the ulnar nerve. The author concluded from these data that the temperature of the index or the little finger is highly representative for the temperature of the sensory area of the median or ulnar nerve, respectively.

Many studies on CTS have used a cold challenge to enhance the thermal contrast between affected fingers. A slow recovery rate after cold exposure is diagnostic for Raynaud's phenomenon [71]. The coincidence of CTS and Raynaud's phenomenon was reported in the literature [72,73].

25.5.1.4 Other Entrapment Syndromes

No clear thermal pattern was reported for the entrapment of the ulnar nerve [65]. A pilot study for the comparison of hands from patients with TOS or entrapment of the ulnar nerve at the elbow found only one out of seven patients with ulnar entrapment who presented with temperature asymmetry of the affected extremity [74]. All patients with TOS who performed provocation test during image recording showed at least in one thermogram an asymmetric temperature pattern.

25.5.2 Peripheral Nerve Paresis

Paresis is an impairment of the motor function of the nervous system. Loss of function of the sensory fibers may be associated with motor deficit, but sensory impairment is not included in the term "paresis." Therefore, most of the temperature signs in paresis are related to impaired motor function.

25.5.2.1 Brachial Plexus Paresis

Injury of the brachial plexus is a severe consequence of traffic accidents and motor cyclers are most frequently affected. The loss of motor activity in the affected extremity results in paralysis, muscle atrophy, and decreased skin temperature. Nearly 0.5% to 0.9% of newborns acquire brachial plexus paresis during delivery [75]. Early recovery of the skin temperature in babies with plexus paresis precede the recovery of motor function as shown in a study from Japan [76].

25.5.2.2 Facial Nerve

The seventh cranial nerve supplies the mimic muscles of the face and an acquired deficit is often named Bell's palsy. This paresis has normally a good prognosis for full recovery. Thermal imaging was used as outcome measure in acupuncture trials for facial paresis [77,78]. Ammer et al. [79] found slight asymmetries in patients with facial paresis, in which hyperthermia of the affected side occurred more frequently than hypothermia. However, patients with apparent herpes zoster causing facial palsy presented with higher temperature differences to the contralateral side than patients with nonherpetic facial paresis [80].

25.5.2.3 Peroneal Nerve

The peroneal nerve may be affected by metabolic neuropathy in patients with metabolic disease or by compression neuropathy due to intensive pressure applied at the site of fibula head. This can result in "foot drop," an impairment in which the patient cannot raise his forefoot. The thermal image is characterized by decreased temperatures on the anterior lower leg, which might become more visible after the patient has performed some exercises [81].

25.6 Complex Regional Pain Syndrome

A temperature difference between the affected and the nonaffected limb equal or greater than 1°C is one of the diagnostic criteria of the complex regional pain syndrome (CRPS) [82]. Ammer conducted a study in patients after radius fracture treated conservatively with a plaster cast [83]. Within 2 h after plaster removal and 1 week later, thermal images were recorded. After the second thermogram, an x-ray image of both hands was taken. The mean temperature difference between the affected and unaffected hand was 0.6 after plaster removal and 0.63 1 week later. In 21 out of 41 radiographs, slight bone changes suspected of algodystropy have been found. Figure 25.4 summarizes the results with respect to the outcome of x-ray images. Figure 25.5 shows the time course of an individual patient.

It was also shown that the temperature difference decrease during successful therapeutic intervention and temperature effect was paralleled by reduction of pain and swelling and resolution of radiologic changes [84].

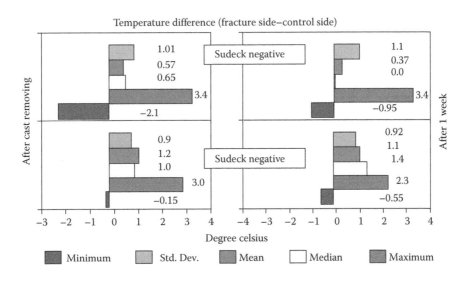

FIGURE 25.4 Diagram of temperatures obtained in patients with positive or negative x-ray images. (From Ammer, K. *Thermol. Österr.*, 1, 4, 1991. With permission.)

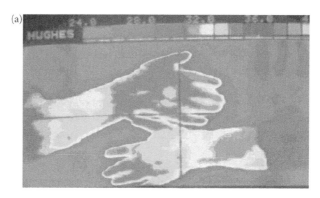

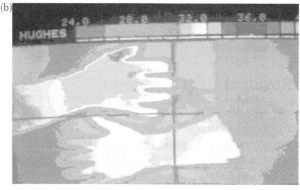

FIGURE 25.5   Early CRPS after radius fracture. (a) Two hours after cast removal; (b) 1 week later.

Disturbance of vascular adaptation mechanism and delayed response to temperature stimuli was obtained in patients suffering from CRPS [85,86]. These alterations have been interpreted as being caused by abnormalities of the autonomic nerve system. It was suggested to use a cold challenge on the contralateral side of the injured limb for prediction and early diagnosis of CRPS. Gulevich et al. [87] confirmed the high diagnostic sensitivity and specificity of cold challenge for the CRPS. Wasner et al. [88] achieved similar results by whole-body cooling or whole-body warming. Most recently, a Dutch study found that the asymmetry factor, which was based on histograms of temperatures from the affected and nonaffected hand, had the highest diagnostic power for CRPS, while the difference of mean temperatures did not discriminate between disease and health [89].

25.7 Thermal Imaging Technique

The parameters for a reliable technique have been described in the past. Reference 90 is a report published in 1978 by a European working group on thermography in locomotor diseases. This paper discusses aspects of standardization, including the need for adequate temperature control of the examination room and the importance of standard views used for image capture. More recently, Ring and Ammer [91] described an outline of necessary considerations for good thermal imaging technique in clinical medicine. This outline has been subsequently expanded to encompass a revised set of standard views, and associated regions of interest for analysis. The latter is especially important for standardization, since the normal approach used is to select a region of interest subjectively. This means that without a defined set of reference points, it is difficult for the investigator to reproduce the same region of interest on subsequent occasions. It is also even more difficult for another investigator to achieve the same, leading to unacceptable variables in the derived data. The aspects for standardization of the technique and the standard views and regions of interest recently defined are the product of a multicentered Anglo-Polish study that is pursuing the concept of a database of normal reference thermograms. The protocol can be found on a British University Research Group's website from University of Glamorgan [10].

25.7.1 Room Temperature

Room temperature is an important issue when investigating this group of diseases. Inflammatory conditions such as arthritis are better revealed in a room temperature of 20°C, for the extremities, and may need to be at 18°C for examining the trunk. This presumes that the relative humidity will not exceed 45%, and a very low airspeed is required. At no time during preparatory cooling or during the examination should the patient be placed in a position where they can feel a draught from moving air. However, in other clinical conditions where an effect from neuromuscular changes is being examined, a higher room temperature is needed to avoid forced vasoconstriction. This is usually performed at 22–24°C ambient. At higher temperatures, the subject may begin to sweat, and below 17°C, shivering may be induced. Both of these thermoregulatory responses by the human body are undesirable for routine thermal imaging.

25.7.2 Clinical Examination

In this group of diseases, it can be particularly important that the patient receives a clinical examination in association with thermal imaging. Observations on medication, range of movement, experience of pain related to movement, or positioning may have a significant effect on the interpretation of the thermal images. Documentation of all such clinical findings should be kept on record with the images for future reference.

References

1. Horvath, S.M. and Hollander, J.L. Intra-articular temperature as a measure of joint reaction. *J. Clin. Invest.*, 13, 615, 1949.
2. Collins, A.J. and Cosh, J.A. Temperature and biochemical studies of joint inflammation. *Ann. Rheum. Dis.*, 29, 386, 1970.
3. Ring, E.F.J. and Collins, A.J. Quantitative thermography. *Rheumatol. Phys. Med.*, 10, 337, 1970.
4. Esselinckx, W. et al. Thermographic assessment of three intra-articular prednisolone analogues given in rheumatoid arthritis. *Br. J. Clin. Pharm.*, 5, 447, 1978.
5. Bird, H.A., Ring, E.F.J., and Bacon, P.A. A thermographic and clinical comparison of three intra-articular steroid preparations in rheumatoid arthritis. *Ann. Rheum. Dis.*, 38, 36, 1979.
6. Collins, A.J. and Ring, E.F.J. Measurement of inflammation in man and animals. *Br. J. Pharm.*, 44, 145, 1972.
7. Bacon, P.A., Ring, E.F.J., and Collins, A.J. Thermography in the assessment of anti rheumatic agents, in *Rheumatoid Arthritis*. Gordon, J.L. and Hazleman, B.L., Eds., Elsevier/North Holland Biomedical Press, Amsterdam, 1977, p. 105.
8. Collins, A.J. et al. Quantitation of thermography in arthritis using multi-isothermal analysis. I. The thermographic index. *Ann. Rheum. Dis.*, 33, 113, 1974.
9. Ammer, K., Engelbert, B., and Kern, E. The determination of normal temperature values of finger joints. *Thermol. Int.*, 12, 23, 2002.
10. Website address, Standard protocol for image capture and analysis, www.medimaging.org.
11. Czerny, V. Eine fokale Malazie des Unterschenkels. *Wien. Med. Wochenschr.*, 23, 895, 1873.
12. Paget, J. On a form of chronic inflammation of bones. *Med. Chir. Transact.*, 60, 37, 1877.
13. Ring, E.F.J. and Davies, J. Thermal monitoring of Paget's disease of bone. *Thermology*, 3, 167, 1990.
14. Tauchmannova, H., Gabrhel, J., and Cibak, M. Thermographic findings in different sports, their value in the prevention of soft tissue injuries. *Thermol. Österr.* 3, 91–95, 1993.

15. Smith, B.L, Bandler, M.K., and Goodman, P.H. Dominant forearm hyperthermia, a study of fifteen athletes. *Thermology*, 2, 25–28, 1986.

16. Fischer, A.A. and Chang, C.H. Temperature and pressure threshold measurements in trigger points. *Thermology*, 1, 212, 1986.

17. Menachem, A., Kaplan, O., and Dekel, S. Levator scapulae syndrome: An anatomic–clinical study. *Bull. Hosp. Jt. Dis.*, 53, 21, 1993.

18. Schmitt, M. and Guillot, Y. Thermography and muscle injuries in sports medicine, in *Recent Advances in Medical Thermography*. Ring, E.F.J. and Philips, J., Eds., Plenum Press, London, 1984, p. 439.

19. Ammer, K. Low muscular activity of the lower leg in patients with a painful ankle. *Thermol. Österr.*, 5, 103, 1995.

20. Kanie, R. Thermographic evaluation of osteoarthritis of the hip. *Biomed. Thermol.*, 15, 72, 1995.

21. Vecchio, P.C. et al. Thermography of frozen shoulder and rotator cuff tendinitis. *Clin. Rheumatol.*, 11, 382, 1992.

22. Ammer, K. et al. Thermography of the painful shoulder. *Eur. J. Thermol.*, 8, 93, 1998.

23. Hobbins, W.B. and Ammer, K. Controversy: Why is a paretic limb cold, high activity of the sympathetic nerve system or weakness of the muscles? *Thermol. Österr.*, 6, 42, 1996.

24. Ring, E.F.J. and Ammer, K. Thermal imaging in sports medicine. *Sports Med. Today*, 1, 108, 1998.

25. Gabrhel, J. and Tauchmannova, H. Wärmebilder der Kniegelenke bei jugendlichen Sportlern. *Thermol. Österr.*, 5, 92, 1995.

26. Devereaux, M.D. et al. The diagnosis of stress fractures in athletes. *JAMA*, 252, 531, 1984.

27. Graber, J. Tendosynovitis detection in the hand. *Verh. Dtsch. Ges. Rheumatol.*, 6, 57, 1980.

28. Mayr, H. Thermografische Befunde bei Schmerzen am Ellbogen. *Thermol. Österr.*, 7, 5–10, 1997.

29. Binder, A.I. et al. Thermography of tennis elbow, in *Recent Advances in Medical Thermography*. Ring, E.F.J. and Philips, J., Eds., Plenum Press, London, 1984, p. 513.

30. Thomas, D. and Savage, J.P. Persistent tennis elbow: evaluation by infrared thermography and nuclear medicine isotope scanning. *Thermology*, 3, 132; 1989.

31. Ammer, K. Thermal evaluation of tennis elbow, in *The Thermal Image in Medicine and Biology*. Ammer, K. and Ring, E.F.J., Eds., Uhlen Verlag, Wien, 1995, p. 214.

32. Devereaux, M.D., Hazleman, B.L., and Thomas, P.P. Chronic lateral humeral epicondylitis—A double-blind controlled assessment of pulsed electromagnetic field therapy. *Clin. Exp. Rheumatol.*, 3, 333, 1985.

33. Ammer, K. et al. Thermographische und algometrische Kontrolle der physikalischen Therapie bei Patienten mit Epicondylopathia humeri radialis. *ThermoMed*, 11, 55–67, 1995.

34. Ammer, K., Schartelmüller, T., and Melnizky, P. Thermography in fibromyalgia. *Biomed. Thermol.* 15, 77, 1995.

35. Ammer, K. Only lateral, but not medial epicondylitis can be detected by thermography. *Thermol. Österr.*, 6, 105, 1996.

36. Hirano, T. et al. Clinical study of shoulder surface temperature in patients with periarthritis scapulohumeralis (abstract). *Biomed. Thermol.*, 11, 303, 1991.

37. Jeracitano, D. et al. Abnormal temperature control suggesting sympathetic dysfunction in the shoulder skin of patients with frozen shoulder. *Br. J. Rheumatol.*, 31, 539, 1992.

38. Biasi, G. et al. The role computerized telethermography in the diagnosis of fibromyalgia syndrome. *Minerva Medica*, 85, 451, 1994.

39. Ammer, K. Thermographic diagnosis of fibromyalgia. *Ann. Rheum. Dis. XIV European League Against Rheumatism Congress, Abstracts*, 135, 1999.

40. Ammer, K., Engelbert, B., and Kern, E. Reproducibility of the hot spot count in patients with fibromyalgia, an intra- and inter-observer comparison. *Thermol. Int.*, 11, 143, 2001.

41. Anbar, M. Recent technological developments in thermology and their impact on clinical applications. *Biomed. Thermol.*, 10, 270, 1990.

42. Uematsu, S. Thermographic imaging of cutaneous sensory segment in patients with peripheral nerve injury. *J. Neurosurg.*, 62, 716–720, 1985.

43. Hoffman, R.M., Kent, D.L., and. Deyo, R.A. Diagnostic accuracy and clinical utility of thermography for lumbar radiculopathy. A meta-analysis. *Spine*, 16, 623, 1991.

44. McCulloch, J. et al. Thermography as a diagnostic aid in sciatica. *J. Spinal Disord.*, 6, 427, 1993.

45. Takahashi, Y., Takahashi, K., and Moriya, H. Thermal deficit in lumbar radiculopathy. *Spine*, 19, 2443, 1994.

46. Kim, Y.S. and Cho, Y.E. Pre- and postoperative thermographic imaging of lumbar disk herniations. *Biomed. Thermol.*, 13, 265, 1993.

47. Zhang, H.Y., Kim, Y.S., and Cho, Y.E. Thermatomal changes in cervical disc herniations. *Yonsei Med. J.*, 40, 401, 1999.

48. Cuetter, A.C. and Bartoszek, D.M. The thoracic outlet syndrome: controversies, overdiagnosism overtreatment and recommendations for management. *Muscle Nerve*, 12, 419, 1989.

49. Schartelmüller, T. and Ammer, K. Thoracic outlet syndrome, in *The Thermal Image in Medicine and Biology*. Ammer, K. and Ring, E.F.J., Eds., Uhlen Verlag, Wien, 1995, p. 201.

50. Schartelmüller, T. and Ammer, K. Infrared thermography for the diagnosis of thoracic outlet syndrome. *Thermol. Österr.*, 6, 130, 1996.

51. Melnizky, P., Schartelmüller, T., and Ammer, K. Prüfung der intra-und interindividuellen Verläßlichkeit der Auswertung von Infrarot-Thermogrammen. *Eur. J. Thermol.*, 7, 224, 1997.

52. Ammer, K. Thermographie der Finger nach mechanischem Provokationstest. *ThermoMed*, 17/18, 9, 2003.

53. Schartelmüller, T., Melnizky, P., and Engelbert, B. Infrarotthermographie zur Evaluierung des Erfolges physikalischer Therapie bei Patenten mit klinischem Verdacht auf Thoracic Outlet Syndrome. *Thermol. Int.*, 9, 20, 1999.

54. Schartelmüller, T. and Ammer, K. Zervikaler Diskusprolaps, Thoracic Outlet Syndrom oder periphere arterielle Verschlußkrankheit-ein Fallbericht. *Eur. J. Thermol.*, 7, 146, 1997.

55. Ammer, K. Diagnosis of nerve entrapment syndromes by thermal imaging, in *Proceedings of the First Joint BMES/EMBS Conference. Serving Humanity, Advancing Technology*, October 13–16, 1999, Atlanta, GA, USA, p. 1117.

56. Atroshi, I. et al. Prevalence of carpal tunnel syndrome in a general population. *JAMA*, 282, 153, 1999.

57. Ammer, K., Mayr, H., and Thür, H. Self-administered diagram for diagnosing carpal tunnel syndrome. *Eur. J. Phys. Med. Rehab.*, 3, 43, 1993.

58. Melnizky, P., Ammer, K., and Schartelmüller, T. Intra- und interindividuelle Verläßlichkeit der elektroneurographischen Untersuchung des Nervus medianus. *Österr. Z. Phys. Med. Rehab.*, 7, S83, 1996.

59. Schartelmüller, T., Ammer, K., and Melnizky, P. Natürliche und postoperative Entwicklung elektroneurographischer Untersuchungsergebnisse des N. medianus von Patienten mit Carpaltunnelsyndrom (CTS). *Österr. Z. Phys. Med.*, 7, 183, 1997.

60. Rosen, H.R. et al. Is surgical division of the carpal ligament sufficient in the treatment of carpal tunnel syndrome? *Chirurg*, 61, 130, 1990.

61. Herrick, R.T. et al. Thermography as a diagnostic tool for carpal tunnel syndrome, in *Medical Thermology*, Abernathy, M. and Uematsu, S., Eds., American Academy of Thermology, Washington, DC, 1986, p. 124.

62. Herrick, R.T. and Herrick, S.K., Thermography in the detection of carpal tunnel syndrome and other compressive neuropathies. *J. Hand Surg.*, 12A, 943–949, 1987.

63. Gateless, D., Gilroy, J., and Nefey, P. Thermographic evaluation of carpal tunnel syndrome during pregnancy. *Thermology*, 3, 21, 1988.

64. Meyers, S. et al. Liquid crystal thermography, quantitative studies of abnormalities in carpal tunnel syndrome. *Neurology*, 39, 1465, 1989.

65. So, Y.T., Olney, R.K., and Aminoff, M.J. Evaluation of thermography in the diagnosis of selected entrapment neuropathies. *Neurology*, 39, 1, 1989.

66. Tchou, S. and Costich, J.F. Thermographic study of acute unilateral carpal tunnel syndromes. *Thermology*, 3, 249–252, 1991.

67. Ammer, K. Thermographische Diagnose von peripheren Nervenkompressionssyndromen. *ThermoMed*, 7, 15, 1991.

68. Hobbins, W.B. Autonomic vasomotor skin changes in pain states: Significant or insignificant? *Thermol. Österr.*, 5, 5, 1995.

69. Ammer, K. et al. The thermal image of patients suffering from carpal tunnel syndrome with a distal latency higher than 6.0 msec. *Thermol. Int.*, 9, 15, 1999.

70. Ammer, K. and Melnizky, P. Determination of regions of interest on thermal images of the hands of patients suffering from carpal tunnel syndrome. *Thermol. Int.*, 9, 56, 1999.

71. Ammer, K. Thermographic diagnosis of Raynaud's phenomenon. *Skin Res. Technol.*, 2, 182, 1996.

72. Neundörfer, B., Dietrich, B., and Braun, B. Raynaud–Phänomen beim Carpaltunnelsyndrom. *Wien. Klin. Wochenschr.*, 89, 131–133, 1977.

73. Grassi, W. et al. Clinical diagnosis found in patients with Raynaud's phenomenon: A multicentre study. *Rheumatol. Int.*, 18, 17, 1998.

74. Mayr, H. and Ammer, K. Thermographische Diagnose von Nervenkompressionssyndromen der oberen Extremität mit Ausnahme des Karpaltunnelsyndroms (abstract). *Thermol. Österr.*, 4, 82, 1994.

75. Mumenthaler, M. and Schliack, H. *Läsionen periphere Nerven*. Georg Thieme Verlag, Stuttgart-New York, Auflage, 1982, p. 4.

76. Ikegawa, S. et al. Use of thermography in the diagnosis of obstetric palsy (abstract). *Thermol. Österr.*, 7, 31, 1997.

77. Zhang, D. et al. Preliminary observation of imaging of facial temperature along meridians. *Chen Tzu Yen Chiu*, 17, 71, 1992.

78. Zhang, D. et al. Clinical observations on acupuncture treatment of peripheral facial paralysis aided by infra-red thermography—A preliminary report. *J. Tradit. Chin. Med.*, 11, 139, 1991.

79. Ammer, K., Melnizky, P. and Schartelmüller, T. Thermographie bei Fazialisparese. *ThermoMed*, 13, 6–11, 1997.

80. Schartelmüller, T., Melnizky, P., and Ammer, K. Gesichtsthermographie, Vergleich von Patienten mit Fazialisparese und akutem Herpes zoster ophthalmicus. *Eur. J. Thermol.*, 8, 65, 1998.

81. Melnizky, P., Ammer, K., and Schartelmüller, T. Thermographische Überprüfung der Heilgymnastik bei Patienten mit Peroneusparese. *Thermol. Österr.*, 5, 97, 1995.

82. Wilson, P.R. et al. Diagnostic algorithm for complex regional pain syndromes, in *Reflex Sympathetic Dystrophy, A Re-appraisal*. Jänig, W. and Stanton-Hicks, M., Eds., IASP Press, Seattle, 1996, p. 93.

83. Ammer, K. Thermographie nach gipsfixierter Radiusfraktur. *Thermol. Österr.*, 1, 4, 1991.

84. Ammer, K. Thermographische Therapieüberwachung bei M.Sudeck. *ThermoMed*, 7, 112–115, 1991.

85. Cooke, E.D. et al. Reflex sympathetic dystrophy (algoneurodystrophy): Temperature studies in the upper limb. *Br. J. Rheumatol.*, 8, 399, 1989.

86. Herrick, A. et al. Abnormal thermoregulatory responses in patients with reflex sympathetic dystrophy syndrome. *J. Rheumatol.*, 21, 1319, 1994.

87. Gulevich, S.J. et al. Stress infrared telethermography is useful in the diagnosis of complex regional pain syndrome, type I (formerly reflex sympathetic dystrophy). *Clin. J. Pain*, 13, 50, 1997.

88. Wasner, G., Schattschneider, J., and Baron, R. Skin temperature side differences—A diagnostic tool for CRPS? *Pain*, 98, 19, 2002.

89. Huygen, F.J.P.M. et al. Computer-assisted skin videothermography is a highly sensitive quality tool in the diagnosis and monitoring of complex regional pain syndrome type I. *Eur. J. Appl. Physiol.*, 91, 516, 2004.

90. Engel, J.M. et al. Thermography in locomotor diseases, recommended procedure. Anglo-dutch thermographic society group report. *Eur. J. Rheumatol. Inflam.*, 2, 299–306, 1979.

91. Ring, E.F.J. and Ammer, K. The technique of infra red imaging in medicine. *Thermol. Int.*, 10, 7, 2000.

26

Functional Infrared Imaging in the Evaluation of Complex Regional Pain Syndrome, Type I: Current Pathophysiological Concepts, Methodology, Case Studies, and Clinical Implications

Timothy D.
Conwell
*Colorado Infrared Imaging
Center*

James Giordano
*Potomac Institute for Policy
Studies*

George Mason University

University of Oxford

26-1

26.1 Introduction

26.1.1 Overview, History, and Contemporary Issues

Complex Regional Pain Syndrome, Type I (CRPS I) is a potentially disabling neuropathic condition characterized by regional pain that is often disproportionate to or occurs in the absence of an identifiable inciting event. This poorly understood disease is the result of a multifactorial interplay between altered somatosensory, motor, autonomic, and inflammatory systems. Peripheral and central sensitization is a common feature in CRPS. The condition is associated with hyperalgesia, allodynia, spontaneous pain, abnormal skin color, changes in skin temperature, abnormal sudomotor activity, edema, active and passive movement disorders, and trophic changes of nails and hair. CRPS I usually begins after minor or major trauma to soft tissue (e.g., strain, sprain, or surgery). Although rare, it can also occur following fracture, visceral trauma, or central nervous system insult (e.g., cerebrovascular accident; CVA). Although similar—if not identical—in presentation, a related syndrome, Type II (CRPS II), occurs following direct insult or injury to peripheral nerve. A complete address of CRPS II is beyond the scope of this chapter (see [1], for review). Sandroni et al. [2] report that CRPS I occurs more frequently than CRPS II (i.e., CRPS I incidence of 5.5/100,000 person years at risk vs. CRPS II incidence of 0.8/100,000 person years at risk), is more prevalent (21/100,000 vs. 4/100,000), and occurs more in females than males with reported ratios of 2:1 or greater [3].

Diagnosis of CRPS I is based upon patient history and evaluation of presenting clinical signs and symptoms. Diagnosis can be complicated by the facts that (1) the severity of the symptoms is characteristically far greater than that of the instigating insult; (2) there is a tendency of the symptoms to spread proximally and in some cases to the contralateral limb, trunk, and face; and (3) although somewhat more rare, all four extremities may be involved [4,5]. In light of these diagnostic issues, a consensus workshop was convened in Orlando in 1993 to posit evaluative criteria and develop a more effective nomenclature for these disorders. From this work, the term "CRPS" was first introduced [7]. Subsequently, The International Association for the Study of Pain (IASP) modified their taxonomy of pain to include these disorders, thus formalizing the use of CRPS I to identify and describe distinct syndromes [6–8]. During the same time period (i.e., circa early 1990s), Veldman et al. developed similar criteria for CRPS [9]. Previously, CRPS I and II were known as reflex sympathetic dystrophy (RSD) and causalgia, respectively, although both syndromes were rather arbitrarily referred to as algodystrophy, Sudeck's atrophy, sympathalgia, and sympathetically maintained pain (SMP). However, it should be noted that SMP remains in use as a diagnostic category for any pain syndrome that involves, and/or is perpetuated by autonomic hyperreactivity, and is identifiable by positive response to sympatholytic intervention.

The IASP diagnostic criteria for CRPS I is based upon nonstandardized signs and symptoms [10–13]. Rigorous discriminant function analyses (DFA) applied to these criteria have revealed problems in reliability and external validity. Results of validation studies have shown that there is significant potential for overdiagnosis due to low specificity [14–16], low interobserver reliability [17,18], and considerable variability in the recognition of relevant clinical signs. Consequently, the maximum potential benefits of the IASP criteria have been unsuccessful. As a result, an international consensus group revised the IASP criteria for CRPS [19]. The schema, known as the "Budapest Criteria" revealed a diagnostic sensitivity of 99%, with an improved specificity of 68%, thereby significantly enhancing the poor specificity (41%) obtained when utilizing the original IASP criteria [20]. Yet, other researchers have expressed a need for more regionally based CRPS diagnostic criteria that differ from the "Budapest Criteria" [21].

Such disparities in diagnostic criteria promote inherent difficulties in the clinical management of CRPS I. Lack of coherence in detection, discernment, and diagnostic acumen arises from a tendency to rely on sympathetic blockade as a diagnostic measure. Reiteratively, it is important to note that while some CRPS I patients will be responsive to sympatholytics (e.g., systemically administered adrenergic antagonists; interventional sympathetic blockade), such SMP does not occur in all CRPS I, and therefore is not an exclusory criterion. Also, given the heterogeneity of signs, symptoms, and overall presentations, there is a level of suspicion among non-subspecialists regarding the validity of CRPS I as a diagnosis. Taken together, these lead to a tendency for either frank undertreatment or excessive utilization of therapeutic interventions that incur significant cost, time, and even health burdens for the patient. Thus, it can be seen that diagnosis is the foundation upon which effective and sound treatment is built. Clearly, more reliable diagnostic tests are needed to advance both our understanding and treatment of CRPS I.

26.1.2 Ethics and Imperatives for Evidence-Based Use of Functional Infrared Imaging in Research and Practice

It is important to emphasize that detection alone is not diagnosis. In formulating any diagnosis, the clinician must utilize distinct domains and types of knowledge to apprehend the objective and subjective features of a disorder as expressed in a particular patient. While this is important in the diagnosis of any malady, it is particularly critical to diagnosing (and treating) pain syndromes. By its nature, pain is subjective, and thus the clinician must utilize patient narrative and history, together with any/ all objective findings to formulate a clinical impression [22]. The "goal" of this initial step of pain medicine is to detect what and how a disorder presents in a particular patient, differentiate these signs and symptoms from other possible disorders, discern the contextual basis of these features, and establish a diagnosis, literally a "…seeing or a knowledge into the problem." This provides the basis for the scope, nature, and extent of subsequent care, and allows for more accurate prognosis (knowledge of what could occur). It is clear that this process—often referred to as the nosological method—is not simply one of applied science, but takes on considerable ethical weight as it names the disorder (and by extension categorizes the patient), frames the patient within a larger community of others, and creates a foundation for the development and implementation of prudent care [23]. The diversity and individuality of signs and symptoms of CRPS I complicates this process. The ethical obligation to treat pain [23] compels research to determine those techniques and technologies that facilitate diagnoses and treatments that are safe and effective, and sustains the utilization of these approaches in clinical practice [24–26].

By classical definition, medicine is dedicated to providing patient care that is technically correct and ethically sound, enacted within the clinical encounter [27]. Research facilitates these ends by affording knowledge that (1) enables the clinician to evaluate the relative value(s), benefits, burdens, and risks of particular diagnostic and therapeutic approaches, so as to ultimately resolve clinical equipoise, and (2) empowers patients to be informed participants in clinical decisions relevant to care, thereby lessening their inherent vulnerability and decreasing the inequalities of knowledge, power, and capacity [25]. The ethical obligations of research and practice are reciprocal: findings from research inform practice, and evaluation of outcomes gained by employing various techniques in practice contributes to a progressive revision and expansion of evidentiary knowledge to instigate and guide further study [26,28]. Thus, we argue that the investigational and clinical use of functional infrared (fIR) imaging is both pragmatically valid and ethically imperative.

26.1.3 Detection, Discernment, and Diagnosis: Bases for Effective Treatment

fIR effectively detects the thermal signature of vasomotor disturbances that are an important factor in establishing a diagnosis of CRPS I. Early detection of CRPS I is essential for successful treatment, yet detection of early-stage CRPS I is difficult because the signs and symptoms mimic other pathologies. Several studies have supported the utility of infrared imaging (thermography) in early detection of CRPS [29–33]. Most studies have utilized quantitative, homologous side-to-side temperature differences as the sole criterion in establishing the diagnosis. These temperature differences have ranged from 0.5°C [29], 0.6°C [34,35], 1°C [32,36–38], 1.5°C [39], and 2°C [33].

Bruehl et al. [34] evaluated thermal asymmetries in patients with CRPS I following equilibration in a 20°C examining room for 20 min, and demonstrated a sensitivity of 60% and specificity of 67% utilizing an asymmetry cutoff of 0.6°C in side-to-side computer-generated temperature differences. Gulevich et al. [38] demonstrated a 93% sensitivity and 89% specificity when three of the following four infrared image categories were present: (1) quantitative homologous side-to-side computer-generated temperature differences of greater than 1.0°C in the region of interest (ROI); (2) the presence of abnormal, disrupted transverse distal thermal gradient lines visualized in the symptomatic extremity; (3) the presence of a "thermal marker" of the symptomatic extremity visualized in the isotherm view; and (4) abnormal warming of the symptomatic extremity secondary to functional cold-water autonomic stress testing of an asymptomatic limb. These contingencies are addressed elsewhere in this chapter (see Section 26.3). Similarly, Wasner et al. [40] showed a sensitivity of 76% and specificity of 93% when patients underwent controlled alteration of sympathetic activity by using thermoregulatory whole-body cooling followed by computer-generated side-to-side temperature differences of homologous body regions. Conwell et al. [41] showed a sensitivity of 72% and specificity of 94%, with a positive predictive value 82% and negative predictive value 90%, when using cold-water autonomic functional stress testing as compared with the modified IASP criteria for CRPS. The use of quantitative temperature differences alone lacks diagnostic specificity because numerous pathologies may present with skin temperature asymmetry. For example, differences in skin temperature occur in focal inflammation and vascular disease. Moreover, thermal asymmetry can result from somatoautonomic vasoconstriction secondary to acute trauma [42], limb immobilization [43], fracture [44], antidromic vasodilatation from small fiber distal neuropathy, and neuropathic pain with sympathetic activity. Infrared imaging (thermography) does not merely measure limb temperature; the infrared thermogram yields a temperature map (IR signature) of an extremity through which a highly trained and experienced examiner can differentiate the IR signatures (thermal patterns) of CRPS I from trauma, inflammation, and vascular disease. The quantitative assessment of the IR signature (vasomotor changes) is instrumental, if not mandatory, for establishing an accurate diagnosis of CRPS I and II [45–47]. Furthermore, the infrared thermographic study is significantly enhanced when the IR data are interfaced with software that enables the examiner to obtain three IR indices [38]: (1) computer-generated side-to-side temperature differences, (2) statistical evaluation of the integrity (i.e., normality vs. abnormality) of distal thermal gradient IR signatures, and (3) responses to cold-water autonomic functional stress testing [41].

An understanding of the capabilities and limitations of fIR technology is critical to establishing its significance in detecting physiological changes that are relevant and meaningful to the diagnosis of CRPS I. Thus, this chapter will present (1) a brief overview of the pathophysiology of CRPS I, with particular emphasis upon those autonomically mediated vasomotor effects that evoke clinically relevant changes in the radiant heat signature; (2) current fIR methodology that has been shown to be capable of effectively detecting such changes; (3) discussion of the current procedures and protocol methods for performing fIR studies of presumed CRPS I (and differentiating this from other pain syndromes); and (4) selected cases of acute trauma, neurogenic inflammation without sympathetic activity, small-caliber fiber distal neuropathy, and acute and chronic CRPS I, to illustrate clinical applications of this technology.

26.2 Pathophysiology of CRPS I

26.2.1 Sympathetic Neural Involvement

Any discussion of CRPS I, II, SMP, and other dysautonomias must address the role of autonomic dysfunction. Ordinarily, sympathetic efferents and sensory afferents are not conjoined [48,49]. However, functional and perhaps structural interactions between nociceptive C-fiber afferents and sympathetic efferent fibers in both the periphery and within the dorsal root ganglion have been elucidated in both animal models of sympathetically mediated pain [50] and in humans with CRPS I [51,52]. This interaction involves the *de novo* expression of alpha-2 adrenoceptors on C-fiber processes, resulting in an increased sensitivity to adrenergic stimulation [52,53]. It remains unclear whether structural synaptogenesis is involved in the periphery or the dorsal root, but clearly the functional properties of C-fiber afferents change, with alteration in excitation threshold(s), after-response duration and frequencies, and perhaps expansion of receptive fields [39,54]. However, while sympathetic contribution to C-fiber activity may be influential and can contribute to the loss of thermoregulatory function (via sensitized alpha-adrenoceptor responses to norepinephrine) that evokes C-fiber-mediated pain and a concomitant axon reflex vasodilatation [55–58], such sympathetically mediated mechanisms are not uniform in all instances of CRPS I, and sympathetic dysregulation does not account for the entire constellation of neurological features of the disorder. Using microneurography to evaluate possible electrophysiological interactions by simultaneously recording single identified sympathetic efferent fibers and C nociceptors while provoking sympathetic neuronal discharges in cutaneous nerves, Campero et al. assessed potential effects of sympathetic activity upon 35 polymodal nociceptors and 19 mechano-insensitive nociceptors in patients with CRPS I and II. These studies failed to reveal activation of nociceptors related to sympathetic discharge, thereby suggesting that sympathetic–nociceptor interactions are the exception [59].

26.2.2 Neurogenic Inflammatory and Central Neural Interactions

Irrespective of sympathetic effects, a neurogenic inflammatory response appears to be perpetuated in CRPS I [60,61,63], thereby supporting the century-old work of Sudeck et al. [62]. This inflammatory response is initiated, at least in part, by substance-P, and involves a nitric oxide-mediated peripheral vasodilatation [64], with enhanced extravasation of serotonin (5-hydroxytryptamine; 5-HT) [65,66], as well a variety of peptides [67] that restimulate C-fiber afferents and elevate local concentrations of proinflammatory cytokines [68–70]. The continued activation of C-fiber afferents, second-order neurons of the spinothalamic tract (STT), and higher supraspinal loci in the neuraxis can evoke functional and perhaps structural changes in the CNS that (1) suppress endogenous pain modulatory mechanisms, (2) decrease pain thresholds, and (3) increase the duration and intensity of pain sensation and perception [24,71]. Such changes have been demonstrated in CRPS I, and may be responsible for "top-down" effects upon the sympathetic system that initiate and/or maintain increased activation of pronociceptive mechanisms [72,73], alter peripheral vaso- and sudomotor control, and may affect the structural integrity of the innervated tissues (e.g., skin, hair, nails, and bone).

26.2.3 Putative Genetic Factors

These plastic changes occur over a variable time course, and it is almost impossible to predict the rapidity of these effects. Thus, there is considerable temporal and individual variation in the presentation of CRPS I. It is not known whether this variability reflects some genetic and/or phenotypic diathesis, a circumstantial effect relative to the provocative insult, or a combination of both. Genetically, particular histocompatibilty complexes have been linked to (certain forms of) CRPS I; Van De Beek and colleagues [74] have shown that spontaneous-onset CRPS I is associated with a newly identified HLA I complex, and increased HLA-DQ1 and HLA-DR13 have been found in CRPS I patients [75,76]. To date, however, it is unclear whether the presence of these histocompatibility complexes are predictive, precipitative, or simply a copresentation (or even epi-phenomenon) of CRPS I. Still, these findings suggest that with further research, HLA testing may become a valuable contribution to the diagnosis of CRPS I. But while genetic testing may be useful to predict or correlate a predisposition to CRPS I, the strength of these findings remains dependent upon other objective measures that detect pathologic changes in support of patients' subjective reports (of pain, sensory and autonomic dysfunction, etc.) and thus reveal the disorder in its pattern of expression. It is in this light that we argue for the utility of fIR as a critical component in the evaluation of CRPS I.

26.3 Methodology: fIR Imaging

26.3.1 Description

Functional infrared thermography (fIR) is a nonionizing, pain-free physiological assessment procedure that has no known adverse biological side effects, is completely safe for (1) women who may become or are pregnant, (2) patients with intractable pain, and (3) patients who cannot tolerate painful invasive procedures. Functional IR imaging is not recommended for patients with casts, bandages, or other technical factors (i.e., patients bound to a wheelchair) that preclude the ability to expose skin to temperature equilibration and imaging.

26.3.2 Brief Historical Context

Measuring human skin temperature as a means of medical assessment began in antiquity with practitioners and scholars who recognized that body temperature is altered during infection and disease. As early as 400 BCE, Hippocrates applied thin layers of moistened clay to various body regions and measured drying times in order to show differences in body temperature, believing that "…in whatever part of the body heat or cold is seated, there is disease" [77]. Some 2000 years later, Galileo Galilee invented the "thermoscope" to detect human body temperature. In the early eighteenth century, Gabriel Fahrenheit invented the mercury thermometer, a device that is strikingly similar to its contemporary counterpart. However, it was not until 1870 that the thermometer was used in medicine. In 1800, William Herschel discovered the infrared spectrum [78], and this ultimately provided the impetus for studies on thermal heat that culminated in the development of imaging of radiant emissions some 200 years later. The incipient use of infrared technology provided a rudimentary depiction of biological heat signatures, and it was not until the mid-1960s that these methods were directly employed as a potentially viable evaluative tool in medicine. In the early 1980s, highly sensitive infrared cameras were interfaced with computer software specifically designed for medical applications. The development of sophisticated computerized electronic infrared detectors that are coupled to the state-of-the-art analytic software has overcome many of the technical and practical limitations of infrared thermography, that now allows accurate real-time, side-to-side homologous temperature differences of the extremities, as well as additional sophisticated computer-enhanced thermal images.

26.3.3 Medical Infrared Equipment

Medical infrared thermographic equipment is able to image and record the radiant infrared emission from skin surface radiation. Optimal detection must be capable of measuring human skin temperature with a thermal energy within the 7.5–13.5 Mm IR waveband. The infrared radiant emission is focused through 25–50 mm lens and detected by an indium antimonide (InSb) focal plane array, third-generation microbolometer (or better) with a 320×240 resolution (or better) and 12–16 bit dynamic range. The thermal resolution should be 60 mK NETD (noise equivalent temperature difference) or less. The IR emission is converted into an electronic signal that is processed by a computer and displayed on a color monitor in real time. Current infrared equipment can detect temperature differences as small as 0.02°C, with high thermal sensitivity and resolution. This high-resolution infrared image is able to detect the subtle temperature differences that are required to evaluate patients with mild sensory and autonomic neuropathies. Advances in medical infrared cameras (focal plane array microbolometer detectors) and sophisticated medical software allows for very accurate computer-generated side-to-side homologous temperature differences of the regions of interest to provide objective discrimination of thermal emissions. The thermal sensitivity range utilized by most clinicians with modern high-resolution IR equipment ranges from 0.2°C to 1.0°C. The thermal range is set according to the anatomic region being studied, as well as the discriminate information relative to the pathology studies. In addition, advanced software is capable of enhancing the thermal image to evaluate autonomic vasomotor tone and maintenance of the sympathetic vasoconstrictor reflex during cold-water autonomic functional stress testing.

26.3.4 Infrared Laboratory Environment

The infrared laboratory environment must be controlled to ensure obtaining accurate readings of the thermal emission that are free of artifact. The laboratory must be maintained at a constant temperature of $20 \pm 1°C$, and may require a 2–4°C reduction in temperature during the hot summer months in order to achieve the necessary ANS arousal (i.e., body cooling) prior to imaging. It is imperative that the subjects' autonomic thermoregulatory function is not affected by the ambient environment. Thus, the laboratory must be free of any ultraviolet rays that may cause aberrant heating of the surface skin temperature, and should be lit with fluorescent light bulbs. Similarly, all windows should be covered to eliminate solar radiation. The laboratory must maintain low humidity to ensure protection from diffuse cooling, and must be as draft-free as possible to prevent cold air blowing on the patient. The laboratory must be carpeted to prevent an inadvertent cooling artifact of the plantar surface of the feet.

26.3.5 Patient Selection: Indication

Patients that are presumed or suspected to have CRPS I and/or II are commonly selected for fIR studies, as are those patients with signs and symptoms of conditions that mimic CRPS I [79,80]. Some of these latter conditions include small fiber distal mononeuropathies, acute trauma, localized inflammation, vascular pathology, peripheral neuropathy, and vasospastic disorders. Follow-up fIR studies may be helpful in determining the effectiveness of sympathetic blockade, sympathectomy, and/or spinal cord stimulator placement. Follow-up fIR studies may also be indicated when evaluating patients' response to treatment, or evaluating progression of the underlying disease state.

26.3.6 Patient Preprocedure Protocol

Patients undergoing fIR testing must follow very specific preprocedure protocols [79,80]. These include discontinuing the use of nicotine and caffeine products 4 h before testing; discontinuing physical therapy and TENS unit the day before testing, and the avoidance of skin lotions, deodorants, moisturizers, liniments, topical OTC medications, skin powders, and makeup the day of testing. Patients are advised not to wear any tight-fitting clothing on the day of the test, and to discontinue the use of braces, bandaging, or neoprene wraps for 24 h before evaluation. Also, patients must not have any form of invasive diagnostic procedures for 24 h before testing. Patients may be required to discontinue certain opioid and nonopioid analgesics, and sympatholytic medications up to 24 h before testing, as these may impact the sympathetic function and alter surface skin temperature. All interventional (sympathetic, Bier, and neurolytic) blocks must be discontinued for a minimum of 3 days before testing.

26.3.7 Patient Protocol

Prior to initiating IR imaging, the patient is required to equilibrate body temperature in a $20 \pm 1°C$ environment for a minimum of 15 min in order to stimulate the thermoregulatory autonomic response. The equilibration time may be prolonged in patients who are carrying a heat load from a warm outside environment or who may have a high basal metabolic rate (BMR). A patient assessment should be performed before equilibration. This includes assessing the ability to tolerate the procedure, evaluation of any contraindications, taking an appropriate medical history, and conducting a physical examination. In addition, mental status, pain levels, symptoms/signs of allodynia, hyperalgesia, hyperpathia, vasomotor or sudomotor findings, and risk of vasomotor instability should be assessed. Documentation of the patient's current medications and therapies, results of any previous thermographic or vascular studies, and results of any previous sympathetic or vascular interventions should also be acquired [79,80].

Before beginning equilibration, the patient is asked to disrobe and put on a loose-fitting cotton gown that covers the breasts and genitalia. The patient is required to stand during the equilibration period and is asked not to scratch, rub, or touch any area of the skin that is going to be imaged. During the equilibration period, the technician should ensure that the patient has followed all preprocedure protocols [79,80].

26.3.8 Study Protocol for fIR Pain Studies

Upper body fIR pain study (qualitative and quantitative images are of the same views):
   Capture IR images of

1. Posterior cervical region
2. Posterior thoracic region that includes posterior arms
3. Anterior thoracic region that includes anterior arms
4. Anterior forearms and palmar hands (preferably in one image)
5. Posterior forearms and dorsal hands (preferably in one image)
6. Radial forearms and hands (preferably in one image)
7. Ulnar forearms and hands (preferably in one image)

Evaluate the following three (3) IR signature indices required for an interpretive impression for presumptive CRPS of the upper extremities:
Index 1: Quantitative side-to-side homologous computer-generated temperature measurements of

1. Anterior forearms and palmar hands (preferably in one image)
2. Posterior forearms and dorsal hands (preferably in one image)
3. Radial and ulnar forearms also

Index 2: Black–white distal thermal gradient IR signatures

1. Palmar hands (preferably in one image)
2. Dorsal hands (preferably in one image)

Index 3: Cold-water autonomic functional stress test IR signatures

1. Distal posterior forearms and dorsal hands (one image)

Lower body fIR pain study (qualitative and quantitative images are of the same views):
   Capture IR images of

1. Posterior lumbar and buttock region
2. Posterior thighs and legs (preferably in one image)
3. Anterior thighs and legs (preferably in one image)
4. Left lateral thigh and leg with medial right thigh and leg (one image)
5. Right lateral thigh and leg with medial left thigh and leg (one image)
6. Dorsal feet (preferably in one image)
7. Plantar feet (preferably in one image)

Evaluate the following IR signature indices required for an interpretive impression for presumptive CRPS of the lower extremities:
Index 1: Quantitative side-to-side homologous computer-generated temperature measurements of

1. Anterior legs (preferably in one image)
2. Posterior legs (preferably in one image)
3. Dorsal feet (preferably in one image)
4. Plantar feet (preferably in one image)

Index 2: Black–white distal thermal gradient imaging of

1. Dorsal feet (preferably in one image)
2. Plantar feet (preferably in one image)

Index 3: Cold-water autonomic functional stress testing of

1. Distal anterior legs and dorsal feet (one image)

26.3.9 Methods for Obtaining Three Specific fIR Indices Required for an Interpretive Impression for Patients with Presumptive CRPS

Index 1: Qualitative/quantitative side-to-side homologous computer-generated temperature views. Both the quantitative and qualitative images are obtained by capturing baseline color qualitative thermal images 0–50°C, with 0.05° accuracy. The user can display the 12-bit data in color or gray scale with software-installed color maps. The images are displayed with an 85–100-color palette and 0.15°C thermal window. Once the qualitative thermal images are captured, the technician outlines the region of interest using the polygon drawing tool that is embedded in the medical software. In an upper body pain study, the technician draws a polygon around the anterior forearms, posterior forearms, palmar hands, and dorsal hands. In a lower body pain study, the technician draws a polygon around the anterior legs, posterior legs, dorsal feet, and plantar feet. The computer software calculates average temperature(s) within each polygon, allowing calculation of side-to-side homologous temperature differences. A side-to-side homologous temperature difference >1.0°C is generally considered indicative of an autonomic abnormality secondary to an underlying pathology. This 1.0°C temperature difference may range from 0.5°C to 1.5°C depending on the laboratory bias.

Index 2: Black–white distal thermal gradient IR signatures. The black–white distal transverse thermal gradient signature views are obtained by imaging the palmar and dorsal surfaces of the hand in an upper body pain study and the dorsal and plantar surfaces of the feet in a lower body pain study. The images are viewed in the black–white mode with a 10-color palette and 0.05°C thermal window.

Index 3: Cold-water autonomic functional stress test IR signatures. Cold-water autonomic functional stress testing is best performed by utilizing real-time dynamic subtraction imaging that is available on most medical IR software programs. Real-time image subtraction is achieved by choosing a starting reference image, then choosing to view only the differences from the reference-to-the-current image. If the individual pixel temperature rises, the difference will be shown in color; if the temperature drops, the image will be displayed in shades of gray. At any time during the imaging process the user can choose to view the reference, delta, or current image. All thermal data have a dynamic range of 12 bits, enabling the user to view 0.05° difference in a 0–50°C temperature range. This testing is performed by imaging the symptomatic and contralateral asymptomatic distal extremity for 5 min while an asymptomatic limb is placed in a 12–16°C cold-water bath. The immersion of a noninvolved limb activates autonomic thermoregulatory function. If autonomic function is intact, there is vasoconstriction in all four extremities due to the central vasoconstrictor reflex. If the autonomic vasoconstrictor reflex is inhibited or there is autonomic failure, then an axon vasodilatation reflex will occur. This reflex will be visualized by a warming of the symptomatic distal extremity, and on occasion the bilateral asymptomatic distal extremity, during the 5-min cold-water autonomic functional stress test.

Conwell et al. [41] have described cold-water autonomic functional stress testing and the sensitivity, specificity, and predictive value in comparing stress test results with modified IASP criteria for CRPS. This IR index, in and of itself, reveals a 72% sensitivity, 94% specificity, positive predictive value 82%, negative predictive value 90%, and kappa statistical analysis of 0.69 (95% confidence interval: 0.55 and 0.83). The authors posited that cold-water autonomic functional stress testing may be helpful in identifying those patients with a sympathetically mediated component that have a high probability to positively respond to sympatholytic intervention(s). Those patients who have an empirically normal cold-water stress test response characteristically do not respond to sympathetic nerve blocks, and are considered to have the sympathetically independent pain (SIP) form of CRPS I.

26.4 Case Studies

26.4.1 Integrating Pathophysiology of CRPS I to Findings from fIR Case Studies

Functional infrared imaging detects the thermal signature produced by changes in cutaneous blood flow regulated by central thermal and respiratory control that affects vasoconstrictor and sudomotor reflexes. This vaso- and sudomotor activity is predominantly, but not exclusively, dependent upon hypothalamic mechanisms. Sympathetic preganglionic neurons project to the paravertebral ganglia and synapse upon postganglionic neurons innervating target organs and cells. Postganglionic sympathetic neurons release norepinephrine (NE) and neuropeptide Y (NPY) to regulate cutaneous blood flow. Studies suggest that the thermoregulatory dysfunction in CRPS I is due to central inhibition of the cutaneous sympathetic vasoconstrictor reflex [81].

In addition to inhibition of efferent sympathetic neurons, afferent neurons also regulate cutaneous blood flow. Sensitization of cutaneous, small-caliber C fibers evokes orthodromic release of the tachykinin substance-P (SubsP) within the dorsal horn to engage the nociceptive neuraxis, and may also cause antidromic release of SubsP, as well as calcitonin gene-related peptide (CGRP) to (1) elicit peripheral vasodilatation, causing extravasation of other pronociceptive and proinflammatory mediators (e.g., serotonin [5-HT], bradykinin, vasoactive intestinal peptide [VIP]), and (2) perpetuate a neurogenic inflammatory response [24,61–65,68,82–85]. This cutaneous vasodilatation may be involved in the clinically observed vasomotor changes in patients with acute CRPS I, and is easily demonstrated with fIR imaging by visualizing the infrared hyperthermic radiation. Wasner et al. revealed that vascular hyperthermic abnormalities seen in acute CRPS I (and visualized via IR signature(s)) are the result of complete inhibition of sympathetic nerve activity, rather than antidromic vasodilatation secondary to activation of nociceptive afferents [86].

26.4.2 Interpretation of fIR Signatures

Gulevich et al. [38] demonstrated 93% sensitivity and 89% specificity with fIR in diagnosing cases presumed to be CRPS I. These results were obtained by evaluating three separate and distinct IR indices:

1. Computer-generated side-to-side quantitative homologous temperature differences of the symptomatic and asymptomatic distal extremities obtained after the patient equilibrated in a controlled temperature environment of ≤20°C for 15 min (see Section 26.3).
2. Black-and-white distal transverse thermal gradient IR signatures evaluated for maintenance or presence of well-defined transverse gradient lines, which represents normal vasomotor presentation or disruption of the transverse gradient lines, which represent normal vasomotor presentation or disruption of the tranverse gradient lines that represent an abnormal vasomotor presentation. The distal thermal gradient patterns are visualized in the hands or feet, paying particular attention to fingers and toes.
3. Responses to cold-water autonomic functional stress testing [38,41,87,88] of the symptomatic and asymptomatic distal extremity to evaluate for autonomic function with concomitant maintenance of the vasoconstrictor reflex. When autonomic function is intact, there is evidentiary cooling of the distal symptomatic extremity due to maintenance of the vasoconstrictor reflex. With autonomic dysfunction, there is warming of the distal symptomatic extremity [38,41]. This warming may be due to inhibition/failure of the vasoconstrictor reflex [38,41,89–92] or adrenergic

sensitization of nociceptors (viz., upregulation and/or hyperaffinity of alpha adrenergic receptors [93]) producing an axon reflex-mediated vasodilatation. This axon reflex-mediated vasodilatation is not suppressed by sympathetic activity due to central inhibition of sympathetic efferent fibers.

26.4.3 Normal fIR Signatures

In an asymptomatic (i.e., healthy, normal control) patient population, the three fIR image indices are entirely normal as demonstrated by (1) symmetrical thermal emission; (2) normal well-defined transverse thermal gradient lines; and (3) normal response (i.e., cooling) to cold-water autonomic functional stress testing [38] (Table 26.1).

IR findings:

- Bilateral thermal symmetry in region of interest (ROI)
- Normal, well-defined and uniform, transverse distal thermal gradient lines
- Cooling of the symptomatic distal extremity during cold-water autonomic functional stress testing

IR indices description:

1. *Quantitative thermal emission (computer-generated side-to-side temperature) image finding*: Uematsu et al. [94] in a pioneering normative study demonstrated that in a normal healthy asymptomatic patient population, the quantitative computer-generated side-to-side temperature differences of homologous body parts is in the range of 0.17–0.45°C, with a human surface temperature symmetry averaging 0.24 ± 0.073°C between homologous sides [95]. The mean standard deviation for repetitive readings over time of computer-generated side-to-side temperature differences of homologous body parts was 30.8 ± 0.032°C [96]. Uematsu's data have been confirmed by numerous authors [38,97,98].

2. *Distal thermal gradient image pattern findings*: In normal, healthy asymptomatic patients, the distal transverse thermal gradient lines visualized in the distal extremities, particularly in the fingers and toes, are well maintained [38]. Normal distal thermal gradient patterns in the fingers and toes are represented by distinct uniform transverse lines that are closely aligned, forming an alternating black–white–black linear pattern. Normal distal thermal gradient IR signatures may be a result of the normal rhythmic cycling of cutaneous blood flow that is seen in healthy asymptomatic individuals [99].

TABLE 26.1   Normal Study

Normal Study	fIR Index 1: ROI delta T°	fIR Index 2: Distal Thermal Gradient IR Signatures	fIR Index 3: Cold-Water Autonomic Functional Stress Test
IR Signature: 1. Quantitative and qualitative thermal symmetry in the ROI	IR Signature: 1. ROI symmetrical IR signature <1.0°C	IR Signature: 1. Maintained distal gradient IR signature	IR Signature: 1. Normal cooling of the symptomatic and asymptomatic distal extremity Putative mechanism: 1. Normal ANS function with maintenance of the vasoconstrictor reflex
Upper body fIR study IR Signatures **Lower body fIR study** IR Signatures			

3. *Cold-water autonomic functional stress test findings:* In normal, healthy asymptomatic patients with intact autonomic function, there is cooling of the distal extremities when a noninvolved extremity is placed in a ≤15°C cold-water bath [38,41] or when subjected to whole-body cooling by perfusion of circulating water into a thermal suit [40,100]. Additional studies have shown that following the arousal stimuli and the cold pressor test, the vasoconstrictor response is observed in asymptomatic patients and is diminished or absent in patients with CRPS I [89–92].

26.4.4 Abnormal fIR Studies

26.4.4.1 Mimics of CRPS I

In symptomatic patients, numerous conditions *mimic* the modified IASP CRPS criteria, and these make differentiation difficult by the nonexpert clinician. Instead, we argue that the following three fIR image indices, with their specific thermal signatures, provides significant aid in differentiating CRPS I from other pain syndromes.

26.4.5 Acute Trauma (Post-Traumatic Injury): fIR Signatures

Following acute trauma, patients without strong clinical signs of sympathetic dysfunction or acute nerve injury may present with symptoms similar to CRPS I (Table 26.2). This patient population shows clinically relevant hypothermia in the symptomatic extremity that is likely due to maintenance of the vasoconstrictor reflex secondary to a peripheral pain generator [101].

Image findings:

- Unilateral hypothermic vascular disturbances in ROI
- Normal well-defined transverse distal thermal gradient lines
- Cooling of the symptomatic distal extremity during cold-water autonomic functional stress testing due to normal autonomic function with maintenance of the vasoconstrictor reflex

TABLE 26.2   Acute Trauma

Acute Trauma	fIR Index 1: ROI delta T°	fIR Index 2: Distal Thermal Gradient IR Signatures	fIR Index 3: Cold-Water Autonomic Functional Stress Test
IR Signature:	IR Signature:	IR Signature:	IR Signature:
1. Quantitative and qualitative thermal asymmetry in the ROI	1. ROI hypothermic IR signature >1.0°C	1. Maintained distal gradient IR signatures	1. Normal cooling of the symptomatic and asymptomatic distal extremity
Putative mechanism:			Putative mechanism:
1. Increased sympathetic vasomotor tone secondary to a peripheral pain generator			1. Normal ANS function with maintenance of the vasoconstrictor reflex
2. No evidence of neurogenic inflammation–antidromic vasodilatation			
Upper body fIR study			
IR Signatures			
Lower body fIR study			
IR Signatures			

Image indices description:

1. *Quantitative thermal emission (computer-generated side-to-side temperature) findings*: In cases of acute trauma with an intact autonomic function, skin temperature cooling is a common finding of the affected symptomatic extremity when compared to the unaffected, contralateral extremity. Cooling in the symptomatic extremity is due to a somatoautonomic reflex secondary to a peripheral pain generator, with a resultant increase in sympathetic vasoconstrictor activity. This normal somatoautonomic reflex vasoconstriction is due to excitation of sensory receptors and/or irritation of a sensory nerve that elicits increased sympathetic tone, resulting in skin cooling. This vasoconstriction is predominantly visualized in the territory in which the afferent stimulus originates [102].
2. *Distal thermal gradient lined patterns*: In post-traumatic states, the distal thermal gradient lines are well maintained. This is believed to be due to maintenance of normal sympathetic vasomotor tone.
3. *Cold-water autonomic functional stress test findings*: In post-traumatic states, autonomic function is intact and elicits cooling of the symptomatic extremity during cold-water autonomic functional stress testing. Several researchers believe that maintenance of the sympathetic vasoconstrictor reflex may be a feature that allows differentiation between normal, post-traumatic states, and CRPS I [41,85,92,103,104].

26.4.6 C-Fiber (Small-Caliber Nociceptive Afferent Excitation) Mononeuropathy: fIR Signatures

Excitation of peripheral sensory C-fiber nociceptive afferents produces both orthodromic and antidromic release of vasodilator substances from the involved nerve terminals (Table 26.3). The vasodilatatory substances include SubsP and CGRP. The (antidromic) vasodilatatory effects induce a rise in surface temperature and increase in the resultant radiant heat signature. This hyperthermia is independent of sympathetic activity, and is localized to the skin territory innervated by the particular C-fiber [83,105,106].

TABLE 26.3   C-Fiber (Small-Caliber Nociceptive Afferent Excitation) Mononeuropathy

Small-Caliber Fiber Distal Sensory Mononeuropathy (ABC Syndrome)	fIR Index 1: ROI delta T°	fIR Index 2: Distal Thermal Gradient IR Signatures	fIR Index 3: Cold-Water Autonomic Functional Stress Test
IR Signature:	IR Signature:	IR Signature:	IR Signature:
1. Vasodilatation (visualized hyperthermic emission) localized in specific skin territory of the involved nerve	1. ROI hyperthermic IR signature >1.0°C localized to the specific skin territory of the involved nerve	1. Disrupted distal gradient IR signature localized to the skin territory of the sensitized peripheral nerve	1. Normal cooling of the symptomatic and asymptomatic distal limb
Putative mechanism:			Putative mechanism:
1. Antidromic vasodilatation unrelated to sympathetic activity			1. Normal ANS function that overrides the antidromic vasodilatation with maintenance of the vasoconstrictor reflex
Upper body fIR study			
IR Signatures			

Image findings:

- Unilateral hyperthermic vascular disturbances isolated to a specific nerve territory
- Disrupted transverse distal thermal gradient lines isolated to a specific nerve territory
- Normal cooling of the symptomatic distal extremity during cold-water autonomic functional stress testing due to normal autonomic function that suppresses vasodilatation induced by antidromic release of pro-inflammatory substances

Image indices description:

1. *Quantitative thermal emission (computer-generated side-to-side temperature) findings*: In these cases there is skin temperature warming (hyperthermia) localized to the skin territory of the involved nerve that is independent of sympathetic activity. There is normal thermal symmetry of the asymptomatic extremity.
2. *Distal thermal gradient lined patterns*: The distal thermal gradient lines are disrupted solely in the skin territory of the involved nerve with loss of the normal transverse lines.
3. *Cold-water autonomic functional stress test findings*: Normal cooling of the symptomatic extremity is observed during cold-water autonomic functional stress testing because the observed vasodilatation is independent of sympathetic activity in this pathology. Therefore, the normal cooling is most likely due to intact autonomic function with maintenance of the vasoconstrictor reflex that suppresses vasodilatation [41,90,94,96–98,107].

26.4.7 Acute CRPS I fIR Signatures

26.4.7.1 Abnormal fIR Signatures Visualized Solely in Symptomatic Limb

The following fIR images were taken from case studies of patients who met both the IASP and Gulevich et al. [34] criteria for CRPS I, which, when taken together, demonstrated a 93% sensitivity and 89% specificity in diagnosing patients with presumed CRPS I (Table 26.4). These fIR cases show hyperthermia of the symptomatic distal extremity visualized by computer-generated side-to-side homologous temperature differences >1.0°C. These cases also show evidence of inhibition of the vasoconstrictor reflex as revealed by abnormal hyperthermic response to cold-water autonomic functional stress testing [41]. This fIR thermal presentation appears to be consistent with the hypothesized central inhibition of sympathetic activity that occurs in acute-stage CRPS I, producing decreased release of NE at the terminal sites, and resulting in vasodilatation and increased cutaneous blood flow [82]. This central sympathetic inhibition results in abnormal warming during cold-water autonomic functional stress testing.

Image findings:

- Unilateral hyperthermic vascular disturbances in ROI
- Disrupted transverse distal thermal gradient lines in ROI
- Abnormal warming of the distal symptomatic extremity (generally seen in the fingers or toes of the involved extremity) during cold-water autonomic functional stress testing

Image indices description:

1. *Quantitative thermal emission (computer-generated side-to-side temperature) findings*: In the acute phase of CRPS I, the affected distal extremity shows abnormal vasodilatation with subsequent skin warming as compared to the contralateral asymptomatic extremity [81,92,108–111]. This vasodilatation is due, in part, to central inhibition of sympathetic activity and a decreased release of NE and NPY at terminal sites of sympathetic neurons [112–114]. Decreased NE activity at the terminal sites results in an increased cutaneous blood flow, which produces the clinical picture of

TABLE 26.4 Acute CRPS I

Acute CRPS I	fIR Index 1: ROI delta T°	fIR Index 2: Distal Thermal Gradient IR Signatures	fIR Index 3: Cold-Water Autonomic Functional Stress Test
IR Signature:	IR Signature:	IR Signature:	IR Signature:
1. Vasodilatation (visualized hyperthermic emission) with global, nondermatomal skin warming in the affected distal extremity	1. ROI hyperthermic IR signature >1.0°C	1. Disrupted distal gradient IR signature in a global, nondermatomal distribution in the symptomatic distal extremity and rarely may be visualized in the asymptomatic contralateral distal extremity	1. Abnormal warming of symptomatic distal extremity with occasional warming of the contralateral side
Putative mechanism:			Putative mechanism:
1. Neurogenic inflammation–afferent axon reflex vasodilatation 2. Central sympathetic inhibition			1. Abnormal ANS function with impairment of the sympathetic vasoconstrictor reflex or sympathetic failure 2. Intense axon reflex vasodilatation secondary to C-fiber sensitivity to circulating NE
Upper body fIR study IR Signatures **Lower body fIR study** IR Signatures			

a warm, discolored extremity. Inhibition of neuronally mediated vasoconstrictor reflexes is not believed to be due to major damage to the peripheral nerve (as in CRPS II), but rather due to a central inhibition of the thermoregulatory function [40,108,111].

2. *Distal thermal gradient line patterns*: In symptomatic patients with thermoregulatory dysfunction, the distal transverse thermal gradient lines visualized in the distal extremities (particularly in the fingers and toes) are disrupted [38]. The disrupted lines are represented by irregular black–white–black gradient lines that are highly irregular without evidence of the normal transverse well-maintained pattern alignment. The disrupted irregular patterns are felt to represent aberrations produced by abnormal sympathetic vasomotor tone.

3. *Cold-water autonomic functional stress test patterns*: In the acute phase of CRPS I, the affected extremity abnormally warms during cold-water autonomic functional stress testing [38,41]. It is hypothesized that this abnormal warming is due to inhibition or complete failure of the vasoconstrictor reflex and concomitant axon reflex-mediated vasodilatation. The axon reflex vasodilatation prevails due to the absence or inhibition of vasoconstrictor reflexes that would normally suppress vasodilatation. This has been illustrated by a case presentation of a patient who developed complete failure of the tonic vasoconstrictor response to cooling within 2 weeks of CRPS I onset [108]. This has been further supported by literature showing diminished sympathetic vasoconstrictor reflexes in the affected limb during the early-acute phases of CRPS I [110,111]. This abnormal warming may be due to adrenergic sensitization of nociceptors [93].

26.4.8 Acute CRPS I fIR Signatures: Bilateral Presentation

26.4.8.1 Abnormal fIR Signatures Visualized in Both Symptomatic and Asymptomatic Limbs

There are reports in the literature of subclinical *contralateral* sympathetic involvement in CRPS I [89,90,99] that are associated with axon reflex vasodilatation (hyperthermia) in both the distal symptomatic and asymptomatic limbs [115]. The following fIR images are taken from case studies of patients who met IASP CRPS I criteria and the criteria of Gulevich et al. [38], which demonstrated 93% sensitivity and 89% specificity in diagnosing patients with presumed CRPS I who demonstrated bilateral vasomotor findings (Table 26.5). These fIR cases show hyperthermia of both the symptomatic and asymptomatic distal extremity, evidenced by computer-generated side-to-side homologous temperature differences of a >1°C hyperthermia of the symptomatic distal extremity. These studies also show evidence of inhibition of the vasoconstrictor reflex subserved by an abnormal axon reflex vasodilatation during cold-water autonomic functional stress testing [41]. The hypothermia is likely due to supersensitivity to circulating catecholamines evoking vasoconstriction [116]. These findings tend to support the notion that CRPS I may also involve central mediation via an increased activation of brainstem or hypothalamic–pituitary–adrenal mechanisms.

Image findings:

- Bilateral distal hyperthermic vascular disturbances with the symptomatic distal extremity (ROI) demonstrating a >1.0°C hyperthermic difference with the contralateral asymptomatic limb
- Bilateral disrupted transverse distal thermal gradient lines in ROI
- Bilateral abnormal warming of the distal extremities (generally seen in the fingers or toes) during cold-water autonomic functional stress testing.

TABLE 26.5 Acute CRPS I (Bilateral Thermal Findings)

Acute CRPS I Bilateral Thermal Findings	fIR Index 1: ROI delta T°	fIR Index 2: Distal Thermal Gradient IR Signatures	fIR Index 3: Cold-Water Autonomic Functional Stress Test
IR Signature:	IR Signature:	IR Signature:	IR Signature:
1. Vasodilatation (visualized hyperthermic emission) with global, nondermatomal skin warming in the affected distal extremity	1. ROI hyperthermic IR signature >1.0°C	1. Bilateral disrupted distal thermal gradient IR signature (seen in the fingers or toes) visualized in a global nondermatomal distribution in the effected distal extremity and asymptomatic contralateral distal limb	1. Abnormal warming of symptomatic and asymptomatic distal extremity
Putative mechanism:			Putative mechanism:
1. Central sympathetic inhibition 2. Neurogenic inflammation–afferent axon reflex vasodilatation			1. Abnormal ANS function with bilateral impairment of the sympathetic vasoconstrictor reflex or sympathetic failure 2. Bilateral axon reflex vasodilatation secondary to C-fiber sensitivity to circulating NE
Upper body fIR study			
IR signatures			

26.4.9 Chronic CRPS I fIR Signatures

The following fIR images are taken from case studies of patients who met the IASP criteria for chronic CRPS I as well as the Gulevich et al. [38] criteria (Table 26.6). These fIR studies show hypothermia of the symptomatic distal extremity visualized by computer-generated side-to-side homologous temperature differences >1.0°C. These case studies also show evidence of inhibition of the vasoconstrictor reflex associated with an axon reflex vasodilatation that is readily visualized in the symptomatic distal extremity, and occasionally the asymptomatic distal extremity during the cold-water autonomic functional stress testing [41].

Image findings:

- Unilateral hypothermic vascular disturbances with the ROI being >1.0°C colder than the contralateral asymptomatic limb
- Unilateral or bilateral disrupted transverse distal thermal gradient lines
- Unilateral or bilateral (warming) abnormal response to cold-water autonomic functional stress testing

Image indices description:

1. *Quantitative thermal emission (computer-generated side-to-side temperature) findings*: In the chronic phase of CRPS I, the affected distal extremity shows abnormal vasoconstriction with subsequent skin cooling of the involved distal limb as compared to the contralateral asymptomatic limb [37,38,40,116,117]. This vasoconstriction is felt to be due to supersensitivity to circulating catecholamines [118]. Numerous mechanisms are responsible for this adrenergic supersensitivity, including, but not limited to, diminished neurotransmitter reuptake, enzyme degradation, increased alpha adrenoceptor binding affinity, and/or density of receptor sites, increased expression of sodium channels and/or sodium–potassium pump inefficiency [119].

TABLE 26.6   Chronic CRPS I

Chronic CRPS I	fIR Index 1: ROI delta T°	fIR Index 2: Distal Thermal Gradient IR Signatures	fIR Index 3: Cold-Water Autonomic Functional Stress Test
IR Signature:	IR Signature:	IR Signature:	IR Signature:
1. Vasoconstriction (visualized hypothermic emission) with skin cooling in the affected distal extremity	1. ROI hypothermic IR signature >1.0°C	1. Bilateral disrupted distal thermal gradient IR signature (seen in the fingers or toes) visualized in a global nondermatomal distribution in the effected distal extremity and asymptomatic contralateral distal limb	1. Abnormal warming of the symptomatic distal extremity with occasional warming of the contralateral asymptomatic extremity
Putative mechanism:			Putative mechanism:
1. Supersensitivity to circulating catecholamines–adrenergic supersensitivity			1. Abnormal ANS function with bilateral impairment of the sympathetic vasoconstrictor reflex or sympathetic failure
2. Minimal or absent sympathetic inhibition			2. Bilateral axon reflex vasodilatation secondary to C-fiber sensitivity to circulating NE

Upper body fIR study

IR signatures

2. *Distal thermal gradient line patterns*: The distal transverse thermal gradient lines in chronic CRPS I are disrupted with a loss of the normal well-defined transverse gradient lines [38]. It is hypothesized that this disruption is the result of vasomotor disturbances evoked by circulating catecholamines.

3. *Cold-water autonomic functional stress test findings*: In the chronic phase of CRPS I, the affected extremity, and occasionally the contralateral unaffected extremity, shows abnormal warming during cold-water autonomic functional stress testing [38,41]. It is hypothesized that this abnormal warming is due to inhibition or complete loss of the vasoconstrictor reflex resulting in an axon reflex-induced vasodilatation.

26.4.10 Clinical Implications and Potential

These findings support our contention that fIR, when coupled to advanced computational software, can effectively detect thermal signatures that reflect particular vaso- and sudomotor disturbances that are important in establishing a differential diagnosis of CRPS I. Obviously, it is important to restate that these objective features cannot be taken in isolation, and thus we do not advocate that fIR be considered or utilized as a stand-alone diagnostic modality. However, it is equally important to recognize that the inherent ambiguities in the presentation of CRPS I (and a persistent reticence to acknowledge the validity of a diagnosis of CRPS I based upon lesser objective findings) fortify the utility of this technology as a part of the diagnostic workup. Functional IR testing, when administered and evaluated by a competently trained professional, can provide reliable data that can contribute to both the diagnosis of a particular patient and a progressive database that can be utilized to develop an objective standard for clinical discernment and diagnoses.

Extant concerns and refutations of the plausibility of IR thermography reflected inadequacies of older technology, and were valid criticisms of the lack of specificity, inappropriate and/or inapt use by untrained personnel, and the entrepreneurial overuse of IR as a "diagnostic" test. However, these biases are no longer applicable as the enjoinment of current technology (based upon declassified military instrumentation) and advanced statistical software has rendered all prior iterations of thermographic detection almost obsolete and established a new benchmark for technical efficiency and accuracy.

This technology will only improve, and thus it is important to both study fIR further to develop new avenues for clinical use and employ these methods in the detection of CRPS I and other diagnostically difficult syndromes.

Dedication

This chapter is dedicated to the memory of Nicholas A. Diakides, D.Sc. Dr. Diakides was a pioneer in the development of knowledge-based databases of standardized IR signatures validated by pathology which was instrumental in setting the groundwork for numerous advances in medical IR imaging. The scientific contribution, by this exceptional human being, to the understanding and potential benefit of medical IR imaging will most certainly play a role in improving the health and well-being of future generations.

Acknowledgments

Contributions to this work were supported in part by Colorado Infrared Imaging Center, Denver Colorado. The fIR case studies were obtained from the center's patient files.

TC gratefully acknowledges Drs. Nicholas Diakides (deceased), William Hobbins, Stephen Gulevich, L. Barton Goldman, Floyd Ring, Richard Stieg, and Neil Rosenberg (deceased) for their support and encouragement over many years.

Contributions to this work were also supported in part by the Institute for Biotechnology Futures, The William H. and Sara Crane Schaeffer Endowment, and Nour Foundation (JG). JG gratefully acknowledges Kim Abramson for assistance on this chapter.

We express our sincere thanks to Maurice Bales, founder and president, Bales Scientific, for his valuable and lifelong dedication to medical infrared imaging. The fIR images in this chapter were captured by a Bales Scientific advanced image acquisition Tip-50 infrared camera interfaced with sophisticated medical computerized analytic system imaging software.

References

1. Hendler, N., Complex regional pain syndrome types I and II, in *Weiner's Pain Management: A Practical Guide for Clinicians*. 7th ed., Boswell, M.V. and Cole, B.E., Eds., CRC Press, Boca Raton, FL, 2005.
2. Sandroni, P. et al., Complex regional pain syndrome type I: Incidence and prevalence in Olmsted County—A population based study, *Pain*, 103, 199, 2003.
3. Allen, G., Galer, B.S., and Schwartz, L., Epidemiology of complex regional pain syndrome: A retrospective chart review of 134 patients, *Pain*, 80, 539, 1999.
4. Harden, R.N., Baron, R., and Janig, W., *Complex Regional Pain Syndrome, Progress in Pain Research and Management*, IASP Press, Seattle, Vol. 22, 2001.
5. Janig, W. and Levine, J.D., Autonomic-endocrine-immune interactions in acute and chronic pain, in *Wall and Melzack's Textbook of Pain*, 5th ed., McMahon, S.P. and Koltzenburg, M., Eds., Churchill Livingston, Edinburgh, 2005.
6. Merskey, H. and Bogduk, N. *Classification of Chronic Pain: Descriptions of Chronic Pain Syndrome and Definitions of Pain Terms*, 2nd ed., IASP Press, Seattle, 1994.
7. Stanton-Hicks, M. et al., Reflex sympathetic dystrophy: Changing concepts and taxonomy, *Pain*, 63, 127, 1995.
8. Janig, W. and Stanton-Hicks, M., Eds. *Reflex Sympathetic Dystrophy: A Reappraisal, Progress and Pain Research and Management*, IASP Press, Seattle, Vol. 6, 1996.
9. Veldman, P.H. et al., Signs and symptoms of reflex sympathetic dystrophy: Prospective study of 829 patients. *Lancet*, 342, 1012, 1993.
10. Bonica, J.J., *The Management of Pain*, Lea and Febiger, Philadelphia, PA, 1953.
11. Kozin, F. et al., Reflex sympathetic dystrophy (RSDS), III: Scintigraphic studies, further evidence for the therapeutic efficacy of systemic corticosteroids and proposed diagnostic criteria, *Am. J. Med.*, 70, 23, 1981.
12. Blumberg, H., A new clinical approach for diagnosing reflex sympathetic dystrophy, in *Proceedings of the VI World Congress on Pain*, Bond, M.R., Charlton, J.E., and Woolf, C.J., Eds., Elsevier, New York, p. 399, 1991.
13. Gibbons, J.J. and Wilson, P.R., RSD score: Criteria for the diagnosis of reflex sympathetic dystrophy and causalgia, *Clin. J. Pain*, 8, 260, 1992.
14. Harden, R.N. and Bruehl, S.P., Diagnostic criteria: The statistical deviation of the four criteria factors, in *CRPS: Current Diagnosis and Therapy*, Wilson, P.R., Stanton Hicks, M., and Harden, R.N., Eds., IASP Press, Seattle, Vol. 32, 2005.
15. Galer, B.S., Bruehl, S., and Harden, R.N., IASP diagnostic criteria for complex regional pain syndrome: A preliminary comparable validation study, *Clin. J. Pain*, 14, 48, 1998.
16. Bruehl, S. et al., External validation of IASP diagnostic criteria for complex regional pain syndrome and proposed research diagnostic criteria, *Pain*, 81, 147, 1999.
17. Van de Beek, W.J. et al., Diagnostic criteria used in studies of reflex sympathetic dystrophy, *Neurology*, 8, 522, 2002.
18. Van den Vusse, A.C. et al., Interobserver reliability of the diagnosis in patients with CRPS, *Eur. J. Pain*, 7, 259, 2003.

19. Harden, R.N. et al., Proposed new diagnostic criteria for complex regional pain syndrome, *Pain Med.*, 8(4), 326, 2007.
20. Harden, R.N. et al., Validation of proposed diagnostic criteria (the "Budapest Criteria") for Complex Regional Pain Syndrome, *Pain*, 150(2), 268, 2010.
21. Sumitani, M. et al., Development of comprehensive diagnostic criteria for complex regional pain syndrome in the Japanese population, *Pain*, 150(2), 243, 2010.
22. Giordano, J., *Pain: Mind, Meaning and Medicine*. Glen Falls, PA: PPM Books, 2009.
23. Giordano, J., Prolegomenon: Engaging philosophy, ethics and policy in, and for pain medicine, in *Pain Medicine: Philosophy, Ethics, and Policy*. Giordano J. and Boswell M.V., Eds., Linton Atlantic Books, Oxon, 2009, pp. 13–20.
24. Giordano, J., The neuroscience of pain and analgesia, in *Weiner's Pain Management: A Practical Guide for Clinicians* 7th ed., Boswell, M.V. and Cole, B.E., Eds., CRC Press, Boca Raton, FL, 2005.
25. Giordano, J., Moral agency in pain medicine: Philosophy, practice and virtue, *Pain Phys.*, 9, 71, 2006.
26. Giordano, J., Good as gold? The randomized controlled trial—Pragmatic and ethical issues in pain research, *Am. J. Pain Manage.*, 16, 68, 2006.
27. Pellegrino, E.D., The healing relationship: The architectonics of clinical medicine, in *The Clinical Encounter: The Moral Fabric of the Patient-Physician Relationship*, Philosophy and Medicine Series 4, Shelp, E.E., Ed. Dordrecht, D., Reidel, 1983.
28. Giordano, J., Pain research: Can paradigmatic revision bridge the needs of medicine, scientific philosophy and ethics? *Pain Phys.*, 7, 459, 2004.
29. Karstetter, K.W. and Sherman, R.A., Use of thermography in initial detection of early reflex sympathetic dystrophy, *J. Am. Podiatr. Med. Assoc.*, 81, 198, 1991.
30. Perelman, R.B., Adler, D., and Humphries, M., Reflex sympathetic dystrophy: Electronic thermography as an aid in diagnosis, *Orthop. Rev.*, 16, 561, 1987.
31. Lightman, H.I. et al., Thermography in childhood reflex sympathetic dystrophy, *J. Pediatr.*, 111, 551, 1987.
32. Lewis, R., Racz, G., and Fabian, G., Therapeutic approaches to reflex sympathetic dystrophy of the upper extremity, *Clin. Issues Reg. Anesth.*, 1, 1, 1985.
33. Uematsu, S. et al., Thermography and electromyography in the differential diagnosis of chronic pain syndromes and reflex sympathetic dystrophy, *Electromyogr. Clin. Neurophysiol.*, 21, 165, 1981.
34. Bruehl, S. et al., Validation of thermography in the diagnosis of reflex sympathetic dystrophy, *Clin. J. Pain.*, 12, 316, 1996.
35. Cooke, E.D. et al., Reflex sympathetic dystrophy (algoneurodystrophy): Temperature studies in the upper limb, *Br. J. Rheumatol.*, 28, 399, 1989.
36. McLeod, J.G. and Tuck, R.R., Disorders of the autonomic nervous system, *Ann. Neurol.*, 21, 419, 1987.
37. Low, P.A. et al., Laboratory findings in reflex sympathetic dystrophy: A preliminary report, *Clin. J. Pain.*, 10, 235, 1994.
38. Gulevich, S.J. et al., Stress infrared telethermography is useful in the diagnosis of complex regional pain syndrome, type 1 (formally reflex sympathetic dystrophy), *Clin. J. Pain.*, 13, 50, 1997.
39. Birklein, F. et al., Neurological findings in complex regional pain syndrome-analysis of 145 cases, *Acta Neurol. Scand.*, 101, 262, 2000.
40. Wasner, G., Schattschneider, J., and Baron, R., Skin temperature side differences: A diagnostic tool for CRPS? *Pain*, 98, 19, 2002.
41. Conwell, T.D. et al., Sensitivity, specificity and predictive value of infrared cold water autonomic functional stress testing as compared with modified IASP criteria for CRPS, *Thermol. Int.*, 20-2, 60, 2010.
42. Schurmann, M. et al., Imaging in early posttraumatic complex regional pain syndrome: A comparison of diagnostic methods, *Clin. J. Pain.*, 23(5), 449, 2007.
43. Terkelsen, A.J. et al., Experimental forearm immobilization in humans induces cold and mechanical hyperalgesia, *Anesthesiology*, 109(2), 297, 2008.

44. Niehof, S.P. et al., Using skin surface temperature to differentiate between complex regional pain syndrome type I patients after a fracture and control patients with various complaints after a fracture, *Anesth. Analg.*, 106(1), 270, 2008.

45. Rommel, O., Habler, H.J., and Schurmann, M., Laboratory tests for complex regional pain syndrome, in *CRPS: Current Diagnosis and Therapy*, Wilson, P.R., Stanton-Hicks, M., Harden, R.N., Eds., IASP Press, Seattle, Vol. 32, 2005.

46. Hooshmand, H., *Chronic Pain: Reflex Sympathetic Dystrophy Prevention and Management*, CRC Press, Boca Raton, FL, 1993.

47. Stanton-Hicks, M., Janig, W., and Boas, R.A., *Reflex Sympathetic Dystrophy*, Kluwer Academic Publications, Norwell, MA, 1990.

48. Baron, R. and Maier C., Reflex sympathetic dystrophy: Skin blood flow, sympathetic vasoconstrictor reflexes and pain before and after surgical sympathectomy, *Pain*, 67, 317, 1996.

49. Wasner, G. et al., No effect of sympathetic sudomotor activity on capsaicin-evoked ongoing pain and hyperalgesia, *Pain*, 84, 331, 2000.

50. McLachlan, E.M. et al., Peripheral nerve injury triggers noradrenergic sprouting within the dorsal root ganglia, *Nature*, 363, 543, 1993.

51. Ali, Z. et al., Intradermal injection of norepinephrine evokes pain in patients with sympathetically maintained pain, *Pain*, 88, 161, 2000.

52. Baron, R. et al., Relation between sympathetic vasoconstrictor activity and pain and hyperalgesia in complex regional pain syndromes: A case control study, *Lancet*, 359, 1655, 2002.

53. Shi, T.S. et al., Distribution and regulation of alpha(2)—Adrenoceptors in rat dorsal root ganglia, *Pain*, 84, 319, 2000.

54. Sieweke, N. et al., Patterns of hyperalgesia in complex regional pain syndrome, *Pain*, 80, 171, 1999.

55. Janig, W. and Habler, H.J., Organization of the autonomic nervous system: Structure and function, in *Handbook of Clinical Neurology*, Appenzeller, O., Ed., Elsevier, Amsterdam, Vol. 74, p. 1, 1999.

56. Janig, W. and Habler, H.J., Neurophysiological analysis of target-related sympathetic pathways from animal to human: Similarities and differences, *Acta Physiol. Scand.*, 177, 255, 2003.

57. Janig, W. and MacLachlan, E.M., Neurobiology of the autonomic nervous system, in *Autonomic Failure: A Textbook of Clinical Disorders of the Autonomic Nervous System*, 4th ed., Mathias, C.J. and Bannister, R., Eds., Oxford University Press, Oxford, p. 3, 1999.

58. Janig, W., The autonomic nervous system and its coordination by the brain, in *Handbook of Affective Sciences, Part II*, Davidson, R.J., Scherer, K.R., Goldsmith, H.H., Eds., Oxford University Press, Oxford, p. 135, 2003.

59. Campero, M., Bostock, H., Baumann, T.K., and Ochoa, J. L., A search for activation of C nociceptors by sympathetic fibers in complex regional pain syndrome, *Clin. Neurophysiol.*, 121(7), 1072, 2010.

60. Oyen, W.J. et al., Reflex sympathetic dystrophy of the hand: An excessive inflammatory response? *Pain*, 55, 151, 1993.

61. Weber, M. et al., Facilitated neurogenic inflammation in complex regional pain syndrome, *Pain*, 91, 251, 2001.

62. Sudeck, P., Über die acute (trophoneurotische) Knochenatrophie nach Entzündugen und Traumen der Extremitüten, *Deutsche Medizinische Wochenschrift*, 28, 336, 1902.

63. Bove, G.M., Focal nerve inflammation induces neuronal signs consistent with symptoms of early complex regional pain syndromes, *Exp. Neurol.*, 219, 223, 2010.

64. Hartrick, C.T., Increased production of nitric oxide stimulated by interferon-gamma from peripheral blood monocytes in patients with complex regional pain syndrome, *Neurosci. Lett.*, 323, 75, 2002.

65. Giordano, J. and Dyche, J., Differential analgesic action of serotonin 5-HT3 receptor antagonists in three pain tests, *Neuropharmacology*, 28, 431, 1989.

66. Giordano, J. and Schultea, T., Serotonin 5-HT3 receptor mediation of pain and anti-nociception, *Pain Phys.*, 7, 141, 2003.

67. Kozin, F. et al., The reflex sympathetic dystrophy syndrome: I. Clinical and histologic studies: Evidence for bilaterality, response to corticosteroids and articular involvement, *Am J. Med.*, 60, 321, 1976.

68. Sufka, K., Schomberg, F., and Giordano, J., Receptor mediation of 5-HT-induced inflammation and nociception in rats, *Pharmacol. Biochem. Behav.*, 41, 53, 1992.

69. Giordano, J. and Sacks, S., Topical ondansetron attenuates capsaicin-induced inflammation and pain, *Eur. J. Pharmacol.*, 354, 13, 1998.

70. Huygen, F.J. et al., Evidence for local inflammation in complex regional pain syndrome type 1, *Mediators Inflamm.*, 11, 47, 2002.

71. Giordano, J., Neurobiology of nociceptive and anti-nociceptive systems, *Pain Phys.*, 8, 277, 2005.

72. Apkarian, A.V. et al., Prefrontal cortical hyperactivity in patients with sympathetically mediated chronic pain, *Neurosci. Lett.*, 311, 193, 2001.

73. Juttonen, K. et al., Altered central sensorimotor processing in patients with complex regional pain syndrome, *Pain*, 98, 315, 2002.

74. Van de Beek, W.J. et al., Susceptibility loci for complex regional pain syndrome, *Pain*, 103, 93, 2003.

75. Kemler, M.A. et al., HLA-DQ1 associated with reflex sympathetic dystrophy, *Neurology*, 53, 1350–1351, 1999.

76. Van Hilten, J.J., Van de Beek, W.J., and Roep, B.O., Multifocal or generalized tonic dystonia of complex regional pain syndrome: A distinct clinical entity associated with HLA-DR13, *Ann. Neurol.*, 48, 113–116, 2000.

77. Galenus, C. (Galen), *Hippocrates Writings*, Franklin Library, Franklin Center, PA, 1979.

78. Clark, R.P., Human skin temperature and its relevance in physiology and clinical assessment, in *Recent Advances in Medical Thermology*, Ring, E.F.J. and Phillips, B., Eds. Plenum Press, New York, 5, 1984.

79. American Chiropractic College of Infrared Imaging a College of the Council on Diagnostic Imaging, *Technical Protocols for High Resolution Infrared Imaging*, American Chiropractic Association, 1999.

80. Schwartz, R.G., Chair, Practice guidelines committee, Guidelines for neuromusculoskeletal thermography, American Academy of Thermology, *Thermology International*, 16, 5, 2006.

81. Wasner, G. et al., Vascular abnormalities in reflex sympathetic dystrophy (CRPS I): Mechanisms and diagnostic value, *Brain*, 124, 587, 2001.

82. Giordano, J. and Gerstmann, H., Patterns of serotonin- and 2-methylserotonin-induced pain may reflect 5-HT3 receptor sensitization, *Eur. J. Pharmacol.*, 483, 267, 2004.

83. Ochoa, J.L. et al., Intrafascicular nerve stimulation elicits regional skin warming that matches the projected field of evoked pain, in *Fine Afferent Nerve Fibers and Pain*, Schmidt, R.F., Schaible, H.G., Vahle-Hinz, C., Eds., VCH Verlagsgesellschaft, Weinheim, Germany, 1987.

84. Holzer, P., Peptidergic sensory neurons in the control of vascular functions: Mechanisms and significance in the cutaneous and splanchnic vascular beds, *Rev. Physiol. Biochem. Pharmacol.*, 121, 49, 1992.

85. Birklein, F., Kunzel, W., and Sieweke, N., Despite clinical similarities there are significant differences between acute limb trauma and complex regional pain syndrome I (CRPS I), *Pain*, 93, 165, 2001.

86. Wasner, G. et al., Vascular abnormalities in acute reflex sympathetic dystrophy (CRPS I): Complete inhibition of sympathetic nerve activity with recovery, *Arch Neurol.*, 56(5), 613, 1999.

87. Hobbins, W.B., Differential diagnosis of painful conditions and thermography, in *Contemporary Issues in Chronic Pain Management*, Norwell, M.A., Ed., Kluwer Academic Publishers, Paris 251, 1991.

88. Edwards, B.E. and Hobbins, W.B., Pain management and thermography, in *Practical Management of Pain*, 2nd ed., Raj, P.P., Mosby-Year Book, St. Louis, p. 168, 1992.

89. Rosen, L. et al., Skin microvascular circulation in the symptomatic dystrophy is evaluated by video-photometric capillaroscopy and laser Doppler fluxmetry, *Eur. J. Clin. Invest.*, 18, 305, 1998.

90. Kurvers, H.J. et al., The spinal component to skin blood flow abnormalities in reflex sympathetic dystrophy, *Arch. Neurol.*, 53, 50, 1996.

91. Schurmann, M., Grab, G., and Furst, H., A standardized bedside test for assessment of peripheral sympathetic nervous function using laser Doppler flowmetry, *Microvasc. Res.*, 52, 157, 1996.

92. Schurmann, M. et al., Assessment of peripheral sympathetic nervous function for diagnosing early post-traumatic complex regional pain syndrome type 1, *Pain*, 80, 149, 1999.

93. Campbell, J.N., Maier, R.A., and Raja, S.N., Is nociceptor activation by alpha-1 adrenoceptors the culprit in sympathetically maintained pain? *APS J.*, 1, 3, 1992.

94. Uematsu, S., Thermographic imaging of the sensory dermatomes, *Soc. Neurosci.*, abstract, 9, 324, 1983.

95. Uematsu, S., Thermographic imaging of cutaneous sensory segments in patients with peripheral nerve injury, *J. Neurosurg.*, 62, 716, 1985.

96. Uematsu, S. et al., Quantification of thermal asymmetry, Part 1: Normal values and reproducibility, *J. Neurosurg.*, 69, 552, 1988.

97. Feldman, F. and Nickoloff, E.L., Normal thermographic standards for the cervical spine and upper extremities, *Skeletal Radiol.*, 12, 235, 1984.

98. Goodman, P.A., Computer-assisted thermography, *Proceedings of the 14th annual meeting of the American Academy of Thermology*, abstract, 36, 1985.

99. Bej, M.D. and Schwartzman, R.J., Abnormalities of cutaneous blood flow regulation in patients with reflex sympathetic dystrophy as measured by laser doppler fluxmetry, *Arch. Neurol.*, 48, 912, 1991.

100. Bini, G. et al., Thermography and rhythm-generating mechanisms governing the sudomotor and vasoconstrictor outflow in human cutaneous nerves, *J. Physiol.*, (London), 206, 537, 1980.

101. Ochoa, J.L. et al., Interactions between sympathetic vasoconstrictor outflow and C-nociceptors-induced antidromic vasodilatation, *Pain*, 54, 191, 1993.

102. Bennett, G.J. and Ochoa, L.J., Thermographic observations on rats with experimental neuropathic pain, *Pain*, 45, 61, 1991.

103. Schurmann, M. et al., Peripheral sympathetic function as a predictor of complex regional pain syndrome type I (CRPS I) in patients with radial fracture, *Autonom. Neurosci.*, 86, 127, 2000.

104. Rosenbaum, R.B. and Ochoa, J.L., Thermography, in *Carpal Tunnel Syndrome and Other Disorders of the Median Nerve*, Rosenbaum, R.B. and Ochoa, J.L., Eds., Butterworth-Heindmann, Boston, p. 185, 1993.

105. Cline, M.A., Ochoa, J.L., and Torebjork, H.E., Chronic hyperalgesia and skin warming caused by sensitized C nociceptors, *Brain*, 112, 621, 1989.

106. Ochoa, J.L., The newly recognize painful ABC syndrome: Thermographic aspects, *Thermology*, 2, 65, 1986.

107. Ochoa, J.L. et al., Antidromic vasodilatation overridden by somatosympathetic reflexes in man-intraneural stimulation and thermography (abstract), *Soc. Neurosci.*, 16, 1280, 1990.

108. Wasner, G. et al., Vascular abnormalities in acute reflex sympathetic dystrophy (CRPS I): Complete inhibition of sympathetic nerve activity with recovery, *Arch. Neurol.*, 56, 613, 1999.

109. Hornyak, M.E. et al., Sympathetic activity influences the vascular axon reflex in the skin, *Acta Physiol. Scand.*, 139, 77, 1990.

110. Kurvers, H.J. et al., Reflex sympathetic dystrophy: Evolution of microcirculatory disturbances in time, *Pain*, 60, 333, 1995.

111. Birklein, F. et al., Sympathetic vasoconstrictor reflex pattern in patients with complex regional pain syndrome, *Pain*, 75, 93, 1998.

112. Drummond, P.D., Finch, P.M., and Smythe, G.A., Reflex sympathetic dystrophy: The significance of differing plasma catecholamine concentrations in affected and unaffected limbs, *Brain*, 114, 2025, 1991.

113. Drummond, P.A. et al., Plasma neuropeptide Y in the symptomatic limb of patients with causalgia pain, *Clin. J. Pain*, 12, 222, 1996.

114. Harden, R.N. et al., Norepinephrine and epinephrine levels in affected versus unaffected limbs in sympathetically maintained pain, *Clin. J. Pain,* 10, 324, 1994.

115. Leis, S. et al., Facilitated neurogenic inflammation in unaffected limbs of patients with complex regional pain syndrome, *Neurosci. Lett.*, 359, 163, 2004.

116. Baron, R. and Maier, C., Reflex sympathetic dystrophy: Skin blood flow, sympathetic vasoconstrictor reflexes and pain before and after surgical sympathectomy, *Pain,* 67, 317, 1996.

117. Birklein, F. et al., Pattern of autonomic dysfunction in time course of complex regional pain syndrome, *Clin. Auto. Res.,* 8, 79, 1998.

118. Cannon, W.B. and Rosenblueth, A., *The Supersensitivity of Denervation Structures: A Law of Denervation*, Macmillan, New York, 1949.

119. Fleming, W.W. and Westphal, D.P., Adaptive supersensitivity, in *Catecholamines I, Handbook of Experimental Pharmacology*, Trendelenburg, U. and Weiner, N., Eds., New York, Springer, Vol. 9, p. 509, 1988.

27

Thermal Imaging in Surgery

Paul Campbell
Ninewells Hospital

Roderick Thomas
Swansea Institute of Technology

27.1 Overview

Advances in miniaturization and microelectronics, coupled with enhanced computing technologies, have combined to see modern infrared imaging systems develop rapidly over the past decade. As a result, the instrumentation has considerably improved, not only in terms of its inherent resolution (spatial and temporal) and detector sensitivity (values ca. 25 mK are typical) but also in terms of its portability: the considerable reduction in bulk has resulted in light, camcorder (or smaller)-sized devices. Importantly, cost has also been reduced so that entry to the field is no longer prohibitive. This attractive combination of factors has led to an ever-increasing range of applicability across the medical spectrum. Whereas the mainstay application for medical thermography over the past 40 years has been with rheumatological and associated conditions, usually for the detection and diagnosis of peripheral vascular diseases such as Raynaud's phenomenon, the latest generations of thermal imaging systems have seen active service within new surgical realms such as orthopedics, coronary by-pass operations, and also in urology. The focus of this chapter relates not to a specific area of surgery per se, but rather to a generic and pervasive aspect of all modern surgical approaches: the use of *energized* instrumentation during surgery. In particular, we will concern ourselves with the use of thermal imaging to accurately monitor temperature within the tissue locale surrounding an energy-activated instrument. The rationale behind this is that it facilitates optimization of operation-specific protocols that may either relate to thermally based therapies or else to reduce the extent of collateral damage that may be introduced when inappropriate power levels, or excessive pulse durations, are implemented during surgical procedures.

27.2 Energized Systems

Energy-based instrumentation can considerably expedite fundamental procedures such as vessel sealing and dissection. The instrumentation is most often based around ultrasonic, laser, or radio-frequency (RF)-current-based technologies. Heating tissue into distinct temperature regimes is required in order to achieve the desired effect (e.g., vessel sealing, cauterization, or cutting). In the context of electrical current heating, the resultant effect of the current on tissue is dominated by two factors: the temperature attained by the tissue and the duration of the heating phase, as encapsulated in the following equation:

$$T - T_0 = \frac{1}{\sigma \rho c} J^2 \delta t \tag{27.1}$$

where T and T_0 are the final and initial temperatures (in degrees Kelvin [K]) respectively, σ is the electrical conductivity (in S/m), ρ is the tissue density, c is the tissue specific heat capacity (J kg^{-1} K^{-1}), J is the current density (A/m^2), and δt is the duration of heat application. The resultant high temperatures are not limited solely to the tissue regions in which the electrical current flow is concentrated. Heat will flow away from hotter regions in a time-dependent fashion given by the Fourier equation

$$Q(r,t) = -k \nabla T(r,t) \tag{27.2}$$

where Q is the heat flux vector, the proportionality constant k is a scalar quantity of the material known as the thermal conductivity, and $NT(r,t)$ is the temperature gradient vector. The overall spatio-temporal evolution of the temperature field is embodied within the differential equation of heat flow (alternatively known as the diffusion equation)

$$\frac{1}{\alpha} \frac{\partial T(r,t)}{\partial t} = \nabla^2 T(r,t) \tag{27.3}$$

where α is the thermal diffusivity of the medium defined in terms of the physical constants, k, ρ, and c as

$$\alpha = k/\rho c \tag{27.4}$$

and the temperature T is a function of both the three dimensions of space (r) and the time t. In other words, high temperatures are not limited to the region specifically targeted by the surgeon, and this is often the source of an added surgical complication caused by collateral or proximity injury. Electrosurgical damage, for example, is the most common cause of iatrogenic bowel injury during laparoscopic surgery and 60% of mishaps are missed, that is, the injury is not recognized during surgery and declares itself with peritonitis several days after surgery or even after discharge from the hospital. This level of morbidity can have serious consequences, in terms of the expense incurred by readmission to hospital or even the death of the patient. By undertaking *in vivo* thermal imaging during energized dissection, it becomes possible to determine, in real time, the optimal power conditions for the successful accomplishment of specific tasks, and with minimal collateral damage. As an adjunct imaging modality, thermal imaging may also improve surgical practice by facilitating easier identification and localization of tissues such as arteries, especially by less experienced surgeons. Further, as tumors are more highly vascularized than normal tissue, thermal imaging may facilitate their localization and staging, that is, the identification of the tumor's stage in its growth cycle. Figure 27.1 shows a typical set-up for implementation of thermography during surgery.

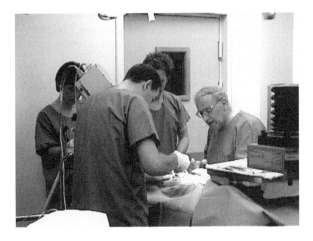

FIGURE 27.1 Typical set-up for a thermal imaging in surgery. The camera is tripod mounted toward the foot of the operating table and aimed at the surgical access site (camera visible over the left shoulder of the nearmost surgeon).

27.3 Thermal Imaging Systems

As skin is a close approximation to an ideal black body (the emissivity, o, of skin is 0.98, whereas that of an ideal black body is o = 1), then we can feel reasonably confident in applying the relevant physics directly to the situation of thermography in surgery. One important consideration must be the waveband of detector chosen for thermal observations of the human body. It is known from the thermal physics of black bodies, that the wavelength at which the maximum emissive power occurs, λ_{max} (i.e., the peak in the Planck curve), is related to the body's temperature T through Wien's law:

$$\lambda_{max}T = 0.002898 \tag{27.5}$$

Thus, for bodies at 310 K (normal human body temperature), the peak output is around 10 μm, and the majority of the emitted thermal radiation is limited to the range from 2 to 20 μm. The optimal detectors for passive thermal imaging of normal skin should thus have best sensitivity around the 10 μm range, and this is indeed the case with many of the leading thermal imagers manufactured today, which often rely on GaAs quantum well infrared photodetectors (QWIPs) with a typical waveband of 8–9 μm. A useful alternative to these longwave detectors involves the use of indium–antimonide (InSb)-based detectors to detect radiation in the mid-wave infrared (3–5 μm). Both these materials have the benefit of enhanced temperature sensitivity (ca. 0.025 K), and are both wholly appropriate even for quantitative imaging of hotter surfaces, such as may occur in energized surgical instrumentation.

27.4 Calibration

While the latest generation of thermal imaging systems are usually robust instruments exhibiting low drift over extended periods, it is sensible to recalibrate the systems at regular intervals in order to preserve the integrity of captured data. For some camera manufacturers, recalibration can be undertaken under a service agreement and this usually requires shipping of the instrument from the host laboratory. However, for other systems, recalibration must be undertaken in-house, and on such occasions, a black body source (BBS) is required.

Most BBS are constructed in the form of a cavity at a known temperature, with an aperture to the cavity that acts as the black body, effectively absorbing all incident radiation upon it. The cavity temperature

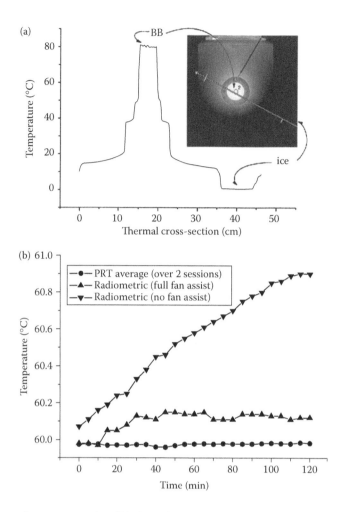

FIGURE 27.2 Thermal cross-section (profile) through the black body calibration source together with equilibrated crushed ice, which acts as a convenient secondary temperature gauge *in situ*. (Insert [left] thermal view with linear region of interest highlighted, and [right] optical view of the black body cavity and beaker of [equilibrated] crushed ice to the lower right.) (b) Radiometric detector drift during start up under two different ambient conditions. The detector readout is centered on the black body cavity source shown in (a), which was itself maintained at a target temperature of 59.97°C throughout the measurements (solid circles). Without fan-assisted cooling of the camera exterior, the measured temperature drifted by 0.8°C over 2 h, hence the importance of calibration under typical operating conditions. With fan-assisted cooling, the camera "settles" within around 30 min of switching on. (Camera: Raytheon Galileo [Raytheon Systems].)

must be measured using a high-accuracy thermometric device, such as a platinum resistance thermometer (PRT), with performance characteristics traceable to a thermometry standard. Figure 27.2b shows one such system, as developed by the UK National Physical Laboratory at Teddington, and whose architecture relies on a heat-pipe design. The calibration procedure requires measurement of the aperture temperature at a range of temperature set-points that are simultaneously monitored by the PRT (e.g., at intervals of 5° between temperature range of 293 and 353 K). Direct comparison of the radiometric temperature measured by the thermal camera with the standard temperature monitored via the PRT allows a calibration table to be generated across the temperature range of interest. During each measurement, sufficient time must be allowed in order to let the programmed temperature set-point equilibrate, otherwise inaccuracies will result. Further, the calibration procedure should ideally be undertaken under similar ambient conditions to those under which usual imaging is undertaken. This may include aspects such as laminar, or even fan-assisted, flow around the camera body which will affect the heat transfer rate from the camera to the ambient and in turn may affect the performance of the detector (Figure 27.2b).

27.5 Thermal Imaging during Energized Surgery

Fully remote-controlled cameras may be ideally suited to overhead bracket mountings above the operating table so that a bird's eye view over the surgical site is afforded. However, without a robotized arm to fully control pitch and location, the view may be restrictive. Tripod mounting, as illustrated in Figure 27.1, and with a steep look-down angle from a distance of about 1 m to the target offers the most versatile viewing without compromising the surgeon's freedom of movement. However, this type of set-up demands that a camera operator be on hand continually in order to move the imaging system to those positions offering best viewing for the type of energized procedure being undertaken.

27.5.1 RF Electrosurgery

As mentioned earlier, the most common energized surgical instrumentation employ a physical system reliant on (high-frequency) electrical current, an ultrasonic mechanism, or else incident laser energy in order to induce tissue heating. Thermal imaging has been used to follow all three of these procedures. There are often similarities in approach between the alternative modalities. For example, vessel sealing often involves placement of elongated forcep-style electrodes across a target vessel followed by ratcheted compression, and then a pulse of either RF current or alternatively ultrasonic activation of the forceps is applied through the compressed tissue region. The latest generations of energized instrumentation may have active feedback control over the pulse to facilitate optimal sealing with minimal thermal spread (e.g., the Valleylab *Ligasure* instrument); however, under certain circumstances, such as with calcified tissue or in excessively liquid environments, the performance may be less predictable.

Figure 27.3 illustrates how thermal spread may be monitored during the instrument activation period of one such "intelligent" feedback device using RF current. The initial power level for each application is determined through a fast precursor voltage scan that determines the natural impedance of the compressed tissue. Then, by monitoring the temperature dependence of impedance (of the compressed tissue) during current activation, the microprocessor-controlled feedback loop automatically maintains an appropriate power level until a target impedance is reached, indicating that the seal is complete. This process typically takes between 1 and 6 s, depending on the nature of the target tissue. The termination of the pulse is indicated by an audible tone burst from the power supply box. The performance of the system has been evaluated in preliminary studies involving gastric, colonic, and small bowel resection [1]; hemorraoidectomy [2]; prostatectomy [3]; and cholecystectomy [4].

FIGURE 27.3 Thermographic sequence taken with the Dundee thermal imaging system and showing (1) energized forceps attached to bowel (white correlates with temperature), (2) detachment of the forceps revealing hot tissue beneath, and (3) remnant hot spot extending across the tissue and displaying collateral thermal damage covering 4.5 mm on either side of the instrument jaws.

Perhaps most strikingly, the facility for real-time thermographic monitoring, as illustrated in Figure 27.3, affords the surgeon immediate appreciation of the instrument temperature, providing a visual cue that automatically alerts to the potential for iatrogenic injury should a hot instrument come into close contact with vital structures. By the same token, the *in situ* thermal image also indicates when the tip of the instrument has cooled to ambient temperature. It should be noted that the amount by which the activated head's temperature rises is largely a function of device dimensions, materials, and the power levels applied together with the pulse duration.

27.5.2 Analysis of Collateral Damage

While thermograms typical of Figure 27.3 offer a visually instructive account of the thermal scene and its temporal evolution, a quantitative analysis of the sequence is more readily achieved through the identification of a linear region of interest (LROI), as illustrated by the line bisecting the device head in Figure 27.4a. The data constituted by the LROI is effectively a snapshot thermal profile across those pixels lying on this designated line (Figure 27.4b). A graph can then be constructed to encapsulate the time-dependent evolution of the LROI. This is displayed as a 3D surface (a function of spatial co-ordinate along the LROI, time, and temperature) upon which color-mapped contours are evoked to represent the different temperature domains across the LROI (Figure 27.4c). In order to facilitate measurement of the thermal spread, the 3D surface, as represented in matrix form, can then be interrogated with a mathematical programming package, or alternatively inspected manually, a process that is most easily undertaken after projecting the data to the 2D coordinate–time plane, as illustrated in Figure 27.4d. The critical temperature beyond which tangible heat damage can occur to tissue is assumed to be 45°C [5]. Thermal spread is then calculated by measuring the maximum distance between the 45°C contours on the planar projection, then subtracting the electrode "footprint" diameter from this to get the total spread. Simply dividing this result by two gives the thermal spread on either side of the device electrodes.

The advanced technology used in some of the latest generations of vessel sealing instrumentation can lead to a much reduced thermal spread, compared with the earlier technologies. For example, with the Ligasure LS1100 instrument, the heated peripheral region is spatially confined to less than 2 mm, even when used on thicker vessels/structures. A more advanced version of the device (LS1200 [*Precise*]) consistently produces even lower thermal spreads, typically around 1 mm (Figure 27.4). This performance is far superior to other commercially available energized devices.

For example, Kinoshita and coworkers [6] have observed (using infrared imaging) that the typical lateral spread of heat into adjacent tissue is sufficient to cause a temperature of over 60°C at radial distances of up to 10 mm from the active electrode when an ultrasonic scalpel is used. Further, when standard bipolar electrocoagulation instrumentation is used, the spread can be as large as 22 mm. Clearly, the potential for severe collateral and iatrogenic injury is high with such systems unless power levels are tailored to the specific procedure in hand and real-time thermal imaging evidently represents a powerful adjunct technology to aid this undertaking.

While the applications mentioned thus far relate to "open" surgical procedures requiring a surgical incision to access the site of interest, thermal imaging can also be applied as a route to protocol optimization for other less invasive procedures. Perhaps the most important surgical application in this regime involves laser therapy for various skin diseases/conditions. Application of the technique in this area is discussed below.

27.6 Laser Applications in Dermatology

27.6.1 Overview

Infrared thermographic monitoring (ITM) has been successfully used in medicine for a number of years and much of this has been documented by Professor Francis Ring (http://www.medimaging.org/),

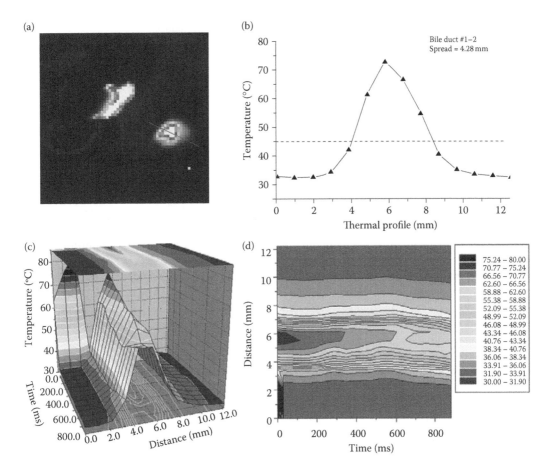

FIGURE 27.4 (a) Mid-infrared thermogram taken at the instant an energized forceps (Ligasure LS1200 "*Precise*") is removed from the surgical scene after having conducted a seal on the bile duct. The hot tips of the forceps are clearly evident in the infrared view (just left of center), as is the remnant hot spot where the seal has occurred on the vessel. By generating a linear region of interest (LROI) through the hot spot, as indicated by the highlighted line in the figure, it is possible to monitor the evolution of the hot spot's temperature in a quantitative fashion. (b) Thermal profile corresponding to the LROI shown in (a). (c) By tracking the temporal evolution of the LROI, it is possible to generate a 3D plot of the thermal profile by simply stacking the individual profiles at each acquisition frame. In this instance, the cooling behavior of the hot spot is clearly identified. Manual estimation of the thermal spread is most easily achieved by resorting to the 2D contour plot of the thermal profile's temporal evolution, as shown in (d). In this instance, the maximal spread of the 45°C contours is measured as 4.28 mm. By subtracting the forcep "footprint" (2.5 mm for the device shown) and dividing the result by 2, we arrive at the thermal spread for the device. The average thermal spread (for six bile-duct sealing events) was 0.89 ± 0.35 mm.

who has established a database and archive within the Department of Computing at the University of Glamorgan, UK, spanning over 30 years of ITM applications. Examples include monitoring abnormalities such as malignancies, inflammation, and infection that cause localized increases in skin temperature, which show as hot spots or as asymmetrical patterns in an infrared thermogram.

A recent medical example that has benefited by the intervention of ITM is the treatment by laser of certain dermatological disorders. Advancements in laser technology have resulted in new portable laser therapies, examples of which include the removal of vascular lesions (in particular port-wine stains [PWS]), and also cosmetic enhancement approaches such as hair (depilation) and wrinkle removal.

TABLE 27.1   Characteristics of Laser Therapy during and after Treatment

General Indicators	Dye Laser Vascular Lesions	Ruby Laser Depilation
During treatment	Varying output parameters	Varying output parameters
	Portable	Portable
	Manual and scanned	Manual and scanned
	Selective destruction of target chromophore (hemoglobin)	Selective destruction of target chromophore (melanin)
After treatment (desired effect)	Slight bruising (purpura)	Skin returns to normal coloring (no bruising)
	Skin retains its elasticity	Skin retains surface markings
	Skin initially needs to be protected from UV and scratching	Skin retains its ability to tan after exposure to ultraviolet light
	Hair follicles are removed	Hair removed

In these laser applications, it is a common requirement to deliver laser energy uniformly without overlapping of the beam spot to a subdermal target region, such as a blood vessel, but with the minimum of collateral damage to the tissue locale. Temperature rise at the skin surface, and with this the threshold to burning/scarring is of critical importance for obvious reasons. Until recently, this type of therapy had not yet benefited significantly from thermographic evaluation. However, with the introduction of the latest generation thermal imaging systems, exhibiting the essential qualities of portability, high resolution, and high sensitivity, significant inroads to laser therapy are beginning to be made.

Historically, lasers have been used in dermatology for some 40 years [7]. In recent years, there have been a number of significant developments, particularly regarding the improved treatment of various skin disorders, most notably the removal of vascular lesions using dye lasers [8–12] and depilation using ruby lasers [13–15]. Some of the general indicators as to why lasers are the preferred treatment of choice are summarized in Table 27.1.

27.7 Laser–Tissue Interactions

The mechanisms involved in the interaction between light and tissue depend on the characteristics of the impinging light and the targeted human tissue [16]. To appreciate these mechanisms the optical properties of tissue must be known. It is necessary to determine the tissue reflectance, absorption, and scattering properties as a function of wavelength. A simplified model of laser light interaction with the skin is illustrated in Figure 27.5.

Recent work has shown that laser radiation can penetrate through the epidermis and basal structure to be preferentially absorbed within the blood layers located in the lower dermis and subcutis. The process is termed selective photothermolysis, and is the specific absorption of laser light by a target

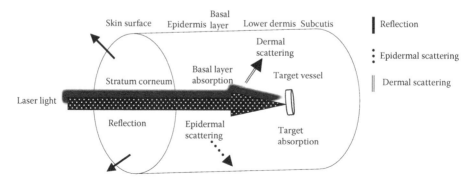

FIGURE 27.5   Passage of laser light within skin layers.

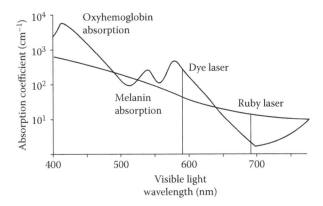

FIGURE 27.6 Spectral absorption curves for human blood and melanin.

TABLE 27.2 Interaction Effects of Laser Light and Tissue

Effect	Interaction
Photothermal	
Photohyperthermia	Reversible damage of normal tissue (37–42°C)
Photothermolysis	Loosening of membranes (odema), tissue welding (45–60°C)
Photocoagulation	
Photocarbonization	Thermal-dynamic effects, micro-scale overheating
Photovaporization	Coagulation, necrosis (60–100°C)
	Drying out, vaporization of water, carbonization (100–300°C)
	Pyrolysis, vaporization of solid tissue matrix (>300°C)
Photochemical	Photodynamic therapy, black light therapy
Photochemotherapy	Biostimulation
Photoinduction	
Photoionization	Fast thermal explosion, optical breakdown, mechanical
Photoablation	shockwave

tissue in order to eliminate that target without damaging surrounding tissue. For example, in the treatment of PWS, a dye laser of wavelength 585 nm has been widely used [17] where the profusion of small blood vessels that comprise the PWS are preferentially targeted at this wavelength. The spectral absorption characteristics of light through human skin have been well established [18] and are replicated in Figure 27.6 for the two dominant factors: melanin and oxyhemoglobin.

There are three types of laser–tissue interaction, namely photothermal, photochemical, and protoionization (Table 27.2), and the use of lasers on tissue results in a number of differing interactions, including photodisruption, photoablation, vaporization, and coagulation, as summarized in Figure 27.7.

The application of appropriate laser technology to medical problems depends on a number of laser operating parameters, including matching the optimum laser wavelength for the desired treatment. Some typical applications and the desired wavelengths for usage are highlighted in Table 27.3.

27.8 Optimizing Laser Therapies

There are a number of challenges in optimizing laser therapy, mainly related to the laser parameters of wavelength, energy density, and spot size. Combined with these are difficulties associated with poor positioning of hand-held laser application that may result in uneven treatment (overlapping spots and/or uneven coverage [stippling] of spots), excessive treatment times, and pain. Therefore, for enhanced

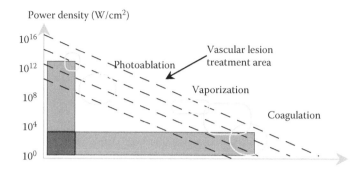

FIGURE 27.7 Physiological characteristics of laser therapy. (From Thomas, R.A., Donne, K.E., Clement, R.M., and Kiernan, M. (2002) Optimised laser application in dermatology using infrared thermography, *Thermosense XXIV, Proceedings of SPIE*, April 1–4, Orlando, USA. With permission.)

TABLE 27.3   Laser Application in Dermatology

Laser	Wavelength (nm)	Treatment
Flashlamp short-pulsed dye	510	Pigmented lesions, for example, freckles, tattoos
Flashlamp long-pulsed dye	585	PWS in children, warts, hypertrophic scars
Ruby single-pulse or Q-switched	694	Depilation of hair
Alexandrite Q-switched	755	Multicolored tattoos, viral warts, depilation
Diode variable	805	Multicolored tattoos, viral warts
Neodymium yitrium aluminum (Nd-YAG) Q-switched	1064	Pigmented lesions; adult port-wine stains, black/blue tattoos
Carbon dioxide continuous pulsed	10600	Tissue destruction, warts, tumors

efficacy, an improved understanding of the thermal effects of laser–tissue interaction benefits therapeutic approaches. Here, variables for consideration include

1. Thermal effects of varying spot size
2. Improved control of hand-held laser minimizing overlapping and stippling
3. Establishment of minimum gaps
4. Validation of laser computer scanning

Evaluation (Figure 27.8) was designed to elucidate whether or not measurements of the surface temperature of the skin are reproducible when illuminated by nominally identical laser pulses. In this case, a 585 nm dye laser and a 694 nm ruby laser were used to place a number of pulses manually on tissue. The energy emitted by the laser is highly repeatable. Care must be taken to ensure that both the laser and

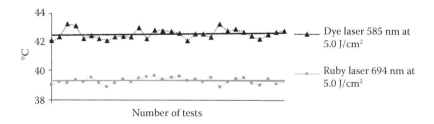

FIGURE 27.8   Repeatability of initial maximum skin temperatures (°C) of two lasers with similar energy density but different wavelengths.

radiometer positions are kept constant and that the anatomical location used for the test had uniform tissue pigmentation.

Figure 27.8 shows the maximum temperature for each of 20 shots fired on the forearm of a representative Caucasian male with type 2 skin*. Maximum temperature varies between 48.90°C and 48.10°C representing a variance of 1°C (±0.45°C). This level of reproducibility is pleasing since it shows that, despite the complex scenario, the radiometer is capable of repeatedly and accurately measuring surface tissue temperatures. In practice, the radiometer may be used to inform the operator when any accumulated temperature has subsided, allowing further treatment without exceeding some damage threshold.

Energy density is also an important laser parameter and can be varied to match the demands of the application. It is normal in the discipline to measure energy density (fluence) in J/cm^2. In treating vascular lesions, most utilize an energy density for therapy of 5–10 J/cm^2 [19]. The laser operator needs to be sure that the energy density is uniform and does not contain hot spots that may take the temperature above the damage threshold inadvertently. Preliminary characterization of the spot with thermal imaging can then aid with fine tuning of the laser and reduce the possibility of excessive energy density and with that the possibility of collateral damage.

27.9 Thermographic Results of Laser Positioning

During laser therapy, the skin is treated with a number of spots, applied manually depending on the anatomical location and required treatment. It has been found that spot size directly affects the efficacy of treatment. The wider the spot size the higher the surface temperature [20]. The type and severity of lesion also determine the treatment required. Its color severity (dark to light) and its position on skin (raised to level). Therefore, the necessary treatment may require a number of passes of the laser over the skin. It is therefore essential as part of the treatment that there is a physical separation between individual spots so that

1. The area is not overtreated with overlapping spots that could otherwise result in local heating effects from adjacent spots resulting in skin damage.
2. The area is not undertreated leaving stippled skin.
3. The skin has cooled sufficiently before second or subsequent passes of the laser.

Figure 27.9 shows two laser shots placed next to each other some 4 mm apart. The time between the shots is 1 s. There are no excessive temperatures evident and no apparent temperature build-up in the

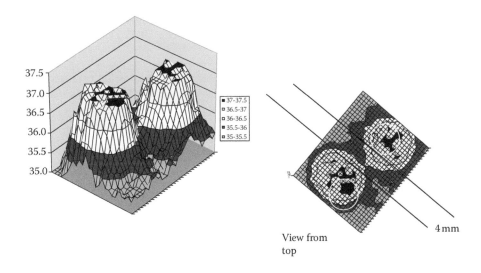

FIGURE 27.9 Two-dye laser spots with a minimum of 4 mm separation (585 nm at 4.5 J/cm^2, 5 mm spot).

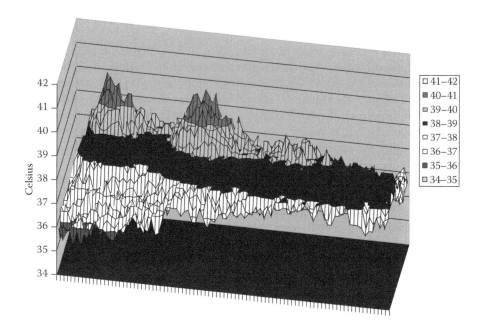

FIGURE 27.10 Three-dye laser spots, 2 s apart with a 5 mm separation (585 nm at 5 J/cm², 5 mm spot).

gap. This result, which concurs with Lanigan [21], suggests a minimum physical separation of 5 mm between all individual spot sizes.

The intention is to optimize the situation leading to a uniform therapeutic and aesthetic result without either striping or thermal build-up. This is achieved by initially determining the skin color (Chromotest) for optimum energy settings, followed by a patch test and subsequent treatment. Increasing the number of spots to 3 with the 4 mm separation reveals a continuing trend, as shown in Figure 27.10. The gap between the first two shots is now beginning to merge in the 2 s period that has lapsed. The gap between shots 2 and 3 remains clear and distinct and there are clearly visible thermal bands across the skin surface between 38–39°C and 39–40°C. These experimental results supply valuable information to support the development of both free-hand treatment and computer-controlled techniques.

27.10 Computerized Laser Scanning

Having established the parameters relating to laser spot positioning, the possibility of achieving reproducible laser coverage of a lesion by automatic scanning becomes a reality. This has potential advantages, which include

1. Accurate positioning of the spot with the correct spacing from the adjacent spots
2. Accurate timing allowing the placement at a certain location at the appropriate lapsed time

There are some disadvantages that include the need for additional equipment and regulatory approvals for certain market sectors.

A computerized scanning system has been developed [13] that illuminates the tissue in a predefined pattern. Sequential pulses are not placed adjacent to an immediately preceding pulse thereby ensuring the minimum of thermal build-up. Clement et al. [13] carried out a trial, illustrating treatment coverage using a hand-held system compared to a controlled computer scanning system. Two adjacent areas (lower arm) were selected and shaved. A marked hexagonal area was subjected to 19 shots using a hand-held system, and an adjacent area of skin was treated with a scanner whose computer control

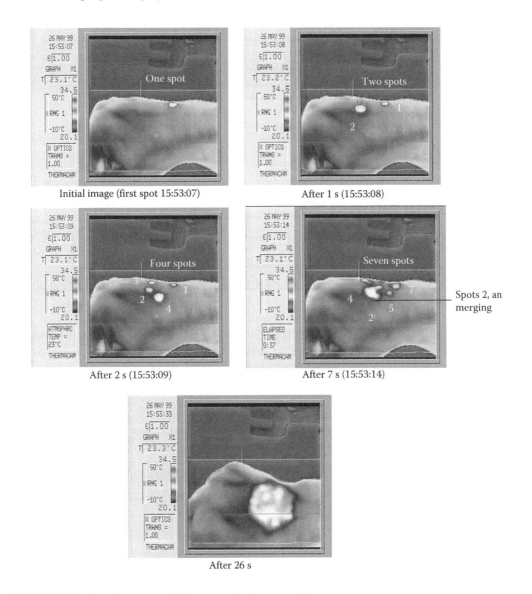

FIGURE 27.11 Sample sequences during computer laser scanning.

is designed to uniformly fill the area with exactly 19 shots. Such tests were repeated and the analyzed statistics showed that, on average, only 60% of area is covered by laser spots. The use of thermography allowed the validation and optimization of this automated system in a way that was impossible without thermal imaging technology. The following sequence of thermal images, Figure 27.11, captures the various stages of laser scanning of the hand using a dye laser at 5.7 J/cm^2. Thermography confirms that the spot temperature from individual laser beams will merge and that both the positioning of spots and the time duration between spots dictate the efficacy of treatment.

27.10.1 Case Study 1: Port-Wine Stain

Vascular nevi are common and are present at birth or develop soon after. Superficial lesions are due to capillary networks in the upper or mid dermis, but larger angiomas can be located in the lower dermis

TABLE 27.4   Vasculature Treatment Types

Treatment Type	Process	Possible Concerns
Camouflage	Applying skin colored pigments to the surface of the skin. Enhancement to this technique is to tattoo skin colored inks into the upper layer of the lesion	Only a temporary measure and is very time consuming. Efficacy dependent on flatter lesions
Cryosurgery	Involves applying supercooled liquid nitrogen to the lesion to destroy abnormal vasculature	May require several treatments
Excision	Common place where the lesion is endangering vital body functions	Not considered appropriate for purely cosmetic reasons. Complex operation resulting in a scar. Therefore, only applicable to the proliferating hemangioma lesion
Radiation therapy	Bombarding the lesion with radiation to destroy vasculature	Induced number of skin cancer in a small number of cases
Drug therapy	Widely used administering steroids	Risk of secondary complications affecting bodily organs

and subcutis. An example of vascular nevi is the PWS often present at birth, which is an irregular red or purple macule which often affects one side of the face. Problems can arise if the nevus is located close to the eye, and in some cases where a PWS involves the trigeminal nerve's ophthalmic division may have an associated intracranial vascular malformation known as Sturge Weber syndrome. The treatment of vascular nevi can be carried out in a number of ways, often dependent on the nature, type, and anatomical and severity of lesion location, as highlighted in Table 27.4.

A laser wavelength of 585 nm is preferentially absorbed by hemoglobin within the blood, but there is partial absorption in the melanin-rich basal layer in the epidermis. The objective is to thermally damage the blood vessel, by elevating its temperature, while ensuring that the skin surface temperature is kept low. For a typical blood vessel, the temperature–time graph appears similar to Figure 27.12. This suggests that it is possible to selectively destroy the PWS blood vessels, by elevating

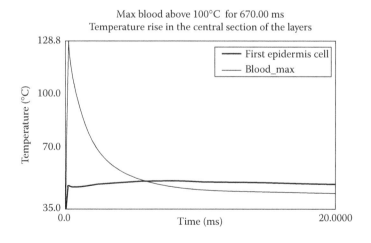

FIGURE 27.12   Typical temperatures for PWS problem, indicating thermal disruption of blood vessel, while skin surface temperature remains low.

them to a temperature in excess of 100°C, causing disruption to the small blood vessels, while maintaining a safe skin surface temperature. This has been proven empirically via thermographic imaging with a laser pulsing protocol that was devised and optimized on the strength of Monte-Carlo-based models [22] of the heat dissipation processes [23]. The two-dimensional Cartesian thermal transport equation is

$$\nabla T^2 + \frac{Q(x,y)}{k} = \frac{1}{\alpha} \frac{\partial T}{\partial t} \tag{27.6}$$

where temperature T has both an implied spatial and temporal dependence, and the volumetric source term, $Q(x,y)$, is obtained from the solution of the Monte-Carlo radiation transport problem [24].

27.10.2 Case Study 2: Laser Depilation

The 694 nm wavelength laser radiation is preferentially absorbed by melanin, which occurs in the basal layer and particularly in the hair follicle base, which is the intended target using an oblique angle of laser beam (see Figure 27.13). A Monte-Carlo analysis was performed in a similar manner to *Case Study 1* above, where the target region in the dermis is the melanin-rich base of the hair follicle. Figure 27.14a and b shows the temperature–time profiles for 10 and 20 J cm² laser fluence [25]. These calculations suggest that it is possible to thermally damage the melanin-rich follicle base while restricting the skin surface temperature to values that cause no superficial damage. Preliminary clinical trials indicated that there is indeed a beneficial effect, but the choice of laser parameters still required optimizing.

Thermographic analysis has proved to be indispensable in this work. Detailed thermometric analysis is shown in Figure 27.15a. Analysis of this data shows that in this case, the surface temperature is raised to about 50°C. The thermogram also clearly shows the selective absorption in the melanin-dense hair. The temperature of the hair is raised to over 207°C. This thermogram illustrates direct evidence for selective wavelength absorption leading to cell necrosis. Further clinical trials have indicated a maximum fluence of 15 J cm² for type III Caucasian skin. Figure 27.15b illustrates a typical thermographic image obtained during the real-time monitoring.

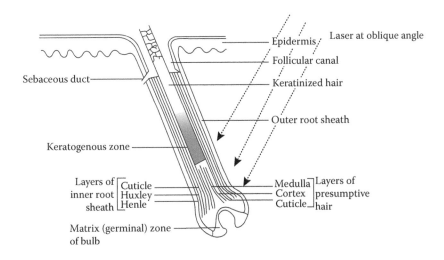

FIGURE 27.13 Oblique laser illumination of hair follicle.

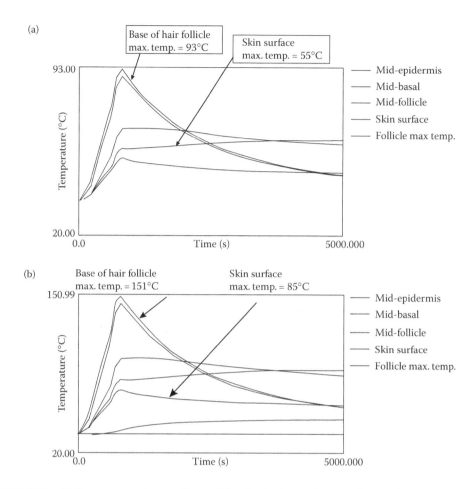

FIGURE 27.14 (a) Temperature–time profiles at 10 J cm² ruby (694 nm), 800 μs laser pulse on Caucasian skin type III. (b) Temperature–time profiles for 20 J cm² ruby (694 nm), 800 μs laser pulse on Caucasian skin type III.

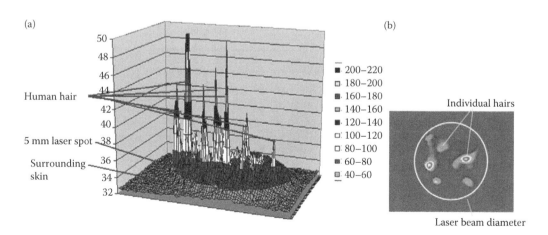

FIGURE 27.15 (a) Postprocessed results of 5 mm. (b) Simplified thermogram diameter 694 nm 20 J cm² 800 μs ruby pulse of ruby laser pulse, with 5 mm spot at 20 J cm².

27.11 Conclusions

The establishment, development, and consequential success of medical infrared thermographic (MIT) intervention with laser therapy is primarily based on the understanding of the following:

1. Problem/condition to be monitored
2. Set-up and correct operation of infrared system (appropriate and validated training)
3. Appropriate conditions during the monitoring process
4. Evaluation of activity and development of standards and protocol

With reference to (1) above, that is, the condition to be monitored, there needs to be a good knowledge as to the physiological aspects of the desired medical process; in laser therapy an understanding as to the mechanisms involved in laser–tissue interaction. A good reference source of current practice can be found in the *Handbook of Optical Biomedical Diagnostics*, published by The International Society for Optical Engineering (SPIE).

In this application, fast data capture (>50 Hz), good image quality (256 × 256 pixels), temperature sensitivity, and repeatability were considered important and an Inframetrics SC1000 Focal Plane Array Radiometer (3.4–5 μm, CMOS PtSi Cooled Detector) with a real-time data acquisition system (Dynamite) was used. There are currently very fast systems available with data acquisition speeds in terms of hundreds of Hertz with detectors that provide excellent image quality. In (2), the critical aspect is training [26]. Currently, infrared equipment manufacturers design systems with multiple applications in mind. This has resulted in many aspects of good practice and quality standards. This is one of the reasons why industrial infrared thermography is so successful. This has not necessarily been the case in medicine. However, it is worth noting that there are a number of good infrared training organizations throughout the world, particularly in the United States. The advantages of adopting training organizations such as these is that they have experience of training with reference to a very wide range and type of infrared thermographic systems, in a number of different applications. This will help in the identification of the optimum infrared technology. In (3), consideration as to the conditions surrounding the patient and the room environment are important for optimum results. In the United Kingdom, for example, Professor Francis Ring, University of Glamorgan, has led the way in the development and standardizations of clinical infrared practice [27]. Finally, (4) the evaluation of such practice is crucial if lessons are to be learnt and protocol and standards are to emerge.

Infrared thermal imaging provides an important tool for optimizing energized surgical interventions and facilitates validation of theoretical models of evolving temperature fields.

References

1. Heniford, B.T., Matthews, B.D., Sing, R.F., Backus, C., Pratt, P., and Greene, F.L. (2001) Initial results with an electrothermal bipolar vessel sealer. *Surg. Endosc.* 15: 799–801.
2. Palazzo, F.F., Francis, D.L., and Clifton, M.A. (2002) Randomised clinical trial of ligasure versus open haemorrhoidectomy. *Br. J. Surg.* 89: 154–157.
3. Sengupta, S. and Webb, D.R. (2001) Use of a computer controlled bipolar diathermy system in radical prostatectomies and other open urological surgery. *ANZ J. Surg.* 71: 538–540.
4. Schulze, S., Krztiansen, V.B., Fischer-Hansen, B., and Rosenberg, J. (2002) Sealing of the cystic duct with bipolar electrocoagulation. *Surg. Endosc.* 16: 342–344.
5. Reidenbach, H.D. and Buess, G. (1992). Anciliary technology: Electrocautery, thermoregulation and laser. In Cuschieri, A., Buess, G., and Perrisat, L. Eds., *Operative Manual of Endoscopic Surgery*. Springer-Verlag, Berlin, pp. 44–60.
6. Kinoshita, T., Kanehira, E., Omura, K., Kawakami, K., and Watanabe, Y. (1999) Experimental study on heat production by a 23.5 kHz ultrasonically activated device for endoscopic surgery. *Surg. Endosc.* 13: 621–625.

7. Wheeland, R.G. (1995) Clinical uses of lasers in dermatology. *Lasers Surg. Med.* 16:2–23.

8. Barlow, R.J., Walker, N.P.J., and Markey, A.C. (1996) Treatment of proliferative haemangiomas with 585 nm pulsed dye laser. *Br. J. Dermatol.* 134: 700–704.

9. Garden, J.M., Polla, L.L., and Tan, O.T. (1988) Treatment of port wine stains by pulsed dye laser—Analysis of pulse duration and long term therapy. *Arch. Dermatol.* 124: 889–896.

10. Glassberg, E., Lask, G., Rabinowitz, L.G., and Tunnessen, W.W. (1989) Capillary haemangiomas: Case study of a novel laser treatment and a review of therapeutic options. *J. Dermatol. Surg. Oncol.* 15: 1214–1223.

11. Kiernan, M.N. (1997) An analysis of the optimal laser parameters necessary for the treatment of vascular lesions, PhD thesis, The University of West of England.

12. Motley, R.J., Katugampola, G., and Lanigan, S.W. (1996) Microvascular abnormalities in port wine stains and response to 585 nm pulsed dye laser treatment. *Br. J. Dermatol.* 135(Suppl. 47): 13–14.

13. Clement, R.M., Kiernan, M.N., Thomas, R.A., Donne, K.E., and Bjerring, P.J. (1999) The use of thermal imaging to optimise automated laser irradiation of tissue. *Skin Research and Technology.* Vol. 5, No. 2, *th Congress of the International Society for Skin Imaging*, July 4–6, 1999, Royal Society London.

14. Gault, D., Clement, R.M., Trow, R.B., and Kiernan, M.N. (1998) Removing unwanted hairs by laser. *Face* 6: 129–130.

15. Grossman et al. (1997) Damage to hair follicle by normal mode ruby laser pulse. *J. Amer. Acad. Dermatol.* 889–894.

16. Welsh, A.J. and van Gemert, M.V.C. (1995) *Optical–Thermal Response of Laser-Irradiated Tissue.* Plenum Press, ISBN 0306449269.

17. Clement, R.M., Donne, K.D., Thomas, R.A., and Kiernan, M.N. (2000) Thermographic condition monitoring of human skin during laser therapy. *Quality Reliability Maintenance, 3rd International Conference*, St Edmund Hall, University of Oxford, March 30–31, 2000.

18. Andersen, R.R. and Parrish, J.A. (1981) Microvasculature can be selectively damaged using dye lasers. *Lasers Surg. Med.* 1: 263–270.

19. Garden, J.M. and Bakus, W. (1996) Clinical efficacy of the pulsed dye laser in the treatment of vascular lesions. *J. Dermatol. Surg. Oncol.* 19: 321–326.

20. Thomas, R.A., Donne, K.E., Clement, R.M., and Kiernan, M. (2002) Optimised laser application in dermatology using infrared thermography, *Thermosense XXIV, Proceedings of SPIE*, April 1–4, Orlando, USA.

21. Lanigan, S.W. (1996) Port wine stains on the lower limb: Response to pulsed dye laser therapy. *Clin. Exp. Dermatol.* 21: 88–92.

22. Wilson, B.C. and Adam, G. (1983) A Monte Carlo model for the absorption and flux distributions of light in tissue. *Med. Phys. Biol.* 1.

23. Daniel, G. (2002) An investigation of thermal radiation and thermal transport in laser–tissue interaction, PhD thesis, Swansea Institute.

24. Donne, K.E. (1999) Two dimensional computer model of laser tissue interaction. Private communication.

25. Trow, R. (2001) The design and construction of a ruby laser for laser depilation, PhD thesis, Swansea Institute.

26. Thomas, R.A. (1999) *Thermography.* Coxmoor Publishers, Oxford, pp. 79–103.

27. Ring, E.F.J. (1995) History of thermography. In Ammer, K. and Ring, E.F.J., Eds., *The Thermal Image in Medicine and Biology.* Uhlen Verlag, Vienna, pp. 13–20.

28

Thermal Signals and Cutaneous Circulation in Physiological Research and Reconstructive Surgery

David D. Pascoe
Auburn University

Louis de Weerd
*University Hospital of
North Norway*

James B. Mercer
University of Tromsø

Joshua E. Lane
*Mercer University School of
Medicine*

*Emory University School of
Medicine*

Sven Weum
Auburn University

28.1 Overview

In the third edition of the *Biomedical Engineering Handbook* (*Physiology of Thermal Signals*), also republished in this edition, we introduced the reader to skin thermal properties in response to stress, regulation of skin blood flow for heat transfers, modeling equations, and objective thermography [1]. The objective thermography explored the efficiency of heat transport in the hand-reduced blood flow, nerve block and laser Doppler mapping, and sports medicine injury applications. Most of the work in this current publication on thermal signals builds on that foundation of information. However, before we go into further detail on the use of IR thermography, it is important to review some basic principles in thermal physiology.

28.1.1 Skin Thermoregulation Models

The physiological research journal articles reporting skin surface temperatures have almost exclusively obtained these measurements through surface skin probes. This is problematic because the attachment of the probe can alter the microenvironment of the skin, and the probe is only measuring one small location that cannot account for the variance in tissue temperature across the surface area. In an attempt to account for the influence of skin thermoregulation on the thermal homeostasis of the individual, formulas combining the various regions of the body have been created. The challenge in acquiring accurate mean skin temperature measures has been related to the large contoured surface area and the assigned contribution of regional areas to the overall thermal status of the individual. In the literature, 20 measurement sites have been identified and numerous formulas varying in the number of reference sites required by each formula have emerged. Should all sites be given equal representation? While the core environment remains fairly constant, the regional peripheral blood flow and temperatures are variable. To obtain an accurate assessment of mean skin temperature, previous research has focused on four basic formula approaches: (1) unweighted average of the sites recorded; (2) weighted formulas based on regional surface averages as defined within population norms; (3) variable weighting based on individually determined surface areas, often determined by the DuBois linear formula; and (4) weighted formulas that incorporate factors that describe both surface area and the thermoregulatory response to thermal stimuli [2–4]. This fourth approach recognized that while the surface areas of the various regions help to explain the exposed surface available for heat exchange (conduction, convection, and radiation), differences in regional blood flow, temperature, and sweat gland distribution due to evaporative cooling may not be adequately expressed.

28.1.2 Skin Surface Temperature Measurements and Infrared Thermography

In the investigative process, noncontact infrared skin surface thermal maps can provide sensitive measurements (0.05°C) of a region based on a large number of individual reference points (pixels) from which the high, low, and mean temperatures can be derived. This approach utilizing IR thermography can be instrumental in our quest to understand and refine our understanding of skin perfusion, heat transfers, and thermoregulation.

Infrared thermography captures the thermal image of the dynamic heat transfer processes of the skin surface that are critical for the maintenance of our body temperature within a survivable range. As a changing, dynamic process, the thermologist needs to be aware of the factors that influence skin heat transfers, skin blood flow, and sources of heat from the core and external environment. An understanding of thermal responses to various stressors that activate the regulation of skin blood flow can allow the physiologist and clinician to "challenge test" for a better understanding of thermoregulation, assess skin blood perfusion, observe and diagnose vascular pathologies, and evaluate thermal therapies. This chapter will introduce the reader to the emergence of infrared thermography in the practice of plastic surgery and tissue transposition. More specifically, the viability of transposed tissue postsurgery is dependent upon the plastic surgeon's ability to identify the underlying blood vessels and take advantage of the elaborate network of blood vessels to perfuse the transferred tissue.

28.2 Skin Thermal Properties

28.2.1 Radiant Heat and Emissivity

Infrared thermography captures a visual thermal map of the skin. Detectors within the imaging system collect the radiant heat that is emitted from an object. An object's ability to radiate heat is termed emissivity and is compared to a blackbody source that is capable of absorbing and radiating all the electromagnetic heat it encounters. Probably the first research efforts that pointed out the usefulness of infrared emission of human skin as a diagnostic aid in medicine were the series of studies published in 1934–1936 by the American physiologists J. D. Hardy and Carl Muschenheim [5–7]. Their work revealed that the human skin, regardless of its color, is a highly efficient radiator with an emissivity of 0.98. This skin value is close to the emissivity of 1.0 (a perfect blackbody source) and the skin provides the human with a very efficient system for dissipating heat from its surface when blood flow is directed toward the outer layers. The determination of the electromagnetic emissivity of an object is a crucial factor in obtaining accurate temperature measurements. When presenting thermographic data, it is a good practice to reference the emissivity and the blackbody reference temperature.

28.2.2 Perfusion and Heat Transfers

Our skin surface serves as an interactive heat transfer medium between the internal heat produced from our metabolic production within the core and the external thermal conditions of our environment. Changes in skin temperature are primarily modulated by cutaneous perfusion. This perfusion is a function of vascular anatomy and vasoactive control by the autonomic nervous system. Under conditions of hyperthermic stress, skin blood flow can be upregulated to provide a larger surface area for transfers of heat from the body. In contrast, when blood flow is shunted from the skin surface, the cutaneous layer serves as an efficient insulator. Infrared imaging provides a visual map of the skin surface temperature as determined by the perfusion to that region. These temperature measurements are indicative of both spatial and temporal changes to the regional temperature distribution.

However, it should be stressed that an IR image cannot quantify measurements of blood flow to the skin tissue. The infrared image provides a thermal map of the skin that is influenced by blood perfusion to the surface area, but these thermal changes may also be influenced by conductive and/or radiant heat provided from an external thermal stressor.

The interpretation of the dynamic process of skin thermal regulation obtained from thermal imaging requires a basic understanding of physiological mechanisms of skin blood flow and factors that influence heat transfers to the skin. Some researchers and practitioners have combined the use of infrared imagery with methodologies and techniques that provide measurements of blood flow to better understand the complex thermal regulation, functions, and responses of the cutaneous tissue layer. For example, the complementary data obtained from infrared thermography with the more direct measure of blood flow from laser Doppler. The laser Doppler can ascertain the direction of skin blood flow but these measurements are restricted to small surfaces and blood flow which is not quantified but reported as changes in arbitrary Doppler [8]. With this understanding, objective data from IR thermography can add valuable information and complement other methodologies in the scientific inquiry and medical practices. Some plastic surgeons have utilized the complementary methods for pre-, intra-, and postoperative evaluations. A further discussion of infrared imaging, complementary blood flow measurements, and techniques that reveal skin tissue structures, rates, and variability of perfusion can be found in Reference 1.

28.2.3 Skin Blood Flow Perfusion Rates

The ability of the skin to substantially increase blood flow, far in excess of the tissue's metabolic needs, alludes to the tissue's role and potential in heat transfer mechanisms. The nutritive need's of skin tissue has been estimated at 0.2 mL/min per cubic centimeter of skin [9], which is considerably lower than the maximal rate of 24 mL/min per cubic centimeter of skin (estimated from total forearm circulatory measurement during heat stress) [10]. If one were to approximate skin tissue as 8% of the forearm, then skin blood would equate to 250–300 mL/100 mL of skin per minute [11]. Applying this flow rate to an estimated skin surface of 1.8 m^2 (average individual) suggests that approximately 8 L of blood flow could be diverted to the skin to dissipate heat at rate of 1750 W to the environment [12,13]. This increased blood flow required for heat transfers from active muscle tissue and skin blood flow for thermoregulation is made available through the redistribution of blood flow (splanchnic circulatory beds, renal tissues) and increases in cardiac output [14]. The increased cardiac output has been suggested to account for two thirds of the increased blood flow needs, while redistribution provides the remaining third [12]. Several good reviews are available regarding cutaneous blood flow, cardiovascular function, and thermal stress [14–20].

In summary, the ability of an object to radiate heat is termed "emissivity" and is compared to a perfect blackbody source. The emissivity determination allows thermal imagers to have accurate measurements of sources of radiant heat. The skin is nearly a perfect radiator with an emissivity of 0.98. Skin surface perfusion can be dramatically altered to accommodate thermal conditions that rely on the control of heat transfers from the skin. The importance of the skin as a primary thermoregulatory organ is clearly recognized by its abundance of blood vessels to the skin surface that far exceed the nutritional needs of the tissue. This skin perfusion and accompanying changes in the surface temperature map make infrared thermography an ideal investigative tool.

28.3 Regulation of Skin Blood Flow

28.3.1 Regional Neural Regulation

The acral regions of the body (palms, plantar surface of feet, nose, and ears) are regulated solely by adrenergic sympathetic vasoconstrictor nerves that alter regional skin blood flow and temperature regulation through adjustments to the vasoconstrictor tone [10,21,22]. Within the acral regions, there are an abundance of arteriovenous anastomoses (AVA). The AVAs are thick-walled, low-resistance vessels that direct blood flow from the arterioles to the venules. Sympathetic adrenergic vasoconstriction controls the arterioles and AVA to modulate flow rates to the skin vascular plexus in accordance to prevailing thermal conditions. The opening and closing of the AVAs within the acral areas can substantially alter skin blood flow responses [23]. During vasoconstriction, blood flow is shunted from the subcutaneous region allowing very little heat loss and when the vasoconstriction is relaxed (effectively causing vasodilation) blood flow is increased to the skin vascular plexus allowing heat transfers to dissipate excess heat. The skin sites where these vessels are found are among those where skin blood flow changes are discernible to IR-thermography. While the AVA are most active during heat stress, their thick walls and high velocity flow rates do not support their significant role in heat transfers to adjoining skin tissue [12].

The nonacral regions possess dual sympathetic control regulation of both adrenergic vasoconstrictor and vasodilator nerve activity. This vasodilator response was evidenced in research from the 1930s that demonstrated skin flushing and skin surface temperature changes of sympathectomized limbs and confirmed in the 1950s by Edholm et al. [24,25] and Rodie et al. [26] using nerve blocks. However, the neurotransmitter for active vasodilatation is not yet known. Current investigations postulate cholinergic sudomotor and cotransmitter activation [27–29]. The nonacral regions do not commonly have AVAs.

For a more in-depth discussion of regional sympathetic reflex regulation, see Charkoudian [10] and Johnson and Proppe [13].

28.3.2 Thermal Homeostasis

The regulation of skin blood flow thermoregulation is controlled by the preoptic anterior hypothalamus of the brain based on the fluctuations of internal core and skin temperature sensory information. The thermal stress from the environment can be exacerbated by the influences of radiant heat, high humidity, forced convective flow, and increased amounts of clothing [30]. Internal thermal stress is mostly influenced by the heat production associated with exercise and increased metabolism. Physiologists utilize the "clo" value that factors both physical activity intensity and the amount of clothing, to assess levels of thermal comfort [31]. In the research and clinical settings, a nonexercising slightly dressed individual is in a thermoneutral environment at 20–25°C. Under these conditions, the skin temperature regulation is controlled by a tonically active sympathetic vasoconstrictor system that maintains stable skin and core temperatures, eliciting perceptual sweating or shivering stress responses. This thermoneutral zone provides the basis for the clinical testing standards for room temperatures being set at 18–25°C during infrared thermographic studies. Controlling room test conditions is important when measuring skin temperature responses as changes in ambient temperatures can alter the fraction of flow shared between the musculature and skin [19]. Under the influence of whole-body hyperthermic conditions, removal of the vasomotor vasoconstrictor tone can account for 10–20% of cutaneous vasodilation, while the vasodilator system provides the remaining skin blood flow regulation [10]. Alterations in the threshold (onset of vascular response and sweating) and sensitivity (level of response) in vasodilation blood flow control can be related to an individual's level of heat acclimation [32], exercise training [25], circadian rhythm [33,34], and women's reproductive hormonal status [10]. Recent literature suggests that some observed shifts in the reflex control are the result of female reproductive hormones. Both estrogen and progesterone have been linked to menopausal hot flashes [10].

28.3.3 Challenge Testing

Under thermoneutral testing conditions, the skin temperature remains stable. However, the thermologist and/or clinician may want to provide a stressor or "challenge" test to assess the functioning of the thermal response. Challenge testing can include cold or hot water immersion, convective air flow, and exercise bouts that are designed to alter the skin blood flow. In some functional ergonomic cases, positioning of a body part at a particular angle can create a nerve impingement that can be detected by changes in the thermal temperature and pattern. Under these thermally challenging conditions, the clinical importance of abnormalities of skin blood flow may be more apparent. However, one must also recognize that as heat is being transferred through the various layers of skin, some of the heat is dissipated into the adjoining tissues. The heat decay as the blood traverses the layers of tissue and its dispersion pattern within the circulatory plexus of the skin may disguise the origin of the tissue producing the abnormal thermal response.

In summary, the regulation of blood flow to the skin surface is different for the acral region (sympathetic vasoconstriction) and nonacral regions (sympathetic vasoconstriction and vasodilation). Additionally, the predominance of AVA blood vessels in the acral region provides a structural mechanism for greatly altering blood flow perfusion to the skin. The increased blood flow needs for thermoregulation are made possible by increases to cardiac output and the redistribution of blood flow from the splanchnic circulatory beds and renal tissues without compromising blood pressure. The stability of the skin surface temperatures during thermoneutral conditions and the alterations in blood flow to challenge testing allow the researcher and clinician the opportunity to evaluate the function and regulatory response of the skin to thermal stressors.

28.4 Heat Transfer Modeling of Cutaneous Microcirculation

28.4.1 Vascular Heat Transfers

The modeling of the microcirculation of the cutaneous tissue is developed around the relationships of heat transfers. Models have been created for whole-body thermal responses [35,36] and localized hyperthermia [37,38]. While all heat transfers within the body occur through conduction and convection, the model equation must account for tissue conductivity, tissue density, tissue specific heat, local tissue temperature, metabolic or external derived heat sources, and blood velocity. The equation must also be able to account for vascular architecture and neural regulatory vascular controls which strongly influence heat transfers and transport. The vascular architecture can be complicated by the size of blood vessels and geometric vessel patterns (plexus, countercurrent vessels, AVAs, etc.). The magnitude of the heat transfer between the blood vessels and tissue are dependent on the vessel size. According to Poiseulle's law, the change in radius alters resistance to the fourth power of the change in radius. For example, a twofold increase in radius decreases resistance by 16-fold! Therefore, vessel resistance is exquisitely sensitive to changes in radius. Neural regulation controls the radius of the perfusion vessels and the "perfusion conductivity" depends on the velocity of the local blood flow and tissue–vessel alignment. Thus, large vessels exchange very little energy with the surrounding tissues. In contrast, the small vessels (precapillary arterioles, capillaries, and venules) and the surrounding tissues are close to thermal equilibrium.

28.4.2 Bioheat Equation and Skin Blood Flow Modeling

In 1948, Pennes performed a series of experimental procedures "to evaluate the applicability of heat flow theory to the forearm in basic terms of the local rate of tissue heat production and volume flow of blood" [9]. The model derived from experimental data from this investigation became known as the "Bioheat equation." The importance of this model is that it accounted for the importance of blood as a carrier of heat. Thus, the Bioheat equation calculates volumetric heat that is equated to the proportional volumetric rate of blood perfusion. Many of the current models are modifications developed from this basic premise. The Pennes' model is developed around the assumptions that thermal equilibrium occurs in the capillaries and venous temperatures were equal to local tissue temperatures. Both of these assumptions have been challenged in more recent modeling research. The assumption of thermal equilibrium within capillaries was challenged by the work of Chato [39] and Chen and Holmes in 1980 [40]. Based on this vascular modeling for heat transfer, thermal equilibrium occurred in "thermally significant blood vessels" that are approximately 50–75 mm in diameter and located prior to capillary plexus. These thermally significant blood vessels derive from a tree-like structure of branching vessels that are closely spaced in countercurrent pairs. For a historical perspective of heat transfer bioengineering, see Chato [39], and for a review of heat transfers and microvascular modeling, see Baish [41].

In summary, the capacity and ability of blood to transfer heat through various tissue layers to the skin can be predicted from models. This modeling literature provides a conceptual understanding of the thermal response of skin when altered by disease, injury, or external thermal stressors to skin temperatures (environment or application of cold or hot thermal sources, convective airflow, or exercise). From tissue modeling, we know that tissue is only slightly influenced by the deep tissue blood supply but is strongly influenced by the cutaneous circulation. With the use of infrared thermography, skin temperatures can be accurately quantified and the thermal pattern mapped. However, these temperatures cannot be assumed to represent thermal conditions at the source of the thermal stress. Furthermore, the thermal pattern only provides a visual map of skin surface in which heat is dissipated throughout the skin's multiple microvascular plexuses.

28.5 Vascular Structure of Skin

28.5.1 Work of Manchot, Spalteholz, Salmon, Taylor, and Palmer

One of the earliest studies of value on the vascular anatomy of the skin is that of the medical student Carl Manchot from Hamburg. In 1889, at the age of 23 and studying at the Kaiser-Wilhelm University Medical School in Strassburg, Manchot published his treatise *Die Hautarterien des Menschlichen Körpers,* in the incredible time of 6 months [42]. He gave a detailed description of the deep cutaneous arteries and assigned them to their underlying source vessels. His ink injection studies on cadavers allowed him to chart the cutaneous vascular territories of these source arteries. Manchot did not have the benefit of radiographic contrast studies since Roentgen was not to make his discovery until several years later; nevertheless, the accuracy of his work has mostly stood the test of time. His work was translated in English and published in 1983 under the title *The Cutaneous Arteries of the Human Body* [43].

Another important study was published in 1893 by Werner Spalteholz of Leizig. Based on the gelatin and pigment cadaveric injection studies, he made a distinction between direct cutaneous arteries whose main purpose is to supply the skin and indirect arteries which supply the deeper tissues, especially the muscles [44].

The French anatomist and surgeon Michel Salmon published in 1936 his eminent work on the cutaneous arteries in his book titled *Les Artères de la Peau.* Salmon's work is a reappraisal of the work of Manchot [45]. His detailed radiographic studies of the skin's vasculature using a lead oxide mixture produced excellent images. Based on these images he could divide the cutaneous circulation of the human body in 80 vascular territories, which is approximately twice the number as described by Manchot. Each vascular territory has its own source artery. Salmon noted a difference in density and size of arteries in different regions of the body. He divided regions into hypo- and hypervascular zones.

Taylor and Palmer's radiographic lead oxide injection studies of the blood supply to the skin and underlying tissue of fresh cadavers are a timely rediscovery and expansion of the works of Manchot and Salmon [46]. Their results made it possible to segregate the human body anatomically into three-dimensional vascular territories called angiosomes. These three-dimensional anatomical territories are supplied by a source vessel and its accompanying vein. Each tissue in an angiosome is supplied by branches from the source artery. These composite blocks of skin, bone, muscle, and other soft tissue fit together like the pieces of a jigsaw puzzle. The individual angiosomes varied in size and each angiosome is linked to its neighbor angiosome at every tissue level. These interconnections are mostly by a reduced caliber choke anastomosis but in some cases by simple anastomotic arterial connections without change in the caliber of the vessel. A similar pattern is seen on the venous side with avascular bidirectional or oscillating veins. The watershed zones that separate the angiosomes can be seen on angiograms of the skin as areas with reduced vessel density. The choke vessels are located in these watershed zones.

28.5.2 Blood Supply to Skin

The anatomy of the microcirculation of the skin is well described by Cormack and Lamberty [47]. The human skin consists of two layers. The outer layer or epidermis is a waterproof layer of keratinizing stratified squamous epithelium. The inner layer, or dermis, supports the epidermis and is a layer of connective tissue. This layer contains among others blood vessels, lymphatic vessels, sensory nerves, and receptors and skin appendages like sebaceous glands and hair follicles. The skin is supplied by two vascular plexi, the subdermal plexus and the superficial plexus. The subdermal plexus is located just underneath the dermis and branches from here feed the superficial plexus. The superficial plexus lies at the junction of the dermis and epidermis. The skin relies for its blood supply on the vascular structure of the underlying tissue. The work of Taylor and Palmer, as well as others, has revealed that the skin receives its blood supply from so-called perforators [46,48,49]. Taylor defines a cutaneous perforator as "any vessel that perforates the outer layer of the deep fascia to supply the overlying subcutaneous tissue and skin" [50]. A perforator consists of an artery and its concomitant vein. Perforators arise from a source artery and its concomitant vein and on their course to the skin they may follow the intermuscular septa or pass through a muscle. A perforator that lies in an intermuscular septum is called a septocutaneous perforator. These are most frequently located on the extremities. A perforator that passes through a muscle is called a musculocutaneous perforator. These perforators are mostly found on the trunk. Taylor and Palmer identified an average of 374 major perforators per cadaver [46]. Perforators can be further divided into direct and indirect perforators. This classification was described originally by Spalteholz and used again by Taylor and Palmer [44]. Direct perforators have a straight course to the skin were they connect with the subdermal plexus. Whether they follow the intermuscular septa, or pierce muscles en route, their main destination is the skin. The indirect perforators also arise from a source artery and concomitant vein. Their main purpose is to supply the muscles and after they have perforated the deep fascia, the deeper tissues.

El-Mrakby and Miller studied specifically the vascular anatomy of the lower abdomen [48,49]. They found that direct perforators had a diameter larger than 0.5 mm and kept a constant diameter throughout their straight course to the skin. These large direct perforators feed the subdermal plexus and supply the superficial fat. This is in contrast to the small indirect perforators that contribute to the formation of the deep subcutaneous vascular plexus at the deep fat level. The venous drainage from the skin occurs via a venous plexus that connects with the concomitant veins of the perforators or by superficial veins.

28.6 Applying Infrared Technology

28.6.1 Static versus Dynamic Infrared Thermography

IR thermography may be static in that a single image is taken. This technique involves the observation of the spatial temperature distribution over the area of interest. The interpretation of such an image mainly depends on identifying the distribution of hot and cold spots and asymmetric temperature distributions. One of the assumptions when using this method is that the distribution of body surface temperature is basically symmetrical [51,52].

Dynamic infrared thermography (DIRT) is another technique recently proposed to better understand IR images. Because of the interference from complex vascular patterns, researchers have proposed to monitor the thermal recovery process after exposure of the area of interest to a thermal stress. The dynamic method is able to detect not only spatial but also temporal behavior of skin temperature. By applying a thermal challenge, the subsequent recovery of the skin temperature toward its thermal equilibrium is evaluated. The images can be analyzed with respect to the rate and pattern of recovery. The dynamic method could be classified as a passive method or as an active method. In the active method, an external thermal stress is applied to the area under investigation. In practice, the skin area being examined is subjected to a thermal stress by fan cooling, water immersion, or by applying cold or warm objects to the skin surface. It is, however, important that the subject's skin is prevented from becoming wet during immersion in water. No external thermal stress is needed in the passive method. In the passive method, the temperature reactions of healthy individuals are compared with those of patients when an internal stress is applied to them. A common internal stress can be seen after and during exercise. A special form of DIRT is the perfusion of tissue with warm or cold perfusate. DIRT has been used for the monitoring of rewarming of skin after a cold challenge, after surgical reperfusion, or for monitoring cooling of organs following perfusion of a cold fluid as used occasionally in cardiac surgery or organ transplantation [53–56].

28.6.2 Infrared Imaging for Pandemic Fever Screening

The skin regulates heat transfers from its surface through changes in blood perfusion.

These heat transfers are conducted with the purpose of maintaining the body's thermal homeostasis. In recent years, infrared thermographic pandemic screening has been introduced as a means of detecting individuals with fevers that may be related to infectious diseases. An elevated core temperature of 1–2°C (1.8–3.6°F) is generally regarded as indicating a febrile response, pyrexia, or fever. It should be noted that the skin temperature recorded by the infrared imaging device is capturing a skin surface temperature and can only be considered as an indirect measurement of core temperature. The relationship between the skin and core temperature measurements can be problematic if the infrared devise operator does not operate the screening procedure under standardized conditions. In a recent publication, Ring and colleagues [57] compared the inner canthus eyes of febrile and nonfebrile children eyes and recommended a threshold temperature of 37.5°C to differentiate those with fever. In pandemic fever screening, it should be recognized that not all infectious diseases develop a fever, some diseases have an infectious period that is not concurrent with a fever, and the identification of a threshold temperature would change the sensitivity and specificity of the screening measurement. The fundamental problem with infrared pandemic screening lies with operators and screening locations that do not abide by the published standards. Failure to follow the published standards undermines the efficacy, specificity, and sensitivity of the infrared screening and detection procedures. For a more complete discussion on the International Electrochemical Commission/International Standards Organization standards for "Thermal Images for Human Febrile Temperature Screening," see [58,59], pandemic screening review articles [60,61], or the chapter in this text on pandemic screening [62].

28.7 IR and Sports Medicine

Physiologist and sports medicine practitioners using infrared thermography must adhere to published infrared imaging practice, procedures, and standards for thermal assessments and research. Additionally, they must possess a basic understanding of skin blood flow responses to thermal conditions, be able to identify thermal abnormalities (cold or hot spots, dermatomes), and potentially incorporate challenge testing. Dermatomes are a localized region of skin with innervation via a single nerve from a single nerve root of the spinal cord. Nerve impairments (impingements, cuts, stretch, compression, and impact injury) to a dermatome region may be evidenced by a cooler thermal response to the skin within that specified surface area. Unlike normal thermal maps of the skin, these dermatome nerve impingements produce a thermal pattern that has very discrete borders. In a recent study, Sefton and colleagues [63] performed therapeutic neck and shoulder massage on patients between Cervical 1 and Thoracic 2 vertebrae (dermatomes C3–C5). Significant increases in temperature (60 min posttreatment) were observed in the treated area (anterior upper chest, posterior neck, upper back) and two adjoining areas (right arm and middle of the back; dermatome C6–C8). Physical therapist, occupational therapist, and athletic trainers have used infrared imaging to help in the diagnosis of injury and to provide evidence of the efficacy of various treatment modalities (ice, ultrasound, massage).

28.8 Plastic Surgery

The basis of plastic surgery is to restore form as well as function in patients with, for example, congenital deformities or with tissue defects caused by trauma, tumor surgery, and pressure sores. The treatment of these deformities and injuries relies on techniques for transposition of tissue from one part of the human body to another. The tissue that is transposed is called a flap. Historically, the word *flap* originated from the sixteenth-century Dutch word "flappe," this being anything that hangs broad and loose, fastened only by one side [47]. A flap can consist of, for example, skin, skin and muscle, or skin, muscle, and bone. The term "flap" is now a general term encompassing both pedicled flaps and free flaps. A pedicled flap is an area of tissue, which is detached from its surroundings except from a bridge of tissue. Through this bridge, also called pedicle, blood enters and leaves the flap. The pedicle may also consist of only an artery that supplies the flap and a vein or veins that drain the flap. In a free flap, the flap has an identifiable pedicle of an artery and a vein. After the flap has been detached from its surroundings, the artery and vein are cut at sufficient lengths and the flap is moved from its donor site, that is, the site where the flap is harvested, and transferred to the recipient site, the site to be reconstructed. Here, the blood circulation of the flap has to be reestablished. Under a microscope and using microsurgical techniques, the artery and vein of the flap are connected to the vessels at the recipient site. This procedure is a critical part in free flap surgery as a flap without blood circulation will not survive.

28.8.1 Perforator Flap in Plastic Surgery

In 1989, Koshima and Soeda introduced the perforator flap in reconstructive surgery [64]. A perforator flap can be defined as a flap of skin and subcutaneous tissue, which is supplied by an isolated perforator vessel. Such a perforator consists of an artery and its accompanying vein. Perforators pass from the source vessel to the skin, either through or between the deep tissues. The inclusion of muscle and fascia, previously thought to be necessary to guarantee flap circulation, are no longer required as illustrated in Figure 28.1. As perforator flaps consist only of skin and subcutaneous tissue, they permit excellent "like to like" replacement with minimal donor site morbidity for defects of skin and subcutaneous tissue. Perforator flaps lend themselves to being used as either pedicled or free flaps.

The main advantage of perforator flaps is their low donor site morbidity. The majority of donor sites can be closed directly and as the underlying muscle and its nerve supply are preserved, donor site morbidity is minimal [65]. Studies also suggest faster recovery, less postoperative pain, and shorter hospital stays with the use of perforator flaps compared to the flaps that include a muscle.

The disadvantages of perforator flaps are largely related to the learning curve. Adequate preoperative planning, meticulous surgical techniques and a thorough understanding of the vascular anatomy are crucial for a successful postoperative result. Inadequate perfusion of the flap is a complication that may occur when using this technique, especially by the inexperienced surgeon. Inadequate perfusion of the flap can lead to partial or even to total flap loss. These complications are a devastating experience for a patient as they clearly influence the final postoperative outcome. Besides the psychological effect such a flap loss may have on a patient, it is also an inefficient use of economical and hospital resources as reoperations are often necessary. Great efforts should therefore be made to reduce the risk of postoperative flap complications.

28.8.2 Infrared Thermography in Plastic Surgery

The use of infrared thermography in flap surgery, where the perfusion and reperfusion of the flap, pedicled or free, is an important predictive parameter for the success of the surgery, is limited to a few articles. Theuvenet et al. were probably one of the first to recognize the potential of the use of infrared thermography in flap surgery [66]. In the preoperative planning, they cooled the area where the flap

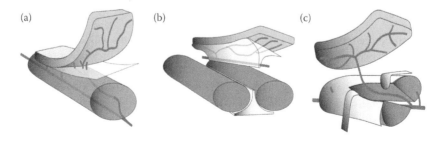

FIGURE 28.1 The evolution from myocutaneous flap (a) to fasciocutaneous flap (b) to perforator flap (c). With the perforator flap, no muscle or fascia is included. In (c), the perforator flap receives its blood supply from a perforator that passes through the muscle and emerges from the source vessel that lies underneath the muscle.

was to be harvested with an ice pack and registered the rewarming of the skin. The locations where hot spots were seen on the infrared images correlated with the locations of perforators as identified during the operation. These flaps were, however, not perforator flaps but musculocutaneous flaps. These musculocutaneous flaps differ from perforator flaps in that a piece of muscle together with a large number of perforators is included in the flap to guarantee flap perfusion. The first to report on the use of infrared thermography in perforator flaps were Itoh and Arai [67]. They illustrated with two clinical cases that a perforator flap could be based on the perforator that was identified by the location of the hot spot on the thermal image during the rewarming of the skin after a cold challenge. Wolff and colleagues used infrared thermography in an experimental study to compare the intraoperative perfusion of different types of flaps [68]. The usefulness of infrared thermography, especially DIRT, has been illustrated in open heart surgery, neurosurgery, and organ transplant surgery [54–56]. We will illustrate in the next section how the use of DIRT can provide the surgeon with valuable information in pedicled and free perforator flap surgery.

28.9 Breast Reconstruction with the Deep Inferior Epigastric Perforator Flap

Breast reconstruction has become an integrated part in the overall treatment of patients diagnosed with breast cancer. The goal of breast reconstruction is to restore a breast mold and to maintain quality of life without affecting the prognosis or detection of recurrence of cancer [69]. Breast reconstruction using tissue from the lower abdomen has become an increasingly popular method after treatment of breast cancer with a mastectomy that is removal of the breast. The lower abdomen is a donor site that remains unmatched in tissue volume, quality, and texture. Patients are specifically pleased with the natural shape, soft consistency, and superior aesthetic results of the reconstructed breast that can be obtained with the use of this donor site [70,71]. An added bonus of an abdominal donor site is for most patients the improved abdominal contour after closure which approximates that of an abdominoplasty or tummy tuck. Although there are several techniques to harvest tissue from the lower abdomen, the technique called the "Deep Inferior Epigastric Perforator" flap or DIEP flap is currently considered the gold standard in breast reconstruction.

The use of the free DIEP flap for breast reconstruction was first described by Allen and Treece in 1994 [72]. The DIEP flap consists of skin and subcutaneous fatty tissue from the lower abdomen. It relies for its blood supply on one of the perforators that arise from the source vessel, the deep inferior epigastric artery and its accompanying vein. After harvest of the DIEP flap on the lower abdomen, the flap is transferred to the thoracic wall. Here, the perforator artery and vein are connected to the internal mammary vessels on the thoracic wall and the blood supply to the flap is reestablished. Finally, a breast is reconstructed. The principle of breast reconstruction with a DIEP flap is illustrated in Figure 28.2.

As with all forms of surgery, breast reconstruction with DIEP flap can be divided into a preoperative phase, an intraoperative phase, and a postoperative phase.

28.9.1 Preoperative Phase

Meticulous planning during the preoperative phase is a prerequisite for successful perforator flap surgery. As the DIEP flap relies for its blood supply on just one perforator, selection of the most suitable perforator to perfuse the flap is crucial for successful surgery. A good perforator can in fact perfuse the whole flap from the lower abdomen.

The perforators of the DIEP flap emerge from the deep inferior epigastric artery and vein, which can be found right underneath the rectus abdominis muscle. While perforators can be located with a handheld Doppler ultrasound probe, studies have shown that this technique is associated with a large number of false-positives [73,74]. The technique is in fact too sensitive and also detects perforators that

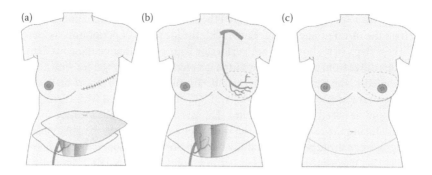

FIGURE 28.2 Breast reconstruction with a DIEP flap. (a) The abdominoplasty flap is harvested from the lower abdomen as a DIEP flap, based on one perforator from the deep inferior epigastric artery and vein. (b) After the flap is transferred to the thoracic wall, its vessels are anastomosed to the internal mammary vessels. (c) A breast is reconstructed and the lower abdomen is closed as an abdominoplasty.

are too small to be used in perforator flap surgery. In 2009, we reported on the value of the use of DIRT in the preoperative planning of the DIEP flap [75]. After the flap had been drawn on the lower abdomen, the locations of arterial Doppler sounds were marked on the skin with a black dot as illustrated in Figure 28.3. All patients were examined in a special examination room with a room temperature of 22–24°C, constant humidity, and air circulation. Typically, the exposed abdomen was subjected to an acclimatization period at room temperature prior to the DIRT examination. A desktop fan was used to deliver the cold challenge by blowing air at room temperature over the abdomen for a period of 2 min. This cold challenge caused visible changes in skin temperature that were well within the physiological range, and had a short recovery period of approximately 5 min. Analysis of the rate and pattern of rewarming of the hot spots allowed a qualitative assessment of all the perforators at the same time. Hot spots that showed a rapid and progressive rewarming could be related to suitable perforators intraoperatively. A rapid rewarming at the hot spot indicates that the perforator is capable of transporting more blood to the skin surface than a hot spot with a low rate of rewarming. A rapid progression of rewarming at the hot spot suggests a better developed vascular network around the hot spot. It appeared that while all first appearing hot spots could be associated with the location of an arterial Doppler sound, not all arterial Doppler sound locations could be related to a hot spot.

The location of the hot spot on the skin could easily be related to the location where the perforator passed through the anterior rectus fascia, although the hot spot as well as the associated Doppler sound were slightly more laterally positioned. The eminent surgeon and anatomist John Hunter (1728–1793)

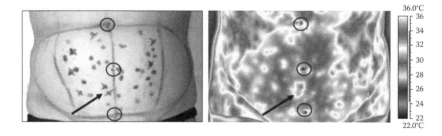

FIGURE 28.3 Left: This photograph of the lower abdomen shows the locations where arterial Doppler sounds were heard and marked with dots. Some of the dots are additionally marked with crosses. The crosses mark arterial Doppler sounds were also associated with hot spots seen in the thermal images shown on the right. The arrows indicate the location where a loud arterial Doppler sound was heard and was associated with a bright hot spot. This perforator could be used to supply a DIEP flap. The circles indicate the positions of small pieces of metal tape used as reference markers.

explained the orientation of vessels as a product of differential growth that had occurred in that area from the stage of fetus to adulthood [76]. Giunta et al. [26] found that the preoperative Doppler location on the skin was located within an average distance of 0.8 cm of that of the exit point of the perforator through the fascia. Interestingly, the selected hot spot on the DIRT images was always associated with an audible Doppler sound and a suitable perforator.

The easiest dissection is reported for those perforators that have a perpendicular penetration pattern through the fascia and a short intramuscular course [73,74,77]. Perforators that are located at the tendineous intersection have these characteristics and are, in addition, larger then average. Interestingly, these perforators were easily identified with DIRT. The short course of the perforator from the source vessel to the skin explains the rapid rewarming of the skin at the hot spot.

In medical infrared thermography, asymmetry on the images has often been associated with pathophysiology [51,78]. The results from our study on the preoperative use of DIRT in the planning of DIEP flaps showed that asymmetry in the distribution and quality of hot spots between both sides on the lower abdomen was more or less a normal finding as can be seen in Figure 28.4. This result is in accordance with other studies related to the preoperative mapping of perforators on the lower abdomen. The use of the modern high-quality IR cameras may be the reason why this nonpathological asymmetry becomes visible now.

Recently, the use of multidetector computed tomography (MDCT) angiography has become increasingly popular for the preoperative planning of perforator flaps, and has become in some hospitals the gold standard [79,80]. The high spatial and temporal resolution achieved with MDCT angiography

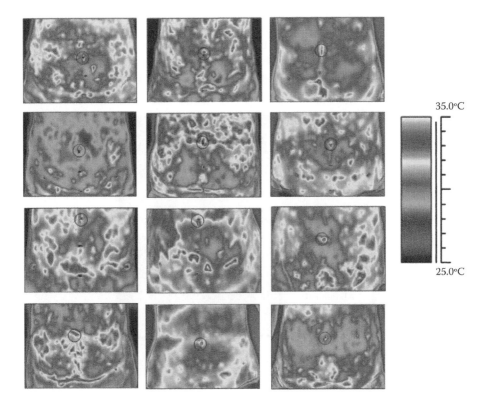

35.0°C

25.0°C

FIGURE 28.4 Preoperative infrared thermal images of the abdominal area of 12 female patients indicating thermal asymmetry. The black circle in each image indicates the position of the navel. The figure illustrates the large variability in the distribution and intensity of hot spots, not only between the left and right side of each individual patient but also between patients.

allows for a precise description of the origin, intramuscular course, and point of fascia penetration of the arterial perforator. MDCT angiography does not provide information on the venous side of the perforator. Such information can be obtained but requires an extra scan. Although DIRT does not show the morphology of the perforators, it identified suitable perforators based on the perforators' physiology.

To adopt MDCT as a routine preoperative imaging modality, the benefit to the patient, for both the reconstructed breast and the abdominal donor site, must outweigh the problems associated with the procedure. These include the risk of intravenous contrast agent, exposure to ionizing radiation, and associated cost. Recent literature cautions against the rising exposure of ionizing radiation to the population due to CT examinations [81,82]. IR thermography has none of these disadvantages.

28.9.2 Intraoperative Phase

During the operation, the DIEP flap is harvested from the lower abdomen (Figure 28.5) and transferred to the thoracic wall where a breast will be reconstructed. On the thoracic wall, the blood circulation to the DIEP flap has to be reestablished. This is a critical part of the whole operation as without blood circulation the flap will not survive. The perforator artery and vein to the DIEP flap are connected to the internal mammary vessels under a microscope and with the use of microsurgical instruments. This connection is called anastomosis. By opening the anastomosis, the flap becomes reperfused. In 2006, we published a study on the intraoperative use of DIRT in free perforator surgery [83]. During the transfer of the DIEP flap from the lower abdomen to the thoracic wall, the flap is a period without blood circulation and as a consequence cools down. After the anastomosis is opened, the flap becomes reperfused and as a consequence rewarms. It showed that the intraoperative use of DIRT allowed to assess perfusion and reperfusion in real time and thus provided the surgeon immediate feedback. The usefulness of DIRT during the period after completion of the anastomosis became readily apparent as partial and total arterial inflow problems were easily detected with analyses of the IR images. Also, venous congestion was easily identified on the IR images. An important advantage of the intraoperative use of DIRT

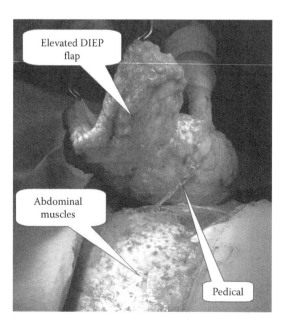

FIGURE 28.5 The skin and subcutaneous tissue of the lower abdomen is elevated as a DIEP flap. The image was taken just prior to transfer of the flap to the thoracic area. The flap receives its blood supply via the pedicle which is the perforator and consists of one artery and one vein. Note no muscle is included.

was that the surgeon could take corrective action when it was most effective, namely during the operation. At the end of the operation, perfusion of the flap could be rapidly and easily evaluated after applying a thermal stress to the skin surface using a cold metal plate and see how the flap rewarms.

28.9.3 Postoperative Phase

During the postoperative phase, the flap is closely monitored for signs of compromised perfusion. Most surgeons rely on clinical observation of skin perfusion using subjective signs such as skin color, capillary refill, skin temperature, and turgor [84]. A typical monitoring protocol includes hourly observations of flaps for the first 24 h and then to extend this to every 4 h for the first three postoperative days. Recognizing the visual cues of a failing flap requires considerable clinical experience. If a flap shows signs of impaired perfusion, whether it is due to an arterial inflow or a venous outflow problem, surgical intervention may be necessary (see Figure 28.6). The success of such secondary microsurgery is inversely related to the time interval between the onset of impaired perfusion and its detection [85,86]. Temperature measurements of skin flaps during the postoperative phase are one of the oldest monitoring methods of free flaps. Measurements of absolute skin surface temperatures of free groin flaps during the postoperative phase was first reported by Baudet et al. in 1976 [87]. Acland promoted the technique of absolute surface temperature monitoring of free flaps and stated that a temperature between 32°C and 30°C was marginal, and a temperature below 30°C was a sign of flap failure [88]. Leonard et al. introduced the concept of differential surface temperature by monitoring

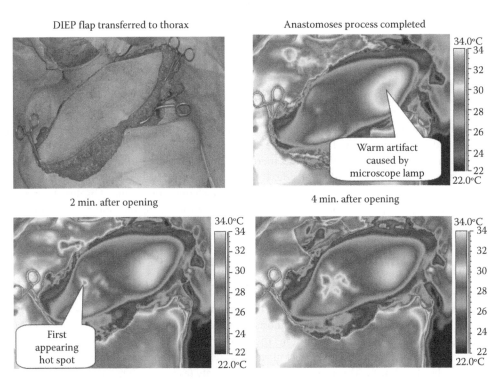

FIGURE 28.6 Digital photograph of a free DIEP flap. The infrared thermal images demonstrate a rewarming of a free DIEP flap after completion of a successful microsurgical anastomosis. The anastomosis was opened in the upper right image. The other images were taken at 2 min intervals after the anastomosis had been opened and the blood flow to the flap was restored. Note the appearance of hot spots that rapidly increase in size and number. A thermal artifact (diffuse warm area) can be seen in the upper left thermal image caused by heating from the microscope lamp. This area cools down after removal of the heat source.

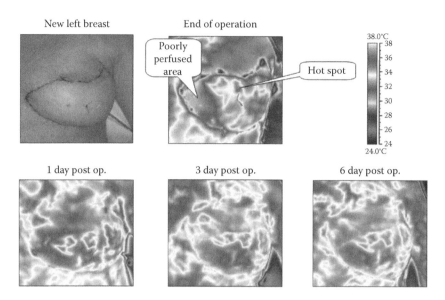

FIGURE 28.7 The photograph on the top left shows a newly reconstructed left breast with a DIEP flap. The infrared thermal images show an improved perfusion over time. Not that part of the newly reconstructed breast is less well perfused at end of operation (indicated by a cooler area) but improves in the days as indicated by the appearance of new hot spots.

the difference between a temperature probe on the flap and a second control temperature probe on adjacent normal skin [89]. Temperature measurements are often included in the protocols of postoperative monitoring of flaps. Recently, Busic et al. criticized the use of temperature measurement in the monitoring of DIEP flaps [90]. We reported on the postoperative use of DIRT on breast reconstruction with a DIEP flap [91]. It appears that the perfusion of DIEP is a dynamic process with a stepwise progression during the first postoperative week. Hot spots are initially seen on the IR images in the area positioned over the location of the entrance point of the perforator in the DIEP flap (Figure 28.7). On subsequent days, the number of hot spots increases, first on the ipsilateral side of the midline and from day three also on the contralateral side. It became clear from this study that the temperature differs from location to location and that the temperature changes over time. Absolute temperature measurements are perhaps less informative then the information on the rate and pattern of rewarming obtained with DIRT.

28.10 Use of Infrared Thermal Imaging to Localize Axial Pedicles in Facial and Nasal Reconstruction

Facial reconstruction is a pivotal component in the treatment of skin cancer as it involves anatomic, functional, and cosmetic challenges. The diagnosis of over one million skin cancers annually exemplifies the increasing prevalence of this condition. The use of Mohs micrographic surgery is the most accurate technique for treatment of all types of skin cancer (melanoma and nonmelanoma) with particular utility on the head and neck. This is due to the high cure rate (exceeding 99%) and superior cosmetic result due to the selective surgical margin that is achieved when performed by fellowship-trained Mohs surgeons.

The Mohs surgical technique yields a surgical defect that must subsequently be repaired. Facial reconstructive techniques encompass a multitude of different options, including primary closure, flaps, grafts, and second intent wound healing. Knowledge of the anatomy, physiology, and flap biomechanics of skin is imperative for successful reconstruction.

There are a multitude of cutaneous flaps in the reconstructive surgeon's armamentarium to repair cutaneous surgical defects. These include primary closures, rotation flaps, advancement flaps, interpolation flaps, allografts, xenografts, and skin grafts, in addition to the numerous variants of each of these. An adequate vascular supply is critical to the success of all of these reconstructive techniques. Reconstruction of the nose specifically mandates a fundamental knowledge of the cutaneous, cartilaginous, and vascular supply for successful reconstruction. A specific reconstruction of larger nasal defects will be utilized to demonstrate the importance of vascular identification and the benefit of adjunctive thermal imaging.

The paramedian forehead flap is an effective surgical technique that can be utilized to reconstruct large surgical defects, primarily of the nasal subunits and/or entire nose [92–94]. This reconstruction relies on a specific arterial supply to maintain patency of the flap. At present, this pedicle flap is identified primarily by the surgeon's knowledge of anatomy and may be confirmed by Doppler ultrasonography.

The use of thermal infrared imaging represents a unique adjunctive technique to allow confirmation of specific arterial sources for cutaneous flaps. Thermal infrared imaging has been shown to provide good sensitivity in the direct imaging of superficial vessels on the face [95]. This fundamental knowledge is applied in the present chapter to a specific artery as it is used in nasal reconstruction.

28.10.1 Anatomy of a Flap

Surgical defects can be repaired with a wide variety of reconstructive techniques. These may include allowing the wound to heal via second intent, primary closure, skin grafts (split thickness and full thickness, allografts, xenografts), and flaps. There are many types of cutaneous flap designs, including transposition, advancement, rotation, and interpolation flaps.

An adequate blood supply is paramount to the survival of a flap. The anatomic vascular characteristics of a flap provide a useful means of classification [96]. Early studies by Manchot and Salmon contributed to the understanding of vascular anatomy, especially in the setting of cutaneous flaps [42,43]. Vascular territories, or angiosomes, are those cutaneous regions that are supplied by a specific vessel or set of vessels.

Proper design of a cutaneous flap must be performed with this angiosome in mind to ensure survival of the flap. In cutaneous flap surgery, arterial supply is characterized as either axial or random pattern [97]. Axial flaps are those that are based on a specific artery, such as the paramedian forehead flap, which is designed around the supratrochlear artery. A random pattern flap is one that is not designed on a specific artery but instead survives on the network of small nonspecified arteries, also known as the subdermal plexus. These vascular territories must be visualized in a three-dimensional view to prevent unintentional transaction of a vessel.

Facial and nasal reconstruction mandates a superior knowledge of anatomy, with particular attention to the vasculature. This allows both the identification of key vessels that are paramount to the survival of a flap in addition to the ability to avoid inadvertent transection of vessels.

The supratrochlear artery traverses the orbital rim between the corrugator and frontalis muscles. It is located above the periosteum as it passes over the medial brow and protected by the corrugator muscle [92–94]. The supratrochlear artery becomes more superficial as it is followed upwards. The artery remains underneath the frontalis muscle until it reaches the mid-forehead, where it then travels through the frontalis muscle and reaches the subdermal position at the superior forehead [93,94]. The supraorbital artery is the dominant artery of the forehead and thus an excellent arterial source for an axial flap [98].

28.10.2 Detection of Vasculature

The surgeon's fundamental knowledge of anatomy is the primary standard for reconstruction. Localization of the supratrochlear artery can be determined by clinical landmarks. This artery is

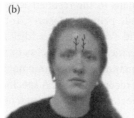
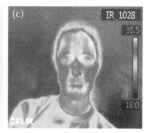

FIGURE 28.8   (a) Digital picture of forehead. (b) The supratrochlear arteries can be localized by clinical land-marks and are typically found originating near the inferior glabellar creases. These near parallel vessels run vertically approximately 1.5–2.0 cm from the midline forehead. (c) Infrared image of forehead.

typically found approximately 1.5–2.0 cm from the midline forehead (see Figure 28.8) [98]. Its origin at the lower forehead can be localized by palpation of the supratrochlear foramen and/or visualization of the inferior glabellar crease [98,99].

Continuous wave Doppler ultrasound is commonly used to confirm and track the location of the supratrochlear artery. This is achieved with a handheld vascular probe that allows audible detection of vessels. This is useful provided the target vessel is of sufficient caliber to allow detection. Handheld Doppler ultrasound is not useful in determining vascular characteristics of random-pattern flaps as these vessels are too small.

The use of thermal infrared imaging marks an additional adjunctive tool to localize specific vessels. This technique allows for real-time visualization of the specific artery and/or a vascular plexus, which can be a tremendous aid in the decision-making process of a reconstruction.

Facial segmentation based on thermal signatures of vascular regions has been utilized to demonstrate vascular versus less vascular angiosomes [95]. The work of Buddharaju and Pavlidis with thermal minutia point extraction shows promising results toward increased thermal detection of cutaneous vasculature [95].

28.10.3 Determinants of Flap Choice

The choice of reconstructive technique is made with consideration of multiple variables. Some of these variables include the location of the surgical defect, the age of the patient, the medical and surgical history of the patient, the type of tumor that was treated, and both functional and cosmetic outcomes from the reconstruction. Additional factors such as the use of tobacco products are significant as nicotine acts as a vasoconstrictor and can decrease arterial flow to a dependent flap [99,100]. The use of surgical delay can be utilized to assist in flaps that are deemed to be risky for a potentially inhibited arterial supply [99,100]. Anatomical differences among various individuals represent another factor that must be considered.

The use of thermal infrared imaging can be used as an adjunctive tool to gauge the vascular health of a cutaneous region. The sensitivity of a camera certainly plays a large part in the amount of information that can be obtained. Imaging for axial versus random pattern flaps has different goals and sensitivities. Imaging for a random pattern flap requires identification of adequate perfusion, while that for an axial flap requires precise localization of a specific vessel. The relatively small caliber vessel in cutaneous flaps makes this more challenging than with the larger vessels due to the decreased thermal signature detected.

28.10.4 Paramedian Forehead Flap

The paramedian forehead flap used today is based on centuries of refinements from surgeons all over the world. In short, this technique classically involves a two-stage transfer of skin from the forehead to

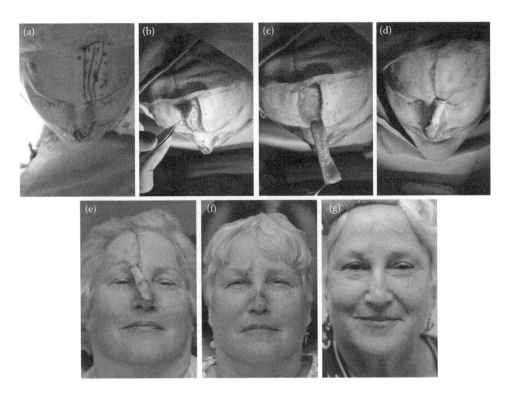

FIGURE 28.9 Paramedian forehead flap reconstruction of a surgical defect following Mohs micrographic surgery for a basal cell carcinoma. The interpolation flap is based on the right supratrochlear artery (a). The flap is elevated superficially at the distal aspect and at the level of the periosteum at the proximal aspect (b, c) to ensure patency of the supratrochlear artery. The flap is subsequently rotated on its proximal base into position to accommodate the surgical defect where it remains for 2–3 weeks (d, e). The second stage of the reconstruction involves taking the flap down (f). The final reconstruction (g) offers a cosmetically acceptable reconstruction.

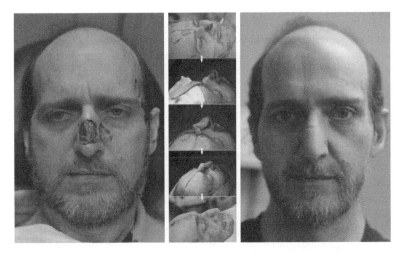

FIGURE 28.10 Paramedian forehead flap reconstruction of a large, infiltrative, morpheaform basal cell carcinoma of the right nasal ala and sidewall treated with Mohs micrographic surgery.

the nose (Figure 28.9). The first English publication of the forehead flap was in 1793; however, this flap is believed to have been used at least since 1440 AD [101,102]. The use of forehead skin to reconstruct the nose is based on its close similarity in both color and texture [92,93]. In addition, the forehead offers a large canvas on which the surgeon can design the flap based on the needed size and thickness.

Multiple optimizations of this flap have been implemented since the initial forehead flap, resulting in a thinner pedicle, less rotation/twisting of the pedicle, and a resultant improved cosmetic outcome [92]. Proper design of the flap with a thinner pedicle requires precise knowledge and localization of the supratrochlear artery. Failure in the identification and/or design of the flap could lead to inadvertent transaction of the artery and subsequent flap necrosis. Basic anatomical landmarks assist in localization of the artery; however, adjunctive techniques for this purpose are always welcome.

Based on the surgical defect and reconstructive requirements, this reconstructive technique may be performed in conjunction with other techniques such as cartilage grafts, nasal mucosal lining flaps and/or grafts, and/or other flaps.

The paramedian forehead flap is well described and thus will only be briefly discussed [103]. As discussed above, proper design of the flap is paramount. Identification of the supratrochlear artery is performed initially based on anatomic landmarks (Figure 28.10) [98,104,105]. This includes multiple different methods such as palpation of foramen and glabellar creases. The use of Doppler ultrasonography is commonly used to localize and/or confirm the course of the supratrochlear artery. Once this axial artery is identified, the flap is designed around the artery based on the necessary dimensions. Incision and elevation of the flap is performed with careful preservation of the supratrochlear artery (Figure 28.4). The flap is incised to the level of the periosteum at the insertion while the level may vary based on surgeon preference at the distal aspect. The interpolation flap is next rotated on its proximal end and placed into position at the surgical defect (Figure 28.3). The flap may be trimmed and tailored for an exact fit and then sutured into position. There are numerous variations on technical aspects of this flap that have been described in detail [92,106]. The donor site on the forehead is repaired as a primary closure, leaving the pedicle intact to maintain an adequate arterial supply to the flap (Figure 28.2d). The flap is typically left in place for 2–3 weeks at which time the flap is divided. This second stage involves removal of the connecting pedicle and fitting the proximal portion of the flap in place (Figure 28.3). There are techniques to perform the paramedian forehead flap in a single stage in addition to three (or more) stages as well.

28.10.5 Utilization of Thermal Imaging as an Adjunctive Tool

The use of thermal infrared imaging offers a safe, real-time visualization of cutaneous vessels. This offers dermatologic, Mohs micrographic, and plastic surgeons an adjunctive tool to assess the vascular characteristics of anatomic regions. The sensitivity of thermal imaging cameras is a critical determinant in this ability to detect the superficial vasculature. This enhanced precision can allow for improvements such as a thinner pedicle to be designed.

Acknowledgments

The schematic diagrams (Figures 28.1 and 28.2) were drawn by Sven Weum, Auburn University, Auburn, Alabama. The images in Figures 28.3 through 28.6 were from patients and volunteers and are courtesy of the Cardiovascular Research Group, Department of Medical Biology, Faculty of Health Sciences, University of Tromsø, Norway and the Department of Hand and Plastic Surgery, University Hospital North Norway, Tromsø, Norway. The infrared thermal images in these examples were taken using an FLIR ThermaCAM®S65-HS uncooled bolometer. The image analysis software used was ThermaCAM researcher ver.2.8(SR2). Examples of IR images (XX) come from the Thermal Lab, Auburn University, Alabama, USA utilizing an FLIR B360. We would like to acknowledge the assistance of Kunal Aswani,

Ryan Ward, and Martha Hart "Ragan" Hart; doctoral students Matt Barberio and David Elmer; and my family for their support and editing (Donna, Corrie, and Annan Pascoe).

References

1. Pascoe D.D., Mercer J.B., DeWeerd L. Physiology of thermal signals. In: *The Biomedical Engineering Handbook; Medical Devices and Systems.* (3rd edition). Joseph D.B. Ed., Boca Raton, FL: CRC Press, Taylor & Francis Group 21.1–21.18, 2006.
2. Nadel E.R., Mitchell J.W., Stowwijk J.A.J. Differential thermal sensitivity in the human skin. *Pflugers Arch.* 340:71–76, 1973.
3. Crawshaw L.I., Nadel E.R., Stolwijk J.A.J., Samford B.A. Effect of local cooling on sweat rate and cold sensation. *Pflugers Arch.* 354:19–27, 1975.
4. Olsen B.W. How many sites are necessary to estimate a mean skin temperature? In: *Thermal Physiology.* Hales J.R.S., Ed., New York: Raven Press, 33–38, 1984.
5. Hardy J.D. The radiation of heat from the human body. III The human skin as a black-body radiator. *J. Clin. Invest.* 13(4): 615–620, 1934.
6. Hardy J.D., Muschenheim C. The radiation of heat from the human body. IV The emission, reflection, and transmission of infra-red radiation by the human skin. *J. Clin. Invest.* 13(5): 817–831, 1934.
7. Hardy J.D., Muschenheim C. Radiation of heat from the human body. V. The transmission of infra-red radiation through skin. *J. Clin. Invest.* 15(1): 1–9, 1936.
8. Ryan T.J., Jolles B., Holti G. *Methods in Microcirculation Studies.* London: H.K. Lewis and Co, Ltd, 1972.
9. Pennes H.H. Analysis of tissue and arterial blood temperatures in resting human forearm. *J. Appl. Physiol.* 1, 93–102, 1948.
10. Charkoudian N. Skin blood flow in adult thermoregulation: How it works, when it does not, and why. *Mayo Clin. Proc.*, 78, 603–612, 2003.
11. Greenfield A.D.M. The circulation through the skin. In: *Handbook of Physiology—Circulation.* Hamilton W.P., Ed., Washington DC: Am. Physiol. Society, Section 3, Vol. II, (Chapter 39), 1325–1351, 1963.
12. Johnson J.M., Brenglemann G.L., Hales J.R.S., Vanhoutte M., Wenger C.B. Regulation of the cutaneous circulation. *Fed. Proc.* 45, 2841–2850, 1986.
13. Johnson J.M., Proppe D.W. Cardiovascular adjustments to heat stress. In: *Handbook of Physiology—Environmental Physiology.* Fregly M.J., Blatteis C.M., Eds., New York/Oxford: Oxford Press, pp. 215–243, 1996.
14. Rowell L.B. Human cardiovascular adjustments to exercise and thermal stress. *Physiol. Rev.* 54, 75–159, 1974.
15. Rowell L.B. Cardiovascular adjustments to thermal stress. In: *Handbook of Physiology—The Cardiovascular System.* Fregly M.J., Blatteis C.M., Eds., Bethesda, MD: Am. Physiol. Society, Section 3, Vol. 5, Part 3, (Chapter 27), 967–1024, 1983.
16. Rowell L.B. *Human Circulation: Regulation during Physiological Stress.* New York: Oxford University Press, 1986.
17. Sawka M.N., Wenger C.B. Physiological responses to acute exercise-heat stress. In: *Human Performance Physiology and Environmental Medicine at Terrestrial Extremes.* KB Pandolf, K.B., Sawka M.N., Gonzales R.R., Eds, Indianapolis, IN: Benchmark Press, pp. 97–151, 1988.
18. Johnson J.M. Circulation to the skin. In: *Textbook of Physiology,* Vol. 2, Patton H.D., Fuchs A.F., Hille B., Scher A.M., Steiner R., Eds, Philadelphia: W.B. Saunders Co., (Chapter 45), 1989.
19. Johnson J.M. Exercise and the cutaneous circulation, *Exercise Sports Sci. Rev.* 20, 59–97, 1992.
20. Charkoudian N. Skin blood flow in adult human thermoregulation: How it works, when it does not, and why. *Mayo Clin. Proc.* 78:603–612, 2003.

21. Gaskell P. Are there sympathetic vasodilator nerves in the vessels of the hands? *J. Physiol.* 131, 647–656, 1956.

22. Fox R.H., Edholm O.G. Nervous control of the cutaneous circulation. *J. Appl. Physiol.* 57, 1688–1695, 1984.

23. Lossius K., Eriksen M., Walloe L. Fluctuations in blood flow to acral skin in humans: Connection with heart rate and blood pressure variability. *J. Physiol.* 460:641–655, 1993.

24. Edholm O.G., Fox R.H., Macpherson R.K. The effect of body heating on the circulation in skin and muscle. *J. Physiol.* 134:612–619, 1956.

25. Edholm O.G., Fox R.H., Macpherson R.K. Vasomotor control of the cutaneous blood vessels in the human forearm. *J. Physiol.* 139:455–465, 1957.

26. Rodie I.C., Shepard J.T., Whelan R.F. Evidence from venous oxygen saturation that the increase in of arm blood flow during body heating is confined to the skin. *J. Physiol.* 134, 444–450, 1956.

27. Kellogg D.L. Jr, Shepard J.T., Whelan R.F. Cutaneous active vasodilation in humans is mediated by cholinergic nerve cotransmission. *Cir. Res.* 77:1222–1228, 1995.

28. Kolka M.A., Stephenson L.A. Cutaneous blood flow and local sweating after systemic atropine administration. *Pflugers Arch.* 410:524–529, 1987.

29. Shastry S., Minson C.T., Wilson S.A., Dietz N.M., Joyner M.J. Effects of atropine and L-NAME on cutaneous blood flow during body heating in humans. *J. Appl. Physiol.* 88:467–462, 2000.

30. Pascoe D.D., Shanley L.A., Smith E.W. Clothing and Exercise I: Biophysics of heat transfer between the individual, clothing, and environment. *Sports Med.* 18(1): 38–54, 1994.

31. Pascoe D.D., Bellinger T.A., McCluskey B.S., Clothing and Exercise II: Influence of clothing during exercise/work in environmental extremes, *Sports Med.* 18(2):94–108, 1994.

32. Roberts M.P., Wenger C.B., Stölwik, Nadel E.R. Skin blood flow and sweating changes following exercise training and heat acclimation *J. Appl. Physiol.* 43, 133–137, 1977.

33. Stephenson L.A., Kolka M.A. Menstrual cycle phase and time of day alter reference signal controlling arm blood flow and sweating *Am. J. Physiol.* 249(2, pt2), R186–R192, 1985.

34. Aoki K., Stephens D.P., Johnson J.M. Diurnal variations in cutaneous vasodilator and vasoconstrictor systems during heat stress. *Am. J. Physiol. Regul. Integr. Comp. Physiol.* 281, R591–R595, 2001.

35. Wissler F.H. Steady State temperature distribution in man. *J. Appl. Physiol.* 16:734–740, 1961.

36. Volpe B.T., Jain R.K. Temperature distribution and thermal responses in humans. I. Stimulation of various modes of whole body hyperthermia in normal subjects. *Med. Physics* 9:506–513, 1982.

37. Chan R.A., Sigelman R.A., Guy A.W. Calculations of therapeutic heat generated by ultrasound in fat-muscle-bone layers. *IEEE Trans. Biomed. Eng.* BME, 21:280–284, 1974.

38. Sekins K.M., Emery A.F., Lehmann J.F., MacDougall J.A. Determination of perfusion field during local hypoerthermia with the adi of finite elements thermal model. *ASME J. Biomech. Eng.* 104:272–279, 1982.

39. Chato J.C. A view of the history of heat transfer in bioengineering. In: *Advances in Heat Transfer*, San Diego/New York/London/Sydney/Boston/Tokyo/Toronto: Academic Press, Vol. 22, pp. 1–19, 1981.

40. Chen M.M., Holmes K.R. Microvascular contributions in tissue heat transfer. In: *Thermal Characteristics of Tumors: Applications in Detection and Treatment.* Jain R.K., Guillino P.M., Eds., Ann N.Y. Acad Sci. 335, 137, 1980.

41. Baish J.W., Microvascular heat transfer, In: *The Biomedical Engineering Handbook.* Bronzino J.D., Ed., Boca Raton, FL: CRC Press/IEEE Press, Vol. II, 98, 1–14, 2000.

42. Manchot C. *Die Hautarterien des Menschlichen Körpers.* Vogel, Leipzig, 1889.

43. Manchot C. *The Cutaneous Arteries of the Human Body.* Springer-Verlag: New York, 1982.

44. Spalteholtz W. Die Vertheilung der Blutgefasse in der Haut. *Arch Anat Entwiecklngs-Gesch (Leipz)* 1:54, 1893.

45. Salmon M. Artères de la peau. Maisson et Cie, Paris, 1936.

46. Taylor G.I., Palmer J.H. The vascular territories (angiosomes) of the body: Experimental study and clinical applications. *Br. J. Plast. Surg.* 40:113–141, 1987.

47. Cormack G.C., Lamberty B.G. The different layers of the integument and functional organization of the microcirculoation. In: *The Arterial Anatomy of Skin Flaps*. Cormack G.C., Lamberty B.G. Eds., New York: Churchill Livingstone, pp. 16–69, 1994.

48. El-Mrakby H.H., Milner R.H. The vascular anatomy of the lower anterior abdominal wall: A microdissection study on the deep inferior epigastric vessels and the perforator branches. *Plast. Reconstr. Surg.* 109:539–543, discussion by Taylor GI, 544–547, 2002.

49. El Mrakby H.H., Milner R.H. Bimodal distribution of the blood supply to lower abdominal fat: Histological study of the microcirculation of the lower abdominal wall. *Ann. Plast. Surg.* 50:165–170, 2003.

50. Taylor G.I. The angiosomes of the body and their supply to perforator flaps. *Clin. Plast. Surg.* 30:331–342, 2003.

51. Jiang L.J., Ng E.Y.K., Yeo A.C.B., Wu S., Pan F., Yau W.Y., Chen J.H., Yang Y. A perspective on medical infrared imaging. *J. Med. Eng. Tech.* 29:257–267, 2005.

52. Wilson S.B., Spence V.A. Dynamic thermography imaging method for quantifying dermal perfusion: Potential and limitations. *Med. Biol. Eng. Comput.* 27:496–501, 1989.

53. Miland Å.O., Mercer J.B. Effect of a short period of abstinence from smoking on rewarming patterns of hands following local cooling. *Eur. J. Appl. Phys.* 98:161–168, 2006.

54. Goetz C., Foertsch D., Schoeberger J., Uhl E. Thermography—a valuable tool to test hydrocephalus shunt patency. *Acta Neurochir. (Wien)* 147:1167–1173, 2005.

55. Garbade J., Ulllmann C., Hollenstein M., Barten M.J., Jacob S., Dhein S., Walther T., Gummert J.F., Falk V., Mohr F.W. Modeling of temperature mapping from quantitative dynamic infrared coronary angiography for intraoperative graft patency control. *J. Thor. Card. Surg.* 131:1344–1351, 2006.

56. Gorbach A., Simonton D., Hale A., Swanson S.J., Kirk A.D. Objective real-time, intraoperative assessment of renal perfusion using infrared imaging. *Am. J. Transplant.* 3:988–993, 2003.

57. Ring E.F.J., Jung A., Zuber J., Rukowski P., Kalicki B., Najwa U. Detecting Fever in Polish Children by Infrared Thermography, *QIRT, Proceedings of 9th International Conference on Quantitative Infrared Thermography*, Krakow Poland, Technical University of Ldz. Institute of Electronics, pp. 125–128, 2008.

58. International Standard IEC 80601-2-59, Medical Electrical Equipment-Part 2–59: Particular requirements for basic safety and essential performance of screening thermographs for human febrile temperature screening IEC Geneva (2008) available at www.webstore.ansi.org RecordDetail.aspx? Sku=IEC+8060 1-2-59+Ed.1.0+b%3a2008, 2008.

59. International Organization for Standards ISO/TR 13154:2009, Medical Electrical Equipment-Deployment, implementation, and operational guidelines for identifying febrile humans using a screening thermograph. ISO Geneva, 2009 www.iso.org/iso/iso_catalogue_tc/catalogue_detail.htm?csnumber=51236, 2009.

60. Mercer, J.B., Ring E.F.J. Fever screening and infrared imaging: Concerns and guidelines. *Thermol. Int.* 19, 67–69, 2009.

61. Pascoe D.D., Ring E.F., Mercer, J.B., Snell J., Osborn D., Hedley-Whyte J. International standards for pandemic screening using infrared thermography ISO TC121/SC3-IEC TC/SC62D/JWG 8, Clinical Thermometers, Project Team 1, Thermal Imagers for Human Febrile Temperature Screening. Medical Imaging 2010: Biomedical Applications in Molecular, Structural, and Functional Imaging, Molthen R.C., Weaver J.B., Eds., *Proc. SPIE*, 7626, 76261Z, 2010.

62. Francis J. Ring, EYK Ng, Infrared thermal imaging standards for human fever detection. In: *Medical Infrared Imaging*. Diakides N.A., Bronzino J.D., Eds, Baton Rouge/London/New York: CRC Press.

63. Sefton J.M., Yarar C., Berry J.W., Pascoe D.D. Therapeutic massage of the neck and shoulder produces changes in peripheral blood flow when assessed with dynamic infrared thermography. *J. Alternative Complimentary Med.* 16(7): 1–10, 2010.

64. Koshima I., Soeda S. Inferior epigastric artery skin flaps without rectus abdominis muscle. *Br. J. Plast. Surg.* 42:645–648, 1989.

65. Blondeel P.N. Soft tissue reconstruction with perforator flaps. In: *Tissue Surgery.* Siemionow M.Z., Ed., London: Springer-Verlag, pp. 87–100, 2006.

66. Theuvenet W.J., Koevers G.F., Borghouts M.H. Thermographic assessment of perforating arteries. A preoperative screening method for fasciocutaneous and musculocutaneous flaps. *Scand. J. Plast. Reconstr. Surg.* 20:25–29, 1986.

67. Itoh Y., Arai K. Use of recovery-enhanced thermography to localize cutaneous perforators. *Ann. Plast. Surg.* 34:507–511, 1995.

68. Wolff K.D., Telzrow T., Rudolph K.H. et al. Isotope perfusion and infrared thermography of arterialised, venous flow-through and pedicled venous flaps. *Br. J. Plast. Surg.* 48:61–70, 1995.

69. Cordeiro P.G. Breast reconstruction after surgery for breast cancer. *N. Engl. J. Med.* 359:1590–1601, 2008.

70. Granzow J.W., Levine J.L., Chiu E.S. et al. Breast reconstruction using perforator flaps. *J. Surg. Oncol.* 94:441–454, 2006.

71. Chevray P.M. Update on breast reconstruction using free TRAM, DIEP, and SIEA flaps. *Sem. Plast. Surg.* 18:97–103, 2004.

72. Allen R.J., Treece P. Deep inferior epigastric perforator flap for breast reconstruction. *Ann. Plast. Surg.* 32:32–38, 1994.

73. Blondeel P.N. One hundred free DIEP flap breast reconstructions: A personal experience. *Br. J. Plast. Surg.* 52:104–111, 1998.

74. Giunta R.E., Geisweid A., Feller A.M. The value of preoperative Doppler sonography for planning free perforator flaps. *Plast. Reconstr. Surg.* 105:2381–2386, 2000.

75. de Weerd L., Weum S., Mercer J.B. The value of dynamic infrared thermography (DIRT) in perforator selection and planning of DIEP flaps. *Ann. Plast. Surg.* 63:274–279, 2009.

76. Hunter J. *A Treatise on the Blood, Inflammation and Gunshot Wounds.* John Richardson, London, 1794.

77. Neligan P.C., Blondeel P.N., Morris S.F., Hallock G.G. Perforator flaps: Overview, Classification, and Nomenclature. In: *Perforator Flaps. Anatomy, Technique & Clinical Applications.* Blondeel P.N., Morris S.F., Hallock G.G., Neligan P.C. Eds., St Louis, Missouri: Quality Medical Publishing, pp. 37–52, 2006.

78. Amalu W.C., Hobbins W.B., Head J.F., Elliott R.L. Infrared imaging of the breast: a review. In: *Medical Infrared Imaging.* Diakides N.A., Bronzino J.D. Eds., Boca Raton: CRC Press, Taylor & Francis Group, pp. 9.1–9.19, 2008.

79. Masia J., Clavero J.A., Larrañaga J.R., Alomar X., Pons G., Serret P. Multidetector-row tomography in the planning of abdominal perforator flaps. *J. Plast. Reconstr. Aesthet. Surg.* 59:594–599, 2006.

80. Casey W., Chew R.T., Rebecca A.M., Smith A.A., Collins J., Pockaj A. Advantages of preoperative Computed Tomography in deep inferior epigastric artery perforator flap breast reconstruction. *Plast. Reconstr. Surg.* 123:1148–1155, 2008.

81. Wiest P.W., Locken J.A., Heintz P.H., Mettler Jr F.A. CT scanning: A major source of radiation exposure. *Sem. Ultrasound MRI* 23:402–410, 2002.

82. Brenner D.J., Hall E.J. Computed Tomography: An increasing source of radiation exposure. *N. Engl. J. Med.* 357:2277–2284, 2007.

83. de Weerd L., Mercer J.B., Setså L.B. Intraoperative dynamic infrared thermography and free-flap surgery. *Ann. Plast. Surg.* 57:279–84, 2006.

84. Disa J.J., Cordeiro P.G., Hidalgo D.A. Efficacy of conventional monitoring techniques in free tissue transfer: An 11-year experience in 750 consecutive cases. *Plast. Reconstr. Surg.* 104:97–101, 1999.

85. Jones N.F. Intraoperative and postoperative monitoring of microsurgical free tissue transfers. *Clin. Plast. Surg.* 19:783–797, 1992.

86. Smit J.M., Acosta R., Zeebregts C.J., Liss A.G., Anniko M., Hartman E.H.M. Early reintervention of compromised free flaps improves success rate. *Microsurgery* 27:612–616, 2007.

87. Baudet J., LeMaire J.M., Guimberteau J.C. Ten free groin flaps. *Plast. Reconstr. Surg.* 57:577–595, 1976.

88. Acland R.D. Discussion of "Experience in monitoring the circulation of free flap transfers". *Plast. Reconstr. Surg.* 68:554–555, 1981.

89. Leonard A.G., Brennen M.D., Colville J. The use of continuous temperature monitoring in the post-operative management of microvascular cases. *Br. J. Plast. Surg.* 35:337–342, 1982.

90. Busic V., Das-Gupta R. Temperature monitoring in free flap surgery. *Br. J. Plast. Surg.* 57:588, 2004.

91. de Weerd L., Miland Å.O., Mercer J.B. Perfusion dynamics of free DIEP and SIEA flaps during the first postoperative week monitored with dynamic infrared thermography (DIRT). *Ann. Plast. Surg.* 62:40–47, 2009.

92. Menick F. *Nasal Reconstruction. Art and Practice.* China: Mosby Elsevier, 2009.

93. Menick F.J. Nasal reconstruction: Forehead flap. *Plast. Reconstr. Surg.* 113, 100e–111e, 2004.

94. Moolenburgh S.E., McLennan L., Levendag P.C., Scholtemeijer M., Hofer S.O.P., Mureau M.A.M. Nasal reconstruction after malignant tumor resection: An algorithm for treatment. *Plast. Reconstr. Surg.* 126, 97–105, 2010.

95. Buddharaju P., Pavlidis I. Physiology-based face recognition in the thermal infrared spectrum. In *Medical Infrared Imaging*. Diakides N.A., Bronzino J.D. Eds., New York: CRC Press, pp. 13-1–13-16, 2008.

96. Taylor G.I., Ives A., Dhar S. Vascular territories. In *Plastic Surgery. Vol. 1 General Principles*. Mathes S.J. Ed., Philadelphia, PA: Saunders Elsevier, pp. 317–363, 2006.

97. Mathes S.J., Hansen S.L. Flap classification and applications. In *Plastic Surgery. Vol. General Principles*. Mathes S.J., Ed., Philadelphia, PA: Saunders Elsevier, pp. 365–481, 2006.

98. Vural E., Batay F., Key J.M. Glabellar frown lines as a reliable landmark for the supratrochlear artery. *Otolaryngol. Head Neck Surg.* 123, 543–546, 2000.

99. Stelnicki E.J., Young V.L, Francel T. Randall P. Vilray P. Blair, his surgical descendents, and their roles in plastic surgical development. *Plast. Reconstr. Surg.* 103, 1990, 1999.

100. Riggio E. The hazards of contemporary paramedian forehead flap and neck dissection in smokers. *Plast. Reconstr. Surg.* 112, 346–347, 2003.

101. Antia N.H., Daver B.M. Reconstructive surgery for nasal defects. *Clin. Plast. Surg.* 8, 535, 1981.

102. Reece E.M., Schaverien M., Rohrich R.J. The paramedian forehead flap: A dynamic anatomical vascular study verifying safety and clinical implications. *Plast. Reconstr. Surg.* 121, 1956–1963, 2008.

103. Baker S.R. Interpolated paramedian forehead flaps. In *Local Flaps in Facial Reconstruction*. Baker S.R. Ed., China: Mosby Elsevier, pp. 265–312, 2007.

104. Shumrick K.A., Smith T.L., The anatomic basis for the design of forehead flaps in nasal reconstruction. *Arch. Otolaryngol. Head Neck Surg.* 118, 373–379, 1992.

105. Ugur M.B., Savranlar A., Uzun L, Küçüker H, Cinar F. A reliable surface landmark for localizing supratrochlear artery: Medial canthus. *Otolaryngol. Head Neck Surg.* 138, 162–165, 2008.

106. Angobaldo J., Malcolm M. Refinements in nasal reconstruction: The cross-paramedian forehead flap. *Plast. Reconstr. Surg.* 123, 87–93, 2009.

29

Infrared Imaging Applied to Dentistry

Barton M. Gratt
University of Washington

29.1 The Importance of Temperature

Temperature is very important in all biological systems. Temperature influences the movement of atoms and molecules and their rates of biochemical activity. Active biological life is, in general, restricted to a temperature range of 0–45°C [1]. Cold-blooded organisms are generally restricted to habitats in which the ambient temperature remains between 0°C and 40°C. However, a variety of temperatures well outside of this occurs on earth, and by developing the ability to maintain a constant body temperature, warm-blooded animals; for example, birds, mammals, including humans have gained access to a greater variety of habitats and environments [1].

With the application of common thermometers, elevation in the core temperature of the body became the primary indicator for the diagnosis of fever. Wunderlich introduced fever measurements as a routine procedure in Germany, in 1872. In 1930, Knaus inaugurated a method of basal temperature measurement, achieving full medical acceptance in 1952. Today, it is customary in hospitals throughout the world to take body temperature measurements on all patients [2].

The scientists of the first part of the twentieth century used simple thermometers to study body temperatures. Many of their findings have not been superseded, and are repeatedly confirmed by new investigators using new more advanced thermal measuring devices. In the last part of the twentieth century, a new discipline termed "thermology" emerged as the study of surface body temperature in both health and in disease [2].

29.2 Skin and Skin-Surface Temperature Measurement

The skin is the outer covering of the body and contributes 10% of the body's weight. Over 30% of the body's temperature-receptive elements are located within the skin. Most of the heat produced within the body is dissipated by way of the skin, through radiation, evaporation, and conduction. The range of ambient temperature for thermal comfort is relatively broad (20–25°C). Thermal comfort is dependent upon humidity, wind velocity, clothing, and radiant temperature. Under normal conditions there is a steady flow of heat from the inside of a human body to the outside environment. Skin temperature distribution within specific anatomic regions; for example, the head vs. the foot, are diverse, varying by as much as ±15°C. Heat transport by convection to the skin surface depends on the rate of blood flow through the skin, which is also variable. In the trunk region of the body, blood flow varies by a factor of 7; at the foot, blood flow varies by a factor of 30; while at the fingers, it can vary by a factor of 600 [3].

It appears that measurements of body (core) temperatures and skin (surface) temperature may well be strong physiologic markers indicating health or disease. In addition, skin (surface) temperature values appear to be unique for specific anatomic regions of the body.

29.3 Two Common Types of Body Temperature Measurements

There are two common types of body temperature measurements that are made and utilized as diagnostic indicators.

1. *Measurement of Body Core Temperature.* The normal core temperature of the human body remains within a range of 36.0–37.5°C [1]. The constancy of human core temperature is maintained by a large number of complex regulatory mechanisms [3]. Body core temperatures are easily measured orally (or anally) with contacting temperature devices including: manual or digital thermometers, thermistors, thermocouples, and even layers of liquid temperature-sensitive crystals, and so on [4–6].
2. *Measurement of Body Surface Temperature.* While body core temperature is very easy to measure, the body's skin surface temperature is very difficult to measure. Any device that is required to make contact with the skin cannot measure the body's skin surface temperature reliably. Since skin has a relatively low heat capacity and poor lateral heat conductance, skin temperature is likely to change on contact with a cooler or warmer object [2]. Therefore, an indirect method of obtaining skin surface temperature is required, a common thermometer on the skin, for example, will not work.

Probably the first research efforts that pointed out the diagnostic importance of the infrared emission of human skin and thus initiated the modern era of thermometry were the studies of Hardy in 1934 [7,8]. However, it took 30 years for modern thermometry to be applied in laboratories around the world. To conduct noncontact thermography of the human skin in a clinical setting, an advanced computerized infrared imaging system is required. Consequently, clinical thermography required the advent of microcomputers developed in the late 1960s and early 1970s. These sophisticated electronic systems employed advanced microtechnology, requiring large research and development costs.

Current clinical thermography units use single detector infrared cameras. These work as follows: infrared radiation emitted by the skin surface enters the lens of the camera, passes through a number of rapidly spinning prisms (or mirrors), which reflect the infrared radiation emitted from different parts of the field of view onto the infrared sensor. The sensor converts the reflected infrared radiation into electrical signals. An amplifier receives the electric signals from the sensor and boosts them to electric potential signals of a few volts that can be converted into digital values. These values are then fed into a computer. The computer uses this input, together with the timing information from the rotating mirrors, to reconstruct a digitized thermal image from the temperature values of each small area within the field of observation. These digitized images are easily viewed and can be analyzed using computer software and stored on a computer disk for later reference.

29.4 Diagnostic Applications of Thermography

In 1987, the *International Bibliography of Medical Thermology* was published and included more than 3000 cited publications on the medical use of thermography, including applications for anesthesiology, breast disease, cancer, dermatology, gastrointestinal disorders, gynecology, urology, headache, immunology, musculoskeletal disorders, neurology, neurosurgery, ophthalmology, otolaryngology, pediatrics, pharmacology, physiology, pulmonary disorders, rheumatology, sports medicine, general surgery, plastic and reconstructive surgery, thyroid, cardiovascular and cerebrovascular, vascular problems, and veterinary medicine [9]. In addition, changes in human skin temperature has been reported in conditions involving the orofacial complex, as related to dentistry, such as the temporomandibular joint [10–25], and nerve damage and repair following common oral surgery [25–27]. Thermography has been shown not to be useful in the assessment of periapical granuloma [28]. Reports of dedicated controlled facial skin temperature studies of the orofacial complex are limited, but follow findings consistent with other areas of the body [29,30].

29.5 Normal Infrared Facial Thermography

The pattern of heat dissipation over the skin of the human body is normally symmetrical and this includes the human face. It has been shown that in normal subjects, the difference in skin temperature from side-to-side on the human body is small, about 0.2°C [31]. Heat emission is directly related to cutaneous vascular activity, yielding enhanced heat output on vasodilatation and reduced heat output on vasoconstriction. Infrared thermography of the face has promise, therefore, as a harmless, noninvasive, diagnostic technique that may help to differentiate selected diagnostic problems. The literature reports that during clinical studies of facial skin temperature a significant difference between the absolute facial skin temperatures of men vs. women was observed [32]. Men were found to have higher temperatures over all 25 anatomic areas measured on the face (e.g., the orbit, the upper lip, the lower lip, the chin, the cheek, the TMJ, etc.) than women. The basal metabolic rate for a normal 30-year-old male, 1.7 m tall (5 ft, 7 in.), weighing 64 kg (141 lbs), who has a surface area of approximately 1.6 m², is approximately 80 W; therefore, he dissipates about 50 W/m² of heat [33]. On the other hand, the basal metabolic rate of a 30-year-old female subject, 1.6 m tall (5 ft, 3 in.), weighing 54 kg (119 lbs), with a surface area of 1.4 m², is about 63 W, so that she dissipates about 41 W/m² of heat [33,34]. Assuming that there are no other relevant differences between male and female subjects, women's skin is expected to be cooler, since less heat is lost per unit (per area of body surface). Body heat dissipation through the face follows this prediction. In addition to the effect of gender on facial temperature, there are indications that age and ethnicity may also affect facial temperature [32].

When observing patients undergoing facial thermography, there seems to be a direct correlation between vasoactivity and pain, which might be expected since both are neurogenic processes. Differences in facial skin temperature, for example, asymptomatic adult subjects (low temperatures differences) and adult patients with various facial pain syndromes (high-temperature differences) may prove to be a useful criterion for the diagnosis of many conditions [35]. Right- vs. left-side temperature differences (termed: delta T or ΔT) between many specific facial regions in normal subjects were shown to be low (<0.3°C) [40], while similar ΔT values were found to be high (>0.5°C) in a variety of disorders related to dentistry [35].

29.6 Abnormal Facial Conditions Demonstrated with Infrared Facial Thermography

29.6.1 Assessing Temporomandibular Joint (TMJ) Disorders with Infrared Thermography

It has been shown that normal subjects have symmetrical thermal patterns over the TMJ regions of their face. Normal subjects had ΔT values of 0.1°C (±0.1°C) [32,36]. On the other hand, TMJ pain patients were found to have asymmetrical thermal patterns, with increased temperatures over the affected TMJ region, with ΔT values of +0.4°C (±0.2°C) [37]. Specifically, painful TMJ patients with internal derangement and painful TMJ osteoarthritis were both found to have asymmetrical thermal patterns and increased temperatures over the affected TMJ, with mean area TMJ ΔT of +0.4°C (±0.2°C) [22,24]. In other words, the correlation between TMJ pain and hyper perfusion of the region seems to be independent of the etiology of the TMJ disorder (osteoarthritis vs. internal derangement). In addition, a study of mild-to-moderate TMD (temporomandibular joint dysfunction) patients indicated that area ΔT values correlated with the level of the patient's pain symptoms [38]. And a more recent double-blinded clinical study compared active orthodontic patients vs. TMD patients vs. asymptomatic TMJ controls, and showed average ΔT values of +0.2, +0.4, and +0.1°C; for these three groups respectively. This study showed that thermography could distinguish between patients undergoing active orthodontic treatment and patients with TMD [39].

29.6.2 Assessing Inferior Alveolar Nerve (IAN) Deficit with Infrared Thermography

The thermal imaging of the chin has been shown to be an effective method for assessing inferior alveolar nerve deficit [40]. Whereas normal subjects (those without inferior alveolar nerve deficit) show a symmetrical thermal pattern (ΔT of +0.1°C [±0.1°C]); patients with inferior alveolar nerve deficit had elevated temperature in the mental region of their chin (ΔT of +0.5°C [±0.2°C]) on the affected side [41]. The observed vasodilatation seems to be due to blockage of the vascular neuronal vasoconstrictive messages, since the same effect on the thermological pattern could be invoked in normal subjects by temporary blockage of the inferior alveolar nerve, using a 2% lidocaine nerve block injection [42].

29.6.3 Assessing Carotid Occlusal Disease with Infrared Thermography

The thermal imaging of the face, especially around the orbits, has been shown to be an effective method for assessing carotid occlusal disease. Cerebrovascular accident (CVA), also called stroke, is well known as a major cause of death. The most common cause of stroke is atherosclerotic plaques forming emboli, which travel within vascular blood channels, lodging in the brain, obstructing the brain's blood supply, resulting in a cerebral vascular accident (or stroke). The most common origin for emboli is located in the lateral region of the neck where the common carotid artery bifurcates into the internal and the external carotid arteries [43,44]. It has been well documented that intraluminal carotid plaques, which both restrict and reduce blood flow, result in decreased facial skin temperature [43–54]. Thermography has demonstrated the ability to detect a reduction of 30% (or more) of blood flow within the carotid arteries [55]. Thermography shows promise as an inexpensive painless screening test of asymptomatic elderly adults at risk for the possibility of stroke. However, more clinical studies are required before thermography may be accepted for routine application in screening toward preventing stroke [55,56].

29.6.4 Additional Applications of Infrared Thermography

Recent clinical studies assessed the application of thermography on patients with chronic facial pain (orofacial pain of greater than 4 month's duration). Thermography classified patients as being "normal" when selected anatomic ΔT values ranged from 0.0°C to ±0.25°C, and "hot" when ΔT values were >+ 0.35°C, and "cold" when area ΔT values were <–0.35°C. The study population consisted of 164 dental pain patients and 164 matched (control) subjects. This prospective, matched study determined that subjects classified with "hot" thermographs had the clinical diagnosis of (1) sympathetically maintained pain, (2) peripheral nerve-mediated pain, (3) TMJ arthropathy, or (4) acute maxillary sinusitis. Subjects classified with "cold" areas on their thermographs were found to have the clinical diagnosis of (1) peripheral nerve-mediated pain, or (2) sympathetically independent pain. Subjects classified with "normal" thermographs included patients with the clinical diagnosis of (1) cracked tooth syndrome, (2) trigeminal neuralgia, (3) pretrigeminal neuralgia, or (4) psychogenic facial pain. This new system of thermal classification resulted in 92% (301 or 328) agreement in classifying pain patients vs. their matched controls. In brief, ΔT has been shown to be within ±0.4°C in normal subjects, while showing values greater than +0.7°C and less than –0.6°C in abnormal facial pain patients [10], making "ΔT" an important diagnostic parameter in the assessment of orofacial pain [35].

29.6.5 Future Advances in Infrared Imaging

Over the last 20 years there have been additional reports in the dental literature giving promise to new and varied applications of infrared thermography [57–63]. While, infrared thermography is promising, the future holds even greater potential for temperature measurement as a diagnostic tool, the most promising being termed dynamic area telethermometry (DAT) [64,65]. Newly developed DAT promises to become a new more advanced tool providing quantitative information on the thermoregulatory frequencies (TRFs) manifested in the modulation of skin temperature [66]. Whereas the static thermographic studies discussed above demonstrate local vasodilatation or vasoconstriction, DAT can identify the mechanism of thermoregulatory frequencies and thus it is expected, in the future, to significantly improve differential diagnosis [66].

In summary, the science of thermology, including static thermography, and soon to be followed by DAT, appears to have great promise as an important diagnostic tool in the assessment of orofacial health and disease.

Acknowledgment

This chapter is dedicated to Professor Michael Anbar, of Buffalo, New York: A brilliant scientist, my thermal science mentor, and "The Father of Dynamic Area Telethermography."

References

1. Grobklaus, R. and Bergmann, K.E. Physiology and regulation of body temperature. In *Applied Thermology: Thermologic Methods*. J.-M. Engel, U. Fleresch, and G. Stuttgen, Eds., Federal Republic of Germany: VCH, 1985, pp. 11–20.
2. *Applied Thermology: Thermologic Methods*. J.-M. Engel, U. Fleresch, and G. Stuttgen, Eds., Federal Republic of Germany: VCH (1985), pp. 11–20.
3. Kirsch, K.A. Physiology of skin-surface temperature. In *Applied Thermology: Thermologic Methods*. J.-M. Engel, U. Fleresch, and G. Stuttgen, Eds., Federal Republic of Germany: VCH (1985), pp. 1–9.
4. Anbar, M., Gratt, B.M., and Hong, D. Thermology and facial telethermography: Part I. History and technical review. *Dentomaxillofac. Radiol.* (1998), 27: 61–67.
5. Anbar, M. and Gratt, B.M. Role of nitric oxide in the physiopathology of pain. *J. Musc. Skeletal Joint Pain* (1997), 14: 225–254.
6. Rost, A. Comparative measurements with an infrared and contact thermometer for thermal stress reaction. In *Thermological Methods*. J.-M. Engel, U. Flesch, and G. Stuttgen, Eds., Weinheim: VCH Verlag (1985), pp. 169–170.

7. Hardy, J.D. The radiation of heat from the human body: I–IV. *J. Clin. Invest.* (1934), 13: 593–620.
8. Hardy, J.D. The radiation of heat from the human body: I–IV. *J. Clin. Invest.* (1934), 13: 817–883.
9. Abernathy, M. and Abernathy, T.B. International bibliography of thermology. *Thermology* (1987), 2: 1–533.
10. Berry, D.C. and Yemm, R. Variations in skin temperature of the face in normal subjects and in patients with mandibular dysfunction. *Br. J. Oral Maxillofac. Surg.* (1971), 8: 242–247.
11. Berry, D.C. and Yemm, R. A further study of facial skin temperature in patients with mandibular dysfunction. *J. Oral Rehabil.* (1974), 1: 255–264.
12. Kopp, S. and Haraldson, T. Normal variations in skin temperature of the face in normal subjects and in patients with mandibular dysfunction. *Br. J. Oral Maxillofac. Surg.* (1983), 8: 242–247.
13. Johansson, A., Kopp, S., and Haraldson, T. Reproducibility and variation of skin surface temperature over the temporomandibular joint and masseter muscle in normal individuals. *Acta Odontol. Scand.* (1985), 43: 309–313.
14. Tegelberg, A. and Kopp, S. Skin surface temperature over the temporo-mandibular and metacarpophalangeal joints in individuals with rheumatoid arthritis. *Odontol. Klin.*, Box 33070, 400 33 Goteborg, Sweden (1986), Report No. 31, pp. 1–31.
15. Akerman, S. et al. Relationship between clinical, radiologic and thermometric findings of the temporomandibular joint in rheumatoid arthritis. *Odontol. Klin.*, Box 33070, 400 33 Goteborg, Sweden (1987), Report No. 41, pp. 1–30.
16. Finney, J.W., Holt, C.R., and Pearce, K.B. Thermographic diagnosis of TMJ disease and associated neuromuscular disorders. *Special Report: Postgraduate Medicine* (March 1986), pp. 93–95.
17. Weinstein, S.A. Temporomandibular joint pain syndrome—The whiplash of the 1980s. *Thermography and Personal Injury Litigation*, Chapter 7. S.D. Hodge, Jr., Ed., New York, USA: John Wiley & Sons (1987), pp. 157–164.
18. Weinstein, S.A., Gelb, M., and Weinstein, E.L. Thermophysiologic anthropometry of the face in home sapiens. *J. Craniomand. Pract.* (1990), 8: 252–257.
19. Pogrel, M.A., McNeill, C., and Kim, J.M. The assessment of trapezius muscle symptoms of patients with temporomandibular disorders by the use of liquid crystal thermography. *Oral Surg. Oral Med. Oral. Pathol. Oral Radiol. Endod.* (1996), 82: 145–151.
20. Steed, P.A. The utilization of liquid crystal thermography in the evaluation of temporomandibular dysfunction. *J. Craniomand. Pract.* (1991), 9: 120–128.
21. Gratt, B.M., Sickles, E.A., Graff-Radford, S.B., and Solberg, W.K. Electronic thermography in the diagnosis of atypical odontalgia: A pilot study. *Oral Surg. Oral Med. Oral Pathol. Oral Radiol. Endod.* (1989), 68: 472–481.
22. Gratt, B.M. et al. Electronic thermography in the assessment of internal derangement of the TMJ. *J. Orofacial Pain* (1994), 8: 197–206.
23. Gratt, B.M., Sickles, E.A., Ross, J.B., Wexler, C.E., and Gornbein, J.A. Thermographic assessment of craniomandibular disorders: Diagnostic interpretation versus temperature measurement analysis. *J. Orofacial Pain* (1994), 8: 278–288.
24. Gratt, B.M., Sickles, E.A., and Wexler, C.E. Thermographic characterization of osteoarthrosis of the temporomandibular joint. *J. Orofacial Pain* (1994), 7: 345–353.
25. Progrell, M.A., Erbez, G., Taylor, R.C., and Dodson, T.B. Liquid crystal thermography as a diagnostic aid and objective monitor for TMJ dysfunction and myogenic facial pain. *J. Craniomand. Disord. Facial Oral Pain* (1989), 3: 65–70.
26. Dmutpueva, B.C., and Alekceeva, A.H. Applications of thermography in the evaluation of the postoperative patient. *Stomatologiia* (1986), 12: 29–30 (Russian).
27. Cambell, R.L., Shamaskin, R.G., and Harkins, S.W. Assessment of recovery from injury to inferior alveolar and mental nerves. *Oral Surg. Oral Med. Oral Pathol. Oral Radiol. Endod.* (1987), 64: 519–526.
28. Crandall, C.E. and Hill, R.P. Thermography in dentistry: A pilot study. *Oral Surg. Oral Med. Oral Pathol. Oral Radiol. Endod.* (1966), 21: 316–320.

29. Gratt, B.M., Pullinger, A., and Sickles, E.A. Electronic thermography of normal facial structures: A pilot study. *Oral Surg. Oral Med. Oral Pathol. Oral Radiol. Endod.* (1989), 68: 346–351.

30. Weinstein, S.A., Gelb, M., and Weinstein, E.L. Thermophysiologic anthropometry of the face in homo sapiens. *J. Craniomand. Pract.* (1990), 8: 252–257.

31. Uematsu, S. Symmetry of skin temperature comparing one side of the body to the other. *Thermology* (1985), 1: 4–7.

32. Gratt, B.M. and Sickles, E.A. Electronic facial thermography: An analysis of asymptomatic adult subjects. *J. Orofacial Pain* (1995), 9: 222–265.

33. Blaxter, K. Energy exchange by radiation, convection, conduction and evaporation. In *Energy Metabolism in Animals and Man.* New York: Cambridge University Press (1989), pp. 86–99.

34. Blaxter, K. The minimal metabolism. In *Energy Metabolism in Animals and Man.* New York: Cambridge University Press (1989), 120–146.

35. Gratt, B.M, Graff-Radford, S.B., Shetty, V., Solberg, W.K., and Sickles, E.A. A six-year clinical assessment of electronic facial thermography. *Dentomaxillofac. Radiol.* (1996), 25: 247–255.

36. Gratt, B.M., and Sickles, E.A. Thermographic characterization of the asymptomatic TMJ. *J. Orofacial Pain* (1993), 7: 7–14.

37. Gratt, B.M., Sickles, E.M., and Ross, J.B. Thermographic assessment of craniomandibular disorders: Diagnostic interpretation versus temperature measurement analysis. *J. Orofacial Pain* (1994), 8: 278–288.

38. Canavan, D. and Gratt, B.M. Electronic thermography for the assessment of mild and moderate TMJ dysfunction. *Oral Surg. Oral Med. Oral Pathol. Oral Radiol. Endod.* (1995), 79: 778–786.

39. McBeth, S.A., and Gratt, B.M. A cross-sectional thermographic assessment of TMJ problems in orthodontic patients. *Am. J. Orthod. Dentofac. Orthop.* (1996), 109: 481–488.

40. Gratt, B.M., Shetty, V., Saiar, M., and Sickles, E.A. Electronic thermography for the assessment of inferior alveolar nerve deficit. *Oral Surg. Oral Med. Oral Pathol. Oral Radiol. Endod.* (1995), 80: 153–160.

41. Gratt, B.M., Sickles, E.A., and Shetty, V. Thermography for the clinical assessment of inferior alveolar nerve deficit: A pilot study. *J. Orofacial Pain* (1994), 80: 153–160.

42. Shetty, V., Gratt, B.M., and Flack, V. Thermographic assessment of reversible inferior alveolar nerve deficit. *J. Orofacial Pain* (1994), 8: 375–383.

43. Wood, E.H. Thermography in the diagnosis of cerebrovascular disease: Preliminary report. *Radiology* (1964), 83: 540–546.

44. Wood, E.H. Thermography in the diagnosis of cerebrovascular disease. *Radiology* (1965), 85: 207–215.

45. Steinke, W., Kloetzsch, C., and Hennerici, M. Carotid artery disease assessed by color Doppler sonography and angiography. *AJR* (1990), 154: 1061–1067.

46. Hu, H.-H. et al. Color Doppler imaging of orbital arteries for detection of carotid occlusive disease. *Stroke* (1993), 24: 1196–1202.

47. Carroll, B.A., Graif, M., and Orron, D.E. Vascular ultrasound. In *Peripheral Vascular Imaging and Intervention.* D. Kim and D.E. Orron, Eds., St. Louis, MO, Mosby/Year Book (1992), pp. 211–225.

48. Mawdsley, C., Samuel, E., Sumerling, M.D., and Young, G.B. Thermography in occlusive cerebrovascular diseases. *Br. Med. J.* (1968), 3: 521–524.

49. Capistrant, T.D. and Gumnit, R.J. Thermography and extracranial cerebrovascular disease: A new method to predict the stroke-prone individual. *Minn. Med.* (1971), 54: 689–692.

50. Karpman, H.L., Kalb, I.M., and Sheppard, J.J. The use of thermography in a health care system for stroke. *Geriatrics* (1972), 27: 96–105.

51. Soria, E. and Paroski, M.W. Thermography as a predictor of the more involved side in bilateral carotid disease: Case history. *Angiology* (1987), 38: 151–158.

52. Capistrat, T.D. and Gumnit, R.J. Detecting carotid occlusive disease by thermography. *Stroke* (1973), 4: 57–65.

53. Abernathy, M., Brandt, M.M., and Robinson, C. Noninvasive testing of the carotid system. *Am. Fam. Physic.* (1984), 29: 157–164.

54. Dereymaeker, A., Kams-Cauwe, V., and Fobelets, P. Frontal dynamic thermography: Improvement in diagnosis of carotid stenosis. *Eur. Neurol.* (1978), 17: 226–234.

55. Gratt, B.M., Halse, A., and Hollender, L. A pilot study of facial infrared thermal imaging used as a screening test for detecting elderly individuals at risk for stroke. *Thermol. Int.* (2002), 12: 7–15.

56. Friedlander A.H. and Gratt B.M. Panoramic dental radiography and thermography as an aid in detecting patients at risk for stroke. *J. Oral Maxillofac. Surg.* (1994), 52: 1257–1262.

57. Graff-Radford, S.B., Ketalaer, M.-C., Gratt, B.M., and Solberg, W.K. Thermographic assessment of neuropathic facial pain: A pilot study. *J. Orofacial Pain* (1995), 9: 138–146.

58. Pogrel, M.A., Erbez, G., Taylor, R.C., and Dodson, T.B. Liquid crystal thermography as a diagnostic aid and objective monitor for TMJ dysfunction and myogenic facial pain. *J. Craniobandib. Disord. Facial Oral Pain* (1989), 3: 65–70.

59. Pogrel, M.A., Yen, C.K., and Taylor, R.C. Infrared thermography in oral and maxillo-facial surgery. *Oral Surg. Oral Med. Oral Pathol. Oral Radiol. Endod.* (1989), 67: 126–131.

60. Graff-Radford, S.B., Ketlaer, M.C., Gratt, B.M., and Solberg, W.K. Thermographic assessment of neuropathic facial pain. *J. Orofacial Pain* (1995), 9: 138–146.

61. Biagioni, P.A., Longmore, R.B., McGimpsey, J.G., and Lamey, P.J. Infrared thermography: Its role in dental research with particular reference to craniomandibular disorders. *Dentomaxillofac. Radiol.* (1996), 25: 119–124.

62. Biagioni, P.A., McGimpsey, J.G., and Lamey, P.J. Electronic infrared thermography as a dental research technique. *Br. Dent. J.* (1996), 180: 226–230.

63. Benington, I.C., Biagioni, P.A., Crossey, P.J., Hussey, D.L., Sheridan, S., and Lamel, P.J. Temperature changes in bovine mandibular bone during implant site preparation: An assessment using infra-red thermography. *J. Dent.* (1996), 24: 263–267.

64. Anbar, M. Clinical applications of dynamic area telethermography. In *Quantitative Dynamic Telthermography in Medical Diagnosis*. Boca Raton, FL: CRC Press, 1994, pp. 147–180.

65. Anbar, M. Dynamic area telethermography and its clinical applications. *SPIE Proc.* (1995), 2473: 3121–3323.

66. Anbar M., Grenn, M.W., Marino, M.T., Milescu, L., and Zamani, K. Fast dynamic area telethermography (DAT) of the human forearm with a Ga/As quantum well infrared focal plane array camera. *Eur. J. Therol.* (1997), 7: 105–118.

30

Laser Infrared Thermography of Biological Tissues

Alexander Sviridov

Institute for Laser and Information Technologies of Russian Academy of Sciences

Andrey Kondyurin

Institute for Laser and Information Technologies of Russian Academy of Sciences

30.1 Introduction

Laser infrared (IR) thermography of biological tissues is a part of active dynamic thermal (ADT) imaging [1], initially developed for noninvasive remote control of materials [2]. It involves heating of a material by some source of energy at controlled conditions and recording of time-dependent, and nonuniform temperature field on a sample surface by thermal imaging camera. Mathematical analysis of the induced temperature field with adequate modeling allows one to assess the physicochemical processes in the material and its characteristics.

Microwaves, ultrasound devices, xenon lamps, and lasers are commonly used as sources of energy for heating of biological tissues. Laser sources have a number of advantages over alternative heating modalities. A light, emitted by commercial lasers, covers a wide spectral range—from ultraviolet to far IR. It allows one to provide desired penetration depths of a laser beam into tissues ranging from a few microns to about 10 cm. The range of light penetration depths in the tissue may be sufficient to analyze thermally induced processes, to control blood circulation, or for thermal tomography. Modern laser sources become compact and relatively cheap. They are often supplied with a processor-based control of output power, allowing to build a computerized laser system with feedback channels. The spatial profile of a laser beam could be precisely managed by optical elements, such as lenses, axicons, optical fibers, or by fast programmable scanning of a narrow beam on a sample surface. It allows one to generate the required temperature field by spatial and temporal control of laser energy supply into the biological tissue.

Development of systems providing the desired laser heating of a given area of biological tissue with IR radiometric feedback control has a practical significance for medical usages such as laser-induced thermolysis and hyperthermia of malignant tissues [3]. The most important problem to solve is the creation of a programmable temperature field in the given spatial and temporal limits with a laser beam of an appropriate intensity profile. Despite many efforts, of different scientific groups, this problem is not yet solved [4]. In fact, laser intensity distribution should be repeatedly corrected during heating procedure depending on the discrepancy between measured and specified temperature fields. Optimal parameters of the feedback system providing minimal discrepancy may be determined from the modeling of three-dimensional (3D) temperature field induced in the tissue by such laser heating taking into account the accuracy of used devices. To be accurate, such models require information about optical and thermophysical properties of tissues, and their temperature behavior which may be determined by using a similar system with an adequate algorithm for solving the inverse thermal problem.

Programmable laser heating of a local area of the biological tissue with IR thermography of surface temperature field may also be used to control thermal processes, stimulated in the tissue. The first approach involves measurements of volume rate and vector of energy dissipation due to blood flow. The controlled parameters may include dynamics of full temperature field on the tissue surface and laser power, when a chosen local region is heated with a given rate (constant, harmonically, or isothermally). By solving numerically a bio-heat transfer equation with perfusion terms, one could evaluate volume rate and the direction of blood flow in the region of interest. Another potential application of programmable laser heating is a remote calorimetry of the processes such as thermally induced chemical reaction and phase transformation of biological tissue components. Such problems can be important in cases of hyperthermia or in thermoplastics of biological tissues.

In some applications, light-emitting diodes (LED) could also be considered as an excitation source for IR thermography of biological tissues and in many respects they are similar to lasers. Therefore, the term "laser" in the title of this chapter should be considered more generally to include photodiodes, as a heating source. Using LEDs instead of laser systems allow for making an excitation system for IR thermography that is more compact and cheap. However, solving the choice between various excitation sources is determined by specific and technical requirements for a control system.

30.2 Programmable Laser Heating of Biological Tissues

A general scheme of the laser system with IR radiometric feedback control for programmable heating of biological tissues and measurement of optical and thermophysical parameters is presented in Figure 30.1.

The laser output power is controlled by a laser internal processor or a programmable external power supply, an actuator which manages the laser beam shape or beam position on the object surface. According to computer commands, IR camera controls the temperature field of the object surface, the data of IR camera are transmitted to the computer, and digitally processed to provide the control data for the actuators of optical system and laser power.

Let us consider a problem of programmable laser heating of a local region of the biological tissue with IR thermography as feedback control. The typical approach for such a type of problem which is widely used in industrial systems is the so-called three-term control, that is, control of the proportional,

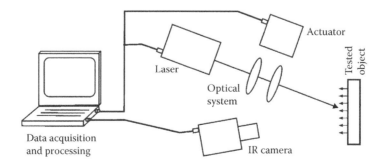

FIGURE 30.1 Block scheme of a laser system with thermovision camera as a feedback control detector.

integral, and derivative values, denoted as *P*, *I*, and *D*, respectively. The weighted sum of these three parameters determines the response of the feedback system to the difference between measured and programmed temperatures at a given point of time. It takes into account the past temperature dynamics, the current temperature, and the rate of temperature change at the current time. Typically, correction of power is performed with some permanent frequency. Depending on the weights of *P*, *I*, and *D* values, one could obtain different temperature time profiles in the region of interest and variations in the laser power, as well. Our goal is to find the optimal set of *P*, *I*, and *D* values to provide a minimal error between measured and programmed temperatures during a whole period of laser operation, also controlling the variations in the laser power. These values are determined by laser beam parameters, as well as thermo-physical and optical characteristics of the sample.

To realize this procedure, let us solve a direct heat conduction problem for a laser-heated sample of a biological tissue under the three-term (PID) control operation. It means that the power of laser will be changed with time as follows:

$$P(t) = K_c e(t) + \frac{K_c}{T_i} \int_0^t e(t) \cdot dt - K_c \cdot T_d \frac{de(t)}{dt} + \sigma_P \cdot NRD, \tag{30.1}$$

where *e(t)* is the difference between measured temperature in the local region (*ValuePoint*) and pro-grammed temperature (*SetPoint*), K_c, K_c/T_i, and $K_c \cdot T_d$ are linearly proportional, integrative and deriva-tive weights, respectively, σ_p is the error of laser power level, a zero-mean normal random distribution (NRD) with a standard deviation of 1. In cylindrical coordinate system, the differential equation to be solved is

$$\frac{\partial T}{\partial t} = \frac{\chi}{r} \frac{\partial}{\partial r}\left(r \frac{\partial T}{\partial r}\right) + \chi \frac{\partial^2 T}{\partial z^2} + f(r,z,t), \tag{30.2}$$

where *z* axis is perpendicular to the sample surface (*z* = 0), *r* is the radial coordinate, *f(r,z,t)* is the heat source function, determined by light absorption. Initial temperature distribution is assumed to be uni-form and equal to the room temperature:

$$T(r,z,t = 0) = T_0 \tag{30.3}$$

According to the Newton's law for forced and free convection between the sample and environment, the following conditions were set at the sample boundary:

$$-\chi \frac{\partial T(r,Z,t)}{\partial z}\bigg|_{z=Z} = \beta \cdot T(r,Z,t) - \mu, -\chi \frac{\partial T(R,z,t)}{\partial r}\bigg|_{r=R} = \beta \cdot T(R,z,t) - \mu, \lim_{r \to 0}\left(r\chi \frac{\partial T}{\partial r}\right) = 0, \tag{30.4}$$

where $\beta = h\rho C_p$, $\mu = \beta T_0$, ρ is density, C_p is specific heat, *h* is convective heat transfer coefficient (W mm^{-2} K^{-1}), and *R*, *Z* are radius and thickness of the sample, respectively. The value of *h* was chosen according to the published data [5]. Parameters χ, α, and C_p are assumed to be independent from the temperature, coordinates, and time.

At first approximation, the radiation intensity inside the sample is described by Beer law. Corre-spondingly, for light beams with Gaussian energy distribution, the function *f(r,z,t)* has been obtained as

$$f(r,z,t) = (1 - R_d)\frac{\alpha P(t)}{C_p \rho \cdot \pi W_L^2} \exp\left[-2 \cdot \left(\frac{r}{W_L}\right)^2\right] \exp(-\alpha z), \tag{30.5}$$

where α is the total attenuation coefficient (cm^{-1}), W_L is the laser beam radius (mm). ρC_p is the specific heat of a unit volume (J/(cm^3 K)), ρ is the density (g/cm^3), t_{imp} is the laser pulse duration (s), E_0 is the light intensity (W/cm^2), W_L is the laser beam radius (mm), and R_d is the dimensionless diffuse reflectance (for the case of cartilage it was measured earlier with an integrating sphere, $R_d = 0.10$ at $\lambda = 1.56$ µm) [6]. The spatial distribution of the laser radiation power density in an absorbing optically inhomogeneous medium is determined by three optical parameters—absorption, scattering coefficients, and scattering anisotropy. Though, strictly speaking, variations of the intensity are not described by the Beer law, the use of the exponential law with a certain generalized index of exponent α to describe the term heat source in an optically inhomogeneous absorbing medium, such as tissues, seems quite reasonable [7]. We refer to this exponent as the total attenuation coefficient.

The induced temperature field was computed using the finite difference method. An implicit difference scheme was constructed that approximates the heat conduction as Equation 30.2 and boundary conditions (Equation 30.4). In the region of continuous variation of coordinates r from 0 to R and z and from 0 to Z, a uniform grid of $I \times J$ elements with a step of $\Delta r = \Delta z = 0.01$ mm was set. Grid functions $T(r_i, z_j, t_m) = T(i, j, m)$ and $f(r_i, z_j, t_m) = \phi(i, j, m)$ were put into correspondence with functions $T(r, z, t)$ and $f(r, z, t)$. Differential equation (Equation 30.1) and boundary conditions (3) are replaced by the difference in analogs [8]. The laser power at a moment m was calculated according to the PID control:

$$P(m) = K_c \left(e(m) + \frac{1}{T_i} \sum_1^m \frac{e(m-1) + e(m)}{2} \Delta t - T_d \frac{e(m) - e(m-1)}{\Delta t} \right), \qquad (30.6)$$

where $e(m)$ was determined as difference $e(m) = SetPoint(m) - ValuePoint(m)$. An error of temperature measurement was taken into account to determine the *ValuePoint(m)*: ValuePoint$(m) = \sigma_T \cdot NRD + u(i_0, j_0, m)$. Here, $u(i_0, j_0, m)$ is the real temperature in the control point (i_0, j_0), σ_T is the error of temperature measurement with thermovision camera ($\sigma_T = 0.05°C$).

The set of optimal PID parameters K_c, T_i, and T_d was found from the solution of the inverse problem, based on minimization of the cumulative square error:

$$\sum_1^M e^2 \left(m; K_c, T_i, T_c \right) \rightarrow \min_{K_c, T_i, T_c} \qquad (30.7)$$

It was performed with the modified Levenberg–Marquard algorithm [9,10]. The influence of laser beam geometric parameters (W_L), thermophysical characteristics of the tissue (χ, ρC_p), effective attenuation coefficient (α), convective heat transfer coefficient (h) on K_c, T_i, and T_d were studied. A number of calculations for different sets of varied parameters were performed. From Equation 30.5 for $f(r,z,t)$, we assume that the coefficient K_c depends mainly on the combination of the listed parameters, namely:

$$K_c = B_{Kc} \cdot \frac{\rho C_p \pi W_L^2}{\alpha}, \qquad (30.8)$$

where B_{Kc} is phenomenological constant to be determined. A number of calculations of laser heating of the sample with the constant rate at $z = 0$, $r = 0$ substantiated this suggestion if χ, ρC_p, and α are varied in the ranges 0.1–0.2 mm^2/s, 2.5–3.5 J/cm^3/K, and 7.5–15 cm^{-1}, respectively. Calculations have shown that optimal values of T_i are independent of ρC_p and α, as well as the optimal value of T_d is equal to zero for all ρC_p and α. Our analysis also has shown that the variation of h in rather a wide range does not affect the parameters of PID controller. However, B_{Kc} and T_i depend on the heating rate. The data on optimal B_{Kc} and T_i, that is, providing minimal cumulative square error of temperature at laser heating of the local area $z = 0$ and $r = 0$ with constant heating rate are presented in Table 30.1.

TABLE 30.1 Values of B_{Kc} and T_i at Different Heating Rates

Heating Rate (°C/s)	$B_{Kc} \times 10^3$ (s^{-1})	T_i (s)
0.1	2.0	1.46
0.3	3.2	0.84
0.4	4.0	0.69
0.5	4.6	0.58
1.0	5.5	0.52
3.0	9.2	0.31
5.0	12.1	0.26

Similarly, the optimal PID parameters that satisfy condition (Equation 30.7) could be found for any heating scenario of a given local region of the sample. Among them harmonic and isothermal scenarios are more interesting for medical diagnostics and treatment. Significance of PID parameters and its optimization is clearly seen from Figures 30.2 and 30.3, presenting the behavior of temperature and laser power as a function of time with nonoptimal and optimal PID parameters. In the case of nonoptimal PID parameters, the temperature curve has some ripples with amplitudes of ~2°C around the straight line (see Figure 30.2a). On the other hand, the temporal behavior of laser power (Figure 30.2b) looks preferably random. Actually it does not allow the use of power as a measured parameter for control. On the contrary, optimal PID parameters provide stable and smooth behavior both for the temperature and the laser power (see Figure 30.3a,b), making laser thermography a promising tool for control of thermal processes in biological tissues and other materials.

Figure 30.4a demonstrates an experimental realization of programmable laser heating of cartilage by glass Erbium-doped fiber laser illuminating at $\lambda = 1.56$ μm (IRE Polus, Moscow, Russia) with a

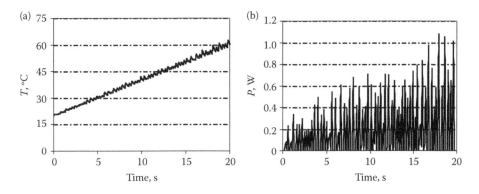

FIGURE 30.2 Temporal behavior of nonoptimal PID parameters: (a) temperature and (b) laser power.

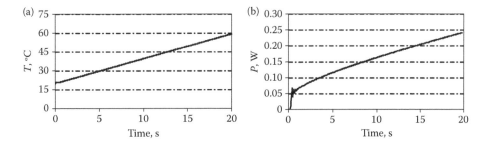

FIGURE 30.3 Temporal behavior of optimal PID parameters: (a) temperature and (b) laser power.

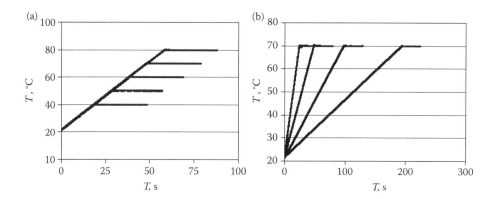

FIGURE 30.4 Examples of experimental realization of programmable laser heating of the cartilage with a given constant rate (1°C/s) up to different temperatures: (a) keeping the rate constant and with different rates up to given temperature (70°C) and (b) keeping the temperature constant.

constant rate (1°C/s) attaining different temperatures and keeping them at a constant level. Similarly, Figure 30.4b demonstrates regimes of programmable laser heating with different rates attaining 70°C and keeping it at constant level. Temperature control was performed with IR thermograph IRTIS-200 (IRTIS, Moscow, Russia).

30.3 Laser Thermography for Control of Energetic Processes

Successive realization of programmable laser heating with thermal control gave a key for the development of remote calorimetry of energetic processes in open thermodynamic systems. Applying it to biological materials and tissues allows one to implement noninvasive measures of the enthalpy of phase transformation or energetic reactions. The idea is based on the registration of laser power temporal function which may be considered as an analog of heat quantity curve in classical thermal analysis such as differential scanning calorimetry (DSC). Thermophysical processes in biological tissues heated by a laser with constant rate should result in compensation of consumption or release of heat by appropriate correction of laser power. Thus, the nonmonotonic temporal behavior of laser power could become an indicator of energetic processes. The dynamics of temperature field in the full area of sample surface in combination with the power dynamics could be an additional indicator of thermal process. To estimate the sensitivity of laser calorimetry let us consider a programmable laser constant rate heating of the sample with optimal parameters of PID controller at the condition of tentative thermal chemical reaction A → B of first order with given Arrhenius parameters and different enthalpies. In such a statement, the direct problem of temperature field may be solved just by adding the term appropriate energy density on the right side of Equation 30.2 with its subsequent application of the approach described previously:

$$\frac{\partial T}{\partial t} = \frac{\chi}{r}\frac{\partial}{\partial r}\left(r\frac{\partial T}{\partial r}\right) + \chi\frac{\partial^2 T}{\partial z^2} + f(r,z,t) + g(T,r,z,t),$$ (30.9)

$$g(r,z,t) = \frac{\Delta_r H}{\rho C_p}\frac{d\xi}{dt}.$$ (30.10)

Here, Δ_r H—specific enthalpy of chemical reaction (J/g), ξ—is the factor of substance A conversion in A → B reaction ($\xi = 0$ before reaction, $\xi = 1$ after reaction), and

$$\frac{d\xi}{dt} = k_0 \cdot \exp\left(-\frac{E_a}{RT}\right) \cdot (1 - \xi) \tag{30.11}$$

For numerical computation of the temperature field $u(i,j,m)$ at every node of spatial–temporal grid, it is necessary to express $g(r,z,t)$ by its grid analogs of differential finite increments:

$$\gamma(i,j,m) = \frac{\Delta_r H}{C_p \rho} \frac{\Delta\xi}{\Delta t} = \frac{\Delta_r H}{C_p \rho} \frac{\left(\xi(i,j,m) - \xi(i,j,m-1)\right)}{\Delta t} \tag{30.12}$$

By integrating Equation 30.11 and making a series of algebraic transformations, one could find the following analytical formulas:

$$\gamma(i,j,m) = \frac{\Delta_r H}{C_p \rho \Delta t}\left(-\exp\left[-I\big(u(i,j,m)\big)\right] + \exp\left[-I\big(u(i,j,m-1)\big)\right]\right), \tag{30.13}$$

where

$$I\big(u(i,j,m)\big) = \left(0.5 k_0 \sum_1^m \left(\exp\left(-\frac{E_a}{R \cdot u(i,j,m-1)}\right) + \exp\left(-\frac{E_a}{R \cdot u(i,j,m)}\right)\right)\Delta t\right). \tag{30.14}$$

Notice, as a first result of temperature field calculations, the feedback control system with laser thermography possesses high stability at optimal PID parameters providing constant heating rate by a smooth correction of laser power despite the occurrence of thermal energetic processes. Figure 30.5 demonstrates laser power variations as a function of time at programmable laser heating of a local area of the sample ($r = 0$, $z = 0$) with the Gaussian light beam ($W_L = 4$ mm, $\lambda = 1.56$ µm) under the condition of a tentative endothermic chemical reaction of first order for different enthalpies. The scattering in the power data is due to an intrinsic error of temperature measurement by thermovision camera (0.1°C) and an error of the power control (10 mJ) by laser actuator. One could see bell-shaped curves of laser power which more distinctly manifested as enthalpy increases. As a criterion of minimal enthalpy

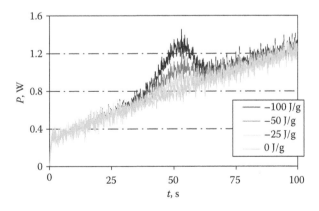

FIGURE 30.5 Temporal behavior of laser power as a function of enthalpy of tentative first-order reaction with Arrhenius-type reaction rate being constant during heating of the local area of the sample with a constant rate 1°C/s. The laser beam has Gaussian intensity distribution $W_L = 4$ mm; $\chi = 0.125$ mm^2/s; $\rho C_p = 3.5$ J cm^3/°C; $\alpha = 10$ cm^{-1}.

measurement, we have chosen the double standard error excess of the power curve which is caused by correction of energetic processes. In this respect the sensitivity of the presented system to the enthalpy is about 25 J/g. This value is rather large, comparing with a sensitivity of the classical DSC (~1–5 J/g). Note that sensitivity of enthalpy measurement with laser thermography may be substantially improved by increasing the accuracy of temperature measurement and laser power control, as well as by improved data processing. A more sophisticated heating mode and an analysis of laser-induced temperature field can potentially increase sensitivity.

30.4 Laser Thermography for Measurements of Optical and Thermophysical Parameters

Since its emergence in the early 1980s, photothermal radiometry (PTR) has become a very powerful tool for the thermophysical characterization and nondestructive evaluation of broad classes of materials [11–14], including homogeneous materials [15] and layered and/or buried structures [16]. This method combines laser heating and radiometric measurement of the sample surface temperature, making it a natural extension of laser thermography.

PTR is based on the fact that the surface temperature temporal variations, measured by the thermovision camera, are determined by the thermal and optical properties of the material that is being heated locally by the laser. To process experimental thermographic data and to calculate the thermal, and optical parameters, the heat conduction equation should be solved. It includes thermal diffusivity Equation 30.2 with the term heat-source (Equation 30.5), characterizing the rate of temperature increase in the irradiated zone. On the one hand, the term heat-source is defined by the radiation characteristics, that is, the wavelength, exposure power, and duration, and, on the other hand, by optical parameters of the medium, that is, the effective absorption coefficient. Hence, it is the set of optical and thermal parameters that define the temperature response of a biomaterial to laser exposure.

Usually, the effect of optical parameters is eliminated by choosing the appropriate laser wavelength in accordance with the absorption spectrum of basic chromophores of the material. The idea is to provide conditions at which penetration depth is much smaller than laser beam diameter. In this case, the temporal variations of the temperature field are defined only by thermal parameters and the temperature evaluation is reduced by comparing the radiometric signal upon laser heating with the theoretically calculated temperature profiles [15,17].

There exist two basic types of time-dependent methods to measure thermal diffusivity: (1) the transient method and (2) the periodic heat flow method. In the former method, a sample is irradiated with a laser pulse and then the temperature evolution is monitored [18,19]. The thermal diffusivity calculation is based on the integral equation expressing the time dependence of the radiometric image $\Delta M(x,y,t)$ by means of the initial 3D distribution of the temperature induced by short laser pulses. Since the pulse duration is <500 ms, the initial 3D temperature distribution can be represented by the product of two independent components, that is, lateral $T(x,y,t = 0)$ and longitudinal $T(z,t = 0)$. In this case, the radiometric image $\Delta M(x,y,t_2)$ at a point of time t_2 is expressed as a convolution of a specific point spread function $K_r(\Delta t = t_2 - t_1)$ depending on χ and the radiometric image $\Delta M(x,y,t_1)$ recorded earlier. Using the Levenberg–Marquardt algorithm to compare a pair of radiometric images, we determine K_r and calculate the thermal diffusivity χ.

In periodic heat-flow method [20–22] a sample of known thickness is irradiated with a harmonic-modulated laser beam launching a thermal wave and the periodic temperature at the front or the back surface of the sample is monitored at several modulation frequencies f. The frequency-dependent thermal diffusion length v is given by

$$v = \sqrt{\frac{\chi}{\pi f}}$$

(30.15)

It is related to the phase shift of the detected temperature wave with respect to the heating source which may be monitored using a lock-in amplifier. The Fourier transform is applied to the measured temperature function, which allows the use of amplitude and phase characteristics instead of temporal characteristics.

Very recently PTR was extended to study the thermal and optical parameters simultaneously. In this work [21], thermal radiometry was applied to characterize both the thermal and optical properties of biomaterials. The dependence of the amplitude and phase on the modulated laser radiation frequency was used to measure the thermal diffusivity $e = (\lambda \rho C_p)^{1/2}$ and the so-called "initial heating coefficient" $g = \mu_{eff}/(\rho C_p)$ where μ_{eff} is the total attenuation coefficient, ρ is the density, and C_p is the specific heat. However, using this method it is impossible to individually and independently determine each of these parameters.

In the paper [23], the noncontact technique to evaluate the thermal and optical properties of biological tissues and materials has been developed. The method is based on a combination of PTR and the solution of the inverse heat conduction problem by the finite difference method. Numerical solution of the heat conduction equation allows rigorous consideration of the heat source distribution over the entire sample volume, which makes it possible to measure the total attenuation coefficient μ_{eff}, characterizing the optical properties of a biomaterial along with the thermal diffusivity χ and specific heat ρC_p.

The calculation of the 3D temperature field induced by IR laser radiation is based on the solution of the classical heat conduction equation, expressed in the cylindrical coordinates, see Equation 30.2.

It has been assumed that the variation of the radiation intensity inside the sample is described by the Beer law; the function $f(r,z,t)$ has been obtained for the light beams with a Gaussian energy distribution (Equation 30.5). The temperature distribution over the sample at the initial time was taken as uniform and equal to the ambient medium temperature T_0. Boundary conditions (Equation 30.4) at the sample surface have been used. The forward heat problem has been solved by the finite-difference method. Similar to Section 30.2, an implicit difference scheme to approximate the heat conduction equation and boundary conditions has been constructed. The determination of the parameters in question ($\chi, \rho C_p$, μ_{eff}) is reduced to conventional minimization of the squares of deviations of the calculated radiometric temperatures from the experimental:

$$\text{SSE}\left(\chi, \rho C_p, \mu_{eff}\right) = \sum_{l=1}^{L} \sum_{m=1}^{M} \left[u_{exp}\left(r_l, t_m\right) - u_{calc}\left(r_l, t_m; \chi, \rho C_p, \alpha\right) \right]^2, \tag{30.16}$$

where $u_{exp}(r_l, t_m)$ and $u_{calc}(r_l, t_m)$ are respectively, temperatures at the point r_l on the sample surface at the moment t, as measured by the thermovision camera and calculated theoretically.

A temporal series of one-dimensional temperature distributions measured on the back surface of the sample was used in calculations, see Figure 30.6a. We have used Erbium glass laser $\lambda = 1.56$ μm with maximal emitting power of 5 W in our experiments. During one laser pulse, the observed temperature increase was up to 6°C. The rate and the shape of the rising temperature front will depend on the distance to the center of the laser beam (Figure 30.6b). The temperature field variations have been recorded as a matrix u_{exp} with size $L \times M$, where L is the number of points with different coordinates r and M is the number of points along the time axis during laser pulse t_{imp}, and some time after the laser is turned off, a substantive cooling of the heated region occurs. The temperature increase at the stage of laser heating depends mostly on the total absorption coefficient α and specific heat ρC_p, while during the cooling phase the temperature is mostly determined by the thermal diffusivity χ. In fact, reliable and accurate measurements of all these parameters proved to be possible, using a few laser pulses and thermovision recordings of the full temperature field induced on the sample surface.

An optimal technique for determining thermal properties should provide both proper accuracy and acceptable speed of calculation. The former depends on both the measurement accuracy of the

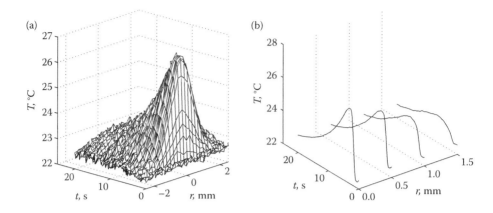

FIGURE 30.6 (a) Variations of the temperature field on the sample surface and (b) examples of temporal profiles of the surface temperature at different distances from the center of the irradiated region.

temperature field and the conditionality of their calculation model [24]. It is necessary to estimate the fraction of the experimental data set that carries the necessary sufficient data for determining all parameters χ, ρC_p, and α with satisfactory accuracy, while keeping the computation time reasonably small. An important stage of testing the technique is also to study the sensitivity of the calculated parameters to the temperature field parameters h and E_0 in the model, which can be a source of certain systematic errors.

The size of the experimental data set is defined by two quantities: L, that is, the number of points of different coordinates r, and M, that is, the number of points along the time axis. To calculate the desired parameters, it is reasonable to use the time dependence of the temperature for points arranged near the center of the irradiated region, since the temperature reaches the maximum exactly in this region (Figure 30.6b). In this case, it is sufficient to use the time dependence of the temperature only for several points from the irradiated region. For these points, the temperature change is several times larger than the temperature measurement error. Quantity L does not affect the calculation time of χ, ρC_p, and α, since the temperature field variations are calculated for the entire sample which is independent of L.

However, the accuracy and computation time of χ, ρC_p, and α depend strongly on the quantity M. The choice of the optimum value of M for calculating the desired parameters consists of three stages [24]:

1. Analysis of the dependences of χ, ρC_p, and α on M;
2. Estimation of the significance of each parameter at various M;
3. Estimation of the accuracy and time for calculating a parameter depending on M.

In addition to the accuracy, quantity M defines the computation time of the desired parameters, which monotonically increases with M. This is because it is necessary to calculate the temperature field variations up to a time point, corresponding to the value of M at each iteration. Thus, computation time of the temperature field variations R_{calc} increases monotonically with M, resulting in an increase of the entire calculation time.

Some arbitrariness in choosing the loss factor h during heat exchange with an ambient medium, which is built into the model for calculating the temperature field variations, is a factor controlling the systematic error of χ, ρC_p, and α measurements. According to the published data, the value of factor h, corresponding to free convection, can differ several times (from 5 to 50 W mm^{-2} K^{-1}) [5]. Therefore, we must determine the sensitivity of the calculated parameters to this parameter. In the paper [23], it has been shown that the dependence of the calculated parameters χ, ρC_p, and α on h can be disregarded if some reasonable value of h, for example, $h = 25$ W mm^{-2} K^{-1} is assumed. It has also been found that optimal M is equal to about 700, providing accuracy of χ, ρC_p, and α measurements in the vicinity of 5%, while $L = 10$.

The error in measurements of the laser radiation intensity E_0 could be the main source of a systematic error in calculating ρC_p, because the quantities E_0 and ρC_p are not independent. In other words, at the same minimum sum of squares error SSE, depending on the value of E_0, one could find the corresponding value of the varied parameter ρC_p. To estimate this factor, it is recommended that the specific heat is to be measured by an independent method such as DSC.

30.5 Tissue-Equivalent Phantoms for Simulation of the Laser-Induced Temperature Fields

Artificial materials, imitating individual physicochemical properties of biological tissues, are more often used to simulate a response caused by various physical interactions. These materials are stable, convenient in applications, allow one to control doses easily and interaction regimes, and simplify the development and tests of the diagnostic medical equipment considerably. Depending on the type of interaction, it is necessary to reproduce various physical properties of biological tissues. At present many phantoms are developed and used to imitate the optical [25–27], acoustic [26–28], electric [29–31], and thermal [29,30,32] properties of biological tissues.

Special interest is of equivalents of biological tissues used to simulate laser-induced heating of tissues. In particular, a moderate heating of biological tissues by laser radiation up to 70–80°C is used for cartilage shaping [33–35], stimulation of regenerative processes, and hyperthermia of malignant tumors. To achieve the therapeutic effect, the temperature field should be reproduced in strictly specified temporal and spatial limits. The spatiotemporal temperature distribution in a biological tissue is determined by the parameters of laser radiation and the absorption, and scattering coefficients specific to heat and thermal diffusivity of the tissue. It is obvious that materials with optical and thermal properties similar to those of biological tissues can be used as their equivalents for simulating laser-induced heating. It is important that all possible thermal and chemical processes would not considerably distort the temperature field and preserve the phase stability.

In the paper [36], the equivalent of biological tissues for simulating temperature fields induced by laser radiation ($\lambda=$) has been developed. The major chromophore, responsible for radiation absorption by biological tissues at wavelength 1.56 µm is water (70–80%). It makes various hydrogels the most convenient basis for obtaining the required tissue equivalent. In the literature different gels were suggested—agar [37], agarose [38], gelatin [39], and polyacrylamide (PAA) hydrogels [29–31]. However, most of these hydrogels easily melt after relatively small heating. Only the PAA hydrogel keeps its elasticity under heating, which makes it possible to synthesize samples of various shapes and to introduce probes and thermocouples into PAA. By synthesizing PAA gels with different monomer concentrations and different degrees of cross-linking, we can obtain a material with thermal and optical properties similar to those of a cartilage tissue.

PAA hydrogels can be easily synthesized by radical copolymerization of acrylamide and N,N'-methylenebisacrylamide in the presence of catalytic amounts of ammonium persulfate and N,N,N',N'-tetramethylethylenediamine [40]. Acrylamide and N,N'-methylenebisacrylamide should be dissolved in distilled water. Then a catalytic amount of $(NH_4)_2S_2O_8$ and N,N,N',N'-tetramethyethylenediamine should be added. At the end of the reaction, the vessel is cooled and hermetically sealed to prevent the evaporation of water.

There are two parameters, which define composition and consequently thermal and optical properties of the PAA hydrogel. The first parameter is the mass content of water in hydrogel. If water is the main absorber of laser radiation, the absorption coefficient of the hydrogel will be proportional to the mass content of water in the hydrogel. The second parameter is the degree of cross-linking, which is equal to the ratio of the initial amounts of N,N'-methylenebisacrylamide and acrylamide and define the value of the scattering coefficient of the hydrogel. PAA hydrogels with water content from 90% to 75% with a degree of cross-linking from 1:9 to 1:24 can be synthesized by the earlier method. On the one hand, the choice of the concentration interval of PAA gels is limited, by the limiting solubility of initial reagents in water, which prevented the preparation of gels with a lower water content, and on the other, by the limiting concentration of initial reagent in water below which gels could not be synthesized.

The thermal analysis of samples performed by the method of DSC shows that the thermal behavior of PAA hydrogels in the temperature range from 25°C to 80°C is stable, and no significant power-consuming processes that could potentially distort the temperature field upon laser heating of the samples, is observed.

30.6 Summary

Feedback control systems combined with a laser as a source of heating, IR radiometer as temperature sensors and PID controller, evaluating and controlling the laser power at a given time could provide programmable heating of the local region of biological tissues with desired temperature variations. Optimal parameters of PID controller could be found by numerically solving a problem of the sample heating under laser radiation. The power of laser is determined by the PID controller to provide minimal integrative error between the current and prescribed temperatures at a given sample area. These errors depend on the optical and thermophysical characteristics of the tissue as well as on the radius of the laser beam. Optimized PID controller provides a smooth and almost linear behavior of the laser power as a function of time when realizing heating of the local sample area with constant rate. Laser power would manifest nonmonotonic behavior if energetic processes are induced in the tissue to compensate its energy consumption/release while the temperature increases linearly. Scattering of laser power data regulated by the PID controller is mainly determined by the errors of temperature measurement by the thermal camera and accuracy of laser power adjustment when performing PID controller commands. Programmable laser heating of tissues could be used to control enthalpies of energetic thermal processes induced in the tissue by the laser.

IR thermography of the temperature field of the tissue surface during postlaser heating allows to express the remote measurement of its total attenuation coefficient, specific heat, and thermal diffusivity simultaneously. The duration of such procedure could be just 10 s while the increment of tissue temperature could be 3–5°C. Since effective thermal diffusivity is strongly related with blood and its volumetric circulation, its measurement may be used for *in vivo* diagnostics of the blood perfusion.

PAA hydrogels may be quite convenient as tissue equivalents to simulate temperature fields induced by the laser radiation in biological tissues as diagnostic and treatment procedures, allowing one to vary the light absorption coefficient by changes in water and organic ratio and light scattering coefficient by changing the concentration of the cross-linking agent. It can also be applied for the calibration of medical devices, based on the interaction of laser light with biological tissues and materials. Laser thermography is proposed to be used as a method to express the characterization of optical and thermophysical properties of such tissue equivalents. Recently, optical and thermophysical equivalents of the cartilage have been developed on the basis of this approach.

References

1. Nowakowski A. (2006). Quantitative dynamic thermal IR-imaging and thermal tomography in medical diagnostic. *Medical Devices and Systems*. Bronzino J. D., Ed. Boca Raton, FL, CRC Press: 22-21-30.
2. Maldague X. P. V. (2001). *Theory and Practice of Infrared Technology for Nondestructive Testing*. New York, J. Wiley & Sons.
3. Zhou J., Chen J. K., Zhang Y. (2010). Simulation of laser-induced thermotherapy using a dual-reciprocity boundary element model with dynamic tissue properties *IEEE Trans. Biomed. Eng.*, 57(2): 238–245.

4. Cheng K.-S., Dewhirst M. W., Stauffer P. F., Das S. (2010). Mathematical formulation and analysis of the nonlinear system reconstruction of the on-line image-guided adaptive control of hyperthermia. *Med. Phys.*, 37(3): 980–994.

5. Orr C. S., Eberhart R. C. (1995). Overview of bioheat transfer. *Optical–Thermal Response of Laser-Irradiated Tissue*. Welch A. J. and van Gemert M. J. C., Ed. New York, Plenum Press, 367–383.

6. Bagratashvili V. N., Bagratashvili N. V., Gapontsev P. V., Makhmutova G. S., Minaev V. P., Omelchenko A. I., Samartsev I. E., Sviridov A. P., Sobol E. N., Tsypina S. I. (1996). Change in the optical properties of hyaline cartilage heated by near-IR laser radiation. *Quantum Electron.*, 31(6): 534–538.

7. Jacques S. L. (1998). Light distributions from point, line and plane sources for photochemical reactions and fluorescence in turbid biological tissues. *Photochem. Photobiol.*, 67(1): 23–32.

8. Samarskii A.A. (1971). *Introduction to the Theory of Differential Schemes*. Moscow, Nauka (in Russian).

9. Marquardt D. W. (1963). An algorithm for least-square estimation of nonlinear parameters. *J. Soc. Indus. Appl. Math.*, 11(2): 431–441.

10. More J. J. (1977). The Levenberg–Marquardt algorithm: Implementation and theory. *Numerical Analysis. Lecture Notes in Mathematics*, Springer. 630: 105–116.

11. Santos R., Miranda L. C. M. (1981). Theory of the photothermal radiometry with solids. *J. Appl. Phys.*, 52(6): 4194–4198.

12. Tom R. D., O'Hara E. P., Benin D. (1982). A generalized model of photothermal radiometry. *J. Appl. Phys.*, 53(8): 5392–5400.

13. Lan T. T., Walther H. G., Goch G., Schmitz B. (1995). Experimental results of photothermal micro-structural depth profiling. *J. Appl. Phys.*, 78(6): 4108–4111.

14. Munidasa M., Funak F., Mandelis A. (1998). Application of a generalized methodology for quantitative thermal diffusivity depth profile reconstruction in manufactured inhomogeneous steel-based materials. *J. Appl. Phys.*, 83(7): 3495–3498.

15. Park H. K., Grigoropoulos C. P., Tam A. C. (1995). Optical measurements of thermal-diffusivity of a material. *Int. J. Thermophys.*, 16(4): 973–995.

16. Salazar A., Sanchez-Lavega A., Terron J. M. (1998). Effective thermal diffusivity of layered materials measured by modulated photothermal techniques. *J. Appl. Phys.*, 84(6): 3031–3041.

17. Fujii M., Park S. C., Tomimura T., Zhang X. (1997). A noncontact method for measuring thermal conductivity and thermal diffusivity of anisotropic materials. *Int. J. Thermophys.*, 18(1): 251–267.

18. Milner T. E., Goodman D. M., Tanenbaum B. S., Anvari B., Nelson J. S. (1996). Noncontact determination of thermal diffusivity in biomaterials using infrared imaging radiometry. *J. Biomed. Opt.*, 1(1): 92–97.

19. Telenkov S. A., Youn J. I., Goodman D. M., Welch A. J., Milner T. E. (2001). Non-contact measurement of thermal diffusivity in tissue. *Phys. Med. Biol.*, 46: 551–558.

20. Munidasa M., Mandelis A. (1994). A comparison between conventional photothermal frequency scan and the lock-in rate window method in measuring thermal-diffusivity of solids. *Rev. Sci. Instrum.*, 65(7): 2344–2350.

21. Gijsbertsen A., Bicanic D., Gielen J. L. W., Chirtoc M. (2004). Rapid, non-destructive and non-contact inspection of solids foods by means of photothermal radiometry, thermal effusivity and initial heating coefficient. *Infrared Phys. Technol.*, 45: 93–101.

22. Wang C. H., Mandelis A. (2007). Characterization of hardened cylindrical C1018 steel rods (0.14%–0.2% C, 0.6%–0.9% Mn) using photothermal radiometry. *Rev. Sci. Inst.*, 78(5).

23. Kondyurin A. V., Sviridov A. P., Obrezkova M. V., Lunin V. V. (2010). Noncontact measurement of thermal and optical parameters of biological tissues and materials using IR laser radiometry. *Russ. J. Phys. Chem.*, 83(8): 1405–1413.

24. Draper N. R., Smith H. (1987). *Applied Regression Analysis*. New York, John Wiley & Sons.

25. Gibson A. P., Hebden J. C., Riley J., Everdell N., Schweiger M., Arridge S. R., Delpy D. T. (2005). Linear and nonlinear reconstruction for optical tomography of phantoms with nonscattering regions. *Appl. Opt.*, 44(19): 3925–3936.

26. Spirou G. M., Oraevsky A. A., Vitkin I. A., Whelan W. M. (2005). Optical and acoustic properties at 1064 nm of polyvinyl chloride-plastisol dor use as a tissue phantom in biological optoacoustics. *Phys. Med. Biol.*, 50(14): 141–153.

27. Devi C. U., Sood A. K. (2007). Measurement of visco-elastic properties of breast-tissue mimicking materials using diffusing wave spectroscopy. *J. Biomed. Opt.*, 12(3): 034035-034031–034035-034035.

28. de Korte C. L., Cespedes E. I., van der Steen A. F. W., Norder B., te Nijenhis K. (1997). Elastic and acoustic properties of vessel mimicking material for elasticity imaging. *Ulrtason. Imaging*, 19: 112–126.

29. Bini M., Ignesti A., Millanta L., Olmi R., Rubino N., Vanni R. (1984). The polyacrylamide as a phantom material for electromagnetic hyperthermia studies. *IEEE Trans. Biomed. Eng.*, BME-31(3): 317–322.

30. Andreuccetti D., Bini M., Ignesti A., Olmi R., Rubino N., Vanni R. (1988). Use of polyacrylamide as a tissue-equivalent material in the microwave range. *IEEE Trans. Biomed. Eng.*, 35(4): 275–277.

31. Surowiec A., Shrivastava P., Astrahan M., Petrovich Z. (1992). Utilization of a multilayer polyacrylamide phantom for evaluation of hyperthermia applications. *Int. J. Hyperthermia*, 8(6): 795–807.

32. Arora D., Cooley D., Perry T., Skliar M., Roemer R. B. (2005). Direct thermal dose control of constrained focused ultrasound treatments phantom and *in vivo* evaluation. *Phys. Med. Biol.*, 50(8): 1919–1935.

33. Sobol E. N., Bagratashvili V. N., Sviridov A. P., Omelchenko A. I., Ovchinnikov A. B., Shechter A. B., Helidonis E. (1994). Laser shaping of cartilage. *Proc. SPIE.* 2128: 43–49.

34. Sobol E. N., Bagratashvili V. N., Sviridov A. P., Omelchenko A. I., Shechter A. B., Jones N., Howdle S., Helidonis E. (1996). Cartilage reshaping with holmium laser. *Proc. SPIE.* 2623: 544–547.

35. Wang Z., Pankratov M. M., Perrault D. F., Shapshay S. M. (1995). Laser-assisted cartilage reshaping: *In vitro* and *in vivo* animal studies. *Proc. SPIE.* 2395: 296–302.

36. Kondyurin A. V., Sviridov A. P. (2008). Equivalent of a cartilage tissue for simulations of laser-induced temperature fields. *Quantum* Electron., 38(7): 641–646.

37. Beck G., Akgun N., Ruck A., Stainer R. (1998). Design and characterisation of a tissue phantom system for optical diagnostics. *Lasers Med. Sci.*, 13: 160–171.

38. Saidi I., Jacques S., Tittel F. (1990). Monitoring neonatal bilirubinemia using an optical patch. *Proc. SPIE.* 1201: 569–578.

39. Hielscher A., Liu H., Chance B., Tittel F., Jacques S. (1996). Time-resolved photon emission from layered turbid media. *Appl. Opt.*, 35(4): 719–728.

40. Tanaka T. (1981). Gels. *Sci. Am.*, 244(31): 110–134.

31

Use of Infrared Imaging in Veterinary Medicine

Ram C. Purohit
Auburn University

Tracy A. Turner

David D. Pascoe
Auburn University

31.1 Historical Perspective

In the mid-1960s and early 1970s, several studies were published indicating the value of infrared (IR) thermography in veterinary medicine [1–3]. In the 1965 research study of Delahanty and George [2], the thermographic images required at least 6 min to produce a thermogram, a lengthy period of time during which the veterinarian had to keep the horse still while the scan was completed. This disadvantage was overcome by the development of high-speed scanners using rotating IR prisms which then could produce instantaneous thermograms.

Stromberg [4–6] and Stromberg and Norberg [7] used thermography to diagnose inflammatory changes of the superficial digital flexor tendons in race horses. With thermography, they were able to document and detect early inflammation of the tendon, 1–2 weeks prior to the detection of lameness using clinical examination. They suggested that thermography could be used for early signs of pending lameness and it could be used for preventive measures to rest and treat race horses before severe lameness became obvious on physical examination.

In 1970, the Horse Protection Act was passed by the United States Congress to ban the use of chemical or mechanical means of "soring" horses. It was difficult to enforce this act because of the difficulty in obtaining measurable and recordable proof of violations. In 1975, Nelson and Osheim [8] documented that soring caused by chemical or mechanical means on the horse's digit could be diagnosed as having a definite abnormal characteristic IR emission pattern in the affected areas of the limb. Even though thermography at that time became the technique of choice for the detection of soring, normal thermography patterns in horses were not known. This prompted the USDA to fund research for the uses of thermography in veterinary medicine.

Purohit et al. [9] established a protocol for obtaining normal thermographic patterns of the horses' limbs and other parts of the body. This protocol was regularly used for early detection of acute and chronic inflammatory conditions in horses and other animal species. Studies at Auburn University vet school used an AGA 680 liquid-cooled thermography system that had a black and white and an accessory color display units that allows the operator to assign the array of 10 isotherms to temperature increments from 0.2°C to 10.0°C. Images were captured within seconds rather than the 6 min required for earlier machines. In veterinary studies at Auburn University, the thermographic isotherms were imaged with nine colors and white assigned to each isotherm that varied in temperature between either 0.5°C or 1.0°C.

In a subsequent study, Purohit and McCoy [10] established normal thermal patterns (temperature and gradients) of the horse, with special attention directed toward thoracic and pelvic limbs. Thermograms of various parts of the body were obtained 30 min before and after the exercise for each horse. Thermographic examination was also repeated for each horse on six different days. Thermal patterns and gradients were similar in all horses studied with a high degree of right to left symmetry in IR emission.

At the same time, Turner et al. [11] investigated the influence of the hair coat and hair clipping. This study demonstrated that the clipped leg was always warmer. After exercise, both clipped and unclipped legs had similar increases in temperature. The thermal patterns and gradients were not altered by clipping and/or exercise [10,11]. This indicated that clipping hair in horses with even hair coats was not necessary for thermographic evaluation. However, in some areas where the hair is long hair and not uniform, clipping may be required. Recently, concerns related to hair coat, thermographic imaging, and temperature regulation were investigated in llamas exposed to the hot humid conditions of the southeast [12]. While much of the veterinary research has focused on the thermographic imaging as a diagnostic tool, this study expanded its use into the problems of thermoregulation in various non endemic species.

Current camera technology has improved scanning capabilities that are combined with computer-assisted software programs. This new technology provides the practitioner with numerous options for image analysis, several hundred isotherms capable of capturing temperature differences in the hundredths of a degree Celsius, and better image quality. Miniaturized electronics have reduced the size of the units, allowing some systems to be housed in portable hand-held units. With lower cost of equipment, more thermographic equipment are being utilized in human and animal veterinary medicine and basic physiology studies.

It was obvious from initial studies by several authors that standards needed to be established for obtaining reliable thermograms in different animal species. The variations in core temperature and differences in the thermoregulatory mechanism responses between species emphasizes the importance of individually established norms for thermographic imagery.

A further challenge in veterinary medicine may occur when animal patient care may necessitate outdoor imaging.

31.2 Standards for Reliable Thermograms

Thermography provides an accurate, quantifiable, noncontact, noninvasive measure and map of skin surface temperatures. Skin surface temperatures are variable and change according to blood flow regulation to the skin surface. As such, IR thermography practitioner must be aware of the internal and external influences that alter this dynamic process of skin blood flow and temperature regulation. While imaging equipment can vary widely in price, these differences are often reflective of the wavelength capturing capability of the detectors and adjunct software that can aid in image analysis. The thermographer needs to understand the limitations of their IR system in order to make appropriate interpretations of their data. There have been some published studies that have not adhered to reliable standards and equipment prerequisites, thereby detracting from the acceptance of thermography as a valuable research and clinical technique. In some cases a simple cause–effect relationship was assumed to demonstrate the diagnosis of a disease or syndrome based on thermal responses as captured by thermographic images.

Internal and external factors have a significant effect on the skin surface temperature. Therefore, the use of thermography to evaluate skin surface thermal patterns and gradient requires an understanding of the dynamic changes which occur in blood flow at systemic, peripheral, regional, and local levels [9,10]. Thus, to enhance the diagnostic value of thermography, we recommend the following standards for veterinary medical imaging:

1. The environmental factors which interfere with the quality of thermography should be minimized. The room temperature should be maintained between 21°C and 26°C. Slight variations in some cases may be acceptable, but room temperature should always be cooler than the animal's body temperature and free from air drafts.
2. Thermograms obtained outdoors under conditions of direct air drafts, sunlight, and extreme variations in temperature may provide unreliable thermograms in which thermal patterns are altered. Such observations are meaningless as a diagnostic tool.
3. When an animal is brought into a temperature-controlled room, it should be equilibrated at least 20 min or more, depending on the external temperature from which the animal was transported. Animals transported from extreme hot or cold environments may require up to 60 min of equilibration time. Equilibration time is adequate when the thermal temperatures and patterns are consistently maintained over several minutes.
4. Other factors affecting the quality of thermograms are exercise, sweating, body position and angle, body covering, systemic and topical medications, regional and local blocks, sedatives, tranquilizers, anesthetics, vasoactive drugs, skin lesions such as scars, surgically altered areas, and so on. As stated prior, the hair coat may be an issue with uneven hair length or a thick coat.
5. It is recommended that the IR imaging should be performed using an electronic non contact cooled system. The use of long wave detectors is preferable.

The value of thermography is demonstrated by the sensitivity to changes in heat on the skin surface and its ability to detect temporal and spatial changes in thermal skin responses that corresponds to temporal and spatial changes in blood flow. Therefore, it is important to have well-documented normal thermal patterns and gradients in all species under controlled environments prior to making any claims or detecting pathological conditions.

31.3 Dermatome Patterns of Horses and Other Animal Species

Certain chronic and acute painful conditions associated with peripheral neurovascular and neuromuscular injuries are easy to confuse with spinal injuries associated with cervical, thoracic, and lumbar-sacral areas [13,14]. Similarly, inflammatory conditions such as osteoarthritis, tendonitis, and other associated conditions may also be confused with other neurovascular conditions. Thus, studies have been done over the past 25 years at Auburn University to map cutaneous and differentiate the sensory–sympathetic dermatome patterns of cervical, thoracic, and lumbosacral regions in horses [13,14]. IR thermography was used to map the sensory–sympathetic dermatome in horses. The dorsal or ventral spinal nerve(s) were blocked with 0.5% of mepevacine as a local anesthetic. The sensory–sympathetic spinal nerve block produced two effects. First, blocking the sympathetic portion of the spinal nerve caused increased thermal patterns and produced sweating of the affected areas. Second, the areas of insensitivity produced by the sensory portion of the block were mapped and compared with the thermal patterns. The areas of insensitivity were found to correlate with the sympathetic innervations.

Thermography was used to provide thermal patterns of various dermatome areas from cervical areas to epidural areas in horses. Clinical cases of cervical area nerve compression provided cooler thermal patterns, away from the site of injuries. In cases of acute injuries, associated thermal patterns were warmer than normal cases at the site of the injury. Elucidation of dermatomal (thermatom) patterns provided location for spinal injuries for the diagnosis of back injuries in horses. Similarly, in a case of a dog where the neck injury (subluxation of atlanto-axis) the diagnosis was determined by abnormal thermal patterns and gradients.

31.4 Peripheral Neurovascular Thermography

When there are alterations in skin surface temperature, it may be difficult to distinguish and diagnose between nerve and vascular injuries. The cutaneous circulation is under sympathetic vasomotor control. Peripheral nerve injuries and nerve compression can result in skin surface vascular changes that can be detected thermographically. It is well known that inflammation and nerve irritation may result in vasoconstriction causing cooler thermograms in the afflicted areas. Transection of a nerve and/or nerve damage to the extent that there is a loss of nerve conduction results in a loss in sympathetic tone which causes vasodilation indicated by an increase in the thermogram temperature. Of course, this simple rationale is more complicated with different types of nerve injuries (neuropraxia, axonotomesis, and neurotmesis). Furthermore, lack of characterization of the extent and duration of injuries may make thermographic interpretation difficult.

Studies were done on horses and other animal species to show that if thermographic examination is performed properly under controlled conditions, it can provide an accurate diagnosis of neurovascular injuries. The rationale for a neurovascular clinical diagnosis is provided in the following Horner's Syndrome case.

31.4.1 Horner's Syndrome

In four horses, Horner's Syndrome was also induced by transaction of vagosympathetic trunk on either left or right side of the neck [15]. Facial thermograms of a case of Horner's Syndrome were done 15 min before and after the exercise. Sympathetic may cause the affected side to be warm by 2–3°C more than the nontransected side. This increased temperature after denervation is reflective of an increase in blood flow due to vasodilation in the denervated areas [15,16]. The increased thermal patterns on the affected side were present up to 6–12 weeks. In about 2–4 months, neurotraumatized side blood flow readjusted to the local demand of circulation. Thermography of both non-neuroectomized and neuroectomized sides looked similar and normal [16]. In some cases, this readjustment took place as early as five days and it was difficult to distinguish the affected side. The intravenous injection of 1 mg of epinephrine in a 1000 lb horse caused an increase in thermal patterns on the denervated side, the same as indicating the presence of Horner's Syndrome. Administration of IV acetyl promazine (30 mg/1000 lb horse) showed increased heat (thermal pattern) on the normal non-neuroectomized side, whereas acetylpromazine had no effect on the neurectomized side. Alpha-blocking drug acetylpromazine caused vasodilation and increased blood flow to normal non-neurectomized side, whereas no effect was seen in the affected neurectomized side due to the lack of sympathetic innervation [16–18].

31.4.2 Neurectomies

Thermographic evaluation of the thoracic (front) and pelvic (back) limbs were done before and after performing digital neurectomies in several horses. After posterior digital neurectomy there were significant increases in heat in the areas supplied by the nerves [17]. Within 3–6 weeks, readjustment of local blood flow occurred in the neurectomized areas, and it was difficult to differentiate between the non-neurectomized and the neurectomized areas. At 10 min after administration of 0.06 mg/kg IV injection of acetylpromazine, a 2–3°C increase in heat was noted in normal non-neurectomized areas, whereas the neurectomized areas of the opposite limb were not affected.

31.4.3 Vascular Injuries

Thermography has been efficacious in the diagnosis of vascular diseases. It has been shown that the localized reduction of blood flow occurs in the horse with navicular disease [11]. This effect was more obvious on thermograms obtained after exercise than before exercise. Normally, 15–20 min of exercise will increase skin surface temperature by 2–2.5°C in horses [10,11]. In cases of arterial occlusion, the area distal to the occlusion in the horses' limb shows cooler thermograms. The effects of exercise or administration of alpha-blocking drugs like acetylpromazine causes increased blood flow to peripheral circulation in normal areas with intact vascular and sympathetic responses [17,18]. Thus, obtaining thermograms either after exercise or after administration of alpha-blocking drugs like acetylpromazine provides prognostic value for diagnosis of adequate collateral circulation. Therefore, the use of skin temperature as a measure of skin perfusion merits consideration for peripheral vascular flow, perfusion, despite some physical and physiological limitations, which are inherent in methodology [19].

Furthermore, interference with the peripheral vascular blood flow can result from neurogenic inhibition, vascular occlusion, and occlusion as a result of inflammatory vascular compression. Neurogenic inhibition can be diagnosed through the administration of alpha-blocking drugs which provide an increase in blood flow. Vascular impairment may also be associated with local injuries (inflammation, edema, swelling, etc.) which may provide localized cooler or hotter thermograms. Thus, evaluation using thermography should note the physical state and site of the injury.

31.5 Musculoskeletal Injuries

Thermography has been used in the clinical and subclinical cases of osteoarthritis, tendonitis, navicular disease, and other injuries such as sprains, stress fractures, and shin splints [10,11,20,21]. In some cases thermal abnormalities may be detected 2 weeks prior to the onset of clinical signs of lameness in horses, especially in the case of joint disease [21], tendonitis [10], and navicular problems [11,20].

Osteoarthritis is a severe joint disease in horses. Normally, diagnosis is made by clinical examination and radiographic evaluation. Radiography detects the problem after deterioration of the joint surface has taken place. Clinical evaluation is only done when horses show physical abnormalities in their gait due to pain. An early sign of osteoarthritis is inflammation, which can be detected by thermography prior to it becoming obvious on radiograms [21].

In studies of standard bred race horses, the effected tarsus joint can demonstrate abnormal thermal patterns indicating inflammation in the joint 2–3 weeks prior to radiographic diagnosis [21]. The abnormal thermograms obtained in this study were more distinct after exercise than before exercise. Thus, thermography provided a subclinical diagnosis of osteoarthritis in this study.

Thermography was used to evaluate the efficacy of corticosteroid therapy in amphotericine-B induced arthritis in ponies [22]. The intra-articular injection of 100 mg of methylprednisolone acetate was effective in alleviating the clinical signs of lameness and pain. It is important to note that when compared with clinical signs of nontreated arthritis, it was difficult to differentiate increased thermal patterns between corticosteroid treated vs. non-treated, arthritis-induced joints. However, corticosteroid therapy did not decrease the healing time of intercarpal arthritis, whereas corticosteroid therapy did decrease the time for return to normal thermographic patterns for tibiotarsal joints. In this study, thermography was useful in detecting inflammation in the absence of clinical signs of pain in corticosteroid treated joints and aiding the evaluation of the healing processes in amphotericin B-induced arthritis [22].

The chronic and acute pain associated with neuromuscular conditions can also be diagnosed by this technique. In cases where no definitive diagnosis can be made using physical examination and x-rays, thermography has been efficacious for early diagnosis of soft-tissue injuries [10,23]. The conditions such as subsolar abscesses, laminitis, and other leg lameness can be easily differentiated using thermography [10,11]. We have used thermography for quantitative and qualitative evaluation of anti-inflammatory drugs such as phenylbutazone in the treatment of physical or chemically induced inflammation. The most useful application of thermography in veterinary medicine and surgery has been to aid early detection of an acute and chronic inflammatory process.

31.6 Thermography of the Testes and Scrotum in Mammalian Species

The testicular temperature of most mammalian species must be below body temperature for normal spermatogenesis. The testes of most domestic mammalian species migrates out of the abdomen and are retained in the scrotum, which provides the appropriate thermal environment for normal spermatogenesis [24,25]. The testicular arterial and venous structure is such that arterial coils are enmeshed in the pampiniform plexus of the testicular veins, which provides a counter current heating regulating mechanism by which arterial blood entering the testes is cooled by the venous blood leaving the testes [24,25]. In the ram, the temperature of the blood in the testicular artery decreases by 4°C from the external inguinal ring to the surface of the testes. Thus, to function effectively, the mammalian testes are maintained at a lower temperature.

Purohit et al. [26,27] used thermography to establish normal thermal patterns and gradients of the scrotum in bulls, stallions, bucks, dogs, and llamas. The normal thermal patterns of the scrotum in all species studied is characterized by right to left symmetrical patterns, with a constant decrease in the thermal gradients from the base to the apex. In bulls, bucks, and stallions, a thermal gradient of 4–6°C from the base to apex with concentric hands signifies normal patterns. Inflammation of one testicle increased ipsilateral scrotal temperatures of 2.5–3°C [26,28] If both testes were inflamed, there was an overall increase of 2.5–3°C temperature and a reduction in temperature gradient was noted.

Testicular degeneration could be acute or chronic. In chronic testicular degeneration with fibrosis, there was a loss of temperature gradient, loss of concentric thermal patterns, and some areas were cooler than others with no consistent patterns [26]. Reversibility of degenerative changes depends upon the severity and duration of the trauma. The IR thermal gradients and patterns in dogs [27] and llamas [27,29] are unique to their own species and the patterns are different from that of the bull and buck.

Thermography has also been used in humans, indicating a normal thermal pattern which is characterized by symmetric and constant temperatures between 32.5°C and 34.5°C [30–33]. Increased scrotal IR emissions were associated with intrascrotal tumor, acute and chronic inflammation, and varicoceles [34,35]. Thermography has been efficacious for early diagnosis of acute and/or chronic testicular degeneration in humans and many animal species. The disruption of the normal thermal patterns of the scrotum is directly related to testicular degeneration. The testicular degeneration may cause transient or permanent infertility in the male. It is well established that increases in scrotal temperature above normal causes disruption of spermatogenesis, affects sperm maturation, and contributes toward subfertile or infertile semen quality. Early diagnosis of pending infertility has a significant impact on economy and reproduction in animals.

31.7 Conclusions

The value of thermography can only be realized if it is used properly. All species studied thus far have provided remarkable bilateral symmetrical patterns of IR emission. The high degree of right-to-left symmetry provides a valuable asset in diagnosis of unilateral problems associated with various inflammatory disorders. On the other hand, bilateral problems can be diagnosed due to changes in thermal gradient and/or overall increase or decrease of temperature, away from the normal established thermal patterns in a given area of the body. Various areas of the body on the same side have normal patterns and gradients. This can be used to diagnose a change in gradient patterns. Alteration in normal thermal patterns and gradients indicates a thermal pathology. If thermal abnormalities are evaluated carefully, early diagnosis can be made, even prior to the appearance of clinical signs of joint disease, tendonitis, and various musculoskeletal problems in various animal species. Thermography can be used as a screening device for early detection of an impending problem, allowing veterinarian institute treatment before the problem becomes more serious. During the healing process post surgery, animals may appear physically sound. Thermography can be used as a diagnostic aid in assessing the healing processes. In equine sports medicine, thermography can be used on a regular basis for screening to prevent severe injuries to the horse. Early detection and treatment can prevent financial losses associated with delayed diagnosis and treatment.

The efficacy of noncontact electronic IR thermography has been demonstrated in numerous clinical settings and research studies as a diagnostic tool for veterinary medicine. It has had a strong impact on veterinary medical practice and thermal physiology where accurate skin temperatures need to be assessed under normal conditions, disease pathologies, injuries, and thermal stress. The importance of IR thermography as a research tool cannot be understated for improving the medical care of animals and for the contributions made through animal research models that improve our understanding of human structures and functions.

References

1. Smith W.M. Application of thermography in veterinary medicine. *Ann. NY Acad. Sci.*, 121, 248, 1964.
2. Delahanty D.D. and George J.R. Thermography in equine medicine. *J. Am. Vet. Med. Assoc.*, 147, 235, 1965.
3. Clark J.A. and Cena K. The potential of infrared thermography in veterinary diagnosis. *Vet. Rec.*, 100, 404, 1977.
4. Stromberg B. The normal and diseased flexor tendon in racehorses. *Acta Radiol.* 305(Suppl.), 1, 1971.
5. Stromberg B. Thermography of the superficial flexor tendon in race horses. *Acta Radiol.* 319(Suppl.), 295, 1972.
6. Stromberg B. The use of thermograph in equine orthopedics. *J. Am. Vet. Radiol. Soc.*, 15, 94, 1974.
7. Stromberg B. and Norberg I. Infrared emission and Xe-disappearance rate studies in the horse. *Equine Vet. J.*, 1, 1–94, 1971.
8. Nelson H.A. and Osheim D.L. Soring in Tennessee walking horses: Detection by thermography. *USDA-APHIS, Veterinary Services Laboratories*, Ames, Iowa, pp. 1–14, 1975.
9. Purohit R.C., Bergfeld II W.A. McCaoy M.D., Thompson W.M., and Sharman R.S. Value of clinical thermography in veterinary medicine. *Auburn Vet.*, 33, 140, 1977.
10. Purohit R.C. and McCoy M.D. Thermography in the diagnosis of inflammatory processes in the horse. *Am. J. Vet. Res.*, 41, 1167, 1980.
11. Turner T.A. et al. Thermographic evaluation of podotrochlosis in horses. *Am. J. Vet. Res.*, 44, 535, 1983.
12. Heath A.M., Navarre C.B., Simpkins A.S., Purohit R.C., and Pugh D.G. A comparison of heat tolerance between sheared and non sheared alpacas (*llama pacos*). *Small Ruminant Res.*, 39, 19, 2001.
13. Purohit R.C. and Franco B.D. Infrared thermography for the determination of cervical dermatome patterns in the horse. *Biomed. Thermol.*, 15, 213, 1995.
14. Purohit R.C., Schumacher J., Molloy J.M., Smith, and Pascoe D.D. Elucidation of thoracic and lumbosacral dermatomal patterns in the horse. *Thermol. Int.*, 13, 79, 2003.
15. Purohit R.C., McCoy M.D., and Bergfeld W.A. Thermographic diagnosis of Horner's syndrome in the horse. *Am. J. Vet. Res.*, 41, 1180, 1980.
16. Purohit R.C. The diagnostic value of thermography in equine medicine. *Proc. Am. Assoc. Equine Pract.*, 26, 316–326, 1980.
17. Purohit R.C. and Pascoe D.D. Thermographic evaluation of peripheral neurovascular systems in animal species. *Thermology*, 7, 83, 1997.
18. Purohit R.C., Pascoe D.D., Schumacher J., Williams A., and Humburg J.H. Effects of medication on the normal thermal patterns in horses. *Thermol. Osterr.*, 6, 108, 1996.
19. Purohit R.C. and Pascoe D.D. Peripheral neurovascular thermography in equine medicine. *Thermol. Osterr.*, 5, 161, 1995.
20. Turner T.A., Purohit R.C., and Fessler J.F. Thermography: A review in equine medicine. *Comp. Cont. Education Pract. Vet.*, 8, 854, 1986.

21. Vaden M.F., Purohit R.C., Mcoy, and Vaughan J.T. Thermography: A technique for subclinical diagnosis of osteoarthritis. *Am. J. Vet. Res.*, 41, 1175–1179, 1980.

22. Bowman K.F., Purohit R.C., Ganjan, V.K., Peachman R.D., and Vaughan J.T. Thermographic evaluation of corticosteroids efficacy in amphotericin-B induced arthritis in ponies. *Am. J. Vet. Res.* 44, 51–56, 1983.

23. Purohit R.C. Use of thermography in the diagnosis of lameness. *Auburn Vet.*, 43, 4, 1987.

24. Waites G.M.H. and Setchell B.P. Physiology of testes, epididymis, and scrotum. In *Advances in Reproductive Physiology*. McLaren A., Ed., London: Logos, Vol. 4, pp. 1–21, 1969.

25. Waites G.M.H. Temperature regulation and the testes. In *The Testis*, Johnson A.D., Grones W.R., and Vanderwork N.L., Eds., New York: Academy Press, Inc., Vol. 1, pp. 241–237, 1970.

26. Purohit R.C., Hudson R.S., Riddell M.G., Carson R.L., Wolfe D.F., and Walker D.F. Thermography of bovine scrotum. *Am. J. Vet. Res.*, 46, 2388–2392, 1985.

27. Purohit R.C., Pascoe D.D., Heath A.M. Pugh D.G., Carson R.L., Riddell M.G., and Wolfe D.F. Thermography: Its role in functional evaluation of mammalian testes and scrotum. *Thermol. Int.*, 12, 125–130, 2002.

28. Wolfe D.F., Hudson R.S., Carson R.L., and Purohit, R.C. Effect of unilateral orchiectomy on semen quality in bulls. *J. Am. Vet. Med. Assoc.*, 186, 1291, 1985.

29. Heath A.M., Pugh D.G., Sartin E.A., Navarre B., and Purohit R.C. Evaluation of the safety and efficacy of testicular biopsies in llamas. *Theriogenology*, 58, 1125, 2002.

30. Amiel J.P., Vignalou L., Tricoire J. et al. Thermography of the testicle: Preliminary study. *J. Gynecol. Obstet. Biol. Reprod.*, 5, 917, 1976.

31. Lazarus B.A. and Zorgiotti A.W. Thermo-regulation of the human testes. *Fertil. Steril.*, 26, 757, 1978.

32. Lee J.T. and Gold R.H. Localization of occult testicular tumor with scrotal thermography. *J. Am. Med. Assoc.*, 1976, 236, 1976.

33. Wegner G. and Weissbach Z. Application of palte thermography in the diagnosis of scrotal disease. *MMW*, 120, 61, 1978.

34. Gold R.H., Ehrilich R.M., Samuels B. et al. Scrotal thermography. *Radiology*, 1221, 129, 1979.

35. Coznhaire F., Monteyne R., and Hunnen M. The value of scrotal thermography as compared with selective retrograde venography of the internal spermatic vein for the diagnosis of subclinical varicoceles. *Fertil. Steril.*, 27, 694, 1976.

32

Standard Procedures for Infrared Imaging in Medicine

Kurt Ammer
*Ludwig Boltzmann
Research Institute for
Physical Diagnostics*

University of Glamorgan

Francis J. Ring
University of Glamorgan

32.1 Introduction

Infrared thermal imaging has been used in medicine since the early 1960s. Working groups within the European Thermographic Association (now European Association of Thermology) produced the first publications on standardization of thermal imaging in 1978 [1] and 1979 [2]. However, Collins and Ring established already in 1974 a quantitative thermal index [3], which was modified in Germany by J.-M. Engel in 1978 [4]. Both indices opened the field of quantitative evaluation of medical thermography.

Further recommendations for standardization appeared in 1983 [5] and 1984, the later related to essential techniques for the use of thermography in clinical drug trials [6]. Engel published a booklet titled *Standardized Thermographic Investigations in Rheumatology and Guideline for Evaluation* in 1984 [7]. The author presented his ideas for standardization of image recording and assessment including some normal values for wrist, knee, and ankle joints. Engel's measurements of knee temperatures were first published in 1978 [4]. Normal temperature values of the lateral elbow, dorsal hands, anterior knee, lateral and medial malleolus, and the first metatarsal joint were published by Collins in 1976 [8].

The American Academy of Thermology published technical guidelines in 1986 including some recommendations for thermographic examinations [9]. However, the American authors concentrated on determining the symmetry of temperature distribution rather than the normal temperature values of particular body regions. Uematsu in 1985 [10] and Goodman in 1986 [11] published the side-to-side variations of surface temperatures of the human body. These symmetry data were confirmed by E.F. Ring for the lower leg in 1986 [12].

In Japan, medical thermal imaging has been an accepted diagnostic procedure since 1981 [13]. Recommendations for the analysis of neuromuscular thermograms were published by Fujimasa et al. in 1986 [14]. Five years later more detailed proposals for the thermal image-based analysis of physiological functions were published in *Biomedical Thermology* [15], the official journal of the Japanese Society of thermology. This chapter was the result of a workshop on clinical thermography criteria.

Recently, the thermography societies in Korea have published a book, which summarizes in 270 pages general standards for imaging recording and interpretation of thermal images in various diseases [16].

As the relationship between skin blood flow and body surface temperature has been obvious from the initial use of thermal imaging in medicine, quantitative assessments were developed at an early stage. Ring developed a thermographical index for the assessment of ischemia in 1980, that was originally used for patients suffering from Raynauds' disease [17]. The European Association of Thermology published a statement in 1988 on the subject of Raynaud's phenomenon [18]. Normal values for recovering after a cold challenge have been published since 1976 [19,20]. A range of temperatures were applied in this thermal challenge test, the technique was reviewed by Ring in 1997 [21].

An overview of recommendations gathered from, the Japanese Society of Biomedical Thermology and the European Association of Thermology was collated and published by Clark and Goff in 1997 [22]. This chapter is based on the practical implications of the foregoing papers taken from the perspective of the modern thermal imaging systems available to medicine.

Finally, a project at the University of Glamorgan, aims to create an atlas of normal thermal images of healthy subjects [23]. This study, started in 2001, has generated a number of questions related to the influence of body positions on accuracy and precision of measurements from thermal images [24,25].

32.2 Definition of Thermal Imaging

Thermal imaging is regarded as a technique for temperature measurements based on the infrared radiation from objects. Unlike images created by x-rays or proton activation through magnetic resonance, thermal imaging is not related to morphology. The technique provides only a map of the distribution of temperatures on the surface of the object imaged.

Whenever infrared thermal imaging is considered as a method for measurement, the technique must meet all criteria of a measurement. The most basic features of measurement are accuracy (in the medical field also named validity) and precision (in medicine reliability). Anbar [26] has listed five other terms related to the precision of infrared-based temperature measurements. When used as an outcome measure, responsiveness, or sensitivity to change is an important characteristic.

32.2.1 Accuracy

Measurements are basic procedures of comparison namely to compare a standardized meter with an object to be measured. Any measurement is prone to error, thus a perfect measurement is impossible. However, the smaller the variation of a particular measurement from the standardized meter, the higher is the accuracy of the measurement or in other words, an accurate measurement is as close as possible to the true value of measurement. In medicine, accuracy is often named validity, mainly caused by the fact, that medical measurements are not often performed by the simple comparison of meter and object. For example, assessments from various features of a human being may be combined to a new construct, resulting in an innovative measurement of health.

32.2.2 Precision

A series of measurements cannot achieve totally identical results. The smaller the variation between single results, the higher is the precision or repeatability (reliability) of the measurement. However, reliability without accuracy is useless. For example, a sports archer who always hits the same peripheral sector of the target, has very high reliability, but no validity, because such an athlete must find the center of the target to be regarded as accurate.

32.2.3 Responsiveness

Both accuracy and precision have an impact on the sensitivity to change of outcome measures. Validity is needed to define correctly the symptom to be measured. Precision will affect the responsiveness also, because a change of the symptom can only be detected if this change is bigger than the variation of repeated measurements.

32.2.4 Sources of Variability of Thermal Images

Table 32.1 shows conditions in thermal imaging that may affect accuracy, precision, and responsiveness.

32.2.5 Object or Subject

As the emittance of infrared radiation is the source of remote temperature measurements, knowledge of the emissivity of the object is essential for the calculation of temperature related to the radiant heat. In nonliving objects emissivity is mainly a function of the texture of the surface.

Seventy years ago, Hardy [27] showed that the human skin acts like an almost perfect black body radiator with an emissivity of 0.98. Studies from Togawa in Japan have demonstrated that the emissivity of the skin is unevenly distributed [28]. In addition, infrared reflection from the environment and substances applied on the skin may also alter the emissivity [29–31]. Water is an efficient filter for infrared rays and can be bound to the superficial corneal layer of the skin during immersion for at least 15 min [32,33] or in the case of severe edema [34]. This can affect the emissivity of the skin.

The hair coat of animal may show a different emissivity than the skin after clipping the hair [35]. Variation in the distribution of the hairy coat will influence the emissivity of the animal's surface [36]. Variation in emissivity will influence the accuracy of temperature measurements.

Homeothermic beings, maintain their deep body (core) temperature through variation of the surface (shell) temperature, and show a circadian rhythm of both the core and shell temperature [37–40]. Repeated temperature registrations not performed at the same time of the day will therefore affect the precision of these measurements.

32.2.6 Camera Systems, Standards, and Calibration

32.2.6.1 The Imaging System

A new generation of infrared cameras has become available for medical imaging. The older systems, normally single element detectors using an optical mechanical scanning process, were mostly cooled by the addition of liquid nitrogen [41–43]. However, adding nitrogen to the system, affects the stability of temperature measurements for a period up to 60 min [44]. Nitrogen-cooled scanners had the effect of limiting the angle at which the camera could be used that restricted operation.

TABLE 32.1 Conditions Affecting Accuracy, Precision, and Responsiveness of Temperature Measures

Condition Affecting	Accuracy	Precision	Responsiveness
Object or subject	X	X	X
Camera systems, standards, and calibration	X	X	X
Patient position and image capture		X	X
Information protocols and resources		X	X
Image analysis	X	X	X
Image exchange	X	X	X
Image presentation	X	X	X

Electronic cooling systems were then introduced, which provided the use of image capturing without restrictions of the angle between the object and the camera. The latest generation of focal plane array cameras can be used without cooling, providing almost maintenance-free technology [45]. However, repeated calibration procedures built inside the camera can affect the stability of temperature measurements [46].

The infrared wavelength, recorded by the camera, will not affect the temperature readings as long as the algorithm of calculation temperature from emitted radiation is correct. However, systems equipped with sensors sensitive in different bands of the infrared spectrum are capable to determine the emissivity of objects [47].

32.2.6.2 Temperature Reference

Earlier reports stipulate the requirement for a separate thermal reference source for calibration checks on the camera [9,48,49]. Many systems now include an internal reference temperature, with manufacturers claiming that external checks are not required. Unless frequent servicing is obtained, it is still advisable to use an external source, if only to check for drift in the temperature sensitivity of the camera. An external reference, which may be purchased or constructed, can be left switched on throughout the day. This allows the operator to make checks on the camera, and in particular provides a check on the hardware and software employed for processing. These constant temperature source checks may be the only satisfactory way of proving the reliability of temperature measurements made from the thermogram [48]. Linearity of temperature measurements which may be questionable in focal plane array equipment, can be checked with two ore more external temperature references. New low-cost reference sources, based on the triple point of particular chemicals, are currently under construction in the United Kingdom [44].

32.2.6.3 Mounting the Imager

A camera stand which provides vertical height adjustment is very important for medical thermography. Photographic tripod stands are inconvenient for frequent adjustment and often result in tilting the camera at an undefined angle to the patient. This is difficult to reproduce, and unless the patient is positioned so that the surface scanned is aligned at 90° to the camera lens, distortion of the image is unavoidable. Undefined angles of the camera view affect the precision of measurements.

In the case of temperature measurements from a curved surface, the angle between the radiating object and the capturing device may be the critical source of false measurements [50–52]. At an angle of view beyond 30° small losses of capturing the full band of radiation start to occur, at an angel of 60° the loss of information becomes critical and is followed by false temperature readings. The determination of the temperature of the same forefoot in different views shows clearly that consideration of the angel of the viewing is a significant task [53]. Unless corrected, thermal images of evenly curved objects lack accuracy of temperature measurements [54].

Studio camera stands are ideal, they provide vertical height adjustment with counterbalance weight compensation. It should be noted that the type of lens used on the camera will affect the working distance and the field of view, a wide-angle lens reduces distance between the camera and the subject in many cases, but may also increase peripheral distortion of the image [55].

32.2.6.4 Camera Initialization

Start up time with modern cameras are claimed to be very short, minutes or seconds. However, the speed with which the image becomes visible is not an indication of image stability. Checks on calibration will usually show that a much longer period from 10 min to several hours with an uncooled system are needed to achieve stable conditions for temperature readings from infrared images [5,46].

32.2.7 Patient Position and Image Capture

Standardized positions of the body for image capture and clearly defined fields of view can reduce systematic errors and increases both accuracy and precision of temperature readings from thermal images recorded in such a manner. In radiography, standardized positions of the body for image capture have been included in the protocol for quality assurance for a long time. Although thermal imaging does not provide much anatomical information compared with other imaging techniques, variation of body positions and the related fields of view affects the precision of temperature readings from thermograms. However, the intra- and inter-rater repeatability of temperature values from the same thermal image was found to be excellent [56].

32.2.7.1 Location for Thermal Imaging

The size of investigation room does not influence the quality of temperature measurements from thermal images, unless the least distance in one direction is not shorter than the distance between the camera and an object of 1.2 m height [57]. Such a condition will result in thermal images out of focus. Other important features of the examination room are thermal insulation and prevention of any direct or reflected infrared radiation sources. Following this proposal will result in an increase of accuracy and precision of measurements.

32.2.7.2 Ambient Temperature Control

This is a primary requirement for most clinical applications of thermal imaging. A range of temperatures from 18 to 25°C should be attainable and held for at least 1 h to better than 1°C. Owing to the nature of human thermoregulation, stability of the room temperature is a critical feature. It has been shown that subjects acclimatized for 40–60 min at a room temperature of 22°C showed differences in surface temperature at various measuring sites of the face after lowering the ambient temperature by 2°C [58]. Whereas the nose cooled on average by 4°C, the forehead and the meatus decreased the surface temperature by only 0.4–0.45%. Similar changes may occur at other acral sites such as tips of fingers or toes, as both regions are highly involved in heat exchange for temperature regulation.

At lower temperatures, the subject is likely to shiver, and over 25°C room temperature will cause sweating, at least in most European countries. Variations may be expected in colder or warmer climates, in the latter case, room temperatures may need to be 1°C to 2°C higher [59].

Additional techniques for cooling particular regions of the body have been developed [60,61]. Immersion of the hands in water at various tempeatures is a common challenge for the assessment of vasospastic disease [21].

Heat generated in the investigation room affects the room temperature. Possible heat sources are not only electronic equipment such as the scanner and its computer, but also human bodies. For this reason the air-conditioning unit should be capable of compensating for the maximum number of patients and staff likely to be in the room at any one time. These effects will be greater in a small room of 2 × 3 m or less.

Air convection is a very effective method of skin cooling and related to the wind speed. Therefore, air-conditioning equipment should be located so that direct draughts are not directed at the patient, and that overall air speed is kept as low as possible. A suspended perforated ceiling with ducts diffusing the air distribution evenly over the room is ideal [62].

A cubicle or cubicles within the temperature-controlled area is essential. These should provide privacy for disrobing and a suitable area for resting through the acclimatization period.

32.2.7.3 Preimaging Equilibration

On arrival at the department, the patient should be informed of the examination procedure, instructed to remove appropriate clothing and jewellery, and asked to sit or rest in the preparation cubicle for a fixed time. The time required to achieve adequate stability in blood pressure and skin temperature is generally considered to be 15 min, with 10 min as a minimum [63–65]. After 30 min cooling, oscillations of the skin temperature can be detected, in different regions of the body with different amplitudes resulting in a temperature asymmetry between left and right sides [64].

Contact of body parts with the environment or with other body parts alters the surface temperature because of the heat transfer by conduction. Therefore, during the preparation the patient must avoid folding or crossing arms and legs, or placing bare feet on a cold surface. If the lower extremities are to be examined, a stool or leg rest should be provided to avoid direct contact with the floor [66]. If these requirements are not met, poor precision of measurements may result.

32.2.7.4 Positions for Imaging

As in anatomical imaging studies, it is preferable to standardize on a series of standard views for each body region. The EAT Locomotor Diseases Group recommendations include a triangular marker system to indicate anterior, posterior, lateral, and angled views [2,67]. However, reproduction of positions for angled views may be difficult, even when aids such as rotating platforms are used [68].

Modern image processing software provides comment boxes that can be used to encode the angle of view which will be stored with the image [69]. It should be noted that the position of the patient for scanning and in preparation must be constant. Standing, sitting, or lying down affect the surface area of the body exposed to the ambient, therefore an image recorded with the patient in a sitting position may not be comparable with one recorded on a separate occasion in a standing position. In addition, blood flow against the influence of gravity contributes to the skin temperature of fingers in various limb positions [70].

32.2.7.5 Field of View

Image size is dependent on the distance between the camera and the patient and the focal length of the infrared camera lens. The lens is generally fixed on most medical systems, so it is a good practice to maintain a constant distance from the patient for each view, in order to acquire a reproducible field of view for the image. If in different thermograms different fields of the same subject are compared, the variable resolution can lead to false temperature readings [71]. However, maintaining the same distance between object and camera, cannot compensate for individual body dimensions, for example, big subjects will have big knees and therefore maintaining the same distance as for a tiny subjects knee is not applicable.

To overcome this problem, the field of view has been defined in the standard protocol at the University of Glamorgan in a twofold way, that is, body position and alignment of anatomical landmarks to the edge of the image [23]. These definitions enabled us to investigate the reproducibility of body views using the distance in pixels between anatomical landmarks and the outline of the infrared images [24–72].

Figure 32.1 gives examples of the views that have been investigated for the reproducibility of body positions. Table 32.2 shows the mean value, standard deviation (SD), and 95% confidence interval (CI) of the variation of body views of the upper and the lower part of the human body. Variations in views of the lower part of the body were bigger than in views of the upper part. The highest degree of variation was found in the view "Both Ankles Anterior," but the smallest variation in the view "Face."

32.2.8 Information Protocols and Resources

Human skin temperature is the product of heat dissipated from the vessels and organs within the body, and the effect of the environmental factors on heat loss or gain. There are a number of further influences that are controllable, such as cosmetics [29], alcohol intake [73–75], and smoking [76–78]. In general terms, the patient attending examination should be advised to avoid all topical applications such as ointments and cosmetics on the day of examination to all the relevant areas of the body [31,47,79,80]. Large meals and above-average intake of tea or coffee should also be excluded, although studies supporting this recommendation are hard to find and the results are not conclusive [81,82].

Patients should be asked to avoid tight-fitting clothing, and to keep physical exertion to a minimum. This particularly applies to methods of physiotherapy such as electrotherapy [83–85], ultrasound [86,87], heat treatment [88–90], cryotherapy [91–94], massage [95–97], and hydrotherapy [31,32,98,99], because thermal effects from such treatment can last for 4–6 h under certain conditions. Heat production by muscular exercise is a well-documented phenomenon [65,100–103].

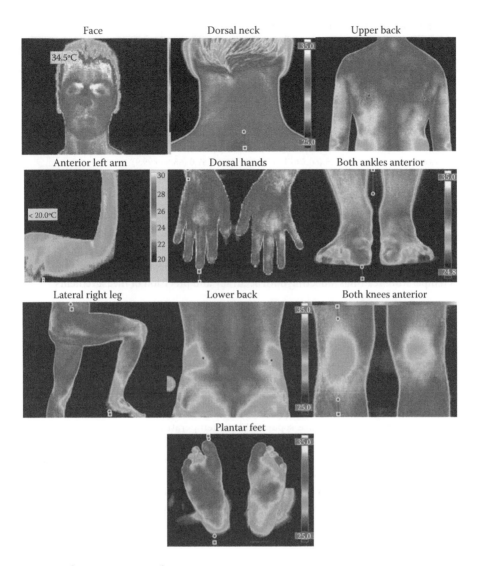

FIGURE 32.1 Body views investigated.

Drug treatment can also affect the skin temperature. This phenomenon was used to evaluate the therapeutic effects of medicaments [6]. Drugs affecting the cardiovascular system must be reported to the thermographer, so that the correct interpretation of thermal images will be given [104–107].

Omitting just one of the aforementioned conditions will result in reduced precision of temperature measurements.

32.2.9 Image Processing

Every image or block of images must carry the indication of temperature range, with color code/temperature scale. The color scale itself should be standardized. Industrial software frequently provides a gray-scale picture and one or more color scales. However, modern image-processing software permits to squeeze the color scale in already-recorded images to increase the image contrast. Such a procedure will affect the temperature readings from thermal images as temperatures outside the compressed

TABLE 32.2 Variation of Positions of All the Investigated Views

View	Upper Edge (Pixel) Mean ± SD (95% CI)	Lower Edge (Pixel) Mean ± SD (95% CI)	Left Side Edge (Pixel) Mean ± SD (95% CI)
Face	0.5 ± 5.3 (−2.2 to 1.9)	4.0 ± 10.9 (−0.03 to 8.2)	
Dorsal neck	−8.4 ± 36.4 (−18.3 to 1.6)	122.6 ± 146.6 (82.6 to 162.6)	
Upper back	4.5 ± 9.9 (0.8 to 8.2)	28.1 ± 22.0 (19.9 to 36.4)	
Anterior left arm	22.4 ± 33.0 (8.7 to 6.0)	15.8 ± 15.4 (9.5 to 22.2)	12.5 ± 16.0 (5.9 to 19.1)
Dorsal hands	41.8 ± 17.8 (35.5 to 48.2)	33.2 ± 22.3 (25.3 to 41.5)	
Both knees anterior	80.7 ± 47.3 (60.7 to 100.7)	84.3 ± 37.0 (68.6 to 99.9)	
Lateral right leg	16.7 ± 21.0 (5.9 to 27.5)	17.2 ± 15.8 (9.0 to 25.3)	
Lower back	17.1 ± 4.2 (8.6 to 25.6)	16.3 ± 4.6 (16.3 to 34.9)	
Both ankles anterior	158.8 ± 12.2 (133.6 to 184.1)	54.9 ± 9.1 (36.1 to 37.8)	
Plantar feet	31.0 ± 24.1 (23.2 to 38.7)	25.7 ± 23.1 (18.3 to 33.1)	

temperature scale will not be included in the statistics of selected regions of interest. This will result in erroneous temperature readings, affecting both accuracy and precision of measurements.

32.2.10 Image Analysis

Almost all systems now use image-processing techniques and provide basic quantitation of the image [108–110]. In some cases, this may be operated from a chip within the camera, or may be carried out through an on-line or off-line computer. For older equipment such as the AGA 680 series several hardware adaptations have been reported to achieve quantitation of the thermograms [111–113].

It has to be emphasized that false color coding of infrared images does not provide means for temperature measurement. If colors are separated by a temperature distance of 1°C, the temperature difference between two points situated in adjacent colors may be between 0.1°C and 1.9°C. It is obvious that false-colored images provide at its best an estimation of temperature, but not a measurement. The same is true for liquid crystal thermograms.

Nowadays, temperature measurements in thermal images are based on the definition of regions of interest (ROI). However, standards for shape, size, and placement of these regions are not available or incomplete. Although a close correlation exists for ROI of different size in the same region [114], the precision of measurement is affected when ROIS of different size and location are used for repeated measurements.

The Glamorgan protocol [23] is the very first attempt to create a complete standard for the definition of regions of interest in thermal images based on anatomical limits. Furthermore, in the view "both knee anterior" the shape with the highest reproducibility was investigated. During one of the Medical Infrared Training Courses at the University of Glamorgan, three newly trained investigators defined on the same thermal image of both anterior knees twice the region of interest in the shape of a box, an ellipsoid or as an hour-glass shape. Similar to the result of a pilot study that compared these shapes for repeatability, the highest reliability was found for temperature readings from the hour-glass shape, followed by readings from ellipsoids and boxes [53]. The repeatability of the regions on the view "Left Anterior Arm," "Both Ankles Anterior," "Dorsal Feet," and "Plantar Feet" were also investigated and resulted in reliability coefficients between 0.7 (right ankle) and 0.93 (forearm). The intraclass correlation coefficients ranged between 0.48 (upper arm) and 0.87 (forearm). Applying the Glamorgan protocol consequently will result in precise temperature measurements from thermal images.

32.2.11 Image Exchange

Most of the modern infrared systems store the recorded thermal images in an own image format, which may not be compatible with formats of thermal images from other manufacturers. However, most of these images can be transformed into established image formats such as TIF, JPEG, GIF, and others. As a thermal image is the pictographic representation of a temperature map, the sole image is not enough unless the related temperature information is not provided. Consequently, temperature measurements from standard computer images derived from thermograms is not possible.

Providing both temperature scale and a scale of gray shades, allows the exchange of thermal images over long distance and between different, but compatible image-processing software [115]. The gray scale must be derived from the original gray shade thermal image. If it has been transformed from a false color image, the resulted black-and-white thermogram may not be representative for the original gray scale gradient as the gray scale of individual colors may deviate from the particular gray shade of the image. This can then result in false temperature readings.

32.2.12 Image Presentation

Image presentation does not influence the result of measurements from thermal images. However, if thermograms are read by eyes, their appearance will affect the credibility of the information in thermal images. This is for instance the case, when thermal images are use as evidence in legal trials [116].

It was stated that for forensic acceptability of thermography standardization and repeatability of the technique are very important features [117]. This supports the necessity of quantitative evaluation of thermal images and standards strictly applied to the technique of infrared imaging will finally result in high accuracy and precision of this method of temperature measurement. At that stage it can be recommended as responsive outcome measure for clinical trials in rheumatology [6,8], angiopathies [107,118], neuromuscular disorders [119], surgery [120], and paediatrics [121].

References

1. Aarts, N.J.M. et al. Thermographic Terminology. *Acta Thermograp.*, 3(Suppl. 2), 1, 1978.
2. Engel, J.M. et al. Thermography in locomotor diseases—Recommended procedure. *Eur. J. Rheum. Inflamm.*, 2, 299, 1979.
3. Collins, A.J. et al. Quantitation of thermography in arthritis using multi-isothermal analysis. *Ann. Rheum. Dis.*, 33, 113, 1974.
4. Engel, J.-M. Quantitative Thermographie des Kniegelenks. *Z. Rheumatol.*, 37, 242, 1978.
5. Ring, E.F.J. Standardisation of thermal imaging in medicine: Physical and environmental factors, in *Thermal Assessment of Breast Health*, Gautherie, M., Albert, E., and Keith, L., Eds., MTP Press Ltd, Lancaster/Boston/The Hague, 1983, p. 29.
6. Ring, E.F.J., Engel, J.M., and Page-Thomas, D.P. Thermologic methods in clinical pharmacology—Skin temperature measurement in drug trials. *Int. J. Clin. Pharm. Ther. Tox.*, 22, 20, 1984.
7. Engel, J.-M. and Saier, U. *Thermographische Standarduntersuchungen in der Rheumatologie und Richtlinien zu deren Befundung.* Luitpold, München, 1984.
8. Collins, A.J. Anti-inflammatory drug assessment by the thermographic index. *Acta Thermograp*, 1, 73, 1976.
9. Pochaczevsky, R. et al. Technical guidelines, 2nd ed. *Thermology*, 2, 108, 1986.
10. Uematsu, S. Symmetry of skin temperatures comparing one side of the body to the other. *Thermology*, 1, 4, 1985.
11. Goodman, P.H. et al. Normal temperature asymmetry of the back and extremities by computer-assisted infrared imaging. *Thermology*, 1, 195, 1986.
12. Bliss, P. et al. Investigation of nerve root irritation by infrared thermography, in *Back Pain—Methods for Clinical Investigation and Assessment*, Hukins, D.W.L. and Mulholland, R.C., Eds., University Press, Manchester, 1986, p. 63.
13. Atsumi, K. High technology applications of medical thermography in Japan. *Thermology*, 1, 79–80, 1985.

14. Fujimasa, I. et al. A new computer image processing system for the analysis of neuromuscular thermograms: A feasibility study. *Thermology*, 1, 221, 1986.

15. Fujimasa, I. A proposal for thermographic imaging diagnostic procedures for temperature related physiologic function analysis. *Biomed. Thermol.*, 11, 269, 1991.

16. Lee, W.-Y. et al. (Eds.) *Practical Manual of Clinical Thermology*, Med Lang, 2004, ISBN 89-954013-04.

17. Ring, E.F.J. A thermographic index for the assessment of ischemia. *Acta Thermograp.*, 5, 35, 1980.

18. Aarts, N.P. et al. Raynaud's phenomenon: Assessment by thermography. *Thermology*, 3, 69, 1988.

19. Acciarri, L., Carnevale, F., and Della Selva, A. Thermography in the hand angiopathy from vibrating tools. *Acta Thermograp.*, 1, 18, 1976.

20. Ring, E.F. and Bacon, P.A. Quantitative thermographic assessment of inositol nicotinate therapy in Raynaud's phenomena. *J. Int. Med. Res.*, 5, 217, 1977.

21. Ring, E.F.J. Cold stress test for the hands, in *The Thermal Image in Medicine and Biology*, Ammer, K. and Ring, E.F.J., Eds., Uhlen-Verlag, Wien, 1995, p. 237.

22. Clark, R.P. and de Calcina-Goff, M. Guidelines for standardisation in medical thermography draft international standard proposals. *Thermol. Österr.*, 7, 47, 1997.

23. Website address, Atlas of Normals, www.medimaging.org.

24. Ammer, K. et al. Rationale for standardised capture and analysis of infrared thermal images, in *Proceedings Part II, EMBEC'02 2. European Medical and Biological Engineering Conference*, Hutten, H. and Krösel, P., Eds. IFMBE, Graz, 2002, p. 1608.

25. Ring, E.F.J. et al. Errors and artefacts in thermal imaging, in *Proceedings Part II, EMBEC'02 2. European Medical and Biological Engineering Conference*, Hutten, H. and Krösel, P., Eds., IFMBE, Graz, 2002, p. 1620.

26. Anbar, M. Recent technological developments in thermology and their impact on clinical applications. *Biomed. Thermol.*, 10, 270, 1990.

27. Hardy, J.D. The radiation of heat from the human body. III. The human skin as a black body radiator. *J. Clin. Invest.*, 13, 615, 1934.

28. Togawa, T. and Saito, H. Non-contact imaging of thermal properties of the skin. *Physiol. Meas.*, 15, 291, 1994.

29. Engel, J.-M. Physical and physiological influence of medical ointments of infrared thermography, in *Recent Advances in Medical Thermology*, Ring, E.F.J. and Phillips, B., Eds., Plenum Press, New York, 1984, p. 177.

30. Hejazi, S. and Anbar, M. Effects of topical skin treatment and of ambient light in infrared thermal images. *Biomed. Thermol.*, 12, 300, 1992.

31. Ammer, K. The influence of antirheumatic creams and ointments on the infrared emission of the skin, in *Abstracts of the 10th International Conference on Thermogrammetry and Thermal Engineering in Budapest 18–20, June 1997*, Benkö, I., Balogh, I., Kovacsics, I., Lovak, I., Eds., MATE, Budapest, 1997, p. 177.

32. Ammer, K. Einfluss von Badezusätzen auf die Wärmeabstrahlung der Haut. *ThermoMed*, 10, 71, 1994.

33. Ammer, K. The influence of bathing on the infrared emission of the skin, in *Abstracts of the 9th International Conference on Thermogrammetry and Thermal Engineering in Budapest 14–16, June 1995*, Benkö, I., Lovak, I., and Kovacsics, I., Eds., MATE, Budapest, 1995, p. 115.

34. Ammer, K. Thermographie in lymphedema, in *Advanced Techniques and Clinical Application in Biomedical Thermologie*, Mabuchi, K., Mizushina, S., and Harrison, B., Eds., Harwood Academic Publishers, Chur/Schweiz, 1994, p. 213.

35. Heath, A.M. et al. A comparison of surface and rectal temperatures between sheared and non-sheared alpacas (*Lama pacos*). *Small Rumin. Res.*, 39, 19, 2001.

36. Purohit, R.C. et al. Thermographic evaluation of animal skin surface temperature with and without haircoat. *Thermol. Int.*, 11, 83, 2001.

37. Damm, F., Döring, G., and Hildebrandt, G. Untersuchungen über den Tagesgang von Hautdurchblutung und Hauttemperatur unter besonderer Berücksichtigung der physikalischen Temperaturregulation. *Z. Physik. Med. Rehabil.*, 15, 1, 1974.

38. Reinberg, A. Circadian changes in the temperature of human beings. *Bibl. Radiol.*, 6, 128, 1975.
39. Schmidt, K.-L., Mäurer, R., and Rusch, D. Zur Wirkung örtlicher Wärme und Kälteanwendungen auf die Hauttemperatur am Kniegelenk. *Z. Rheumatol.*, 38, 213, 1979.
40. Kanamori, T. et al. Circadian rhythm of body temperature. *Biomed. Thermol.*, 11, 292, 1991.
41. Friedrich, K.H. Assessment criteria for infrared thermography systems. *Acta Thermograp.*, 5, 68, 1980.
42. Alderson, J.K.A. and Ring, E.F.J. "Sprite" high resolution thermal imaging system. *Thermology*, 1, 110, 1985.
43. Dibley, D.A.G. Opto-mechanical systems for thermal imaging, in *The Thermal Image in Medicine and Biology*, Ammer, K. and Ring, E.F.J., Eds., Uhlen-Verlag, Wien, 1995, p. 33.
44. Plassmann, P. Advances in image processing for thermology, *Presented at Int. Cong. of Thermology*, Seoul, June 5–6, 2004, p. 3.
45. Kutas, M. Staring focal plane array for medical thermal imaging, in *The Thermal Image in Medicine and Biology*, Ammer, K. and Ring, E.F.J., Eds., Uhlen-Verlag, Wien, 1995, p. 40.
46. Ring, E.F.J., Minchinton, M., and Elvins, D.M. A focal plane array system for clinical infrared imaging. *IEEE/EMBS Proceedings*, Atlanta 1999, p. 1120.
47. Hejazi, S. and Spangler, R.A. A multi-wavelength thermal imaging system, in *Proceedings of the 11th Annual International Conference IEEE Engineering in Medicine and Biology Society*, II, 1989, p. 1153.
48. Ring, E.F.J. Quality control in infrared thermography, in *Recent Advances in Medical Thermology*, Ring, E.F.J. and Phillips, B., Eds., Plenum Press, New York, 1984, p. 185.
49. Clark, R.P. et al. Thermography and pedobarography in the assessment of tissue damage in neuropathicand atherosclerotic feet. *Thermology*, 3, 15, 1988.
50. Clark, J.A. Effects of surface emissivity and viewing angle errors in thermography. *Acta Thermograp.*, 1, 138, 1976.
51. Steketee, J. Physical aspects of infrared thermography, in *Recent Advances in Medical Thermology*, Ring, E.F.J. and Phillips, B., Eds., Plenum Press, New York, 1984, p. 167.
52. Wiecek, B., Jung, A., and Zuber, J. Emissivity-Bottleneck and challenge for thermography. *Thermol. Int.*, 10, 15, 2000.
53. Ammer K. Need for standardisation of measurements, in *Thermal Imaging in Thermography and Lasers in Medicine*, Wiecek, B., Ed., Akademickie Centrum Graficzno-Marketigowe Lodart S.A, Lodz, 2003, p. 13.
54. Anbar, M. Potential artifacts in infrared thermographic measurements. *Thermology*, 3, 273, 1991.
55. Ring, E.F.J. and Dicks, J.M. Spatial resolution of new thermal imaging systems, *Thermol. Int.*, 9, 7, 1999.
56. Melnizky, P., Schartelmüller, T., and Ammer, K. Prüfung der intra-und interindividuellen Verlässlichkeit der Auswertung von Infrarot-Thermogrammen. *Eur. J. Thermol.*, 7, 224, 1997.
57. Ring, E.F.J. and Ammer, K. The technique of thermal imaging in medicine. *Thermol. Int.*, 10, 7, 2000.
58. Khallaf, A. et al. Thermographic study of heat loss from the face. *Thermol. Österr.*, 4, 49, 1994.
59. Ishigaki, T. et al. Forehead–back thermal ratio for the interpretation of infrared imaging of spinal cord lesions and other neurological disorders. *Thermology*, 3, 101, 1989.
60. Schuber, T.R. et al. Directed dynamic cooling, a methodic contribution in telethermography. *Acta Thermograp.*, 1, 94, 1977.
61. Di Carlo, A. Thermography in patients with systemic sclerosis. *Thermol. Österr.*, 4, 18, 1994.
62. Love, T.J. Heat transfer considerations in the design of a thermology clinic. *Thermology*, 1, 88, 1985.
63. Ring, E.F.J. Computerized thermography for osteo-articular diseases. *Acta Thermograp.*, 1, 166, 1976.
64. Roberts, D.L. and Goodman, P.H. Dynamic thermoregulation of back and upper extremity by computer-aided infrared imaging. *Thermology*, 2, 573, 1987.
65. Mabuchi, K. et al. Development of a data processing system for a high-speed thermographic camera and its use in analyses of dynamic thermal phenomena of the living body, in *The Thermal Image in Medicine and Biology*, Ammer, K. and Ring, E.F.J., Eds., Uhlen-Verlag, Wien, 1995, p. 56.

66. Cena, K. Environmental heat loss, in *Recent Advances in Medical Thermology*, Ring, E.F.J. and Phillips, B., Eds., Plenum Press, New York, 1984, p. 81.

67. Engel, J.-M. Kennzeichnung von Thermogrammen, in *Thermologische Messmethodik*, Engel, J.-M., Flesch, U., and Stüttgen, G., Eds., Notamed, Baden–Baden, 1983, p. 176.

68. Park, J.-Y. Current development of medical infrared imaging technology, *Presented at Int. Cong. of Thermology*, Seoul, June 5–6, 2004, p. 9.

69. Plassmann, P. and Ring, E.F.J. An open system for the acquisition and evaluation of medical thermological images. *Eur. J. Thermol.* 7, 216, 1997.

70. Abramson, D.I. et al. Effect of altering limb position on blood flow, O_2 uptake and skin temperature. *J. Appl. Physiol.*, 17, 191, 1962.

71. Schartelmüller, T. and Ammer, K. Räumliche Auflösung von Infrarotkameras. *Thermol. Österr.*, 5, 28, 1995.

72. Ammer, K. Update in standardization and temperature measurement from thermal images, *Presented at Int. Cong. of Thermology*, Seoul, June 5–6, 2004, p. 7.

73. Mannara, G., Salvatori, G.C., and Pizzuti, G.P. Ethyl alcohol induced skin temperature changes evaluated by thermography. Preliminary results. *Boll. Soc. Ital. Biol. Sper.*, 69, 587, 1993.

74. Melnizky, P. and Ammer, K. Einfluss von Alkohol und Rauchen auf die Hauttemperatur des Gesichts, der Hände und der Kniegelenke. *Thermol. Int.*, 10, 191, 2000.

75. Ammer, K., Melnizky, P., and Rathkolb, O. Skin temperature after intake of sparkling wine, still wine or sparkling water. *Thermol. Int.*, 13, 99, 2003.

76. Gershon-Cohen, J., Borden, A.G., and Hermel, M.B. Thermography of extremities after smoking. *Br. J. Radiol.*, 42, 189, 1969.

77. Usuki, K. et al. Effects of nicotine on peripheral cutaneous blood flow and skin temperature. *J. Dermatol. Sci.*, 16, 173, 1998.

78. Di Carlo, A. and Ippolito, F. Early effects of cigarette smoking in hypertensive and normotensive subjects. An ambulatory blood pressure and thermographic study. *Minerva Cardioangiol.*, 51, 387, 2003.

79. Collins, A.J. et al. Some observations on the pharmacology of "deep-heat," a topical rubifacient. *Ann. Rheum. Dis.*, 43, 411, 1984.

80. Ring, E.F. Cooling effects of Deep Freeze Cold gel applied to the skin, with and without rubbing, to the lumbar region of the back. *Thermol. Int.*, 14, 64, 2004.

81. Federspil, G. et al. Study of diet-induced thermogenesis using telethermography in normal and obese subjects. *Recent Prog. Med.*, 80, 455, 1989.

82. Shlygin, G.K. et al. Radiothermometric research of tissues during the initial reflex period of the specific dynamic action of food. *Med. Radiol. (Mosk)*, 36, 10, 1991.

83. Danz, J. and Callies, R. Infrarothermometrie bei differenzierten Methoden der Niederfrequenztherapie. *Z. Physiother.*, 31, 35, 1979.

84. Rusch, F., Neeck, G., and Schmidt, K.L. Über die Hemmung von Erythemen durch Capsaicin. 3. Objektivierung des Capsaicin-Erythems mittels statischer und dynamischer Thermographie, *Z. Phys. Med. Baln. Med. Klim.*, 17, 18, 1988.

85. Mayr, H., Thür, H., and Ammer, K. Electrical stimulation of the stellate ganglia, in *The Thermal Image in Medicine and Biology*, Ammer, K. and Ring, E.F.J., Eds., Uhlen-Verlag, Wien, 1995, p. 206.

86. Danz, J. and Callies R. Thermometrische Untersuchungen bei unterschiedlichen Ultraschallintensitäten. *Z. Physiother.*, 30, 235, 1978.

87. Demmink, J.H., Helders, P.J., Hobaek, H., and Enwemeka, C. The variation of heating depth with therapeutic ultrasound frequency in physiotherapy. *Ultrasound Med. Biol.*, 29, 113–118, 2003.

88. Rathkolb, O. and Ammer, K. Skin temperature of the fingers after different methods of heating using a wax bath. *Thermol Österr.*, 6, 125, 1996.

89. Ammer, K. and Schartelmüller, T. Hauttemperatur nach der Anwendung von Wärmepackungen und nach Infrarot-A-Bestrahlung. *Thermol. Österr.*, 3, 51, 1993.

90. Goodman, P.H., Foote, J.E., and Smith, R.P. Detection of intentionally produced thermal artifacts by repeated thermographic imaging. *Thermology*, 3, 253, 1991.

91. Dachs, E., Schartelmüller, T., and Ammer, K. Temperatur zur Kryotherapie und Veränderungen der Hauttemperatur am Kniegelenk nach Kaltluftbehandlung. *Thermol. Österr.*, 1, 9, 1991.

92. Rathkolb, O. et al. Hauttemperatur der Lendenregion nach Anwendung von Kältepackungen unterschiedlicher Größe und Applikationsdauer. *Thermol. Österr.*, 1, 15, 1991.

93. Ammer, K. Occurrence of hyperthermia after ice massage. *Thermol. Österr.*, 6, 17, 1996.

94. Cholewka, A. et al. Temperature effects of whole body cryotherapy determined by thermography. *Thermol. Int.*, 14, 57, 2004.

95. Danz, J., Callies, R., and Hrdina, A. Einfluss einer abgestuften Vakuumsaugmassage auf die Hauttemperatur. *Z. Physiother.*, 33, 85, 1981.

96. Eisenschenk, A. and Stoboy, H. Thermographische Kontrolle physikalisch-therapeutischer Methoden. *Krankengymnastik*, 37, 294, 1985.

97. Kainz, A. Quantitative Überprüfung der Massagewirkung mit Hilfe der IR-Thermographie. *Thermol. Österr.*, 3, 79, 1993.

98. Rusch, D. and Kisselbach, G. Comparative thermographic assessment of lower leg baths in medicinal mineral waters (Nauheim Springs), in *Recent Advances in Medical Thermology*, Ring, E.F.J. and Phillips, B., Eds., Plenum Press, New York, 1984, p. 535.

99. Ring, E, F.J., Barker, J.R., and Harrison, R.A. Thermal effects of pool therapy on the lower limbs. *Thermology*, 3, 127, 1989.

100. Konermann, H. and Koob, E. Infrarotthermographische Kontrolle der Effektivität krankengymnastischer Behandlungsmaßnahmen. *Krankengymnastik*, 27, 39, 1975.

101. Smith, B.L., Bandler, M.K., and Goodman, P.H. Dominant forearm hyperthermia: A study of fifteen athletes. *Thermology*, 2, 25, 1986.

102. Melnizky, P., Ammer, K., and Schartelmüller, T. Thermographische Überprüfung der Heilgymnastik bei Patienten mit Peroneusparese. *Thermol. Österr.*, 5, 97, 1995.

103. Ammer, K. Low muscular acitivity of the lower leg in patients with a painful ankle. *Thermol. Österr.*, 5, 103, 1995.

104. Ring, E.F., Porto, L.O., and Bacon, P.A. Quantitative thermal imaging to assess inositol nicotinate treatment for Raynaud's syndrome. *J. Int. Med. Res.*, 9, 393, 1981.

105. Lecerof, H. et al. Acute effects of doxazosin and atenolol on smoking-induced peripheral vasoconstriction in hypertensive habitual smokers. *J. Hypertens.*, 8, S29, 1990.

106. Tham, T.C., Silke, B., and Taylor, S.H. Comparison of central and peripheral haemodynamic effects of dilevalol and atenolol in essential hypertension. *J. Hum. Hypertens.*, 4, S77, 1990.

107. Natsuda, H. et al. Nitroglycerin tape for Raynaud's phenomenon of rheumatic disease patients—An evaluation of skin temperature by thermography. *Ryumachi*, 34, 849, 1994.

108. Engel, J.M. Thermotom- ein Softwarepaket für die thermographische Bildanalyse in der Rheumatologie, in *Thermologische Messmethodik*, Engel, J.-M., Flesch, U., and Stüttgen, G., Eds., Notamed, Baden–Baden, 1983, p. 110.

109. Bösiger, P. and Scaroni, F. Mikroprozessor-unterstütztes Thermographie-System zur quantitativewn on-line Analyse von statischen und dynamischen Thermogrammen, in *Thermologische Messmethodik*, Engel, J.-M., Flesch, U., and Stüttgen, G., Eds., Notamed, Baden–Baden, 1983, p. 125.

110. Brandes, P. PIC-Win-Iris Bildverarbeitungssoftware. *Thermol. Österr.*, 4, 33, 1994.

111. Ring, E.F.J. Quantitative thermography in arthritis using the AGA integrator. *Acta thermograp.*, 2, 172, 1977.

112. Parr, G. et al. Microcomputer standardization of the AGA 680 M system, in *Recent Advances in Medical Thermology*, Ring, E.F.J. and Phillips, B., Eds., Plenum Press, New York, 1984, pp. 211–214.

113. Van Hamme, H., De Geest, G., and Cornelis, J. An acquisition and scan conversion unit for the AGA THV680 medical infrared camera. *Thermology*, 3, 205, 1990.

114. Mayr, H. Korrelation durchschnittlicher und maximaler Temperatur am Kniegelenk bei Auswertung unterschiedlicher Messareale. *Thermol. Österr.*, 5, 89, 1995.

115. Plassmann, P. On-line Communication for Thermography in Europe, *Presented at Int. Cong. of Thermology*, Seoul, June 5–6, 2004, p. 50.

116. Ring, E.F.J. Thermal imaging in medico-legal claims. *Thermol. Int.*, 10, 97, 2000.

117. Sella, G.E. Forensic criteria of acceptability of thermography. *Eur. J. Thermol.*, 7, 205, 1997.

118. Hirschl, M. et al. Double-blind, randomised, placebo controlled low level laser therapy study in patients with primary Raynaud's phenomenon. *Vasa*, 31, 91, 2002.

119. Schartelmüller, T., Melnizky, P., and Engelbert, B. Infrarotthermographie zur Evaluierung des Erfolges physikalischer Therapie bei Patienten mit klinischem Verdacht auf Thoracic Outlet Syndrome. *Thermol. Int.*, 9, 20, 1999.

120. Kim, Y.S. and Cho, Y.E. Pre- and postoperative thermographic imaging in lumbar disc herniations, in *The Thermal Image in Medicine and Biology*, Ammer, K. and Ring, E.F.J., Eds., Uhlen-Verlag, Wien, 1995, p. 168.

121. Siniewicz, K. et al. Thermal imaging before and after physial exercises in children with orthostatic disorders of the cardiovascular system. *Thermol. Int.*, 12, 139, 2002.

33

Storage and Retrieval of Medical Infrared Images

Gerald Schaefer
Loughborough University

33.1 Introduction

Advances in camera technologies and reduced equipment costs have led to an increased interest in the application of thermal imaging in the medical fields [7]. Medical infrared images are typically recorded and stored in digital form, and computerized image processing techniques have been used in acquiring and evaluating medical thermal images [13,26] and proved to be important tools for clinical diagnostics. Yet, these tools rely on the digital images to be in a certain format. Unfortunately, manufacturers of medical infrared cameras have their own proprietary image formats with little or no possibility of data interchange between suppliers. There is therefore a need to develop a standardized format for storing and processing thermograms. In the first part of this chapter, we will show that the DICOM (Digital Imaging and Communications in Medicine) medical imaging standard [12] can be adopted for this purpose.

Obviously, the more images are captured the more attention has to be put on necessary resources such as storage space and bandwidth. For example, the images for one person captured according to Reference 14, which suggests 27 standard views, requires, assuming a 12-bit thermal camera with 680×512 resolution, more than 13 megabytes of disk space. The application of compression methods is therefore often a necessary step to reduce these storage requirements. For images, there are two kinds of compression methods: lossless compression which preserves all of the original information and lossy compression which sacrifices some of the visual quality to gain in terms of compression rate. While approaches for lossy compression of medical infrared images have been presented [19,20], clinicians often prefer lossless algorithms to ensure no information is lost. Also, in some countries it is forbidden by law to lossy compress images used for medical diagnosis.

In the second part of the chapter, we evaluate several "standard" lossless image compression algorithms for compressing medical infrared images. Lossless JPEG [10], JPEG-LS [4], JPEG2000 [5], PNG [28], and CALIC [27] are compared on an image set comprising more than 380 thermal images organized into 20 groups according to Reference 14.

Thermograms are typically stored for archival and legal purposes only and are not being retrieved again once a successful diagnosis has been made. In the final part of this chapter, we show how these images of past cases can be used to aid in the diagnosis of new ones. Our approach is based on the concept of content-based image retrieval (CBIR) which has been an active research area in image processing for many years [23]. The principal aim is to retrieve digital images based not on textual annotations but on features derived directly from image data. These features are then stored alongside the image and serve as an index. Retrieval is often performed in a query by example (QBE) fashion where a query image is provided by the user. The retrieval system is then searching through all images in order to find those with the most similar indices which are returned as the candidates most alike to the query.

We consider the application of content-based image retrieval as a generic approach for the analysis and interpretation of medical infrared images. CBIR allows the retrieval of visually similar and hence usually relevant images based on a predefined similarity measure between image features derived directly from the image data. In terms of medical infrared imaging, images that are visually similar to a given sample will also be likely to have medical relevance. These known cases together with their medical reports should then provide a valuable asset for diagnostic purposes.

While the first introduced approach is based on the extraction of image features from the raw (uncompressed) images, we furthermore show that content-based retrieval of medical infrared images can also be performed directly on compressed image data. In particular, we show how retrieval based on wavelet image descriptors allows the retrieval of similar thermograms.

Finally, we also introduce an approach that allows the browsing of whole databases of thermograms, again based on visual features. Utilizing a feature dimensionality reduction method, all thermal images are shown on the computer screen so that images which are close by are also visually similar to provide an alternative method of navigating through these image collections as opposed to standard methods which are restricted to searching by patient name or similar attributes.

The remainder of the chapter is organized as follows. Section 33.2 stresses the need for a standard format for thermograms. Section 33.3 evaluates the performance of several compression algorithms on thermograms, both in terms of compression speed and in terms of compression ratio. Section 33.4 details our methods of applying content-based image retrieval techniques to the domain of thermal medical images and introduces a system of visually navigating collections of thermograms. Section 33.5 concludes the chapter.

33.2 Toward a Standardized Thermogram Format

Various camera suppliers are competing for their share in the market of medical infrared imaging. Unfortunately, each of these also stores the captured images in its own, proprietary file format. This fact makes it hard to impossible to share medical infrared images between users of different systems, despite the urgent need for this facility in the light of increase of telemedicine and other emerging technologies. Thermal imaging packages such as CTHERM [13] allow the capture from various types of cameras and store images in a simple common format; yet this approach is only a step toward a suitable solution.

What is needed is a recognized standard for storage and interchange of thermograms. Clearly this standard needs to be supported by suppliers of both cameras and software packages. Most importantly, such a standard must support the preservation of the original radiometric information. What this also means is that it must support a variety of spatial and radiometric resolutions both of which vary from camera model to camera model. Apart from the storage of the actual image information in digital form, a useful standard format will provide various other properties. Patient identification and the addition of patient information as well as attaching information on the clinicians and treatments should be supported as well as the addition of other information items deemed useful for interpreting the thermogram.

Looking at other medical fields that deal with storage and exchange of digital images, DICOM [12] has emerged as the major standard and is in common use for many imaging modalities such as MRI or CT scans. We therefore want to briefly investigate whether the DICOM standard can be adopted for storing thermograms.

DICOM supports arbitrary image resolutions and various bit depths of data. Saving the radiological information of thermograms accurately will therefore be ensured, though the exact format will need to be specified. Storage of patient and medical information is also provided by default as is the possibility to provide annotations and extra information in the form of tags. This feature can be employed to save information on which part of the anatomy is captured and the storage of region of interest information as suggested in Reference 14 and can hence be used to integrate the efforts of standardization there into a common file format. Furthermore, DICOM supports not only storage of the original image data but also compression thereof which will be investigated in further detail in the next section of this chapter. Overall, DICOM supports the main requirements of a standard for storage and communication of medical thermograms while its application and enforcement will have a major positive impact on the community of users providing them with an effective and efficient way of sharing and interpreting medical infrared images.

33.3 Compression of Thermograms

In this section, we investigate the use of lossless compression algorithms for storing medical infrared images in a more compact manner.

33.3.1 Image Dataset

In order to provide a useful comparison of the performance of compression algorithms, one requires an image set that reflects the diversity of types of thermal images that are typically captured. We have therefore compiled such a data set which follows the standard views introduced in Reference 14. There, 27 standard poses are defined which are designed to capture every view possible necessary for composing an atlas based on infrared imaging. Of these 27 views, we omitted 7 poses which are very similar to some of the other ones due to symmetry reasons (either left/right or anterior/dorsal). Of each of the remaining 20 image groups about 20 images were collected using CTHERM [13]; details regarding each group are given in Table 33.1. With a few exceptions, all images are of size 680×512, and the image bit depth is 7 bits.

33.3.2 Compression Algorithms

We evaluated five popular compression algorithms of which four have also been adapted as international imaging standards. Below, we briefly characterize the algorithms and implementations we used:

- Lossless JPEG—former JPEG committee standard for lossless image compression [10]. The standard describes predictive image compression algorithm with Huffman or arithmetic entropy coder. JPEG is supported by the DICOM standard. The results are reported for the predictor function SV 2, which resulted in the best average compression ratio for the dataset, and Huffman coding.
- JPEG-LS—standard of the JPEG committee for lossless and near-lossless compression of still images [4]. The standard, which is based on the LOCO-I algorithm [25], describes low-complexity predictive image compression algorithm with entropy coding using modified Golomb–Rice family. JPEG-LS is supported by the DICOM standard.
- JPEG2000—a recent JPEG committee standard describing an algorithm based on wavelet transform image decomposition and arithmetic coding [5]. Apart from lossy and lossless compressing and decompressing of whole images, it delivers many interesting features (progressive transmission, region of interest coding, etc.) [1]. JPEG2000 is supported by the DICOM standard.

TABLE 33.1   Image Groups in the Dataset

Abbreviation	Description	Number of Images
ABD	Abdomen, anterior view	19
BAA	Both ankles, anterior view	21
BHD	Both hands, dorsal view	16
BKA	Both knees, anterior view	18
CA	Chest, anterior view	22
DF	Dorsal feet	15
FA	Face	23
LAD	Left arm, dorsal view	15
LB	Lower back, dorsal view	17
LLA	Lower legs, anterior view	20
LLD	Lower legs, dorsal view	19
LRL	Right leg, lateral view	19
ND	Neck, dorsal view	23
PF	Plantar feet	23
TA	Thighs, anterior view	20
TBA	Total body, anterior view	19
TBD	Total body, dorsal view	16
TBR	Total body, right view	17
TD	Thighs, dorsal view	18
UB	Upper body, dorsal view	22
Total		382

- PNG—standard of the WWW Consortium for lossless image compression [28]. PNG is a predictive image compression algorithm using the LZ77 [29] algorithm and Huffman coding. The results are reported for the *sub* predictor function (filter), which resulted in the best average compression ratio for the dataset.
- CALIC—a relatively complex predictive image compression algorithm using arithmetic entropy coder, which because of its usually high compression ratios, is commonly used as a reference for other image compression algorithms [27].

33.3.3 Compression Results

Before detailing the experimental procedure, it should be stressed that, although all evaluated algorithms are lossless, they are only able to preserve the data that is originally presented and which therefore has to be ensured to be radiometrically sound.

Experimental results were obtained on an HP Proliant ML350G3 computer equipped with two Intel Xeon 3.06 GHz (512 kB cache memory) processors, Windows 2003 operating system, and the algorithms were compiled using Intel C++ 8.1 compiler. To minimize effects of the system load and the input–output subsystem, the compressors were run several times. The time of the first run was ignored while the collective time of other runs (executed for at least 1 s, and at least 5 times) was measured and then averaged. The time measured is hence the sum of time spent by the processor in application code and in kernel functions called by the application, as reported by the operating system after application execution.

In Tables 33.2 and 33.3, we show the results obtained by the compression algorithms introduced in Section 33.3.2 for the image dataset described in Section 33.3.1. Results are given in terms of compression speed, expressed in megabytes per second (MB/s), where 1 MB = 2^{20} bytes in Table 33.2 and compression ratio defined as the ratio of the file size of the original image and that of the compressed file, in Table 33.3. The numbers are calculated as an average for all images contained in the group; since not all

TABLE 33.2 Compression Speeds, Given in MB/s

Images	L-JPG	JPEG-LS	JPEG2000	PNG	CALIC
ABD	11.9	17.7	3.8	3.3	4.1
BAA	11.8	18.2	4.0	3.4	4.3
BHD	11.8	17.9	3.8	3.2	4.3
BKA	11.7	14.5	3.5	2.8	3.7
CA	11.6	14.2	3.4	2.6	3.8
DF	11.8	17.6	3.9	3.3	4.2
FA	12.0	20.2	4.1	3.8	4.7
LAD	12.1	22.4	4.3	4.0	5.0
LB	11.6	13.7	3.4	2.6	3.6
LLA	12.2	23.6	4.4	4.3	4.9
LLD	12.0	21.6	4.4	4.0	4.6
LRL	11.9	20.3	4.0	3.8	4.8
ND	11.6	15.3	3.6	2.9	3.8
PF	11.7	17.7	3.9	3.2	4.4
TA	11.8	19.0	4.0	3.7	4.4
TBA	12.4	29.7	4.7	4.9	5.7
TBD	12.5	28.8	4.6	4.9	5.5
TBR	12.5	30.2	4.8	4.8	5.7
TD	12.0	19.6	4.1	3.6	4.5
UB	11.7	14.6	3.5	2.7	3.7
All	11.9	19.6	4.0	3.6	4.5

groups contain the same number of images, the average results for all images may be slightly different from the average of all groups.

Among the tested algorithms the JPEG-LS is clearly the best when we consider the compression speeds listed in Table 33.2. All other algorithms are noticeably slower—from 39% (Lossless JPEG) to 82% (PNG). The speeds of JPEG2000, PNG, and CALIC are similar (CALIC is faster than the remaining two algorithms by 11% and 20%, respectively). Both CALIC and JPEG2000 use an arithmetic entropy coder in contrast to PNG which is based on faster techniques (LZ77 and the Huffman coding). Based on this, we expected PNG to obtain compression speeds close to Lossless JPEG with Huffman coding. Therefore, the low speed of PNG is probably due to the implementation (we used NetPBM). Looking at the results on a group by group basis, the best compression speeds were obtained for groups TBA, TBD, and TBR that is, those groups where the total body is captured. The worst compression speeds were achieved for groups BKA, CA, LB, and UB.

Considering the compression ratios from Table 33.3, we see that the ratios for JPEG-LS and CALIC are higher than those for JPEG2000 which in turn performs better than PNG and Lossless JPEG. While for various continuous tone grayscale images, JPEG2000 has been reported as close to or little worse than JPEG-LS [16], our results indicate that for medical infrared images, JPEG2000 is significantly worse than JPEG-LS which is the best-performing algorithm. CALIC performs slightly worse than JPEG-LS but is also computationally much more complex. Looking at the results for each image group, the highest compression ratios are achieved for the three body groups TBA, TBD, and TBR while images of groups BKA, CA, LB, and UB are least compressible. Correlating this with the compressions speeds from Table 33.2, we see that there is a direct link between efficiency and efficacy.

Overall, it is clear that JPEG-LS is the best-performing algorithm for lossless compression of medical infrared images. Not only does it provide the highest compression ratios, it is also the fastest of the tested methods. Furthermore, JPEG-LS is already included in the DICOM standard and can hence be readily employed.

TABLE 33.3 Compression Ratios

Images	L-JPG	JPEG-LS	JPEG2000	PNG	CALIC
ABD	2.83	3.86	3.48	3.06	3.76
BAA	3.03	4.12	3.73	3.28	4.04
BHD	2.86	3.81	3.41	3.05	3.75
BKA	2.50	3.11	2.85	2.52	3.10
CA	2.41	3.06	2.78	2.55	3.01
DF	3.03	4.04	3.68	3.20	3.99
FA	3.18	4.73	4.23	3.70	4.57
LAD	3.48	5.23	4.68	4.12	4.98
LB	2.36	2.92	2.65	2.41	2.88
LLA	3.75	5.71	5.09	4.27	5.52
LLD	3.64	5.34	4.79	4.06	5.18
LRL	3.24	4.71	4.19	3.66	4.56
ND	2.55	3.33	3.04	2.78	3.28
PF	2.94	3.96	3.59	3.27	3.89
TA	3.14	4.40	3.98	3.44	4.28
TBA	4.21	7.35	6.36	5.42	6.96
TBD	4.20	7.19	6.22	5.36	6.84
TBR	4.26	7.36	6.39	5.64	6.92
TD	3.21	4.49	4.08	3.46	4.36
UB	2.53	3.24	2.95	2.65	3.20
All	3.04	4.21	3.79	3.35	4.11

33.4 Retrieval of Thermograms

While medical infrared images are typically stored for archival purposes, in this section, we demonstrate that a database of images can also be usefully employed for retrieving similar medical cases. For this, we employ concepts initially developed for querying general-purpose image collections.

33.4.1 Content-Based Retrieval

33.4.1.1 Moment Invariant Features

In content-based image retrieval techniques, each image is characterized by a set of features that serve as an index into the database. The features we propose to store as indices for thermal images are invariant combinations of moments of an image. Two-dimensional geometric moments m_{pq} of order $p + q$ of a density distribution function $f(x,y)$ are defined as

$$m_{pq} = \int\limits_{-\infty}^{\infty} \int\limits_{-\infty}^{\infty} x^p y^q f(x, y)\mathrm{d}x\mathrm{d}y \tag{33.1}$$

In terms of a digital image $g(x,y)$ of size $N \times M$, the calculation of m_{pq} becomes discretized and the integrals are hence replaced by sums leading to

$$m_{pq} = \sum_{y=0}^{M-1}\sum_{x=0}^{N-1} x^p y^q g(x, y) \tag{33.2}$$

Rather than m_{pq}, often central moments

$$\mu_{pq} = \sum_{y=0}^{M-1} \sum_{x=0}^{N-1} (x - \overline{x})^p (y - \overline{y})^q \, g(x, y) \tag{33.3}$$

with

$$\overline{x} = \frac{m_{10}}{m_{00}} \quad \overline{y} = \frac{m_{01}}{m_{00}}$$

are used, that is, moments with the center of gravity moved to the origin (i.e., $\mu_{10} = \mu_{01} = 0$). Central moments have the advantage of being invariant to translation.

It is well known that a small number of moments can characterize an image fairly well; it is equally known that moments can be used to reconstruct the original image [3]. In order to achieve invariance to common factors and operations such as scale, rotation, and contrast, rather than using the moments themselves, algebraic combinations thereof, known as moment invariants, are used that are independent of these transformations. It is a set of such moment invariants that we use for the retrieval of thermal medical images. In particular, the descriptors we use are based on Hu's original moment invariants given by [3]

$$
\begin{aligned}
M_1 &= 4\mu_{20} + \mu_{02} \\
M_2 &= (\mu_{20} - \mu_{02})^2 + 4\mu_{11}^2 \\
M_3 &= (\mu_{30} - 3\mu_{12})^2 + 3(\mu_{21} + \mu_{03})^2 \\
M_4 &= (\mu_{30} + \mu_{12})^2 + (\mu_{21} + \mu_{03})^2 \\
M_5 &= (\mu_{30} - 3\mu_{12})(\mu_{30} + \mu_{12})[(\mu_{30} + \mu_{12})^2 - 3(\mu_{21} + \mu_{03})^2] \\
&\quad + (3\mu_{21} - \mu_{03})(\mu_{21} + \mu_{03})[3(\mu_{30} + \mu_{12})^2 - (\mu_{21} + \mu_{03})^2] \\
M_6 &= (\mu_{20} - \mu_{02})[(\mu_{30} + \mu_{12})^2 - (\mu_{21} + \mu_{03})^2] + 4\mu_{11}(\mu_{30} + \mu_{12})(\mu_{21} + \mu_{03}) \\
M_7 &= (3\mu_{21} - \mu_{03})(\mu_{30} + \mu_{12})[(\mu_{30} + \mu_{12})^2 - 3(\mu_{21} + \mu_{03})^2] \\
&\quad + (\mu_{30} - 3\mu_{12})(\mu_{21} + \mu_{03})[3(\mu_{30} + \mu_{12})^2 - (\mu_{21} + \mu_{03})^2]
\end{aligned}
\tag{33.4}
$$

Combinations of Hu's invariants can be found to achieve invariance not only to translation and rotation but also to scale and contrast [11]

$$
\begin{aligned}
\beta_1 &= \frac{\sqrt{M_2}}{M_1} \\[4pt]
\beta_2 &= \frac{M_3 \mu_{00}}{M_1 M_2} \\[4pt]
\beta_3 &= \frac{M_4}{M_3} \\[4pt]
\beta_4 &= \frac{\sqrt{M_5}}{M_4} \\[4pt]
\beta_5 &= \frac{M_6}{M_1 M_4} \\[4pt]
\beta_6 &= \frac{M_7}{M_5}
\end{aligned}
\tag{33.5}
$$

33.4.1.2 Similarity Metric

Each thermal image is characterized by the six moment invariants from Equation 33.5 which form a vector $\Phi = \{\beta_i, i = 1\ldots6\}$. It should be noted that in the context of the DICOM standard, these features can be integrated into the thermogram through a set of annotations and can hence be shared by different users.

As similarity metric or distance measure between two thermograms, we use the Mahalanobis norm which takes into account different magnitudes of different components in Φ. The Mahalanobis distance between two invariant vectors $\Phi(I_1)$ and $\Phi(I_2)$ computed from two thermal images I_1 and I_2 is defined as

$$\mathrm{d}(I_1, I_2) = \sqrt{(\Phi_1 - \Phi_2)^T C^{-1} (\Phi_1 - \Phi_2)} \tag{33.6}$$

where C is the covariance matrix of the distribution of Φ.

33.4.1.3 Retrieval through Query by Example

One main advantage of using the concept of content-based image retrieval is that it represents a generic approach to the automatic processing of images. Rather than employing specialized techniques which will capture only one type of image or pose (or one kind of disease or defect), image retrieval, when supported by a sufficiently large medical image database, will provide those cases that are most similar to a given one. The QBE method whereby an image is provided to the system and corresponding images from the database are retrieved and returned in order of decreasing similarity, is perfectly suited for this task. Typically, it is sufficient to restrict the attention to the top 20 retrieved images.

The moment invariant descriptors described in Section 33.4.1.1 were used to index an image database of several hundred thermal medical images provided by the University of Glamorgan [8]. An example of an image of an arm was used to perform QBE retrieval on the whole dataset. The result of this query is given in Figure 33.1 which shows those 20 images that were found to be closest to the query (sorted by

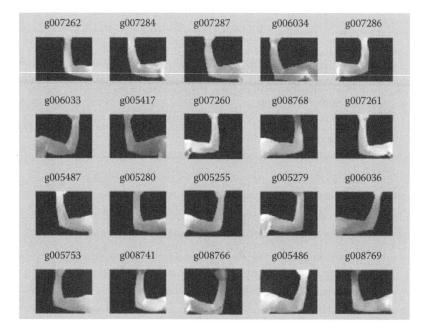

FIGURE 33.1 Example of retrieving thermal images based on a query of an *arm* image. The query image is the one on the top left; the other images are those most similar to the query ordered in decreasing similarity (from left to right, top to bottom).

descending similarity from left to right, top to bottom). As can be seen, all 20 retrieved images are of the same category, that is, arm images. The QBE method can hence be effectively employed to retrieve similar thermograms from an existing medical image repository. If coupled with patient data and related medical records, a powerful information retrieval process supporting clinical diagnosis can be initiated.

33.4.1.4 Comparison with Database of Normals

The QBE paradigm is well suited for checking an existing database of thermal images for similar cases, for example, for cases of the same disease or similar manifestations of a disease. Provided a sufficiently comprehensive database is available, it will hence provide useful results to be used for medical diagnosis. However, in many cases this prerequisite on the database is not realistic. In such instances, an alternative approach based on dissimilarities rather than similarities can be taken. Based on the assumption that a dysfunction will show on the taken thermogram, it can be compared, based on the features and similarity measure described above, to a dataset of healthy subjects captured in the same pose (such a database is described in Reference 14). If the average distance to the healthy subjects is much larger than the intercluster distances within this group of normals, this provides an indication of a certain dysfunction that needs further investigation by the clinician.

Figure 33.2 shows three infrared images of hands. Two of the images depict healthy subjects. However, the third one shows a clear inflammation on one of the fingers and generally cooler hands. Following the strategy laid out above, all hand images in the database were taken and their intercluster distance, that is, the average distance (again, based on the moment invariant features) between any two hand images, calculated. This was then compared with the average distance of the third image to all "healthy" hand thermograms. While the average distance between two images of healthy individuals was 3.53,

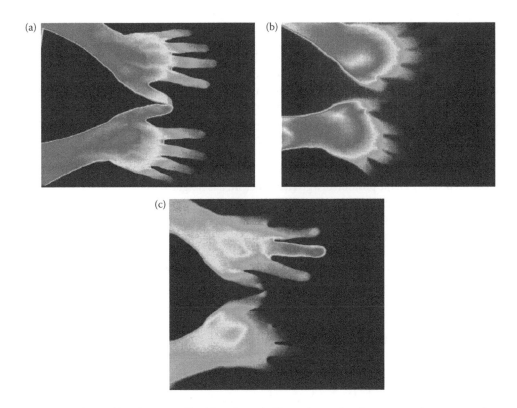

FIGURE 33.2 Thermal hand images of healthy subject (a, b) together with a thermogram showing an inflammation (c).

the distance of the image with the inflammation was 15.78. The difference between the thermogram showing an inflammation and images of healthy subjects is hence calculated as about 4 times the average difference among the images of normals. The presented image indexing techniques can therefore also be employed when no comprehensive database of pathological cases is present. Thermal images falling outside the range of thermograms recorded from healthy subjects can thus be automatically detected and should give a clear indication to the clinician to instigate further investigation.

33.4.2 Compressed-Domain Retrieval

While thermograms are typically stored in uncompressed form, as mentioned earlier in this chapter, image compression techniques can be applied to reduce storage space and resources. For performing content-based retrieval of compressed thermograms, the images would normally need to be uncompressed to allow the calculation of features such as the image moments discussed above, which clearly leads to a computational overhead. Through careful crafting of image descriptors, it is however also possible to perform retrieval directly in the compressed domain of images, for example, for images compressed using a wavelet-based techniques such as JPEG2000 [5].

In wavelet-based compression algorithms, an image is first decomposed using an M-level wavelet transform. The transform of level 1 is obtained in the following way: We start with the original image and apply one-dimensional wavelet low- and highpass filters to the rows of the image. As a result, the image is transformed into low- and high-frequency bands. Then, we apply the transform to rows of bands transforming each band into its respective low- and high-frequency band. The wavelet transform of level 1 produces four bands, each of them of resolution equal to 1/2 of the resolution of the original image. The output band of (double) lowpass filtering is a low-frequency band of level 1 (hence, actually a reduced-size original image), others are high-frequency bands representing fine details oriented at different directions. To obtain a transform of level $N + 1$, we apply the transform of level 1 to the low-frequency band of level N. As a result, we get $3M + 1$ bands of wavelet coefficients in M resolution levels (see Figure 33.3).

In the fast multiresolution image quering method (FMIQ) [6], properties of wavelet coefficients are used directly for the retrieval task. In both variants of the transform, the top-left coefficient represents the average intensity of the image, while the remaining coefficients represent image details of various sizes and orientations (coarsest near the top left corner and finer as we get more distant from there). After decomposition, we threshold the coefficients and only a small number N of wavelet coefficients (we usually choose 100) of largest absolute values (i.e., representing the most important details of image) are used as a "fingerprint of the image." It is also possible to design a special metric that takes into account only these N nonzero wavelet coefficients, thanks to which efficient comparison of fingerprints is possible. Also, depending on the position of a given coefficient, it is weighted using one of six weights (for six ranges of distance from top left corner to coefficient's position). Weights permit adjusting the algorithm to given image types. Images of low resolutions (e.g., 128×128), or resized to low resolution, were found sufficient for generating fingerprints.

In Figure 33.4, we show an example of this approach, based on the same image dataset that we have used earlier in this chapter. It can be seen that all 10 retrieved images are similar (i.e., of the same category) to the query image.

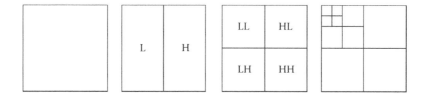

FIGURE 33.3 Wavelet transform (three levels).

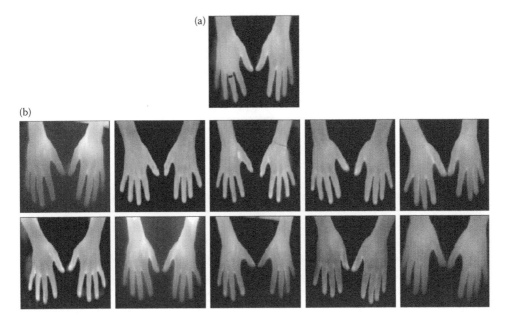

FIGURE 33.4 Example of compressed-domain retrieval showing the top 10 retrieved images based on the query image at the top.

33.4.3 Browsing of Thermal Image Database

When storing thermograms (or other medical images), often little sought is given to the actual organization and management of these image collections. While in typical medical image databases data records are accessed by patient name or other similar attributes, here we present a method that allows access to and navigation through the database based on visual cues. All images in the collection are projected onto a two-dimensional space in such a way that images that are similar to each other are located close to one another in this configuration. The result is then displayed on screen and provides a picture of the complete dataset. That visually similar images are located close to each other (i.e., clustered) is fundamental to this approach as it provides an intuitive and easy-to-understand interface for the user.

33.4.3.1 Features and Similarity Measure

Since the browsing approach taken in this chapter is again based on visual similarity between images, we make use of the features and the similarity metric we introduced and described in Sections 33.4.1.1 and 33.4.1.4, that is, moment invariants and Mahalanobis distance. Doing so allows the same features to be used for both retrieval and browsing without any additional overhead.

33.4.3.2 Multidimensional Scaling

Multidimensional scaling (MDS) [9] expresses the similarities between different objects in a small number of dimensions, allowing for a complex set of interrelationships to be summarized in a single figure. MDS can be used to analyze any kind of distance or similarity/dissimilarity matrix created from a particular dataset.

In this chapter, we follow ideas from Reference 15 and apply MDS for thermal medical image database display and navigation. By applying MDS based on the distances between moment invariant vectors, we produce a way of not only visualizing the retrieved images in terms of decreasing similarities but also according to their common similarities. All images are implanted (based on their similarities) in a two-dimensional Euclidean space where the original distances are preserved as closely as possible.

In detail, first a distance matrix which contains all pairwise distances between the medical images in the databases need to be obtained. As mentioned above, we define as the distance d between two images the Mahalanobis distance between their moment invariant vectors. Euclidean distances $\hat{d}_{i,j}$ are calculated and initially compared using Kruskal's stress formula [9]

$$\text{Stress} = \frac{\sum_{i,j}(\hat{d}_{i,j} - d_{i,j})}{\sum_{i,j} d_{i,j}^2} \tag{33.7}$$

which expresses the difference between the distances d and the Euclidean values $\hat{d}$ between all images. The aim of nonmetric MDS is to assign locations to the input data so that the overall stress is minimal.

Typically, an initial configuration is found through principal component analysis (PCA). While the degree of goodness of fit after this is in general fairly high, it is not optimal. To move toward a better solution, the locations of the points are updated in such a way as to reduce the overall stress. If, for instance, the distance between two specific samples has been overestimated, it will be reduced to correct this deviation. It is clear that this modification will have implications for all other distances calculated. Therefore, the updating of the coordinates and the recalculation of the stress is being performed in an iterative way, where during each iteration the positions are slightly changed until the whole configuration is stable and the algorithm has converged into a minimum where the distances between the projected samples correspond accurately to the original distances. Several termination conditions can be applied such as an acceptable degree of goodness of fit, a predefined maximal number of iterations or a threshold for the overall changes in the configuration. Once the calculation is terminated, the images can then be plotted at the calculated coordinates on the screen.

33.4.3.3 Browsing

Navigation through the image database starts typically with a global display of the entire dataset with images positioned in relation to how similar they are to all others. From here the user has the ability to zoom into certain regions of interest to enlarge for further querying. For each localized visualization occurrence, the images selected in the area have their distance matrix recalculated and MDS reapplied.

FIGURE 33.5   Global visualization view of complete thermal image database.

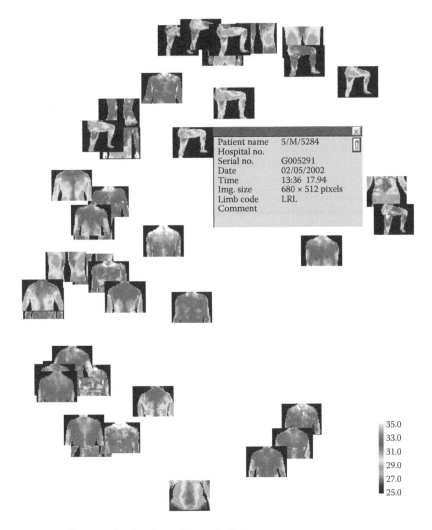

Patient name   5/M/5284
Hospital no.
Serial no.   G005291
Date   02/05/2002
Time   13:36 17.94
Img. size   680 × 512 pixels
Limb code   LRL
Comment

35.0
33.0
31.0
29.0
27.0
25.0

FIGURE 33.6 Zoomed-in area showing leg and upper body images.

The same image database as employed for the retrieval experiments has been used to test our browsing method. The triangular distance matrix between all invariant vectors was obtained using the Mahalanobis distance from Equation 33.6. Based on this distance matrix, multidimensional scaling was applied as explained in Section 33.4.3.2. The resulting global view is shown in Figure 33.5 where all images are projected (although some are occluded by others) onto the visualization plane which is then displayed on screen. It can be see that similar images are indeed placed close to each and that clusters or groups are formed. Typically, the user will then zoom into one of the clusters or areas of interest to narrow down the search as was done to produce Figure 33.6 which displays a zoomed-in area showing mainly *legs* and *upper body* images. MDS is reapplied to the images in the localized area to provide a less cluttered view. Also shown in Figure 33.6 is the possibility to retrieve further information on a certain image which is displayed in a popup window. Clearly, additional medical information such as the patient's medical history, medication, or current diagnosis information can be added to this view which hence provides an intuitive interface.

33.5 Conclusions

In this chapter, we have looked at issues regarding the storage and retrieval of medical infrared images. First, we stressed the need of a standard format for storage and communication of thermograms and suggested the adoption of the DICOM standard for this purpose.

A number of lossless compression algorithms were then evaluated on a large database of thermal medical images. It was shown that, depending on the type of image, compression ratios of up to 1:4 are possible. Furthermore, it was shown that JPEG-LS seems the most suitable algorithm to employ as it provides both the best compression performance and is also the fastest of the tested algorithms. As JPEG-LS is also certified as an international standard and is further integrated into the DICOM standard for the exchange of medical imagery, it could be recommended as a standard for storing and exchanging thermal medical images.

We have also proposed the application of content-based image retrieval to the domain of thermal medical images. Each image can be characterized by a set of moment invariants which are independent to translation, scale, rotation, and contrast. Alternatively, information on wavelet coefficients can be employed which in turn allows for retrieval also directly in the compressed domain. Retrieval is performed by returning those images whose descriptors are most similar to the ones of a given query image. In addition to retrieval, the same features can be used to provide an intuitive interface for browsing a collection of thermograms. Using multidimensional scaling, all images are projected onto a visualization plane so that images that are visually similar are also located close to each other. Experimental results for both techniques have been presented and demonstrated the usefulness of the proposed techniques.

Acknowledgments

This chapter is based on the work published in References 17, 18, 21, 22, 24. The author would like to thank Roman Starosolski (Silesian Technical University), Shao Ying Zhu, and Brian Jones (University of Derby) for their contributions as well as the Medical Computing Research Group at the University of Glamorgan for providing the test image dataset.

References

1. C. Christopoulos, A. Skodras, and T. Ebrahimi. The JPEG2000 still image coding system: An overview. *IEEE Transactions on Consumer Electronics*, 46(4):1103–1127, 2000.
2. Consultative Committee for Space Data Systems. Lossless data compression. *CCSDS Recommendation for Space System Data Standards, CCSDS 121.0-B-1, Blue Book*, 1997.
3. M.K. Hu. Visual pattern recognition by moment invariants. *IRE Transactions on Information Theory*, 8(2):179–187, February 1962.
4. ISO/IEC. Lossless and near-lossless compression of continuous-tone images—Baseline. *ISO/IEC International Standard 14495-1*, 1999.
5. ISO/IEC. JPEG2000 image coding system: Core coding system. *ISO/IEC International Standard 15444-1*, 2002.
6. C.E. Jacobs, A. Finkelstein, and D.H. Salesin. Fast multiresolution image querying. In *SIGGRAPH 95*, pp. 277–286, 1995.
7. B.F. Jones. A reappraisal of infrared thermal image analysis for medicine. *IEEE Transactions on Medical Imaging*, 17(6):1019–1027, 1998.
8. B.F. Jones. EPSRC Grant GR/R50134/01 Report, 2001.
9. J.B. Kruskal and M. Wish. *Multidimensional Scaling*. Beverly Hills, CA: Sage Publications, 1978.
10. G. Langdon, A. Gulati, and E. Seiler. On the JPEG model for lossless image compression. In *2nd Data Compression Conference*, pp. 172–180, 1992.
11. S. Maitra. Moment invariants. *Proceedings of the IEEE*, 67:697–699, 1979.

12. National Electrical Manufacturers Association. Digital Imaging and Communications in Medicine (DICOM). *Standards Publication PS 3.1-2004*, 2004.

13. P. Plassmann and E.F.J. Ring. An open system for the acquisition and evaluation of medical thermological images. *European Journal of Thermology*, 7:216–220, 1997.

14. E.F.J. Ring, K. Ammer, A. Jung, P. Murawski, B. Wiecek, J. Zuber, S. Zwolenik, P. Plassmann, C. Jones, and B.F. Jones. Standardization of infrared imaging. In *26th IEEE International Conference on Engineering in Medicine and Biology*, pp. 1183–1185, 2004.

15. Y. Rubner, L. Guibas, and C. Tomasi. The earth mover's distance, multi-dimensional scaling, and color-based image retrieval. In *Image Understanding Workshop*, pp. 661–668, 1997.

16. D. Santa-Cruz and T. Ebrahimi. A study of JPEG2000 still image coding versus other standards. In *10th European Signal Processing Conference*, pp. 673–676, 2000.

17. G. Schaefer and R. Starosolski. A comparison of two methods for retrieval of medical images in the compressed domain. In *30th IEEE International Conference Engineering in Medicine and Biology*, pp. 402–405, 2008.

18. G. Schaefer, R. Starosolski, and S.Y. Zhu. An evaluation of lossless compression algorithms for medical infrared images. In *27th IEEE International Conference Engineering in Medicine and Biology*, pp. 1125–1128, 2005.

19. G. Schaefer and S.Y. Zhu. Compressing thermal medical images. In *UK Symposium on Medical Infrared Thermography*, 2004. Abstract.

20. G. Schaefer and S.Y. Zhu. Lossy compression of medical infrared images. In *3rd European Medical and Biological Engineering Conference*, 2005.

21. G. Schaefer, S.Y. Zhu, and B. Jones. Retrieving thermal medical images. In *International Conference on Computer Vision and Graphics*, Computational Imaging and Vision. Springer, 2006.

22. G. Schaefer, S.Y. Zhu, and S. Ruszala. Visualisation of medical infrared image databases. In *27th IEEE International Conference Engineering in Medicine and Biology*, pp. 1139–1142, 2005.

23. A.W.M. Smeulders, M. Worring, S. Santini, A. Gupta, and R.C. Jain. Content-based image retrieval at the end of the early years. *IEEE Transactions on Pattern Analysis and Machine Intelligence*, 22(12):1349–1380, 2000.

24. G. Schaefer, J. Huguet, P. Plassmann, S.Y. Zhu, and F. Ring. Adopting the DICOM standard for medical infrared images, *28th IEEE Int. Conference on Engineering in Medicine and Biology*, pp. 236–239, 2006.

25. M.J. Weinberger, G. Seroussi, and G. Sapiro. The LOCO-I lossless image compression algorithm: Principles and standardization into JPEG-LS. *IEEE Transactions on Image Processing*, 9(8):1309–1324, 1996.

26. B. Wiecek, S. Zwolenik, A. Jung, and J. Zuber. Advanced thermal, visual and radiological image processing for clinical diagnostics. In *21st IEEE International Conference on Engineering in Medicine and Biology*, 1999.

27. X. Wu and N. Memon. Context-based adaptive lossless image codec. *IEEE Transactions on Communications*, 45(4):437–444, 1997.

28. WWW Consortium. PNG (Portable Network Graphics) specification. Version 1.0, 1996.

29. J. Ziv and A. Lempel. A universal algorithm for sequential data compression. *IEEE Transactions on Information Theory*, 32(3):337–343, 1977.

34

Ethical Obligations in Infrared Imaging Research and Practice

James Giordano
Potomac Institute for Policy Studies

George Mason University

University of Oxford

Kim Abramson
Institute for BioTechnology Futures

34.1 Introduction

34.1.1 A Perspective on Ethical Issues in the Use of Infrared Technology in Medicine

For the past two decades, technological advancement has led to increased capability and efficacy of functional infrared imaging (fIR). Developments in computerized image acquisition, data processing, and interpretation derived and directly incorporated from formerly classified military applications have increased and solidified the medical utility of fIR. Previous and residual concerns and criticisms of IR as being inadequate to effectively detect thermal signatures that are important to establishing particular differential diagnoses have been assuaged and refuted by the functional sophistication produced through advanced image acquisition and computerized analytic systems. There is recurrent contention surrounding the inapt use of fIR by untrained personnel, and the overuse of IR as an improperly administered or interpreted "diagnostic" test. The aforementioned progress in the technology domains of the field has deepened concerns over the ethical use of fIR, and compels a need for increased stringency in the education, training, and certification of professionally qualified, competent clinicians and technicians to be the sole providers of this technology.

In this chapter, we address the ethical obligations that compel and sustain the use of fIR in medical research and practice. We base this discussion on the premise that medicine is not merely applied science, but rather is a profession that mandates the use of scientific (i.e., theoretical) information in ways that sustain and are resonant to the humanitarian essence of the interaction between clinician and patient (Dell'Oro, 2005). To be sure, there are areas of medicine (e.g., basic science research, epidemiology/public health, etc.) that are not explicitly focused upon the clinical encounter. Yet, their relevance to the clinician–patient relationship, while indirect, implicitly reinforces the scope, tenor, and basis of patient care. This is perhaps most evident in basic sciences' research; while the proximate goal of the basic sciences is the contribution of theoretical knowledge (i.e., "knowledge for knowledge's sake"), this theoretical information contributes to a larger body of epistemic capital that is applied within the context of clinical practice as a fundamental domain of relevant knowledge (Feinstein, 1967). It is in this light that we discuss the imperative to conduct and advance research that is aimed at elucidating the capacities, enhancing the efficacy, and evaluating the outcomes of IR technology as a component of clinical paradigm of detection, discernment, and ultimately diagnosis.

But how this knowledge is utilized is as important as what this knowledge entails, and thus a brief overview of the domains of knowledge that maintain the intellectual virtues important for clinical medicine is provided. For if the information gained from basic sciences' studies of IR is to be the bedrock on which its use in medical practice is to be built, then we must recognize the moral obligations to utilize such knowledge and technical acumen in prudent ways that support the humanitarian dimensions that define medicine as an interpersonal act. This is particularly true when regarding IR technology as a viable method of detection to facilitate diagnosis, as the values in and of diagnosis ultimately establish the subsequent construct(s), content, and the context of care.

Thus, we argue that the use of IR (or any technique or technology) in the medical milieu must be explicitly relevant to and consistent with (if not wholly supportive of) the moral obligations and ethical integrity of clinical encounter between clinical and patient. The clinical encounter is, in essence, a nexus for (1) the multidimensional aspects of medicine as a science, engaging research findings within the context of a complex practice and (2) the enactment of medicine as a humanitarian endeavor, in which the conjoinment of participatory agents who maintain the expectation that the ends, or *telos*, of technically right and morally good care is enabled and provided (Pellegrino, 1983). Therefore, the capacities and limitations of IR must be recognized and explicated, so as to ensure (technically and ethically) appropriate use within the medical fiduciary.

34.1.2 On the Need for Continuing and Progressive Research

Research facilitates the defined *telos* of medicine by affording knowledge that (1) enables the clinician to evaluate the relative value(s), benefits, and risks of particular diagnostic and therapeutic approaches, and ultimately resolve equipoise, and (2) empowers patients to be informed participants in clinical decisions relevant to care, thereby lessening their inherent vulnerability and decreasing the inequalities of knowledge, power, and capacity (Freedman, 1987; Giordano, 2006a). Of course, such research must be methodologically rigorous and ethically sound (Fried, 1974; see Levine, 1986, for review). The relationship of research and practice is reciprocal. Findings from research inform practice, and evaluation of outcomes gained by employing various techniques in practice contributes to a progressive revision and expansion of an evidentiary base of knowledge to instigate and guide further study (Giordano, 2004, 2006a, 2009).

In both cases, the imperative for developing, implementing, and promoting ongoing research to assess and evaluate the capabilities and clinical application(s) of fIR is grounded upon the moral obligation of beneficence (Frankena, 1982). Irrespective of the ethical system that is utilized to guide medical research or care (e.g., employment of *prima facie* principles, casuistic approaches, ethics of care, virtue ethics, etc.), the primacy of patient benefit is critical, if not fundamental to morally sound conduct of clinical research and practice (Giordano, 2007). Pellegrino and Thomasma (1988) have defined beneficence in the clinical encounter as fourfold, encompassing the biomedical (i.e., technical) good, the good as relevant to the choices of the patient, the good for the patient as a (*n* autonomous) person, and the existential good, in context. Obviously, research is instrumental in determining the technical or practical "good" of a particular technique or technology, such as fIR. But even at this level, application as a meaningful biomedical "good" requires translational studies that evaluate the broader implications of use-in-practice to determine viability in real-world medical settings. In recognizing and achieving the biomedical good the somewhat more passive maxim of non-harm can be concomitantly realized. By understanding the actions, effects and capabilities of fIR, we can also recognize, weigh, and may be able to compensate for (1) inherent limitations of the technology in various applications; (2) potential risks of use or nonuse; and (3) possible burdens incurred to the patient, as well as the medical system.

This knowledge serves to illustrate the potential viability of fIR as a tool within the armamentarium of evaluative techniques that contribute to accurate diagnosis. When coupled to an understanding of the mechanisms of particular pathologies, such knowledge allows the clinician to address whether fIR can, and indeed should, be employed in specific determinative clinical situations. If properly employed, fIR can fortify the informational base necessary toward the resolution of equipoise, and thus allow treatment that is appropriate to the pathologic state, as well as the choices and overall integrity of a particular patient. Ultimately, this process, together with the prudent, practical wisdom of the clinician (i.e., *phronesis*) distinguishes what care *should* be rendered, from what care *can* be rendered (Davis, 1997).

Information fuels this decisional process, and as this volume well illustrates, the technical sophistication of current iterations of fIR enables its utility as a tool in detecting thermal changes that reflect pathologic processes of several disorders (e.g., complex regional pain syndrome, certain vasculopathies, tumorigenesis, etc.). Still, there is equivocal discussion that questions the value and necessity of continuing fIR research. The majority of contention is focused on the incapability of fIR to accurately detect thermal variation with sufficient sensitivity to be relevant to the clinical discernment of particular pathophysiological states. From this assumption would arise the issue that fIR research is impractical, or worse, that time and effort spent pursuing clinical application of fIR are pragmatically, temporally, and economically wasteful and therefore constitute an unethical exploitation of research resources or patient trust and expectation. We do not feel that these criticisms are valid based on the following grounds: first, to reiterate, the wedding of advanced image acquisition systems to state-of-the-art computerized systems using complex statistical data analyses has allowed for increased image clarity and specificity of signal-to-noise discrimination (Kakuta et al., 2002). The technical advancement of current fIR has rendered previous iterations of this technology obsolete (Irvine, 2002). Second, it must be borne in mind that the applied goals of basic (and clinical) research are to address, identify, and develop methods of overcoming problematic issues that affect or impede medical care (Goodman, 2003). But research is not merely a means to these ends; rather it is an end unto itself, being a moral enterprise that must sustain the obligations to patients (as research subjects) and to society as a public good (Giordano, 2006b; Pedroni, 2006). As a public good it should: (1) be effective in achieving goals consistent with stated ends (in this case, of medicine); (2) not be burdensome to individual patient subjects or society (May, 2003); and (3) be relatively self-advancing—that is, maintain a reciprocally translational focus that contributes to the knowledge base that infuses and advances the benefits of patient care (see also Giordano, 2004). Contemporary fIR research satisfies each of these requirements, and thus we argue that contentions against ongoing research in this area are ill informed, improperly directed and hence, unjustified.

But if the knowledge acquired through well-designed and rigorously conducted research studies is to be of any meaningful value, it must be appropriately and effectively used within the clinical encounter. How such knowledge gained from research is to be incorporated within the larger fabric of clinically relevant information is important to both its practical and ethical use, and reflects the intellectual skills necessary to the act of medicine (Pellegrino and Thomasma, 1993; Toulmin, 1975).

34.2 Domains of Knowledge in Clinical Medicine

Arguably, medicine as a science relies on theoretical knowledge that affords understanding of mechanisms of physiological process(es), disease, and diagnostic and therapeutic interventions. Such knowledge is based not only in part on *a priori* understanding of the workings of the world as natural phenomena that are logically consistent, but also on the acquisition of new knowledge that modifies our basic concepts and, as such, provides a somewhat changing epistemological capital of "truth(s)" (Fuller, 2003; Kuhn, 1962). Philosophically, this reflects the essence of science as being self-critical and self-revising as new information is acquired and incorporated into the accepted fund of knowledge

(Nagel, 1961). Theoretical knowledge is gained through research, both as an observational and experimental undertaking; this criticality of epistemic knowledge supports the importance of research to science and medicine, in general, and in the present case, undergirds the importance of progressive research to advance an understanding of the capabilities of fIR.

Given the position that medicine is not just applied science, we believe that its practice involves far more than a theoretical knowledge of physiological systems, pathologies, and the mechanics of particular tests, tools, and interventions. We hold that knowing "what should be done" mandates understanding of "how," "when," and "why" something is done (Giordano, 2006b, 2009). Specifically, knowing whether fIR represents an appropriate detective protocol for a particular clinical case requires understanding not only of the technology (and its capacities and limitations) but also the suspected pathology, the physiological changes it incurs, the viability of fIR to detect such changes, as well as how the technology (of both image acquisition conditions and data analyses) should be employed to meet the particular circumstances. This necessitates experience in differing situations and across time. Experiential knowledge can be gained through an expanding body of research findings, and may be acquired through training and direct activity (Davis, 1997; Giordano, 2006b). The necessity of experiential knowledge in clinical contexts further endorses the need for continued research in mechanisms and applications of fIR, and additionally supports the continued utilization of fIR in practice so as to contribute to a growing database of clinically relevant applications and possibilities. Ultimately, theoretical and experiential knowledge work synergistically when relating current cases to paradigmatic examples (i.e., the casuistic approach), and in this way theory and experience are conjoined to determine if, how and why fIR can and should be utilized as a step in the process of diagnosing particular pathology in a specific patient.

In this situation, contextual knowledge allows the clinician to

1. Utilize technical and experiential knowledge of fIR as a tool in detecting the thermal signatures of certain pathophysiological states.
2. Use this theory and experience to determine whether fIR represents an appropriate evaluative method.
3. Utilize theoretical and experiential understanding to formulate and implement the right conditions for use and interpretation of fIR in a given case.

Taken together, contextual knowledge fuses research to practice (Giordano, 2009), and allows the clinician to avoid a simplistic "one-size-fits-all" approach to assessment of those pathologies for which fIR has been shown to be a valid and effective evaluative tool. Further, fIR provides objective assessment of otherwise subjective diagnostics. Objective measures increase the accuracy of diagnosis and, by extension, the effectiveness of treatment, and are therefore emphasized within most medical disciplines in order to validate subjective information (Johnston et al., 2007; Reinhard et al., 2007; Zhang & Zhou, 2007).

Experiential and contextual knowledge rely on a theoretical foundation, and both require and infuse skill. However, it is important to note that there is an abstract dimension to medicine. Indeed, medicine has been described as both skill and art (i.e., *tekne*) that establishes the distinction between simply applied science and the finesse that empowers its subjective, hermeneutic character (Edelstein, 1967; Owens, 1977; Svenaeus, 2000). The critical intellectual, practical, and moral step in medical practice occurs as description, detection, and differentiation coalesce into the process of diagnosis (Wulff, 1976). It is from this point of determining "what is wrong" with a particular patient that the possibilities for and selection of subsequent treatment can be ascertained (Pellegrino, 1979). Indeed, this is the point at which individual and clinical equipoise are reconciled and resolved on grounds that provide the best choice(s) for ethically good patient care. This process of using theoretical, experiential, and contextual knowledge, together with any and all tools and methods that are nonburdensome, manifest low risk, and which effectively produce reliable data to generate an evidence-based approach, is essential to clinical decision making. *Phronesis*, the intellectual virtue of practical wisdom, is imperative to determine (1) which tools, tests, and approaches best contribute to the diagnostic process; (2) the meaning and relative value(s) of findings; and (3) how these findings facilitate diagnosis (Giordano, 2006a, b, 2007, 2009; see also Davis, 1997). As well, *phronesis* allows the clinician to intuit the nature and power in and of diagnosis as an act of naming the disorder, claiming particular "truths" about the disorder and its care, and framing the patient within a population of those with this disorder and establishing the trajectory of subsequent clinical interaction(s) (Giordano, 2006b, 2007, 2009). The power of this step is exceptional, and as such the inherent responsibilities must be borne by individuals who are capable of integrating multidimensional knowledge, weighing the complex choices arising from diverse circumstances, prudently determining the reasons to guide right action(s), and recognizing and accepting consequences of those actions (Pellegrino, 1979). For fIR to be accurately, effectively, and thus ethically employed, it must be administered and interpreted by individuals who are not simply repeating a task by rote (even if safely, and efficiently), but rather by individuals who exercise practical wisdom in the professional responsibilities of the diagnostic process (i.e., by one who utilizes and embodies *phronesis* by virtue of character—a *phronimos*). By Aristotelian definition, a *phronimos* is concerned not just with knowing the right things, but in using that knowledge in the right ways to achieve the good(s) intrinsic to the practice (Aristotle, 1999; Edelstein, 1967).

34.3 The Contribution of fIR Detection to Diagnostic Value(s)

Assessment, detection, and discernment are not diagnosis; fIR cannot and should not be considered or utilized as a singular method of detection or explicitly diagnostic technique. However, the use of IR evaluation by a competently trained professional, acting prudently to weigh the value of information gained, can serve as an important diagnostic asset. The word diagnosis is derived from the Greek *diagignoskein*, meaning "to distinguish." This distinguishing quality reflects the abilities of the clinician to utilize the aforementioned domains of knowledge to apprehend the features of a disorder as expressed in a particular patient, differentiate these signs and symptoms from other possible disorders, discern the contextual basis of these features and in so doing establish diagnosis, literally as a "... seeing or a knowledge into the problem" (Mainetti, 1992, p. 256). Any and all contributory methods and acts of diagnosis should be oriented toward (1) simplification, (2) ongoing reevaluation and interpretation, (3) synergizing theoretical and contextual understanding and moral intention, and (4) the primacy of the patients' best interest(s) (Sadler, 2004, 2005). Each of these criteria can be met by fIR, for as this volume illustrates, the technical advances that have been made over the past 20 years have allowed fIR to be used to (1) facilitate detection and thus simplify particular diagnoses, (2) afford the ability for (re)evaluation and interpretation of pathological advancement or effects of therapeutic intervention, and (3) incur low/minimal risk and thus be nonburdensome to the best interest(s) of the patient. In this latter regard, it should be noted that as part of a more expansive diagnostic protocol, fIR may lessen the risks and burdens of medical care through earlier detection of particular pathologies, to prompt better diagnosis and care that ultimately could reduce the extent, duration, as well as negative (personal and financial) impact of subsequent intervention(s). It is clear that this process takes on considerable ethical weight as it provides the basis for prognosis (a knowledge of what could occur), and creates a foundation for the development and implementation of prudent care (Giordano, 2009). Thus, diagnosis also relies at least in part on *phronesis*, as its technical rectitude and moral gravity must be oriented to and consistent with the *telos* of medicine. These telic obligations compel both research to determine those techniques and technologies that enhance diagnoses, as well as the utilization of these approaches in clinical practice (Giordano, 2006b).

34.3.1 Ethical (and Legal) Obligations for Education and Training

If research is to inform practice, and the actions of morally sound practice are reliant on the phronetic integration of distinct types and domains of knowledge, then how can this relationship be fortified to ensure that fIR is aptly utilized? It is notable that the more befouling criticisms of fIR have centered upon its use by poorly trained "technicians" or nonprofessionals, and on unethical claims for financial reimbursement for improper characterization and employment as a "diagnostic" technique. Both of these denunciations reflect inadequacies or improprieties in education and training. Thus, we argue that the technical understanding of fIR technology, use and applications must be provided by stringent programs of didactic pedagogy, practical training, and evaluation that reflect the complexity and advancement of the field, acknowledge the technical and ethical power of this technology in medical diagnosis and care, and uphold a high standard of professionalism. This is not only an ethical question, but a legal one. Improperly or inadequately trained personnel risk negligence resulting in malpractice suits, increasing medical costs, hampering patient trust, and damaging the reputation of the medical provider. Whether personnel simply appear negligent, based on a lack of professionalism, which potentially results in unsuccessful lawsuits, or actually perform an act of malpractice that results in a successful claim against the provider, legal as well as ethical consequences can be dire. We therefore impugn those approaches and programs that claim to "certify" individuals as fIR "technicians" or experts after a course of brief home study, weekend seminar and/or superficial "examination." Such programs are wholly iniquitous, depreciate the field and threaten patient care and the integrity of the medical fiduciary. Rather, we propose that fIR be incorporated into extant training programs of medical imaging, either as an additional curricular focus, or as a specialized program of study. Similarly, we propose that fIR training be provided to licensed medical professionals (i.e., physicians, veterinarians) through continuing education and postgraduate programs that (1) afford thorough information on the technology, conditions and constraints of use, limitations and delimiting factors in clinical application, and (2) mandate practical training and experience in fIR use and methods of data analysis and outcomes evaluation. Progress in technological development should not be ethically undermined by capricious, unprofessional utilization engendered through improper education, training, and certification; and this is sound legal practice, as well, to clearly demonstrate that professional and ethical standards have been instituted.

To be sure, these insurances are demanding, and require revision and ongoing monitoring of research, education, training, and application(s) in practice. Yet, as technology progresses, so too do the responsibilities to utilize and apply such technology in ways that are consistent with an expanding epistemology and are adherent to moral obligations and ethical affirmations for good and nonharm (Jonas, 1973).

34.3.2 The Role of Policy

But while these incentives may affect change on a grass-roots level, meaningful change in the "culture" of medicine (and the economic strata that support it) cannot be fully actualized without "top-down" implementation. This is the role of policy. If positive change is to occur, it must reflect the technological progress in the field, identify and be consonant with the purpose of those individuals who embrace and utilize these developments for identifiably good ends, and enact a process that engages each of the tiers affected (i.e., research, education, training, and application; Light, 2000). The voices of the research, clinical, and patient communities must reflect valuation of fIR as a meaningful clinical tool that requires further study, development, and application, thus necessitating policy that provides enablement and economic support of these endeavors. In this way "bottom-up" values and interests will encourage "top-down" enactment of public policy programs for subsidizing these enterprises. How could this be realistically achieved? First, it is important to recognize that learning must precede change—the vectors for change agency can only be empowered when the potential and possibilities that change incurs are generally appreciated as positive and nonthreatening to some real or perceived status quo. Input from the affected communities, including patients, medical providers and technology developers, is crucial, to identify authentic and perceived risks, fears, and concerns, as well as potential education (for both medical providers and for the general population) and outreach opportunities. By demonstrating the progress in fIR and pragmatically addressing the potential, possibilities, and implications that such progress may afford to healthcare, a pediment to learning can, and hopefully has, been achieved. This is critical for dispelling prior and extant misconceptions, and engendering favorable reappraisal. But "bottom-up" efforts must also reflect a "house cleaning" of sorts, to fortify the field from within, ensuring professionalism that adheres to, and reflects strict ethical standards, as previously described. Without such, any viability of fIR as a technology and technique will be overshadowed by ethical improbity in application, and the value of policy to support further research and clinical use will be abrogated.

The fusion of "bottom-up" and "top-down" change to advance the medical use of fIR requires a multipartite effort of (1) integrating research into formalized education and training, (2) instilling and ensuring high ethical standards for the use of fIR by competent professionals, and (3) instituting public policy to assure ongoing administrative and economic support. The work described by this volume certainly serves both as a cornerstone to such efforts and a catalyst for progress in the ethical use of this technology.

Acknowledgments

This work was supported, in part, by funding from the Institute for BioTechnology Futures (JG and KA), the William H. and Sara Crane Schaefer Endowment (JG), and the Nour Foundation (JG).

References

Aristotle. *The Nicomachean Ethics* (T. Irwin, trans.). Hackett Publishing, Indianapolis, 1999.

Davis FD. Phronesis, reasoning and Pellegrino's philosophy of medicine. *Theoret. Med.*, 1997; 18: 173–198.

Dell'Oro R. Interpreting clinical judgment: Epistemological notes on the praxis of medicine. In: Viafora C. (ed.) *Clinical Bioethics—A Search for Foundations.* Springer, Dordrecht, 2005.

Edelstein L. The professional ethics of the Greek physician. In: Temkin, O. and Temkin, C.L. (eds.) *Ancient Medicine: The Selected Papers of Ludwig Edelstein.* Johns Hopkins Press, Baltimore, MD, 1967.

Feinstein AR. *Clinical Judgment.* Williams & Wilkins, Baltimore, MD, 1967.

Frankena W. Beneficence in an ethics of virtue. In: Shelps, E. (ed.) *Beneficence and Health Care.* D. Reidel, Doredrecht, 1982.

Freedman B. Equipoise and the ethics of clinical research. *N. Engl. J. Med.*, 1987; 317: 141–145.

Fried C. *Medical Experimentation: Personal Integrity and Social Policy.* North-Holland, Amsterdam, 1974.

Fuller S. *Kuhn vs. Popper: The Struggle for the Soul of Science.* Columbia University Press, New York, 2003.

Giordano J. Pain research: Can paradigmatic revision bridge the needs of medicine, scientific philosophy and ethics? *Pain Phys.*, 2004; 7: 459–463.

Giordano J. Good as gold? The randomized control trial—Paradigmatic revision and ethical responsibility in pain research. *Am. J. Pain Manage.*, 2006a; 16: 68–71.

Giordano J. Moral agency in pain medicine: Philosophy, practice and virtue. *Pain Phys.*, 2006b; 9: 71–76.

Giordano J. Pain, the patient and the physician: Philosophy and virtue ethic in pain medicine. In: Schatman M.E. (ed.) *Ethical Issues in Chronic Pain Management.* CRC/Taylor-Francis, Boca Raton, FL, 2007; 1–18.

Giordano J. *Pain, Mind, Meaning and Medicine.* PPM Books, Glen Falls, PA, 2009.

Goodman KW. *Ethics and Evidence-Based Medicine: Fallibility and Responsibility in Clinical Science.* Cambridge University Press, Cambridge, 2003.

Irvine IM. Targeting breast cancer detection with military technology: Applicability of automated target recognition technology to early screening for breast cancer. *Eng. Biol. Med.*, 2002; 21: 36–40.

Johnston DW, Propper C, and Shields MA. Comparing subjective and objective measures of health: Evidence from hypertension for the income/health gradient. Discussion paper. Bonn, Germany: Forschungsinstitut zur Zukunft der Arbeit (Institute for the Study of Labor), 2007.

Jonas H. Technology and responsibility. *Social Res.*, 1973; 40: 31–54.

Kakuta N, Yokoyama S, and Mabuschi K. Human thermal models for evaluating infrared images: Comparing infrared images under various thermal environmental conditions through normalization of skin surface temperature. *Eng. Biol. Med.*, 2002; 21: 63–72.

Kuhn T. *The Structure of Scientific Revolutions.* University of Chicago Press, Chicago, IL, 1962.

Levine RJ. *Ethics and Regulation of Clinical Research*, 2nd ed. Yale University Press, New Haven, CT, 1986.

Light DW. The sociological character of health care. In: Albrecht, G.L., Fitzpatrick, R., and Scrimshaw, S.C. (eds.) *Healthcare Markets, Handbook of Social Studies in Health.* Sage Publications, London, 2000.

Mainetti JA. Embodiment, pathology, diagnosis. In: Pesert J.L. and Gracia D. (eds.) *The Ethics of Diagnosis.* Philosophy and Medicine Series, Vol. 40, Kluwer, Boston, MA, 1992.

May WF. Contending images of the healer in an era of turnstile medicine. In: Walter J.K. and Klein E.P. (eds.) *The Story of Bioethic: From Seminal Works to Contemporary Explorations.* Georgetown University Press, Washington, DC, 2003.

Nagel E. *The Structure of Science: Problems in the Logic of Scientific Explanation.* Harcourt, Brace and World, New York, 1961.

Owens J. Aristotelian ethics, medicine and the changing nature of man. In: Spicker S and Engelhardt HT (eds.) *Philosophical Medical Ethics—Its Nature and Significance*, Vol. 3. D. Reidel, Dordrecht, 1977.

Pedroni J. Going off the gold standard—Pain research beyond the randomized controlled trial. *Am. J. Pain Manage.*, 2006; 16: 61–65.

Pellegrino ED. The anatomy of clinical judgments: Some notes on right reason and right action. In: Engelhardt H.T., Spicker S. and Towers B. (eds.) *Clinical Judgment: A Critical Appraisal.* D. Reidel, Dordrecht, 1979.

Pellegrino ED. The healing relationship: Architectonics of clinical medicine. In: Shelp E (ed.) *The Clinical Encounter: The Moral Fabric of the Patient–Physician Relationship.* Philosophy and Medicine Series, Vol. 4, D. Reidel, Dordrecht, 1983.

Pellegrino ED and Thomasma DC. *For the Patient's Good: The Restoration of Beneficence in Health Care.* Oxford University Press, New York, 1988.

Pellegrino ED and Thomasma DC. *The Virtues in Medical Practice.* Oxford University Press, New York, 1993.

Reinhard MJ, Hinkin CH, Barclay TR, Levine AJ, Mario S., Castellon SA, Longshore D et al. Discrepancies between self-report and objective measures for stimulant drug use in HIV: Cognitive, medication adherence and psychological correlates. *Addictive Behavior*, 1997; 32(12): 2727–2736.

Sadler JZ. Diagnosis/antidiagnosis. In: Radden J (ed.) *The Philosophy of Psychiatry: A Companion.* Oxford University Press, New York, 2004.

Sadler JZ. *Values and Psychiatric Diagnosis.* Oxford University Press, New York, 2005.

Svenaeus F. *The Hermeneutics of Medicine and the Phenomenology of Health: Steps Toward a Philosophy of Medical Practice*, Kluwer, Dordrecht, 2000.

Toulmin S. Concepts of function and mechanism in medicine and medical science (Hommage a Claude Bernard). In: Engelhardt HT and Spicker S (eds.) *Evaluation and Explanation in the Biomedical Sciences.* Philosophy and Medicine Series, Vol. 1, D. Reidel, Dordrecht, 1975.

Wulff HR. *Rational Diagnosis and Treatment.* Blackwell, Oxford, 1976.

Zhang L and Zhou Z. Objective and subjective measures for sleep disorders. *Neuroscience Bulletin*, 2007; 23(4).

Index